SOCIOLOGY

SOCIOLOGY

Seventh Edition

Jon M. Shepard

Virginia Polytechnic Institute
and State University

Wadsworth

I(T)P® An International Thomson Publishing Company

Belmont, CA • Albany, NY • Boston • Cincinnati • Johannesburg • London • Madrid • Melbourne
Mexico City • New York • Pacific Grove, CA • Scottsdale, AZ • Singapore • Tokyo • Toronto

Publisher: Eve Howard
Development Editor: Robert Jucha
Assistant Editor: Barbara Yien
Marketing Manager: Chaun Hightower
Editorial Assistant: Angela Nava
Project Editor: Jerilyn Emori
Print Buyer: Karen Hunt
Permissions Editor: Robert Kauser
Design: Hespenheide Design
Production: Hespenheide Design
Copy Editor: Kathy Pruno
Illustrator: Randy Miyake/Hespenheide Design
Cover Design: Hespenheide Design
Cover Image: Christopher Brown, *Station,* 1993.
  Fine Arts Museums of San Francisco, Achenbach
  Foundation for Graphic Arts. Crown Point Press Archive,
  Crown Point Press, gift of Kathan Brown, 1995.
Composer: Hespenheide Design
Printer: World Color Book Services

Library of Congress Cataloging-in-Publication Data

Shepard, Jon M.
  Sociology / Jon M. Shepard. — 7th ed.
    p. cm.
  Includes bibliographical references and index.
  ISBN 0-534-55578-0 (pbk.)
  1. Sociology.   I. Title
HM51.S5175 1998   98-9938
301—dc21

COPYRIGHT © 1999 by Wadsworth Publishing Company
A Division of International Thomson Publishing Inc.
I(T)P® The ITP logo is a registered trademark under license.

Printed in the United States of America
 3 4 5 6 7 8 9 10

For more information, contact Wadsworth Publishing Company, 10 Davis Drive, Belmont, CA 94002,
or electronically at http://www.wadsworth.com

International Thomson Publishing Europe
Berkshire House 168-173
High Holborn
London, WC1V 7AA, England

International Thomson Publishing Asia
60 Albert Street
#15–01 Albert Complex
Singapore 189969

Thomas Nelson Australia
102 Dodds Street
South Melbourne 3205
Victoria, Australia

International Thomson Publishing Japan
Hirakawacho Kyowa Building, 3F
2-2-1 Hirakawacho
Chiyoda-ku, Tokyo 102, Japan

Nelson Canada
1120 Birchmount Road
Scarborough, Ontario
Canada M1K 5G4

International Thomson Publishing
Southern Africa
Building 18, Constantia Park
240 Old Pretoria Road
Halfway House, 1685 South Africa

International Thomson Editores
Campos Eliseos 385, Piso 7
Col. Polanco
11560 México D.F. México

Dedicated
To Sam and Jon

Contents in Brief

Contents

Preface

A Note to Students From the Author

As a college student on a football scholarship, my goal in life was to become a high school football coach. When the young woman I was dating (now my wife) suggested I take a course in introductory sociology with her, I did—mainly because it allowed us to be together and it happened to fit into my schedule.

The issues that came up in that class were ones jocks don't usually spend a lot of time on: Is divorce more likely or less likely when people have the same social-class background? Are some races inferior to others? What is the social significance of Darwinism?

Suddenly, I began to see human behavior in a different light. I discovered that Richard Wright's classics *Native Son* and *Black Boy* were not just stories about black youth but autobiographical reflections of the black experience in America. Prejudice and discrimination were not characteristics of individuals; they were built into society. I learned that the game I was playing to pay my way through college was actually as much a business as a sport. It became apparent that the fraternity I was about to join was not only a brotherhood but also part of the campus social hierarchy.

I began to see social relationships as essential for human survival. And if the world is a stage and its inhabitants are players, we generally deliver our social lines and act out our parts as if they were rehearsed and with a definite flair for mimicry. Yet the action that sociologists have labeled "social structure" depends less on the conscious learning of appropriate attitudes, beliefs, and behavior than on *unreflective* acceptance of one's culture and society. In one sense, we are puppets responding to tugs on the strings that bind us to essential social relationships.

People, I came to understand, do not usually behave randomly and do not always behave merely as individuals. People often think, feel, and behave in rather predictable ways because of what they are taught and because of the social pressures on all sides. At the same time, individuals create their own understandings of a situation as they interact with others. We are not really puppets, for human beings not only are capable of bucking tradition, we also are active, thinking creatures even when we are conforming. Society was demystified for me. I came to value sociology as a tool for understanding the world around me. In fact, this experience led me to major in sociology in college and subsequently to obtain my Ph.D. in sociology. America lost a football coach, but it gained a committed teacher. I have never regretted my choice.

You probably have never shared my aspiration to be a football coach. You might not major in sociology, though I hope to show you the life-long benefits of studying sociology. You can, however, enjoy this course and take from it the unique slant on social life that sociology provides.

DISTINCTIVE STUDY AIDS

Students through six editions continue to express enthusiasm for the unique study features central to this text. Via these features you become an active learner, comprehending and retaining the material more fully.

SQ3R: A Format for Study

This text is designed throughout with the "SQ3R" study format at its core. Research demonstrates that this approach will help you identify significant ideas, understand these ideas rapidly, remember important points, and review effectively for exams. As a result, you will learn more about sociology more easily while performing better on tests.

The letters in SQ3R symbolize five steps in effective reading and learning: survey (S), question (Q), and read, recite, and review (3R). The steps in the SQ3R method are built into each chapter.

1. *Survey.* Read the outline, the introduction, and the summary that opens each *Review Guide* at the end of each chapter before beginning a careful reading of the chapter. This survey, which will give you an overall picture of the chapter content, should take only a few minutes.

2. *Question.* Third-order headings are phrased as questions to help you select and concentrate on the important points. For example, instead of seeing subtopic headings such as "Working Women," you will find such questions as "Why are more women working today?" and "How does the increasing size of the female labor force affect the family?" Increased comprehension will result from focusing on each question as you read the material that answers it. More will be said about this step in the following section on critical thinking.

3. *Read.* Read only the material that takes you through a third-order heading posed as a question. Although each major section has several questions within it, take step 4 before proceeding to read about the next subtopic.

4. *Recite.* In this phase, try to answer each question heading immediately after reading the relevant material. You may wish to state the answer to yourself or to write it down in note form. Keeping a set of notes with questions and brief answers will be helpful. If you are unable to answer a question, examine the material until you find the answer. As a final check, answer the self-test questions appearing in the *Feedback* section at the end of every major topic. (See sample

Feedback below.) Attempt to answer these *Feedback* questions without looking at the correct answers that appear beneath them. If you cannot answer a self-test question, note the correct answer given under the questions and look back at the text material to find out why this is the right answer. Then proceed to the next major section and follow the same procedure. Repeat until you have completed the chapter. This process will prevent you from deluding yourself into believing that you understand material when, in fact, you do not. The recitation dimension of the SQ3R method is designed to replace surface recognition of material with true understanding and comprehension.

5. *Review.* After completing a chapter according to the procedure just outlined, attempt to answer briefly the questions posed in the third-order headings. If possible, have another person ask you the questions to see if you can state the answers in your own words. You can complete your review of a chapter by using the *Review Guide* that appears at the end of each chapter.

Review Guide

Each chapter concludes with a *Review Guide*. Each guide begins with a chapter summary followed by the chapter learning objectives. Next is a concept review feature composed of about one-half of the concepts covered in the chapter to which you are asked to match a set of definitions. This is an opportunity for you to test your grasp of key concepts. Several critical thinking questions follow the concept review. These broad questions encourage the application of critical thinking skills to sociological questions and provide practice for essay tests. A set of multiple choice questions is then provided as a mini self-test. The feedback review consists of a sample of questions taken directly from the *Feedback* questions throughout the chapter. The *Review Guide* in most chapters includes a graphic review feature, which gives you the opportunity to test your understanding of a particular table or figure in the chapter by answering

two open-ended questions. An answer key is provided at the end of each *Review Guide*.

Critical Thinking: A Questioning Mind

Critical thinking is aimed at challenging conventional wisdom. Through the critical thinking process, one is able to ask incisive questions, select and properly consider relevant evidence, form judgments based on reason rather than emotion, and formulate solutions to problems (Halpern, 1997). An inclination to question commonly held assumptions is bound to be an important result of thinking critically about social phenomena.

Critical thinking is important for today's college students. First, the tradition of liberal education, of which sociology is a vital part, is the tradition of critical thought (Pelikan, 1992). Second, as the nature of work continues to move from physical labor to mental activities, the facility for critical reasoning becomes an increasingly valuable asset on the job. Third, it doesn't take a rocket scientist to figure out the need for critical thought among all American voters; behavior of our political leaders makes this point almost daily. Finally, not the least use of critical thinking comes in more personal contexts such as family life, decision making outside of work, and personal enrichment. By thinking critically we can make our own decisions, free from the media, advertising, and many other assaults on personal choice that are part of daily life. Critical thinking can be liberating because it empowers us to take control of our own decisions by making informed decisions (Waller, 1994).

Critical thinking is incorporated in this edition in four ways:

1. **Critical Thinking Within the SQ3R Method** The promotion of critical thinking lies at the heart of the "question" step of the SQ3R method. For any given topic, initial questions are typically informational. Follow-up questions, designed to take you more deeply into the topic, are critical thinking questions.
2. **Critical Thinking Questions** As mentioned earlier, a set of four to six critical thinking

FEEDBACK

1. _____ is the scientific study of social structure.
2. Match the social sciences listed below with the examples of research projects beside them.

____ a. sociology	(1) a study of how children learn to talk
____ b. anthropology	(2) a study of the impact of taxation on consumer spending
____ c. psychology	(3) a study of African American family structure during the slavery era
____ d. economics	(4) a study of village ruins
____ e. political science	(5) a study of presidential power
____ f. history	(6) a study of drug use patterns among high school students

Answers: 1. Sociology 2. a. (6) b. (4) c. (1) d. (2) e. (5) f (3) questions are included in the *Review Guide* at the end of each chapter. These wide-ranging questions provide you the opportunity to think critically and creatively about the ideas within a chapter. Sometimes you will apply these ideas to a particular aspect of social life. Other times you will use sociological ideas to analyze and understand events and experiences in your own life.

3. **Critical Feedback**   Each *Doing Research* (a feature to be described shortly) closes with a series of critical feedbacks, designed to help you better understand the piece of sociological research involved and to probe below the surface. For example, these questions are posed following the description of E. E. LeMasters' study of blue-collar workers and middle-class lifestyle: "Do you believe that the working class constitutes a subculture?" "Do you believe that the lifestyle distinctions between the working and middle class will continue to exist?"

4. **Questioning Conventional Social Wisdom** The existence of society, of course, depends upon members' acceptance of many ideas that are passed from generation to generation. At the same time, it is easy to fall into a pattern of nonreflection about prevailing ways of thinking, feeling, and behaving. The "sociological imagination," to be discussed in the next section, promotes critical thinking about the nature of society.

UNIFYING THEMES

Sociological Imagination

False ideas about social life survive mainly because so many of us rely on nonsociological, nonscientific sources of knowledge. Although knowledge exists—and we are often exposed to it—we tend to reject or ignore it because it contradicts what we have been conditioned to believe. The study of sociology encourages critical thinking about conventional wisdom (social myths thought generally to be true) through the development of the *sociological imagination*—the mindset that enables individuals to see the relationship between events in their personal lives and events in their society. By questioning conventional wisdom, one is in a better position to make a decision or judgment based on reality rather than on socially accepted false beliefs that obscure the truth. Personal freedom from social restrictions based on erroneous information is one happy consequence of this form of critical thinking.

Each chapter of this edition opens with a question about some aspect of social life. The answer to each

question contradicts a popular or commonsense belief. Sometimes the question will focus on a result that even sociologists may have doubted until a sufficient amount of convincing research had been done. The correct answer to each question is given at the beginning of the chapter and is elaborated on within the chapter itself.

Doing Research: Sociologists as Scientists

This feature presents the theory, methods, conclusions, and implications of significant sociological studies. Several criteria guided the selection of these research studies. Some studies are sociological landmarks. Others are included because they reinforce a major point in a chapter. Still other studies are included because they illustrate the imaginative use of a major research method.

Given these criteria, it is hardly surprising that many of the studies included are sociological "classics": they have had a lasting influence on the field and are continuously being cited by other researchers. Like classics in all fields, these pieces of research generally have high interest value. They are innovative in approach and explore important topics in ingenious ways. If read carefully, these detailed accounts of significant sociological studies cannot fail to pique your interest in social research and stimulate your sociological imagination.

Culture: A Global Perspective

Using the sociological perspective in relation to other societies is becoming increasingly important. The world is becoming a "global village." Events in one part of the world increasingly carry political, social, and economic consequences for other parts. Will the United States send troops to restore to power a country's democratically elected president ousted by the military? Will Americans continue to have a peace-keeping presence in Bosnia? Will further military action have to be taken against Saddam Hussein in Iraq? The repercussions of the breakup of the former Soviet Union, the economic development of China, and the economic unification of Western Europe are as uncertain as they are sure to be significant (Schaeffer, 1997).

The economic watchword is "globalization of the economy." A few countries, such as the United States and England, no longer control the economic tides; they live in a world without economic borders (Schnitzer, 1997). Japan, a relatively new economic power player, is now feeling the challenge from the so-called Little Dragons of South Korea, Taiwan, Singapore, and Hong Kong, not to mention other parts of the world. Large American corporations, as well as many small and medium-size companies, are becoming increasingly intertwined in pervasive global economic competition and cooperation. Companies fail to develop a global perspective at their peril.

The employees of these companies must also be prepared for a new economic reality in which there will be "no such organization as an 'American' (or British or French or Japanese or West German) corporation, nor any finished good called an 'American' (or British, French, Japanese, or West German) product" (Reich, 1993: 110). American companies are sending increasing numbers of their domestic employees to work in other countries, while simultaneously hiring foreign workers in large numbers. Forty percent of IBM's world employees, for example, are now foreign. Also, foreign investment in the American economy is staggering. A substantial proportion of America's manufacturing assets are owned by economic interests based outside the country, and foreign companies in the United States employ nearly one-tenth of America's manufacturing workers.

As you move into a world without borders, the sociological perspective will help you participate in it effectively. This perspective has an important contribution to make toward one's success in this globalized world.

You will begin in Chapter 1 with an introduction to the sociological perspective. This unique perspective is an integral part of one's exposure to the liberal arts, vital to the development of an educated person.

SOCIOLOGY INTO THE 21ST CENTURY

Sociology in the News

Sociology in the News is a unique feature new to this edition. All chapters contain a boxed insert briefly summarizing a CNN news report that your instructor can show the class from a tape provided by the publisher of this text. Each news report amplifies a point of interest in the chapter. Two questions are provided in the boxed insert asking you to relate some aspect of the chapter to the news report.

Internet Link

Also new to this edition is a feature called *Internet Link*. In nearly all chapters, one table or graphic presenting important information on a topic will contain an Internet address that will give you electronic access to additional information.

Connections: Interactive Sociology

Connections is an interactive CD-ROM for introductory sociology. The CD-ROM includes multimedia overviews of standard topics, Hyper-Soc, a detailed tutorial of key concepts, a glossary, simulations, and graphs. *Connections* is an excellent resource for student self-study.

ACKNOWLEDGMENTS

Several colleagues have provided thoughtful and helpful reviews for the seventh edition. Many thanks to the following individuals:

- Robert Anwyl, Miami Dade Community College North
- Jerome Crane, Rock Valley Community College
- Ray Darville, Stephen F. Austin University (Texas)
- Mary Donahy, Arkansas State University
- Lois Easterday, Onondage Community College (New York)
- Jesse Frankel, Pace University
- Larry Frye, St. Petersburg Junior College
- Roberta Goldberg, Trinity College, Washington, D.C.
- Thomas Harlach, Orange County Community College (New York)
- Joseph Kotarba, University of Houston
- Jerry M. Lewis, Kent State University
- Jieli Li, Ohio University
- Doris Miga, Utica College
- Mary Steward, University of Nevada at Reno

Some of my colleagues listed above have also reviewed earlier editions. In addition to those individuals, the following colleagues have provided critiques of previous editions: G. William Anderson, Paul J. Baker, Melvin W. Barber, Jerry Bode, Patricia Bradley, Ruth Murray Brown, Brent Bruton, Victor Burke, Bruce Bylund, David Caddell, Albert Chabot, Stephen Childs, Carolie Coffey, Kenneth Colburn, Jr., Alline DeVore, Mary Van DeWalker, Susan B. Donohue, M. Gilbert Dunn, Mark G. Eckel, Irving Elan, Ralph England, K. Peter Etzkorn, Mark Evers, Joseph Faulkner, Kevin M. Fitzpatrick, Ramona Grimes, James W. Grimm, Rebecca Guy, Penelope J. Hanke, Cynthia Hawkins, Kenneth E. Hinze, Carla Howery, Gary Kiger, James A. Kitchens, Irving Krauss, Mark LaGory, Raymond P. LeBlanc, Roger Little, Richard Loper, Roy Lotz, Scott Magnuson-Martinson, R. Lee McNair, Edward V. Morse, Charles Mulford, Bill Mullin, Daniel F. O'Connor, Jon Olson, Charles Osborn, Thomas R. Panko, Margaret Poloma, Carol Axtell Ray, Ellen Rosengarten, William Roy, Josephine Ruggiero, Steven Schada, Paul M. Sharp, James Skellinger, Robert P. Snow, Robert F. Szafran, Ralph Thomlinson, Charles M. Tolbert, David Waller, Michael E. Weissbuch, Carol S. Wharton, Douglas L. White, Paul Wozniak, David Zaret, Wayne Zatopek.

I wish to thank several people for their contributions to this edition. Bob Jucha, my developmental editor, has guided me since the fourth edition. His substantive and artistic contributions over these years have been invaluable. Our close collaboration has been both productive and enjoyable. Sandy Crigger has also worked with me over the last four editions. The transformation from the illegible pages I gave her to the flawless text she returns is a miracle. I have come to depend very heavily on her ability to understand what needs to be done and to often do it before I can even make a request. Dinah Akers once again cheerfully transformed extremely rough ideas into an attractive graphic format. The willingness of Melissa Kessinger to fill in for Sandy at times is greatly appreciated. Craig VanSandt's attention to detail and perseverance in the library and on the Internet were critical to the updating of this edition. He exceeded my expectations regularly.

Hespenheide Design performed the amazing feat of producing a visually attractive book while switching from a four-color format in the last edition to a two-color layout in the current one. Kathy Pruno, who copyedited this edition as well as the sixth edition, again displayed her meticulous attention to detail, accuracy, and consistency.

My wife, Kay, once again displayed her superb judgment throughout this revision. She always knows just what to say in the most constructive and supportive way. The marriage of our son, Jon, to Samantha Williams, made the workload much lighter.

Chapter One

The Sociological Perspective

After careful study of this chapter, you will be able to:
- Define sociology and illustrate its unique perspective from both the micro and macro levels of analysis.
- Describe three uses of the sociological perspective.
- Distinguish sociology from other social sciences.
- Outline the distinctive contributions of the major pioneers of sociology.
- Summarize the beginnings of sociological development in the United States.
- Differentiate the major theoretical perspectives in sociology today.
- Compare two recent developments in symbolic interactionism.

SOCIOLOGICAL IMAGINATION

Why do people commit suicide?   Answers to this question immediately come to mind: prolonged illness, loss of a lover, public disgrace, depression, heavy financial loss. Our explanations assume that suicide is due solely to personal motivation. Emile Durkheim demonstrated that suicide is, in fact, affected not only by a personal crisis, but also by social forces that increase the likelihood of a lone individual jumping from a bridge, deliberately taking an overdose of drugs, or driving a car into a wall. Specifically, he found that variations in suicide rates are related to the extent to which people are tied to their social groups. Highly socially embedded categories of people—married, females, Catholics—exhibit lower rates of suicide. More socially isolated categories of persons—unmarried, males, Protestants—show higher suicide rates. After one hundred years, research continues to support Durkheim's findings and conclusions.

Durkheim is one of the pioneers of sociology who will be profiled in this chapter. Before turning to these pioneers, however, we will discuss the sociological perspective that Durkheim so uniquely identified and developed.

THE SOCIOLOGICAL PERSPECTIVE

Truly, one's perspective matters. Babies are usually brighter and better looking to their parents, and people in love nearly always find their lovers infinitely more attractive than do their friends.

More fundamentally, we all interpret what is happening around us through the particular perspectives we have acquired. It can hardly be any other way. We normally do not realize the extent to which our view of reality is determined by our perspectives. Only when our perspectives are challenged are we jarred into realizing how much we take our worldview for granted.

This was aptly reflected in a Native American's speech to the white men in eighteenth-century Virginia who had offered to "properly educate" some of his young men at the College of William and Mary in Williamsburg, Virginia.

> We know that you highly esteem the kind of learning taught in . . . [your] colleges. . . . But you, who are wise, must know that different nations have different conceptions of things; and you will not therefore take it amiss, if our ideas of this kind of education happen not to be the same with yours. We have had some experience of it; several of our young people were formerly brought up at the colleges of the northern provinces; they were instructed in all your sciences; but, when they came back to us, they were bad runners, ignorant of every means of living in the woods, unable to bear either cold or hunger, knew neither how to build a cabin, take a deer, nor kill an enemy, spoke our language imperfectly, were therefore neither fit for hunters, warriors, nor councellors; they were totally good for nothing.
>
> We are however not the less obligated by your kind offer, though we decline accepting it; and, to show our grateful sense of it, if the gentlemen of Virginia will send us a dozen of their sons, we will take care of their education, instruct them in all we know, and make men of them.

To the surprise of the colonists, the benefits of a white gentleman's education were not desired by the tribal elders. In fact, from the Native American perspective, white young men would be much better off to leave William and Mary and pursue an education in the tribe. Because of their own perspective, no William and Mary students transferred.

Sociology has its own unique perspective that never takes the individual as an object of investigation. The first section of this chapter is devoted to an elaboration of sociology's unique focus, a focus that always remains above the level of the individual.

A Definition of Sociology

As a novice to the field, you may wish at first to view sociology as the study of human groups. This inclination has the advantage of tapping ideas you already possess. However, as you progress through Chapter 3 ("Culture") and Chapter 5 ("Social Structure and Society"), you will acquire a more precise understanding of **sociology** as the scientific study of social structure and come to realize that a group is only one type of social structure. In the meantime, your understanding of the concept of social structure will be understandably vague. For now it is sufficient for you to become familiar with two key dimensions of the sociological perspective: (1) the interaction between individuals and social structures; and (2) the idea of levels of analysis.

Social Structure and the Individual: A Two-Way Street

The starting point for sociology and central to the understanding of the sociological perspective is the assumption of predictability and recurrence in social relationships. Sociologists are interested not in the idiosyncratic behavior of individuals but in patterned social relationships referred to as **social structures**. The person on the street tends to explain human behavior in individualistic terms—a young man goes to war because he hates the enemy, a woman divorces her husband to develop her potential, a recently retired person commits suicide to escape depression. On the other hand, sociology attempts to explain these same events without relying on individualistic factors—young men go to war because they have been taught by their society to be patriotic; women may divorce because of the social trend toward sexual equality; retired persons commit suicide because society has suddenly separated them from their work, a primary source of meaning in life. Sociologists do not speak of a young man, a married woman, or a retired person. They focus on categories of people—young men, married women, retired persons.

To cite another example, college students do not all behave exactly alike in class—some attempt to write down everything their professors say, some use tape recorders instead of taking notes, some just listen to the lecture, some daydream, some write letters home. Yet, if you visit almost any college or university, you will find patterned relationships—students remain in their seats whereas professors lecture, professors give examinations whereas students take them. Although the individual characteristics of students and professors may vary from school to school, students and professors do relate in similar patterned ways. It is the patterned interaction of people and the social structures created by such interaction that capture the attention of sociologists.

Sociologists, then, assume that social relationships are not determined by the idiosyncratic characteristics of the individuals involved. Emile Durkheim (1966, originally published in 1895), a pioneering nineteenth-century sociologist, developed this aspect of the sociological perspective. He argued, for example, that we do not attempt to explain bronze from its component parts (lead, copper, and tin). Instead, we consider bronze an alloy, a unique metal produced by the synthesis of several distinct metals. The hardness of bronze is not predictable from its components, each of which is soft and malleable. Durkheim reasoned that if a combination of certain metals produces a unique metal, some similar process might happen in groups of people. He argued that people's behavior within a group cannot be predicted from the characteristics of individual group members. Something new is created when individuals come together as a collective. To use an extreme example, rioters in South Central Los Angeles, following the acquittal of the officers charged with police brutality in the Rodney King case in 1992, behaved differently than they would have as individuals on their own.

The decision to attend college is not merely a choice made by an isolated individual. Social relationships with teachers, parents, peers, and others influence a person's decision to become a college student. Sociologists study the various patterned social relationships referred to as social structures.

Individualistic explanations of group behavior are inadequate because human activities are influenced by social forces that individuals have not created and cannot control. We live in groups that range in size from a family to an entire society and they all encourage conformity. Thus, people who belong to similar groups tend to think, feel, and behave in similar ways. American, Russian, and Chinese citizens have distinctive eating habits, types of dress, religious beliefs, and attitudes toward family life.

Conformity within a group occurs partly because most of its members believe that their group's ways of thinking, feeling, and behaving are the best; they have been successfully taught to value their group's ways. Group members also tend to conform even when their personal preferences are not the same as their group's. For example, beer drinking is common in college fraternities and sororities; drinking with one's brothers or sisters is expected. Because of social pressures, some fraternity and sorority members drink alcoholic beverages even though, as individuals, they would prefer not to. Similarly, because of peer pressure, members of campus religious groups may not drink alcohol despite their preference for it.

Whether because members value their group's ways or because they yield to social pressures of the moment, behavior within a group is not usually predictable from knowledge about its individual members. A group is not equal to the sum of its parts.

Sociologists pursue a variety of activities in attempts to understand, explain, and predict these often hidden processes that permit successive generations to carry out relatively predictable and orderly lives without having to forge anew the rules for their social life. Yet because each generation usually is spared the trouble of creating new rules and roles by which to conduct their social lives, its members fail to ask: Why are things the way they are? How do they change? Sociologists, in contrast, constantly wrestle with these basic questions. Even if you do not plan to become a professional sociologist, you can share in the excitement of trying to answer these questions.

If this description of sociology causes you to believe that sociology is interested only in the effects of existing social structures on individuals, it has been misleading. It is necessary to begin with this emphasis because without some considerable degree of conformity there can be no group life. This necessary emphasis, however, should not obscure the effects of individuals on social structures. The interplay between individuals and their social structures is a two-way street—people are affected by social structures, and people change their social structures. Symbolic interactionism, one of the major theoretical perspectives covered in the following chapter, underscores the dynamic, creative construction of social interaction by individuals. Consequently, you

should not conclude that all human behavior in groups is determined by preexisting social structures.

An important distinction is implied in our discussion of the sociological perspective thus far. A careful reading reveals the interest of sociology in both the interaction of people "within" social structures (microsociology) and the "intersection" of social structures per se (macrosociology). The distinction between microsociology and macrosociology as distinct levels of analysis is a second integral dimension of the sociological perspective.

Levels of Analysis: Microsociology and Macrosociology

Microsociology and macrosociology should be thought of as levels of analysis, because they determine which aspects of social structure a sociologist wishes to examine. Whereas *micro* means small, as in microscopic, *macro* means large. At the micro level, social relationships are investigated at close range (Scheff, 1990). The macro level of analysis considers social structure at a distance, or at a higher level (Helle and Eisenstadt,1985a, 1985b; Hawley, 1992).

Suppose for the moment that social structures are directly observable as tangible objects and that we are cruising at 20,000 feet at the controls of *Sociologist One*. On a clear day at this altitude we would be able to observe definite social structures just as we can observe the lay of the land from an airplane. At 20,000 feet we will see no movement in the form of people interacting. We are at the macro level. Suppose we wish to get a closer look and start descending. As we descend, we will begin to see people interacting. When we focus on these relationships, we are at the micro level. Formal definitions of microsociology and macrosociology will make more sense against this general background.

Microsociology is concerned with the study of people as they interact in daily life. Consider the practice of knife fighting among street gangs. At the micro level, a sociologist attempts to explain participation in gang knife fighting via the social relationships involved. Street-gang leaders, for example, may feel they must fight either a member of their own gang who is challenging their position or the leader of another gang in order to validate periodically their right to leadership. As implied in this example, at the micro level it is not the individuals per se that are important; it is their interpretations of the parts they play as participants in social relationships. Politicians are expected to compromise to achieve their goals, business people are expected to keep a constant eye on the balance sheet, and professors are expected to grade students fairly. In turn, we have ideas about how we should behave in social relationships. We see nothing inappropriate in asking a politician for a favor; we know that wariness is in order when a used-car dealer delivers a pitch on that low-mileage, one-owner car at a giveaway price; we believe that professors will grade impartially. As will be seen later in this chapter, and throughout this book, individuals, being the active, thinking beings they are, do not simply follow established social scripts. Individuals interpret events and construct their behavior as they participate in social structures.

Macrosociology focuses on relationships among social structures without reference to the interaction of the people involved. Some advocates of macro level analysis restrict their analysis to entire societies (Tilly, 1978, 1986; Wallerstein, 1979a, 1979b, 1984; Skocpol, 1985; Lenski, 1988; Lenski, Nolan, and Lenski, 1995). We will use macrosociology to refer to the study of both societies as a whole as well as the relationships among social structures within societies. The following are sample research questions that can be asked at the macro level of analysis: What are the patterned relationships between the defense industry and the federal govern-

ment? How does the economy (for example, preindustrial or industrial) affect the stability of the family? In what ways do different governmental structures promote conflict between societies?

Because both the micro and the macro levels of analysis focus on social structure, they obviously are complementary; their combined use will tell us more about social structure than either one used alone (Alexander et al., 1987). To understand gang warfare, a sociologist would want to know about the social relationships involved—for example, the relationships between gang leaders and followers or between gang members and the police on the beat. To supplement this understanding of gangs derived from the micro level, the macro level can be used to examine the larger social structures of which gangs are a part. At the macro level, a sociologist might study those aspects of a society producing the poverty, the lack of education, and the joblessness that are the breeding grounds of organized delinquency.

USES OF THE SOCIOLOGICAL PERSPECTIVE

Why study sociology? To learn more about what it means to be a professional sociologist? To satisfy the desire for knowledge that springs from an intrinsic love of ideas? These are perfectly legitimate motivations for studying sociology. But there are additional reasons.

The sociological perspective has three primary uses. First, it encourages development of the "sociological imagination." Second, sociological theory and research can be applied to important public issues such as crime and health care. Third, the study of sociology can sharpen skills useful in many occupations.

The Sociological Imagination

Interplay of self and society. In his famous essay, *On Liberty*, nineteenth-century philosopher John Stuart Mill wrote:

It will probably be conceded that it is desirable people should exercise their understandings, and that an intelligent following of custom, or even occasionally an intelligent deviation from custom, is better than blind and simple mechanical adhesion to it (Mill, 1993:68–69, originally published in 1859).

Knowledge of the ways social forces affect our lives, the study of sociology tells us, prevents us from being prisoners of those forces. C. Wright Mills (1959) called this personal use of sociology the **sociological imagination**—the set of mind that enables individuals to see the relationship between events in their personal lives and events in their society. Events affecting us as individuals, Mills pointed out, are closely related to the ebb and flow of our society. Decisions, both minute and momentous, are not just isolated, individual matters. For example, American society has shown a strong bias against married couples remaining childless or having only one child. Such couples have been considered selfish, and only children have been labeled as spoiled (Benokraitis, 1996). This myth dates back to the societal need for the species to survive. Later, large families were needed both because of the high infant mortality rate and because of the need for labor on family farms. People generally responded to society's pressure by having large families. Only now, as the need for large families has disappeared, are we reading of the benefits of one-child families—to the child, to the family, and to society. The sociological imagination provides the intellectual perspective needed to understand the effects of social trends on our daily lives. With this understanding, we are in a better position to make autonomous

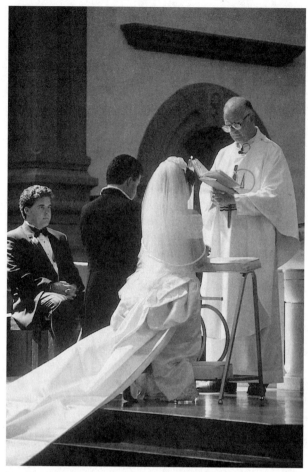

For this couple, their marriage, and the particular ceremonial form it is taking, was the result of their decision as individuals. To sociologists, the decision to marry and to do so in a particular way is socially channeled.

decisions rather than succumb to the barrage of societal pressures to which we are subjected (Farley, 1996; Peck and Hollingsworth, 1996; Game and Metcalfe, 1996).

This broadened social awareness enables us to read the newspaper with a better understanding of the implications of social events for individuals. We can see a letter to the editor opposing welfare payments not merely as a reflection of someone who cares nothing for hungry children or elderly people but as an expression of the importance Americans place on independence and self-help (Lee, 1990).

Ashley Montagu has coined the term *psychosclerosis* to describe a mental condition that deprives people of the ability to accept new ideas. What are the symptoms of psychosclerosis, or "hardening of the mind"?

> *Well, a certain lack of ability to see 360 degrees, to see all the way behind, sideways, as well as forward, to pick up the small and subliminal signals. To be wedded to fixed, stereotyped ideas that this is the right way and the only way* (Montagu, 1977:50).

Because the sociological imagination expands our intellectual horizons, it may be considered an antidote for psychosclerosis.

The debunking theme. Despite the availability of accurate information and explanations, people tend to cling to myths and false ideas about social life that are passed from generation to generation. Plato's "Allegory of the Cave" comes to mind. In *The Republic*, Plato describes a cave in which people have been chained from childhood so they cannot move. Forced to look only straight ahead, these prisoners cannot see behind them the blazing fire and a raised walkway in front of the fire. Men walk along the wall carrying all types of objects. The shadows on the wall in front of the prisoners made by the walking men and the fire are the only things the prisoners have ever seen. Their interpretations of these shadows constitute reality to them. Sociology attempts to replace common misconceptions about social life (shadows on the cave wall) with accurate information and explanations (Ruggiero, 1995).

Because the task of sociology is to reveal the nature of social life, it often leads to questioning what most people take for granted. What people in a society take for unassailable truth, may, under scientific examination, prove to be false. This aspect of the sociological imagination, then, involves investigating accepted interpretations of social life in order to see reality. Recognition of this benefit from sociology is reflected in the question and answer that appears at the beginning of each chapter. For additional description of this feature, see "Preface: A Note to Students from the Author" preceding this chapter.

Intellectual liberation. Like all liberal arts courses—anthropology, history, literature, and philosophy—sociology helps free people from the confines of the time, location, and situation in which they grew up (Bierstedt, 1963; Brouilette, 1985). You can, for example, learn that the sacrificial murder of infants, a practice repugnant to most people, is a normal and legitimate aspect of some cultures and that those who participate in this practice are not evil and despicable. Among the Winnebago Indians of Wisconsin, for example, it was socially acceptable for parents to kill children who became too great a liability (Radin, 1953). Among the Toda of southern India, female infants were frequently killed at birth, and if twins of different sexes were born, the female was always killed (Murdock, 1935). The Winnebago and Toda, like all humans, had been taught the ways of their group. Through the sociological imagination, you can come to understand that a society under extreme population pressure, living on the margin of subsistence, might adopt infanticide as part of its way of life.

Sociology provides a window to the social world that lies outside ourselves. It allows us to see the many social forces that shape our own lives. Sociology thus complements rather than replaces other ways of viewing human behavior. The more we seek to apply a range of different perspectives in attempting to interpret social forces and their meanings, the greater will be our understanding of our own behavior as well as the behavior of others. With such understanding comes the potential for greater personal freedom from social pressures. Consider divorce, for example. Social research reveals many factors in modern society that promote marital failure: Divorce is more likely to occur among couples who have low levels of education, marry early, have had a premarital pregnancy, or live with their parents after marriage. As predictive as these characteristics are, they cannot be used to force couples not to marry. However, knowing in advance the kinds of pressures that such marriages are inevitably going to experience may convince couples under these circumstances to either not marry or at least anticipate and prepare more effectively for such problems. The successful application of the sociological imagination enables us to begin to share Somerset Maugham's insight that "tradition is a guide and not a jailer."

Applied Sociology

Should social scientists be social fixers? Do sociologists *as scientists* have a moral responsibility to speak against and attempt to change aspects of social life they believe to be wrong? From the time of its appearance on the American academic scene in the late 1800s, sociology has steadily attempted to move from its origins as a social problem-solving discipline to a nonsocially involved science. During the intervening years, disagreement has periodically surfaced on the compatibility of these two viewpoints (Lee, 1978). A central argument against the involvement of social scientists in the eradication of social ills is that science is "value neutral"—that is, the conduct of scientific research is supposed to be directed at discovering what *actually* exists, with no room for personal judgments as to what *ought* to exist. The idea of value-neutrality—that scientists are not supposed to permit their own value judgments (ideas of what is good and bad) to influence their scientific work—has dominated sociological thought for a long time. And although the idea of science as a value-neutral enterprise remains strong among sociologists, those favoring the interjection of standards of good and bad into scientific activity are on the increase (Bickman and Rog, 1997). If the debate regarding the moral responsibility of the social sciences to improve society is not a new one, it is an issue that has once again gained considerable prominence in the form of **humanist sociology**—the theoretical perspective that places human needs and goals at the center of sociology (Lee, 1978; Scimecca, 1987; Giddens, 1997).

Sociological research in fact does influence public policies and programs. For example, social scientists contributed to the 1954 Supreme Court decision outlawing separate but equal schools for blacks and whites. This is one of the earliest examples of the United States court system accepting social science research as legitimate evidence for making a legal decision. Subsequent research has explored a variety of related topics, including the effects of school quality on student performance, the impact of social environment on measured IQ, the influence of school desegregation on the performance of black and white students, the effectiveness of busing in desegregating schools, and the relationship between busing and the movement of white middle-class families to suburbs. An illustration of the concern of sociology for application is provided by the closing words of William Julius Wilson's 1990 presidential address to the American Sociological Society:

With the reemergence of poverty on the nation's public agenda, researchers have to recognize that they have the political and social responsibility as social scientists to ensure that their findings and theories are interpreted accurately by those in the public who use their ideas. They also have the

The debate over whether sociology should be concerned with solving social problems dates back to its inception. Today, there is a strong current in sociology advocating concern for social ills. Humanist sociologists wish to apply sociological theories and research findings to problems such as those that occupied Nobel Peace Prize winner Jane Addams—the urban poor in general and the plight of poor working women in particular.

intellectual responsibility to do more than simply react to trends or currents of public thinking. They have to provide intellectual leadership with arguments based on systematic research and theoretical analyses that confront ideologically driven and short-sighted public views (Wilson, 1991:12).

Some sociologists advocate contributions beyond policy-related research. Following a long-neglected lead (Wirth, 1931), clinical sociologists are now using sociological theories, principles, and research to promote change. Clinical sociologists may redesign the social environment of cancer patients to help them better adapt to their situation, work as marital and family therapists, or design intervention programs to reduce juvenile delinquency. Clinical sociologists provide help for individuals or serve as agents for change in organizations, communities, and even entire societies (Black, 1984; Clark, 1984; Fritz, 1985; Glaser and Erez, 1988; Lee, 1990; Fritz and Clark, 1989, 1991; Rebach and Bruhn, 1991; Straus, 1994).

Sociology and Occupational Skills

The predominant trend in higher education at the undergraduate level has been toward professions, such as business and engineering. This emphasis on vocationalism is now being questioned by critics who believe that American higher education has lost its way. Many who are concerned about the condition of higher education say that in the rush to provide a curriculum that leads to employment, American institutions of higher education are failing to provide students with an adequate background in the humanities and social sciences.

Students have a legitimate concern about their future employment. The point is not to eliminate professional, specialized training. At issue is how to provide job skills and a broad education at the same time. Although no easy and general solution for this problem exists, students can increase their job skills while receiving a broader education. That is, exposure to liberal arts courses actually provides the opportunity for the development of skills required for employment in almost any field. This is because liberal arts courses such as sociology teach what is now being called higher order thinking skills, a phrase created to describe the process of learning how to learn. Higher order thinking skills include the ability to see structure in apparent disorder, to handle complex structures and ideas, and to be able to develop multiple solutions to problems (Rankin, 1991). Critical thinking, a synonym for higher order thinking skills, is a necessity to compete successfully in a modern economy. For Robert Reich, author of *The Work of Nations,* and former Secretary of Labor in the Clinton administration, there are four dimensions of critical thinking essential for employees in the new workplace: the capability to see patterns (abstraction), the ability to discern cause-and-effect relationships, the desire for experimentation, and the social skills necessary for teamwork (Reich, 1993).

Employers are interested in three types of functional skills: the capacity for working well with others, the ability to write and speak fluently, and capabilities for problem solving and data analysis. The enhancement of these functional skills is important. Computers, for example, have revolutionized the office. Data analysis skills are becoming much more important to managers at all levels in all types of organizations. In addition, the increasing complexity of work demands greater critical-analysis and problem-solving skills for the decision-making process. As noted above, the levels of each of these functional skills can be improved through the study of sociology (Ruggiero and Weston, 1986; Billson and Huber, 1993; Miller, 1994; American Sociological Association, 1997). In addition to these general functional skills, specific sociology subfields offer preparation for fairly specialized jobs. Consider some examples:

1. Training in race relations is an asset for working in the human resources department of a firm with a culturally diverse workforce, a hospital, or a day-care center.
2. Background in urban sociology can be put to good use in an urban planning agency.
3. Courses focusing on gender and race serve as valuable background for the affirmative action field.

The modern world has entered the computer age. Job requirements faced by this woman at her computer, however, go beyond technical skills. She will be required to exercise higher order thinking skills such as seeing pattern in apparent disorder, handling complex ideas, and developing alternative solutions to problems.

1. The _____ is the set of mind that enables individuals to see the relationship between events in their personal lives and events in their society.

2. Which of the following is *not* one of the uses of the sociological imagination?
   a. Seeing the interplay of self and society.
   b. Capacity for creating new aspects of culture not thought of by others.
   c. Ability to question aspects of social life most people take for granted.
   d. Capability of understanding the social forces that shape daily life.

3. _____ sociology places human needs and goals at its center.

Answers: 1. sociological imagination 2. b 3. Humanist

4. Training in criminology is sought by agencies dealing with criminal justice, probation, and juvenile delinquency.
5. Courses in social psychology, by providing knowledge of human motivation and attitudes, are valuable for sales, marketing, and advertising, as well as for work involving counseling.

These jobs only scratch the surface; there are many other careers that students of sociology are well trained to pursue. Consider this selected list of occupations: manager, executive, college placement officer, community planner, employment counselor, foreign service worker, environmental specialist, guidance counselor, health planner, journalist, labor relations specialist, marketing researcher, public relations supervisor, research analyst, writer, and editor (American Sociological Association, 1997).

It is my hope that what has been said so far has created an awareness on your part for the need to learn more about sociology. In addition to introducing you to the unique nature of sociology, the next section should make the uses of sociology even more clear to you.

THE NATURE OF SOCIAL SCIENCE

I have attempted to provide you an initial understanding of the nature of sociology by elaborating on some of the unique dimensions and uses of the sociological perspective. Another way to promote your understanding of sociology is to distinguish it from the other social sciences.

The identification of what we now call the social sciences is a relatively recent occurrence. It was not until the nineteenth century that Auguste Comte, the founder of sociology, formulated the term *sociology*. Comte's ideas about sociology as the one comprehensive social science were embedded in his philosophical ideas about psychology, theology, economics, ethics, and politics.

Gradually, scholars have developed the various social sciences to the point that they can be legitimately distinguished. Despite genuine differences among the social sciences, however, they share enough in common to overlap with one another at many points.

Anthropology

The social science most related to sociology is anthropology. Historically, anthropologists have concentrated on the study of the nonliterate or "primitive" societies, whereas sociologists have focused on the complex, industrial societies of Western civilization. Because anthropologists are interested primarily in small nonindustrial societies, they tend to study a culture in its entirety. This holistic approach cannot be used by sociologists studying large modern societies. Rather, sociologists investigate various aspects of modern societies, such as political revolutions or the status of women in the workplace vis-à-vis men.

Psychology

Psychology focuses on the development and functioning of mental-emotional processes in human beings. Whereas sociology and anthropology both concentrate on human groups, psychology takes as its domain the individual. Sociology and psychology share some common interest in an aspect of behavioral science called social psychology, a field that takes interactions among individuals and groups as its focus. A social psychologist, for example, would be interested in the ways in which groups promote conformity among their individual members.

Economics

Economics is the study of the ways in which people produce, distribute, and consume goods and services. It deals with such matters as the impact of interest rates on the flow of money and the effects of taxation on consumption. Economics uses highly developed mathematical models to predict changes in economic indicators. Although these economic models are not always correct, they are sophisticated, and they provide economists with greater predictive power than is possessed by most other social scientists. Sociology and economics merge in an area of study called economic sociology, which concentrates on the interrelationships among the economic and noneconomic aspects of social life. An economic sociologist would be interested, for example, in the relationship between the extent of industrialization in a particular society and the degree of workers' involvement in union activity.

Political Science

Political science traditionally has studied the organization, administration, history, and theory of government. Political scientists have been concerned, for example, with voting patterns and participation in political parties. In recent years, however, political science has moved away from its preoccupation with the workings of government toward the study of political behavior in its broadest sense. This trend has inevitably brought closer together the interests of political scientists and sociologists specializing in a subfield known as political sociology. Both political scientists and political sociologists share an interest in the social interaction that occurs within the political institution and between the political institution and other institutions, including economic and educational institutions. Both conduct research on the location and use of power, the process of political socialization, the functioning of special interest groups, and the workings of political protest.

History

The academic discipline of history is concerned with the study of past events of human beings. As a field of study, history currently is sharply divided into two camps (Landes and Tilly, 1971). Humanistic historians believe that history cannot use the methods of the social sciences. They view any attempts at scientific generalization as the simplification of complex, unique human experiences rooted in specific times and places. Description, they argue, is the only legitimate function of the historian. Social scientific historians, in contrast, assume the existence of uniform, recurring patterns of human behavior that transcend time and place. The role of the historian, they contend, is to identify and verify these patterns. In recent years, the relationship between sociology and history has become more mutually beneficial. Sociology is currently more open to the use of historians' work in their attempts to document recurring patterns and provide explanations of historical reality. More historians are now committed to the collection

and analysis of primary sources with an eye to establishing recurrent and universal historical patterns.

Social Sciences and the Family

Examining a specific topic from the viewpoint of each of these social science disciplines will further clarify each discipline's unique approach. The family can serve as an illustration.

- *Sociologists.* Some sociologists investigating the family focus on the social causes of divorce. Factors thought to contribute to higher divorce rates in American society include early marriage, poor mate selection, economic prosperity, increased participation of women in the labor force, and the weakening of the stigma formerly associated with divorce. (See Chapter 11, "Family.")
- *Anthropologists.* An interest of some anthropological researchers is whether the nuclear family of married biological parents and their legitimate offspring exists universally in human societies. This topic can be studied by examining documented evidence on as many societies as possible and by attempting to find a society or societies without a nuclear family form (Leach, 1982).
- *Psychologists.* Some psychologists are interested in studying familial retardation—cases of mental retardation that tend to appear in poor families with no known organic basis. Familial retardation is traced largely to an impoverished environment, including inferior nutrition and medical care and the relative absence of early emotional support and intellectual stimulation (Coon, 1997).
- *Economists.* Economists use the family as the unit of analysis for the distribution of income. For example, currently, the bottom 60 percent of American families receive less than one-third of the nation's total income (Baumol and Blinder, 1996).
- *Political scientists.* The relationship between voting and the family is a topic of research for political scientists. Political scientists, for example, have found that although voting on election day is done by private ballot, voting is primarily a group phenomenon. The family is the most important influence in forming political party identification and determining the voting behavior of its members. Political attitudes and behaviors are shaped, often unintentionally, by family members; and members of the same family exist in similar economic, religious, social class, and geographical environments (Patterson, 1993).
- *Historians.* The relationship between slavery and the family has been the subject of research by

historians. Using slavery in the United States as a case study, a controversial study by historian Henry Gutman (1983) presents evidence from twenty-one urban and rural communities in the South between 1865 and 1880 to challenge the long-standing assumption that slavery destroys the family structure. For example, Gutman concluded that a majority of African American households during the slavery era had a father and mother present. In short, African American slave families tended to be stable and intact.

EARLY EUROPEAN ORIGINS OF SOCIOLOGY

Sociology is a relatively new science, emerging as a distinct area of study in the late nineteenth century (Turner, 1993). Moreover, it is only during the past half century that sociology has taken its place as a legitimate social science.

Nineteenth-century Europe was torn by social change and controversy. The old order of the Middle Ages—which had been based on social position, land ownership, church, kinship, community, and autocratic political leadership—was crumbling under the social and economic influences of the Industrial Revolution and the French Revolution. Philosophers had long been speculating about human society and the human species, but the phenomena of industrialism and democracy forced them to take a less abstract look at the problems surrounding them. The change from farming to factory life, and the loss of the stabilizing ties people had had with their churches and small communities caused intellectuals of that time to focus more closely on social order and social changes. In an important sense, the rise of sociology was a conservative reaction to the social chaos of the nineteenth century. This social disorder and its consequences led Auguste Comte and others to grapple with the restoration of a sense of community, a sense of social order and predictability (Nisbet, 1966; Mazlish, 1993). This signaled the beginnings of sociology—a field of study distinguishable from philosophy, natural sciences, and other social sciences such as economics, psychology, anthropology, political science, and history. The central ideas of the major pioneers of sociology, all of whom were European, will shed further light on the origins of the field (Hughes, Martin, and Sharrock, 1995; Camic, 1998).

Auguste Comte

Auguste Comte (1798–1857) was a Frenchman born to a government bureaucrat. Comte had some difficulty balancing his genuine interest in school and his rebellious nature. In fact, his protest against the examination procedures at the *Ecole Polytechnique* (of the stature of M.I.T.) led to his dismissal. A year later Comte became secretary to the famous philosopher Henri Saint-Simon. Most of Comte's significant ideas were reflected in the writings he and Saint-Simon did. Differences between Comte and his mentor slowly eroded their relationship, which was finally broken in an argument over whose name was to appear on a new publication.

What were Comte's major ideas?   Comte, generally recognized as the founder of sociology, took the betterment of society as his main concern. Like many philosophers of his day, Comte was both disturbed and fascinated by the social disorder created by the French Revolution.

If societies were to advance, Comte believed, social life had to be studied scientifically. Because no science of society existed, Comte attempted to create one himself. One of his enduring contributions is the idea that sociology should rely on **positivism;** that is, it should use observation and experimentation, methods of the physical sciences, in the study of social life. When Comte wrote that sociology should rely on positivism, he meant that sociology should be a science based on facts, on things about which we can be "positive."

According to Comte, society tends to evolve through three stages: the theological, the metaphysical, and the positivistic. Sociologist-priests would be the rulers during the positivistic stage. He also distinguished between **social statics**—the study of stability and order—and **social dynamics**—the study of change. This distinction is still being debated by modern sociologists.

FEEDBACK

1. Match the following fields of study and aspects of family life.

____	a. anthropology	(1)	Distribution of income and the family.
____	b. sociology	(2)	Effects of slavery on family stability.
____	c. history	(3)	Relationship between voting and the family
____	d. political science	(4)	Effects of early marriage on divorce
____	e. economics	(5)	Link between early childhood emotional support and familial retardation.
____	f. psychology	(6)	Universality of the nuclear family.

Answers: 1. a. (6) b. (4) c. (2) d. (3) e. (1) f. (5)

Comte was almost prevented from writing his masterwork, *Positive Philosophy,* as a result of his bitter breakup with Saint-Simon. (It was in this work that Comte elucidated his principle of "cerebral hygiene." To prevent others from polluting his thoughts, Comte stopped reading.) After parting ways with Saint-Simon, Comte was never able to find another well-paid position and was forced to live by tutoring and testing in mathematics, along with various other activities. It was only when a few of his followers invited him to prepare some private lectures that Comte came to present his ideas to the world.

Comte died before many beyond his private circle came to appreciate his work. Once recognized, however, his belief that sociology could promote social progress was widely adopted by other European scholars whose own societies were feeling the impact of change unleashed by industrialism and the democratic revolution. Many nineteenth-century scholars came to embrace Comte's ideas, but others did not.

Harriet Martineau

Although the role of women in the history of sociology has traditionally been excluded, the current feminist movement in the United States has opened a crack in the patriarchal foundation of the field. Through that crack is coming overdue acknowledgment of women who were also present at the creation of sociology. Harriet Martineau (1802–1876) is one of those women who deserves to be included in the history of sociological thought. Some believe her to be the first woman sociologist (Rossi, 1973; Hocker-Drysdale, 1994); others believe she should be acknowledged as a founding mother of sociology (Riedesel, 1981).

Harriet Martineau was an Englishwoman born into a solidly middle-class home. Her writing career, which included fiction as well as sociological work, began in 1825 after the Martineau's family textile mill was lost to a business depression. Without the family income, and with few immediate prospects for marriage following a broken engagement, Martineau was forced to seek a dependable source of income for self-support (Webb, 1960; Fletcher, 1974; Pichanik, 1980). She was a popular writer of celebrity stature, whose work outsold Charles Dickens's.

What did Martineau contribute to the early development of sociology? Martineau is best known among sociologists for her translation (and condensation) of Comte's *Positive Philosophy.* Accomplished with Comte's approval, Martineau's translation remains the most readable one (Turner, Beeghley, and Powers, 1989). Familiarity through this avenue should not overshadow her own original contributions in the areas of research methods, political economy, and feminist theory.

Her book *How to Observe Manners and Morals* (1838) is the first book on methodology in sociology (Lipset, 1962). The research principles she emphasized were the use of a theoretical framework to guide social observation, the development of predetermined questions to be used for gathering information, objectivity, and representative sampling.

Martineau followed her own advice while conducting a comparative study of European and American society. Although not as well known as Alexis de Tocqueville's *Democracy in America* published at about the same time (1835), Martineau's *Society in America* (1837) remains her most widely read book today. Martineau compared America favorably with England, but believed that America had not lived up to its promise of democracy and freedom for all its people. Slavery and the domination of women revealed to her a significant gap between American ideals and social practices (Terry, 1983).

In *Society in America,* Martineau established herself as a pioneering feminist theorist who, like other early feminists, saw a link between slavery and the oppression of women. Consequently, she was a strong and outspoken supporter of the emancipation of both women and slaves. Through her penetrating analysis of the subordination of American women, she stands as an important precursor of contemporary feminist theory. Martineau identified economic dependency as the linchpin of female subordination. Doors to the economic institution were closed. The oppression of women, in short, is due to sociological forces; subordination of women is part of the structure of society.

Harriet Martineau (1802–1876)

Herbert Spencer

Herbert Spencer (1820–1903), the sole survivor of nine children, was born to an English schoolteacher. Due to continual ill health, Spencer was taught exclusively by his father and uncle, mostly in mathematics and the natural sciences. Because of his poor background in Latin, Greek, English, and history, Spencer did not feel qualified to enter Cambridge University, his father's alma mater. His subsequent career was a mixture of engineering, drafting, inventing, journalism, and writing.

What do we remember most about Herbert Spencer? Although he publicly denied Comte's influence on his work, Herbert Spencer also developed theories on social statics and social change. He used an organic analogy to explain social stability: Like humans, a society is composed of interrelated parts that work together to promote its well-being and survival. An individual has a brain, a stomach, a nervous system, limbs; a society has an economy, a religion, a state, a family form. Just as the eyes and the heart make essential contributions to the functioning of the human body, the legal and educational systems are crucial for a society's functioning.

Spencer's theory of change was based on a combination of this organic analogy and Charles Darwin's theory of evolution. An advocate of evolutionary change, Spencer thought that social change led to progress, provided that people did not interfere. Why? Because societies, like people, pass through natural stages of growth from birth to decay. If left alone, natural selection ensures that inferior social arrangements perish and that superior ones survive. On these grounds, Spencer opposed social reform. The poor deserve to be poor, the rich to be rich. Society profits from allowing individuals to find their own social class level without help or hindrance; to interfere with the existence of poverty—or the result of any natural process—would be harmful to society.

When Spencer visited America in 1882, he was warmly greeted, particularly by the captains of industry. After all, his ideas of noninterference with natural processes and survival of the fittest provided a respectable justification for their selfish, ruthless, and often shady economic practices. As advocacy for government intervention to cushion social ills increased, however, Spencer's ideas began to slip out of fashion. Spencer reportedly died with a sense of failure.

Karl Marx

The most influential of the nineteenth-century intellectuals was Karl Marx (1818–1883), who was born in Germany to parents whose lineage included long lines of rabbis on both sides. Although he did not consider himself a sociologist (his doctorate was in philosophy), his ideas have had an enormous influence in shaping the field. Preferring social activism to the abstractness of philosophy, Marx's life was guided by his conviction that social scientists should seek to change the world rather than merely observe it. In part through his long association with Friedrich Engels, the son of an industrialist, Marx's concern for democracy and humanism was channeled toward a concern for the poverty and inequality suffered by the working class. Forced by political pressures, Marx moved from Germany to France and finally settled in London, where he devoted the remaining decades of his life to a systematic analysis of capitalism.

Unfortunately, Marxism is equated in the minds of many today with communism, a burden his work should not have to bear. It is likely that he would be as discouraged with the practice of communism in the twentieth century as he was with the utopian communists of his time. In fact, at one point Marx was so unhappy with others' erroneous interpretations of his work that he disavowed being a Marxist himself.

What is the legacy of Marx? Herbert Spencer and Karl Marx had different conceptions of society and social change. Spencer depicted society as a set of interrelated parts that promoted its own welfare. Marx described society as a set of conflicting groups with different values and interests; the selfishness and ruthlessness of capitalists harmed society. Spencer saw progress coming from only noninterference with natural, evolutionary processes. Marx, too, believed in an unfolding, evolutionary pattern of social change. He envisioned a linear progression of modes of production from primitive communism through slavery, feudalism, capitalism, and communism. However, he was also convinced that the transformation from capitalism to communism could be speeded up through planned revolution. His political objective was to explain the workings of capitalism in order to hasten its fall through revolution, but he believed capitalism would eventually self-destruct anyway because of its inherent contradictions.

Although recognizing the presence of several social classes in nineteenth-century industrial society—farmers, servants, factory workers, craftspeople, owners of small businesses, moneyed capitalists—Marx predicted that all industrial societies ultimately would contain only two social classes—the **bourgeoisie** (those who own the means for producing wealth in industrial society) and the **proletariat** (those who labor for the bourgeoisie at subsistence wages). For Marx, the key to the unfolding of history was **class conflict**—conflict between those controlling the means for producing wealth and those laboring for them. Just as the slave owners had been overthrown by the slaves and the landed aristocracy revolted against by the peasants, the

Karl Marx (1818–1883)

From Judaic family training and an intimate environment Durkheim gained a deep and permanent concern for universal moral law and the problems of ethics, a concern that was not combined with any indulgent sense of humor. Indeed, he was eminently without humor and somewhat "heavy-handed" (Simpson, 1963:1).

Always a studious individual, Durkheim eventually became the first French academic sociologist.

What were Durkheim's foremost contributions?
According to Durkheim, social order exists because of a broad consensus on the nature of values and institutions (government, family, religion) among members of a society. This consensus is especially characteristic of nonliterate societies based on **mechanical solidarity**—social unity that comes from a consensus of values and norms, strong social pressures for conformity, and dependence on tradition and family. Durkheim, witnessing the social upheaval brought on by the industrial and democratic revolutions, attempted to describe how social order was achieved in complex, industrial society. In modern society, he contended, social order is based on **organic solidarity**—social unity based on a complex of highly specialized roles that makes members of a society dependent on one another. For example, people in industrial society depend on one another to provide goods and services.

Although Comte and other early sociologists emphasized the need to make sociology scientific, their own research methods were not very scientific. They did, however, influence later sociologists to replace arm-chair speculation with careful observation, to engage in the collection and classification of data, and to use data for formulating and testing social theories. One of the most prominent of these later sociologists was Durkheim, who first introduced the use of statistical techniques in the study of human groups in his groundbreaking research on suicide. (See *Sociological Imagination* at the beginning of this chapter and *Doing Research* in Chapter 2, "Emile Durkheim—The Study of Suicide.")

In his study, Durkheim demonstrated that suicide involves more than individuals acting alone. By showing that suicide rates vary according to group characteristics—the suicide rate is lower among Catholics than Protestants and lower among married than single persons—Durkheim convincingly supported the idea that social life must be explained by social factors rather than by individualistic ones.

capitalists would fall to the wage workers. Out of the conflict would emerge a classless society without exploitation of the powerless by the powerful.

According to the principle of **economic determinism** (an idea often associated with Marx), the nature of a society is based on the society's economy. A society's economic system determines the society's legal system, religion, art, literature, and political structure. Marx himself did not use the term *economic determinism;* the term was applied to his ideas by others, no doubt a consequence of his concentration on the economic sphere in capitalist society. The mistake interpreters have made is to assume that because Marx perceived the economic institution as having primacy in capitalist society, he believed that all societies operated according to the same principles. Moreover, Marx recognized that even in capitalist society, economic and noneconomic institutions mutually affect each other. Marx even wrote that sometimes the economy "conditions" rather than "determines" the historical process in capitalist society.

Emile Durkheim

Emile Durkheim (1858–1917) was born into the home of a French rabbi. Durkheim originally intended to follow the family tradition of becoming a rabbi. Later experiences led him to become an agnostic, although he retained an intellectual interest in religion throughout his life. In fact, one of Durkheim's major concerns was social and moral order. This emphasis was doubtless related to his upbringing in the home of a rabbi:

Max Weber

Max Weber (1864–1920) was the eldest son of a well-to-do German lawyer and politician. Weber's father was a man rooted in the pleasures of this world. His mother, in stark contrast, was a strongly devout Calvinist who rejected the pleasure seeking of her husband. While Weber's father concentrated on life on earth, Weber's mother was preoccupied with her future salvation. Weber was affected psychologically by the resulting family discord; he eventually suffered a complete mental breakdown from which he recovered to do some of his best work. The influence of his mother is clearly reflected in his work, especially in the sociology of religion and in one of his most famous books, *The Protestant Ethic and the Spirit of Capitalism* (Ritzer, 1996).

What did Weber contribute to the development of sociology? As a university professor trained in law and economics, Weber wrote on a wide variety of topics, including the relationship between capitalism and Protestantism, the nature of power, the development and nature of bureaucracy, the religions of the world, and the nature of social classes. Through the quality of his work and the diversity of his interests, Weber has been credited with being the single most important influence on the development of sociological theory.

Unlike Durkheim, Weber did not favor studying human beings as though they were physical things. Weber's approach was a subjective one—humans act on the basis of their own understanding of a situation. Consequently, sociologists must discover the personal meanings, values, beliefs, and attitudes underlying behavior. Understanding the subjective intentions of people could be accomplished through what Weber called the method of *verstehen*—understanding the behavior of others by putting oneself mentally in their place. This perspective was apparently rooted in Weber's personal nature. According to Weber's wife, Weber "showed throughout his life an extraordinary appreciation for the problems other people faced and for the shades of mood and meaning that characterized their outlook on life" (Bendix, 1962:8).

Since Weber and Marx were not intellectual contemporaries, and since Weber disagreed with Marx on many points, it is often said that Weber was debating with Marx's ghost (Wiley, 1987). It was out of honest differences on issues, not disrespect, that Weber took his positions against some of Marx's ideas. Although Marx was, at times, openly political in advocating his ideas, Weber stressed objectivity. He strongly counseled sociologists to conduct **value-free research**—research in which personal biases are not allowed to affect its conduct and outcome. Marx saw religion as retarding social change, whereas Weber believed religion could promote change. Also, whereas Marx emphasized the role of the economy in social stratification, Weber's analysis involved several dimensions.

One should not conclude from these differences, however, that Weber's approach to sociology is antagonistic to Marx's. Actually, Weber may be thought of as

Max Weber (1864–1920)

the most prominent of the second generation of German scholars concerned with power and conflict in society. Although there are clearly significant differences in detail, both Marx and Weber took the problem of capitalism as their unifying theme (Collins, 1994).

SOCIOLOGY IN AMERICA

Although the early development of sociology occurred in Europe, the maturation of sociology has taken place largely in the United States. Because sociology has become a science largely through the efforts of American sociologists, it is not surprising that the majority of all sociologists are American. Sociological writings in English are used by sociologists throughout the world, reflecting the tremendous influence American sociologists have had globally since World War II.

In 1893, the first department of sociology was established at the University of Chicago. A few years later, in 1905, the American Sociological Society (since renamed the American Sociological Association) was founded after a split from the American Economic Association. It was not until 1930 that sociology became a department at Harvard University. Presently, almost all colleges and universities have sociology departments.

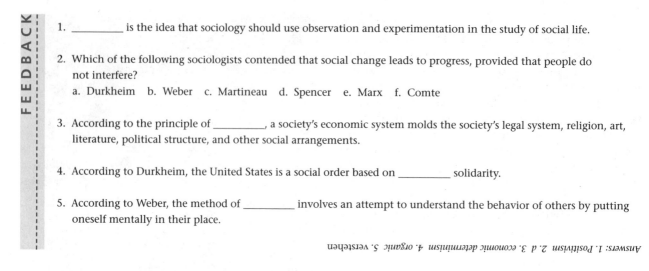

1. _____ is the idea that sociology should use observation and experimentation in the study of social life.

2. Which of the following sociologists contended that social change leads to progress, provided that people do not interfere?
   a. Durkheim   b. Weber   c. Martineau   d. Spencer   e. Marx   f. Comte

3. According to the principle of _____, a society's economic system molds the society's legal system, religion, art, literature, political structure, and other social arrangements.

4. According to Durkheim, the United States is a social order based on _____ solidarity.

5. According to Weber, the method of _____ involves an attempt to understand the behavior of others by putting oneself mentally in their place.

Answers: 1. Positivism 2. d 3. economic determinism 4. organic 5. verstehen

American sociologists have been heavily influenced by the European originators of sociology. This is, of course, partly because early American sociologists had to rely on the writings of Comte, Martineau, Spencer, Marx, Durkheim, Weber, and others. Where would they have turned in their efforts to develop sociology but to the existing body of literature published by the Europeans? There is, however, another reason for the strong European influence. Just as sociology in Europe was born during a time of rapid, disruptive change, sociology emerged in the United States during the period of rapid urbanization and industrialization following the Civil War. It is little wonder that Lester Ward (1841–1913), the founder of American sociology, picked up on Comte's conviction that the study of sociology could promote social progress. Like Comte, Ward believed that industrial, urban society could be improved through sociological analysis.

From its founding in 1893 to World War II, the sociology department at the University of Chicago stood at the forefront of American sociology. Housed at the University of Chicago were several exceptional minds. George Herbert Mead, John Dewey, and William I. Thomas pioneered the study of human nature and personality. Urban social problems such as prostitution, slums, and crime were studied extensively by Robert E. Park and Ernest Burgess. The study of social and cultural change was championed by William F. Ogburn. On the whole, the so-called Chicago School became closely linked with the idea of social reform (Bulmer, 1984; Lengermann, 1988; Fine, 1995).

After World War II, eastern universities such as Harvard and Columbia, midwestern universities such as Wisconsin and Michigan, and western universities such as Stanford and the University of California at Berkeley emerged as leading institutions fostering a diversity of sociological emphases. Sociologists such as George A. Lundberg underscored the application of the research methods of physical scientists in order to make sociological research more sophisticated and scientific. Lundberg was known as a neo-positivist for his revival of Comte's concern that sociology be a science. A sociologist at Harvard, Talcott Parsons, returned to the development of general theories of society à la Weber, Durkheim, and other European scholars. Parsons has been one of the leading proponents of functionalism (discussed later in the chapter), one of several dominant theoretical perspectives in sociology today. Robert K. Merton, a Columbia University sociologist and a student of Parsons, advocated "theories of the middle range," or more specific theories, over general theories and stressed the importance of empirical research. By the 1970s, the number of practicing sociologists grew dramatically. As a result, talented and influential sociologists are no longer located in only a few elite colleges and universities.

In their quest toward developing general sociological theories and establishing sociology as a science, sociologists between World War II and the 1960s nearly lost sight of the idea that sociology could help solve social problems. Social reform emerged again during the turbulent sixties and remains an important part of sociology today. "Humanist" sociologists, who owe much to sociologist C. Wright Mills, today believe that sociology cannot be ethically neutral with respect to important social issues. Rather, they argue, sociologists have an inherent obligation to be critical of existing social arrangements and to work for social transformation (Lee, 1978; Scimecca, 1987; Giddens, 1987, 1997).

THEORETICAL PERSPECTIVES

In *Beyond Good and Evil*, nineteenth-century philosopher Friedrich Nietzsche wrote,

It is more comfortable for our eye to react to a particular object by producing again an image it has often produced before

than by retaining what is new and different in an impression (Nietzsche, 1987:97, originally published in 1886).

This fundamental point about the power of a perspective to blind one to other possibilities is graphically illustrated in two drawings psychologists often use to illustrate the concept of perception. One is a picture of an old witchlike woman, the other a picture of a rabbit. If you stare at the old woman long enough, she becomes a beautiful young woman with a feather boa around her neck. If you stare at the rabbit long enough, a duck appears. If you continue to look, both drawings return to their original form. You cannot, however, see the old woman and the young woman or the rabbit and the duck at the same time.

Which is real, the old woman or the young woman, the rabbit or the duck? It depends on your focus. So it is with any perspective. The perspective you take influences what you see. One perspective emphasizes certain aspects of an event; another perspective accents different aspects of the same event. When a perspective highlights certain parts of something, it necessarily places other parts in the background. (For a firsthand experience with the difference a perspective makes, study Figure 1.1.)

In science, a more technical term for a **theoretical perspective** is "paradigm": a set of assumptions (in this case regarding the workings of society) that is accepted as true by its advocates and that stimulates coherent traditions of scientific research (Kuhn, 1996). Theoretical perspectives or paradigms are broader than **theories**: logically interrelated sets of propositions that can be tested against reality. Although a theoretical proposition generated by a paradigm can be tested, the assumptions of a paradigm cannot. A paradigm is

accepted as true and is used as a way of viewing selected aspects of society. Paradigms, however, are used to generate theoretical propositions whose accuracy can be tested scientifically. For example, one of the assumptions of the conflict perspective (to be discussed shortly) is that a society rests on the constraint of some of its members by others. Advocates of the conflict perspective do not attempt to test the truth of this assumption; they accept its basic legitimacy. An advocate of the conflict perspective might, however, examine an aspect of social life with this assumption in mind. He or she might, for example, attempt to test the theoretical proposition that those who hold power over others will exercise that power by demonstrating through research that prison guards coerce prisoners to perform illegal acts or that a strong interest group such as the National Rifle Association uses its political power to influence the votes of members of Congress on a piece of gun-control legislation.

Sociology has three fundamental theoretical perspectives: functionalism, conflict, and symbolic interactionism. Each of these perspectives provides a different slant on human behavior in groups. The exclusive use of any one of them prevents one's seeing other aspects of group life, just as one cannot see the old woman and the young woman at the same time. All three perspectives together, however, allow us to see most of the important dimensions of social life that are of interest in sociology.

These three perspectives can be placed within the context of the macrosociology-microsociology distinction. Functionalism and the conflict perspective are best viewed from the macro level. Symbolic interactionism deals with social phenomena at the micro level.

 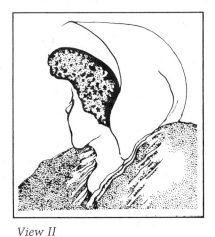 

View I   *View II*   *View III*

Figure 1.1 The Importance of Perspectives
These three pictures illustrate the importance of perspective. That is, our perception of things is heavily influenced by the perspective we bring to them. Decide whether you want to look at View I or View II (cover all other views). After examining View I or View II,

look at View III. If you decided to look at View I, you should see the old woman in View III. If you decided to look at View II, you should see the young woman in View III. Can you see both? Try it on your friends.

Macrosociology: Functionalism

Functionalism emphasizes the contributions (functions) made by each part of a society. It focuses on social integration, stability, order, and cooperation. "Parts" of a society include the family, the economy, and religion. For example, the family contributes to society by providing for the reproduction and care of its new members. The economy contributes by dealing with the production, distribution, and consumption of goods and services. Religion contributes by helping people focus on beliefs and practices related to sacred things.

Who were the originators?   One of the major pioneers of this perspective was, of course, Spencer, who compared living organisms to societies. Although sociologists no longer use the organism-society analogy, they still see society as a system of interrelated parts. Durkheim, a contemporary of Spencer's, greatly contributed to the development of this theoretical perspective. Some of the strongest supporters of functionalism have been American sociologists Talcott Parsons and Robert Merton.

What are the assumptions of functionalism?   One assumption of functionalism views the parts of a society as organized into an integrated whole. Consequently, a change in one part of a society leads to change in other parts. A major change in the economy, for example, may change the family. This is precisely what happened as a result of the Industrial Revolution. The need for a large farm labor force (fulfilled by having many children) disappeared as industrialization proceeded, and family size decreased.

Functionalists realize that societies are not perfectly integrated. A certain degree of integration is necessary for the survival of a society, but the actual degree of integration varies.

According to a second assumption of functionalism, societies tend to return to a state of stability or equilibrium after some upheaval has occurred. A society may undergo change over time, but functionalists believe that it will return to a state of stability by incorporating these changes so that the society will again be similar to what it was before any change occurred.

Because a society both changes and maintains most of its original structure over time, functionalists refer to a **dynamic equilibrium**—a constantly changing balance among its parts. The student unrest on college and university campuses during the late 1960s illustrates the concept of dynamic equilibrium. The activities of student radicals did create some changes: The public no longer accepts all American wars as legitimate, universities are now more responsive to students' needs and goals, and the public is more aware of the importance of environmental protection. These changes, however, have not revolutionized American society. They have been absorbed into it, leaving it only somewhat different from the way it had been before the student unrest.

According to functionalism, most aspects of a society have evolved to promote the society's survival and welfare. It is for this reason that all complex societies have economies, families, governments, and religions. If these elements did not contribute to a society's well-being and survival, they would not survive.

According to Robert Merton (1996), some functions—**manifest functions**—are intended and recognized and others—**latent functions**—are unintended and unrecognized. Moreover, not all elements of a society make a positive contribution; those that have negative consequences result in **dysfunction.** One of the manifest functions of reliance on rules in a bureaucracy is the creation of an environment espousing equitable treatment for everyone. A latent (and positive) function of bureaucratic rules is the promotion of equal occupational opportunity for employees. Dysfunctions of bureaucratic rules include rigidity, inefficiency, and impersonality.

Of course, what is negative for one part of a society may be positive for another. Poverty obviously has negative consequences for the poor, but Herbert Gans (1971) believes that poverty has some benefits for other parts of society. Without the poor, who would buy substandard products such as day-old bread and overripe fruit, who would hire over-the-hill lawyers and doctors, and where would social workers and penologists find work?

According to functionalism, there is among most members of a society a consensus on values and interests. Most Americans, for example, agree on the desirability of democracy, success, and equal opportunity. This high degree of consensus, say the functionalists, accounts for the great degree of cooperation found in any society.

One of the major criticisms of functionalism is its neglect of social change and its conservative tendency to legitimatize the status quo. The conflict perspective, in contrast, takes social change as a focal point.

Macrosociology: The Conflict Perspective

The **conflict perspective** emphasizes conflict, competition, change, and constraint within a society (Schellenberg, 1982; Giddens, 1987, 1997). Although this theoretical perspective was not very popular among most American sociologists until the 1960s, its roots go back as far as those of functionalism.

Who were the originators?   Karl Marx, Max Weber, and Georg Simmel (1858–1918) are three German sociologists whose work underlies the conflict perspective. As noted earlier, Marx contended that the nature of a society is based on the way people meet their basic needs and that class conflict is inevitable in all capital-ist economies. Modern conflict advocates such as C. Wright Mills, Ralf Dahrendorf, and Randall Collins do not limit themselves to class conflict. They broaden Marx's insights to include conflict among any segments of a society, a point Weber had made much earlier (Collins, 1994). For example, conflict exists between Republicans and Democrats, unions and management, industrialists and environmentalists.

Understanding the conflict perspective is easy when you understand functionalism, because the assump-tions behind these two perspectives are often the reverse of each other. (See Table 1.1.)

What is the role of conflict and constraint in the con-flict perspective?   Assuming a basic agreement on val-ues and interests within a society, functionalism emphasizes the ways people cooperate to reach com-mon goals. The conflict perspective focuses on the inevitable disagreements among various segments of a society or between societies. Segments compete and conflict as they attempt to preserve and promote their own special values and interests.

Advocates of the conflict perspective, then, depict social life as a contest. Their central question is, Who gets what? Their answer: Those with the most **power**— the ability to control the behavior of others even against their will—get the largest share of whatever is considered valuable in a society. Those with the most power have the most wealth, the most prestige, and the most privileges. Because some segments have more power than others, they are able to constrain the less powerful to do as they wish. Income tax laws, for exam-ple, are manipulated most often to benefit the wealthy and powerful. Citizens for Tax Justice, a citizens' watch-dog organization, reported that forty large American corporations paid no federal taxes in 1984. In fact, these forty firms, earning a combined $10.4 billion in

Conflict theorists can see no function for rural poverty such as this. Poor people, they contend, end up at the bottom of the strati-fication structure, in part, because they lack power to influence social policy.

profits, received a total of $657 million in tax refunds. In 1989, 1,081 Americans with incomes over $200,000 paid no federal income taxes whatever (Ehrenreich, 1994). Although the lowest 20 percent family income category in the United States gained 5 percent in after-tax income between 1980 and 1995, the top 5 percent gained 46 percent. (See Table 1.2.)

Tax reform in 1986 plugged some tax loopholes, but examples of wealthy and powerful interests getting their way are numerous (Birnbaum and Murray, 1988). In 1985, the House of Representatives and the Senate passed legislation continuing a federal sugar subsidy that placed an average of $250,000 in the pockets of some 12,000 American sugar growers. As a result, American consumers had to pay among the highest

TABLE 1.1	Assumptions of Functionalism, the Conflict Perspective, and Symbolic Interactionism	
Functionalism	**The Conflict Perspective**	**Symbolic Interactionism**
1. A society is a relatively integrated whole.	1. A society experiences inconsistency and conflict everywhere.	1. Human beings act according to their own interpretations of reality.
2. A society tends to seek relative stability or dynamic equilibrium.	2. A society is continually subjected to change everywhere.	2. Subjective interpretations are based on the meanings we learn from others.
3. Most elements of a society con-tribute to the society's well-being and survival.	3. Elements of a society tend to con-tribute to the society's instability.	3. Human beings are constantly inter-preting their own behavior as well as the behavior of others in terms of learned symbols and meanings.
4. A society rests on the consensus of its members.	4. A society rests on the constraint and coercion of some of its mem-bers by others.	

Source: The assumptions of functionalism and the conflict perspective are based on the work of Ralf Dahrendorf, "Toward a Theory of Social Conflict," *Journal of Conflict Resolution 2* (June 1958):174. Adapted by permission of Sage Publications, Inc., ©1958.

TABLE 1.2	Average Income Before and After Tax, 1980 and 1995 (in constant 1995 dollars, adjusted for changes in family size)					
Family Income Category	**Pretax Incomes**			**After-Tax Incomes**		
	1980	1995	% Change	1980	1995	% Change
Lowest 20%	11,872	11,265	−5%	7,382	7,783	+5%
Second 20%	25,843	25,955	0%	17,477	18,161	+4%
Middle 20%	39,041	40,637	+4%	26,968	28,365	+5%
Fourth 20%	54,164	59,457	+10%	37,816	41,337	+9%
Highest 20%	91,196	119,453	+31%	61,168	77,140	+26%
Top 5%	129,642	204,863	+58%	84,972	123,839	+46%

Source: U.S. Bureau of the Census. "Historical Income Tables–Families; Table F-3; Mean Income Received by Each Fifth and Top 5 Percent of Families, by Race and Hispanic Origin of Householder: 1966 to 1995." April, 1997. At http://www.census.gov/hhes/income/histinc/f03.html.

Internet Link: To learn more about historical patterns of income distribution in the United States, you can visit the website of the U.S. Bureau of the Census at http://www.census.gov/.

prices in the world for their sugar—a quantity of sugar costing 85 cents in Washington, D.C., sold for half that amount in Ottawa, Canada. According to evidence, members of the House of Representatives were likely to vote for the bill in direct proportion to the amount of sugar-lobby campaign contributions they had received (Stern, 1988).

How does the conflict perspective explain social change? Because many conflicting groups exist and the balance of power among groups may shift, social change is constant, according to the conflict perspective. For example, the women's movement is attempting to change the balance of power between the sexes. As this movement continues to progress, we will see larger numbers of women in important occupations. More women will be making or influencing decisions in business, politics, medicine, and law. Gender relations will change in other ways as well. More women will choose to remain single, to marry later in life, to have fewer children or none, to divide household tasks with their husbands or roommates. According to the conflict perspective, these changes are the cause and the result of women's gaining true power.

Which theoretical perspective—functionalism or the conflict perspective—is better? Neither; there is no one better macro perspective. Each of these perspectives sheds light on certain aspects of social life. The advantages of one perspective are the disadvantages of the other. Functionalism explains much of the consensus, stability, and cooperation within a society. The conflict perspective explains much of the constraint, conflict, and change.

Because each perspective has captured an essential side of society's nature, their combination or synthesis is a reasonable next step. Some attempts to combine func-

tionalism and the conflict perspective have already been made (Dahrendorf, 1958b; van den Berghe, 1963, 1978). One of the most promising is the attempt to specify the conditions under which conflict and cooperation occur. Gerhard Lenski (1984; Lenski, Nolan, and Lenski, 1995) contends that people cooperate—even share the fruits of their labors—when scarcity threatens their survival. But conflict, competition, and constraint are likely to occur when there is more than enough for everyone. Thus, as a society moves from a subsistence economy to an affluent one, conflict, competition, and constraint increase. Or to cite another example, the more a deprived group questions the legitimacy of its condition, the more likely it is to conflict with privileged groups (Coser, 1987). The civil rights and women's movements are recent examples of groups questioning the legitimacy of the inequalities they live with. (For a more detailed discussion of a theoretical synthesis, see Chapter 18, "Social Change in Modern Society.")

Microsociology: Symbolic Interactionism

Who were the originators? Both functionalism and the conflict perspective deal with large social units and broad social processes—the state, the economy, evolution, class conflict. At the close of the nineteenth century, some sociologists began to alter their focus on the social world. Instead of being totally preoccupied with larger structures, they began to recognize the importance of the ways people interact and relate to one another. Weber and Simmel were the earliest contributors to the theoretical perspective on interactionism that attempts to understand social life from the viewpoint of the individuals involved. Later sociologists, such as Charles Horton Cooley, George Herbert Mead, W. I. Thomas, Erving Goffman, Harold Garfinkel, and Herbert Blumer, developed in greater detail the insight that groups can exist only because their members influence one another's behavior. They created symbolic interactionism, the most influential approach to interactionism.

What is symbolic interactionism? **Symbolic interactionism** is the theoretical perspective that focuses on interaction among people based on mutually understood symbols. A **symbol** is something that stands for or represents something else. Symbols can take an infinite variety of forms, including words, concepts, sounds, facial expressions, and body movements. Symbols are not determined by the things they represent; they are determined by those who create and use them.

> *One cannot tell by looking at an X in an algebraic equation what it stands for; one cannot ascertain with the ears alone the symbolic value of the phonetic compound si; one cannot tell merely by weighing a pig how much gold he will exchange for; one cannot tell from the wave length of a color whether it stands for courage or cowardice, "stop" or "go" (White, 1969:26).*

Because the meanings of things are not inherent in themselves, the meanings must be determined by the people who use them. What does it mean when members of an audience whistle? In one culture whistling may symbolize approval; in another it may signify disapproval. In Latin America, whistling at a sporting event is equivalent to booing in the United States. A bullfighter in Mexico who does not please the crowd will draw derisive whistling.

If people in a group do not share the same meanings for a given symbol, confusion results. If some members of a community interpret the red light on a traffic signal to mean they can proceed whereas others interpret the green light to mean they can proceed, chaos and violence will reign in the streets.

What are the basic assumptions of symbolic interactionism? Herbert Blumer (1969a, 1969b), who coined the term *symbolic interactionism,* has outlined three assumptions central to this perspective. (Refer to Table 1.1.) First, we act according to our own interpretations of reality. That is, we act toward things according to the meanings those things have for us. Is shoulder-length hair more appropriate for women than men? Although in some historical periods long hair has been considered appropriate for women only, in other periods long hair has been the rule for men. During the 1960s in the United States, most American fathers disliked long hair on their sons because of the sloppiness and effeminacy it represented to them. Today, many American fathers wear their hair long and are unhappy with their sons who have short hair with designs shaved in it. Fathers in different eras are not responding to hair length per se but to what it represents to them.

Second, subjective interpretations are based on the meanings we learn from others. We learn the meaning of something by the way we see others acting toward it. American musicians in a Latin American country might misinterpret whistling at the end of their performance and deliver an encore. The audience would probably let them know that their response was inappropriate. Once these musicians had learned this different meaning of whistling, they would be able to respond appropriately the next time an audience whistled.

Third, we are constantly interpreting our own behavior as well as the behavior of others in terms of the symbols and meanings we have learned. Because we can carry on mental conversations with ourselves, we can imagine how others will respond to us before we act. This enables us to guide our actions according to the behavior we expect from others. Thus, our behavior is continually being created on the spot as we interact with others.

What are some more recent developments in symbolic interactionism? In an attempt to understand the interaction of participants in social structures, Erving Goffman introduced **dramaturgy,** an approach that depicts social life as theatre—people are viewed as performers on the stage (Goffman, 1961a, 1963, 1974, 1979, 1983; Lemert and Branaman, 1997). Humans are seen as actors stage-managing their conduct to control the attitudes and responses of others toward them. **Presentation of self,** or "impression management," refers, then, to the ways we, in a variety of social situations, attempt to create a favorable evaluation of ourselves in the minds of others.

Symbolic interactionists are interested in interaction among people based on mutually understood symbols. The two teenage females above are interpreting their own behavior, as well as the behavior of the teenage male, in terms of symbols (facial expressions, words, body language) whose meanings they have learned to share. The teenage male is doing the same thing.

DOING RESEARCH

Frances Fox Piven and Richard Cloward—Conflict Theory and Poor People's Movements

Frances Fox Piven and Richard Cloward researched four poor people's social movements, two originating during the Great Depression of the 1930s and two spawned after World War II. During the 1930s, the Workers' Alliance of America emerged from the protest of the unemployed, and the Congress of Industrial Organizations rose from the insurgency of industrial workers. The post–World War II era brought two poor people's movements— the civil rights movement and the movement of welfare recipients that created the National Welfare Rights Organization.

Little work has been done on the movement of the unemployed in the Great Depression. Analysis of the civil rights movement was based on the direct knowledge of and involvement in the movement on the part of Piven and Cloward, along with the few existing studies of the movement.

All of these protest movements were researched from a conflict perspective. Piven and Cloward contend that power is held by those who control the means of physical coercion and the means of creating wealth. Because institutions, which are in the hands of the elite who hold the power, are extremely efficient in maintaining political docility, Piven and Cloward argue, the lower classes have the chance to push for their own class interests only on rare occasions. Thus, on those rare occasions, the forms the protests take and the effects they can have are largely shaped by the elites. This is what Piven and Cloward mean when they depict social

protest as a socially structured phenomenon. This influence of the larger social context through elites, the researchers argue, normally limits the size and force of the poor people's protests.

According to Piven and Cloward, leaders of lower class protest movements have generally relied on the creation of formal organizations with a mass membership composed of the lower classes themselves. This approach, contend the authors, diminishes the probability of success for two reasons, one external to movements and one internal to them.

Externally, the creation of organizations does not successfully force the elites to make the concessions necessary to generate the resources needed to sustain protest organizations over a sufficiently extended period of time. Elites, because of fear of the excessive political energy of the masses, seem to side with the insurgents, asking for their views and encouraging them to state their griev-

Think about impression management as a beginning college student. Most students come to college concerned about their acceptance by the strangers they are to meet. Consequently, decisions are made (consciously or not) about how students want others to view them. Choices in clothing are made; some students wear blue jeans and T-shirts, whereas others choose a preppie look. Posters and magazines may be prominently displayed to present the desired image. Personal information such as athletic, musical, or sexual prowess will be appropriately leaked.

Performances, according to Goffman's theatre analogy, may have a front and a back stage. College students may behave one way with dates and let down when alone with other campus classmates. Students may behave as expected when with their parents on campus, only to later tell their roommates what they really think. Behavior changes as it is managed in various settings.

Goffman's theoretical approach has stimulated the research of other sociologists. Dramaturgical symbolic interactionists have documented the drama and ritual within big-time college football (Deegan and Stein,

1978) as well as within other areas of social life, including singles' bars and the effects of corruption in politics, athletics, and religion on the American public's view of their social world (Young, 1990).

Ethnomethodology—the study of processes people develop and use in understanding the routine behavior expected of themselves and others in daily life—is another fairly recent theoretical development in microsociology (Garfinkel, 1984; Heritage, 1984). Because ethnomethodology involves both a theory and associated research methods, a discussion of this perspective is reserved for Chapter 2 ("Sociologists Doing Research").

What are the limitations of symbolic interactionism?
By virtue of being a microsociological theoretical perspective, symbolic interactionism sometimes fails to take the larger social picture into account. Social interaction in everyday life is sometimes affected by societal forces beyond the control (or even awareness) of individuals. Anthony Giddens (1979, 1982, 1984, 1987, 1997), in his *structuration theory,* attempts to overcome

ances to appropriate governmental bodies. The elites, say Piven and Cloward, are not actually responding to the protest organization but are responding rather to the fear of insurgency itself. Once insurgency subsides, as it usually does, most of the organizations based on it disappear. The protest-based organizations that do survive are those that are useful to the elites who control the resources on which their existence depends. Surviving organizations must abandon their politics of opposition.

Unwittingly, then, protest organizers help the elites channel the protest of the insurgent masses into normal political channels. As a result, two things tend to occur. After the heat of insurgency has subsided, either the organizations fade away for lack of elite-supplied resources, or they survive because of their increasing usefulness to the elites on whom they are dependent for resources. The poor, of course, lose with either alternative.

Not only are protest organizations not as successful as they might be because of the influence of elites, assert Piven and Cloward, but the energy spent on building organizations channels resources away from activities that would produce better results. Instead of mobilizing the protest sentiments of the poor to create major change, leaders of the movement are preoccupied with organization building. As a result, the potential momentum of the poor people's willingness to protest is diminished. Piven and Cloward put it this way:

When workers erupted in strikes, organizers collected dues cards; when tenants refused to pay rent and stood off marshals, organizers formed building committees; when people were burning and looting, organizers used that "moment of madness" to draft constitutions (Piven and Cloward, 1979:xii).

Piven and Cloward see a lesson for leaders of future poor people's protests. Future organizers and leaders should understand that past protest efforts have failed because their leaders did not understand the influence of existing social structure on the forms of political activity available to the poor. Future lower class protest mobilizations should not be squandered out of ignorance of political and social tendencies.

Critical feedback

1. Explain why you think the conflict perspective might underlie this research.
2. Are there alternative theoretical perspectives that could have been used? Why or why not?
3. Place the study in the context of any one of the theoretical perspectives discussed in this chapter.

this potential limitation. Although Giddens recognizes that social structures influence the individual—a central point of both functionalism and conflict theory—he emphasizes the effects of individuals on social structures through their recognition of these social forces and their subsequent actions based on this recognition. Despite these limitations, symbolic interactionism is one of the most influential theoretical perspectives in sociology today (Fine, 1993).

What are the contributions of the theoretical perspectives?   Because these three theoretical perspectives have some contradictory assumptions and each is incomplete in itself, you may be somewhat confused. Where do perspectives in sociology now stand? The answer to this question can be direct: These three perspectives tend to complement one another because they focus on different aspects of social life.

Each of the perspectives could be used, for example, to better understand such deviant behavior as crime, juvenile delinquency, and mental illness. Functionalism emphasizes integration, stability, and consensus, so it accents the negative consequences of deviance. It underscores the cost of crime, including the money spent on police departments and prisons, violence in the streets, and the threat to public safety in inner cities and suburbs. The conflict perspective views deviance quite differently. Because various groups compete to influence the passage and enforcement of legislation serving their interests, many laws are created and enforced by those with political power. Consequently, those with the least influence—prostitutes, drug addicts—are the most likely to be classified as deviants. Interactionism focuses on the relationships within a group of deviants. It looks, for example, at the processes by which members of a drug-using group influence others to join them. It studies the ways drugs are bought from pushers, the methods for raising money to pay the pushers, and the ways people who use drugs interpret the user's activities.

The unique contribution of each of the three perspectives can be illustrated by examining another aspect of group life. American women currently earn about seventy-seven cents for every dollar earned by men. An earn-

SOCIOLOGY IN THE NEWS

Effects of Divorce

This CNN report focuses on the many sides of divorce in modern American society. One of the most devastating effects of divorce is the psychological pain usually experienced by children of divorce. This pain extends into adulthood. Divorce is an excellent topic on which to assess your understanding of the major theoretical perspectives in sociology.

1. State how functionalists would view the high divorce rate in the United States.

_____

_____

2. Focusing on self-esteem, identify some problems faced by children of divorce as a symbolic interactionist would.

_____

_____

Divorce is usually painful for everyone involved. It is always difficult for children who are least able emotionally to handle the situation.

ing gap exists even between men and women in the same occupations with the same educational qualifications. How would functionalists, conflict advocates, and symbolic interactionists view this piece of social reality? Functionalists claim income inequality between men and women is a dysfunction, or negative consequence, of the broader social condition of sexual inequality. Conflict advocates attribute income inequality to the power men exercise in controlling women's level of income. Symbolic interactionists would point out that income inequality has existed for a very long time. It is only in recent years that many Americans have come to define income inequality as an unfair condition in need of eradication. In other words, income inequality now has a different symbolic meaning than it has had in the past.

Because each of these perspectives emphasizes different aspects of social life, one cannot be said to be inherently superior to the others. Except for sheer personal preference, the old witchlike lady is no worse than the beautiful young woman, the rabbit no better than the duck. By the same token, no theoretical perspective is innately superior to another. Each of the three major perspectives has its own worth, because each tells us something different about group life.

FEEDBACK

1. A _____ perspective is a set of assumptions accepted as true by its advocates.
2. Indicate whether the following statements represent functionalism (F), the conflict perspective (C), or symbolic interactionism (S).
   ____ a. Societies are in dynamic equilibrium.
   ____ b. Power is one of the most important elements in social life.
   ____ c. Religion helps hold a society together morally.
   ____ d. Symbols are crucial to social life.
   ____ e. A change in the economy leads to a change in the family structure.
   ____ f. People conduct themselves according to their subjective interpretations of reality.
   ____ g. Many elements of a society exist to benefit the powerful.
   ____ h. Different segments of a society compete to achieve their own self-interests rather than cooperate to benefit others.
   ____ i. Social life should be understood from the viewpoint of the individuals involved.
   ____ j. Most members of a society agree that democracy is desirable.
   ____ k. Social change is constantly occurring.
   ____ l. Conflict is harmful and disruptive to society.
3. According to the _____ approach, humans use impression management to control the attitudes and responses of others toward them.

Answers: 1. theoretical 2. a. (F) b. (C) c. (F) d. (S) e. (F) f. (S) g. (C) h. (C) i. (S) j. (F) k. (C) l. (F) 3. dramaturgical

SUMMARY

1. Sociology is the scientific study of social structure. It maintains a group rather than an individual focus by emphasizing the patterned and recurrent social relationships among group members and by using social conditions to explain behavior in groups. As levels of analysis, microsociology and macrosociology are crucial to understanding the sociological perspective.

2. Sociology benefits both the individual and the public. First, through the sociological imagination, individuals can better understand the relationship between what is happening in their personal lives and the events occurring in their society. The sociological imagination also promotes questioning of conventional, and often misleading, ways of thinking. In addition, the sociological imagination provides a vision of social life that extends far beyond the often narrow confines of one's limited personal experience. Second, sociological research contributes to public policies and programs in such areas as school desegregation, poverty, treatment of the mentally ill, and the relationship between drug treatment and crime. Third, sociology contributes to the development of occupational skills.

3. It was not until the nineteenth century that the process of differentiating the various social sciences seriously began. Five of the major social sciences—anthropology, psychology, economics, political science, and history—are clearly distinguishable from sociology. Although these are separate fields in their own right, they share some areas of interest with sociology.

4. Sociology is rather young. It was born out of the social upheaval created by the French Revolution and the Industrial Revolution. In an attempt to understand the social chaos of their time, early sociologists emphasized social order and social change. Sociology received its start primarily from the writings of European scholars Auguste Comte, Harriet Martineau, Herbert Spencer, Karl Marx, Emile Durkheim, and Max Weber.

5. Auguste Comte, the generally acknowledged father of sociology, believed that society could advance only if studied scientifically. He advocated use of the research methods of the physical sciences to study social life. Harriet Martineau, probably the first woman sociologist, made original contributions to research methods and to theory in political economy and feminism. According to Herbert Spencer, social progress would occur if people did not intervene. Spencer's opposition to social reform was based on a survival-of-the-fittest philosophy. In contrast to Spencer, Karl Marx argued that only planned change brings social progress. History unfolds, he wrote, according to the outcome of conflict between social classes. In capitalist societies, the conflict is between the ruling bourgeoisie and the ruled proletariat.

6. Durkheim shared a concern for social order with the earlier pioneers. Two of his major contributions were the nonpsychological explanation of social life and the introduction of statistical techniques in social research. Max Weber's major contribution was also methodological. His method of *verstehen* assumed an understanding of social life that comes from mentally putting oneself in the place of others.

7. American sociology has been heavily influenced by early European sociologists, in part because it too was born during a time of social upheaval (following the Civil War). From its founding in the 1800s to World War II, the hotbed of American sociology was the University of Chicago. After World War II, sociology departments in the East and Midwest rose to prominence.

8. A theoretical perspective is a set of assumptions considered true by its advocates. Two major sociological perspectives on the macro level are functionalism and conflict theory. On the micro level is the perspective of symbolic interactionism.

9. According to functionalism, a society is an integrated whole, seeks a dynamic equilibrium, is composed of elements promoting its well-being, and is based on the consensus of its members. The conflict perspective contradicts each of these assumptions. A society experiences conflict at all times, is constantly changing, contains elements contributing to its change, and rests on the domination of some of its members by other members.

10. Human beings, according to symbolic interactionism, behave on the basis of their own interpretations of reality. These subjective interpretations are derived from the meanings learned from others. Humans use learned symbols and meanings to interpret their own behavior as well as the behavior of others.

11. No single sociological perspective is all-encompassing. Each perspective makes a unique contribution; combined, these perspectives shed considerable light on social life.

REVIEW GUIDE

LEARNING OBJECTIVES REVIEW

After careful study of this chapter, you will be able to:

- Define sociology and illustrate its unique perspective from both the micro and macro levels of analysis.
- Describe three uses of the sociological perspective.
- Distinguish sociology from other social sciences.
- Outline the distinctive contributions of the major pioneers of sociology.

- Summarize the beginnings of sociological development in the United States.
- Differentiate the major theoretical perspectives in sociology today.
- Compare two recent developments in symbolic interactionism.

CONCEPT REVIEW

Match the following concepts with the definitions listed below them.

____ a. economic determinism
____ b. mechanical solidarity
____ c. positivism
____ d. dynamic equilibrium
____ e. social structure

____ f. bourgeoisie
____ g. macrosociology
____ h. sociology
____ i. *verstehen*
____ j. symbol

____ k. latent function
____ l. conflict perspective
____ m. presentation of self
____ n. social dynamics
____ o. theoretical perspective

1. A set of assumptions accepted as true by its advocates.
2. The theoretical perspective that emphasizes conflict, competition, change, and constraint within a society.
3. An unintended and unrecognized consequence of some element of a society.
4. The ways people attempt to create a favorable evaluation of themselves in the minds of others.
5. The study of social change.
6. The method of understanding the behavior of others by putting oneself mentally in another's place.
7. Patterned, recurring social relationships.
8. The scientific study of social structure.
9. The use of observation, experimentation, and other methods of the physical sciences in the study of social life.
10. Something that stands for or represents something else.
11. The idea that the nature of a society is based on the society's economy.
12. Social unity based on a consensus of values and norms, strong social pressure for conformity, and dependence on tradition and family.
13. The assumption on the part of functionalists that a society both changes and maintains most of its original structure over time.
14. Members of industrial society who own the means for producing wealth.
15. The level of analysis that focuses on relationships among social structures without reference to the interaction of the people involved.

CRITICAL THINKING QUESTIONS

----

1. Apply the sociological imagination to your decision to come to college. Identify the social forces (family, society, peers) that led you to your present situation.

   _____

   _____

2. Discuss the possible benefits of sociology to your future personal and work life.

   _____

   _____

3. Think about some social change (for example, increased drug use or later age at first marriage) that you think is particularly significant. Is this change better explained by conflict theory or functionalism?

   _____

   _____

4. Could any of your behavior over the past week be described by the dramaturgical perspective? Explain, using personal examples.

_____

_____

MULTIPLE CHOICE QUESTIONS

1. According to C. Wright Mills, the _____ is the set of mind that allows individuals to see the relationship between events in their personal lives and events in their society.
   a. sociological awareness
   b. macrosociology
   c. sociological imagination
   d. microsociology
   e. moral imperative

2. Shepard defines sociology as
   a. the scientific study of socialism and other political-economic systems.
   b. the opposite of psychology.
   c. the scientific arm of social work.
   d. the study of how culture affects the personality.
   e. the scientific study of social structure.

3. Microsociology focuses on
   a. class conflicts.
   b. relationships among social structures.
   c. the way individuals interact with each other in social relationships.
   d. material very similar to psychology.
   e. emotional-mental development.

4. Macrosociology focuses on
   a. relationships between individuals within groups.
   b. political systems.
   c. various systems of producing, distributing, and consuming goods and services.
   d. how individuals interpret and react to their own culture.
   e. relationships among social structures without concern for the interaction of the individuals involved.

5. The sociological perspective that places human needs and goals as the primary concern of sociology is referred to as
   a. the sociological imagination.
   b. qualitative sociology.
   c. the ministerial perspective.
   d. humanist sociology.
   e. pure sociology.

6. Positivism is defined as
   a. the approach that should be used in dealing with the soft sciences.
   b. a religion that started in the early 1900s in France.
   c. a mental attitude designed to ensure success in life.
   d. the use of observation and experimentation in research.
   e. the balance of power between groups.

7. Karl Marx saw class conflict as
   a. one of the major economic consequences of World War I.
   b. the product of insecurities of the Cold War.
   c. one of the social and economic consequences of the social chaos of nineteenth-century Europe.
   d. conflict between those controlling the means for producing wealth and those laboring for them.
   e. conflict among those controlling the means for producing wealth.

8. Karl Marx's political objective was to
   a. end the British Monarchy.
   b. obtain a seat in the British Parliament.
   c. explain the workings of capitalism in order to hasten its fall through revolution.
   d. promote the use of religion as the opium of the masses.
   e. identify the distinctive characteristics of monogamy in order to enhance family stability.

9. According to Emile Durkheim, the social unity based on a complex of highly specialized roles that makes members of a society dependent on one another is
   a. mechanical solidarity.
   b. organic solidarity.
   c. positivism.
   d. value freedom
   e. dialectic materialism.

10. The theorist who wrote *The Protestant Ethic and the Spirit of Capitalism* was
   a. Emile Durkheim.
   b. Alexis de Tocqueville.
   c. Max Weber.
   d. Lester Ward.
   e. Robert Merton.

11. Functionalism is defined as
   a. the theoretical perspective that emphasizes the contributions made by each part of society.
   b. the type of society characterized by weak family ties, competition, and impersonal social relationships.
   c. the form of control by which authority is split evenly between husband and wife.
   d. a system in which strategic industries are owned and operated by the state.
   e. a sense of identification with the goals and interests of the members of one's social class.

12. _____ emphasizes conflict, competition, change, and constraint within a society.
   a. The functionalist perspective
   b. Critical theory
   c. The symbolic interactionist perspective
   d. Social revolutionist theory
   e. The conflict perspective

13. "Impression management" is a central concept of which of the following theoretical perspectives?
   a. functionalism
   b. conflict theory
   c. symbolic interactionism
   d. critical theory
   e. exchange theory

FEEDBACK REVIEW

True-False

1. Microsociology focuses on the relationships among social structures without reference to the interaction of the people involved. *T or F?*

Fill in the Blank

2. _____ explanations of group behavior are inadequate because human activities are influenced by social forces that individuals have not created and cannot control.

3. According to Durkheim, the United States is a social order based on _____ solidarity.

4. The _____ is the set of mind that enables individuals to see the relationship between events in their personal lives and events in their society.

5. A _____ perspective is a set of assumptions accepted as true by its advocates.

Matching

6. Match the following fields of study and aspects of family life.
   ____ a. anthropology   (1) distribution of income and the family
   ____ b. sociology   (2) effects of slavery on family stability
   ____ c. history   (3) relationship between voting and the family
   ____ d. political science   (4) effects of early marriage on divorce
   ____ e. economics   (5) link between early childhood emotional support and familial retardation
   ____ f. psychology   (6) universality of the nuclear family

7. Indicate whether the following statements represent functionalism (F), the conflict perspective (C), or symbolic interactionism (S).
   ____ a. Religion helps hold a society together morally.
   ____ b. Symbols are crucial to social life.
   ____ c. A change in the economy leads to a change in the family structure.
   ____ d. Many elements of a society exist to benefit the powerful.
   ____ e. Social life should be understood from the viewpoint of the individuals involved.
   ____ f. Social change is constantly occurring.

Multiple Choice

8. Which of the following is *not* one of the uses of the sociological imagination?
   a. seeing the interplay of self and society
   b. capacity for creating new aspects of culture not thought of by others
   c. ability to question aspects of social life most people take for granted
   d. capability of understanding the social forces that shape daily life

9. Which of the following sociologists contended that social change leads to progress, provided that people do not interfere?
   a. Durkheim
   b. Weber
   c. Martineau
   d. Spencer
   e. Marx
   f. Comte

GRAPHIC REVIEW

Table 1.2 contains data on average income (before and after taxes) in America by income category. Answering the following questions will test your understanding of this table:

1. State what you think is the most important generalization you could make from these data.

   _____

   _____

2. State the likely interpretation conflict theorists would make of Table 1.2.

   _____

   _____

3. Would functionalists agree with the interpretation of conflict theorists? Why or why not?

   _____

   _____

   _____

ANSWER KEY

Concept Review	Multiple Choice	Feedback Review
a. 11	1. c	1. F
b. 12	2. e	2. Individualistic
c. 9	3. c	3. organic
d. 13	4. e	4. sociological imagination
e. 7	5. d	5. theoretical
f. 14	6. d	6. a. (6)
g. 15	7. d	b. (4)
h. 8	8. c	c. (2)
i. 6	9. b	d. (3)
j. 10	10. c	e. (1)
k. 3	11. a	f. (5)
l. 2	12. e	7. a. F
m. 4	13. c	b. S
n. 5		c. F
o. 1		d. C
		e. S
		f. C
		8. b
		9. d

Chapter Two

Sociologists Doing Research

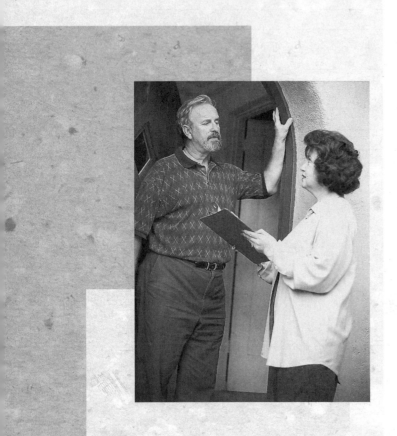

LEARNING OBJECTIVES

After careful study of this chapter, you will be able to:
- Identify major nonscientific sources of knowledge and explain why science is a superior source of knowledge.
- Apply the concept of causation and the controlled experiment to the logic of science.
- Differentiate the major quantitative research methods used by sociologists.
- Describe the major qualitative research methods used by sociologists.
- Explain the steps in the model sociologists use to guide their research.
- Describe the place of ethics in research.
- State the place of concern for reliability, validity, and replication in social research.

SOCIOLOGICAL IMAGINATION

What is the relationship between church attendance and juvenile delinquency?   The finding of a statistical link between church attendance and delinquency (delinquency increases as church attendance decreases) meets the test of common sense. One can easily speculate on why this would be the case. An observed relationship between these two events, however, should not lead us to conclude that one causes the other. In fact, delinquency increases as church attendance decreases because of a third factor—age. Age is related to both delinquency and church attendance. Older adolescents both go to church less often and are more likely to be delinquents. The apparent relationship between church attendance and delinquency, then, is actually produced by a third factor—age—that affects both of the original two factors.

Such mistaken ideas as a causal relationship between lower church attendance and juvenile delinquency can survive in large part because people so often rely on sources of knowledge not grounded in the use of reason and the search for reality. One of the major benefits of sociological research lies in its challenge of commonly held false beliefs and its attempt to replace such beliefs with more accurate knowledge. Before turning to the logic of scientific research and the methods of sociological research, it will be helpful to examine some major nonscientific sources of knowledge. This will provide a context for the appreciation of the utility of a scientific approach.

SOURCES OF KNOWLEDGE

How do we know what we know? Four major nonscientific sources of knowledge are intuition, common sense, authority, and tradition.

Nonscientific Sources of Knowledge

Intuition is quick and ready insight that is not based on rational thought. To intuit is to have the feeling of immediately understanding something because of insight from an unknown inner source. For example, the decision against buying a particular house because "it feels wrong" is a decision based on intuition.

Common sense refers to opinions that are widely held because they seem so obviously correct. The problem with commonsense ideas is that they are often wrong. Some claim common sense tells us, for example, that property values almost always decline when African Americans move into a white, middle-class neighborhood. Research demonstrating that property values do not necessarily decline when African Americans move in is hard for many people to accept because it goes against common sense. As philosopher Alasdair MacIntyre writes, "because common sense is never more than an inherited amalgam of past clarities and past confusions, the defenders of common sense are unlikely to enlighten us" (MacIntyre, 1997: 117).

An *authority* is someone who is supposed to have special knowledge that we do not have. A king believed to be ruling by divine right is an example of an authority. Reliance on authority is often appropriate. It is more reasonable to accept a doctor's diagnosis of an illness than to rely on information from a neighbor whose friend had the same symptoms (although even a single doctor's diagnosis should not be accepted uncritically). In other instances, however, authority can obscure the truth. Astrologers who advise people to guide their lives by the stars are an example of a misleading authority.

The fourth major nonscientific source of knowledge is *tradition*. It is traditional to believe that only children are all self-centered and socially inept. Despite evidence to the contrary, Americans generally still wish to have two or more children in order to avoid these alleged personality traits (Sifford, 1989). Barriers to equal opportunity for women persist in industrial societies despite evidence that traditional negative ideas about the capabilities of women are fallacious.

What is the major problem with nonscientific sources of knowledge?   The major problem with nonscientific sources of knowledge is that they often provide misleading or false information. In fact, nonscientific sources can lead to completely opposite conclusions. One person's intuition tells her to buy oil stocks, whereas another person's intuition tells him to avoid all energy stocks. One person's commonsense conclusion may be that the Equal Rights Amendment will destroy most sexual differences, although it seems perfectly obvious to someone else that this will not be the case.

Because these sources of knowledge are often accepted at face value, most people seldom challenge the information obtained through them. Consequently, reality can be distorted for a long time. Science is a more reliable method for obtaining knowledge because it is based on the principles of objectivity and verifiability.

Science as a Source of Knowledge

What is objectivity?   According to the principle of **objectivity,** scientists are expected to prevent their personal biases from influencing their results and their interpretation of those results. A male, antifeminist biologist investigating the intelligence levels of males and females, for example, is supposed to guard against any unwarranted tendency to conclude that males are more intelligent than females. Researchers must interpret their data solely on the basis of merit; the outcome they personally prefer is supposed to be irrelevant. This is what Max Weber (1946b, originally published in 1918) meant by value-free research. (Refer to Chapter 1, "The Sociological Perspective.")

Can scientists really be objective?   Inevitably, scientists' personal views do affect their work. In exceptional cases, this is deliberate. A promising young Harvard Medical School physician, John R. Darsee, admitted faking data in an experiment on heart attack prevention ("Harvard Delays in Reporting Fraud," 1982). A National Institute of Mental Health investigatory panel ruled that Stephen Breuning, an assistant professor of child psychiatry, deliberately and repeatedly engaged in deceptive research practices (Brand, 1987).

Sometimes scientists unintentionally let their personal biases influence their work. For example, pioneering sex researcher, Alfred Kinsey, has been accused of being both a homosexual and a masochist, characteristics said to unduly influence his research. In his recent book, James Jones (1997) makes this charge and presents evidence that Kinsey, who revolutionized popular thinking about sex in America in the 1950s, was a man with an ideological agenda whose research methods undermine his claim to objectivity. Jones refers to scientific critics who point out that Kinsey generalized about

When forming opinions about the risk of AIDS, many Americans rely on nonscientific sources. The protestors in this picture are concerned about the harm done to children in an environment void of scientifically based knowledge.

the American population on the basis of data gathered largely from volunteers, including disproportionate numbers of male prostitutes, gays, and prison inmates.

Scientists cannot possibly be completely objective. But if subjectivity in research cannot be eliminated, it can be reduced.

How can subjectivity be reduced?   If researchers are aware of their biases, they can consciously take them into account. They can be more careful in designing research instruments, selecting samples, choosing statistical techniques, and interpreting results. According to Swedish economist Gunnar Myrdal (1969), personal recognition of biases is insufficient; public exposure of them is essential. Personal values, Myrdal contends, should be explicitly stated so that those who read a research report can be aware of the author's biases.

What is verifiability?   **Verifiability** means that any study can be duplicated by other scientists. This is possible because scientists report in detail how they conducted their research. Verifiability is important because it exposes a piece of research to scientists' critical examination, retesting, and revision by colleagues. If researchers repeating a study produce results at odds with the original study, the original findings will be questioned. Erroneous theories, findings, and conclusions cannot survive in the long run in this type of system (Begley, 1997).

1. Intuition is quick and ready insight based on rational thought. *T or F?*
2. The major problem with nonscientific sources of knowledge is that such sources often provide erroneous information. *T or F?*
3. Define *objectivity* and *verifiability* as used in science.
4. According to Gunnar Myrdal, it is enough that scientists themselves recognize their biases. *T or F?*

Answers: 1. F 2. T 3. Objectivity exists when an effort is made to prevent personal biases from distorting research. Verifiability means that any given piece of scientific research done by one scientist can be duplicated by other scientists. 4. F

CAUSATION AND THE LOGIC OF SCIENCE

The Nature of Causation

In science it is assumed that an event occurs for a reason. According to the concept of **causation,** events occur in predictable, nonrandom ways, and one event leads to another. Why does this book remain stationary rather than slowly rising off your desk, past your eyes, to rest against the ceiling? Why does a ball thrown into the air return to the ground? Why do the planets stay in orbit around the sun? According to Aristotle, heavier objects fall faster than lighter ones. In the late 1500s, Galileo contended that all objects fall with the same acceleration (change of speed) unless slowed down by air resistance or some other force. It was not until the late 1600s that Isaac Newton developed the theory of gravity. We now know that objects fall because the earth has a gravitational attraction for objects near it. The planets remain in orbit around the sun because of the gravitational force that the sun creates.

In sociology, religious affiliation, political preference, educational achievement, child-rearing practices, and divorce rates can be predicted, in large part, on the basis of social class membership. Because science is based on causation, one of its main goals is to discover cause-and-effect relationships. Scientists attempt to discover the factors—there is usually more than one—that cause events to happen.

Why multiple causation?   Leo Rosten, noted author and political scientist, once wrote, "If an explanation relies on a single cause, it is surely wrong." Events in the physical or social world are generally too complex to be explained by any single factor. For this reason, scientists are guided by the principle of **multiple causation,** which states that an event occurs as a result of several factors operating in combination. What, for example, causes crime? Cesare Lombroso, a nineteenth-century Italian criminologist, believed that the predisposition to crime was inherited and that criminals could be identified by certain primitive physical traits (large jaws, receding foreheads). Modern criminologists reject Lombroso's (or anyone's) one-factor explanation of crime. They now cite numerous factors that contribute to crime, including subcultures of violence turned against society; rapid social change and economic development; excessive materialism; hopeless poverty in slums; and overly lax, overly strict, or erratic child-rearing practices.

How does the concept of variable fit into a discussion of causation?   A **variable** is a characteristic (age, education, social class) that occurs in different degrees. Some materials have a greater specific gravity than others; some people have higher incomes than others; the average level of education is higher in developed countries than in developing countries. Each of these is a *quantitative* variable, a variable numerically measured. Because differences can be measured numerically, individuals, groups, objects, or events can be pinpointed at some specific point along a continuum. A *qualitative* variable consists of categories rather than numerical units; it measures differences in kind rather than numerical degree. Sex, marital status, and group membership are three qualitative variables often used by sociologists: People are male or female; they are single, married, widowed, separated, or divorced; they are sorority members or they are not.

Variables that cause something to occur are **independent variables.** Variables in which a change (or effect) can be observed are **dependent variables.** Marital infidelity is an independent variable (although, of course, not the only one) that can cause the dependent variable of divorce. The independent variable of poverty is one of several independent variables that can produce change in the dependent variable of hunger. Whether a variable is dependent or independent varies with the context. The extent of hunger may be a dependent variable in a study of poverty; it may be an independent variable in a study of crime.

If a causal relationship exists between an independent variable and a dependent variable, the variables must be correlated. A **correlation** exists when a change in one variable is associated with a change (either positively or negatively) in the other. Establishment of causation, however, is much more complicated than establishment of a correlation between two variables.

What are the criteria for establishing a causal relationship? Three standards are commonly used for establishing causality (Lazarsfeld, 1955; Hirschi and Selvin, 1973). These standards can be illustrated using the mistaken assumption that lower church attendance causes higher juvenile delinquency discussed at the beginning of this chapter (see "Sociological Imagination").

1. *Two variables must be correlated.* Some researchers found that juvenile delinquency increases as church attendance declines (Stark, Kent, and Doyle, 1982). Does the existence of this negative correlation mean that lower church attendance causes higher delinquency? To answer this question, the second criterion of causality must be met.

2. *All possible contaminating factors must be controlled.* Although all cause-and-effect relationships involve a correlation, the existence of a correlation does not necessarily indicate a causal relationship. Just because two events vary together does not mean that one causes the other. Two totally unrelated variables may have a high correlation. In fact, the correlation between lower church attendance and delinquency is known as a **spurious correlation**—an apparent relationship between two variables that is actually produced by a third variable that affects both of the original two variables.

   The negative relationship between church attendance and delinquency occurs because age is related to both church attendance (older adolescents attend church less frequently) and delinquency (older adolescents are more likely to be delinquents). Thus, before we could be sure that a causal relationship exists between church attendance and delinquency, we would need to control for all variables relevant to the relationship. In this instance, controlling for age revealed that the relationship between church attendance and delinquency is not a causal one.

   A major problem in establishing causality lies in the control of all relevant variables. Normally, such control is impossible. Researchers are usually not aware of all possible factors that might affect the relationship between an independent variable and a dependent variable, and even if they were, it is often not feasible to control for all of them. Discovering and controlling for contaminating factors is one of the greatest challenges in science.

3. *A change in the independent variable must occur before a change in the dependent variable can occur.* Does lack of church attendance precede delinquency, or vice versa? Logically, either one could precede the other, or they could occur simultane-

ously. Thus, if the original correlation between church attendance and delinquency was maintained after controlling for possible contaminating factors, causality between these two variables could not be established, because it cannot be said which is temporally prior to the other.

Although the successful use of these criteria of causation is not always complete, the criteria are important standards for which scientists continue to strive. Moreover, research results—even if they meet these criteria of causation—require theory to make empirical data meaningful.

The Controlled Experiment as a Model

A description of the controlled experiment provides an excellent means to illustrate causation. Though used infrequently by sociologists, the controlled experiment provides insight into the nature of all scientific research because it is grounded in the concept of causation.

A **controlled experiment** takes place in a laboratory and attempts to eliminate all possible contaminating influences on the variables being studied. The basic idea of the controlled experiment is to rule out the effects of extraneous factors to see the effects (if any) of an independent variable on a dependent variable. According to the logic of the controlled experiment, if the dependent variable changes when the experimental (independent) variable is introduced but does not change when it is absent, the change must have been caused by the independent variable.

The basic ingredients of a controlled experiment are a pretest, a posttest, an experimental (independent) variable, an experimental group, and a control group. Suppose a researcher wants to study experimentally the effects of providing information on drug use to junior high school students. After selecting a class of eighth graders, the researcher could first measure the teenagers' attitudes toward drug use (pretest). Then, at a later time a film demonstrating the harmful effects of drug use might be shown to the class (experimental variable). After the movie, the students could again be questioned about their attitudes toward drug use (posttest). Any changes in their attitudes toward drug use that took place between the pretest and the posttest could be attributed to the experimental variable. Such a conclusion might be wrong, however, because the change could have been due to factors other than the experimental variable—a student in the school might have died from an overdose of drugs, a nationally known rock singer might have publicly endorsed drug use, or a pusher might have begun selling drugs to the students.

The conventional method for controlling the influence of contaminating variables is to select a control group as well as an experimental group. In the preceding example, half of the eighth-grade class could have been assigned to the **experimental group**—the group exposed to the experimental variable—and half to the **control group**—the group not exposed to the experimental variable. Assuming that the members of each group had similar characteristics and that their experiences between the pretest and the posttest had been the same, any difference in attitudes toward drug use between the two groups could safely be attributed to the students' exposure or lack of exposure to the film.

How can experimental and control groups be made comparable? The standard ways of making experimental and control groups comparable in all respects except for exposure to the experimental variable are through *matching* or *randomization.* In matching, participants in an experiment are matched in pairs according to all factors thought to affect the relationship being investigated and members of each pair are then assigned to one group or the other. In randomization, which is preferable to matching, subjects are assigned to the experimental and control groups on a random or chance basis. Assignment to one group or the other can be determined by flipping a coin or by having subjects draw numbers from a container. Whether matching or randomization is used, the goal is the same: to form experimental and control groups that are alike with respect to all relevant characteristics except the experimental variable. If this requirement has been met, any significant change in the experimental group as

BASIC STATISTICAL MEASURES

The trend in sociology today is toward more complicated and sophisticated statistical measures. However, the statistics you will encounter in this textbook and in the sources you are likely to read later, such as *The Wall Street Journal, Time, Newsweek,* and *The Economist,* are easily comprehended. Among the most basic statistical measures are averages (modes, means, medians) and correlations.

An *average*—a measure of central tendency—provides a single number representing the way numerical values are distributed. Consider the following hypothetical salary figures for the nine highest-paid Major League baseball players by position:

$3,300,000 (Catcher)	$4,500,000 (First Base)
$3,600,000 (Second Base)	$4,900,000 (Starting Pitcher)
$3,600,000 (Third Base)	$5,300,000 (Left Field)
$4,200,000 (Center Field)	$6,100,000 (Right Field)
$4,300,000 (Shortstop)	

There are three averages that can be used to make these numerical values more manageable and meaningful. Each of these three measures of central tendency gives a different picture. When any one measure of central tendency is misleading, researchers usually present two or more.

The *mode*—in this case $3,600,000—is the numerical value that occurs most frequently. If a researcher were to rely on the mode alone in a report of these Major League salaries, readers would be misled, because no mention is made of the wide range of salaries ($3,300,000 to $6,100,000). The mode is appropriate only when the objective is to indicate the most popular value.

In common usage, something that is average lies somewhere in the middle of a range. The *mean* is the measure of central tendency closest to the everyday meaning of the term *average.* The mean of the salary figures above—

$4,422,222—is calculated by adding all of the figures together and dividing by the number of figures ($39,800,000 ÷ 9). The mean, unlike the mode, takes all of the figures into account, but it is distorted by the extreme figure of $6,100,000. Although one player earns $6,100,000, most players make considerably less—the highest-paid player earns nearly twice as much as the lowest-paid player in this elite category. The mean distorts when there are extreme values at either the high or low end of a scale; it is more accurate when extremes are not widely separated.

The *median* is the number that divides a series of values in half; half of the values lie above it, half below. In this example, the median is $4,300,000—half of the salaries are above $4,300,000, half are below. Should there be an even number of values in a series, the median would be the mean of the two middle figures. The advantage of the median is that it is not distorted by extremes.

Measures of central tendency describe a single set of values, whereas a *correlation coefficient* indicates the strength of the relationship between two variables. A correlation coefficient of zero indicates that two variables are absolutely unrelated, as in the death rate in South Africa and the number of victories in a Los Angeles Dodgers's season. A perfect *positive correlation*—as in the case of the rate of descent of a parachutist and the earth's gravitational pull—has a value of +1.0. A perfect *negative correlation*, expressed numerically as −1.0, exists when the occurrence of one variable always leads to the absence of another. A perfect negative correlation exists between sunlight and darkness. Because correlations in sociological research are seldom perfect, judgments must be made about the strength of relationships. Correlation coefficients of plus or minus 0.4 and up are considered respectable in most sociological research, although sociologists have much more confidence in correlation coefficients above 0.6.

1. Match the following concepts and statements:

_____ a. causation
_____ b. multiple causation
_____ c. variable
_____ d. quantitative variable
_____ e. qualitative variable
_____ f. independent variable
_____ g. dependent variable
_____ h. correlation
_____ i. spurious correlation

(1) something that occurs in varying degrees
(2) the variable in which a change or effect is observed
(3) a change in one variable associated with a change in another variable
(4) the idea that an event occurs as a result of several factors operating in combination
(5) a factor that causes something to happen
(6) the idea that the occurrence of one event leads to the occurrence of another event
(7) a factor consisting of categories
(8) when a relationship between two variables is actually the result of a third variable
(9) a variable consisting of numerical units

2. A _____ attempts to eliminate all possible contaminating influences on the variables being studied.

3. The group in an experiment that is not exposed to the experimental variable is the _____ group.

4. Experimental and control groups are made comparable in all respects except for exposure to the experimental variable through _____ or _____.

Answers: 1. a.(6) b.(4) c.(1) d.(9) e.(7) f.(5) g.(2) h.(3) i.(8) 2. controlled experiment 3. control 4. matching or randomization

compared to the control group can be attributed with considerable confidence to the experimental variable. That is, a causal link will have been established between the independent and dependent variables.

QUANTITATIVE RESEARCH METHODS

Because sociologists find it difficult to create controlled situations, they tend to rely more on other research methods, classified either as quantitative or qualitative. About 90 percent of the research published in major sociological journals is based on surveys, so this approach is discussed first among major sociological research methods.

Survey Research

A **survey,** in which people are asked to answer a series of questions, is the most widely used research method among sociologists because it is ideal for studying large numbers of people. In survey research, care must be taken in the selection of respondents and in formulating the questions to be asked (Weisberg and Krosnick, 1996).

A **population** consists of all those people with the characteristics a researcher wants to study. A population could be all college sophomores in the United States, all former drug addicts now living in Connecticut, or all current inmates of the Ohio State Penitentiary. Most populations are too large and inaccessible to permit the collection of information on all members. For this reason, for example, the U.S. Bureau

of the Census has asked Congress for approval to change its method from an attempted survey of the entire American population to some limited use of scientific sampling (McAllister, 1997). A **sample,** of course, is a limited number of cases drawn from the larger population. A sample must be selected carefully if it is to have the same basic characteristics as the population. If a sample is not representative of the population from which it is drawn, the survey findings cannot be used to make generalizations about the entire population (Winship and Mare, 1992).

How can a representative sample be drawn? A **random sample**—a sample selected on the basis of chance so that each member of a population has an equal opportunity of being selected—is the standard way of selecting a representative sample. A random sample can be selected by assigning each member of the population a number and then drawing numbers from a container after they have been thoroughly scrambled. An easier and more practical method, particularly with large samples, involves the use of a table of random numbers in which numbers appear without pattern. After each member of the population has been assigned a number, the researcher begins with any number in the table and goes down the list until enough subjects have been selected.

If greater precision is desired, a *stratified random sample* can be drawn. This is accomplished by dividing the population into strata (categories such as sex, race, age, or any other relevant variable) and then selecting a random sample from each category. The proportion of persons in a given category, or stratum, should equal their proportion in the population at large.

CLOSED-ENDED AND OPEN-ENDED QUESTIONS

Closed-ended Questions

Please tell me whether you strongly agree, agree, disagree, or strongly disagree with each of the following statements:

	Strongly Agree	Agree	Disagree	Strongly Disagree
a. Most school teachers really don't know what they are talking about.	1	2	3	4
b. To get ahead in life, you have to get a good education.	1	2	3	4
c. My parents encouraged me to get a good education.	1	2	3	4
d. By the time children are sixteen years old, they should be ready to leave school.	1	2	3	4
e. Too much emphasis is put on education these days.	1	2	3	4
f. My parents thought that going to school was a waste of time.	1	2	3	4

Open-ended Question

In your own words, please describe your views on the education of your children.

_____

_____

_____

How is information gathered in surveys? In surveys, information is obtained either through a questionnaire or an interview. A *questionnaire* is a written set of questions that survey participants fill out by themselves; in an *interview*, a trained interviewer asks questions and records answers. Questionnaires or interviews may be composed of either closed-ended or open-ended questions. (See "Closed-Ended and Open-Ended Questions.") Closed-ended questions are those for which a limited, predetermined set of answers is possible. Because participants must choose from rigidly predetermined answers, closed-ended questions sometimes fail to elicit the participants' real attitudes and opinions. On the positive side, closed-ended questions make answers easier to quantify and compare. Open-ended questions ask for answers in the respondents' own words. Answers to open-ended questions, however, are not easy to quantify. And interviewers make the comparison of answers even more difficult when they change the meaning of questions by rephrasing them.

What are the advantages and disadvantages of survey research? Surveys—especially those based on structured questions—have the advantage of precision and comparability of responses. They permit the use of statistical techniques, a feature they have in common with experiments. Statistical techniques can be used because of still other advantages in survey research. Surveys per-

mit the collection of large samples, which in turn permit more detailed analysis; surveys include a large number of variables; variables in surveys can be quantified.

These employees of the U.S. Census Bureau are entering data from one of this government agency's many surveys. The results of these surveys, considered to be representative of the United States population, are widely utilized for decision making by private individuals, business organizations, and political leaders.

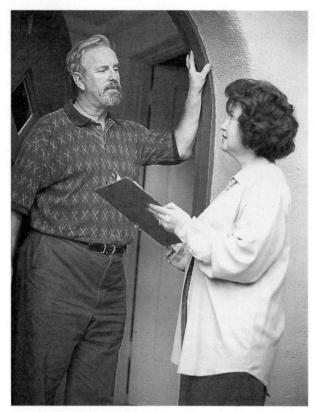

The survey is the most widely used research method for collecting data in sociology. Surveys are usually conducted in person, although use of the telephone is becoming much more common. One of the advantages of the survey is that it permits the gathering of information on a large number of people.

The survey research method has several disadvantages, however. First, surveys tend to be expensive because of the large samples that are usually involved. Second, because survey questions are predetermined, interviewers cannot always include important unanticipated information, although they are encouraged to write such information in the margin or on the back of the interview form. Third, the response rate—particularly in mailed questionnaires—is often low. A respectable return rate is about 50 percent, although researchers make an effort to obtain a return rate of 80 percent or higher. Even in interviews, some people are not available and some refuse to answer the questions. Because nonresponses can make the sample unrepresentative, surveys may be biased. Fourth, the phrasing of survey questions may also introduce bias. For example, negatively phrased questions are more likely to receive a no answer than neutrally phrased questions. It is better to ask, "Are you in favor of abortion?" than "You aren't in favor of abortion, are you?" Respondents also interpret the same question differently. If asked about the extent of their drug use, some respondents may include alcohol in their answers, others may not. As in experiments, there is a tendency for respondents

to give answers that they think the interviewer wants to hear or that they think are socially acceptable. Fifth, surveys cannot probe deeply into the context of the social behavior being studied; they draw specific bits of information from respondents, but they cannot capture the total social situation. Finally, survey researchers must be on guard for the **Hawthorne effect**—when unintentional behavior on the part of researchers influences the results they obtain from those they are studying (Roethlisberger and Dickson, 1964, originally published in 1939). As researchers and survey participants interact, participants detect cues regarding what the researchers are trying to find. The participants, depending on the circumstances, may subsequently attempt to please the researcher or frustrate the researcher's goals.

Precollected Data

The use of information already collected by someone else for another purpose is a well-respected method among sociologists, known as **secondary analysis.** In fact, the first sociologist to use statistics in a sociological study—Emile Durkheim—relied on precollected data. (See *Doing Research.*)

What are the major types of precollected data?   The sources for precollected data are as varied as government data, company records, voting records, prison records, and reports of research done by other social scientists.

One of the most important sources of precollected data for sociologists is the census bureau. Countries collect various types of information from their populations. The U.S. Bureau of the Census collects a wealth of information on the total population every ten years and conducts countless specific surveys each year. Because of this, detailed information exists on such topics as income, education, race, sex, age, marital status, occupation, and death and birth rates. Other government agencies collect additional information. The U.S. Department of Labor regularly collects information on the nation's income and unemployment levels across a variety of jobs. The U.S. Department of Commerce issues monthly reports on various aspects of the economy's health.

What are the advantages and disadvantages of precollected data?   Precollected data provide sociologists with inexpensive, quality information. Existing sources of information also permit the study of a topic over a long period of time. With census data, for example, we can trace changes in the relative income levels of blacks and whites since the war on poverty began in the 1960s. Also, because the data have been collected by others, the researcher cannot influence answers to questions he or she is using.

SCRUTINIZING POPULAR REPORTS OF SOCIAL SCIENCE RESEARCH

We are being bombarded daily with such a mass of new information that it is difficult to process it adequately. Consequently, becoming a critical, selective, and informed consumer of information is increasingly important. Discussed below are several means for better evaluating reports on social science research that you may encounter in the media.

Maintain a Skeptical Attitude   Be skeptical, because the media have a tendency to sensationalize and distort. For example, the media may report that a university researcher spent $500,000 to find out that love keeps families together when, in fact, this was only one small aspect of the larger research project. Moreover, chances are the media have oversimplified even this part of the researcher's conclusions.

Consider the Source of Information   For example, find out whether a study on the relationship between cancer and smoking has been sponsored by the tobacco industry or by the American Cancer Society. Representatives of tobacco companies deny the existence of any research linking throat and mouth cancer with dipping snuff. A medical researcher contended that putting a "pinch between your cheek and gum" has, in the long run, led to cancer in humans. Whom do you believe? At the very least you want to know the background of the source of information before making a judgment about scientific conclusions. This caution is especially relevant to the Internet, which is now a new major source of information. Because this information varies widely in its accuracy and reliability, sources must be evaluated with particular care.

Determine Whether a Control Group Has Been Used   Knowing whether a control group has been used in the research may be important. For instance, increases in self-esteem and physical energy may be reported in a study of participants in a meditation program. Was this because of the respect and attention they were given during the training period or because of the meditation techniques themselves? Or a study may report that the productivity of a group of workers in an office increased dramatically because the workers were allowed to participate in work-related decisions. Was the productivity increase due to the employees' being involved in something new and exciting or because of the participation in decision making itself? Without one or more control groups, you cannot be certain of what caused the changes in the meditation participants or in the office workers.

Do Not Mistake Correlation for Causation   A correlation between two variables does not necessarily mean that one caused the other. For example, at one time the percentage of Americans who smoked was increasing at the same time life expectancy was increasing. Did this mean that smoking caused people to live longer? Actually, a third factor—improved health care—accounts for the increased life expectancy. Do not assume that two events are related causally just because they occur together.

The use of precollected data also has disadvantages. The existing information may not exactly suit the researcher's needs, because it was collected for different purposes. Also, those people who collected the data may have been biased. Finally, sometimes precollected data are too old to be currently valid.

QUALITATIVE RESEARCH METHODS

Surveys and analysis of precollected data have been adopted by sociology in an attempt to emulate the quantification of the physical sciences. Another

FEEDBACK

1. Match the following terms and statements:

    ____ a. population
    ____ b. representative sample
    ____ c. random sample
    ____ d. sample
    ____ e. survey

    (1) selected on the basis of chance so that each member of a population has an equal opportunity of being selected
    (2) all those people with the characteristics the researcher wants to study within the context of a particular research question
    (3) a limited number of cases drawn from the larger population
    (4) a sample that has basically the same relevant characteristics as the population
    (5) the research method in which people are asked to answer a series of questions

2. Use of company records would be an example of using _____ data.

Answers: 1. a. (2) b. (4) c. (1) d. (3) e. (5) 2. precollected

DOING RESEARCH

Emile Durkheim— The Study of Suicide

Emile Durkheim, the first person to be formally recognized as a sociologist and the most scientific of the pioneers, conducted a study that stands as a research model for sociologists today. His investigation of suicide was, in fact, the first sociological study to use statistics. In *Suicide* (1964, originally published in 1897), Durkheim documented his contention that some aspects of human behavior—even something as allegedly individualistic as suicide—can be explained without reference to individuals.

Like all of Durkheim's work, *Suicide* must be viewed within the context of his concern for social integration (Collins, 1994). Durkheim wanted to see whether suicide rates within a social entity (for example, a group, organiza-

tion, or society) are related to the degree to which individuals are socially involved (integrated and regulated). Durkheim described three types of suicide: egoistic, altruistic, and anomic. *Egoistic* suicide is promoted when individuals do not have sufficient social ties. Because single (never married) adults, for example, are not heavily involved with family life, they are more likely to commit suicide than are married adults. *Altruistic* suicide, on the other hand, is more likely to occur when social integration is too strong. The ritual suicide of Hindu widows on their husbands' funeral pyres is one example. Military personnel, trained to lay down their lives for their country, provide another illustration.

Durkeim's third type of suicide—*anomic* suicide—increases when the social regulation of individuals is disrupted. For example, suicide rates increase during economic depressions. People suddenly without jobs or hope of finding them are more prone to kill themselves. Suicide may also increase during periods of prosperity. People may loosen their social

ties by taking new jobs, moving to new communities, or finding new mates.

Using data from the government population reports of several countries (much of it from the French government statistical office), Durkheim found strong support for his line of reasoning. Suicide rates were higher among single than married people, among military personnel than civilians, among divorced than married people, and among people involved in nationwide economic crises.

Durkheim's primary interest, however, was not in the empirical (observable) indicators he used, such as suicide rates among military personnel, married people, and so forth. Rather, Durkheim used the following indicators to support several of his contentions: (1) social behavior can be explained by social rather than psychological factors; (2) suicide is affected by the degree of integration and regulation within social entities; and (3) because society can be studied scientifically, sociology is worthy of recognition in the academic world (Ritzer, 1996).

---

research approach assumes that some aspects of social reality can be reached only by using qualitative, or nonquantitative, research methods. Qualitative research methods include *field research* and the *subjective approach* (Schwandt, 1997).

Field Research

Field research is used for studying aspects of social life that cannot be measured quantitatively and that are best understood within a natural setting. The world of prostitution, the inner workings of a Mafia family, and events during a riot are examples of field research.

The most often used approach to field research is the **case study**—a thorough investigation of a small group, an incident, or a community. Case studies are accomplished primarily through intensive observation, information obtained from informants, and informal interviews. Newspaper files, formal interviews, official

records, and surveys can be used to supplement these techniques.

This method assumes that the findings in one case can be generalized to other situations of the same type. The conclusions of a study on prostitution in Chicago, for example, should apply to other large cities as well. It is the researcher's responsibility to indicate factors that might make one situation different from similar situations in other places. Researchers conducting case studies often use the technique of *participant observation*.

What is participant observation? In **participant observation,** a researcher becomes a member of the group being studied. A researcher may join a group with or without informing its members that he or she is a sociologist. A compelling account of the use of covert participant observation appears in *Black Like Me,* a book written by John Howard Griffin (1961), a white journalist who dyed his skin to study the life of blacks in

The self-inflicted burning on the part of the Reverend Quang Duc, a Vietnamese Buddhist monk, is a prime example of altruistic suicide.

Durkheim was successful on all three counts. If Auguste Comte told us that sociology *could* be a science, Durkheim showed us *how* it could be a science.

Critical feedback

1. Do you believe that Durkheim's study of suicide supported his idea that much social behavior cannot be explained psychologically? State why or why not.
2. Durkheim used precollected data in researching suicide. Referring to the other major methods discussed in this chapter, indicate one or more other ways this problem could be studied.
3. The functionalist, conflict, and symbolic interactionist perspectives were discussed in Chapter 1. In which of these theoretical traditions does Durkheim seem to belong? Support your choice by relating his study to the assumptions of one of the perspectives.

the South. Although he had visited the South as a white man, the behavior of southern whites looked quite different to him through the eyes of a black man.

Sociologists sometimes identify themselves as researchers who want to observe firsthand a group's way of life. Elliot Liebow's study of two dozen lower-class black men who hung around a corner in Washington, D.C., illustrates the open approach to participant observation. Even though he was a white outsider, Liebow was allowed to participate in the daily activities of the men: "The people I was observing knew that I was observing them, yet they allowed me to participate in their activities and take part in their lives to a degree that continues to surprise me" (Liebow, 1967:253).

What are the advantages and disadvantages of field studies? Field studies can produce a depth and breadth of understanding unattainable with experi-

ments and surveys. They cannot be matched in their ability to reveal the meanings of a social situation from the angle of the people involved. Adaptability is another advantage. Once a survey has begun, it is not practical to make significant changes when new insights or oversights are discovered. But because it is unstructured, a field study can easily be altered. Field studies are especially valuable for situations in which quantitative research either is impossible or would yield biased results, as in a study of skid-row derelicts or organized crime. Because of these advantages, field studies may produce insights and explanations not likely to be unearthed through quantitative research.

Disadvantages do exist, however. The findings from one case may not be generalizable to similar situations. One mental hospital or community may be quite unlike any other mental hospital or community. If the possible bias of the sample is a major problem, so is the potential bias of the researcher. In the absence of more

READING TABLES AND GRAPHS

Tables and graphs are often confusing even though they are intended to present information concisely and unambiguously. Because of an inability to read tables and graphs, many people either misinterpret them or rely on an author's summary of what the data mean. However, another person's interpretation of a table or graph may be deliberately biased, accidentally misleading, or incomplete. Tables and graphs have a lot of information packed into them, but if they have been properly organized, you can easily understand them by following certain steps (Wallis and Roberts, 1962:195–207). The steps outlined below are keyed to Table 2.1 and Figure 2.1.

1. **Begin by reading the title of the table or graph carefully; it will tell you what information is being presented.** Table 2.1 shows median annual incomes in the United States by sex, race, and education.
2. **Find out the source of the information.** You will want to know whether the source is reliable, whether its techniques for gathering and presenting data are sound. The figures originated from the U.S. Bureau of the Census, a highly trusted source. If you know the source of data, you can investigate further on your own.
3. **Read any notes accompanying the table or graph.** Not all tables and graphs have notes, but if they do, the notes should be read for further information about the data. The notes in Table 2.1 and in Figure 2.1 explain that all the data refer to the total money income of full-time and part-time workers, ages 25 and over, in a March 1995 survey.
4. **Examine any footnotes.** Footnotes in Table 2.1 and Figure 2.1 indicate that the data are categorized by the highest grade actually completed. Although you may have assumed this correctly, years of schooling could have referred to the total number of years in school, regardless of the grade level attained.

5. **Look at the headings across the top and down the left-hand side of the table or graph.** To observe any pattern in the data, it is usually necessary to keep both types of headings in mind. Table 2.1 and Figure 2.1 show the median annual income of black and white males and females for several levels of education.
6. **Find out what units are being used.** Data can be expressed in percentages, hundreds, thousands, millions, billions, means, and so forth. In Table 2.1 and Figure 2.1, the units are dollars and years of schooling.
7. **Check for trends in the data.** For tables, look down the columns (vertically) and across the rows (horizontally) for the highest figures, lowest figures, trends, irregularities, and sudden shifts. If you read Table 2.1 vertically, you would be able to see how income varies by race and sex within each level of education. If you read the table horizontally, you could see how income varies with educational attainment for white males, black males, white females, and black females. A major advantage of graphs is that the sudden shifts, trends, irregularities, and extremes are easier to spot than they are in tables.
8. **Draw conclusions from your own observations.** Table 2.1 and Figure 2.1 show that although income tends to rise with educational level for both blacks and whites, it increases much less for black men and for women of both races than for white men. At each level of schooling, black men earn less than white men. In fact, white male high school dropouts have incomes only $485 below black male high school graduates; white male high school graduates earn nearly $2,000 more than black males with some college but no degree. White women appear to improve their earning power through college education to a greater extent than do black women.

---

precise measuring devices, the researcher has to rely on personal judgment and interpretation. Because of personal blind spots or because of emotional attachment to the people being studied, the researcher may not accurately see what is actually happening. Moreover, the lack of objectivity and standardized research procedures makes it difficult for another researcher to duplicate or replicate a field study. Because of these disadvantages, many sociologists regard the results of field studies as insights to be investigated further with more precise methods.

The Subjective Approach

The subjective approach to research has a long and honorable place in sociology. Recall from Chapter 1 Max Weber's method of *verstehen,* in which the subjective intentions of people are to be discovered by an attempt to imagine ourselves in their place. The **subjective approach,** then, studies some aspect of social structure through an attempt to ascertain the interpretations of the participants themselves. A prominent example of the subjective approach is *eth-*

TABLE 2.1	Median Annual Income by Sex, Race, and Education					
Demographic Group	Overall Median Income	Years of Schooling*				
		Less than 9	9–11	12	13–15	16 or More
White males	$30,409	$13,995	$18,403	$26,135	$30,293	$45,228
Black males	$21,531	$11,791	$16,323	$18,888	$24,161	$35,122
White females	$17,784	$ 9,338	$ 9,883	$15,133	$17,385	$28,492
Black females	$16,754	$ 9,730	$ 9,416	$14,017	$17,757	$27,280

Note: These figures include the total money income of full-time and part-time workers, ages 25 and over, as of March 1995.

* In terms of highest grade completed.

Source: U.S. Bureau of the Census, unpublished data: "Table 15. Educational Attainment—Total Money Earnings in 1995 of Persons 25 Years Old and Over, by Age, Race, Hispanic Origin, Sex, and Work Experience in 1995."

nomethodology, a relatively recent development in microsociology that attempts to uncover taken-for-granted social routines.

How does ethnomethodology work?

Ethnomethodology is the study of processes people develop and use in understanding the routine behaviors expected of themselves and others in everyday life. Ethnomethodologists assume that people share the meanings that underlie much of their everyday behavior. Through observing others and a process of trial and error in social situations, people develop a sense of appropriate ways of behaving. This understanding prevents them from making silly or serious social errors and saves them from having to decide constantly how they should behave in particular situations. Predictable, patterned behavior is a product of this process (Handel, 1982; Sharrock and Anderson, 1986; Livingston, 1987; Atkinson, 1988; Hilbert, 1990; Pollner, 1991).

How can ethnomethodologists discover what is going on in the minds of individuals as they construct a mental sense of social reality?

Because they are not mind readers, ethnomethodologists have had to seek other solutions. Harold Garfinkel, a prominent advocate of ethnomethodology, believes that the best way to understand how people construct social reality is to deprive them momentarily of their mental maps of daily routines. If people are deprived of their previous definitions of expected behaviors, they reconstruct a coherent picture of social reality. Ethnomethodologists can then learn by observing this process of reconstruction.

Garfinkel writes of situations that his students have created in order to observe what people do when

deprived of their taken-for-granted social routines. The following passage describes a situation in which an

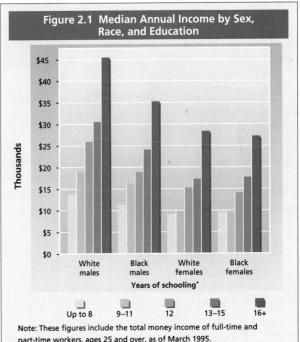

Figure 2.1 Median Annual Income by Sex, Race, and Education

Note: These figures include the total money income of full-time and part-time workers, ages 25 and over, as of March 1995.

*In terms of highest grade completed.

Source: U.S. Bureau of the Census, unpublished data: "Table 15. Educational Attainment—Total Money Earnings in 1995 of Persons 25 Years Old and Over, by Age, Race, Hispanic Origin, Sex, and Work Experience in 1995."

Internet Link: To learn more about income in the United States, visit the U.S. Bureau of the Census web site at http://www.census.gov. On its home page, select "Subjects A–Z," then select "I," "Income," and then your choice of topics.

experimenter (E) is attempting to deprive a subject (S) of his sense of expected routine by asking for more detailed information than is normally required in everyday situations. In the context of watching television, the experimenter first asks, "How are you tired? Physically, mentally, or just bored?"

(S): I don't know, I guess physically, mainly.
(E): You mean that your muscles ache or your bones?
(S): I guess so. Don't be so technical.
(After more watching)
(S): All these old movies have the same kind of old iron bedstead in them.

(E): What do you mean? Do you mean all old movies, or some of them, or just the ones you have seen?
(S): What's the matter with you? You know what I mean.
(E): I wish you would be more specific.
(S): You know what I mean! Drop dead! (Garfinkel, 1984:43).

The researcher continues this type of conversation until the subject is disoriented and can no longer respond within a previously developed frame of reference. The researcher can then observe the subject creating a new definition of what the expected or "normal" pattern of social interaction should be.

FEEDBACK

1. Field studies are best suited for situations in which _____ measurement cannot be used.
2. A _____ is a thorough investigation of a small group, an incident, or a community.
3. In _____, a researcher becomes a member of the group being studied.
4. According to the _____ approach, some aspect of social structure is best studied through an attempt to ascertain the interpretations of the participants themselves.
5. _____ is the study of the processes people develop and use in understanding the routine behaviors expected of themselves and others in everyday life.

Answers: 1. quantitative 2. case study 3. participant observation 4. subjective 5. Ethnomethodology

A MODEL FOR DOING RESEARCH

In an effort to obtain accurate knowledge, sociologists, like other scientists, use a model that involves the application of several distinct steps to any research problem. These steps, regularly referred to as the "scientific method," include identifying a problem, reviewing the literature, formulating hypotheses, developing a research design, collecting data, analyzing data, and stating conclusions (Hoover and Donovan, 1995).

Identify the Problem

Research begins with determining the object of investigation. A research question may be chosen because it interests the researcher. Or it may be pursued because it addresses a current social problem, attempts to test a major theory, or responds to a government agency wishing to support the research.

Review the Literature

Once the object of study has been identified, it is then defined within the context of relevant theories and previous research findings. For example, a sociologist investigating suicide will probably develop an approach by relating it to the classic study of suicide by Emile Durkheim (see *Doing Research* in this chapter) and to other sociologists who have done research on the topic.

Formulate Hypotheses

From a careful examination of relevant theory and previous findings, a sociologist is able to state one or more **hypotheses**—tentative, testable statements of relationships among variables. These variables must be defined precisely enough to be measurable. One hypothesis might be "The longer couples are married, the less likely they are to divorce." The independent variable (length of marriage) and the dependent variable (divorce) must be defined and measured. Scientists measure variables through the use of **operational definitions**—definitions of abstract concepts in terms of simpler, observable procedures. Divorce could be defined operationally as the legal termination of marriage. Measurement of divorce would be qualitative—the couple is either legally married or not. Length of marriage would be measured quantitatively—for example, the number of years a couple has been legally married. Other operational definitions may involve defining poverty for a family of four at some dollar level ($16,183 in 1996) or determining social class level by a combination of occupational, educational, and income levels.

Develop a Research Design

A research design defines the procedures for collecting and analyzing data. Will the study be a survey or a case study? If it is a survey, will data be collected from a cross section of an entire population, such as the Harris and

Gallup polls, or will a sample be selected from only one city? Will simple percentages or more sophisticated statistical methods be used? These and many other questions must be answered while the research design is being developed.

Collect Data

There are three basic ways of gathering data in sociological research: asking people questions, observing behavior, and analyzing existing materials and records. Sociologists interested in studying the harmony in interracial marriages could question couples directly about how well they get along. They could locate an organization with a large number of interracially married couples and observe the couples' behavior. Or they could compare the divorce rate among interracially married couples to the divorce rate of the population as a whole.

Analyze Data

Once the data have been collected and classified, they can be analyzed to determine whether the hypotheses were supported. This is not as easy or automatic as it sounds, because results are not always obvious. Because the same data can be interpreted in several ways, judgments have to be made. Guarding against personal biases is especially important in this phase of research.

State Conclusions

After analyzing the data, a researcher is ready to state the conclusions of the study. It is during this phase that the hypotheses are formally accepted, rejected, or modified. The conclusions of the study are related to the theory or research findings on which the hypotheses are based, and directions for further research are suggested. Depending on the findings, the original theory itself may have to be altered. Whether the statement of conclusions appears in a scientific journal, a book, or a mimeographed report, it includes a description of the methods used. By making the research procedures public, scientists make it possible for others either to duplicate the research, conduct a slightly different study, or go in a very different direction.

Some sociologists believe that this model is too rigid to capture spontaneous, subjective, and changeable social behavior. They prefer to discover what exists rather than to bias their observations with preconceived hypotheses and an inflexible research design. Even sociologists who generally follow the steps outlined above usually do not do so mechanically. They may conduct exploratory studies prior to stating hypotheses and developing research designs. Or they may alter their hypotheses and research designs as their investigations proceed.

FEEDBACK

1. Listed below are the steps in the research model. Beside these steps are some concrete examples related to the sociability of the only child. Indicate the appropriate example for each step number.

____ Step 1: identify the problem
____ Step 2: review the literature
____ Step 3: formulate hypotheses
____ Step 4: develop a research design
____ Step 5: collect data
____ Step 6: analyze data
____ Step 7: state conclusions

a. Read past theory and research on the sociability of only children.
b. From previous research and existing theory, a researcher states that only children appear to be more intelligent than children with siblings.
c. A researcher collects data on only children from a high school in a large city.
d. A researcher writes a report giving evidence that only children are more intelligent than children with brothers or sisters.
e. A researcher decides to study the intelligence level of only children.
f. A researcher classifies and processes the data collected in order to test a hypothesis.
g. A researcher decides on the data needed to test a hypothesis, the methods for data collection, and the techniques for data analysis.

Answers: 1. Step 1: e. Step 2: a. Step 3: b. Step 4: g. Step 5: c. Step 6: f. Step 7: d.

ETHICS IN SOCIAL RESEARCH

The Issue of Ethics

Sociological research is a distinctly human endeavor. Although there are canons for conducting research, such as objectivity and verifiability, scientists don't always live up to them. As for people in other occupations, among scientists there is sometimes a discrepancy between the rules of work and the actual performance of work. Conducting scientific research requires ethical values as surely as it requires theoretical and methodological skills (Kuznik, 1991; Hessler, 1992).

Unfortunately, there is a long list of examples that call into question the ethical standards of researchers. During the Nuremberg trials, twenty Nazi doctors were convicted of conducting sadistic experiments on concentration camp inmates. From 1932 to 1972, the Public Health Service of the U.S. government deliberately did not treat approximately four hundred syphilitic black sharecroppers and day laborers so that biomedical researchers could study the full evolution of the disease (Jones, 1993). Ethical questions have been raised upon disclosure that researchers at Germany's University of Heidelberg had, for twenty years, used human corpses, including children, in high-speed automobile crash tests (Fedarko, 1993). Federal investigators in the United States have documented over ten years of fraud in some of the most important breast cancer research ever done, including a study that sanctioned lumpectomy as a safe operation (Crewdson, 1994).

Several social scientists have been criticized for conducting what many scientists view as unethical research. In each case, subjects were placed in stressful situations without being informed of the true nature of the experiments (Milgram, 1963, 1965, 1974; Zimbardo, Anderson, and Kabat, 1981). These and other studies have created great interest in a code of ethics. There is, in fact, a formal code of ethics for professional sociologists (American Sociological Association, 1997).

A Code of Ethics in Sociological Research

The formal code of ethics for sociologists covers a variety of important areas beyond research, including relationships with students, employees, and employers (American Sociological Association, 1997). In broad terms, the code of ethics is generally concerned with maximizing the benefits of sociology to society and minimizing the harm sociological work might create. Of importance in the present context are the research-related aspects of the code.

In this regard, sociologists are committed to objectivity, adherence to the highest technical research standards, accurate public exposure of their findings and methods, and protection of the rights, privacy, integrity, dignity, and autonomy of the subjects of their research. Because most of these topics have already been covered in this chapter, the focus in the present section is on the rights, privacy, integrity, dignity, and autonomy of participants in sociological research.

Sociologists routinely protect the rights of participants and avoid deceiving or harming them, so it is normally only the violators of the code of ethics that are publicized. Occasionally, adherence to the code is documented. Mario Brajuha, a graduate student at a major American university, kept detailed field notes while engaging in a participant observation study of restaurant work (Brajuha and Hallowell, 1986). Because of suspected arson after a fire at the restaurant where he was employed as a waiter, his field notes became the object of interest of the police, the district attorney, the courts, and some suspects. By refusing to reveal the contents of his field notes, Brajuha protected the rights of those individuals described in his notes. He did so in the face of a subpoena, threats of imprisonment, and the specter of personal harm to himself, his wife, and his children. The case was finally dropped after two difficult years.

Though infinitely rarer, much can be learned about ethics in sociological research from examination of a negative case. A case study of homosexuals conducted by sociologist Laud Humphreys (1979) provides a background against which to examine further the code of ethics.

Humphreys studied homosexual activities in men's public bathrooms ("tearooms"). By acting as a lookout to warn the homosexuals of approaching police officers, he was able to observe their activities closely. After the men left the tearooms, Humphreys recorded their license plate numbers to obtain their addresses for subsequent personal interviews. Humphreys waited a year so that any memory the men had of him would have faded, and then he falsely presented himself to them as a survey researcher to obtain additional information.

Did Humphreys violate the code of ethics as a covert participant observer? Yes, Humphreys violated the privacy of these people. Most did not want their sexual activities known, and Humphreys did not give them the opportunity to refuse to participate in the study. Humphreys also deceived the men by misrepresenting himself in both the tearooms and their homes. Finally, by recording his observations, Humphreys placed these people in jeopardy of public exposure, arrest, or loss of employment. (Actually, because of his precautions, none of the subjects was injured as a result of his research. In fact, to protect their identities, Humphreys even allowed himself to be arrested.)

Good scientific research is difficult from both a financial and a technical viewpoint. Ethical concerns make it even harder. Still, it is the researcher's responsibility to decide when a particular action crosses an ethical line—a decision not always easy to make, because moral lines are often blurred. Moreover, the researcher must balance a concern for the rights and protection of those being studied with the need to use certain methods to obtain knowledge. Kai Erikson is one of the most sensitive and outspoken critics of disguised observation, but he has defended it on the grounds that it is sometimes the only way to obtain information.

Some of the richest material in the social sciences has been gathered by sociologists who were true participants in the group under study but who did not announce to other members that they were employing this opportunity to collect research data. . . . It would be absurd, then, to insist as a point of ethics that sociologists should always introduce themselves as investigators everywhere they go and should

inform every person who figures in their thinking exactly what their research is about (Erikson, 1967:368).

Balance is the key to the issue of ethics. Subjects—whether in experiments, surveys, or field studies—above all should be protected from social, financial, psychological, or legal damage.

1. Three situations involving ethics in social research are cited below (Babbie, 1995:475). Match each situation with the appropriate aspects of the social science code of ethics for research on human subjects.

____ (1) concern for participants' privacy
____ (2) avoidance of deception
____ (3) obligation not to harm participants

   a. After a field study of deviant behavior during a riot, law enforcement officials demand that the researcher identify those people who were observed looting. Rather than risk arrest as an accomplice after the fact, the researcher complies.

   b. A research questionnaire is circulated among students as part of their university registration packet. Although students are not told they must complete the questionnaire, the hope is that they will believe they must—thus ensuring a higher completion rate.

   c. Researchers obtain a list of right-wing radicals they wish to study. They contact the radicals with the explanation that each has been selected "at random" from among the general population to take a sampling of "public opinion."

2. Match the concepts on the left side with the definitions on the right side.

____ a. reliability   (1) when a measurement technique yields consistent results on repeated applications
____ b. validity   (2) the duplication of the same study to ascertain its accuracy
____ c. replication   (3) when a measurement technique actually measures what it is designed to measure

Answers: 1. a. (3) b. (2) c. (1) 2. a. (1) b. (3) c. (2)

SOCIOLOGY IN THE NEWS

Research in Advertising

This CNN report illustrates the role research now plays in the retail industry. For example, studies of women and men reveal that their patterns of shopping behavior are quite different. This information is used by business to encourage more purchasing by both genders. Such market research strikes some sociologists as deceptive and manipulative.

1. Name research methods that would be suitable to explore this area of human behavior.

   _____
   _____

2. Do you see any ethical issues for a sociologist conducting market research on shopping behavior? Why or why not?

   _____
   _____

Suppose that you were sent to a department store to investigate the attitudes of men and women about shopping. What questions would you ask to determine any differences between the genders?

A FINAL NOTE

Reliability, Validity, and Replication

Researchers can be guided by all the important research considerations we have discussed in this chapter and still not conduct a good study. They can be mindful of objectivity, sensitive to the criteria of causation, careful in the selection of the most appropriate method (survey, precollected data, field study), and still fail to produce knowledge superior to that yielded by intuition, common sense, authority, or tradition.

What else must a researcher do?   To practice good social science, sociologists must pay careful attention to the quality of measurement (Babbie, 1995). Consequently, they must emphasize *reliability* and *validity* in the creation and evaluation of the measuring devices they use for the variables they wish to investigate.

What is reliability?   **Reliability** exists when a measurement technique yields consistent results on repeated applications. Reliability is tested by repeated administration of a measurement technique, such as a questionnaire, to the same subjects to ascertain whether the same results occur each time. Suppose a researcher, after deciding to study satisfaction with day care among parents, designed a questionnaire. If, on repeated applications, the level of satisfaction with day care on the part of the sample of parents remained consistent, then confidence in the reliability of the measurement device rises. Should, on the other hand, the level of satisfaction from one administration of the questionnaire to the next vary over a period of time, then we would doubt that satisfaction with child care is actually being measured.

The problem of reliability is involved in qualitative research also. Suppose that our researcher is also interested in satisfaction with day care among the children. If different conclusions about the level of satisfaction among the children, arrived at by asking them questions or observing their behavior, seemed different each day to the researcher, then doubt is raised about the reliability of the measurement technique being used.

Although a measurement technique may be reliable when used in a study, it still may not produce scientifically sound results. This is because a measurement technique must be not only reliable, but also valid.

What is validity?   **Validity** exists when a measurement technique actually measures what it is designed to measure. Thus, a technique intended to measure parental satisfaction with day care may yield consistent results on repeated applications to a sample of parents, but not really be measuring satisfaction at all. The measurement device might be tapping parental need to view day care positively in order to mask guilt feelings about permitting someone else to be the care-provider during working hours. Children at a day-care center may appear satisfied to the visiting researcher because they are neglected during the day and welcome his or her attention or because the children have been coached by the day-care provider to appear satisfied. A measurement technique, in short, may be consistently measuring something very different from what it purports to measure.

What is the relationship among reliability, validity, and replication?   In the first part of this chapter, attention was drawn to the importance of verifiability in science. Verifiability, we stated, is crucial to science as a superior source of knowledge due to its contribution to the self-corrective nature of research. Verifiability depends on the process of **replication**—the duplication of the same study to ascertain its accuracy. Replication is closely linked to both reliability and validity in that reliability and validity problems unknown to original researchers are likely to be revealed as subsequent social scientists repeat their research. It is partially through replication that scientific knowledge accumulates and changes over time.

A major goal of scientific research is to generate knowledge that is more reliable than can be obtained from such nonscientific sources as intuition, common sense, authority, and tradition. Through efforts to be objective and to make their research subject to replication by others, researchers attempt to portray reality as accurately as possible. The methods of research presented in this chapter are the specific tools sociologists use to create knowledge of social life that is as accurate as possible at the time.

However, empirical results obtained through the use of research methods are not the final goal of science. As Gerhard Lenski has stated, "Science is more than method: *its ultimate aim is the development of a body of 'verified' general theory*" (Lenski, 1988: 163). For this reason, there is constant interaction between sociological theory and research methods. Theory is used to develop hypotheses capable of being supported or falsified through testing. These results, in turn, may support existing theory, alter it, or lead to its ultimate rejection and the creation of a new theory. One of Lenski's major points is that, divorced from research methods, "sociological theory has more in common with seminary instruction in theology and biblical studies" (1988:165) than it does with the natural sciences model that sociology is emulating. Theory is trustworthy and useful only to the extent that it has been tested and found to be valid.

SUMMARY

1. People tend to get information from such nonscientific sources as intuition, common sense, authority, and tradition. Generally speaking, these sources are inadequate for obtaining accurate knowledge about social life. The advantage of scientific knowledge is its grounding on the principles of objectivity and verifiability.

2. Complete objectivity is impossible because sociologists, like all scientists, have values, beliefs, attitudes, and prejudices that affect their work to some extent. Subjectivity can be minimized, however, if researchers make themselves aware of their biases and make their biases public when presenting their findings.

3. The concept of causation—the idea that the occurrence of one event leads to the occurrence of another event—is central to science. All events have causes, and scientists attempt to discover the factors causing the events.

4. Three criteria must be met before a cause-and-effect relationship can be said to exist. First, two variables must be correlated. That is, change in the independent variable (the causal factor) must be associated with a change in the dependent variable (effect). Second, the correlation must not be spurious, that is, due to the effects of a third variable. Third, it must be shown that the independent variable always occurs *before* the dependent variable. Scientists think in terms of multiple causation because events are usually caused by several factors, not simply by a single factor.

5. Although sociologists rarely use the controlled experiment, they must understand this research method because it is based on the idea of causation. Sociologists generally employ nonexperimental research methods in attempting to establish causality. This is dictated by the difficulty of controlling relevant variables in the world outside the laboratory.

6. Two major quantitative research methods in sociology are the survey and precollected data. Surveys can draw on large samples, are quantitative, include many variables, are relatively precise, and permit the comparison of responses, but this method must take care to collect representative samples. Use of precollected data permits sociologists to do high-quality research at reasonably low cost and to trace changes in variables over an extended period of time.

7. Field studies are best used when some aspect of social structure cannot be measured quantitatively, when interaction should be observed in a natural setting, and when in-depth analysis is needed. The case study is the popular approach to field research. Some sociologists have adopted a subjective approach in which emphasis is on ascertaining the subjective interpretations of the participants themselves.

8. A research model involves several distinct steps: identifying the problem, reviewing the literature, formulating hypotheses, developing a research design, collecting data, analyzing data, and stating conclusions. These steps are a model for scientific research, but it is not necessary that they always be strictly followed.

9. Researchers have an ethical obligation to protect participants' privacy and to avoid deceiving or harming participants. Preserving the rights of subjects is sometimes weighed against the value of the knowledge to be gained. Most of the time these compromises are harmless, but they sometimes place the subjects in jeopardy.

LEARNING OBJECTIVES REVIEW

After careful study of this chapter, you will be able to:

- Identify major nonscientific sources of knowledge and explain why science is a superior source of knowledge.
- Apply the concept of causation and the controlled experiment to the logic of science.
- Differentiate the major quantitative research methods used by sociologists.
- Describe the major qualitative research methods used by sociologists.

- Explain the steps in the model sociologists use to guide their research.
- Describe the place of ethics in research.
- State the place of a concern for reliability, validity, and replication in social research.

REVIEW GUIDE

CONCEPT REVIEW

Match the following concepts with the definitions listed below them:

____ a. participant observation   ____ f. independent variable   ____ k. field research
____ b. controlled experiment   ____ g. objectivity   ____ l. case study
____ c. verifiability   ____ h. correlation   ____ m. survey
____ d. subjective approach   ____ i. population   ____ n. replication
____ e. experimental group   ____ j. sample

1. The group in an experiment exposed to the experimental variable.
2. A statistical measure in which a change in one variable is associated with change in another variable.
3. A research approach for studying aspects of social life that cannot be measured quantitatively and that are best understood within a natural setting.
4. A thorough, recorded investigation of a small group, incident, or community.
5. All those people with the characteristics a researcher wants to study within the context of a particular research question.
6. The principle of science stating that scientists are expected to prevent their personal biases from influencing their results and their interpretation of the results.
7. A variable that causes something to happen.
8. The type of field research technique in which a researcher becomes a member of the group being studied.
9. A principle of science by which any given piece of research can be duplicated (replicated) by other scientists.
10. A research method in which people are asked to answer a series of questions.
11. The duplication of the same study to ascertain its accuracy.
12. A limited number of cases drawn from a population.
13. A laboratory experiment that attempts to eliminate all possible contaminating influences on the variables being studied.
14. A research method in which the aim is to understand some aspect of social reality through the study of the subjective interpretations of the participants themselves.

CRITICAL THINKING QUESTIONS

1. Suppose that on a break from college you return home and a noncollege friend insists that you are wasting your time because the experience gained from the "university of hard knocks" is all she needs to know the truth. What arguments would you use to defend science as a better source of knowledge?

2. In class, your sociology professor reports on his recent study showing that men are generally better managers in business than women. If you were concerned about a possible lack of objectivity on his part, what questions would you ask him in order for you to place more confidence in his results?

3. The controlled experiment is the research model for investigating causal relationships. What is there about the nature of causation and the design of experiments that supports this claim?

4. Do you think that selecting a sample of three thousand individuals would produce an accurate picture of the U.S. population? Why or why not?

5. Pretend that you are a sociologist studying the relationship between the receipt of welfare payments and commitment to working. Describe the research method you would use and show why it is the most appropriate to this topic.

MULTIPLE CHOICE QUESTIONS

1. The concept of intuition refers to
   a. quick and ready insight that is not based on rational thought.
   b. opinions that are widely held because they seem so obviously correct.
   c. someone who is supposed to have special knowledge that we do not have.
   d. the fourth major nonscientific source of knowledge.
   e. a variable that causes something to happen.

2. According to your text, causation can be asserted when
   a. going from particular instances to general principles.
   b. there are only a limited number of cases taken from society.
   c. events occur in a predictable, nonrandom way, and one event leads to another.
   d. people develop and use routine behaviors expected of themselves and others in everyday life.
   e. a change in one variable is often accompanied by a change in another variable.

3. Several factors have been shown to influence crime rates in poor neighborhoods. This illustrates the principle of
   a. the poverty/crime hypothesis.
   b. multiple causation.
   c. verifiability.
   d. criminology.
   e. variance.

4. A variable that causes something else to occur is a/an
   a. dependent variable.
   b. correlation variable.
   c. causation variable.
   d. independent variable.
   e. qualitative variable.

5. The term *correlation* is defined as
   a. a change in one variable associated with a change in the other.
   b. an apparent relationship between two variables that is actually produced by a third variable that affects both of the original two variables.
   c. an event that occurs as a result of several factors operating in combination.
   d. something that occurs in different degrees among individuals, groups, objects, and events.
   e. a research method in which people are asked to answer a series of questions.

6. All of the following are criteria for establishing a causal relationship *except:*
   a. All possible contaminating factors must be controlled.
   b. A relationship representing a spurious relationship must exist.
   c. The independent variable must occur before the dependent variable.
   d. Two variables must be correlated.

7. All of the following statements about controlled experiments are true *except:*
   a. A description of the controlled experiment provides an excellent means to illustrate causation.
   b. A controlled experiment provides insight into the nature of all scientific research.
   c. Controlled experiments take place in a laboratory.
   d. The basic idea of the controlled experiment is to rule out the effect of extraneous factors to see the effects of an independent variable on a dependent variable.
   e. Controlled experiments do not need a control group because of the controlled atmosphere the laboratory provides.

8. The experimental group is exposed to the experimental variable; the group that is not exposed to the experimental variable is a/an
   a. natural group.
   b. experiential group.
   c. control group.
   d. dependent group.
   e. independent group.

9. The standard ways of making experimental and control groups comparable in all respects except for exposure to the experimental variable are through
   a. qualifying or quantifying.
   b. matching or randomizing.
   c. pretesting or posttesting.
   d. testing or retesting.
   e. verification or replication.

10. What do we call a written set of questions that survey participants are asked to fill out by themselves?
    a. survey
    b. interview
    c. questionnaire
    d. survey research
    e. independent variable

11. Use of data from the U.S. Bureau of the Census is an example of
    a. primary analysis.
    b. population sampling.
    c. the Hawthorne effect.
    d. secondary analysis.
    e. a case study.

12. What type of research is used for studying aspects of social life that cannot be measured quantitatively and that are best understood in a natural setting?
    a. field research
    b. survey research
    c. participant observation
    d. analysis of precollected data
    e. content analysis

13. **Ethnomethodologists assume that**
    a. the subjective approach relies too much on intuition.
    b. the behavior of people is random.
    c. the underlying factor explaining human behavior is ethnicity.
    d. questionnaires need to be tightly structured.
    e. people share the meanings underlying much of their everyday behavior.

14. **What is defined as a tentative, testable statement of a relationship among variables?**
    a. hypothesis
    b. operational definition
    c. formal argument
    d. correlation
    e. conclusion

15. **In Laud Humphreys's study of homosexual activities occurring in men's public bathrooms ("tearooms"), what ethical standard did he violate?**
    a. He studied homosexuals.
    b. He acted as a participant observer.
    c. He violated the privacy of the participants.
    d. He used the research to become famous.
    e. He did not violate any ethical standards.

FEEDBACK REVIEW

True-False

1. The major problem with nonscientific sources of knowledge is that such sources often provide erroneous information. *T or F?*
2. According to Gunnar Myrdal, it is enough that scientists themselves recognize their biases. *T or F?*

Fill in the Blank

3. The group in an experiment that is not exposed to the experimental variable is the _____ group.
4. Field studies are best suited for situations in which _____ measurement cannot be used.
5. In _____, a researcher becomes a member of the group being studied.
6. A _____ attempts to eliminate all possible contaminating influences on the variables being studied.
7. Use of company records would be an example of using _____ data.
8. According to the _____ approach, some aspects of social structure are best studied through an attempt to ascertain the interpretations of the participants themselves.

Matching

9. Listed below are the steps in the research model. Beside these steps are some concrete examples related to the sociability of the only child. Indicate the appropriate example for each step number.

    ____ Step 1: identify the problem
    ____ Step 2: review the literature
    ____ Step 3: formulate hypotheses
    ____ Step 4: develop a research design
    ____ Step 5: collect data
    ____ Step 6: analyze data
    ____ Step 7: state conclusions

    a. Read past theory and research on the sociability of only children.
    b. From previous research and existing theory, a researcher states that only children appear to be more intelligent than children with siblings.
    c. A researcher collects data on only children from a high school in a large city.
    d. A researcher writes a report giving evidence that only children are more intelligent than children with brothers or sisters.
    e. A researcher decides to study the intelligence level of only children.
    f. A researcher classifies and processes the data collected in order to test a hypothesis.
    g. A researcher decides on the data needed to test a hypothesis, the methods for data collection, and the techniques for data analysis.

10. Three situations involving ethics in social research are cited below. Match each situation with the appropriate aspect of the social science code of ethics for research on human subjects.

    ____ (1) concern for participants' privacy
    ____ (2) avoidance of deception
    ____ (3) obligation not to harm participants

    a. After a field study of deviant behavior during a riot, law enforcement officials demand that the researcher identify those people who were observed looting. Rather than risk arrest as an accomplice after the fact, the researcher complies.
    b. A research questionnaire is circulated among students as part of their university registration packet. Although students are not told they must complete the questionnaire, the hope is that they will believe they must, thus ensuring a higher completion rate.
    c. Researchers obtain a list of right-wing radicals they wish to study. They contact the radicals with the explanation that each has been selected "at random" from among the general population to take a sampling of "public opinion."

GRAPHIC REVIEW

Table 2.1 displays the median annual income in the United States by sex, race, and education. Demonstrate your understanding of the information in this table by answering the following questions:

1. State briefly what this table tells us about the relationship among sex, race, and education in the United States.

_____

_____

2. Identify the demographic group that enjoys the greatest economic benefits of education.

_____

_____

3. Identify the demographic group that benefits the least economically from higher levels of education.

_____

_____

ANSWER KEY

Concept Review	Multiple Choice	Feedback Review
a. 8	1. a	1. T
b. 13	2. c	2. F
c. 9	3. b	3. control
d. 14	4. d	4. quantitative
e. 1	5. a	5. participant observation
f. 7	6. b	6. controlled experiment
g. 6	7. e	7. precollected
h. 2	8. c	8. subjective
i. 5	9. b	9. Step 1: e
j. 12	10. c	Step 2: a
k. 3	11. d	Step 3: b
l. 4	12. a	Step 4: g
m. 10	13. e	Step 5: c
n. 11	14. a	Step 6: f
	15. c	Step 7: d
		10. 1. c
		2. b
		3. a

Chapter Three

Culture

LEARNING OBJECTIVES

After careful study of this chapter, you will be able to:
- Explain the relationship between culture and the genetic heritage of humans.
- Describe and illustrate the interplay between language and culture.
- Classify various aspects of culture into one of its three major dimensions.
- Discuss how cultural diversity is promoted within a society.
- Describe and illustrate the relationship between cultural diversity and ethnocentrism.
- Outline the advantages and disadvantages of ethnocentrism and discuss the role of cultural relativism in combating ethnocentrism.
- Explain the existence of certain aspects of culture that are shared around the world.

SOCIOLOGICAL IMAGINATION

Is there a common denominator among cultures around the world?   In recent years, there has been a growing trend toward believing that the existence of cultural diversity rules out the existence of universally accepted aspects of culture. Adoption of the perspective of cultural relativism has caused some Americans to conclude that aspects of one culture are neither superior nor inferior to those of another culture. Other Americans, in their enthusiastic endorsement of multiculturalism, have even stopped looking for similarities among cultures. This is a mistaken view. Social scientists have, in fact, documented a wide variety of "cultural universals" shared by all cultures.

This chapter focuses on the concept of culture as sociologists view it. After distinguishing between culture and society, this chapter will examine the relationship between culture, which is created by humans, and the biological makeup inherited by all people.

CULTURE, SOCIETY, AND GENETIC HERITAGE

Culture and Society
- -

Culture consists of the material objects as well as the patterns for thinking, feeling, and behaving that are passed from generation to generation among members of a society. On the physical, or *material,* side, culture in America includes skyscrapers, McDonalds restaurants, spaceships, and cars. On the *nonmaterial* side, American culture includes such things as beliefs, rules, customs, a family system, and a capitalist economy. More concisely, culture is a society's total way of life. Although culture and society are tightly interwoven—one cannot exist without the other—they are not identical. Culture is a society's total way of life; a **society** is composed of people living within defined territorial borders who share a common culture.

To understand sociology as the study of social structure, we must realize first that culture helps to explain human behavior. What people do, prohibit, like, dislike, believe, disbelieve, value, and discount are all based, in large part, on culture. Culture provides the blueprints people in a society use to guide their social relationships. It is from culture that teenage girls learn to fight with their peers for a piece of the clothing of a current rock star, and it is from culture that teenage boys come to believe that "pumping iron" is the gateway to popularity.

If human behavior is based on culture, which is created by humans, it cannot be biologically inherited. Human behavior, which is based on culture, must be learned. The concepts presented in this chapter substantiate this point. It is best to begin with an exploration of the interrelationships among culture, biology, and human nature.

Since people's ways of thinking, feeling, and behaving are based on learning rather than biology, there is much variation in human expression. These Shilluk (in the Sudan) learned to celebrate the thatching of their king's huts through this unique ritual.

Culture and Genetic Heritage

Instincts are genetically inherited, complex patterns of behavior that always appear among members of a particular species under appropriate environmental conditions. Though nonhuman animals are programmed for action through instincts, human infants cannot go very far on the basis of their genetic heritage alone. They lack immediate and automatic solutions to the problems they face. And later, without instincts to determine the type of shelter to build, the type of food to eat, the time of year to have children, or the type of mating pattern to follow, humans are forced to create and learn ways of thinking, feeling, and behaving—that is, they are forced to rely on culture.

If humans had instincts, they would all behave in the same way with respect to those instincts. If, for example, women had an instinct for mothering, then all women would want children and all women would love and protect their children. In fact, some women do not want to have children, and some women who do give birth abuse or abandon their children.

Is genetic heritage without influence on human behavior? Because humans are without instincts does not mean that they are unaffected by their genetic nature. Research on twins reared together and apart has examined the relative influence of heredity and environment on personality traits. According to a recent study, about 50 percent of the diversity in measured personality characteristics is attributable to genetic heritage (Tellegen et al., 1993).

In addition, humans have **reflexes**—simple, biologically inherited, automatic reactions to physical stimuli. A human baby, for example, cries when burned, blinks its eyes in response to a foreign particle, and contracts and dilates its pupils as lighting conditions change. Humans also have biologically inherited **drives,** or impulses, to reduce discomfort. They want to eat, drink, sleep, associate with others, and have sexual relations.

Genetically inherited personality traits, reflexes, and drives, however, do not "determine" how humans behave. Instead, culture channels the expression of these biological characteristics (Good, et al., 1994). Boys in "Old American" culture, for example, are taught not to cry in response to pain, whereas boys in Jewish and Italian families learn to pay attention to and to be more expressive regarding physical discomfort (Zborowski, 1952, 1969). To cite another example, humans have inherited the capacity for love. Awareness of this fact, however, does not allow us to predict the precise ways that different groups of people will express the ability to love. The people of one culture may believe that being one of several husbands is the most natural form of marriage, whereas members of another culture may endorse monogamous marriage.

Some social and biological scientists are challenging the traditional sociological perspective on human nature. This new perspective is called *sociobiology.*

Sociobiology

What is sociobiology? **Sociobiology** is the study of the biological basis of human behavior. Sociobiologists argue that, like physical characteristics, social behaviors are shaped through the evolutionary process. Thus, sociobiology is the application of Darwinian natural selection and modern genetics to behavior. According to Darwin's theory of evolution, all organisms evolve through the process of natural selection—those organisms best suited to an environment survive and reproduce themselves, and the rest perish. Sociobiologists assume that behavior enhancing the chances for survival is retained in an organism's genetic code to be passed from generation to generation (Degler, 1991).

If human behavior is based on millions of years of evolution, say the sociobiologists, then most human behavior is basically self-protective (Dawkins, 1976). Those behaviors that improve a species' chances for survival are said to be genetically retained. Parental affection and care, friendship, sexual reproduction, and the education of children must therefore be biologically based because they contribute to the survival of the human species. Some sociobiologists even contend that there may be specific genes for such things as aggression, religion, homosexuality, group loyalty or altruism, the creation of hierarchies, and the incest taboo.

Sociobiologists believe that the sharp line usually drawn between human and nonhuman animals is inappropriate. Nonhuman animals not only learn, they contend—as when a group of primates spontaneously begin to use long sticks to ferret out ants from an anthill for a meal—and transmit their knowledge, but also appear to be capable of using language. Many nonhuman animals, assert sociobiologists, exhibit intelligence of a kind formerly thought to be unique to humans (Begley, 1993; Linden, 1993a). In fact, pioneer sociobiologist Edward Wilson (1978, 1986) contends that the study of human behavior must begin with the genetic heritage of humans.

What do the critics of sociobiology say? Some critics of sociobiology fear its use to label some races inferior; others fear that it will be used to argue for the superiority of the male (Andersen, 1997). Still other critics fear that the sociobiological perspective will bolster rational choice theory and its basic assumption of innate human selfishness. Marxists object to sociobiology on the grounds that it justifies the existence of Western capitalism.

Critics join in common objection to the attempt by sociobiologists to explain human behavior genetically.

There is, they contend, too much social diversity around the world to explain human behavior biologically. The human brain and the unique human capacity for using language and creating social life have allowed humans to overcome any contribution to behavior that might come from their genes. Because of the size of their cerebral cortex—which permits, among other things, abstract thinking—humans possess an almost limitless capacity for creating and learning a variety of ways of thinking, feeling, and behaving. Nonhuman species, which either have no cerebral cortex or have an undeveloped cerebral cortex, behave similarly because of a genetic code. Birds do not walk south for the winter, salmon do not fly upstream, and lions do not prefer ferns to fresh meat. Humans, liberated from the confines of their genes by a large, well-developed cerebral cortex, on the other hand, create and transmit a dazzling array of ways of thinking, feeling, and behaving. Critics believe that sociobiology should not sidetrack sociologists from their efforts to understand and explain human behavior from the vantage point of the sociological perspective.

Since the early days of the rise of sociobiology, some common ground has developed between sociologists and sociobiologists. Anthropologist Marvin Harris (1980) agrees that there is some biological continuity between human and nonhuman animals and that there is a basic human nature. Several sociologists now contend that human biology and the human capacity for creating a nearly infinite variety of ways of thinking, feeling, and behaving are two sides of the same coin. The genetic heritage of humans, they argue, shapes and limits human nature and social life.

Consequently, they contend, sociologists should stop ignoring the complex relationship between genetic heritage and the human capacity for creating social life (Wozniak, 1984; Lenski, 1988; Lopreato, 1990).

LANGUAGE AND CULTURE

Culture is the social heritage of humans. This heritage is altered by each generation and must be learned by new members of society. Both the creation and the transmission of culture depend heavily on the human capacity to develop and use symbols, the most significant of which make up language. The following discussion relies on concepts central to symbolic interactionism with its emphasis on subjective interpretations of reality based on meanings learned from others.

Symbols, Language, and Culture

What are symbols? In Lewis Carroll's *Through the Looking Glass,* Humpty Dumpty says to Alice with some finality, "When I use a word, it means just what I choose it to mean—neither more nor less." So it is with **symbols**—things that stand for or represent something else. Symbols range from physical objects to sounds, smells, and taste.

The meaning of a symbol is not dictated by the physical characteristics of the thing it represents. There is nothing naturally good about the sound created by applauding. Although applause warms the heart of an entertainer, a politician, or a professor in the United

FEEDBACK

1. _____ consists of all the material objects as well as the patterns for thinking, feeling, and behaving that are passed from generation to generation among members of a society.
2. A _____ is composed of a people living within defined territorial borders who share a common culture.
3. A genetically inherited, complex pattern of behavior that always appears among members of a particular species under appropriate environmental conditions is a(n)
   a. reflex   b. instinct   c. drive   d. need
4. Scientists agree that humans have instincts for self-preservation, motherhood, and war. *T or F?*
5. Indicate which of the following are drives (D), which are reflexes (R), which are instincts (I), and which are human creations (H).
   ____ a. eye blinking in dust storm
   ____ b. need for sleep
   ____ c. reaction to a loud noise
   ____ d. socialism
   ____ e. sex
   ____ f. racial inequality
6. According to sociology, culture totally determines the nature of human behavior. *T or F?*
7. _____ is the study of the biological basis of human behavior.
8. Sociologists now agree that genetic heritage plays no part in the shaping and limiting of social life. *T or F?*

Answers: 1. Culture 2. society 3. b 4. F 5. a. (R) b. (D) c. (R) d. (H) e. (D) f. (H) 6. F 7. Sociobiology 8. F

SOCIOLOGY IN THE NEWS

The Symbolic River

In this clip, CNN reports on the symbolic meaning of the Ganges River in India. Although its symbolic meaning is religiously based (see *Sociology in the News*, Chapter 14), in the present context you should view its symbolic meaning much more broadly. Hindu attitudes and beliefs regarding the Ganges can be used to explore the general relationship between language and culture.

Symbols stand for or represent something else. For these Asian Indians, the water of the Ganges River is sacred.

1. Use the symbolic nature of the Ganges River to illustrate the relationship between culture and language.

    _____

    _____

2. Suppose you were going to explore the Sapir-Whorf hypothesis by focusing on the symbolic value of the Ganges. Indicate the kinds of things you would like to explore.

    _____

    _____

States, in Latin America it symbolizes disapproval. Of course, once meaning has been assigned to applause and we learn to associate it with approval or disapproval, it seems as though the appropriate meaning is determined by the applause itself. The seven-starred Confederate flag is an emotionally laden symbol that still divides African Americans and whites, particularly in the South (Pedersen, 1996). A red ribbon insignia has become the controversial symbol for AIDS supporters (Peyser, 1993). For some examples of new symbols in the computer age, see Table 3.1.

What is the relationship between language and culture? Language frees humans from the limits of time and place. It allows them to create culture. The Wright brothers' victory over gravity did not come from their personal attempts to fly from a hilltop with a complex arrangement of sticks and cloth attached to their backs. They constructed their airplane according to aerodynamic principles known for some time before their first successful flight. They could read, discuss, and recombine existing ideas and technology to accelerate their work.

Equipped with language, humans can transmit their experiences, ideas, and knowledge to others. Children can be taught things without any actual experience on their part. Although it may take some time and repetition, children can be taught the dangers of fire and heights without being burned or toppling from stairs. This principle of learning, of course, applies to other

cultural patterns, such as exhibiting patriotism, consuming food, or staying awake in church.

The Sapir-Whorf Hypothesis

According to Edward Sapir (1929) and Benjamin Whorf (1956), language is our guide to reality; our view of the world depends on the particular language we have learned. Our perception of reality is at the mercy of the words and grammatical rules of our language. And because our perceptions are different, our worlds are different. This is known as the *hypothesis of linguistic relativity.*

When something is important to people, the language of those people will contain many words to describe the thing. The importance of time in American culture is reflected in the many words the American language has that refer to time—*period, era, moment, interim, recurrent, century, afternoon, semester, eternal, annual, meanwhile, regularly,* just to name a few. When something is unimportant to people, such people may have no word for the thing. When Christian missionaries went to the Orient, they were dismayed because the Chinese language contained no word for—and no concept of—sin. Other missionaries were no less dismayed to learn that Africans and Polynesians had no way to express the idea of a single God.

The Sapir-Whorf hypothesis of linguistic relativity has been extended into areas as diverse as visual perception, audio perception, and ideas about privacy and

TABLE 3.1	New Symbols in the Computer Age
Beam	To transfer (software) of a file electronically.
Biff	To notify someone of incoming mail.
Casters-up mode	Synonym for "broken" or "down."
Dogwash	A project of minimal priority, undertaken as an escape from more serious work.
Fascist	Said of a computer system with excessive or annoying security barriers, usage limits, or access policies.
Hack together	To throw something together so it will work.
Pencil and paper	An archaic information storage and transmission device that works by depositing smears of graphite on bleached wood pulp.
Silly walk	A ridiculous procedure required to accomplish a task.
Snail-mail	Paper mail, as opposed to electronic.
Snark	A system failure.
Tired iron	Hardware that is perfectly functional but far enough behind the state of the art to have been superseded by new products.
User-friendly	Generally used by hackers in a critical tone, to describe systems that hold the user's hand so obsessively that they make it painful for the more experienced and knowledgeable to get any work done.
Wave a dead chicken	To perform a ritual in the direction of crashed software or hardware that one believes to be futile but is nevertheless necessary so that others are satisfied that an appropriate degree of effort has been expended.
Zipperhead	A person with a closed mind.

Source: Eric S. Raymond, *The New Hacker's Dictionary* (Cambridge, MA: The MIT Press, 1991).

space. The Tzeltal language in southern Mexico recognizes five basic colors—white, black, red, green, and yellow. The language of the Jalé of New Guinea has terms only for black and white (Berlin and Kay, 1969). The Japanese use paper walls as sound barriers and are unbothered by noise in adjacent rooms. Westerners staying at a Japanese inn during a party feel they are being bombarded with noise because they have not been mentally trained to screen out sound (Hall, 1966). Germans feel very protective about their privacy. Problems arise in subsidiaries of American firms in Germany because American executives leave their doors open and German executives have a closed-door policy. The German preference for physical barriers to protect their privacy is reflected in the prevalence of double doors in German hotels and public and private buildings (Hall, 1976, 1979).

Are people forever prisoners of their language? As far as we know, all people have the genetic capacity to learn any language. Thus, although the principle of linguistic relativity states that a person's view of the world is colored by his or her language, it does not state that people are forever trapped by their language. Exposure to another language can alter a person's perception of the world.

People can begin to view the world differently as they learn a new language. Most people, however, confine themselves to their native language. Consequently, most people interpret their environment through the words and structure of their language. This limits their ability to perceive and understand. Language, then, so necessary for the creation and transmission of culture, is also constricting.

DIMENSIONS OF CULTURE

If you wanted to describe and analyze a culture, what would you look for? How could you begin to classify the components of a people's way of life? A handy classification system provides three major dimensions of culture: the normative, the cognitive, and the material. An elaboration of these dimensions will help you understand the nature of culture. Note the relationships between sociology's major theoretical perspectives and the dimensions of culture.

1. _____ are things that stand for or represent something.
2. Because of _____, culture can be created and transmitted.
3. According to the hypothesis of linguistic relativity, words and the structure of a language cause people to live in a distinct world. *T or F?*
4. The language we learn determines forever the ways we see the world. *T or F?*

Answers: 1. Symbols 2. language 3. T 4. F

The Normative Dimension

The normative dimension of culture, which consists of the standards for appropriate behavior of a group or society, is heavily tied into functionalism with its emphasis on social integration, stability, and consensus. The most important aspects of the normative dimension are norms, sanctions, and values.

What are norms?   The poor in India can be found dead of starvation beside perfectly healthy and edible cattle (Harris, 1974). A Far Eastern woman may have her head severed for going to bed with a man before marriage. Roman emperors sent relatives to small isolated islands for life for disgracing the family. Each of these instances is a result of cultural **norms**—rules defining appropriate and inappropriate ways of behaving for specific situations. Norms explain why people act similarly in similar circumstances.

William Graham Sumner (1906), the early sociologist who wrote perceptively about norms, stated that anything can be considered appropriate when norms exist that approve of it. This is because once norms are learned, members of a society use them to guide their daily activities. Norms are so ingrained that they guide our activities without our awareness. In fact, we may not even be consciously aware of a norm until it has been violated. For instance, we do not think about standing in line for concert tickets as a norm until someone attempts to step in front of us. Then it immediately registers that taking one's turn in line is expected behavior and that someone is violating that norm.

Norms vary widely from society to society. While visiting the Persian King Gelese in the nineteenth century, Sir Richard Francis Burton witnessed the "Ku to Man" ceremony performed in his honor. The dramatic climax of this ceremony was the ritual execution of twenty-three resigned and terrified slaves (Kruschwitz and Roberts, 1987). Mark Shivas produced *The Borgias,* the story of a powerful family of the late fifteenth and early sixteenth century for the British Broadcasting Corporation. Shivas has been quoted as saying, "We did the orgy last week, perhaps the best-documented orgy in history. They had these naked women picking up chestnuts with their teeth. They called it 'The Chestnut Orgy,' and we shot it, you'll be sorry to hear, in 'good

taste.' " These organized activities occurred under one of the Borgias, named Pope Alexander VI, who distributed prizes after the orgies to those males who made love with the women involved the greatest number of times (Manchester, 1993). In recognition of the variability in cultural norms, a sanitized version of an orgy was shot for the countries involved in the production of *The Borgias.*

Norms also change within the same society. Premarital sexual intercourse among eighteen-year-old American women has increased dramatically since the 1950s. A group of high school boys in California, known as the "Spur Posse," tallied their sexual conquests in a competitive contest. One of the boys, subsequently arrested with several of his friends, claimed a

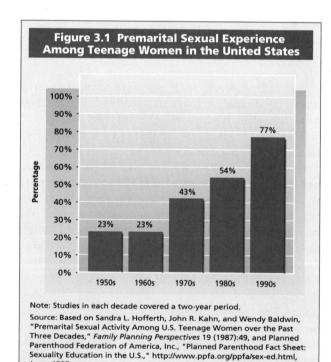

Figure 3.1 Premarital Sexual Experience Among Teenage Women in the United States

Note: Studies in each decade covered a two-year period.

Source: Based on Sandra L. Hofferth, John R. Kahn, and Wendy Baldwin, "Premarital Sexual Activity Among U.S. Teenage Women over the Past Three Decades," *Family Planning Perspectives* 19 (1987):49, and Planned Parenthood Federation of America, Inc., "Planned Parenthood Fact Sheet: Sexuality Education in the U.S.," http://www.ppfa.org/ppfa/sex-ed.html, June, 1995.

Internet Link: The Internet has an extensive list of sources that pertain to premarital sexual activity. Several of these provide useful data for the sociologist researching in this field. To begin, go to Campaign for Our Children at http://www.cfoc.org/

"score" of sixty-three points, each point earned by having an orgasm with a different girl (Smolowe, 1993a) Figure 3.1 illustrates the precipitous increase in premarital sexual intercourse from the 1960s to the 1990s in the United States.

Norms range from insignificant rules, such as applause after a performance, to important ones, such as laws against stealing. Sumner identified three basic types of norms: folkways, mores, and laws. These three types of norms vary in their importance within a society, and their violation is tolerated to different degrees.

What are folkways? Rules that cover customary ways of thinking, feeling, and behaving but lack moral overtones are called **folkways.** For example, whether one sleeps in a bed or on the floor is not a moral issue; one may eat six meals a day without being accused of immorality. Other folkways include eating breakfast, shaking hands when introduced, and drinking coffee during study breaks.

Because folkways are not considered vital to the welfare of the group, disapproval of violators is not very great. Those who consistently violate folkways—say, by persistently talking loudly in quiet places, wearing shorts with a suit coat and tie, or wearing a different-colored sock on each foot—might be considered odd and may be avoided, but are not considered wicked or immoral.

Some folkways are more important than others, and the social reaction to their violation is more intense. Failure to offer a woman a seat on a crowded bus draws little reaction, but obnoxious behavior at a party after excessive drinking may bring a severe negative response from others.

What are mores? Mores (pronounced MOR-ays; singular is *mos*) are norms of great moral significance and are thought to be vital to the well-being of a society. Conformity to mores elicits significant social approval; violation of this type of norm evokes strong disapproval. For example, Americans subscribe to a long-standing mos requiring able-bodied men to work for a living. Able-bodied men who do not work are stigmatized (Waxman, 1983).

Although conformity to folkways is generally a matter of personal choice, conformity to mores is required of all members of a society. As in the case of folkways, some mores are more vital to a society than others. Failure to respect the American flag while the national anthem is being played is not as serious a violation of American mores as using profanity aloud during a church service, and choosing to live on welfare is a more serious breach of appropriate conduct than either of these.

There are some mores that are more serious than others. A **taboo** is a norm so strong that its violation is

Norms, rules specifying appropriate behavior, vary from the optional to the mandatory. Norms against incest are very strong and demand compliance. Although failure to shake hands upon meeting someone is considered poor etiquette, failure to do so is not a moral issue in American society. Norms also vary widely from society to society. To bow in greeting is the Japanese equivalent of the American handshake.

thought to be punishable by the group or society or even by some supernatural force. Although definitions of incest vary from society to society, the incest taboo is generally regarded as the only taboo existing in all societies. There are, however, taboos other than the prohibition against having sexual relations with relatives. The mother-in-law taboo existing in some societies prohibits or severely restricts social contact between a husband and his wife's mother. Taboos existing in Western societies prohibit transvestism and bestiality (having sexual relations with animals). As Christie Davies (1982) points out, taboos such as those against transvestism and bestiality cannot have a psychological or biological origin, because some societies do not have such taboos.

What are laws? Law is a third type of norm. **Laws** are norms that are formally defined and enforced by designated persons. Folkways and mores emerge slowly and are often unconsciously created, but laws are consciously created and enforced.

Mores are an important source of laws. At one time in human history, the norm against murder was not written down. But as civilization advanced, the norm prohibiting murder became formally defined and was enforced by public officials. But not all mores become laws. For example, it is not against the law to cheat on a college examination. Nor have all laws been mores at one time—fines for overtime parking and laws against littering have never been mores.

Mores and laws sometimes overlap. Although private citizens may show their disapproval of the father of an illegitimate child, public officials can do little with regard to the father's responsibilities to support the child unless the mother is willing to start legal action. If the mother does start legal proceedings, however, it is the duty of the appropriate public officials to handle the situation. If a jury decides that the accused man is the child's biological father, compensation takes the form of financial support geared to the needs of the child and the mother and to the father's ability to pay. The refusal to provide compensation leaves the father open to a jail sentence for contempt of court.

Laws often remain on the books for a long time after the mores of a society have changed. It is illegal in Minnesota to hang male and female undergarments on the same clothesline; New York prohibits card playing on trains; elephants in Natchez, Mississippi, cannot legally drink beer; and it is against the law to wear roller skates in public bathrooms in Portland, Oregon.

Because norms must be learned and accepted by individuals, conformity to them is not automatic. Groups must have some means for teaching norms and encouraging conformity to them. This is done, in part, through the application of sanctions.

How are norms enforced? **Sanctions** are rewards and punishments used to encourage socially acceptable behavior. **Formal sanctions** may be given only by officially designated persons, such as judges, executioners, or college professors. Formal sanctions range widely in their severity. From the Middle Ages to the Protestant Reformation, the unpardonable sin was buying wholesale to sell at a profit. This civil crime was punishable on the third offense by public humiliation and social and economic ruin (Cockburn, 1994). More recently, and less severely, courts across the United States are once again handing down sentences involving public shaming, from permitting a woman attacked by her former husband to spit in his face to requiring child molesters to place signs in front of their residences describing their crimes (El Nasser, 1996). In 1997, Latrell Sprewell, National Basketball Association star player for the Golden State Warriors, had his $32 million, four-year contract revoked and was suspended by the NBA for one year after physically attacking his coach, P. J. Carlesimo.

Formal sanctions can also be positive. A soldier may be awarded the Congressional Medal of Honor for heroism in battle, and a professor may reward some students with A's.

Informal sanctions—which can be applied by most members of a group—can also be positive or negative. They include anything from thanking someone for pushing your car out of a snowbank to looking harshly at someone who is cheating on a test. The star of the television show "Roseanne," then Roseanne Barr Arnold, was booed and heavily criticized for her shrill rendition of the national anthem and parody of baseball players (at the conclusion of the song, she grabbed herself between the legs and spat) at a 1990 Major League baseball game.

Sanctions are not used randomly. Specific sanctions are associated with specific norms. Children who do not come home when their parents call are not supposed to be beaten and locked in a closet. A convicted murderer is not invited to the White House for a special celebration. Instead, a society or group develops appropriate sanctions for following or failing to follow specific norms (Horwitz, 1990).

Sanctions do not have to be used most of the time. After we reach a certain age, most of us conform without being pressured by others. We may conform because we believe that the behavior expected of us is appropriate, because we wish to avoid guilt feelings, or because we fear social disapproval. In other words, if we have been properly socialized, we will sanction ourselves mentally before doing something deviant.

The severity of the sanction for a deviant act varies from one society to another and from one time to another. A convicted thief in the United States might be sentenced to only a few months in jail. A convicted thief in North Yemen might have to pick up his amputated hand and salute the judge with it. In 1993, five Somalian women accused of adultery were stoned to death and a sixth woman was lashed one hundred times by a mob of Muslim fundamentalists. During the 1960s, possession of an ounce of marijuana could result in several years in prison. Today, possession of a small amount of marijuana in most states is a misdemeanor, and in several states it is equivalent to a minor traffic violation.

Sometimes, informal sanctions are illegally imposed. A soccer sportscaster was shot in the knees in Avellino, Italy, by irate fans. The rash of "dowry deaths" in India not too long ago is also an example of illegal sanctioning. According to police statistics, almost four hundred women were set afire in New Delhi in 1981. Although some of these deaths were accidental and some were suicides, it is suspected that many of them were murders. Some of the mothers-in-law, upset over the small size of their son's wife's dowry, poured gasoline over their daughters-in-law and threw matches on them.

So far we have discussed norms and sanctions as aspects of normative culture. Though norms and sanctions are relatively specific and concrete, the next major component of the normative dimension of culture—*values*—is rather broad and abstract. Values are much more general than norms, which offer specific guidelines for behavior.

What are values? **Values** are broad cultural principles embodying ideas about what most people in a society consider to be desirable. Values are so general that they do not specify appropriate ways of thinking, feeling, and behaving. Thus, different societies or different groups within the same society can have quite different norms based on the same value. For instance, the value of freedom has not been expressed through the same norms in America and the former Soviet Union. In the Soviet Union, as Robin Williams (1970) notes, freedom was expressed in the right to such things as employment, medical care, and education. Americans have different norms based on the value of freedom—the right to free speech and assembly, the right to engage in private enterprise, the right to a representative government, the right to change where they live and work. Patriotism for some citizens might involve dying for their country in combat. For others, refusing to fight in a war they consider illegal and immoral might be patriotic. Identical values do not result in identical norms.

Values have a tremendous influence on daily life because norms are based on them. A society that values democracy will have norms insuring personal freedom; a society that values humanitarianism will have norms providing for its most unfortunate members; a society that values hard work will have norms against laziness. Values are also important because they are so general that they tend to permeate most aspects of daily life. In America, for example, the influence of the value of freedom goes beyond political life. It affects such diverse areas as relationships in the family, treatment within the legal system, the functioning of organizations, and religious affiliation.

What are the basic values in American society?   Any attempt to state the basic values of American society is risky. Because America has so many diverse groups and because it is constantly changing, any one set of values is unlikely to receive unanimous support. Despite these problems, Robin Williams (1970) has done an excellent job of outlining fifteen major values guiding the daily lives of most Americans. Whether or not his classification is complete, it provides a picture of many of the major values influencing Americans. These values include achievement and success, activity and work, humanitarianism, efficiency and practicality, progress, material comfort, equality, freedom, democracy, individuality, science and rationality, external conformity, group (racial, ethnic, religious) superiority, morality, and patriotism. Following is a sample of the values identified by Williams.

- *Achievement and success.* Americans emphasize achievement, especially in the world of work. Success is supposed to be based on effort and competition; success is viewed as a reward for performance; wealth is a symbol of success and personal worth; and quantity often represents quality.
- *Activity and work.* We tend to stress action over inaction in almost all instances. For most Americans, continuous and regular work is an end in itself, advancement is to be based on merit rather than favoritism, and all citizens are supposed to have the opportunity to perform at their best.
- *Efficiency and practicality.* Americans pride themselves on getting things done by the most rational means. We are always searching for better ways of doing things, praising good workmanship, judging work performance by its practical consequences, and depending on science and technology.
- *Equality.* Americans advocate equality for all citizens. In interpersonal activities, we tend to treat one another as equals, defend everyone's legal rights, and favor equal opportunity for everyone.

Achievement is one of America's most basic values. The expressions on these parents' faces reflect happiness and pride over the significant accomplishment of their daughter.

- *Democracy.* Americans emphasize that all citizens are entitled to equal rights. We all should have equal rights and equal opportunity under the law, and power should not be concentrated in the hands of an elite.
- *Group superiority.* Despite their concern for individual personality and equality of opportunity, Americans tend to place value on people according to the race, ethnic group, social class, or religious group they belong to.

These values clearly are interrelated. The values of achievement and success and activity and work are related to the values of efficiency and practicality. Equally obvious is the conflict among some values: Americans value group superiority while stressing equality and democracy.

Are these values as prominent in American society today? This is a legitimate question because Williams identified these major values approximately thirty years ago. Although these values have remained remarkably stable over the years, some have changed somewhat. For example, there is less emphasis on group superiority in America than in the past, as can be seen in the apparent decline of racist attitudes and behavior. Still, due to the durability of values, racism remains part of the fabric of American culture. Actually, it is usually norms and behavior rather than values that change radically. In other words, because of the passage of laws forbidding discrimination and other related changes, many Americans are currently less likely to make overt racist statements and discriminate against minority members even though they may still subscribe to the value of group superiority. They are simply

Although Americans stress equality of opportunity and individual merit, they also evaluate others on the basis of ascribed characteristics such as social class, race, and ethnic group. Members of this Chinese subculture in the United States experience prejudice and discrimination based on their physical and cultural characteristics.

behaving in a way that is consistent with changed norms. The norms related to hard work and activity have apparently also changed in recent years. According to Lionel Lewis (1982), many Americans now work as hard at their leisure (for example, long-distance running and mountain climbing) as they do at their jobs.

Although Williams's analysis of major American values remains basically valid today, it is not surprising that some, quite validly, believe that Williams's list is incomplete. George and Louise Spindler (1983), for example, wish to add optimism, honesty, and sociability to the list of major American values.

Although functionalism is the basis of the normative dimension of culture, symbolic interactionism is related to the cognitive and material dimensions. Be alert to these theoretical relationships.

The Cognitive Dimension

Cognition is the mental process that enables humans to think, remember, recognize, and imagine. The most important aspect of the cognitive dimension of culture is **beliefs**—ideas concerning the nature of reality. Actually, beliefs can be true or false. The Romans believed that Caesar Augustus was a god; the Tanala, a hill tribe of Madagascar, believed that the souls of their kings passed into snakes; and many Germans believed that pictures of Hitler on their walls would prevent the walls from crumbling during bombing raids. On the American frontier, it was believed that a dog howling in the distance foretold a death in one's family, that anyone who brought a shovel into a cabin would carry out a coffin, and that a bird on one's windowsill or inside one's home meant sorrow in the future. The Black Death, which killed twenty-five million Europeans during the fifteenth century, was believed to be God's punishment for the sins of the people. More recently, there are those who believe that the Nazis under Hitler's command did not murder six million Jews (Shapiro, 1993) and that abductions by extraterrestrials do occur (Willwerth, 1994). There is little reason to consider any of these beliefs true, but for other beliefs—such as the belief that the human eye can distinguish over seven million colors or the belief that the earliest known ancestors of humans lived three to four million years ago (Begley, 1994; Lemonick, 1994)—there is factual evidence. Beliefs are important, however, because people accept them as being true and base their activities on them.

The Material Dimension

Material culture consists of the concrete, tangible objects within a culture—automobiles, chairs, highways, birth-control pills, art, Guess? jeans. Artifacts or physical objects have no meaning or use apart from the meanings people give them. Consider newspaper and pepper as physical objects. Of course, each of these two things has various meanings for you, but can you think of a use for them in combination? Some midwives have used pepper and newspaper in a process known as nettling. An elderly medical doctor tells the story of his first encounter with nettling:

> The ink of my medical license was hardly dry, and as I was soon to find out, my ears would not be dry for some time. I had never delivered a baby on my own and faced my maiden voyage with some fear.
>
> Upon entering Mrs. Williamson's house, I found a local midwife and several neighbors busily at work preparing for the delivery. My fear caused me to move rather slowly and my happiness over my reprieve prompted me to tell the women that they were doing just fine and to proceed without my services.
>
> Having gotten myself off the hook, I watched the ladies with a fascination that soon turned to horror.
>
> At the height of Mrs. Williamson's labor pains, one of the neighbors rolled a piece of newspaper into a funnel shape. Holding the bottom end of the cone she poured a liberal amount of pepper into it. Her next move was to insert the sharp end of the cone into Mrs. Williamson's nose. With the cone in its "proper" place, the neighbor inhaled deeply and blew the pepper from the cone into the inner recesses of Mrs. Williamson's nose—if not her mind.
>
> Suddenly alert, Mrs. Williamson's eyes widened as her senses rebelled against the pepper. With a mighty sneeze, I was introduced to nettling. The violence of that sneeze reverberated through her body to force the baby from her womb in a skittering flight across the bed. An appropriately positioned assistant fielded the baby in midflight and only minor details of Orville's rite of birth remained.

Before the doctor was introduced to nettling, this particular combination of newspaper and pepper had no meaning for him. And until nettling was devised, the combination of the objects was without meaning for anyone, even though the separate physical objects existed as part of the culture.

Physical objects do not all have the same meanings and uses in all societies. Although it is conventional to use a 747 jet for traveling, it is possible that a crashed 747 in a remote jungle region of the world could be used as a place of worship, a storage bin, or a home. In the United States, out-of-service buses, trains, and trolley cars have been converted to restaurants.

Clearly, the meaning of physical objects is not determined by the physical characteristics of the objects. People use newspaper and pepper during childbirth or make a temple of a downed 747 jet because of cognitive and normative definitions. The meanings of physical objects are based on the beliefs, norms, and values people use with regard to them. This is readily apparent when new meanings of a physical object are considered. At one time, only pianos and organs were used in church services; guitars, drums, and trumpets were considered too secular to accompany a choir. Many churches today use these "worldly" instruments regularly in their worship activities. These instruments have not changed physically, but the cultural meanings placed on them have.

Ideal and Real Culture

A gap sometimes exists between cultural guidelines and actual behavior. This gap is captured in the concepts of ideal and real culture. **Ideal culture** refers to aspects of culture publicly embraced by members of a society. **Real culture** refers to subterranean patterns for thinking, feeling, and behaving. These subterranean patterns are aspects of culture guiding behavior that are publicly denied because they conflict with the ways people profess to behave. One aspect of America's ideal culture is honesty. Yet many taxpayers annually violate both the letter and spirit of existing tax laws, many people in business engage in dishonest business practices, many students cheat on exams, and many college athletes do the "high $500" handshake (during which a team booster leaves illegal money in their palms). These are not isolated instances. These are real cultural patterns passed on from generation to generation. Keep in mind that we are not referring to cases of individual deviance such as people who murder, rape, and rob. These types of behavior violate even real culture.

Does this make ideal culture meaningless? Absolutely not. In an imperfect world, ideal culture provides high standards. These ideals become targets that most people attempt to reach most of the time. Otherwise, social chaos would prevail. Ideal culture also permits the detection of deviant behavior. The sanctioning of individuals who carry the real cultural pattern too far preserves the ideal pattern from extinction.

Culture as a Tool Kit

According to the dominant conception, culture guides behavior by providing the values or ends toward which behavior is directed. Ann Swidler, a contemporary critic of the culture-as-a-way-of-life approach, thinks culture should be viewed as a " 'tool kit' of symbols, stories, rituals, and world views, which people may use in varying configurations to solve different kinds of problems" (Swidler, 1986:273). In this view, culture provides a range of choices to be applied in defining and solving the problems of living. Because the content of any culture is not fully consistent, the "tools" that are chosen vary among individuals and situations. This view of

FEEDBACK

1. _____ are rules defining appropriate and inappropriate ways of behaving.
2. _____ are rewards and punishments used to encourage desired behavior.
3. Indicate whether the following are formal sanctions (F) or informal sanctions (I).
   ____ a. A mother spanks her child.
   ____ b. A professor fails a student for cheating on an exam.
   ____ c. A jury sentences a person to life in prison for espionage.
   ____ d. A husband separates from his wife after she has an affair.
4. _____ are broad cultural principles embodying ideas about what most people in a society consider to be desirable.
5. _____ are ideas concerning the nature of reality.
6. Indicate whether these statements best reflect a belief (B), folkway (F), mos (M), law (L), or value (V).
   ____ a. conception that God exists
   ____ b. norm against cursing aloud in church
   ____ c. norm encouraging the eating of three meals daily
   ____ d. idea of progress
   ____ e. norm against burning a national flag
   ____ f. norm encouraging sleeping in a bed
   ____ g. norm prohibiting murder
   ____ h. norm against overtime parking
   ____ i. idea that the earth is elliptical
   ____ j. idea of freedom
7. _____ culture consists of the concrete, tangible objects within a culture.
8. _____ culture refers to aspects of culture publicly embraced by members of a society.

Answers: 1. Norms 2. Sanctions 3. a. (I) b. (F) c. (F) d. (I) 4. Values 5. Beliefs 6. a. (B) b. (M) c. (F) d. (V) e. (M) f. (F) g. (L) h. (L) i. (B) j. (V) 7. Material 8. Ideal

culture is gaining visibility. Its eventual prominence in the sociological approach to culture remains to be seen (Hays, 1994).

CULTURAL DIVERSITY AND SIMILARITY

Theoretical Perspectives and Culture

Inherent in the discussion thus far is the relationship between sociology's major theoretical perspectives and some central cultural concepts. Symbolic interactionism is intimately connected with the role of language in culture and with the cognitive and material dimensions of culture. The assumptions of functionalism underlie the normative dimension of culture. As you view cultural diversity and similarity, you will again be aware of the appropriate theories. Conflict theory, with its emphasis on inconsistency, conflict, and change, undergirds any discussion of cultural diversity, whereas functionalism is the basis for a sociological analysis of cultural similarity.

Cultural Diversity

In ancient Rome, among the upper classes, adultery was a commonly accepted practice and divorce was granted upon agreement of the couple. In modern Italy, adultery is frowned on and divorce has only recently been legalized. Under Saudi Arabian religious laws, adulterers are stoned to death in public; some American married couples advertise their "swinging" behavior in newspapers and magazines. Eskimos used to offer their wives as sexual favors to visitors—even strangers—and felt offended if the offer was refused. These examples reflect the almost endless cultural variations existing in the world. Not only do different societies have different beliefs, norms, values, and sanctions, but various groups within the same society have their own cultural patterns. Even in a simple society, cultural diversity makes it impossible for all members to participate in all aspects of culture. In modern societies, cultural diversity is staggering. Given the biological similarity among humans, this diversity must be explained by nongenetic factors.

How is cultural diversity promoted? Cultural diversity exists in all societies in part because of the presence of **social categories**—persons who share a social characteristic such as age, sex, or religion. Members of social categories are expected to participate in aspects of culture unique to them. As one's age changes, different kinds of behavior are in order; the sexes have traditionally been assigned distinctly different roles.

Cultural diversity also stems from the existence of groups that are somewhat separate from the larger culture. These groups participate in the larger culture—they may speak the language, work in regular jobs, eat and dress like others, attend recognized churches, and so on. But despite this participation, these groups have some ways of thinking, feeling, and behaving that set them apart from the dominant culture. These groups—known as *subcultures* and *countercultures*—are more prevalent in large, complex societies.

A **subculture** is a group that is part of the dominant culture but differs from it in some important respects. One of the clearest illustrations of a subculture is an ethnic minority concentrated in one location, such as San Francisco's Chinatown. Early Chinese immigrants brought much of their native culture with them to America and have attempted to retain it by passing it from generation to generation. Although Chinese residents of Chinatown have been greatly affected by American culture, they have retained many cultural patterns of their own, such as language and family structure.

Residents of southern Appalachia can be considered another example of a subculture. These people subscribe to ways of thinking, feeling, and behaving that are said to be quite different from those of what might be called middle-class America. According to Jack Weller (1965), middle-class Americans believe in such things as freedom to determine one's destiny, progress, planning ahead, status seeking, education, and social participation. In contrast, Weller contends, southern Appalachians are fatalistic, present oriented, not status seeking, and nonparticipative. The designs for living of the southern Appalachians may be different from middle-class American patterns, but they cannot be considered to be in opposition to the larger culture. Their ways of living neither were created in reaction to nor are sustained by opposition to American cultural patterns. Rather, Weller argues, the existence of the southern Appalachian subculture represents an adaptation to the reality of living a deprived and frustrating existence. (See also Rodman, 1963; Rainwater, 1967; Ball, 1968; and Lewis, 1978. Weller's interpretation is merely one of many attempts to describe the Appalachian subculture. Many sociologists take issue with his particular interpretation.)

Some other subcultures in America are adolescents, the poor, drug addicts, and the wealthy, as well as people in prisons, mental hospitals, convents, and universities. (For other excellent examples, see Kephart and Zeller, 1994; Fine, 1996.)

A **counterculture** is a subculture that is deliberately and consciously opposed to certain central aspects of the dominant culture. A counterculture can be understood only within the context of its underlying opposition to the dominant culture. (See *Doing Research.*) For example, a counterculture was formed on American college and university campuses in the 1960s.

What Edward Suchman (1968) called the "hang-loose" ethic rejected such central aspects of American culture as material success, achievement, hard work, efficiency, authority, premarital chastity, and the nuclear family. Delinquent gangs, motorcycle gangs, certain types of drug groups, and revolutionary political or religious groups may also form countercultures (Yinger, 1984; Cushman, 1995; Roszak, 1995; Zellner, 1995).

What is the relationship between cultural diversity and ethnocentrism? Once members of a society learn their culture, they tend to become strongly committed to it. In fact, they tend to become so committed that they cannot conceive of any other way of life. If members of other groups seem different, they are considered strange. Cultural differences may even become indications of inferiority. This tendency to judge others in terms of one's own cultural standards is referred to as **ethnocentrism.**

People who spend most of their lives with others culturally similar to themselves—who hardly ever have to deal with people different from themselves—will almost inevitably use their own cultural standards to judge those who are different. The ethnocentric eye sees those who are different as inferior, ignorant, crazy, or immoral. Christian missionaries were appalled at the sexual and religious habits of Polynesian peoples. We are shocked by a people such as the Ik, who willingly

These members of a street gang are participants in a counterculture. They are aware of the beliefs, values, and norms of the larger culture but consciously participate in a lifestyle that opposes it.

The effects of judging others in terms of one's own cultural standards, known as ethnocentrism, can be disastrous. The recent genocide in Bosnia is only one contemporary example.

abandon their children at three years of age (Turnbull, 1972). It is hard to understand that the Yanomamö women of Venezuela and Brazil measure the affection of their husbands by how many beatings they get (Chagnon, 1997).

Less exotic examples of ethnocentrism between and within societies are plentiful. Members of all societies tend to offer themselves as exemplary models of what the gods really had in mind for human beings. The Olympic Games are much more than an arena for young men and women to engage in healthy and exuberant competition. In addition to this stated purpose of the games, they are also an expression of ethnocentrism. Influential political and nationalistic undercurrents run through the Olympics. A country's final ranking in this athletic competition for gold, silver, and bronze medals is frequently taken as a reflection of the country's worth and status on the world stage.

Consider some intrasocietal instances of ethnocentrism. Boston is said by some (mostly Bostonians) to be the hub of the universe. Some inhabitants of the Northeastern Seaboard have been known to express the belief that only a wasteland separates them from the Pacific Ocean; some more enlightened Easterners are willing to concede the addition of California to the East Coast in describing all that is good in the United States. Of course, this long-standing ethnocentric viewpoint has been subject to attack by those in Middle America who have always felt that virtue actually resides with them. Finally, the members of country clubs, churches, and schools all over America feel that their particular ways of living are eminently suitable for social and cultural duplicating.

What are advantages and disadvantages of ethnocentrism? To this point, ethnocentrism has been portrayed as either villainous or ridiculous. It may be both.

But it is inevitable. And its inevitability is partly rooted in the advantages it offers to social life. Imagine the expense and effort required to create from scratch the integration, high morale, loyalty, and stability within a social structure that ethnocentrism—based on natural cultural learning—provides. Few things draw people closer together than the shared conviction of their superiority vis-à-vis others. Such a certainty makes people feel good about themselves and their fellow members. The fires of nationalism and patriotism—prominent forms of social loyalty—cannot be kindled by propaganda proclaiming national inferiority. Indeed, because of ethnocentrism, life itself is the loyal patriot's freely offered gift for the common cause. Finally, social stability is promoted because the need for change is seldom entertained by those armed with the conviction that truth and beauty are already theirs.

Ethnocentrism not only encourages group integration, loyalty, morale, and stability, but also provides a justification for the script by which group members can make their way in their social orbits. Released from the need for attending to routine matters, group members have more time and energy for creativity in handling either novel situations or more pressing and enduring problems. That excessive ethnocentrism can transform this asset into a liability is an observation that will be elaborated on shortly.

People are not ethnocentric simply because they consciously realize the advantages of ethnocentricity for social life. No. Ethnocentrism would be pervasive even if it did not offer these advantages. Ethnocentrism is inevitable largely because of successful socialization. We are taught the rightness of our culture from a variety of sources (home, church, peers, schools, mass media), and thus it is only natural that this "rightness" becomes the yardstick for evaluating the ways of thinking, feeling, and behaving of social structures other than our own. Judgments of all kinds are nearly always alloyed with socially derived convictions regarding right and wrong. Ethnocentrism, then, is a predictable by-product of the transmission and learning of culture.

That ethnocentrism carries certain advantages has just been noted. However, a price may have to be paid for the integration, morale, loyalty, stability, and cultural justification it supplies. Extreme ethnocentrism has ill effects within societies as well as between them. On the intrasocietal side, extreme ethnocentrism may create such a high degree of integration and stability that innovation is hampered. Societies whose members are too firmly convinced of their righteousness may experience a choking off of internal exploration for new solutions to persistent problems. Further, they may reject, without examination, solutions that might be gleaned from the experience of other societies. If heightened ethnocentrism impedes both the internal creation and the external borrowing of ideas for solving old problems, it may create an even greater disadvantage in meeting the challenge of new problems and unfamiliar circumstances.

Although extreme ethnocentrism may contribute to rigid integration within a society, it may also promote injurious intrasocietal fragmentation. A society fragmented by ethnocentrically induced barriers separating subcultures or social categories also suffers from the damper placed on innovation. Thus, within societies, extreme ethnocentrism, whether it fosters societal integration or fragmentation, works counter to the general societal welfare.

On the intersocietal front, conflict is nearly always the result of extreme ethnocentrism. Global peace and welfare become secondary goals in a world in which ethnocentrism intensifies intersocietal conflicts based on power struggles for economic superiority and/or military supremacy.

We have already seen some of the negative consequences of ethnocentrism for a society. Ethnocentrism has harmful effects on individuals as well. **Culture shock**—the psychological and social stress we may experience when confronted with a radically different cultural environment—is one such negative consequence. Even cultural anthropologists, who are trained professionals, may experience culture shock. Napoleon Chagnon gives this account of his reaction to first contact with the Yanomamö:

My heart began to pound as we approached the village and heard the buzz of activity within the circular compound. Mr. Barker commented that he was anxious to see if any changes had taken place while he was away and wondered how many of them had died during his absence. I felt into my back pocket to make sure that my notebook was still there and felt personally more secure when I touched it. Otherwise, I would not have known what to do with my hands.

The entrance to the village was covered over with brush and dry palm leaves. We pushed them aside to expose the low opening to the village. The excitement of meeting my first Indians was almost unbearable as I duck-waddled through the low passage into the village clearing.

I looked up and gasped when I saw a dozen burly, naked, filthy, hideous men staring at us down the shafts of their drawn arrows! Immense wads of green tobacco were stuck between their lower teeth and lips making them look even more hideous, and strands of dark-green slime dropped or hung from their noses. We arrived at the village while the men were blowing a hallucinogenic drug up their noses.

One of the side effects of the drug is a runny nose. The mucus is always saturated with the green powder and the Indians usually let it run freely from their nostrils. My next discovery was that there were a dozen or so vicious, underfed dogs snapping at my legs, circling me as if I were going to be their next meal. I just stood there holding my notebook, helpless and pathetic. Then the stench of the decaying vegetation and filth struck me and I almost got sick. I was horrified. What sort of a welcome was this for the person who came here to live with you and learn your way of life, to become friends with you? They put their weapons down when they recognized Barker and returned to their chanting, keeping a nervous eye on the village entrances (Chagnon, 1997:5).

If trained anthropologists can experience culture shock, it is easy to understand why others are stunned by exposure to cultural practices foreign to their own. In fact, culture shock is a problem for American workers with overseas assignments (Wagster, 1993).

What can be done to reduce the negative personal effects of ethnocentrism? Awareness of its existence and the harm it can cause is a necessary first step. An important second step is to understand that values, norms, beliefs, and attitudes are not in themselves correct or incorrect, desirable or undesirable. They simply exist within the total cultural framework of a people and should be evaluated in terms of their place within the larger cultural context of which they are a part rather than according to some alleged universal standard that applies across all cultures. This perspective is known as **cultural relativism** and gives us a unique window through which to observe cultural variations.

Offering one's mate for sexual dalliance with an overnight guest is not allowed within most societies. There are, however, some societies in which trading spouses is not only acceptable but expected behavior. In traditional Eskimo society, it was a serious personal affront to the husband if a guest refused to "laugh" (have sexual intercourse) with his wife. Below is a passage from a novel based on Eskimo life. Some time has elapsed since Ernenek's rage caused him to accidentally kill the guest who had refused to have sexual relations

Bennett Berger—The Counterculture of a Commune

It took Bennett Berger several years to develop his ideas about the meaning of his research on communes, but he succeeded in writing a book, *The Survival of a Counterculture.* This book provides an excellent example of a sociologist's ability to shed light on the nature and importance of shared understandings within human groups.

During the period of his research, Berger never lived in communes as a permanent resident, but he and his research associates visited many of them. One commune—described in Berger's book as "The Ranch"—was visited repeatedly by Berger and provided the major source of information for Berger's research.

The members of The Ranch were ambivalent about Berger's role. They liked Berger as a person, but they had a low regard for sociology, which they viewed as "cold and objective—part and parcel of the bureaucratic spirit of modern life that they regarded as the spirit of death" (Berger, 1981:41). They were willing to tolerate Berger's presence, but they were distrustful of the real reasons for his being there.

Such communes usually have short lifespans, but The Ranch had survived for more than ten years before Berger's research began. Berger offers no single explanation for the commune's success, but he believes that the inhabitants' shared belief system and capacities for "remedial ideological work" provide a partial explanation.

Remedial ideological work refers to the process of altering beliefs when they "do not seem to be very effective in serving the interests of believers—or when routine behavior is in apparent contradiction with professed belief" (Berger, 1981:15). Members of The Ranch, for example, were very concerned with ecological ideals and distrustful of modern technology. For these reasons, they refused to allow their land to be used for commercial logging (thereby foregoing considerable income), and despite problems with insects, they would not use commercial insecticides. Yet they bought a chain saw for cutting firewood. The argument about the chain saw illustrates their ability to modify their thinking when necessary, that is, to do remedial ideological work:

The argument in favor of the chain saw was based on its efficiency and the situational requirements for fuel in winter; the argument against its use involved a more abstract ideological feeling against technology: the ugliness of the sound, the need for more gasoline, making them more

with his wife Asiak. Ernenek and Asiak are talking with a potential ally in the matter, a white man whose life Ernenek had saved.

"You have saved my life, Ernenek," the white man said, "and I wish to straighten things out so that you need no longer fear my companions. But you will have to stand before a judge. I will help you explain things."

"You are very kind," said Ernenek, happy.

"You said the fellow you killed provoked you?"

"So it was."

"He insulted Asiak?"

"Terribly."

"Presumably he was killed as you tried to defend her from his advances?"

Ernenek and Asiak looked at each other and burst out laughing.

"It wasn't so at all," Asiak said at last.

"Here's how it was," said Ernenek. "He kept snubbing all our offers although he was our guest. He scorned even the oldest meat we had."

"You see, Ernenek, many of us white men are not fond of old meat."

"But the worms were fresh!" said Asiak.

"It happens, Asiak, that we are used to foods of a quite different kind."

"So we noticed," Ernenek went on, "and that's why, hoping to offer him at last a thing he might relish, somebody proposed him Asiak to laugh with."

"Let a woman explain," Asiak broke in. "A woman washed her hair to make it smooth, rubbed tallow into it, greased her face with blubber and scraped herself clean with the knife, to be polite."

"Yes," cried Ernenek, rising. "She had purposely groomed herself! And what did the white man do? He turned his back to her! That was too much! Should a man let his wife be so insulted? So somebody grabbed the scoundrel by his miserable little shoulders and beat him a few times against the wall—not in order to kill him, just wanting to crack his head a little. It was unfortunate it cracked a lot."

"Ernenek has done the same to other men," Asiak put in helpfully, "but it was always the wall that went to pieces first."

The white man winced, "Our judges would show no understanding for such an explanation. Offering your wife to other men!"

dependent on outside sources, the psychological association of the saw with the hated lumber companies. The clinching argument on behalf of the chain saw was the one that denied that the use of the chain saw constituted giving in to "technology." A chain saw was not technology; it was a "tool" (Berger, 1981:115–116).

Like all settlers, inhabitants of The Ranch used tools on a daily basis. Classifying a chain saw as "tool" rather than "technology" therefore removed it from the category of "things to be despised." Such remedial ideological work is common, but it did not weaken the group's shared understandings, because the basic belief system was left intact.

On many occasions, on the other hand, individuals were expected to alter their behavior to conform to the group's norms. For example, when Vickie and Paul, two long-time members of the commune who were also lovers, had a quarrel because Paul had slept with another woman, the other commune members sided with Paul. They did so because of their belief that members of the commune have no right to be possessive of one another. Rather than allowing remedial ideological work to be done on the nonpossessiveness rule, they encouraged the rejected parties to focus on their relationship to the group. Ranch members

. . . tell of sitting with a brother or sister while his former lover is making love with his or her new partner nearby: holding, rocking, succoring through the tears, the trembling, the nausea, the sweat, and the waves of goosepimples on the skin. There is a strong element of ideological comfort as well: lovers come and go, but brothers and sisters (kinship) are forever (Berger, 1981:148).

Berger used a long and sometimes difficult period of research to show that the members of The Ranch shared a set of cultural understandings that set them apart from many segments of society but added to their internal strengths as a social entity.

Critical feedback

1. Berger claims to have experienced problems of panic and emotional involvement while conducting his research. In your opinion, could such problems interfere with the validity of his conclusions? Why or why not?

2. Communes like The Ranch usually have short lifespans. What factors allowed The Ranch to survive?

3. What does Berger mean by "remedial ideological work"? How does this concept contribute to an understanding of change in cultural systems?

"Why not? The men like it and Asiak says it's good for her. It makes her eyes sparkle and her cheeks glow."

"Don't you people borrow other men's wives?" Asiak inquired.

"Never mind that! It isn't fitting, that's all."

"Refusing isn't fitting for a man!" Ernenek said indignantly. "Anybody would much rather lend out his wife than something else. Lend out your sled and you'll get it back cracked, lend out your saw and some teeth will be missing, lend out your dogs and they'll come home crawling, tired— but no matter how often you lend out your wife she'll always stay like new" (Ruesch, 1959:86–88).

It is clear from this conversation that cultural blindness and ethnocentrism are making communication between the Eskimos and their white friend quite difficult. The white man automatically assumed that Ernenek's rage was for an attempted sexual assault on Asiak. This interpretation by the white man provides amusement for the Eskimos, who go on to explain that not only did the dead man refuse to "laugh" with Asiak but he rejected the delicacy of old meat containing the freshest worms. Ernenek plays a game of cultural one-upmanship by responding to the white man's explanation to Asiak that his own people are accustomed to a different kind of food with a sly "So we noticed." Upon comprehending the final provocation for murder, their white friend immediately tells them of the hopelessness of their case in a white man's court. In fact, ethnocentrism, if not mild culture shock, overtakes their white friend who attempts to shut the case with a dogmatic "It isn't fitting, that's all." But to Ernenek and Asiak it is by no means unfitting to lend wives. With resentment fired by ethnocentrism, Ernenek forcefully explains that of all the possessions a man owns, only a wife stays like new with use. And if you ask Asiak, use brought on by "laughing" adds to her value.

Ernenek's heated justification for wife lending is an excellent illustration of the cultural relativistic perspective—the Eskimos' possessions were handmade, difficult to replace, subjected to hard use, and easily destroyed. Within this context, the lending of wives served a positive function and made good sense. As a perspective, cultural relativism directs that any judgment regarding the goodness or badness of one or more of the ways of thinking, feeling, and behaving in a particular society be based on how well they fit with other aspects of its culture. In this perspective, the test of goodness is whether an aspect of culture is sufficiently consistent with other aspects of culture to be considered acceptable within a given group, subculture, or society. Thus, cultural relativism can, in an important sense, be viewed as a partial antidote to ethnocentrism. It is impossible to view aspects of a group or culture ethnocentrically while applying the principle of cultural relativism.

Practicing cultural relativism does not mean that you have to accept the ways of others as your own. It does not mean that you should abandon your child at three years of age as some Ik do or lend out your wife as Eskimos did. You can practice cultural relativity without abandoning your own moral standards or culture.

Cultural relativism, moreover, can remove barriers between you and others who are culturally different from you. It will help you to adjust more quickly and easily when you meet new people or enter new situations. For example, your adjustment to college roommates from another country, another region of the United States, or even another social class will be enhanced if you attempt to understand your roommates from the viewpoint of their cultural background rather than your own. If significant cultural differences exist, cultural relativism can help you adjust to the person you marry as well as to his or her family. It may even help you at work, where you will very likely be required to cooperate with people who do not think, feel, and act like you. And given the growing importance of America's international business ties, someday you may well be practicing cultural relativism with people from Japan, Germany, Egypt, or Saudi Arabia.

Cultural Similarity

If you had not read the *Sociological Imagination* at the beginning of this chapter, the discussion of cultural diversity might have left the false impression that there are no cultural similarities from one culture to another. Anthropologist George Murdock (1945), in attempting to identify the common denominator of cultures, listed seventy-odd **cultural universals,** or general cultural traits, thought to exist in all known cultures. These included athletic sports, cooking, courtship, division of labor, education, etiquette, funeral rites, family, government, hospitality, housing, incest taboos, inheritance rules, joking, language, law, medicine, marriage, mourning, music, obstetrics, property rights, religious rituals, sexual restrictions, status differences, and tool making. When each of these universals is examined more closely, the similarity among cultures becomes even more apparent.

Just because societies share a cultural universal does not mean, however, that they will express it in the same way. Quite different ways of handling the same general problem are usually developed by each culture. These may be thought of as **cultural particulars.** In the United States, though it is now changing, the division of labor has traditionally been based on gender roles—women have traditionally been assigned domestic tasks and men have traditionally worked outside the home. Among the Manus of New Guinea, on the other hand, the man is completely in charge of child rearing. Among the Mbuti pygmies, the Lovedu of Africa, and the Navajo and Iroquois Indians, men and women share equally in domestic and economic tasks (Little, 1975).

Why do cultural universals exist? The biological similarity shared by all humans helps to account for the presence of cultural universals. If a society is to survive, children must be born and cared for and some type of family structure must exist. Groups that have deliberately eliminated the family—such as the Shakers—have disappeared. And although a third sex might well introduce some useful—even interesting—dimensions to family life, the existence of two sexes limits the type of marital relationships that can exist: one male and one female (monogamy), several females and one male (polygyny), one female and several males (polyandry), several males and females (group marriage), and two people of the same sex (homosexual marriage). Because people become ill, there must be some sort of medical care. Because people die, there must be funeral rites, mourning, and inheritance rules. Because babies are born, some form of obstetrics is required. Because food is necessary, cooking must exist. A list of human biological similarities and their influence on culture could go on and on.

A second source of cultural universals is the physical environment. For example, the awesomeness of nature and people's inability to explain physical phenomena—such as the eclipse of the sun or the creation of the universe—helped to encourage the development of religion. Because humans cannot survive in extreme climates without artificial protection, some form of housing must be created. Conflicts often occur over natural territorial borders such as rivers and mountains, so some provision must be made for maintaining order within as well as between societies.

Finally, cultural universals exist because all societies face many of the same problems in maintaining social life. If a society is to survive, certain social provisions must be made: New members must be socialized; goods and services must be produced and distributed; means of dealing with the supernatural must be devised; tasks must be assigned; work must be accomplished.

1. Indicate which of the following are social categories (SC), subcultures (S), or countercultures (C).

    ____ a. Chinatown in New York City   ____ d. female
    ____ b. motorcycle gang   ____ e. revolutionary political group
    ____ c. Catholics   ____ f. the super rich

2. The tendency to judge other societies or groups according to one's own cultural standards is known as _____.

3. The psychological and social stress one may experience when confronted with a radically different cultural environment is called _____.

4. _____ is the idea that any given aspect of a particular culture should be evaluated in terms of its place within the larger cultural context of which it is a part rather than in terms of some alleged universal standard that applies across all cultures.

5. _____ are general cultural traits thought to exist in all known cultures.

6. Which of the following is *not* one of the reasons that cultural universals exist?

    a. physical environment   c. cultural predetermination
    b. genetic heritage of humans   d. problems of maintaining social life

Answers: 1. a. (S) b. (C) c. (SC) d. (SC) e. (C) f. (SC) 2. ethnocentrism 3. culture shock 4. Cultural relativism 5. Cultural universals 6. c

SUMMARY

1. Culture consists of all the material objects as well as the patterns for thinking, feeling, and behaving passed from generation to generation within a society. A society is composed of a people living within defined territorial borders who share a culture.

2. Contrary to popular belief, humans do not have instincts. Most behavior among nonhuman animals is instinctual, but human behavior is not the sole product of genetic heritage. Human behavior is learned. Even genetically inherited reflexes and drives do not determine how humans will behave, because people are heavily influenced by culture. Although culture does not determine human nature, it does condition it significantly. Sociobiologists are now arguing for recognition of the role of biology in human behavior.

3. Because humans are capable of creating and communicating arbitrary symbols such as language, they have the ability to create and transmit culture. Language is also important because it organizes people's view of reality. According to the hypothesis of linguistic relativity, people actually live in different worlds because their languages make them aware of different aspects of their environment. People are not forever trapped by their language, but they are limited by it unless they learn to see the world from another linguistic viewpoint.

4. Three broad dimensions to culture are normative, cognitive, and material. The normative dimension of culture is composed of norms, sanctions, and values. Norms are rules defining appropriate and inappropriate ways of behaving. There are several types of norms. Folkways are not considered vital to a group and may be violated without significant consequences. When mores—norms considered essential—are violated, social disapproval is strong. Some norms become laws. Sanctions—positive and negative, formal and informal—are used to encourage confor-

mity to norms. Values are broad cultural principles defining the desirable. Norms and values are not the same thing—the same value can be expressed in radically different norms. In complex societies, some values conflict.

5. Language is an important aspect of the cognitive dimension of culture. Beliefs, another important aspect of the cognitive dimension, are ideas concerning the nature of reality. Whether or not they are actually true, beliefs have a great influence on members of a society.

6. Material culture is composed of the concrete, tangible aspects of a culture. Aspects of material culture—desks, trucks, cups, money—have no inherent meanings. Material objects have meanings only when people assign meanings to them.

7. Wide cultural variation exists from one society to another as well as within a single society. Because all humans are basically the same biologically, cultural diversity must be explained by nongenetic factors. Cultural diversity within societies is promoted by the existence of social categories, subcultures, and countercultures.

8. Ethnocentrism—judgment of another group in terms of one's own cultural standards—is a major consequence of cultural diversity. The consequences of ethnocentrism are both positive and negative. The principle of cultural relativism—the idea that any given aspect of a culture should be evaluated in terms of its place within the larger cultural context of which it is a part—helps to combat ethnocentrism.

9. Some cultural universals are found in all societies. Their expression in specific practices, however, varies widely from one society to another. Cultural universals exist because of the biological similarity of humans, common limitations of the physical environment, and the common problems of sustaining social life.

LEARNING OBJECTIVES REVIEW

After careful study of this chapter, you will be able to:

- Explain the relationship between culture and the genetic heritage of humans.
- Describe and illustrate the interplay between language and culture.
- Classify various aspects of culture into one of its three major dimensions.
- Discuss how cultural diversity is promoted within a society.
- Describe and illustrate the relationship between cultural

diversity and ethnocentrism.
- Outline the advantages and disadvantages of ethnocentrism and discuss the role of cultural relativism in combating ethnocentrism.
- Explain the existence of certain aspects of culture that are shared around the world.

CONCEPT REVIEW

Match the following concepts with the definitions listed below them.

____ a. cultural universals   ____ f. cognition   ____ k. ethnocentrism

____ b. sociobiology   ____ g. society   ____ l. mores

____ c. sanctions   ____ h. cultural particulars   ____ m. subculture

____ d. informal sanctions   ____ i. laws   ____ n. formal sanctions

____ e. real culture   ____ j. taboo   ____ o. culture shock

1. The study of the biological basis of human behavior.
2. General cultural traits thought to exist in all known cultures.
3. Subterranean patterns for thinking, feeling, and behaving.
4. Rewards and punishments that may be applied by most members of a group.
5. Norms that are formally defined and enforced by designated persons.
6. A norm so strong that its violation is thought to be punishable by the group or society or even by some supernatural force.
7. The widely varying, often distinctive ways societies may use to handle common problems presented by the existence of cultural universals.
8. The psychological and social stress one may experience when confronted with a radically different cultural environment.

9. The mental process that enables humans to think, remember, recognize, and imagine.
10. Norms of great moral significance thought to be vital to the well-being of a society.
11. Rewards and punishments used to encourage desired behavior.
12. People who live within defined territorial borders and who share a common culture.
13. Rewards and punishments that may be given only by officially designated persons.
14. The tendency to judge others in terms of one's own cultural standards.
15. A group that is part of the dominant culture but differs from it in some important respects.

CRITICAL THINKING QUESTIONS

1. Sometimes you hear it said that people are just "naturally" selfish. Do you agree or disagree that humans are at bottom only capable of pursuing their own self-interest? Defend your position.

_____

_____

2. The ability to use language is cited by sociologists as a characteristic separating humans from nonhuman animals. Discuss the relationship between language and culture. Provide examples.

_____

_____

3. Distinguish the normative, cognitive, and material aspects of culture. Cite illustrations to show an understanding of the differences. _____

_____

_____

4. Functionalists and conflict theorists would be expected to have different views of countercultures. Identify these differences and defend the position you prefer in terms of the theoretical perspective you support.

_____

_____

5. Discuss the relationship between cultural relativism and ethnocentrism. Give one or more examples to make the connection.

_____

_____

MULTIPLE CHOICE QUESTIONS

1. **In your text, culture is defined as**
   a. a number of persons who share a social goal.
   b. rules defining appropriate and inappropriate ways of behaving for specific situations.
   c. the material objects as well as the patterns for thinking, feeling, and behaving that are passed from generation to generation among members of a society.
   d. the idea that any given aspect of a particular culture should be evaluated in terms of its place within the larger cultural context of which it is a part.
   e. people who live within defined territorial borders and who share common ties.

2. **Simple, biologically inherited, automatic reactions to physical stimuli are known as**
   a. reflexes.
   b. impulses.
   c. instincts.
   d. drives.
   e. needs.

3. **_____ is the study of the biological basis of human behavior.**
   a. Anthropology
   b. Ethnomethodology
   c. Sociolinguistics
   d. Sociobiology
   e. Social psychology

4. **The Sapir-Whorf hypothesis holds that**
   a. people are forever prisoners of their language.
   b. our view of the world depends on the particular language we have learned.
   c. Germans are less concerned about privacy than are Americans.
   d. exposure to another language can cause a person to become marginalized in their own culture.
   e. functionalism is more plausible than conflict theory.

5. **Shepard distinguishes between three major dimensions of culture—the normative, the material, and the**
   a. cognitive.
   b. ecological.
   c. ethical.
   d. genetic.
   e. noncognitive.

6. **_____ are rules defining appropriate and inappropriate ways of behaving for specific situations.**
   a. Folkways
   b. Values
   c. Mores
   d. Norms
   e. Laws

7. **_____ are applied only by officially designated persons, such as judges, executioners, or college professors.**
   a. Sanctions
   b. Formal sanctions
   c. Laws
   d. Mores
   e. Statutes

8. **The example in the text of a doctor's introduction to nettling demonstrates that**
   a. material aspects of culture have the same meanings and uses in all cultures.
   b. it is conventional to use a 747 jet for traveling.
   c. separate physical objects existing as part of a culture may be combined to have meaning for some people and no meaning for others.
   d. newspaper and pepper are an ineffective way of delivering babies.
   e. laws regulating midwifery are unnecessary.

9. **Which of the following statements is *not* consistent with the "tool kit" view of culture?**
   a. Culture provides a range of choices to be applied in the definition and solution of the problems of living.
   b. It is a means for "fixing" social problems.
   c. It proposes that culture should be viewed as a group of symbols, stories, rituals, and worldviews.
   d. Because the content of any culture is not fully consistent, the symbols, stories, rituals, and worldviews that are chosen vary among individuals and situations.

10. **Societies experience cultural diversity in part because of**
    a. social groups.
    b. social organizations.
    c. social categories.
    d. social classes.
    e. aggregates.

11. **In your text, a counterculture is referred to as**
    a. a subculture that is deliberately and consciously opposed to certain central aspects of the dominant culture.
    b. the material objects as well as the patterns for thinking, feeling, and behaving that are passed from generation to generation among members of a society.
    c. general cultural traits thought to exist in all known cultures.
    d. aspects of culture publicly embraced by members of a society.
    e. a group that is part of the dominant culture but differs from it in some important respects.

12. **Negative consequences of extreme ethnocentrism include all of the following *except***
    a. culture shock.
    b. intrasocietal fragmentation.
    c. intersocial conflict.
    d. resistance to innovation.
    e. feelings of relative deprivation.

13. **The idea that any given aspect of a particular culture should be evaluated in terms of its place within the larger cultural context of which it is a part rather than in terms of some alleged universal standard that applies across all cultures is known as**
    a. cultural universalism.
    b. ethnocentrism.
    c. ecological adaptation.
    d. cultural relativism.
    e. pragmatism.

14. **_____ are general traits thought to exist in all cultures.**
    a. Cultural particulars
    b. Cultural universals
    c. Ecological adaptations
    d. Material adaptations
    e. Cultural adaptations

FEEDBACK REVIEW

True-False

1. According to the hypothesis of linguistic relativity, words and the structure of a language cause people to live in a distinct world. *T or F?*
2. Sociologists now agree that genetic heritage plays no part in the shaping and limiting of social life. *T or F?*

Fill in the Blank

3. _____ consists of all the material objects as well as the patterns for thinking, feeling, and behaving that are passed from generation to generation among members of a society.
4. Because of _____, culture can be created and transmitted.
5. The tendency to judge other societies or groups according to one's own cultural standards is known as _____.
6. _____ is the idea that any given aspect of a particular culture should be evaluated in terms of its place within the larger cultural context of which it is a part rather than in terms of some alleged universal standard that applies across all cultures.
7. _____ are broad cultural principles embodying ideas about what most people in a society consider to be desirable.

Matching

8. Indicate whether these statements best reflect a belief (B), folkway (F), mos (M), law (L), or value (V).

_____ a. conception that God exists

_____ b. norm against cursing aloud in church

_____ c. norm encouraging the eating of three meals daily

_____ d. idea of progress

_____ e. norm against burning a national flag

_____ f. norm encouraging sleeping in a bed

_____ g. norm prohibiting murder

_____ h. norm against overtime parking

_____ i. idea that the earth is elliptical

_____ j. idea of freedom

9. Indicate whether the following are formal sanctions (F) or informal sanctions (I).

_____ a. A mother spanks her child.

_____ b. A professor fails a student for cheating on an exam.

_____ c. A jury sentences a person to life in prison for espionage.

_____ d. A husband separates from his wife after she has an affair.

10. Indicate which of the following are social categories (SC), subcultures (S), or countercultures (C).

_____ a. Chinatown in New York City

_____ b. motorcycle gang

_____ c. Catholics

_____ d. female

_____ e. revolutionary political group

_____ f. the super rich

GRAPHIC REVIEW

The trend toward increased premarital sexual experience among teenage females in the United States is graphically illustrated on Figure 3.1. The questions below ask you to relate this trend to the normative dimension of culture.

1. What type of norm did the prohibition against premarital sex among teenage females represent in the United States during the 1950s?

_____

2. What type of norm regarding premarital sex among teenage females exists in America today?

_____

3. Compare and contrast the types of formal and informal sanctions used to deal with premarital sexual activity among American female teenagers during the 1950s and today.

_____   _____

_____

ANSWER KEY

Concept Review	Multiple Choice	Feedback Review	
a. 2	1. c	1. T	9. a. I
b. 1	2. a	2. F	b. F
c. 11	3. d	3. Culture	c. F
d. 4	4. b	4. language	d. I
e. 3	5. a	5. ethnocentrism	10. a. S
f. 9	6. d	6. Cultural relativism	b. C
g. 12	7. b	7. Values	c. SC
h. 7	8. c	8. a. B	d. SC
i. 5	9. b	b. M	e. C
j. 6	10. c	c. F	f. SC
k. 14	11. a	d. V	
l. 10	12. e	e. M	
m. 15	13. d	f. F	
n. 13	14. b	g. L	
o. 8		h. L	
		i. B	
		j. V	

Chapter Four

Socialization Over the Life Course

LEARNING OBJECTIVES

After careful study of this chapter, you will be able to:

- Discuss the contribution of the socialization process to human development.
- Describe the contribution symbolic interaction makes to our understanding of socialization, including the concepts of the self, the looking-glass self, significant others, and role taking.
- Compare and contrast the life-course theories of Freud, Erikson, and Piaget.
- Distinguish among the concepts of desocialization, resocialization, and anticipatory socialization.
- Outline the major sources of influence on the socialization of the young.
- Describe the stages of development in adulthood.
- Discuss the primary demands of socialization encountered in late adulthood.
- Compare and contrast the approaches of functionalism and conflict theory to the mass media's effects on socialization.

SOCIOLOGICAL IMAGINATION

Does violence on television lead to violent acts in real life?   Until fairly recently, Americans strongly disagreed about the spillover effect of violence from television to the real world. In fact, social scientists traditionally have hesitated to claim a causal link between viewing television violence and unleashing violence on others. No longer. After hundreds of studies, researchers now find common ground in citing the connection between televised aggression and personal aggressiveness.

Finding a link between watching televised violence and committing real violence is yet another illustration that most human behavior is culturally transmitted. Human beings can (and do) adopt patterns of behavior through learning the culture around them. This learning process begins at birth and continues into old age; it is called *socialization*.

THE IMPORTANCE OF SOCIALIZATION

One of the major points of Chapter 3, "Culture," is that human beings at birth are helpless and without knowledge of their society's ways of thinking, feeling, and behaving. If a human infant is to learn how to participate in social life, much cultural learning has to take place.

Nearly all the social activities we consider natural and normal are learned. How, for example, is one supposed to behave at a cocktail party? Certainly an individual is expected to have a drink. But what will the drink be, white wine or Perrier? Asking seriously for a Bloody Mary with fresh cow's blood probably would result in some signs of disgust and social disapproval. Yet, in some African tribes a glass of cow's blood would be relished (Douglas, 1979). A preference for cow's blood or martinis is, like nearly all aspects of social life, learned during the process of **socialization**—learning to participate in group life through the acquisition of culture.

Learning about the countless aspects of social life in a particular society begins at birth and continues throughout life. For example, infants in most American homes are taught to eat certain foods, to sleep at certain times, and to smile at certain sounds made by their parents. But socialization is not limited to the early years; it is a lifelong process. Successful socialization enables people to fit into all kinds of social groups. Socialization must occur if presidents of the United States are to govern, if plebes are to survive at West Point, and if residents of a home for the aged are to adjust successfully to their new life. Would-be executives who prefer warm cow's blood to dry white wine will have to keep their tastes a secret or abandon their occupational aspirations.

Although socialization is a lifelong process—and this point is well developed in this chapter—the most important socialization occurs early in life. In fact, many of the characteristics associated with being human do not appear in individuals who are deprived in their early years of prolonged and intensive interaction with others. Cases of extreme social isolation, as will be shown, reveal that the absence of early prolonged and intensive social contact leaves children without the facility for such basics as walking, talking, and loving. In the absence of socialization, a human

Socialization is the process of learning to participate in group life through acquiring culture. These alums are visiting the college at which they spent formative years learning aspects of their culture in preparation for the adult lives they now lead.

infant cannot develop a **personality**—the relatively organized complex of attitudes, beliefs, values, and behaviors associated with an individual.

The main themes of this chapter have just been outlined. In this first section, the importance of socialization for human development is documented, followed by a discussion of how theoretical perspectives, especially symbolic interactionism, approach socialization. Socialization as it occurs throughout the life cycle, even to death, occupies the next two sections of the chapter. The last section of the chapter applies the functionalist and conflict theories to the American mass media.

Evidence on the Importance of Socialization

According to an enduring popular belief, the basic good nature of humans is spoiled by civilization. Society corrupts. Jean-Jacques Rousseau's "noble savage" has been a frequent theme in romantic novels, and writers have even contended that individuals raised in complete social isolation would be the masterpieces Adam and Eve were before their corruption and downfall. Underlying this idea is the belief in a fundamental human nature that exists prior to social contact, a belief now rejected by social scientists.

How can the effects of socialization be assessed? Ideally, assessing the effects of socialization requires an experiment with a control group of normally socialized infants and an experimental group of socially isolated infants. Assuming that these two groups of children were biologically the same, differences between them at the end of the experimental period could be attributed to the process of socialization. For obvious reasons, there is no such evidence on human infants. There has, however, been some experimental research with monkeys and some nonexperimental evidence from studies of socially isolated children.

How do monkeys react to social isolation? A series of experiments by psychologist Harry Harlow (Harlow and Zimmerman, 1959; Harlow and Harlow, 1962; Harlow, 1967) has shown the negative effects of social isolation among rhesus monkeys. In one experiment, a group of infant monkeys separated from their mothers at birth was exposed to two surrogate or artificial mothers—wire dummies of the same approximate size and shape as a real adult monkey. One of the substitute mothers had an exposed wire body; the other was covered with soft terry cloth. Free to choose between them, the infant monkeys consistently spent more time with the soft, warm mother. Even when the exposed wire surrogate became the only source of food, the terry cloth mother remained their preference. To the surprise of the researchers, closeness and comfort were more important

to these monkeys than food. When upset by a mechanical toy bear or a rubber snake, these infant monkeys consistently ran to their cloth mothers for security and protection. Apparently, infant monkeys need intimacy, warmth, physical contact, and comfort. Harlow has also shown that infant monkeys raised in isolation become distressed, apathetic, withdrawn, hostile adult animals. They never exhibit normal sexual patterns, and mothers either reject or ignore any babies they may have, sometimes even abusing them physically.

Can we generalize from monkeys to humans? Generalizing research findings from nonhumans to humans is risky, because people are not monkeys. Nevertheless, many experts on human development believe that a human infant's need for affection, intimacy, and warmth—like Harlow's monkeys—is as important as the infant's need for food, water, and physical protection. Babies denied close human contact usually have difficulty forming emotional ties with others. Touching, holding, stroking, and communicating appear to be essential to normal human development. Although Harlow's findings with monkeys cannot be applied directly to humans, similar findings on human children have given them considerable credibility. Lawrence Casler (1965), for example, found that the developmental growth rate of institutionalized children can be improved by only twenty minutes of extra touching a day. Cases of socially isolated children provide additional support.

Do extremely isolated children develop human characteristics? There are documented cases of children who have been kept alive physically but deprived socially, emotionally, and mentally. Fortunately, documentation includes observations of the changes that occur when these children are placed in an environment designed to socialize them (Curtiss, 1977; Pines, 1981). Case histories of two unrelated girls, fictitiously named Anna and Isabelle, have been documented in a classic article by Kingsley Davis (1947).

Anna was the second illegitimate child born to her mother. Because they lived with Anna's grandfather, who was incensed by the latest evidence that his daughter did not measure up to his strict moral code, the mother and child were forced to move out of the house. After several months of moving from one place to another, repeated failures at adoption, and with no further alternatives, Anna was returned to her grandfather's house. Anna's mother so feared that the sight of the child would anger her father that she confined both herself and Anna to an atticlike room on the second floor of the farmhouse. Anna was given little physical attention beyond the exclusive diet of milk that kept her alive until she was discovered at the age of five. Barely alive, she was extremely emaciated and under-

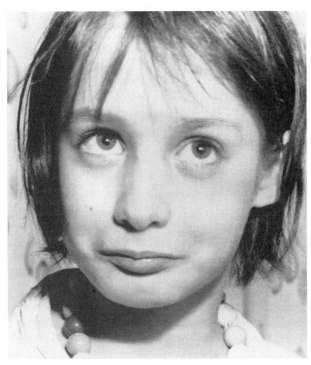

Normal human social and personality development requires intensive and prolonged human interaction in the formative years. Lack of adult warmth and care can produce many enduring social and psychological problems. Loneliness and difficulty in interacting with others, which are among these problems, are easy to sense in this young girl.

nourished, with skeletonlike legs and a bloated stomach. Apparently, Anna had seldom been moved from one position to another, and her clothes and bed were filthy. Positive emotional attention was unfamiliar to her. When she was found, Anna could exhibit no signs of intelligence, nor could she walk or talk.

During the first year and a half after being found, Anna made some developmental progress in a county home for children. Among other things, she learned to walk, understand simple commands, eat by herself, tend to personal neatness somewhat, and recall people she had seen. But her speech was that of a one-year-old. Anna was then transferred to a school for retarded children, where she made some further progress. But still, at the age of seven her mental age was only nineteen months, and her social maturity was that of a two-year-old. A year later she could bounce and catch a ball, participate as a follower in group activities, eat normally (although with a spoon only), attend to her toilet needs, and dress herself aside from handling buttons and snaps. Significantly, she had acquired the speech level of a two-year-old. By the time of her death at age ten, Anna had made some additional progress. She could carry out instructions, identify a few colors, build with blocks, wash her hands, brush her teeth, and try to help other children. Her developing capacity for emotional

attachment was reflected in the love she had developed for a doll.

Nine months after Anna was found, Isabelle was discovered. She, too, was an illegitimate child kept in isolation for fear of social disapproval. Isabelle's mother, a deaf-mute since the age of two, stayed with the child in a dark room secluded from the rest of the family. When found at the age of six and a half, Isabelle was physically ill from an inadequate diet and lack of sunshine. Her legs were so bowed that when she stood the soles of her shoes rested against each other, and her walk was a skittering movement. Unable to talk except for a strange croaking sound, Isabelle communicated with her mother by means of gestures. On seeing strangers—especially men—her fear and hostility caused her to act like a wild animal. Some of her actions were those of a six-month-old infant.

In spite of the interpretation that Isabelle was feeble-minded—her first IQ score was near the zero point and her social maturity was at the level of a two-and-a-half-year-old—an intensive program of rehabilitation was begun. After a slow start, Isabelle progressed through the usual stages of learning and development, much like any normal child progressing from ages one to six. Although the pace was faster than normal, the stages of development were in their proper order. It took her only two years to acquire the skills mastered by a normal six-year-old. By the time she was eight and a half, Isabelle was on an educational par with children her age. By outward appearances, she was an intelligent, happy, energetic child. At age fourteen, she participated in all the school activities normal for other children in her grade. To Isabelle's good fortune, she, unlike Anna, benefited from intensive training at the hands of trained professionals. Her progress may also have been due to the presence of her mother during the period of isolation.

A more recent case did not end as happily as Isabelle's. Susan Curtiss (1977), a psycholinguist, attempted to rehabilitate Genie, a thirteen-year-old girl who had been kept isolated in a locked room by her father from the time she was nearly two. When found, much of Genie's behavior was subhuman. Because her father severely punished her for making any vocal sounds whatever, she was completely silent; in fact, she did not even sob when she cried or speak when in a fit of rage. Because she had hardly ever worn clothing, Genie did not notice changes in temperature. Never having been given solid food, she could not chew. Because she had spent her entire life strapped in a potty chair, Genie could not stand erect, straighten her arms or legs, or run. Her social behavior was primitive. She blew her nose on whatever was handy or into the air when nothing was available. Without asking, she would take from people things that attracted her attention.

Attempts to socialize Genie over a four-year period were not highly successful. At the end of Curtiss's efforts, Genie could not read, spoke primarily in short phrases, and had just begun to control some of her feelings and behavior. Genie paid a high price—her full development as a human being—for the isolation, abuse, and lack of human warmth she experienced in the environment created by her father.

What other disruptions in social contact retard human development? The implication of the cases of Anna, Isabelle, and Genie is unmistakable: The personal and social development associated with being human is acquired through intensive and prolonged social contact with others. But cases of extreme social isolation are not the only evidence for this generalization. Children can be affected adversely when the degree of contact with others is limited or when emotional attachments cannot be formed.

In pioneering research, René Spitz (1946a, 1946b) compared the infants in an orphanage with those in a women's prison nursery. After two years, some of the children in the orphanage were retarded, and all were psychologically and socially underdeveloped for their age. By the age of four, a third of them had died. No such problems were observed among the prison nursery infants. Not one of them died during this period, even though the physical environment was not as clean as that of the orphanage. Spitz traced the difference between the two groups of infants to the emotional, physical, and mental stimulation offered in each setting. In the prison nursery, the infants' mothers were with them for the first year of their lives. The mothers were not present in the foundling home, and the infant-nurse ratio there was seven or eight to one. Also, in contrast to the prison nursery, the infants in the orphanage lacked the stimulation normally provided by toys and were isolated from the other children. The conclusion to be drawn is that social isolation need not be extreme—as it was for Anna, Isabelle, and Genie—in order to damage social and personality development.

Some other research supports Spitz's conclusions. William Goldfarb (1945), for example, compared a group of children reared in an infant home from shortly after birth to three years of age with children who had spent nearly all their lives in foster homes. He found that the institutionalized children suffered personality

defects from their early years of deprivation that persisted even after they had been placed in foster homes. Compared with the children who had known only foster homes, the institutionalized children had lower scores on IQ tests, were more immature, had more difficulty caring about others, and were more passive and apathetic.

Lytt Gardner (1972) attributed a condition known as deprivation dwarfism to emotional deprivation. He conducted an intensive study of six "thin dwarfs"—children who are underweight and short for their age and who have retarded skeletal growth. Gardner found that these children had come from families missing the normal emotional attachment between parents and children. When such children were removed from their hostile family environments, they began to grow physically; their growth was stunted once again on returning home. According to Gardner, deprivation dwarfism is a concrete example—an experiment of nature, so to speak—demonstrating "the delicacy, complexity and crucial importance of infant-parent interactions" (Gardner, 1972).

In another study, some infants adopted from a Lebanese orphanage were compared with children who remained in the orphanage. Initially, all of these children existed in stark cribs, lying on their backs in a barren room. Human touch was limited to diaper changing. After one year, the children's intellectual, motor, and social development was on a six-month-old level. Those children left in the orphanage continued to be underdeveloped, whereas the adopted infants made many developmental strides. In another orphanage study, some preschool-age children (average IQ 64) were placed in an institution for retarded adults where each was "adopted" by an older woman and received much attention from patients and staff members. Children who remained in the orphanage (or another orphanage) lost an average of 20 IQ points over three years. An average gain of 28 points was experienced among the "adopted" orphans (Ornstein and Ehrlich, 1989).

Thus, the process of socialization permits us to develop the basic characteristics we associate with being human. It is also through socialization that we learn culture and learn how to participate in social structures. We turn next to the processes of socialization.

FEEDBACK

1. _____ is the process through which people learn to participate in group life through the acquisition of culture.
2. According to sociologists, no fundamental human nature exists prior to social contact. *T or F?*
3. Thanks to recent breakthroughs, research findings on the need of infant monkeys for warmth and affection can easily be applied to humans. *T or F?*
4. The cases of Anna, Isabelle, and Genie indicate that the personal and social development associated with being human is acquired through intensive and prolonged social contact with others. *T or F?*
5. Social isolation need not be extreme in order to damage social and personality development. *T or F?*

Answers: 1. Socialization 2. T 3. F 4. T 5. T

PROCESSES OF SOCIALIZATION

The Theoretical Perspectives and Socialization

Each of the three major theoretical perspectives—functionalism, conflict theory, symbolic interactionism—sheds light on the processes of socialization. The contribution of functionalism to our understanding of the socialization process is more implied than explicit. That is, the very concept of socialization is based on the idea that people fit into groups by virtue of their learning the culture of their society. People take their places in society by learning and accepting what society expects of them. The process of socialization assumes continuity, stability, and equilibrium. In fact, if it were otherwise, the existence of society would not be possible.

Like functionalism, the contribution of the conflict perspective to our knowledge of the socialization process is not obvious. The conflict perspective views the learning of roles and acceptance of statuses as a way of perpetuating the status quo. When people are socialized to accept their family's social class, for example, they are involved in the perpetuation of the existing class structure. People are socialized into an acceptance of their fate before they have enough self-awareness to realize what is happening. Once social class socialization has taken place, it is very difficult to overcome. Consequently, socialization serves to maintain the social, political, and economic advantages of the higher social classses. People who do not question their lot in life can never mount a revolution against the class structure.

The contribution of symbolic interactionism to our understanding of the socialization process is the most fully developed approach, as will be seen from the discussion that follows. Symbolic interactionism helps us understand the development of the self-concept, the role of symbols and language in interpreting the social environment, the process of learning and assuming roles, and the social antecedents of human nature.

Symbolic Interactionism and Socialization

The symbolic interactionist perspective approaches socialization through social interaction based on symbols, especially language. (Refer to Chapter 1, "The Sociological Perspective.") Charles Horton Cooley and George Herbert Mead, the originators of symbolic interactionism, challenged the belief, prominent in their day, that human nature is biologically determined. To symbolic interactionists, human nature is a social product.

Symbolic interactionism, as applied to socialization, involves a number of key concepts, including those of the self-concept, the looking-glass self, significant others, role taking, the imitation stage, the play stage, the game stage, and the generalized other.

What is the looking-glass self?   Charles Horton Cooley (1864–1929) made some observations about the development of the **self-concept**—an image of oneself as an entity separate from other people—from watching his own children at play. Cooley (1902) noted the many ways that children learn to get attention by interpreting the reactions of others toward them. For example, they learn quickly that causing some disturbance when adult company is in the house will focus attention on themselves rather than on visitors. From such insights, children learn to judge themselves in terms of how they imagine others react to them. Thus, others serve as mirrors for the development of the self.

Cooley called this way of learning the **looking-glass self**—self-concept based on our perceptions of others' judgments of us. We use others as a mirror reflecting their reactions to us. According to Cooley, the looking-glass self is the product of a three-stage process that is constantly taking place. First, we imagine how we appear to others. Next, we imagine the reaction of others to our imagined appearance. Finally, we evaluate ourselves according to how we imagine others have judged us. The result of this process is positive or negative feelings about ourselves. Of course, this is not a conscious process, and the three stages occur in rapid succession in any given instance.

Consider this example of the looking-glass process. Suppose that a college student is going to the spring dance with someone she really wants to impress. In preparation for the dance, she gets a tan, has her hair done, and buys a stunning new dress. As she dresses for the dance, she has an image of how she is going to look (stage 1). As she walks down the stairs to meet her date, however, she thinks she senses disappointment in his eyes, and he fails to compliment her (stage 2). Because she wants her date to be impressed with her appearance, she feels bad about herself (stage 3).

Is the looking glass distorted?   Note that the looking-glass process takes place in the mind. Because the looking-glass self is the product of our imagination, the mirrors we use may be distorted; the looking glass may not accurately reflect what other people really think of us (Gecas, 1982). In the preceding example, the girl's date may have been so stunned and dazzled by her appearance that he felt inadequate and could not show her how he really felt. If this was the case, she misread his lack of expression and silence. Parents, to take another example, may punish a child for some deed that in fact endears the child to them. A child sent to her room for using obscene language may not know how hard it was for her parents to conceal their amusement. An anorexic teenager may continue to threaten her health by not eating because she believes that others see her as fat.

Although we may misinterpret others' perceptions of us, this does not diminish the effectiveness of the looking-glass process. Even when we erroneously believe that, say, a date or our parents dislike us, the consequences are as real to us as if they actually did have that opinion. E. L. Quarantelli and Joseph Cooper (1966) found the self-concept of dental students to be nearer to the evaluations they thought their instructors had given them than to the evaluation their instructors had actually given them. Despite the possibility of distortion in the looking glass, the relationship between self and others is well established.

Can we go beyond the looking glass?   Although not denying that our self-evaluations are heavily influenced by our perception of others' evaluations of us, Viktor Gecas and Michael Schwalbe (1983) think that other factors also affect self-concept. They suggest going "beyond the looking glass" to consider other sources of influence on the development of self-concept.

According to Gecas and Schwalbe, the looking-glass self-orientation depicts human beings as oversocialized, passive conformists. That is, self-concepts are assumed to be based solely on the real or imagined opinions of others. (This is ironic, because Cooley and other symbolic interactionists view individuals as active and creative.) To correct this passive view of human beings, Gecas and Schwalbe developed the argument that individuals' self-concepts are also derived from personal judgments regarding how effectively they control their environments. Individuals with power on the job, for example, tend to have higher self-esteem than powerless individuals (Kanter, 1993). True, part of the response is due to social recognition (or lack of it) from others; part of this effect, however, is attributable to the presence or absence of control over one's fate. A feeling of self-efficacy, in short, contributes to self-esteem.

Do we use all others equally as mirrors?   One's self-concept is not equally influenced by all persons. As George Herbert Mead (1863-1931) pointed out, some people are more important to us than others (Mead, 1934). Those whose judgments are most important to our self-concept are called **significant others.** A new babysitter who sends the children to their rooms for misbehaving will probably not influence the children's self-concepts very much. A mother or father is more likely to have influence. For a child, significant others are likely to include mother, father, grandparent, teachers, and playmates. Teenagers place heavy reliance on their peers. The variety of significant others is greater for adults, ranging from parents and friends to ministers and employers.

Others serve as mirrors for the development of our self-concepts. The looking-glass self is based on our perceptions of others' judgments of us. Part of what this young woman is thinking involves her imagined reactions of others to her appearance.

What is role taking?   Because humans have language and the capacity for thinking, we can carry on mental conversations. That is, we can mentally indicate something to ourselves and respond internally to it. This facility is crucial for anticipating the behavior of others. Through internal conversation we can imagine the thoughts, emotions, and behavior of others in a given situation. As Mead noted, the ability to have internal conversations enables humans to engage in **role taking**—the process of mentally assuming the viewpoint of another individual and then responding to oneself from that imagined viewpoint.

Role taking is a mental process that permits us to avoid the trial-and-error method, which would be necessary if we could not mentally anticipate the behavior of others. Say, for example, that you wish to ask your employer for a raise. If you could not mentally put yourself in your boss's place, you would have no idea about the objections that might be raised. But by engaging in a mental conversation that anticipates possible objections, you can be ready to defend your request.

How does the ability for role taking develop?
According to Mead, the ability for role taking is the product of a three-stage process, which he called the imitation, play, and game stages. In the **imitation**

stage, which begins around age one and a half to two years, the child imitates the physical and even the verbal behavior of a significant other without comprehending the meaning of what is being imitated. This is the first step in developing the capacity for role taking. At about the age of three or four, children begin to assume the statuses of specific individuals and to act out the behavior they associate with them. A young child can be seen playing at being mother, father, police officer, teacher, or astronaut. This play involves acting and thinking as a child imagines another person would. This is what Mead called the **play stage**—the stage during which children take on the roles of individuals, one at a time.

The third phase in the development of role taking Mead labeled the **game stage,** which enables children to engage in more sophisticated role taking. After a few years in the play stage, children begin to be able to consider the roles of several people at the same time. Games involve several participants, and there are specific rules designed to ensure that the behavior of the participants fits together. All participants in a game must know what they are supposed to do as well as what is expected of all others in the game. Imagine the confusion if young first-base players have not yet mastered the idea that the ball hit to a teammate will usu-

The ability for role taking begins with children imitating significant others. As this young child "helps" her father sweep the sidewalk, she is laying the foundation for her ability to assume mentally the viewpoint of another person and then respond to herself from that imagined viewpoint.

ally be thrown to them. In the play stage, a child may pretend to be a first-base player one moment and pretend to be a base runner the next. In the game stage, however, first-base players who drop their gloves and run to second base when an opposition player hits the ball will not remain in the game for very long. It is during the game stage that children learn to gear their behavior to the group.

During the game stage, a child's self-concept, attitudes, beliefs, and values gradually come to depend less on individuals and more on a generalized referent. Being an honest person is no longer merely a matter of pleasing such significant others as one's mother, father, or minister. Rather, it just begins to seem wrong to be dishonest. As this change takes place, a **generalized other**—an integrated conception of the norms, values, and beliefs of one's community or society—emerges.

What is the self?   According to Mead, the self is composed of two analytically separable parts: the "me" and the "I." The **"me"** is the part of the self formed through socialization. Because it is socially derived, the "me" accounts for predictability and conformity. Yet much human behavior is spontaneous and unpredictable. A rejected lover may, for example, spontaneously and unaccountably strike the person he loves most. Afterward, his dazed reaction may be, "I don't know what came over me. How could I do that to the person I cared for most?" To account for this spontaneous, unpredictable, often creative part of the self, Mead wrote of another dimension of the self—the **"I."**

The "I" operates not just in extreme situations of rage or excitement but constantly interacts with the "me" as we conduct ourselves in social situations. According to Mead, the first reaction of the self comes from the "I," but before we act, the initial impulse is directed in socially acceptable channels by the "me." Thus, the "I" normally takes the "me" into account before acting. However, the uniqueness and unpredictability of much human behavior demonstrates that the "me" does not always control the innovative, unpredictable dimension of the self.

LIFE-COURSE THEORIES

The approach of symbolic interactionism to socialization, although it contains some life-course dimensions, was covered separately because it is distinctly sociological in nature and because it is much broader than a life-course theory. At this point, several psychologically oriented life-course theories are presented. They range from the psychoanalytic theory of Sigmund Freud to the psychosocial perspective of Erik Erikson and the cognitive perspective of Jean Piaget.

1. In its approach to socialization, _____ emphasizes social interaction based on symbols.
2. According to the looking-glass process, our _____ is based on how we think others judge the way we look and act.
3. Those individuals whose judgments of us are the most important to our self-concept are _____.
4. _____ is the process of mentally assuming the viewpoint of another individual and then responding to oneself from that imagined viewpoint.
5. According to George Herbert Mead, children learn to take on the roles of individuals, one at a time, during the _____ stage.
6. In Mead's theory of the self, the _____ is the unsocialized side and the _____ is the socialized side.

Answers: 1. symbolic interactionism 2. self-concept 3. significant others 4. Role taking 5. play 6. "I"; "me"

A Psychoanalytic Perspective

Aside from his theories of the unconscious, Sigmund Freud's (1856–1939) greatest contribution was his work regarding the influence of infancy and early childhood experiences on personality development. Experiences within the family during the first few years of life, Freud contended, largely shape future psychological and social functioning.

What is the composition of the human personality? According to Freud, the personality has three parts: the id, the ego, and the superego. The **id** is made up of biologically inherited urges, impulses, and desires. It is selfish, irrational, impulsive, antisocial, and unconscious. The id operates on the *pleasure principle*—the principle of having whatever feels good. Newborn infants are said to be totally controlled by the id—they want their every desire fulfilled without delay. Very early in life, members of society—usually parents—interfere with the pleasure of infants. Infants gradually learn to wait until it is time to eat, to control their bowels and bladder, and to hold their tempers. In many ways, they must face the fact that others are not around merely to satisfy their impulses.

To cope with this denial of pleasure, children begin to develop an **ego**—the conscious, rational part of the personality that thinks, plans, and decides. The ego is ruled by the reality principle, which allows us to delay action until a time when the gratification of our desires is more likely. Thus, the ego mediates between the biological, unconscious impulses of the id and the denying social environment. But the ego is not itself sufficient to control the id. Nor need it be.

At about four or five years of age, the **superego**—roughly the conscience—begins to develop. It contains all the ideas about what is right and wrong that we have learned from those close to us, particularly our parents. By incorporating their ideals into our personality, we develop what may be thought of as an internal parent. Parents are no longer the only source of punishment for wrongdoing; we punish ourselves through guilt feelings.

At the same time, we feel good about ourselves when we live up to the standards contained in our superego. Through this internal monitoring system, we learn to channel our behavior in socially acceptable ways and to repress socially undesirable thoughts and actions.

What are the relationships among the id, the ego, and the superego? Freud did not see the id, ego, and superego as separate regions of the brain or as little people battling and negotiating inside our heads. Rather, he saw them as separate, interacting, and conflicting processes within the mind. The id demands satisfaction; the superego prohibits it. The ego supplies rational information in this conflict; it attempts to gain satisfaction within the limits set by the superego and the social environment. The following example, although an oversimplification, should clarify the relationships among these three parts of the personality.

> Let's say you are sexually attracted to someone. The id clamors for immediate satisfaction of its sexual desires, but is opposed by the superego (which finds the very thought of sexual behavior shocking). The id says, "Go for it!" The superego icily replies, "Never even think that again!" And what does the ego say? The ego says, "I have a plan!" (Coon, 1997:460).

What will the ego do in this situation? If the ego has difficulty controlling the id, then some sexually related experience—sexual intercourse, masturbation, even rape—may follow. Should the superego be dominant, sexual desires may be sublimated into other activities, such as working, dancing, studying, or stamp collecting. If the ego is not overwhelmed by the id or superego, it will encourage more socially acceptable behavior, such as dating or marriage.

Psychosocial Development

Psychoanalyst Erik Erikson (1902–1994) has extended the work on human development of his mentor, Sigmund Freud. Like his teacher, Erikson (1964) emphasized the role of the ego as the mediator between the individual and society and wrote of "crises" accompanying each stage in human development. Erikson, however, differed from Freud in his specific ideas on the relationship between self and socialization as well as in his underlying assumptions. To Freud, the human personality is almost totally determined in early childhood. Erikson, in contrast, believed that personality may change at any time in life. Whereas Freud attributed most human behavior to universal instincts, Erikson emphasized cultural variations. Although Freud contended that parents are the most influential forces in personality development, Erikson considered the impact of peers, friends, and spouses to be significant as well.

According to Erikson, all individuals pass through a series of eight developmental stages from infancy to old age. Each of these developmental stages involves a psychosocial crisis or developmental task. (See Table 4.1.) Personal identity is firmly established or impaired, depending on how successfully an individual handles each turning point. The effects of successfully or unsuccessfully meeting these crises are cumulative—successful management of a crisis at an earlier stage increases the chances of mastering later developmental tasks; those who fall behind in one stage will have increasingly greater difficulty in the following stages. Those individuals whose egos adequately meet the psychosocial crisis at each developmental stage are the most mature and the happiest and have the most stable personal identities.

Several additional points about Erikson's eight stages of psychosocial development should be kept in mind. First, Erikson does not use the word *crisis* to mean a threat or catastrophe. Rather, it is a turning point, a crucial period during which growth or maladjustment may occur, depending on how the crisis is handled. Second,

although the way in which a crisis in one developmental period is handled affects the management of later crises, no crisis is necessarily solved permanently. Thus, a sense of trust developed early in life may, because of social experiences, turn to distrust later, and a sense of self-doubt may later be changed to self-confidence through relationships with others. Third, as implied above, each crisis must be solved in interaction with others, and characteristics developed at earlier stages of development (trust, autonomy, industry, and so forth) require the support of others if they are to be maintained. Fourth, each crisis is solved within a particular social setting—family group, school, peer group, neighborhood, or workplace.

Cognitive Development

Jean Piaget (1896–1980) was concerned primarily with the development of intelligence, or cognitive abilities— thinking, knowing, perceiving, judging, and reasoning. According to Piaget (1950; Piaget and Inhelder, 1969), children gradually develop cognitive abilities through interaction with their social setting through the process of socialization. Piaget contended that children are not merely passive recipients of social stimuli; they are actively engaged in interpreting their environment as they attempt to adjust to it. Cognitive ability, in Piaget's theory, advances in stages. Learning to solve problems in one stage must precede problem solving at later stages. Piaget has isolated four stages of cognitive development: the sensorimotor stage, the preoperational stage, the concrete operational stage, and the formal operations stage.

What cognitive development occurs in the sensorimotor stage? The *sensorimotor stage,* where the basis for

TABLE 4.1	Erik Erikson's Stages of Psychosocial Development	
Stage	**Crisis**	**Favorable Outcome**
1. First year of life	Trust versus Mistrust	Faith in the environment and in others
2. Ages two to three	Autonomy versus Shame and Doubt	Feelings of self-control and adequacy
3. Ages four to five	Initiative versus Guilt	Ability to begin one's own activities
4. Ages six to twelve	Industry versus Inferiority	Confidence in productive skills; learning how to work
5. Adolescence (ages twelve to eighteen)	Identity versus Role Confusion	Integrated image of oneself as a unique person
6. Early adulthood (ages eighteen to thirty-five)	Intimacy versus Isolation	Ability to form bonds of love and friendship with others
7. Middle adulthood (ages thirty-five to sixty)	Generativity versus Stagnation	Concern for family, society, and future generations
8. Late adulthood (over age sixty)	Integrity versus Despair	Sense of dignity and fulfillment; willingness to face death

thought is laid, begins at birth and lasts until the age of eighteen months to two years. It is called the sensorimotor stage precisely because most of its activities are associated with learning to coordinate body movements with information obtained through the senses—touching, hearing, seeing, feeling. Only gradually do children come to realize that they are an object separate from other things. Initially, they do not realize that they can cause things to happen. A baby shaking bells on its crib does not realize who has caused the sound.

One of the most significant developments during this stage is the development of a sense of "object permanence"—a sense that objects exist even when they cannot be seen. Once this level of thought is reached, a child will, for example, pursue a ball that has rolled under a bed. Later a child can anticipate the reappearance of an object. Instead of looking into the entrance of a tunnel that an electric train has entered, a child will watch for the train's reappearance at the other end of the tunnel. During this stage, children come to see their world as an understandable and predictable place.

How do children think during the preoperational stage? Between the ages of two and seven, during the *preoperational stage,* children learn to think symbolically and to use language. Initially, they have difficulty distinguishing between what they call something (a symbol) and the object itself; an object and the symbol used to represent it are the same to them. If you throw away a milk carton that has been used as a child's toy truck, you will have a difficult time convincing the child that it was only a milk carton.

Another dominant characteristic of children during the preoperational stage is self-centeredness. Children's extreme egocentrism is clearly reflected in their inability to see things from others' points of view. Many children believe that the sun and the moon follow them when they are taking a walk. In one type of experiment, children have been placed opposite adults with a two-sided mirror between them. When asked what the adults see in their side of the mirror, the children answer "me!"

In the preoperational stage, thoughts and operations are not reversible. Consider this conversation researcher John Phillips (1975) had with a four-year-old boy:

"Do you have a brother?"
"Yes."
"What's his name?"
"Jim."
"Does Jim have a brother?"
"No."

What are the characteristics of the stage of concrete operations? The dominant characteristic of thinking in the stage of *concrete operations,* which spans the ages

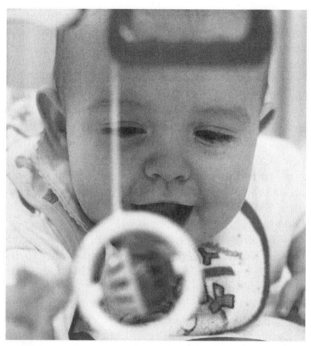

This toy is helping this baby develop the capacity to distinguish itself from other objects. If the toy makes a sound, the infant is not aware of having caused it.

from seven to eleven, is increasing abstractness. A child begins to be able to think logically about time, quantity, and space; handle arithmetic operations; and place things logically in categories. In addition, children can reverse thoughts and operations during this stage. The reversibility of thought allows an older child to recognize that A-B-C is the same as C-B-A or to recognize that if he has a brother, his brother also has one. A certain degree of abstract thinking in children of this age is reflected in the children's ability to imagine themselves in the place of others and to gear their behavior to others.

Despite the movement toward abstract thinking that occurs during the stage of concrete operations, the child still has difficulty if concrete objects are not involved. Logical thinking is tied to action, to the observation or manipulation of concrete symbols and objects. For example, children at this stage of development can master the concept of the conservation of quantity by watching someone pour liquids into different-sized glasses. In the preoperational stage, if children see water poured from a slender glass into a wide glass, they assume that the wider glass has less water in it because the level of the water appears to be lower. During the stage of concrete operations, however, children come to realize that the amount of water is the same in both glasses. They could not grasp this idea without an understanding of concrete operations. It is only in the next stage that completely abstract thought becomes possible.

What is unique about thinking in the stage of formal operations? A fundamental change in cognitive ability begins to occur sometime after the age of eleven. Children learn to think without the aid of concrete objects and manipulations; they can begin to think in terms of abstract ideas and principles. Gradually the adolescent can reason about hypothetical matters. For example, if you ask younger children, "What would life be like if people had wings?" they would say something like "That's silly" or "People don't have wings." But during the stage of *formal operations,* children become willing to speculate on such matters. Children also learn during this stage to consider relationships that are logical, even if they are ridiculous. If younger children are asked to complete the sentence "If day is dark, night is _____," they will argue that day is not dark. Older children will answer that night is light. It is during this stage, then, that the capacity for adult thinking is developed.

SOCIALIZATION AND THE LIFE COURSE

Desocialization, Resocialization, and Anticipatory Socialization

The concepts of *desocialization, resocialization,* and *anticipatory socialization,* associated with symbolic interactionism, can help us understand that over the life course old ways must be abandoned, new ways adopted, and preparation made for transitions from one period in life to another. Erving Goffman's work on

"total institutions" provides an excellent illustration of these concepts because of the clarity they assume in the extreme setting he describes.

Goffman believes that people are not always free to manage their own lives but are players under the direction of others. In *Asylums,* Goffman (1961b) writes about places such as mental hospitals and prisons as **total institutions**—places in which residents separated from the rest of society are controlled and manipulated by those in charge. The control and the manipulation exercised by the staff are used deliberately to change residents. The first step is **desocialization**—the process of relinquishing old norms, values, attitudes, and behaviors. In extreme situations such as mental hospitals and prisons, which Goffman describes, desocialization entails an attempt to obliterate the resident's old self-concept along with the norms, values, attitudes, and behaviors associated with the personal identity the resident had upon entering the institution. This is accomplished in many ways. Replacing personal possessions with standard-issue items promotes sameness among the residents and deprives them of the personal effects (long hair, hair brushes, jeans, T-shirts) they have been using to present themselves as unique individuals. Serial numbers and the loss of privacy also contribute to the breaking down of past identity.

Once the self-concept has been fractured, **resocialization**—the process of learning to adopt new norms, values, attitudes, and behaviors—can begin. The staff, using an elaborate system of rewards and punishments, attempts to create a new sense of self for residents. Rewards for conformity to the new identities and

1. Listed below are the three major parts of the personality according to Freud. Match these parts with the examples.
   ____ a. id
   ____ b. ego
   ____ c. superego
   (1) A student decides not to cheat on an examination because the professor seems to be watching the class closely.
   (2) A student does not cheat on an examination because she feels too guilty about it.
   (3) A lover kills his girlfriend in a fit of jealous rage.
2. The major role of the _____ in Freud's theory of the personality is to mediate between innate impulses and the conscience.
3. Whereas Freud believed that the human personality is almost totally determined in early childhood, Erikson held that personality can change _____.
4. According to Erikson, the successful management of a psychosocial crisis at an earlier stage increases the chances of mastering later crises. This means that the effects of success or lack of success at earlier crisis points are _____.
5. Match Jean Piaget's stages of cognitive growth with the statements.
   ____ a. sensorimotor stage
   ____ b. concrete operations stage
   ____ c. formal operations stage
   ____ d. preoperational stage
   (1) A child thinks without the aid of concrete objects and manipulations.
   (2) A child recognizes that 3 + 2 is the same as 2 + 3.
   (3) A child learns that a cat that runs under a couch still exists.
   (4) A child believes that a box is a spaceship.
6. The ability to imagine oneself in the place of others develops during the stage of _____.

Answers: 1. a.(3) b.(1) c.(2) 2. ego 3. at any time 4. cumulative 5. a.(3) b.(2) c.(1) d.(4) 6. concrete operations associated ways of thinking and behaving that the staff tries to impose on residents include extra food or special periods of privacy. Punishments for nonconformity to the new identities involve loss of special privileges, physical punishment, or physical isolation.

The concepts of desocialization and resocialization were developed by Goffman to analyze social processes in extreme situations. They apply to other social settings, ranging from torture by the enemy in wartime to basic training in the U.S. Marine Corps and plebe year at the United States Military Academy. Happily, these concepts also illuminate changes undergone in life by those fortunate enough to have avoided any such extreme exposures. Desocialization and resocialization occur during the transition into the adolescent subculture, when young adults begin their occupational careers, and among the elderly as they move into retirement or widowhood.

Anticipatory socialization—the process of preparing oneself for learning new norms, values, attitudes, and behaviors—does not generally occur in the extreme social settings represented by total institutions. This is because anticipatory socialization involves the idea of people looking forward to a change they wish to occur. This might be viewed as "voluntary" resocialization. Because teenagers want to resemble and to be liked by those of their own age, they willingly abandon many of the norms, values, attitudes, and behaviors learned previously. Consequently, preteens begin very early to observe the ways of their new **reference group**—a group we use to evaluate ourselves and to acquire attitudes, values, beliefs, and norms. Graduating seniors, normally seen on campus only in jeans and oversized sweatshirts, suddenly are wearing tailored suits and much more serious expressions. They are talking with friends who have already graduated and with representatives of companies in order to begin the transition into the adult world of work. By looking ahead to the new environment they are about to enter, graduating seniors are preparing themselves for the process of resocialization they know awaits them.

These stages of the life course as we know them are unknown in other societies. A *midlife crisis* could not have occurred in preindustrial society, because people died early. Even as late as 1900, life expectancy in the United States was just over forty-nine years (Atchley, 1997). The concept of childhood as we know it did not exist in medieval society. The problems of adolescence are a relatively recent development. Among the Truks of Micronesia, the spectre of death looms at forty.

In Truk, they rely basically on bread, fruit and fish. In order to get fish a young man has to be strong and agile; he has to be a good paddler of a canoe, a good navigator to go out on the reef and gather fish. When the Trukese reaches about forty years of age, his strength begins to decline; he also does

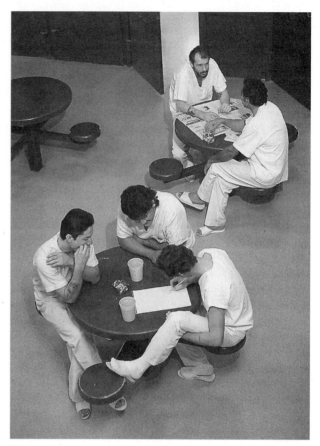

Total institutions are places in which residents separated from the rest of society are controlled and manipulated by those in charge. These inmates in a Texas jail share the experience of life in a total institution.

not climb trees as well as he used to. And when his strength begins to wane, he begins to feel that his life is ebbing away and he begins to prepare himself for death (Kübler-Ross, 1975:28).

Thus, the stages in the life cycle are not strictly defined by biology or psychology; they are also socially and culturally defined (Kübler-Ross, 1986, 1993).

Socialization of the Young

What are the major effects of the family? The child's first exposure to the world occurs within the family. Being dependent and highly impressionable, the child is virtually defenseless during the first few years of life. By the time the child develops some independence and judgment, much of the socializing work of the family has been accomplished. Personality development does not end at age five, but a foundation for later development has been firmly established. Through close interaction with a small number of people—none of whom the child has selected—the child learns to think and speak; internalizes norms, beliefs, and values; forms

In addition to teaching academic subjects, schools have a "hidden curriculum"—informal and unofficial aspects of culture taught to children to prepare them for life in the larger society. These early American immigrants are clearly being taught, among other things, loyalty and patriotism.

some basic attitudes; develops a capacity (or incapacity) for intimate and personal relationships; and acquires a self-image (Handel, 1990).

The impact of the family reaches far beyond its direct effects on the personal and social development of the child. One's family of birth largely determines one's place in society. At birth, our family bestows on us social characteristics that significantly affect what we think of ourselves and how others treat us. Jean Evans offers an illustration of this in the case of Johnny Rocco, a twenty-year-old living in a city slum.

> *Johnny hadn't been running the streets long when the knowledge was borne in on him that being a Rocco made him "something special"; the reputation of the notorious Roccos, known to neighbors, schools, police, and welfare agencies as "chiselers, thieves, and trouble-makers" preceded him. The cop on the beat, Johnny says, always had some cynical smart crack to make…. Certain children were not permitted to play with him. Wherever he went—on the streets, in the neighborhood, settlement house, at the welfare agency's penny milk station, at school, where other Roccos had been before him—he recognized himself by a gesture, an oblique remark, a wrong laugh (Evans, 1954:11).*

The impact of social class usually extends into adulthood. For example, in a major study of occupational attainment, Peter Blau and Otis Duncan make this observation about the relationship between one's family and subsequent work life that is still accurate today: "The family into which a man is born exerts a profound influence on his career, because his occupational life is conditioned by his education, and his education depends to a considerable extent on his family" (Blau and Duncan, 1967:330).

What is the place of the school in socialization?   In school, all children are under the care and supervision of adults who are not relatives. The first year of school involves a transition from an environment saturated with personal relationships to one that is more impersonal. Rewards and punishments are based more on performance than on personality. Although a mother may accept any picture that her child draws as good work, a teacher evaluates all members of the class by the same standards and informs students who are not meeting these standards. Slowly, children are taught to be less dependent emotionally on their parents. In addition, the school ties children to the broader society by creating feelings of loyalty and allegiance to something beyond their families.

The socialization process in school involves more than teaching skills (reading, writing) or subject matter (mathematics, English). Underlying the formal goals of the school is what has come to be known as the *hidden curriculum*—the informal and unofficial things that children are taught to prepare them for life in the larger society. The hidden curriculum teaches children such matters as discipline, order, cooperativeness, and conformity—skills thought to be needed for success in modern bureaucratic society whether the child becomes a doctor, college president, secretary, assembly-line worker, or professional athlete.

According to educational critic John Holt (1967), the schools use a variety of means to prepare children for twentieth-century civilization. Life in schools is, like the "real world," run by the clock. Whether or not a student really understands something he or she has been working on and whether or not the child is psychologically ready to switch to a completely different subject, a bell signals that all children must move to the

Harry Gracey—Kindergarten as Academic Boot Camp

Sociologists view education as an important agency for socialization. To both classical and contemporary sociologists, schooling is a vital mechanism for the survival of society. Schools transmit aspects of culture considered essential for a society's perpetuation. These aspects of culture, as shown in Chapter 12 ("Education"), extend well beyond academic studies and skills. Schools are one means for molding the young into acceptable members of society.

Harry Gracey, using a case study approach, researched the ways in which the kindergarten attempts to socialize preschoolers not only for the first grade but also for life beyond school. The rigidity of the kindergarten's structure he observed led Gracey to compare preschool training with military boot camp. Gracey's reasons for choosing this analogy will become obvious.

Kindergarten children, contends Gracey, learn to go unquestioningly through routines and follow orders even when those routines and orders make no sense to them. They become disciplined to do what an authoritative person tells them without much argument. For example, young physically active children must wonder why they need to do calisthenics in class. Yet they do them when told to do so. Similarly,

pledging allegiance to the flag and singing "America"—everyday rituals—must go over the heads of the children performing them.

Student conformity and discipline, Gracey argues, exist because of the rigid social structure created by the teacher. This structure allows the teacher to control the behavior of most of the students most of the time. A prescribed set of daily routines and rituals seems to be the basis for the inflexible control structure. Each day for the children, Gracey observed, was broken into six unvarying segments: Serious Time, Sharing Time, Play Time, Work Time, Clean-up Time, and Rest Time. Programmed activities existed for each segment. Moreover, spontaneous interests or activities on the part of the students were never developed by the teacher. When spontaneous interest was generated among students,

next scheduled event. Getting through a predetermined set of activities within a given time period often becomes more important than learning. There are rules and regulations to cover almost all activities—how to dress, how to wear one's hair, which side of the hall to walk on, when to speak in class, when to go to the bathroom. Teachers reward children with praise and acceptance when the children recite the "right" answers, behave "properly," or exhibit "desirable" attitudes. (See Chapter 12, "Education.")

How do peer groups contribute to socialization? The family and the school are both agencies of socialization organized and operated by adults. The **peer group**—composed of individuals of roughly the same age and interests—is the only agency of socialization that is not controlled primarily by adults. Children usually belong to several peer groups. A child may belong to groups composed of neighborhood children, schoolmates, Girl or Boy Scouts, and church friends.

The peer group contributes to socialization by providing children with experiences that are unlikely to be provided within the family (Corsaro and Rizzo, 1988, 1990). Because children are subordinated to adults in the family, the peer group provides them an opportunity to engage in give-and-take relationships usually not

possible at home. Children learn to engage in exchange and in conflicting, competitive, and cooperative relationships with others. The peer group also gives children experience in self-direction. Children can begin to make their own decisions; experiment with new ways of thinking, feeling, and behaving; and engage in activities that involve self-expression. Independence from adults is also promoted by the peer group because the peer group introduces the child to a world that is often in conflict with the adult world. Children learn to be different from their parents in certain ways, a step that contributes to the development of self-sufficiency. These experiences cannot easily be obtained within the more rigidly organized family and school environments. A capacity for intimacy can also be enhanced within the peer group because the peer group provides an opportunity for children to develop close ties with friends they choose, including members of the opposite sex. At the same time that children are making close friends with a few individuals, they are learning to get along with large numbers of people, many of whom are quite different from themselves. This helps develop the social flexibility needed in a mobile, rapidly changing society. The ability to participate effectively in adult life is also cultivated through learning to act in accordance with the peer group's unwritten web of rules.

it was quickly squelched, writes Gracey, because of the teachers' need to maintain control. Because self-expression and creativity take children into their own world and away from the adult-structured world, they were purposefully discouraged. Thus, the children's school environment was imposed on them rather than created by them.

Children who conform and identify with the structure become the school's idea of a "good student." Those who submit to the discipline but do not identify with it are adequate students who find their identities outside school. Students who rebel become "bad students" and "problem children." These children are often turned over to school counselors or clinical psychologists.

Thus, kindergarten is a year during which children learn the role of student. The discipline, routines, rituals, and conformity prepare children for the first grade where the role of student is reinforced.

There is, Gracey concludes, a further function of this training in kindergarten and later grades. This function—preparation for work life in a bureaucratic society—is stated very effectively by Gracey himself:

Once out of the school system, young adults will more than likely find themselves working in large-scale bureaucratic organizations, perhaps on the assembly line in the factory, perhaps in the paper routines of the white collar occupations, where they will be required to submit to rigid routines imposed by "the company" which may make little sense to them. Those who can operate well in this situation will be successful bureaucratic functionaries. Kindergarten, therefore, can be seen as preparing children not only for participation in the bureaucratic organization of large modern school systems, but also for the large-scale occupational bureaucracies of modern society (Gracey, 1977:226).

Critical feedback

1. Gracey's study seems to involve both the conflict perspective and functionalism. Explain why this seems to be true or argue that it is not the case.

2. State the biggest flaw you see in Gracey's research approach. Are you convinced of the validity of his analysis? State why or why not.

The rise of formal education has contributed immensely to the emergence of a peer world that is not only separate from adults but also beyond their control. Children are isolated from adult society by being set apart in school for most of their preadult lives. Because they are separated from the adult world for such a long time, young people are forced to depend on one another for social life.

Another factor contributing to the dependence of adolescents on each other is the distribution of population in advanced industrial societies. The majority of Americans now live in either urban or suburban areas. The prevalence of the dual-employed family in American society is further accentuating this situation. Both parents may commute many miles to work; if so, they spend much of their time away from home. Consequently, once children reach the upper levels of grade school, they spend more time with their peers than they do with their parents. Urie Bronfenbrenner (1970) notes that peer groups fill the vacuum created in the lives of children who receive an insufficient amount of attention from their parents.

What role does the mass media play in socialization?
The **mass media** are those means of communication that reach large heterogeneous audiences without any personal interaction between the senders and the receivers of messages. Television, radio, newspapers, magazines, movies, books, the Internet, tapes, and discs are the major forms of mass communication. It is not the technology per se that makes these "mass" media. Newspapers existed in the American colonies, but they were limited in circulation and aimed at more local audiences (DeFleur and Ball-Rokeach, 1982). Technologies become mass media only when their audiences include any average person who wishes exposure to them. Sociologists agree that the mass media are powerful socializing agencies.

One of the primary functions of the mass media is introducing children to their culture. From the mass media, children learn of the behavior expected of individuals in certain social statuses. Despite the fact that these popular images are usually highly distorted—detective and police work are not as exciting and glamorous as depicted in books, in movies, and on television—they nevertheless introduce children to certain aspects of their culture. If parents or others do not counteract these false ideas—if a child's view of the world is shaped by the mass media without corrective guidance—these ideas become real to the child unless altered by personal experience.

Though often stereotyped, the mass media also display role models that children can imitate. They usually

While these teenagers are having fun, they are also engaged in the serious process of learning to participate in group life through their peer group. In a rapidly changing world, these young people will continue to be "socialized" throughout their lives.

present characters in such simple, one-sided forms that it is easy to recognize behavior suitable for men, women, heroes, and villains. Learning these role models helps to integrate the young into society.

The mass media, by their content alone, teach many of the ways of the society. This is evident in the behavior we take for granted—the duties of the detective, waitress, or sheriff; the functions of the hospital, advertising agency, and police court; behavior in hotel, airplane, or cruise ship; the language of the prison, army, or courtroom; the relationship between nurses and doctors or secretaries and their bosses. Such settings and relationships are portrayed time and again in films, television shows, and comic strips; and all "teach"—however misleadingly—norms, status positions, and institutional functions (Elkin and Handel, 1991:189).

The mass media give children ideas about the values in their society. They provide a child with a slick, idealized picture of the importance of such things as success, money, consumption, sex, youth, and good looks in American society.

The only activity American children engage in more than watching television is sleeping. An average American watches television for over seven hours each day. Any medium with which people spend so much time must contribute significantly to the socialization process.

Consider the relationship between violence on television and real-life violence. Social scientists traditionally have been reluctant to conclude that research demonstrates a causal connection between television violence and real-life violence. However, after hundreds of studies involving over ten thousand children, one can fairly conclude that watching aggressive behavior on television significantly increases the expression of

aggressive impulses. There is now a broad consensus among researchers that a causal relationship exists between television violence and a subsequent increase in aggressive behavior (U.S. Department of Health and Human Services, 1982a, 1982b; Joy, Kimball, and Zabrack, 1986; Levinger, 1986; Schutte et al., 1988; Strasburger, 1995).

As discussed in *Sociological Imagination*, researchers now generally agree that at times, the effect of television is direct, concrete, and dramatic. A nine-year-old girl, for example, was gang-raped with a beer bottle by three teenage girls and a boy three days after they had seen a similar incident on a television movie. At least twenty-nine Americans have shot themselves while imitating the Russian roulette scene from *The Deer Hunter*. A two-year-old girl died when her older brother, age five, set the house on fire with matches while imitating behavior he had seen on "Beavis and Butt-Head." Television's effects, of course, are usually more hidden, subtle, and long term. In fact, extreme pressure from a variety of sources has caused the major television networks to institute ratings for their programs (Farhi, 1997). Most critics of violence on television saw this self-policing move as an evasion tactic by an industry under public attack. Among other things, they strongly recommend the "V chip," which a parent who buys a new television set starting in 1998 can preset to block any program rated V-for-violent. More will be said about the effects of the mass media in the socialization process in the closing section of this chapter.

Are there other agencies of socialization during childhood and adolescence? The family, school, peer group, and mass media are the major agencies of socialization during childhood and adolescence, but they are not the only ones. Although religion does not have the

SOCIOLOGY IN THE NEWS

Television and Children

The mass media have become like a "third parent" to America's children. Because children watch television so much (over seven hours daily), it is one of the primary agents of socialization. Consequently, children learn a great deal about their culture from watching television. As this CNN report shows, television may have either negative or positive effects.

1. If you watched television for a day as an alien, how would you describe what American children are learning about their culture?

   _____

   _____

2. What would you do as a parent with respect to the socializing effects of television on your children?

   _____

   _____

Conflict theories emphasize the media's role in socializing the masses. Since 98 percent of all households have at least one television set, and those in the average household watch television over seven hours daily, it follows that television is one of the most powerful influences on our values and ways of thinking, feeling, and behaving.

same degree of influence in all societies, it can affect the moral outlook of young people even in secular societies such as the United States. Athletic teams teach young children to compete, win and lose, cooperate with others, follow rules, make friends with others, and handle their dislike of some peers. Other potentially significant agencies of socialization are youth organizations such as the Cub Scouts, Girl Scouts, and YMCA.

Do these agencies of socialization sometimes conflict?
In complex societies, conflict among agencies of socialization is inevitable. Some families work at cross-purposes with the schools; the church and the peer group may place conflicting demands on adolescents; parents may believe that the mass media are undermining the values they have taught their children. Conflict may even occur within a single agency of socialization. One of a child's parents may emphasize materialism, whereas the other minimizes the importance of possessions. Little League baseball coaches may talk of fair play but encourage their players to win at all costs.

Early and Middle Adulthood Socialization

The discussion of socialization appropriately has focused on childhood and adolescence. Without a doubt, socialization during our preadult years lays a foundation that affects our self-concept and ability to participate in social life for as long as we live. This does not mean, however, that the socialization process sud-

denly ends at age eighteen. Although indispensable, childhood and adolescent socialization does not completely prepare us to meet the demands of adult life. If adults are to mature and to participate successfully in their society, they must continue to undergo socialization (George, 1993). And with the extension of the life span, the elderly confront new situations and roles (Clausen, 1986; Hogan and Astone, 1986; Cox, 1988; Barrow, 1996; Atchley, 1997).

There are several models of the stages of development in adulthood (van Gennep, 1960; Erikson, 1964; Neugarten, 1968; Gould, 1975; Levinson, 1978, 1986). Despite differences in detail, it is possible to paint a portrait of socialization in adulthood reflecting commonalities among these models.

What are the stages of development in adulthood?
Early adulthood begins toward the end of the teen years and extends into the late twenties. (See Figure 4.1 for a graphic presentation of a model of the stages of development in adulthood.) This period involves moving beyond adolescence and making a preliminary step into adulthood. It ends when the individual has made a life within the adult world. Young people forge a temporary link between themselves and the adult world that involves, among other things, choosing an initial occupation and establishing a new family through marriage. Toward the end of the twenties and early thirties, some of the provisional choices made earlier may be reevaluated. Adults may reassess earlier decisions with

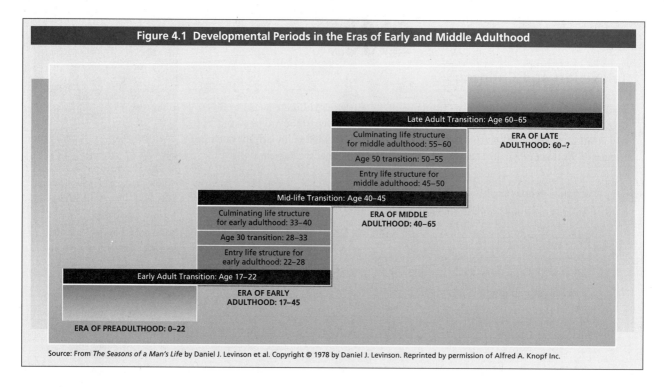

Figure 4.1 Developmental Periods in the Eras of Early and Middle Adulthood

Late Adult Transition: Age 60–65

Culminating life structure for middle adulthood: 55–60

ERA OF LATE ADULTHOOD: 60–?

Age 50 transition: 50–55

Entry life structure for middle adulthood: 45–50

Mid-life Transition: Age 40–45

Culminating life structure for early adulthood: 33–40

ERA OF MIDDLE ADULTHOOD: 40–65

Age 30 transition: 28–33

Entry life structure for early adulthood: 22–28

Early Adult Transition: Age 17–22

ERA OF EARLY ADULTHOOD: 17–45

ERA OF PREADULTHOOD: 0–22

Source: From *The Seasons of a Man's Life* by Daniel J. Levinson et al. Copyright © 1978 by Daniel J. Levinson. Reprinted by permission of Alfred A. Knopf Inc.

the realization that if life changes are to be made, it is time to do so before it is too late. Marriages are in jeopardy, divorce is a possibility, and extramarital affairs are not uncommon. Between ages thirty and forty, a settling-down period tends to occur, during which adults, more conscious of their mortality, attempt to make a place in life for themselves and their families. Central to this period is an interest in advancement in terms of money, prestige, recognition, contribution to society, and establishment of a solid family life.

Next comes the transition from early to *middle adulthood.* This is the period when new questions about one's place in the world may arise. Adults may wonder what they have done with their lives and what they wish to do with the rest of their working years. Sometime during the forties choices are made to continue either the path already taken or to form a somewhat different life. If it is the latter, a number of things may occur, including such drastic steps as an occupational change, a divorce, an extramarital affair, or a move to a new community. Typically, more subtle changes take place. For example, family life improves or deteriorates, work becomes more important, or thoughts turn toward retirement. Between the midforties and retirement age, adults wind up the tasks of middle adulthood. Acceptance of one's level of achievement occurs. Tension over achievement of one's self and one's children is replaced by concern with personal relationships and the small pleasures of life, including children and grandchildren. (*Late adulthood,* beginning after sixty years of age, will be explored shortly.)

Does this model of stages of adult development apply to women? The research on which this model is based is on samples of males. Because insufficient research has been done explicitly on females, it remains to be seen how well women fit the developmental periods. Preliminary work, however, indicates that the model depicts reality for women less than it does for males (Rosenfeld and Stark, 1987). Research tells us that contemporary women have socialization experiences unique to them.

What are some of the socialization experiences unique to women? In early adulthood, women may experience conflict between the ideal wife-mother model they learned as children and the modern roles they now value. Men often reinforce the traditional domestic roles by expecting their mates, who have demanding jobs outside the home, to perform most of the household work (Pendleton, Poloma, and Garland, 1980; Skinner, 1983; Hochschild, 1997). Many women now experience the extra pressure in a society that expects more of them than it does of men.

Midlife brings some additional difficulties for women related to socialization. For those women who spent most of their adult lives as full-time wives and mothers, middle adulthood presents some special problems. Those women who based their identity and self-worth on their home, husbands, and children may experience a void when grown children don't require as much attention and husbands at the peak of their careers are preoccupied with work. Also, economically dependent women who experience the death of a husband or divorce may experience the multiple shocks associated with a declining level of living, breaking into the job market, and returning to school.

Although signs of early aging such as wrinkles, weight gains, and loss of body shape come to men as well as women, it is females who feel the greatest negative repercussions. This is in large part due to the double standard of aging. Many of the characteristics associated with masculinity—aggressiveness, competitiveness, ambitiousness, and decisiveness—are heightened by physical maturity. In contrast, traditionally held images of femininity such as seductiveness, sensuality, and good looks, are diminished in midlife.

The current generation of more highly educated and occupationally successful young women will be in a much stronger position to handle the socialization difficulties faced by today's women in middle adulthood. Actually, some evidence indicates that women in middle adulthood are faring better than the popular image suggests. The loss of family responsibilities is opening new doors of opportunity for many women at the midpoint of their lives (Baruch and Brooks-Gunn, 1984; Mirowsky and Ross, 1986; Coleman et al., 1987; Rosenfield, 1989). Many women in midlife are initially entering the labor force, resuming earlier careers, or pursuing higher education.

Late Adulthood Socialization

What are the major demands of late adulthood?
Length of life is closely linked with the nature of a society. A study of the Bushmen, a hunting and gathering people, shows that 40 percent of them died before the age of fifteen. Even a century ago, the life expectancy of most people did not exceed thirty-five to forty years. In modern societies, adults can now expect to live beyond seventy. This longer life span exposes aging people to demands unique in human history. The major demand for older people in American society is withdrawal from participation in certain major aspects of social life (Achenbaum, 1978; Benokraitis, 1996). Roles are lost because statuses are lost. Two of the most important lost statuses are worker and spouse. One cannot act like a worker if no coworkers or bosses are present; one cannot act like a wife if no husband exists.

In modern societies, older people are expected to retire from work. There are many cues. Workmates begin to encourage retirement, a high level of job performance is no longer expected, deserved promotions are given to younger people, rewards for work diminish, duties are reduced, demotions occur, retraining programs are not made available. Moreover, retirement is considered by most people to be the right thing to do.

Although older people are socialized to retire, there is less emphasis on preparation for life in retirement

In American society, a significant proportion of the population lives into the retirement years. While not all older people adjust well to separation from meaningful work roles, this couple appears to be adjusting nicely to the role requirements of late adulthood.

(George and Maddox, 1979; Plank, 1989). Older workers are simply cut off at an arbitrary age to do as they wish—to be creative and active or to stagnate and be passive. They are forced to assume what has been called a *roleless status* (Shanas, 1968; Hooyman and Kiyak, 1996). The transition from work to retirement may carry negative consequences for social identity and self-esteem (Rosnow, 1985).

An individual may intellectually apprehend the processes of compulsory retirement policies and yet be emotionally unable to accept them as beyond control. If the older worker incorporates the negative societal stereotypes, a gradual shift in self-image will occur, one that is difficult to overcome in the retirement years. When the negative appraisal summarized in "We see you as a bumbling old fool" becomes internalized to "I am a bumbling old fool," a downward spiral is set in motion that is difficult to break (Hendricks and Hendricks, 1986:332).

Most older people must also face the loss of a spouse. There is little preparation for a single life before the death of one's husband or wife, and very few single people sixty-five and older remarry, particularly women. There is, however, intense socialization immediately following the death of a wife or husband. Those who have lost their spouses are expected to be as active as they were before the death occurred. Those who resume normal functioning and establish their independence are praised, and those who mourn too long and withdraw are criticized. However, many members of the community—doctors, social workers, friends, ministers, relatives—are available during the mourning period to help the widowed person learn to establish an independent life. This is particularly important in a society where the elderly frequently live alone rather than with relatives. Like retirement, however, there is almost no cultural definition of what a widowed person should do; widowhood is another roleless status. (For a further discussion of the problems faced by the elderly, see Chapter 16, "Population and Urbanization.")

How do older people face death?   Americans are not very well socialized when it comes to handling death (Moller, 1995a, 1995b). Death is simply not a subject very much discussed in the United States. In fact, Americans do not even die; they "pass on," "go to their just reward," "go to a better world," "leave us," "buy the farm," or just "go." Sixty-six euphemisms for the process of dying have been identified (DeSpelder and Strickland, 1991). Americans do not like to talk about death, particularly to the dying. Relatives often attempt to keep the dying person from knowing that he or she is dying, and dying people sometimes keep their own prognosis secret (Kearl, 1989; Leming and Dickinson, 1990; Atchley, 1997). Consequently, dying Americans do not know how to fulfill the status of a dying person. The status of a dying person is, like retirement and widowhood, a status almost devoid of roles. Dying Americans face conflicting advice. One person may advise a dying person to accept death after a long and fruitful life. Another person may suggest Dylan Thomas's advice: "Do not go gentle into that good night, Old age should burn and rave at close of day; Rage, rage against the dying of the light."

Death is handled very differently in other cultural traditions (Cleiren, 1992; Kübler-Ross, 1975, 1993; Kertzer and Laslett, 1995). Among Alaskan Indians, the dying participate in the timing of and planning for their deaths; they play a very active role in their own dying and death. According to Lois Grace, a deacon of the Kawaiahao Church in Hawaii, the Hawaiian people believe that death is a constant companion, whether they are hunting, partying, or staying at home. When someone dies, a big luau among friends and relatives follows a wailing procession. Children are prominently included in the funeral so that they can learn early that death is a part of life. Among the Japanese, death is not something to fear. A dead person "goes to the pure land, the other shore, a place often described as beautifully decorated pools and silver, gold and lapis lazuli" (Kübler-Ross, 1975:30). Memorial services held by ministers are more for the family than for the deceased.

The American aversion to the topic of death has, until lately, prevented us from gaining much knowledge about death and dying. Partly because of the gradual aging of the American population, death and dying have recently become topics of great interest to researchers. Consequently, we are beginning to learn how dying people face death. We know, for example,

that people are more accepting of death if they are permitted to talk openly and freely about it (Martocchio, 1985; Sanders, 1992). The dying in **hospices**—organizations designed to provide support for the dying and their families—appear to be better socialized to dying in part because they can share their feelings with others who are either dying or expecting death soon (Mor et al., 1988; DeSpelder and Strickland, 1991).

Elisabeth Kübler-Ross (1993) has identified five stages that the dying, at least in the United States, go through after learning that they are terminally ill. In the first stage, *denial and isolation,* there is a refusal to accept the fact that one is going to die; some individuals even attempt to convince themselves that they have been suddenly cured. In the second stage, *anger* is the dominant emotion. After the dying person has accepted the fact that this is actually happening to him or her, the reaction becomes "Why me?" Patients in this stage may express their anger by saying something emotional, like "I am alive, don't forget that. You can hear my voice, I am not dead yet." The third stage involves *bargaining*. Most bargains are made with God and involve asking for more time. Postponement is the basis of bargaining—"If only I could see my son married" or "If only I could see my grandchild born." The fourth stage, *depression,* begins when the terminally ill patient can no longer deny the reality of dying. The predominant emotion during this stage is a deep sense of loss for oneself and for one's family. In the *acceptance,* or last, phase, the dying person is able to think about death with a degree of quiet expectation. Although this is not a happy stage, it is, as one patient expressed it to Kübler-Ross, a time for "the final rest before the long journey."

Whether the five stages in the dying process identified by Kübler-Ross apply to other cultures besides the United States has yet to be determined. Even for American society, others have pointed out that this process does not follow these stages strictly in all cases. People may move back and forth among the stages and may exhibit characteristics of more than one stage at the same time. It is dangerous to view these stages in the dying process as an inevitable progression.

The Sociology of the Life Course

Some qualifications to this description of development during early, middle, and late adulthood need to be noted. First, as you have already seen, the model of early and middle adulthood is based primarily on samples of males. Second, although this model may describe a general pattern, not all people go through each of these stages. Some individuals handle developmental tasks in unique ways, and some skip periods of growth and change. Third, the ages given for each period are only approximate; individuals may enter stages at different ages. Although these qualifications do nothing to diminish the value of attempting to examine adulthood from a developmental perspective, they do alert us to the existence of sociological factors that differentially affect the socialization experiences of both women and men. Obviously, complete reliance on this model for predicting what will happen to people and how they will behave masks variations traceable to sociological factors that need to be discussed (Rossi, 1974; George, 1993). The model omits, for example, important social influences of gender, social class, race, and ethnicity. For purposes of illustration, let's consider the effects of gender and social class on retirement and widowhood, the two major adjustments discussed regarding late adulthood.

How does social class influence retirement? **Social class** refers to a segment of a population whose members have a relatively similar share of the desirable things and who share attitudes, values, norms, and an identifiable lifestyle. (Social class is explored in detail in Chapter 8, "Social Stratification.") Two desirable things associated with social class are economic resources and occupational status. Each of these sociological factors affects the experience of retirement.

Financial security is one of the most important determinants of satisfaction with life among retirees (Atchley and Miller, 1983; Calasanti, 1988). Occupational status also ranks high as a predictor of satisfaction with retirement. First, retirees who held higher status jobs are generally more financially secure than those with lower status jobs. Second, higher status occupations have characteristics promoting better adjustment to retirement. For example, higher status occupations involve more interpersonal contact and

greater participation in a variety of structured activities. The social skills and more positive attitudes toward involvement in group activities are carried into retirement in the form of more satisfying nonwork activities (Kohn and Schooler, 1983). One positive consequence is less social isolation and loneliness.

How do gender differences affect widowhood? Widowhood negatively influences the quality of life of both elderly men and elderly women. Gender, however, appears to exercise differential effects. The major sociological factors associated with gender are the degree of financial security (gender and social class interact) and the extent of social involvement.

Adjustment to widowhood appears to be related to financial resources (Brubaker, 1991). And older widows are usually at a disadvantage financially, especially if they have been economically dependent on their husbands and are not covered by their husbands' pension plan (Choi, 1992). The situation is even worse for economically dependent women who are under sixty years of age because they do not qualify for Social Security benefits. Even if they qualify for Social Security benefits (under $400 per month), they live well below the poverty line (Watterson, 1990). Thus, poverty or near poverty is the fate of a substantial proportion of widows in the United States (Thomas, 1994).

Although women are at a disadvantage in adjusting to widowhood financially, they benefit from their greater capabilities for emotional expression and for attaining and maintaining a network of social support involving friends, relatives, and other widows. Widowers, on the other hand, are more socially isolated, have more difficulty displaying grief, express greater loneliness, and recover more slowly from the loss of their spouses (Anderson, 1984; Burgess, 1988; Morgan, 1989).

FEEDBACK

1. The informal and unofficial things that are taught children in school to prepare them for life in the larger society are called the _____.
2. The _____ provides children with needed experiences they are unlikely to obtain within the family.
3. Because there is almost no cultural definition of what retired and widowed persons should do, they are in what has been called a _____ status.
4. Match the following stages of dying with the accompanying statements.

    ____ a. denial and isolation   (1) "Why me?"
    ____ b. anger   (2) "What will Martha do without me?"
    ____ c. bargaining   (3) "I would be fine, if I could just see Julia graduate."
    ____ d. depression   (4) "I am going to be fine, you'll see; just leave me alone for a while."
    ____ e. acceptance   (5) "My marriage has been a good one. There is little left for me to do."

Answers: 1. *hidden curriculum* 2. *peer group* 3. *roleless* 4. a.(4) b.(1) c.(3) d.(2) e.(5)

MASS MEDIA AND THE THEORETICAL PERSPECTIVES

Socialization through the family, schools, and religion are covered later in separate chapters. Because the mass media are playing an increasingly important role in the socialization process in American society, additional coverage is appropriate. This coverage involves application of the functionalist and conflict theories to the mass media.

Functions of the Mass Media

A major problem with identifying the functions of the mass media is the separating of their unique functions from those of other institutions that use the mass media for their own purposes. Although isolating the societal functions of the mass media is somewhat difficult, some successful attempts have been made (Lasswell, 1948; Wright, 1960, 1974; McQuail, 1987).

What are the functions of the mass media?   The media have consequences for societies, groups, and individuals. We focus here only on societal consequences.

1. *The mass media provide information regarding events inside and outside a society.* Warnings of natural disasters or military invasions are delivered by the mass media. The mass media also provide data necessary for the operation of other institutions—information can be gathered regarding political candidates, stock market trends, or consumer products. In addition, the media promote social control by making information about deviant behavior public.
2. *Social continuity and integration are promoted by the mass media.* The mass media accomplish this by

exposing the entire population to the society's dominant beliefs, values, and norms. The media socialize the young and continue the process of socialization into adulthood. New cultural developments are also transmitted to the population by the mass media.
3. *Entertainment is supplied by the mass media.* By providing amusement, diversion, and relaxation, the media contribute to the reduction of social tension. Thus, people are better able to perform their social functions and society is made more stable.
4. *The mass media explain and interpret meanings of events and information.* This process tends to support established cultural forms and enhances the building of consensus among members of a society. Obviously, the media are involved in the process of socialization as they attempt to translate meanings of events and information for the population.
5. *The mobilization of society can be accomplished through the mass media.* A society can be galvanized for war, great numbers of people can be motivated to join a humanitarian movement, or citizens can be moved to protest governmental policies through the mass media.

The functions just outlined are positive ones. Functional analysis, as you know, also involves the identification of negative consequences, or dysfunctions.

What are some dysfunctions of the mass media?   The media may foster panic while delivering information, increase social conformity, legitimate the status quo and impede social change while promoting social continuity and integration, divert the public from serious long-term social issues through a constant menu of trivial entertainment and short-term emphasis, shape public views through editorializing as they "interpret"

This NBC news studio contributes to the functioning of society. Among other things, its newscasts contribute to the socialization of the population.

the meanings of events and information, and create violence in mobilization of the public.

The Conflict Perspective and the Media

As you have come to expect, conflict theory underscores much more fundamental drawbacks to the mass media. This section examines the mass media from the conflict perspective, beginning with the neo-Marxian model and concluding with the power-elite model.

What is the Marxian view of the media in America? Marxists view the media as a means of producing wealth. Moreover, they consider the media to be monopolized by the ruling class for the purposes of earning a profit and bolstering their power. In the process, workers are exploited by being paid less than they deserve and consumers are overcharged, which places excessive profits in the hands of the ruling class. Finally, the media are said to be necessary to disseminate the **ideology** of the ruling class (that is, the set of ideas they use to justify and defend their interests and actions), and to reduce the impact of competing ideas. By controlling the marketplace of ideas, the ruling class can reduce the chances of the development of class consciousness and subsequent political action on the part of the workers. The media, then, are seen as a tool of manipulation by which the ruling class maintains its power.

There is a non-Marxian theory that also sees a few elite individuals in control of American society, and, of course, in control of the mass media. This is the theory of the **power elite**—a unified coalition of top military, corporate, and government leaders (the executive branch in particular). The power-elite theory is covered in greater depth in Chapter 13 ("Political and Economic Institutions"). For now, we will concentrate on this theory and the mass media.

Do members of the power elite control the mass media? Thomas Dye (1990) explores three lines of indirect evidence: the organizational concentration of the media, the agenda-setting power of the media, and the media's ability to socialize the population.

The mass media are increasingly becoming concentrated in fewer corporate hands. Despite the growth of cable television, American television is still controlled primarily by three major networks: the American Broadcasting Companies (ABC); CBS, Incorporated (CBS); the National Broadcasting Company (NBC). This control is still intact despite the recent competition of the Fox network, a fledgling network making some bold moves to become a big player (Reibstein and Hass, 1994: Zoglin, 1994). Most of America's local television stations must affiliate with one of these four networks because they cannot afford to produce all of their own programming. Each of the networks is permitted to

own local stations, a combination that covers over one-third of households in the United States. Although some ground has been lost to cable television—over 65 percent of all households in the United States are cable subscribers—the four networks held about 48 percent of all television viewing in 1994–95, down from 64 percent in 1986–87 (*The World Almanac and Book of Facts,* 1996, 1997). Moreover, all four networks are now part of larger corporations, a move that greatly magnifies their power. ABC, for example, is now part of The Walt Disney Company, a corporation that owns, in addition to its famous movie studio and theme parks, the ABC television network, eight television stations, six radio networks, nineteen radio stations, four daily and thirty-five weekly newspapers, numerous magazine and book publications, and such cable television holdings as ESPN (an all-sports network), the Arts and Entertainment Network, and Lifetime Cable Network (*Directory of Corporate Affiliations,* 1996).

Of the four networks, Fox is the only one located outside the United States. News Corp. LTD, an Australian-based corporation, in 1996 also owned Fox Television Stations, Inc., Harper Collins Publishers, TV Guide Magazine, The Boston Herald, News American Publishing Company, News LTD of Australia, and Twentieth Century-Fox Film Corporation (*Directory of Corporate Affiliations,* 1996). News Corp. LTD's principal activities are the printing and publishing of newspapers, magazines, and books; commercial printing; television broadcasting; film production and distribution; and motion picture studio operations.

Newspapers and radio stations receive most of their national and international news from two wire services: the Associated Press (AP) and United Press International (UPI). Newspaper ownership is increasingly becoming monopolized as major newspaper chains absorb local papers. Consider the concentration of power in the hands of the New York Times Company, which owns not only the *New York Times* but also many other daily newspapers along with radio and television stations. Time Warner, Incorporated, owns *Time, Life, Sports Illustrated, Fortune, People,* and *Money* magazines, along with a number of television stations, a host of cable television companies, Home Box Office, the Book of the Month Club, movie companies, and several book and magazine publishers (*Directory of Corporate Affiliations,* 1996).

The ability of the media to set the political agenda is also underscored by advocates of the power-elite model. *Agenda setting* in the context of the media refers to the capability of selecting for the public which issues are important for consideration (Turner and Killian, 1987). The media can create, publicize, and dramatize an issue to the point that it becomes a topic for debate among both political leaders and the public. Or, the media can downplay an issue or event so that it disappears from public view. Although agenda setting is not accomplished just by television, television is the most powerful force: "TV is the Great Legitimator. TV confers reality. Nothing happens in America, practically everyone seems to agree, until it happens on television" (Henry, 1981:134). Not only can the media attempt to tell us "what" to think, as the functionalists suggest, they also are said to be extremely successful in determining what we think "about."

Conflict theorists also underscore the media's role in socializing the population. It is through television that most Americans stay in touch with what is happening in their country and in the world. Ninety-eight percent of all households in the United States have at least one television set; the average number of sets in a household is two; those in an average household view television over sixteen hours per week (*The World Almanac and Book of Facts,* 1996, 1997). America's favorite television shows are viewed by an average of almost 20 percent of households each week. (See Table 4.2).

Because television is the most widely shared experience in the United States, it has the dominant role in portraying what the media consider to be the most important beliefs, norms, attitudes, and values for people to hold. Through entertainment programming, news coverage, and advertising, Americans are constantly being exposed to a particular view of the most desirable and appropriate ways to think, feel, and behave.

The foregoing analysis does not prove that the mass media serve primarily the interests of the elite. It does,

TABLE 4.2	Percentage of American Households Viewing the Most Popular Network Programs, 1995–1996
Programs	**Percentage of TV Households**
E.R.	22.0
Seinfeld	21.2
Friends	18.7
Caroline in the City	17.9
NFL Monday Night Football	17.1
Single Guy	16.7
Home Improvement	16.2
Boston Common	15.6
60 Minutes	14.2
NYPD Blue	14.1
Frasier	13.6
20/20	13.6
Grace Under Fire	13.2
Coach	12.9
NBC Monday Night Movies	12.9

Source: *The World Almanac and Book of Facts,* Mahwah, NJ: K–111 Reference Corporation, 1996, p. 295.

however, suggest the potential for the media to be doing so. Conflict theorists, of course, contend that this potential is being realized.

FEEDBACK

1. The _____ are those means of communication that reach large heterogeneous audiences without any personal interaction between the senders and the receivers of messages.
2. Which of the following was *not* discussed in the text as a function of the mass media?
   a. provision of information
   b. creation of consensus
   c. mobilization of society
   d. promotion of social discontinuity and disintegration
3. Which of the following was *not* presented as indirect evidence for the power-elite model of the mass media?
   a. foreign infiltration into the media industry
   b. organizational concentration of the media industry
   c. agenda setting
   d. socialization of the population

Answers: 1. mass media 2. d. 3. a.

SUMMARY

1. Socialization, the process of learning to participate in group life through the acquisition of culture, is one of the most important social processes in human society. Without it, we would not be able to participate in group life and would not develop many of the characteristics we associate with being human. Evidence of the importance of socialization has been shown in studies of isolated monkeys and children. Socially isolated primates—including humans—do not develop as we would expect.

2. Several major theories emphasize the importance of socialization for human development. Although functionalism and conflict theory bear on the socialization process, the most fully developed sociological approach to socialization comes from symbolic interactionism. Key concepts include the self-concept; the looking-glass self; significant others; role taking; the imitation, play, and game stages; and the generalized other.

3. Several major life-course theories emphasize the importance of socialization for human development from a psychological viewpoint. Sigmund Freud accentuated the interaction between the biological nature of human beings and the social environment. Erik Erikson, a student of Freud's, described a series of developmental stages that occur from infancy to old age. Each of these developmental stages is accompanied by a psychosocial crisis, or developmental task. In Erikson's view, socialization and personality development do not stop in childhood; they are lifelong processes. According to Jean Piaget, the ability to think, know, and reason develops through interaction with others; if we are to mature cognitively, Piaget contends, each of us must pass through four identifiable stages in the proper developmental sequence.

4. Symbolic interactionism contributes to our understanding that socialization is a lifelong process. The concepts of total institutions, desocialization, resocialization, anticipatory socialization, and reference group are particularly important in this regard.

5. During childhood and adolescence, the family, school, peer group, and mass media are the major agencies of socialization. The family makes a tremendous impression on the child both because it is the first agency of socialization and because the child is dependent and highly impressionable.

6. The school is generally the first agency of socialization that is controlled by nonrelatives. In school, children are exposed to new standards of performance applied to everyone, are encouraged to develop loyalties beyond their own families, and are prepared in a number of ways for adult life. In addition, by exposing children to such skills as reading and writing and such subject matter as mathematics and English, schools train children to be disciplined, orderly, cooperative, and conforming.

7. The peer group is the first agency of socialization that is not controlled by adults and consequently provides young people with experiences they cannot easily obtain elsewhere. In peer groups, young people learn to deal with others as equals, gain experience in self-direction, establish some degree of independence from their parents and other adults, develop close relationships with friends of their own choosing, and learn to associate with various types of people.

8. Specific effects of the mass media are difficult to document. Still, we know that television, radio, newspapers, and other mass media play a significant part in the socialization process by introducing children to various aspects of their culture.

9. Socialization in infancy, childhood, and adolescence makes a profound contribution to personality development and one's ability to participate in social life. There is, however, an increasing recognition that socialization continues throughout life, even to old age. There is much interest now in viewing adulthood from a developmental perspective.

10. The extension of life expectancy through medical science has created problems of adjustment unparalleled in human history. At an arbitrary age, most older workers are deprived of a meaningful work life. They are forced to retire whether or not they are still productive. Just as there is little preparation for retirement, there is hardly any prior socialization for losing a husband or wife. Like retirement, widowhood is a roleless status. Unlike people in some other cultures, Americans are not well socialized to face the prospect of dying. Despite this relative lack of socialization, Americans tend to go through several distinguishable stages in the process of dying.

11. The mass media encompass those means of communication that reach large heterogeneous audiences without any personal interaction between the senders and the receivers of messages. The major forms of mass communication are television, radio, newspapers, magazines, movies, books, tapes, discs, and now the Internet.

12. Although the functions of the media are intertwined with other institutions, it is possible to isolate their major functions. The media provide information, promote social continuity and integration, supply entertainment, explain and interpret events and information, and can mobilize the society when necessary. Some negative consequences, or dysfunctions, are also associated with the mass media.

13. Both the neo-Marxist and power-elite views have been applied to the mass media. Proponents of the power-elite model point to the organizational concentration of the media industry, the agenda-setting power of the media, and the media's ability to socialize the population as at least indirect evidence of a ruling class in America.

LEARNING OBJECTIVES REVIEW

After careful study of this chapter, you will be able to:

- Discuss the contribution of the socialization process to human development.
- Describe the contribution symbolic interaction makes to our understanding of socialization, including the concepts of the self, the looking-glass self, significant others, and role taking.
- Compare and contrast the life-course theories of Freud, Erikson, and Piaget.
- Distinguish among the concepts of desocialization, resocialization, and anticipatory socialization.

- Outline the major sources of influence on the socialization of the young.
- Describe the stages of development in adulthood.
- Discuss the primary demands of socialization encountered in late adulthood.
- Compare and contrast the approaches of functionalism and conflict theory to the mass media's effects on socialization.

CONCEPT REVIEW

Match the following concepts with the definitions listed below them.

____ a. significant others
____ b. ideology
____ c. looking-glass self
____ d. personality
____ e. total institution

____ f. anticipatory socialization
____ g. resocialization
____ h. hospices
____ i. social class
____ j. peer group

____ k. "I"
____ l. reference group
____ m. ego

1. The process of preparing oneself for learning new norms, values, attitudes, and behaviors.
2. The conscious, rational part of the personality.
3. Organizations designed to provide support for the dying and their families.
4. The spontaneous and unpredictable part of the self.
5. The set of ideas used to justify and defend the interests of those in power in a society.
6. One's self-concept based on perceptions of others' judgments.
7. A group composed of individuals of roughly the same age and interests.
8. A relatively organized complex of attitudes, beliefs, values, and behaviors associated with an individual.

9. A group we use to evaluate ourselves and to acquire attitudes, beliefs, values, and norms.
10. The process of learning to adopt new norms, values, attitudes, and behaviors.
11. Those persons whose judgments are the most important to an individual's self-concept.
12. A segment of the population whose members have a relatively similar share of the desirable things and who share attitudes, values, norms, and an identifiable lifestyle.
13. A place in which residents separated from the rest of society are controlled and manipulated by those in charge.

CRITICAL THINKING QUESTIONS

1. Defend the proposition that human nature is more a matter of nurture than nature. Cite evidence in your argument.

_____

_____

2. What do you think is the greatest contribution symbolic interactionism has made to our understanding of the socialization process? Why?

_____

_____

3. You are now a college student. Have you undergone (are you currently undergoing) desocialization, resocialization, or anticipatory socialization? Provide examples from your own experience.

_____

_____

4. Evaluate the statement that socialization ends after childhood. Be specific in your line of argument.

_____

_____

5. Discuss the pros and cons of the mass media from the viewpoint of sociologists. Use specifics in developing your thoughts.

_____

_____

MULTIPLE CHOICE QUESTIONS

1. _____ is the process through which people learn to participate in group life through the acquisition of their culture.
   a. Imprinting
   b. Behavior modification
   c. Socialization
   d. Imitation
   e. Acculturation

2. **The relatively organized system of attitudes, beliefs, values, and behaviors associated with an individual is known as**
   a. personality.
   b. internalization.
   c. humanization.
   d. socialization.
   e. dualism.

3. **René Spitz's study of infants in an orphanage and infants in a women's prison nursery found that**
   a. the infants in both the orphanage and the prison nursery developed normally.
   b. social isolation need not be extreme in order to damage social and personality development.
   c. social isolation must be extreme in order to damage social and personality development.
   d. infants in both the orphanage and the women's prison died at early ages.
   e. children in prison nurseries are worse off than children in orphanages.

4. _____ allows for the greatest understanding of the socialization process.
   a. Functionalism
   b. Conflict theory
   c. Symbolic interactionism
   d. Sociobiology
   e. Game theory

5. **The _____ is based on our perceptions of others' judgments of us.**
   a. significant other self
   b. alternative self

   c. distorted self-concept
   d. other-directed self
   e. looking-glass self

6. **The process of mentally assuming the viewpoint of another individual and then responding to oneself from that imagined viewpoint is referred to as**
   a. role reversal.
   b. play acting.
   c. gaming.
   d. role taking.
   e. role distancing.

7. **According to George Herbert Mead, children learn to take on the roles of other individuals, one at a time, during the _____ stage.**
   a. game
   b. imitation
   c. work
   d. play
   e. Oedipal

8. **According to Freud, the id, ego, and superego should be thought of as**
   a. separate regions of the brain.
   b. separate, interacting, and conflicting processes within the mind.
   c. a mass of impulses, urges, and desires.
   d. multiple personalities.
   e. little people negotiating inside our heads.

9. **According to Erikson, all individuals pass through a series of eight developmental stages from infancy to old age, each of which involves a _____ crisis or _____ task.**
   a. psychosocial; developmental
   b. sociobiological; cognitive
   c. developmental; psychopathic
   d. psychosocial; cognitive
   e. developmental; cognitive

10. Jean Piaget was concerned primarily with the development of _____, which is the ability to think, know, perceive, judge, and reason.
    a. psychosocial development
    b. cognitive ability
    c. self-concept
    d. socialization
    e. role taking

11. _____ are most likely to participate in anticipatory socialization.
    a. New prison inmates
    b. Teenagers
    c. Office workers
    d. New mental patients
    e. Cab drivers

12. The group composed of individuals of roughly the same age and interests is known as a
    a. subculture.
    b. culture cohort.
    c. peer group.
    d. resocialization group.
    e. society.

13. Because there is almost no cultural definition of what retired and widowed persons should do, they occupy what has been called a/an _____ status.
    a. undefinable
    b. casual
    c. master
    d. limbo
    e. roleless

14. According to the Marxian view, the media in America are
    a. a threat to social conformity.
    b. the "mouthpiece" of organized labor.
    c. a means of production.
    d. a tool of manipulation by which the ruling class maintains its power.
    e. a threat to the proletarian revolution.

15. The power elite is composed of
    a. top university, corporate, and government elite.
    b. those in direct control of the mass media.
    c. top military, corporate, and government leaders.
    d. top religious, university, and corporate leaders.
    e. those who place controlling others over economic success.

FEEDBACK REVIEW

True-False

1. According to sociologists, no fundamental human nature exists prior to social contact. *T or F?*
2. Thanks to recent breakthroughs, research findings on the need of infant monkeys for warmth and affection can easily be applied to humans. *T or F?*

Fill in the Blank

3. In its approach to socialization, _____ emphasizes social interaction based on symbols.
4. According to the looking-glass process, our _____ is based on how we think others judge the way we look and act.
5. Whereas Freud believed that the human personality is almost totally determined in early childhood, Erikson held that personality can change _____.
6. According to Erikson, the successful management of a psychosocial crisis at an earlier stage increases the chances for mastering later crises. This means that the effects of success or lack of success at earlier crisis points are _____.
7. The informal and unofficial things that are taught children in school to prepare them for life in the larger society are called the _____.
8. Because there is almost no cultural definition of what retired and widowed persons should do, they are in what has been called a _____ status.

Matching

9. Match Jean Piaget's stages of cognitive growth with the statements.
    ____ a. sensorimotor stage
    ____ b. concrete operations stage
    ____ c. formal operations stage
    ____ d. preoperational stage

    (1) A child thinks without the aid of concrete objects and manipulations.
    (2) A child recognizes that 3 + 2 is the same as 2 + 3.
    (3) A child learns that a cat that runs under a couch still exists.
    (4) A child believes that a box is a spaceship.

10. Match the following stages of dying with the accompanying statements.

_____ a. denial and isolation   (1) "Why me?"
_____ b. anger   (2) "What will Martha do without me?"
_____ c. bargaining   (3) "I would be fine, if I could just see Julia graduate."
_____ d. depression   (4) "I am going to be fine, you'll see; just leave me alone for a while."
_____ e. acceptance   (5) "My marriage has been a good one. There is little left for me to do."

ANSWER KEY

Concept Review	Multiple Choice	Feedback Review
a. 11	1. c	1. T
b. 5	2. a	2. F
c. 6	3. b	3. symbolic interactionism
d. 8	4. c	4. self-concept
e. 13	5. e	5. at any time
f. 1	6. d	6. cumulative
g. 10	7. d	7. hidden curriculum
h. 3	8. b	8. roleless
i. 12	9. a	9. a. 3
j. 7	10. b	b. 2
k. 4	11. b	c. 1
l. 9	12. c	d. 4
m. 2	13. e	10. a. 4
	14. d	b. 1
	15. c	c. 3
		d. 2
		e. 5

Chapter Five

Social Structure and Society

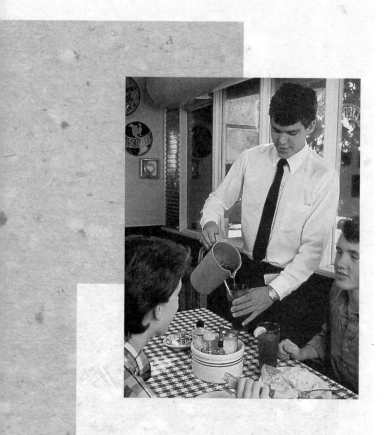

OUTLINE

Sociological Imagination
The Nature of Social Structure

Sociology in the News: Role Conflict in Blended Families

Society

Doing Research: Arlie R. Hochschild—The Social Management of Feelings

LEARNING OBJECTIVES

After careful study of this chapter, you will be able to:

- Explain what sociologists mean by social structure.
- Distinguish and illustrate the basic building blocks of social structure.
- Define the concept of society.
- Describe the basic solution to the problem of survival employed by the major types of preindustrial societies.
- Discuss the basic characteristics of industrial society.
- Compare and contrast preindustrial, industrial, and postindustrial societies.

SOCIOLOGICAL IMAGINATION

Are humans genetically selfish?   Many would say that, by their nature, people are basically selfish. If humans, then, as a species, shared the trait of selfishness, we would not expect to find a society populated by cooperative members. Evidence, on the contrary, indicates that economic relationships in hunting and gathering societies are, in fact, grounded in the practice of sharing their surplus with others. Absent a conception of private property or ownership, members of hunting and gathering societies view even thrift as an indication of selfishness. If genetically programmed to be selfish, members of hunting and gathering societies (or any society for that matter) would be incapable of engaging in behavior oriented toward concern for others. They could not, to follow the example above, exist without a strong sense of personal ownership and act on the belief that being thrifty is selfish.

Members of hunting and gathering societies, like members of all social groups, know what is expected of them and what they can expect from others, and they engage in the same basic social patterns time after time. Consequently, all groups have patterned and predictable social relationships that are passed from generation to generation. These patterned social relationships are referred to as *social structure,* which is the basic subject matter of sociology.

THE NATURE OF SOCIAL STRUCTURE

It is in this chapter that your understanding of the basic nature of sociology will jell. You learned in the last chapter that culture helps to explain human behavior in groups because, in the absence of biological pro-

gramming, culture has to provide the raw material for thinking, feeling, and behaving. In the absence of culture, humans would have no blueprint for social living. This chapter will lead you through a set of concepts sociologists use to demonstrate how culture is translated into social relationships called social structures.

The Concepts of Social Structure

Social structure refers to patterned, recurring social relationships. In *As You Like It,* William Shakespeare captured the essence of this concept in these well-known lines:

> *All the world's a stage. And all the men and women merely players; They have their exits and their entrances; And one man in his time plays many parts.*

Members of any group have parts they are expected to play. Students are expected to attend class, listen to the instructor, fulfill class requirements, and take examinations. Professors are expected to hold class, teach, make assignments, and grade examinations. If you were to observe any class on an American college campus, you would find similar relationships between students and faculty members. Student and faculty behavior is orderly and predictable because of its underlying structure. If you, as a new college student, visited one class in which the students shouted the

A status is a position within a social structure. Each of the Clintons has the ascribed status of their respective genders by virtue of birth. President Clinton and Hillary Clinton each occupy the achieved status of parent. Through graduation from high school and entrance into Stanford University, Chelsea has achieved the status of college student.

professor down, another in which the custodian contributed to the class discussion, and another in which students did all the teaching, you would miss the presence of structure. You would look for order and predictability. Otherwise, you would feel that the educational process was in chaos, and on a personal level, you would be apprehensive about relating to the professors in your new classes.

Fortunately, upon entering a new group, we are usually spared such confusion because we bring some knowledge of the ways people normally relate to one another, even if we have never been in the group before. In other words, we usually carry a social map in our minds for various group situations, and we have mental images—however unconscious and hazy—of the structure of the groups in which we want to participate. This knowledge permits us to engage in the patterned social relationships existing within various groups without undue personal embarrassment or social disruption. This underlying pattern of social relationships is called social structure.

The mental maps of social structure are not acquired automatically; they must be learned from others. In the process, we learn statuses and roles—major elements of social structure.

What do sociologists mean by status? People in modern societies may refer to themselves as students, doctors, welders, secretaries, mothers, or sons. Each of these labels refers to a **status**—a position a person occupies within a social structure. Some social statuses are acquired at birth; a newly born female instantly becomes a child and a daughter. From then on, she assumes an increasingly larger number and variety of statuses.

There are two basic types of social statuses. An **ascribed status** is neither earned nor chosen; it is assigned to us. At birth, an infant is either a male or a female. Except in instances of sex-change operations, our gender is beyond our power to select or refuse. Age is another prominent example of ascribed social status. In some societies, religion and social class are ascribed on the basis of the family into which individuals are born. If you were born into a lower-class home in some societies, for example, you would not be permitted to rise to a higher social class.

An **achieved status** is one that can be earned; people have some degree of control and choice. In most modern societies, for example, an individual can decide whether to become a spouse or a parent. Occupations are also achieved statuses in modern societies; people have some latitude in deciding the type of work they will pursue. Thus, plumber, electrician, sales representative, nurse, executive, lawyer, and doctor are achieved statuses in modern societies.

Sociologists are interested in the relationships between social statuses. A sociologist investigating prostitution would focus on the status of the prostitute in relation to the statuses of the police officer, judge, customer, pimp, and pusher. Figure 5.1 illustrates the status of a social worker as related to various other statuses.

In addition to these relationships, any particular social worker will occupy various other statuses that may be totally unrelated to that of social worker. A **status set** is the totality of statuses that an individual occupies at any particular time. One social worker may be a wife, mother, author, and church choir director; another may be a bachelor, Boy Scout leader, and jazz musician. Each of these statuses would be part of another network of sta-

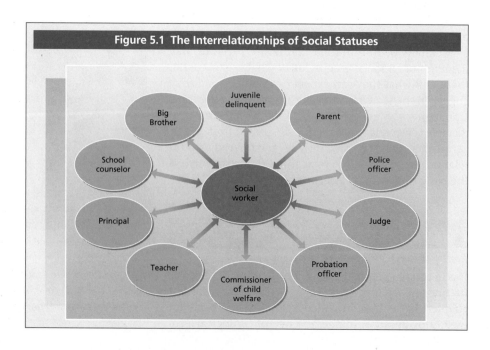

Figure 5.1 The Interrelationships of Social Statuses

tuses. Assume, for example, that in addition to being a social worker, an individual is also a part-time jazz musician. In this status, he would interact with, say, the statuses of nightclub owner, dancer, fellow musician, and representative of the musician's union.

Thus far, social statuses have been described as separate positions that people occupy at different stages of their lives or at different times in their daily lives. Single social statuses, as you have seen, are also interrelated.

Some statuses are more important than others. **Master statuses** are important because they influence most other aspects of a person's life. Master statuses may be achieved or ascribed. In industrial societies, occupations, achieved for the most part, are master statuses. One's occupation strongly influences such important matters as where one lives, how well one lives, and how long one lives. "Criminal" or "drug addict" are also achieved master statuses whose effects permeate the rest of one's life.

Ascribed master statuses are no less important for a person's life than achieved ones. A person who acquires immune deficiency syndrome through a blood transfusion in a hospital becomes an AIDS victim who will likely suffer prejudice and discrimination in employment, housing, and social relationships. The physically handicapped often have similar experiences.

Age, gender, race, and ethnicity are also ascribed master statuses because they significantly affect the likelihood of achieving other social statuses. Age categories, gender, and various racial and ethnic minorities do not normally constitute social classes because all social classes contain males, females, the young, the old, and persons of various racial and ethnic backgrounds. But they are especially influential in social class placement because they affect other social statuses that a person may occupy. When will the United States have a female president? How many women doctors, lawyers, and executives are there in America? Would you let a nineteen- or ninety-year-old handle your case in court or remove your appendix? What about the absurdity of an African American Ku Klux Klan Grand Dragon, Cyclop, or Kleagle? Or, less frivolously, why have so many African Americans turned to sports and show business to make it?

Social statuses are similar to the parts performers play on the stage. Prostitute, pimp, police officer, and judge are parts that individuals play in real life. The behavior of individuals depends largely on the part (or status) such individuals hold. This is because parts in real life are based on culturally defined roles.

What are roles?   **Roles** are the culturally defined rights and obligations attached to statuses; they indicate the behavior expected of individuals holding them. Any status carries with it a variety of roles. The roles of a modern doctor include keeping abreast of new med-

As the occupant of the achieved status of talk show host, Oprah Winfrey has a variety of roles. For example, she is expected to prepare for a show, ask the questions, and keep the proceedings entertaining.

ical developments, keeping office appointments, diagnosing illnesses, and prescribing treatment.

Roles can be thought of as the glue that holds a network of social statuses together—the roles of one status are matched with the roles of other statuses through rights and obligations. **Rights** inform individuals of behavior they can expect from others. **Obligations** inform individuals of the behavior others expect from them. The rights of one status correspond to the obligations of another. Doctors, for example, are obligated to diagnose their patients' illnesses. Correspondingly, patients have the right to expect that their doctors will try to diagnose their physical problems. Patients have an obligation to keep their office appointments, and doctors have a right to expect that they will do so.

To continue with the stage analogy, roles can be viewed as the script that indicates the beliefs, feelings, and actions people holding a certain status are supposed to adopt. Just as a playwright or screenwriter specifies the content of a performer's part, culture underlies the parts played in real life.

What are role performance and social interaction?
Statuses and roles provide the basis for group life. But roles are activated only when people in statuses engage in role performance. This can occur only through the process of social interaction.

As noted in Chapter 1 ("The Sociological Perspective"), Erving Goffman, with his dramaturgical approach, deserves the most credit for developing the analogy between the stage and social life. In a sense, Goffman's life's work was devoted to the distinction between culturally prescribed roles and the fulfillment of those roles. **Role performance** is the actual conduct involved in activating a role. Although some role performance can occur in isolation, as when a student studies alone for an examination, most of it involves social interaction. **Social interaction** exists when two or more persons mutually influence each other's behavior. For example, before two boys begin to fight, they have probably gone through a process of insult, counterinsult, and challenge. Although most social interaction is not as negative and violent, the same process of mutual influence is involved.

Not all social interaction is expected role performance. When patients are told to disrobe for a medical examination, they are not supposed to take this as a cue from the doctor for sexual activity, nor should doctors expect this instruction to evoke sexual cues from their patients. If such sexual episodes occur, they are violations of expected role performance for doctors and patients.

If statuses are analogous to the parts of a play and roles to the script, then social interaction is similar to the give-and-take of cues in a performance and role performance is the actual performance. Table 5.1 illustrates this analogy.

Can we go beyond the stage analogy?   Although the stage analogy is a valid one, there is a danger in taking analogies at face value. The two situations being compared are seldom exactly the same. The analogy between the stage and group life is no exception. The stage analogy can be useful, but insights can also be gained by looking at the differences between group life and the theater.

First, "delivery of the lines" in life is not the conscious process that we see in performers on the stage. Most role performance occurs without much forethought—we usually act in ways that seem to be natural and correct. In fact, these "natural" and "correct" ways have been unconsciously adopted through observation, imitation, and socialization.

Second, there is considerably more of a discrepancy between roles and role performance in group life than between a stage script and the play's performance. Performers may substitute lines for momentarily forgotten ones, deliberately change lines to suit themselves, and introduce a little "business" here and there, but they must stick closely to the script. Differences between a role and role performance in real life are neither as easy to detect nor as easy to control as departures from a script.

Third, on the stage there is a programmed and predictable relationship between cues and responses. One performer's line is a cue for a very specific response from another actor. In life, we can choose from a variety of cues and responses. A student may decide to tell a professor that her examinations are the worst he has ever

TABLE 5.1	The Stage Analogy
Stage	**Social Life**
Parts	Statuses
Script (lines)	Roles
Cues	Social interaction
Actual performance	Role performance

encountered. On hearing this, the professor may tell the student that it is not his place to judge, or the professor may ask for further explanation so that possible improvement may be made. In other words, the professor can choose from several roles. And the student can choose from a variety of responses to the professor's behavior. Should the professor tell the student he is out of line, the student may demand to be heard, or he may report the matter to the head of the professor's department. This process of role selection prior to behavior occurs in nearly all instances of social interaction.

However, the range of acceptable responses is not limitless. Only certain responses are considered legitimate. It is not an appropriate response for the professor to bodily eject the student from her office, and the student is not supposed to pound in protest on the professor's desk.

How are culture and social structure related?   Figure 5.2 shows how culture and social structure are linked. Starting at the top of this figure, the first major bond between culture and social structure is found in the concept of role. A position that a person occupies within a group is a status, the second link in this conceptual chain. Roles are attached to each status, and it is through these roles that culture enters the picture. Yet people do not always follow roles exactly. The manner in which roles are actually carried out is role performance, the third link in the conceptual chain. It is through social interaction that role performance occurs. This is the fourth link between culture and social structure. Social interaction that is based on roles is observable as patterned relationships, which make up social structure. In turn, existing social structure affects the creation of and changes in culture. (Note the two-headed arrow in Figure 5.2.)

Are all these terms necessary?   Concepts are as necessary for sociology as they are for physics. Both sociologists and physicists deal with abstractions, with unobservable phenomena whose existence must be assumed from the existence of things that can be seen. Before 1970, physicists had never directly observed individual atoms. Instead, they used indirect methods, such as photographs of X rays beamed at a small crystal of rock salt, to infer the existence of atoms. Just as

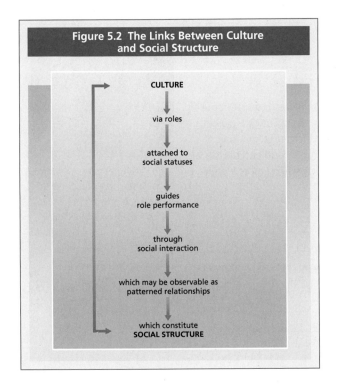

Figure 5.2 The Links Between Culture and Social Structure

CULTURE

↓

via roles

↓

attached to
social statuses

↓

guides
role performance

↓

through
social interaction

↓

which may be observable as
patterned relationships

↓

which constitute
SOCIAL STRUCTURE

physicists have used indirect methods, sociologists use the observation of patterned relationships (something concrete) to infer the existence of social structure (something abstract). If the concepts and relationships outlined in Figure 5.2 were not known, it would be as difficult for sociologists to establish the existence of social structure as it would have been for the ancient Greeks to document the existence of atoms. Concepts and their relationships are the necessary building blocks of any area of scientific study.

Role Conflict and Role Strain

The existence of statuses and roles permits social life to be predictable, orderly, and recurrent. At the same time, each status involves many roles, and each individual holds many statuses. This diversity invites conflict and strain. **Role conflict** exists when the performance of a role in one status clashes with the performance of a role in another one. A Catholic priest, for example, upon hearing a criminal confession, may experience role conflict. His role as priest demands confidentiality; his role as a citizen obligates him to inform the police.

Role strain occurs when the roles of a single status are inconsistent. Basketball coaches, for example, have to do their recruiting for the following season while attempting to achieve a winning season. University professors who are required to teach, spend time with students, and publish their research findings experience strain. Each of these roles is time consuming, and the fulfillment of one role may interfere with the per-

formance of the others. If your expectations as a college student require you to perform well academically, join a Greek organization, pursue a sport, date, and participate in other college activities, you will probably experience role strain.

How can role conflict and strain be handled? Role conflict and strain may lead to discomfort and confusion. To feel better and to have smoother relationships with others, we often solve role dilemmas by setting priorities. When roles clash, we decide which role is more important to us and act accordingly. Priests must decide whether to honor their vows of confidentiality or perform as good citizens. A university professor may have to decide to emphasize either teaching or research. By ranking incompatible roles in terms of their importance to us, we can reduce role conflict and strain.

Another way to minimize role conflict and strain is to segregate roles. This is especially effective for reducing the negative effects of conflicting roles. Priests may convince themselves they can be good citizens without revealing confidential information. Thus, they act one way as priests and another way as citizens. A member of the Mafia may reduce role conflict by segregating his criminal activities from his roles as church member and loving father.

Can conformity to roles ever be complete? The presence of role conflict and strain indicates that complete conformity to all our roles is virtually impossible, but this lack of complete conformity poses no problem as long as role performance occurs within accepted tolerance limits. Professors at research-oriented universities may be permitted to emphasize teaching over research if they do not totally neglect research. Coaches may accent fair play, character building, and scholarship over winning if their records are acceptable or if their school administrators have little interest in winning teams. Professors at research universities who do too little publishing or coaches who win too few games, however, usually will not be rewarded for very long. At some point they will have exceeded the acceptable limits for deviation from expected role performance and will be considered deviants.

Tolerance limits for deviance vary according to the individuals and the roles involved. Individual characteristics allow some people greater latitude for nonconformity. For example, the president of a major motion picture company, because of his past success, was reinstated even after he admitted embezzling $60,000 from his company. Other individuals—usually the poor—often receive several years in prison for stealing.

According to Erving Goffman, **role distancing**—the process of mentally separating oneself from a role—is another obstruction to total role conformity. A child too old to be sitting on Mother's lap will grin as if to say,

Role Conflict in Blended Families

The topic of the CNN report is the *blended family*—a remarried couple with one or more children from a previous marriage. The complexity of this family arrangement is obvious in the report. For one thing, several new statuses are created—stepfather, stepmother, stepchild, newlywed. Occupants of these new statuses encounter role conflict. This is a growing problem because by the year 2000 it is expected that there will be more blended families than traditional ones.

1. Describe the major role conflicts identified in the CNN report.

   _____

   _____

2. If you were a parent in a blended family, how would you attempt to handle role conflict?

   _____

   _____

In an earlier time, it would have been natural to assume this to be a picture of a mother and her biological daughter. Today, this could well be a stepmother and stepchild or a single mother with an adopted child.

"This is not the real me." An inexperienced teenager horseback riding at a public park may make light of riding (riding sidesaddle while the horse is walking, pretending to post) in order to protect herself from the inadequacies she feels, especially in the company of experienced riders. Goffman reports on a chief surgeon who, after being accidentally cut by an assistant surgeon, relieved tension in the operating room by stepping outside his formal leadership role to say: "If I get syphilis I'll know where I got it from and I'll have witnesses" (1961a:122). A professor may use improper English (as in "If it ain't broke, don't fix it") to appear down to earth and nonprofessorial.

Theoretical Perspectives and Social Structure

If you stop to think about it for a moment, you will realize that the concepts of role, social status, and patterned relationships reflect the concern of functionalism with stability, order, and consensus. The concepts of role behavior and social interaction have the strong

Many people experience role conflict—they find themselves performing a role in one status that clashes with a role in another status. The smile on this male Latino college student's face belies the conflict he feels because he is attempting to attend school and earn money to pay for that schooling simultaneously.

FEEDBACK

1. _____ refers to the patterned relationships among individuals and groups.
2. A _____ is a position a person occupies within a group.
3. _____ are culturally defined rights and obligations attached to statuses.
4. The rights of one social status correspond to the _____ of another one.
5. Match the following concepts with the examples listed beside them:

____ a. ascribed status   (1) Basketball coach, mother, author, daughter, professor.
____ b. achieved status   (2) A mother is expected to take care of her children.
____ c. master status   (3) Person who gets AIDS through a blood transfusion.
____ d. role   (4) A husband and wife discuss the disciplining of one of their children.
____ e. status set   (5) Occupation, gender, race, ethnicity.
____ f. role performance   (6) A university president hands out diplomas at a graduation ceremony.
____ g. social interaction   (7) The President of the United States.

6. Which of the following is *not* one of the differences between the stage and social life?
   a. There is considerably more discrepancy between roles and role performance in social life than between a stage script and its performance.
   b. Unlike the stage, there are no cues and responses in real life.
   c. Role performance in real life is not the conscious process that actors go through on the stage.
   d. In social life, the cues and responses are not as programmed and predictable as on the stage.
7. _____ exists when the performance of a role in one status clashes with the performance of a role in another one.
8. _____ occurs when the roles of a single status are inconsistent.

Answers: 1. Social structure 2. status 3. Roles 4. obligations 5. a. (5) b. (3) c. (7) d. (2) e. (1) f. (6) g. (4) 6. b 7. Role conflict 8. Role strain

flavor of symbolic interactionism. Also present, but not as readily apparent perhaps, is the place of the conflict perspective in the study of social structure. We can easily observe change and conflict in social structures. For change, consider colleges and universities where administrators no longer see themselves as substitute parents for students and where some cigarette machines have been converted to condom dispensers. Illustrations of conflict are the struggles between street gangs and law enforcement officials and the clashes between members of high school youth subcultures and their parents.

SOCIETY

Thanks to recent publicity about certain sociological concepts, the term *society* is at least vaguely familiar to most of us. Most Americans no longer associate this term with "high society," with the life of the rich and the super rich. Still, most people do not have a firm idea of what society is.

A **society** is composed of a people living within defined territorial borders who share a common culture. It is the largest and most nearly self-sufficient social structure in existence. A society theoretically is independent of all outsiders. It contains enough smaller social structures—family, economy, government, religion—to fulfill all the needs of its members.

Preindustrial societies could be independent and totally self-sufficient. Modern societies, although capable of taking care of most of their members' needs, must also have political, military, economic, cultural, and technological ties with many other societies. In fact, as the fate of all societies becomes linked with the fate of the world, there is a movement toward a global society.

Societies have been classified by anthropologists and sociologists in various ways. Because all societies must find ways to fulfill the need for food, one important classification system is based on the way the problem of subsistence is solved. The evolution from preindustrial society (including hunting and gathering, horticultural, and agricultural societies) to industrial society and postindustrial society has been based on the solution to the subsistence problem (Plog and Bates, 1990). The way in which a society solves the problem of subsistence is heavily reflected in the society's culture and social structures.

Preindustrial Societies

What is unique about a hunting and gathering society? The **hunting and gathering society** survived by hunting animals and gathering edible foods such as wild fruits and vegetables. This is the oldest solution to the subsistence problem. In fact, it was only about nine thousand years ago that other methods of solving the subsistence problem emerged.

DOING RESEARCH

Arlie R. Hochschild—The Social Management of Feelings

Sociologist Arlie Hochschild grew up as the daughter of diplomat parents in the U. S. Foreign Service. She believes that the roots of her interest in the management of emotions, captured in her book *The Managed Heart* (1983), lie in her observation as a preteenager of the behavior of her parents' official guests. Hochschild transported her fascination with the meanings of body language—handshakes, smiles, glances—into her sociological research.

Using student responses to questionnaire items and personal observation and interviews with Delta Airlines officials, flight attendants, and bill collectors, Hochschild formulated the view that the management of emotions is based on latent "feeling rules." These feeling rules guide us in our daily efforts to understand and honor social expectations attached to roles and ensuing social relationships:

What gives social pattern to our acts of emotion management? I believe that when we try to feel, we apply latent feeling rules. . . . We say, "I shouldn't feel so angry at what she did," or "given our agreement, I have no right to feel jealous." Acts of emotion management are not simply private acts; they are used in exchanges under the guidance of feeling rules. Feeling rules are standards used in emotional conversation to determine what

is rightly owed and owing in the currency of feeling. Through them, we tell what is "due" in each relation, each role. We pay tribute to each other in the currency of the managing act. In interaction we pay, overpay, underpay, play with paying, acknowledge our dues, pretend to pay, or acknowledge what is emotionally due another person (Hochschild, 1983:18).

Hochschild calls this work emotional labor. Emotional labor involves making an effort to call up or suppress feelings in order to create an appearance that produces the socially appropriate reaction in others with whom we are engaging in role performance. For example, flight attendants know their jobs have been done well when customers appear to be content. Producing this contentment requires the delivery of service in a particular way. Airline passengers do not want fake

---

Although not all hunting and gathering societies are exactly alike, they do share some basic features (Beals, Hoijer, and Beals, 1977; Service, 1979). Hunting and gathering societies are usually nomadic—they must move from place to place as the food supply and seasons change. Because they must carry all their possessions with them on each move, these nomads accumulate few material goods. Hunting and gathering societies also tend to be very small—usually with fewer than fifty people—with their members scattered over a relatively wide area. Because the family is the only real institution in hunting and gathering societies, it takes care of nearly all the needs of its members. On a day-to-day basis, association is typically limited to one's immediate family. In fact, these societies are organized either as self-sufficient family groups or, more typically, as loose combinations of families, or bands. Thus, hunting and gathering societies are tied together by kinship. Most members are related by blood or marriage, although marriage is usually limited to those outside one's family or band.

As indicated in the *Sociological Imagination* opening this chapter, economic relationships within hunting and gathering societies are based on cooperation—members share their surplus with other members. Most members of this type of society are kin, so sharing takes

place without the implication that a return must be made. Members of hunting and gathering societies seem simply to give things to one another. In fact, the more scarce something is, the more freely it is shared without strings attached. Generosity and hospitality are valued; thrift is considered a reflection of selfishness. Because the obligation to share goods is one of the most binding aspects of their culture, members of hunting and gathering societies have little or no conception of private property or ownership.

Without a sense of private ownership and with few possessions for anyone to own, hunting and gathering societies have no social classes. They have no rich and poor. These societies lack differences in status based on political authority because they have no political institutions; there is nobody to organize and control activities. When the Eskimos, for example, wanted to settle disputes, they had dueling songs. Those in disagreement prepared and sang songs to express their side of the issue. They were accompanied by their families as choruses, and the victor was chosen by the applause of those listening to the duel (Hoebel, 1983).

The division of labor in these societies is limited to the sex and age distinctions found in most families. Men and women are assigned separate tasks, and cer-

smiles or insincere courtesy. They want flight attendants who appear to love their jobs. Emotional labor is required to create this appearance. This emotional labor is defined as part of the job and leads to actually trying to love work and to enjoy the customers. Successful emotional labor—that is, feeling that one does, in fact, love the job—helps flight attendants to perform their job.

According to Hochschild, emotion management is not evenly distributed throughout the population; it is not practiced to the same extent by members of all social categories. As a matter of tradition, she concludes, women are better acquainted with the management of feeling and are more likely to engage in it as one of the assets they trade for economic support in marriage. Especially women of the middle

and upper classes perform the job of shaping the emotional atmosphere of social occasions. They share in the joy of others opening presents at Christmas, birthdays, and baby showers; attempt to ensure the surprise at birthday parties; and exhibit fear at the sight of a mouse or snake. Those men who are good at the management of emotions, Hochschild asserts, share less in common with other men than women who are good at it do with other women.

Various social classes also engage in the management of emotions to different degrees. Members of the middle and upper classes are more prone to perform emotional work. This is partly because working-class people work more with things, whereas middle-class and upper-class people spend more of their time working with people. Working women of these

higher social classes are more likely than men to deal with people as a job.

Both women in general and men and women of the middle and upper classes, then, sell the management of emotions as labor. This is what Hochschild means by the "commercialization of human feeling."

Critical feedback
1. Discuss Hochschild's research within the context of the most appropriate of the three major theoretical perspectives.
2. If you were going to test further Hochschild's conclusions, identify the research method(s) you would use. Defend your choice.
3. Apply the concept of role strain to men and women who engage in extensive emotional labor in a corporate environment.

tain tasks are given to the old, the young, and young adults. This scant division of labor exists because there are no institutions beyond the family.

What is a horticultural society?   According to archaeological evidence, **horticultural societies** solved the subsistence problem primarily through the domestication of plants. This type of society came into being about nine thousand years ago. The transition from hunting and gathering to horticultural societies was not an abrupt one; it occurred over several centuries. In fact, some hunters and gatherers in the Middle East were harvesting wild grains with stone sickles ten thousand years ago, and early horticultural societies also hunted and gathered (Lenski, Nolan, and Lenski, 1995).

The gradual shift from hunting and gathering to horticulture led to more permanent settlements. Subsistence no longer required people to move frequently in order to take advantage of available food; people could instead work a particular land area for extended periods of time. This relative stability in even simple horticultural societies—and particularly in the permanent settlements in the Middle East and southeastern Europe—permitted the growth of large societies with somewhat greater population densities.

The family is even more basic to social life in horticultural societies than in those based on hunting and gathering. In hunting and gathering societies, the survival of the band has top priority. In horticultural societies, primary emphasis is on the provision of a livelihood for household members. But producing a living in horticultural societies does not require the labor of all family members. With more labor than is needed to survive, households can be more self-sufficient and independent: "The family is to the tribal economy as the manor to medieval European economy, or the industrial corporation to modern capitalism: each of these is the central production-institution of its time" (Sahlins, 1968:75).

Hunters and gatherers seldom engage in interband conflict raids, but there is considerable intervillage conflict in horticultural societies. Much of the fighting, however, resembles modern contact sports more than warfare. Satisfaction is often gained from defeating the enemy in games rather than from killing them. Although conflict does sometimes lead to heated battle and numerous deaths, religious beliefs and rituals keep slaughter within bounds (Plog and Bates, 1990). Still, Napoleon Chagnon (1997) estimates that at least 24 percent of all male deaths among the Yanomamö are

Horticultural societies solved the subsistence problem primarily through the domestication of plants. These Indonesians are threshing grain by hand in the rice terrace at harvest season.

due to warfare, a figure consistent with those for other tribes that constantly feud with their neighbors.

What are the characteristics of an agricultural society? The transition from horticultural to **agricultural society** was made possible largely through the invention of the plow (Lenski, Nolan, and Lenski, 1995). The longer the horticulturalists worked the soil with their digging sticks, hoes, and spades, the greater the disadvantage. With these crude implements, horticulturalists could not keep pace with the weeds, and they could not dig deeply enough to reach the soil's nutrients, which recede farther below the root line each year of cultivation. The plow, which appeared about five or six thousand years ago, did not solve these problems completely, but it was effective enough to permit the permanent cultivation of land. The plow not only allowed the control of weeds but also turned the weeds into fertilizer by burying them under the soil. By digging deeply into the ground, the plow brought back to the surface nutrients that had sunk below root level.

Moreover, the plow permitted a shift from human to animal energy. Only humans could use a hoe or spade, but oxen could pull a plow. This new technology increased productivity because larger areas could be cultivated with fewer people. As a result, more people were released from the land to engage in noneconomic activities. This led in turn to the establishment of cities; the development of more complex economic specialization; and the emergence of separate political, economic, and religious institutions.

Although family ties remained important, the state replaced the kinship group as the basic force holding agricultural societies together. Advanced agricultural societies were monarchical, headed by a king or an emperor who desired as much control as possible. Distinct social classes appeared for the first time in advanced agricultural societies. Wealth and power were based on land ownership, which was controlled by the governing elites. The elites enjoyed the benefits of the economic surplus, but the peasants did most of the work. Urban merchants were considered below the governing class because they, too, worked for their living. The economy, involving considerable trade and a monetary system, began to emerge as an identifiable institution during the agricultural era. The agricultural economy developed two basic sections: a rural agricultural sector and an urban commercial and handicraft sector. Institutional specialization was also reflected in the increasing separation of political and religious elites. Although rulers were believed to be divinely chosen, few of them doubled as religious leaders.

Industrial Society

The Industrial Revolution was "a kind of discontinuous leap in human history; a leap as important as that which had lifted the first . . . [horticultural] settlements above the earlier hunting communities" (Heilbroner, 1993:102). The Industrial Revolution created **industrial society**—a society whose subsistence is based primarily on the application of science and technology to the production of goods and services.

The problem of subsistence is solved in industrial society primarily through the application of science and technology to the production of goods and provision of services. The technology available earlier in the development of industrial society was crude, as shown in this cotton mill in North Carolina in 1908. Still, the power looms used to weave cotton cloth was an important technological advance in the rise of industrial society.

What are the basic features of industrial society? Neil Smelser (1976) has identified some basic structural changes that occur in societies undergoing the shift from an agricultural to an industrial base. As implied in the definition of industrial society, industrialism brings with it a change from simple, traditional technology toward the application of scientific knowledge. With this change comes the discovery and adoption of new, complex technological devices. Examples from the past include the steam engine and electrical power. More recent technological developments include nuclear energy and aerospace-related inventions. These technological devices are used in factories and in agriculture. In industry, the trend is away from human and animal power toward the use of power-driven machines run by wage earners who produce goods for sale on the market. In agriculture, the shift is away from subsistence farming toward the selling of agricultural goods on the open market. Farmers, through the use of new non-human-powered technological devices and the hiring of farm labor, are able to produce a surplus sufficient to support not only themselves but also many others. This food surplus supports the movement of people from farms and villages toward gigantic urban concentrations. Industrial society tends to be urban society as well, with vast concentrations of people located in and around many urban centers.

The concept of structural differentiation also helps to describe industrial society. **Structural differentiation** occurs when one social structure divides into two or more social structures that operate more suc-

cessfully separately than the one alone would under the new historical circumstances (Smelser, 1976). Under the domestic form of production, for example, the family performs a variety of economic roles. In the change from domestic to factory production, the economic activities are moved from the family to the factory. Economic and familial activities are separated into two institutions rather than lodged in one. Familial and economic functions are increasingly segregated.

Other kinds of changes occur through structural differentiation. The education of the young moves from the home to the specialized institution of the school. Because an industrial society requires an educated labor force, education extends from the elite to the masses. Kinship declines in importance as the immediate family begins to separate socially and physically from the larger family of relatives. Personal choice and love replace arranged marriages. Women, through their entrance into the workforce, become less subordinate to their husbands. Religion no longer pervades all aspects of social life. Various institutions—political, economic, familial, educational—become more and more independent of religious influence. Secularization accompanies the rise of science and technology. Individual mobility increases dramatically with industrialization. Mobility is based more on occupational achievement than on assignment of social position according to family of birth. This reflects the differentiation of the occupational status of the adult from his or her social standing at birth.

Definite differences exist among the three preindustrial societies, and even greater differences can be seen between preindustrial societies and industrial society.

How do preindustrial and industrial societies differ?
Ferdinand Tönnies (1957, originally published in 1887), an early German sociologist, distinguished between **gemeinschaft** (community) and **gesellschaft** (society). The former type—closely approximating preindustrial society—is based on tradition, kinship, and intimate social relationships. The latter—representing industrial society—is characterized by weak family ties, competition, and less personal social relationships.

Shortly after Tönnies formulated this distinction, Emile Durkheim (1964a, originally published in 1893) made a similar one distinguishing between the nature of the social solidarity existing in the two types of societies. Durkheim contended that a society remains integrated because of the cultural elements its members share; a society is held together through commonly held beliefs, values, and norms. This he termed *social solidarity.* The nature of social solidarity in a society is based on the nature of the society's division of labor. In societies in which the division of labor is simple—in which most people are doing the same type of work—**mechanical solidarity** is the foundation for social unity. A society based on mechanical solidarity, according to the French sociologist, achieves social unity through a consensus of beliefs, values, and norms; strong social pressures for conformity; and dependence on tradition and family. In this type of society, best observed in small, nonliterate societies, people tend to behave, think, and feel in much the same ways, to place the group above the individual, and to emphasize tradition and family.

Modern industrial societies, in contrast, involve a very complex and differentiated division of labor. Because people perform very specialized jobs, they are dependent on others and must cooperate with them. Although a farmer in a simple society is largely self-sufficient, a factory worker in an industrial society depends on a wide variety of people to fulfill his or her needs—barbers, bakers, manufacturers, and suppliers of services of all kinds. A society based on **organic solidarity,** then, achieves social unity through a complex of highly specialized statuses that makes members of a society dependent on one another. This dependence and need for cooperation replaces the homogeneity of beliefs, values, and norms characteristic of simpler societies. The term *organic solidarity* is based on an analogy with biological organisms. If a biological organism composed of highly specialized parts is to survive, its parts must work together. Similarly, the parts of a society based on organic solidarity must cooperate if the society is to survive. In modern societies, individualism tends to replace strict group conformity; people do not

share all ways of thinking, feeling, and behaving in common; and the importance of kinship declines.

In this same vein, anthropologist Robert Redfield (1941) described a folk and an urban society. A **folk society** rests on tradition, cultural and social consensus, family, personal ties, little division of labor, and an emphasis on the sacred. In **urban society,** social relationships are impersonal and contractual; the importance of the family declines; cultural and social consensus is diminished; economic specialization becomes even more complex; and secular concerns outweigh sacred ones.

Clearly, these sociologists were attempting to accomplish the same thing—to isolate the central features differentiating preindustrial from industrial society. Contemporary sociologists generally agree that they were successful.

Postindustrial Society

Some social analysts believe that advanced industrial societies, particularly the United States, have passed beyond industrial society into a new form known as *postindustrial society.*

What are the major features of postindustrial society?
Daniel Bell (1976) is the earliest predictor of the rise of postindustrial society. He identifies five major features of postindustrial society:

1. *For the first time, the majority of the labor force is employed in services rather than in agriculture or manufacturing.* Today, the United States is the only country in the world in which more than half of all employment is in services—trade, finance, transportation, health, recreation, research, and government. In fact, about 75 percent of all employed workers in the United States are engaged in services.
2. *In industrial society, most workers are unskilled, semiskilled, or skilled blue-collar workers.* The shift to a service economy in postindustrial society—with an emphasis on office work—leads to a preponderance of white-collar workers, who outnumbered blue-collar workers in the United States for the first time in 1956. The gap between the numbers of blue-collar and white-collar workers is still increasing. The most rapid growth among white-collar workers has been in professional and technical employment.
3. *Theoretical knowledge is the key organizing feature in postindustrial society.* Knowledge is used for the creation of innovations (computers lead to more sophisticated defense systems) as well as for the formulation of government policy (computers

Theoretical knowledge is the key organizing feature in postindustrial society. Contrast the relatively high theoretical knowledge requirements of this plant worker at his computer with the minimal theoretical knowledge required of the young girl and woman in the early twentieth-century North Carolina cotton mill shown in the preceding photograph.

a technological innovation are not assessed before the introduction of the technology into an economic system. The pesticide DDT was introduced as a benefit to agriculture before it had been determined that it is harmful to living things. The internal combustion engine contributes to our affluence and economic growth, but it contaminates the environment. Technological assessment will permit us to consider the effects—good and bad—of an innovation before it is introduced.

5. *A new intellectual technology dominates human affairs in postindustrial society, much as production technology has dominated industrial society for the past 150 years.* Intellectual technology is the replacement of human judgment with mathematically based problem-solving rules. With modern computers, it is possible to take into account a large number of interacting variables at the same time. This capability allows us to manage the large-scale organizations that prevail in postindustrial society. Intellectual technology enables complex organizations—including government at national, state, and local levels to set rational goals and to identify the means for reaching them.

permit economic forecasting so that various theories can be tested to see their probable effects if actually applied to a real economic system). As theoretical knowledge becomes more important, so do educational, research, and intellectual institutions.

4. *Through new means of technological forecasting, postindustrial society can plan and control technological change.* In industrial society, technological change is uncontrolled. That is, the effects of

These are some of the most general features of postindustrial society. Considerable detail on the nature of capitalism and work in postindustrial society is presented in Chapter 13 ("Political and Economic Institutions").

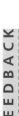

1. A/An _____ is the largest and most nearly self-sufficient group in existence.
2. The _____ society is the oldest known solution to the subsistence problem.
3. The _____ society solved the subsistence problem primarily through the domestication of plants.
4. The transition from horticultural to agricultural society was made possible largely through the invention of the

   _____.
5. _____ occurs when one social structure divides into two or more social structures that operate more successfully separately than the one alone would under new historical circumstances.
6. A society based on _____ solidarity achieves social unity through a complex of highly specialized roles that makes members of a society dependent on one another.
7. Which of the following is *not* one of the major features of postindustrial society?
   a. emphasis on theoretical knowledge
   b. employment of the majority of the labor force in service industries
   c. reliance on intellectual technology
   d. increased dependence on skilled blue-collar workers
   e. shift toward the employment of white-collar workers

Answers: 1. *society* 2. *hunting and gathering* 3. *horticultural* 4. *plow* 5. *Structural differentiation* 6. *organic* 7. *d.*

SUMMARY

1. Relationships among individuals and groups are patterned. The underlying pattern of social relationships is called social structure. Social scientists have developed the concepts of social status, role, role performance, and social interaction to describe social structure.

2. A social status is a position that a person occupies within a social structure. Individuals occupying interrelated statuses usually behave toward one another in orderly and predictable ways. Some statuses are ascribed, or assigned, to people; other statuses are achieved, or earned. Still other statuses are so important, that they affect most other aspects of a person's life.

3. Roles—the glue that binds together a network of social statuses—are the culturally defined rights and obligations attached to social statuses. Rights inform one person of the behavior that can be expected from another person; obligations inform individuals of the behavior others expect from them. The rights of one status correspond to the obligations of another status.

4. Role performance occurs when roles are put into action. Role performance takes place because people influence one another's behavior because they engage in social interaction. Role conflict exists when the performance of a role in one status clashes with the performance of a role in another status. Role strain occurs when the roles of a single position are inconsistent. Role conflict and strain,

which may lead to discomfort and confusion, can be reduced by setting priorities and segregating roles. Because of role conflict, role strain, and role distancing, an individual's conformity to all roles is impossible. This causes no problem as long as roles are followed within certain tolerance limits.

5. A society is composed of a people living within defined territorial borders who share a common culture. The way in which a society solves the problem of subsistence heavily influences the society's culture and social structure. Historically, societies have become larger and more complex as the means for solving the subsistence problem have improved. The major types of societies are hunting and gathering, horticultural, agricultural, and industrial.

6. Although both the horticultural and agricultural revolutions brought significant change, the greatest transformation came with the Industrial Revolution. The leap from preindustrial to industrial society was so great that many scholars have been able to isolate the basic characteristics distinguishing these two types of societies.

7. Postindustrial society is on the rise. This type of society has a predominately white-collar labor force concentrated in service industries, is based on technical knowledge, relies on technical forecasting, and depends on computer technology for organizing and making decisions.

LEARNING OBJECTIVES REVIEW

After careful study of this chapter, you will be able to:

- Explain what sociologists mean by social structure.
- Distinguish and illustrate the basic building blocks of social structure.
- Define the concept of society.
- Describe the basic solution to the problem of survival employed by the major types of preindustrial societies.

- Discuss the basic characteristics of industrial society.
- Compare and contrast preindustrial, industrial, and postindustrial societies.

CONCEPT REVIEW

Match the following concepts with the definitions listed below them:

____ a. society
____ b. urban society
____ c. formal organization
____ d. industrial society
____ e. organic solidarity

____ f. achieved status
____ g. rights
____ h. hunting and gathering society
____ i. social structure
____ j. role strain

____ k. master status
____ l. structural differentiation
____ m. gesellschaft
____ n. folk society
____ o. role performance

1. A slot within a social structure occupied because of an individual's efforts.
2. A group deliberately created for the achievement of one or more specific goals.
3. A society that solves the subsistence problem through hunting animals and gathering edible fruits and vegetables.
4. Roles informing individuals of the behavior that can be expected from others.
5. When one social structure divides into two or more social structures that operate more successfully separately than the one alone would under the new historical circumstances.
6. A society that rests on tradition, cultural and social consensus, family, personal ties, little division of labor, and an emphasis on the sacred.
7. The actual conduct involved in putting roles into action.
8. Occurs when the roles of a single status are inconsistent with one another.

9. Patterned recurring social relationships among individuals and groups.
10. People who live within defined territorial borders and who share a common culture.
11. A status that affects most other aspects of a person's life.
12. Tönnies's term for the type of society characterized by weak family ties, competition, and impersonal social relationships.
13. A society whose subsistence is based primarily on the application of science and technology to the production of goods and services.
14. Social unity based on a complex of highly specialized statuses that makes members of a society dependent on one another.
15. A society in which social relationships are impersonal and contractual, kinship is deemphasized, cultural and social consensus are not complete, the division of labor is complex, and secular concerns outweigh sacred ones.

CRITICAL THINKING QUESTIONS

1. Suppose that a college friend of yours wants to know the meaning of the term *social structure*. Use the stage analogy to develop an understandable answer. Use examples.

_____
_____
_____

2. Have you experienced role conflict and/or role strain lately? If so, describe the situation. If not, explain why you have been immune to role conflict and strain by making clear the meaning of the concepts.

_____
_____
_____

3. Discuss some of the basic distinguishing features of preindustrial and industrial societies. Illustrate your answer.

_____
_____
_____

4. In what ways will your work life be different in a postindustrial society than in an industrial society?

_____
_____

MULTIPLE CHOICE QUESTIONS

1. _____ is defined as patterned, recurring social relationships.
   a. Informal structure
   b. Social structure
   c. Constructive structure
   d. Anomic structure
   e. Symbolic interaction

2. The type of status that is neither earned nor chosen, but is assigned to us is called
   a. achieved status.
   b. ascribed status.
   c. acquired status.
   d. assigned status.
   e. symbolic status.

3. The president of a corporation is an example of a/an
   a. achieved status.
   b. ascribed status.
   c. acquired status
   d. assigned status.
   e. symbolic status.

4. _____ are culturally defined rights and obligations attached to statuses that indicate the behavior expected of individuals holding them.
   a. Norms
   b. Values
   c. Mores
   d. Roles
   e. Beliefs

5. Rights inform individuals of behavior they can expect from others; _____ inform individuals of the behavior others can expect from them.
   a. exchange principles
   b. roles
   c. statuses
   d. cues
   e. obligations

6. _____ exists when the performance of a role in one status clashes with the performance of a role in another one.
   a. Role conflict
   b. Role strain
   c. Role distancing
   d. Role conformity
   e. Role deformity

7. By ranking incompatible roles in terms of their importance to us, we can reduce role _____ and _____.
   a. conflict; strain
   b. conflict; behavior
   c. conflict; ambiguity
   d. strain; ambiguity
   e. ambiguity; distancing

8. A/An _____ is composed of a people living within defined territorial borders who share a common culture.
   a. society
   b. institution
   c. community
   d. social group
   e. formal organization

9. The _____ society solved the subsistence problem through the application of science and technology to the production of goods and services.
   a. hunting and gathering
   b. horticultural
   c. agricultural
   d. industrial
   e. postindustrial

10. Emile Durkheim used the term _____ to describe social unity that is achieved through a consensus of beliefs, values, and norms; strong social pressure for conformity; and a dependence on tradition and family.
    a. organic solidarity
    b. mechanical solidarity
    c. gesellschaft society
    d. gemeinschaft society
    e. traditional society

11. Which of the following is *not* one of the major features of postindustrial society?
    a. emphasis on theoretical knowledge
    b. reliance on intellectual technology
    c. increased reliance on skilled blue-collar workers
    d. shift toward the employment of white-collar workers
    e. shift toward greater employment in the service industries

FEEDBACK REVIEW

FILL IN THE BLANK

1. _____ refers to the patterned relationships among individuals and groups.
2. _____ are culturally defined rights and obligations attached to statuses.
3. The rights of one social status correspond to the _____ of another one.
4. _____ occurs when the roles of a single status are inconsistent.
5. A _____ is the largest and most nearly self-sufficient group in existence.
6. The transition from horticultural to agricultural society was made possible largely through the invention of the _____.
7. A society based on _____ solidarity achieves social unity through a complex of highly specialized roles that makes members of a society dependent on one another.

MULTIPLE CHOICE

8. Which of the following is *not* one of the differences between the stage and social life?
   a. There is considerably more discrepancy between roles and role performance in social life than between a stage script and its performance.
   b. Unlike the stage, there are no cues and responses in real life.
   c. Role performance in real life is not the conscious process that actors go through on the stage.
   d. In social life, the cues and responses are not as programmed and predictable as on the stage.
9. Which of the following is *not* one of the major features of postindustrial society?
   a. emphasis on theoretical knowledge
   b. employment of the majority of the labor force in service industries
   c. reliance on intellectual technology
   d. increased dependence on skilled blue-collar workers
   e. shift toward the employment of white-collar workers

MATCHING

10. Match the following concepts with the examples listed beside them:
    ____ a. ascribed status   1. Basketball coach, mother, author, daughter, professor.
    ____ b. achieved status   2. A mother is expected to take care of her children.
    ____ c. master status   3. Person who gets AIDS through a blood transfusion.
    ____ d. role   4. A husband and wife discuss the disciplining of one of their children.
    ____ e. status set   5. Occupation, gender, race, ethnicity.
    ____ f. role performance   6. A university president hands out diplomas at a graduation ceremony.
    ____ g. social interaction   7. The President of the United States.

ANSWER KEY

Concept Review	Multiple Choice	Feedback Review	
a. 10	1. b	1. Social structure	10. a. 3
b. 15	2. b	2. Roles	b. 7
c. 2	3. a	3. obligations	c. 5
d. 13	4. d	4. Role strain	d. 2
e. 14	5. e	5. society	e. 1
f. 1	6. a	6. plow	f. 6
g. 4	7. a	7. organic	g. 4
h. 3	8. a	8. b.	
i. 9	9. d	9. d.	
j. 8	10. b		
k. 11	11. c		
l. 5			
m. 12			
n. 6			
o. 7			

Chapter Six

Groups and Formal Organizations

LEARNING OBJECTIVES

After careful study of this chapter, you will be able to:
- Define the concept of the group and differentiate it from social categories and social aggregates.
- Outline the basic characteristics of primary and secondary groups.
- Describe the five major types of social interaction.
- State the nature of social processes in groups relative to task accomplishment, leadership styles, and decision making.
- Define the concept of formal organization and identify the major characteristics of bureaucracy.
- Discuss the advantages and disadvantages of bureaucracy.
- Distinguish between formal and informal organization.
- Describe the iron law of oligarchy and demonstrate its importance with some examples.
- Compare and contrast the major features of Western organizations and Japanese organizations.

SOCIOLOGICAL IMAGINATION

Are there benefits to conflict?   Most people assume that the disruptive effects of conflict warrant avoiding it whenever possible. Despite the negative consequences of conflict, it does have some social benefits. Conflict may divide some groups, for example, but it can also strengthen the bonds among members of other groups. The conflict arising from the movement for racial equality led by Martin Luther King in the 1960s united blacks and many whites in opposition to institutionalized racism. This coalescence later led to the extension of some very important legal and social rights to African Americans.

Because groups tend to draw lines around themselves, there are insiders (whites) and outsiders (blacks). Some types of groups have tighter, more definite boundaries than others. Boundaries between blacks and whites in the South of the 1960s were so rigid that a movement had to be mounted for blacks to be able to do such simple things as drink from the same water fountains, use the same restrooms, and eat at the same restaurants as whites. Group boundaries may also change, as they have between blacks and whites in the United States since the 1960s.

Many different types of groups exist within a society. They range from the small and informal to the large and highly formalized. All groups are important elements of society because many of the shared understandings and social relationships that provide the basis for social structure and culture emerge and are sustained in group settings. As this chapter shows, both small, informal groups and large, formal organizations play a part in the development of understandings and relationships. Both are also common settings for conflict, as the civil rights movement in the South demonstrates.

CONCEPT OF THE GROUP

Americans are socialized to think in individualistic terms. In fact, one of their primary values, as noted in Chapter 4 ("Socialization Over the Life Course"), is individuality. This individualistic orientation is often a stumbling block to those first introduced to the concept of the group. Perhaps it will ease the transition into thinking of groups rather than individuals by focusing on what is thought of as an individualistic activity but which, in fact, is a group endeavor. Art is a likely candidate for this purpose.

Think of painting and you are likely to have images of Rembrandt, Renoir, or Picasso. The art of ballet will call to mind Nureyev, Baryshnikov, or Kirkland. It is Gershwin, Sondheim, or Webber who comes to the fore when the subject of musical achievement on Broadway is the topic. In short, when we think of art, we think of individual artists. Without denigrating the talent, hard work, and achievements of individual artists whose works are precious to us, Howard Becker wants to underscore the fact that even such an allegedly individualistic pursuit as art "involves the joint activity of a number, often a large number of people" (Becker, 1982:1).

Becker begins his book with this story told by Anthony Trollope, the famous nineteenth-century English novelist:

It was my practice to be at my table every morning at 5:30 a.m.; and it was also my practice to allow myself no mercy. An old groom, whose business it was to call me, and to whom I paid £5 a year extra for the duty, allowed himself no mercy. During all those years at Waltham Cross he was never once late with the coffee which it was his duty to bring me. I do not know that I ought not to feel that I owe more to him than to any one else for the success I have had. By beginning at that hour I could complete my literary work before I dressed for breakfast.

Becker's point is that Trollope's groom is one member of the complex network of social relationships that permitted this artist to accomplish his work. Moreover, the activities of both Trollope and his groom had to mesh with those of many other people (printers, publishers, and editors, to mention only a few) if the literary works of this Victorian writer were ever to become available to the public. Some of these relationships are personal and face-to-face; others are not. This chapter elaborates on the variety of group relationships and the contexts in which they appear throughout social life. A **group** is composed of several people who are in contact with one another; share some ways of thinking, feeling, and behaving; take one another's behavior into account; and have one or more interests or goals in common. Because of their characteristics, groups play an important role in the lives of the people who are part of them and often influence the societies in which they exist.

Human accomplishments, even those attributed to individuals such as Shakespeare or pianist Vladimir Horowitz, rest on the support of groups. The school newspaper on which these students are working is produced by a complex web of group relationships.

TYPES OF GROUPS

Categories, Aggregates, and Groups

A group should be distinguished from a **social category**—a number of people who share a social characteristic. Taxpayers, women, and college graduates belong to social categories. A group is also sometimes confused with a **social aggregate**—a number of people who happen to be in the same place at the same time, such as passengers waiting for the subway. Although neither categories nor aggregates are groups, some of their members under certain circumstances may form groups. Witnesses of a disaster (an aggregate), such as residents of a town devastated by a flood or tornado, may respond by working together to handle the emergency. Members of a state (a social category) may band together in an organized tax revolt. If such people begin to interact regularly; share ways of thinking, feeling, and behaving; take one another's behavior into account; and share one or more common interests or goals, they have formed a group.

Charles Horton Cooley, one of the founders of symbolic interactionism, was interested in socialization and formulated the concept of the looking-glass self. In addition to having an interest in the processes by which humans develop a self-concept, Cooley was interested in the context in which this development occurs. He used the term *primary group* to describe the groups that form the most important context for socialization. He felt that such groups as the family, neighborhood, and childhood play groups are primary because they are the first groups an infant experiences and they are extremely important in the development of self and personality. Clearly, people participate in primary groups not just in childhood but also throughout life. Cliques of close friends in high school and college, neighbors who keep an eye on one another's children, and older men who gather in poolrooms or country stores are examples of primary groups.

Symbolic interactionism underlies the sociological analysis of groups, in part because of Cooley's influ-

ence. George Herbert Mead, the other founder of symbolic interactionism, also contributed to the sociological study of groups. Mead's imitation stage and play stage occur in primary groups, as do the development of the "I" and the "me" and the ability to take the role of the other. And, of course, a child's parents become the child's first significant others.

It is useful to keep symbolic interactionism in mind when reading about the sociological approach to the study of groups.

Primary Groups

Primary and secondary groups are two principal types of groups. The characteristics of these two types of groups—and the relationships that occur within them—may be thought of as opposites in extreme cases. However, it is more useful to think of groups as falling at different points along a continuum from primary to secondary.

What is a primary group?   A **primary group** is composed of people who are emotionally close, know one another well, seek one another's company because they enjoy being together, and have a "we" feeling. Primary groups are characterized by relationships that are intimate, personal, based on genuine concern for another's personality, and fulfilling in themselves. Such relationships are termed **primary relationships.**

What conditions promote the development of primary groups?   A number of conditions favor the development of primary groups and primary relationships. Although primary relationships may occur in their absence, the likelihood of having a primary relationship is increased if these conditions exist.

- **Small group size.** It is difficult for the members of large groups to spend enough time together to develop close emotional ties; the chances of knowing everyone fairly well are far greater in small groups. Thus, members of a bridge foursome are more likely to develop primary relationships than members of a metropolitan chamber of commerce, even when both groups meet regularly.
- **Face-to-face contact.** People have maintained primary relationships despite separation by war, prison, or residential changes and have even established long-distance primary relationships through letter writing or telephone conversations, but primary relationships occur more easily when interaction is face-to-face. People who can see each other and who can experience such nonverbal communication as facial expressions, tone of voice, and touch are much more likely to develop close ties.
- **Continuous contact.** The probability of developing a primary relationship also increases with prolonged contact. Intimacy and concern for the total personality rarely develop in a short period of time. In spite of the occasional reports of love at first sight, most of us require repeated social contact for development of a close, personal relationship.
- **Proper social environment.** The development of primary relationships is also affected by the social environment in which interaction occurs. If individuals are expected to relate to one another strictly on the basis of status or role, primary relationships are unlikely to develop. Total personalities are usually not considered, and personal concern for one another does not seem appropriate. Lawyers who see judges in court face-to-face, in the presence of a small number of people, and over a long period of time are nevertheless unlikely to develop primary relationships with them. Forming primary relationships when unequal statuses are involved is always difficult. This is why primary relationships do not usually develop between students and professors, bosses and employees, or judges and lawyers.

What are the functions of primary groups? First, primary groups provide emotional support through caring, personal, and intimate relationships. During World War II, the German army refused to disintegrate despite years of being outnumbered, undersupplied, and outfought. Although these conditions should have led to desertion and surrender, they did not. According to Edward Shils and Morris Janowitz (1948), strong primary relationships within German combat units accounted for the Germans' stability and resilience against overwhelming odds. Although social cohesion

People in primary groups are emotionally close, know each other well, enjoy each other's company, and share a feeling of "we-ness." The family is the most important primary group to which we belong.

developed among American soldiers during the Korean conflict, it was not the same as during World War II. Because of special combat conditions and a personnel rotation system, "the basic unit of cohesion was a two-man relationship rather than one that followed squad or platoon boundaries" (Janowitz, 1974:109).

Second, primary groups contribute to the socialization process. For infants, the family is the primary group that provides the emotional support necessary for the development of an integrated personality and a sense of self. The family also conveys information about culture so that children learn how to participate in social life. In addition, primary groups promote adult socialization as individuals adjust to new and changing social environments—as they change schools, take new jobs, change social classes, marry, and retire. Primary groups make these adjustments less painful because membership in them helps people fit into new social situations. Frequent interaction with people who demonstrate genuine concern for others aids adjustment to a new social setting.

Finally, primary groups promote conformity and contribute to social control. The stability and perpetuation of a society depend on the members' acceptance of the society's norms and values. Unless a majority of the members of a society supports the norms and values, that society cannot continue to exist in its present form. Primary groups teach new members to accept norms and values and provide group pressure to promote conformity to them. William F. Whyte's (1993) study of an Italian slum gang illustrates social control within primary groups. Whyte found that rank within this gang determined how well the members performed in sports. One member of the gang, a former semiprofessional

SOCIOLOGY IN THE NEWS

Primary Relationships in Blended Families

A *blended family* is formed when a remarriage occurs involving at least one child from a previous marriage. Because a family is characterized by close, personal, and intimate relationships, it is a *primary* group. Members of a primary group usually like to be together and have a feeling of unity. As this CNN report demonstrates, conflict within blended families often prevents the development of such feelings.

1. Identify a problem the blended family poses for the socialization of children.

   _____

   _____

2. Does a blended family that is not blending well actually constitute a secondary group? Why or why not?

   _____

   _____

An important function of primary groups is the provision of emotional support for its members. Marital conflict, such as in this blended family, can threaten the psychological well-being of its members.

---

baseball player, was one of the gang's best players. Yet, after losing status in the gang, he could no longer play very well. According to Whyte, he indicated that he would play baseball better when gang members were not involved. Whyte also reported that bowling scores corresponded with status in the gang—the higher the rank, the higher the score. If a lower-ranked member began to bowl better than those above him, verbal remarks—"You're bowling over your head" or "How lucky can you get?"—were used to remind him that he was stepping out of line.

Secondary Groups

What is a secondary group?   A **secondary group** is impersonal and task oriented and involves only a segment of the lives and personalities of its members. Work groups in organizations, volunteers who work together during disasters, and environmentalists who participate in political demonstrations are examples of secondary groups.

The relationships in secondary groups are **secondary relationships**—impersonal, involving only limited parts of a person's personality, and existing to accomplish a specific purpose beyond the relationship itself. Interactions between clerks and customers, employers and workers, and dentists and patients are examples of secondary relationships.

As these examples indicate, secondary relationships are not necessarily unpleasant. In fact, members of secondary groups may enjoy pleasant interaction and friendship and identify with one another, but the pur-

pose of the group is to accomplish a task, not to enrich friendships or provide close, emotional relationships. In fact, if the enrichment of friendships becomes more important than the task to be achieved, a secondary group may lose its effectiveness. If the members of a high school basketball team become more interested in the emotional relationships among themselves or with their coach than in playing their best basketball, their play on the court will probably suffer.

Do secondary groups ever have primary relationships? Although primary relationships are more likely to occur in primary groups and secondary relationships in secondary groups, there are many exceptions. Many secondary groups are settings for some primary relationships. Members of work groups relate to one another in a manner that is personal, demonstrate genuine concern for one another as total personalities, and have relationships that are fulfilling in themselves. Likewise, members of a primary group occasionally engage in secondary interaction. One family member may, for example, lend money to another member of the family at a given interest rate with a specific repayment schedule.

Reference Groups

What is a reference group?   We use groups to evaluate ourselves and to acquire attitudes, values, beliefs, and norms. Groups used in this way are called **reference groups.** Your family, college classmates, student government leaders, a Greek campus organization, professors, the Dallas Cowboys, or a rock group may each be a reference group.

Our self-esteem and behavior are significantly affected by our reference groups. Failure to make good grades can make you feel bad about yourself and motivate you to spend more time studying because of your imagined reaction on the part of your parents and teachers.

We need not be a member of a reference group; we may only aspire to be a member. For example, we do not need to be in high school to have high school students as a reference group. We need only to evaluate ourselves in terms of their standards and to subscribe to many of their beliefs, values, and norms. Junior high school girls, for example, may imitate high school girls by using makeup, wearing jeans, and valuing good looks. Junior high school boys may copy high school boys by drinking alcoholic beverages, bragging about their sexual exploits, and valuing athletic prowess.

Reference groups need not be positive in nature. Observing the behavior of some group you dislike may reinforce a previously formed preference for certain ways of acting, feeling, and deciding. Or, rejection of a "negative" reference group may help you as you attempt to develop your own preferences. For example, a strongly felt rejection of a neighborhood violent gang may provide you a model for what you do not wish to be.

In-Groups and Out-Groups

What are in-groups and out-groups? In-groups and out-groups can only exist in relation with each other; one is the flip side of the other. An **in-group** has one's intense identification and loyalty. The level of identification and loyalty is sufficiently intense to entail the conscious exclusion of nonmembers. Thus, an **out-group** is a group toward which one feels opposition, antagonism, and competition. It is with regard to these types of groups that "we" and "they" and "in-siders" and "out-siders" derive their social meaning (Rose, 1980). In-groups and out-groups form around campus organizations, athletic teams, cheerleading squads, marching bands, racially and ethnically divided neighborhoods, and countries at war internally or externally. An in-group–out-group mentality may even develop between generations. Baby boomers—Americans born between 1946 and 1964—are being accused of bashing the so-called Generation X—America's twenty somethings—as a bunch of lazy whiners who want the good life handed to them in order to feel better about themselves (Giles, 1994).

The very existence of in-groups and out-groups depends on the establishment and protection of group boundaries. If no lines distinguish "us" from "them," then there can be no "ins" and "outs." Because bound-

aries remind members of the in-group that movement over the line places them too on the outside, loyalty and conformity to group values, norms, attitudes, and beliefs are promoted. At the same time, boundaries act as an entrance barrier to outsiders.

How are group boundaries maintained? Maintenance of something as important as group boundaries requires effort. Unfortunately, this effort may involve clashes with outsiders. Skinheads, German neo-Nazis, periodically smash windows or faces of Jews. Members of an urban gang may injure or kill a member of another gang who has entered its territory. Boundaries are also maintained through group symbols. Symbols may be as benign as a right-to-life bumper sticker or as intentionally threatening as having all one's hair shaved off. Groups also may adopt certain words and rituals to demarcate themselves. This is easily recognized at a meeting of the Ku Klux Klan or at a college Greek organization initiation ceremony.

As this description of some ways to maintain group boundaries suggests, in-group and out-group relationships can have serious consequences. Let's consider some of these repercussions.

What is the sociological importance of in-groups and out-groups? Membership in an in-group has both positive and negative aspects. Those on the inside benefit from the heightened self-esteem and sense of belonging that come with a shared belief in superiority over those outside the circle. However, in-group members may suffer from the false picture they have formed of themselves and others. Falsely inflating one's worth by deflating that of others—a distortion of reality in-groups deliberately create—is self-deluding. If such self-delusion becomes central to the identity of in-group members, genuine psychological and social development is retarded. Moreover, they may be personally damaged on realizing the falseness of their basis for self-esteem.

Harm is not confined to in-group members. In fact, out-group members experience many more negative effects. The increase in self-esteem enjoyed by in-group members usually comes at the expense of outsiders. Even more serious is the physical damage that may be done by extremist groups such as the German neo-Nazis noted above or religious factions around the world willing to kill others in the name of their deity (O'Brien and Palmer, 1993). Powerful racial or ethnic majorities have been effective in keeping minorities under their control, denying them everything from a sense of worth to equal opportunity (Schaefer, 1993). Males have used a similar pattern of dominance against females (Andersen, 1997).

In-group–out-group relationships, of course, generally do not take such extreme forms. Spirited competition among sororities for new pledges and friendly

Each of the participants in this mass office party is part of a social network. Not all these people know each other or interact with each other; some do. For each person, their relationships in this organization are merely one part of their broader social network, which may include family members, friends, golf partners, and psychotherapy groups.

athletic rivalries between towns are more typical expressions of these types of groups. The relative harmlessness of most cases, however, should not mask the serious consequences of in-groups and out-groups, consequences related to conflict, power, and self-respect.

Social Networks

What are social networks?   As individuals and as members of primary and secondary groups, we interact with a large number of people and groups. The typical American has between 500 and 2,500 people he or she knows, many of whom know each other (Milgram, 1967). All of an individual's social relationships constitute that person's **social network**—a web of social relationships that joins a person directly to other people and groups and, through those individuals and groups, indirectly to additional parties. This total set of social relationships—this social network—includes family members, work colleagues, classmates, physicians, church members, close friends, car mechanics, and store clerks. Social networks tie us to hundreds or thousands of people within our communities, throughout the country, and even around the world.

Are social networks groups?   Although a person's social network involves groups, it is not a group per se. A social network, being looser and more diffuse than a group, does not involve close or continuous interaction among all members. Members of a social network do not all experience a feeling of membership in the network, because too many of the relationships are sporadic and indirect for a sense of togetherness to

develop. This characteristic of a social network does not preclude close ties, such as college fraternity brothers who maintain contact via telephone, computers, and reunions. Still, most ties in a social network are secondary in nature.

How strong are the ties in a social network?   Social relationships within a network, which vary in intensity, have been characterized as having both strong and weak ties (Granovetter, 1973; Freeman, 1992). A clear parallel exists between strong and weak ties and primary and secondary relationships. Strong ties, such as those with one's parents or spouse, are emotionally close and intimate and are based on a genuine concern for the other person(s). Weak ties, including those with salespersons, distant relatives, and most coworkers, are more impersonal and goal oriented.

The number and type of strong ties are heavily influenced by level of education. More highly educated people have more strong ties, a greater variety of strong ties, and more strong ties outside the family. Not unexpectedly, older people have fewer strong ties, and urban dwellers have more strong ties than do rural residents. Although women have as many strong ties as do men, more of these ties are within the family (Fischer, 1982; Marsden, 1987).

What are the functions of social networks?   Social networks serve several important functions. First, networks provide a sense of belonging. Second, they are a source of social support in the form of help and advice. Finally, networking—getting to know people who may be useful to one's career by frequenting the right places, joining the right organizations, and living in the right neighborhood—is a prime tool in today's labor market.

The use of networks to locate new places of employment deserves special attention because of gender differences. Although the social networks of females and males are equally extensive, men are more enmeshed in "good-old-boy networks" containing many higher status contacts (Lin, Ensel, and Vaughn, 1981). Women tend to have more ties with family members and fewer higher status ties. This situation is changing, however, as more women are seeing networks as a pathway to higher status and more powerful positions (Speizer, 1983).

SOCIOLOGICAL ANALYSIS OF GROUPS

Types of Social Interaction

Social interaction is a crucial process in group life. Robert Nisbet (1970) has described five types of social interaction basic to group life: cooperation, conflict, social exchange, coercion, and conformity.

What is cooperation? **Cooperation** is a form of interaction in which individuals or groups combine their efforts to reach some common goal. Cooperation usually occurs when the chances of reaching a goal are slight unless resources and efforts are pooled. The survivors of a plane crash in a snowy mountain range must cooperate if they are to survive. Victims of floods, mudslides, tornadoes, droughts, or famines must help one another to survive their crisis. This is not to say that cooperation exists only during emergencies; without some degree of cooperation, social life could not exist. Children who agree to a set of rules for a game, couples who agree to share household duties, and people who organize car pools are all cooperating.

What is conflict? Benefits and rewards in any society are limited. Individuals or groups who work together to obtain certain valued things are cooperating. Those who work against others to obtain a larger share of the valuables are in **conflict.** In conflict, defeating the opponent is considered essential. Consequently, defeating the opponent may become more important than achieving the goal; conquering the opponent may bring more joy than winning the prize.

FEEDBACK

1. A _____ is composed of several people who are in contact with one another; share some ways of thinking, feeling, and behaving; take one another's behavior into account; and have one or more interests in common.
2. A _____ group is composed of people who are emotionally close, who know one another well, who seek one another's company because they enjoy being together, and who have a "we" feeling.
3. Primary groups have little impact on the development of self and personality. *T or F?*
4. Listed below are some examples of primary and secondary relationships. Indicate which examples are most likely to be primary relationships (P) and which are most likely to be secondary relationships (S).
   ____ a. a marine recruit and his drill instructor at boot camp
   ____ b. a married couple
   ____ c. a manager and his professional baseball team
   ____ d. professors and students
   ____ e. car salespersons and their prospective customers
5. Which of the following is *not* a condition that promotes the development of primary groups?
   ____ a. small group size   ____ c. continuous contact
   ____ b. face-to-face contact   ____ d. interaction on the basis of status or role
6. State the three functions of primary groups. _____
   _____
   _____
7. Primary relationships can occur in secondary groups. *T or F?*
8. Match the following terms and statements:
   ____ a. group   (1) skinheads
   ____ b. in-group   (2) spectators of a Fourth of July fireworks display
   ____ c. social aggregate   (3) orchid growers
   ____ d. social category   (4) teenage subculture
   ____ e. reference group   (5) members of a regular Saturday night poker game
9. A _____ is a web of social relationships that joins a person directly to other people and groups and, through those individuals and groups, indirectly to additional parties.

Answers: 1. group 2. primary 3. F 4. a. (S) b. (P) c. (S) d. (S) e. (S) 5. d. 6. provide emotional support, contribute to socialization, promote conformity 7. T 8. a. (5) b. (1) c. (2) d. (3) e. (4) 9. social network

Cooperation may occur wherever individuals or groups decide to work together toward a common objective. Cooperation is especially important when only pooled efforts will accomplish a task. The cleanup effort following the bombing of the Murrah Federal Building in Oklahoma City in 1995 illustrates cooperation.

As was pointed out in the *Sociological Imagination* opening this chapter, conflict is usually considered disruptive to social life and therefore a form of interaction to be minimized or eliminated. A cooperative, peaceful society is assumed to be better than one containing conflict. Although conflict unquestionably does have some undesirable consequences, this type of interaction can also be socially beneficial.

What are the benefits of conflict? According to Georg Simmel (1858–1918), one of the major contributions of conflict is the promotion of cooperation and unity within opposing groups. Conflict may lead to separation of groups, but it strengthens the relationships among the groups' members. Thus, "the United States needed its revolutionary war; Switzerland, the fight against Austria; the Netherlands, rebellion against Spain" (Simmel, 1964:100). Similarly, a labor union becomes more integrated during collective bargaining for a new contract; police officers attempting to stop a husband from physically abusing his wife may find themselves under attack by both mates. According to historian Stephen Oates (1977), the "outside meddling" of Stephen Douglas and other Democrats united the Illinois Republican party and secured the U.S. Senate nomination for Abraham Lincoln.

Another positive consequence of conflict is the revival of norms and values that may have slipped into the background. Equality of opportunity, freedom, and democracy have always been important values to Americans. But until the nonviolent civil rights demonstrations, the urban violence in large northern cities, and the abuse experienced by civil rights supporters—all of which occurred during the past few decades—the poverty, inequality, and lack of opportunity existing among America's minorities were largely invisible to white society. Once these problems were forcibly called to the public's attention, some actions were taken to reaffirm these basic values. Among these actions were Lyndon Johnson's War on Poverty, the Civil Rights Act of 1964, the Equal Employment Opportunity Act of 1972, and affirmative action programs (Fineman 1992).

Conflict may also be beneficial when it alters norms, beliefs, and values. Domestic opposition to the Vietnam War in the 1960s—which involved unparalleled conflict and violence on college and university campuses—ended the military draft, brought into question the legitimacy of our political and military intervention in other countries, enlarged the American value of patriotism to include opposition to certain wars, and encouraged colleges and universities to be more responsive to students' needs. Events such as the upheaval over the Vietnam War are both cataclysmic and rare. Such events are important because a cataclysmic upheaval is often required to alter norms, beliefs, and values, a lesson not lost on African Americans during the 1992 riots in Los Angeles.

A study by Douglas Maynard (1985) concludes that conflict has benefits even for young children. In a study of first-grade children, Maynard observed a sense of social structure emerging from disputes and arguments. Authority and friendship patterns were also the product of social conflict.

What is social exchange? In *The Nicomachean Ethics,* Aristotle observed:

All men, or most men, wish what is noble but choose what is profitable; and while it is noble to render a service not with an eye to receiving one in return, it is profitable to receive one. One ought, therefore, if one can, to return the equivalent of services received, and to do so willingly.

In this passage, Aristotle touches on **social exchange,** in which one person voluntarily does something for another expecting a reward in return. If one person helps a friend paint her house, expecting that some equivalent help will be returned, the relationship is one of exchange. In an exchange relationship, it is the benefits derived rather than the relationship itself that are central. An individual who does something for someone obligates that person. According to Peter Blau (1964), this obligation can be repaid only by doing something in return. Thus, the basis of an exchange relationship is reciprocity, the idea that you should do for others as they have done for you.

Exchange relationships need not be exploitative; many exchange relationships are based on trust, gratitude, and affection. You may help a classmate move out of a dormitory into an apartment because she has been friendly to you. Of course, you expect some response, whether it be to help when you move later or a special show of appreciation for your efforts. We are not always conscious of these expectations and often think that we expect nothing in return. Nevertheless, we usually do expect a certain response. To say, for example, "I really got ripped off in that relationship" implies an unequal exchange.

Are not exchange and cooperation the same thing? True, both cooperation and social exchange involve people working together. But there is a significant difference between these two types of interaction. In cooperation, individuals or groups work together to achieve a goal that is valued by all involved. Reaching this common goal, however, may or may not benefit those who are cooperating. And although individuals or groups may profit from cooperating, that is not their main objective. For example, a group can work toward building and maintaining an adequate supply of blood for a local blood bank without thought of benefit for itself. If, however, the primary objective of the group is to guarantee that blood is available for its own members, then it has an exchange relationship with the blood bank. In cooperation, the question is, How can we reach our goal? In exchange relationships, the background question is, What is in it for me?

What is coercion? **Coercion** is the type of social interaction in which an individual is compelled to behave in certain ways by another individual or a group. The central element of coercion is the domination of some people by other people. In a sense, coer-

cion is an unequal exchange: Because of superior power, someone can get something from someone else without repaying. Domination may occur through physical force, such as imprisonment, torture, or even death. But coercion is generally expressed in more subtle ways. Social pressures of all kinds—ridicule, rejection, ostracism, denial of love or recognition—are the most widely used and most effective ways of coercing others in the long run. Coercion is, in fact, a part of many of our social relationships. Parents coerce children by imposing an early curfew; employers coerce employees by threatening dismissal; and ministers coerce parishioners into making contributions to a church building fund by appealing to their obligation to God's work. Obviously, conflict theory underlies this type of social interaction.

What is conformity? **Conformity** is the type of social interaction in which an individual behaves toward others in ways expected by the group. When we conform, we adapt our behavior to fit the behavior of those around us. Some individuals are more conforming than others, but most people are conforming to the expectations of some group most of the time. Social life—with all its uniformity, predictability, and orderliness—simply could not exist without this type of social interaction. Without conformity, there could be no churches, families, universities, or governments; without conformity there could be no culture or social structure.

Other Social Processes in Groups

Nineteenth-century social philosopher Georg Simmel was one of the first to recognize the importance of groups for society. He argued that group members face a dilemma. Belonging to a group requires that members submit to social control, losing some personal freedom in the process. If the members remain outside the group, however, they lose such possible benefits of membership as economic security and social acceptance (Simmel, 1964). For this reason, most groups experience both cooperation and conflict. Because groups experience pressures that might interfere with their smooth functioning, social scientists have devoted considerable attention to task accomplishment within groups, various leadership styles for groups, and group decision making.

What is the nature of task accomplishment within groups? Most groups are intended to accomplish certain tasks, which can range from planning leisure activities to manufacturing a product. At the same time, members of groups have self-images and desires that are not necessarily identical to the goals of the group as a whole. Beginning with Robert Bales (1950), social scientists have tried to understand the ways in which task accomplishment within groups could expedite both the achievement of the group's goals and the needs of individual group members.

Bales developed a scheme, labeled "interaction process analysis," to classify group interactions into several categories. He identified twelve separate categories, which can be understood in terms of two basic problems that, according to Bales, must be solved by any social system. First, *instrumental problems* must be solved. Instrumental problems are those that are directly related to achieving the goals of the group. A group trying to elect a political candidate must solve problems regarding fund-raising, advertising, arrangements for speeches and rallies, charges from the opposing candidate, and so on. Second, *social-emotional problems* must receive attention. These are the problems involving the maintenance of individual satisfaction, the handling of disagreements, and the handling of related matters that inevitably arise whenever individuals try to coordinate their activities.

A number of laboratory studies conducted by Bales and his associates have further indicated that both instrumental and social-emotional processes are important.

> . . . if the group is going to do its job, (1) task-oriented interaction must be high; (2) more answers must be given than questions asked; (3) there must be more interaction contributing to the solidarity of the group (positive social-emotional) than detracting from solidarity (negative social-emotional). In other words, there must be, as Bales's theory suggests, some degree of balance to keep the individuals together as a group (Wilson, 1978:66–67).

What is the nature of leadership in groups? It was once believed that leaders are born and that people had to have certain traits to be effective as leaders. It is now known that no leadership trait is absolutely necessary for effective leadership, but Ralph Stogdill has suggested that successful leaders are likely to have the following combination of traits:

> The leader is characterized by a strong drive for responsibility and task completion, vigor and persistence in pursuit of goals, venturesomeness and originality in problem solving, drive to exercise initiative in social situations, self-confidence and sense of personal identity, willingness to accept consequences of decision and action, readiness to absorb interpersonal stress, willingness to tolerate frustration and delay, ability to influence other persons' behavior, and capacity to structure social interaction systems to the purpose at hand (Stogdill, 1974:81).

Such traits are likely to increase the effectiveness of leaders only if they are translated into behavior, and appropriate leader behavior has been the subject of con-

Conformity is the type of social interaction in which people behave as others expect them to behave. Without conformity, society could not exist. Imagine the chaos that would ensue if these people shopping in a grocery store did not remain in line at the checkout counter.

siderable research. Studies by Rensis Likert (1961, 1967) led Likert to emphasize the importance of *supportive behavior* in which leaders try to provide direction in a way that maintains others' sense of personal worth and importance. Similarly, Ralph White and Ronald Lippitt (1960) found that *democratic* leaders (who try to build consensus about decisions) are more effective than *authoritarian* ones (who give orders without considering the preferences of group members). Both, however, were found to be more effective than *laissez faire* leaders (who make little or no attempt to organize group activities). Subsequent studies of leadership have led to the development of approaches that suggest that appropriate leadership behavior depends on the situation presented by the group to be led. Fred Fiedler (1967), for example, has argued that leaders should be more authoritarian if they face a situation in which they have very high or very low amounts of control. In intermediate situations, leaders should be more democratic and concerned with the quality of human relationships.

If Fiedler's theory is correct, the popular leader of an infantry platoon (high control) or the disliked supervisor of an ineffective work crew (low control) should be more authoritarian than the well-liked director of a research team facing an unstructured task (intermediate control). Fiedler's theory has been criticized on several grounds, but it is an indication of the increasing tendency for social scientists to understand leadership in its situational context (Ashour, 1973; Schriesheim and Kerr, 1977).

Can groups make better decisions than individuals? Groups are used in many settings for the solution of a variety of problems. In some situations, group decisions can be shown to be superior to the decisions that would be made by individuals acting alone. For example, researchers have found groups to be a useful setting for solving intellectual tasks because they allow people to hear several points of view and they provide settings in which the merits of various ideas can be discussed (Kelley and Thibaut, 1969).

However, group pressure can also harm the quality of decisions by leading to excessive conformity and loss of individuality. The Spanish writer José Rodó has written that "every society to which you remain bound robs you of a part of your essence, and replaces it with a speck of the gigantic personality which is its own." Actually, the same thing could be said of each group to which an individual is strongly committed. For the sake of group membership, we may relinquish part of our individuality and independence.

The tendency to conform to group pressure has been dramatically illustrated in an experiment by Solomon Asch (1955) in which many participants publicly denied their own senses in order not to deviate from majority opinion. Asch assembled numerous groups of seven to nine male college students, ostensibly for experiments in visual judgment. The task was to compare the length of lines printed on two cards. (See Figure 6.1.) One card showed a single vertical line, which was to be used as a standard for comparison with three lines of different lengths on a second card. Each participant was asked in turn to identify the line on the second card that matched the length of the line on the first card. In each group, all but one of the subjects had been instructed by Asch to choose a line that obviously did not match. The naive subject—the only member of each group unaware of the real nature of the experiment—was forced either to select the line he actually thought matched the standard line or to yield to the

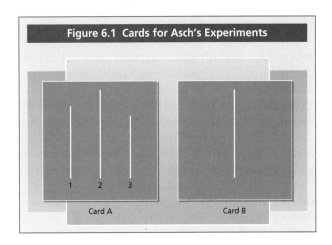

Figure 6.1 Cards for Asch's Experiments

Card A   1   2   3   Card B

unanimous opinion of the rest of the group. In earlier tests of individuals in isolation, Asch had found that the error rate in matching the lines was only 1 percent. Under group pressure, however, the naive subjects accepted the majority's wrong opinion over one-third of the time. If this large a proportion of naive subjects yielded to group pressure in the absence of primary relationships, it is not difficult to imagine the overconformity and loss of individuality that occurs in groups to which people are emotionally committed.

Because of such tendencies in groups, Irving Janis (1982) has argued that many group decisions are likely to be the product of **groupthink**, a situation in which pressures toward uniformity discourage members of the group from expressing their reservations about group decisions. During the Kennedy administration, for example, a decision to launch the ill-fated Bay of Pigs invasion against Cuba was made only because several top advisers to the president failed to reveal their reservations about the plan. A more recent example of groupthink is found in the Reagan administration's

---

FEEDBACK

1. Match the following types of social interaction with the examples beside them:
   ____ a. cooperation   (1) paid blood donors
   ____ b. conflict   (2) students do what a professor assigns
   ____ c. social exchange   (3) Saddam Hussein invades Kuwait
   ____ d. coercion   (4) flood victims help each other
   ____ e. conformity   (5) employees are forced to work overtime or be fired

2. According to Robert Bales, interaction within groups must be designed to solve both _____ problems and _____ problems.

3. Social scientists agree that leaders must have certain traits if they are to be effective. *T or F?*

4. According to Ralph White and Ronald Lippitt, which of the following leadership styles is most likely to be effective?
   ____ a. authoritarian   ____ b. democratic   ____ c. laissez faire

5. Solomon Asch's experiment demonstrates the positive consequences of group pressure. *T or F?*

6. A situation in which pressures toward uniformity discourage members of a group from expressing their reservations about group decisions is called _____.

Answers: 1. a. (4) b. (3) c. (1) d. (5) e. (2) 2. instrumental, social-emotional 3. F 4. b 5. F 6. groupthink plan to sell arms to Iran in exchange for the release of American hostages. The small number of designers of this plan—including Lieutenant Colonel Oliver North—minimized opposition by concealing information from officials in the administration who did not share their convictions. Such hesitancy to destroy the illusion of "assumed consensus" often leads to group decisions that have not been subjected to critical evaluation. Research indicates that groupthink can be avoided only when leaders or group members make a conscious effort to see that all group members participate actively and that points of disagreement and conflict are tolerated.

We have examined several important aspects of groups that are relatively small social units. Next we turn to larger social units known as formal organizations. It will become even more evident that groups are very much a part of any formal organization.

TABLE 6.1	Largest Industrial Corporations by Sales, Assets, and Profits, 1993		
Corporate Size	**Percentage of Total**		
	Sales	**Assets**	**Profits**
Top 100	71%	75%	78%
101–200	14	13	19
201–300	7	6	3
301–400	5	4	(2)
401–500	3	2	2

Source: U.S. Bureau of the Census, *Statistical Abstract of the United States: 1996*, Washington, DC: U.S. Government Printing Office, 1996, p.554.

Internet Link: There is a wealth (pardon the pun) of information on the characteristics of large American corporations. For example, the *1996 Statistical Abstract on the Internet* often offers even more data than is found in the book *Statistical Abstract of the United States*. To access this data, go to http://www.census.gov/stat_abstract/

FORMAL ORGANIZATIONS

Even in the early part of this century, Cooley foresaw a shift away from primary groups toward secondary groups. Until the 1930s, most Americans lived on farms or in small towns and villages. Almost all of their daily lives were spent in primary groups—families, neighborhoods, churches. As industrialization and urbanization advanced, however, more and more Americans became enmeshed in secondary groups and relationships. Born in hospitals, educated in large schools, employed by huge corporations, regulated by government agencies, living in old-age homes, and buried by funeral establishments, Americans—like members of all industrialized societies—find themselves conducting many of their daily activities within the type of secondary group known as a *formal organization*. This trend has been one of the major changes accompanying the shift from a preindustrial to an industrial society, and the trend is continuing. Table 6.1 shows that sales, assets, and profits are concentrated in the very largest corporations. People have become highly dependent on such organizations, not only for jobs but also for many major products, services, and even ideas. To understand organizations is to take an important step toward understanding industrial and postindustrial society (Blau and Meyer, 1987).

Theoretical Perspectives and Formal Organizations

Although it is not always explicit, functionalism and conflict theory are intimately involved in the sociological study of formal organizations. Bureaucracy, for example, with its emphasis on structure, stability, rules, and hierarchy, is based squarely on functionalism. When sociologists examine the dynamics of organizations, they don't abandon functionalism, but conflict theory does play a predominant role. The presence of conflict and change within organizations is spotlighted when the concepts of informal organization, power, and organizational environment are brought to bear on the study of the dynamics of organizations.

The Structure of Formal Organizations

As stated in Chapter 5 ("Social Structure"), a **formal organization** is a social structure deliberately created for the achievement of one or more specific goals. Examples of formal organizations are colleges, corporations, government agencies, hospitals, and high schools.

Most formal organizations in the United States and other industrialized societies are **bureaucracies**—a type of formal organization based on rationality and efficiency. Although bureaucracies are popularly thought of as monuments to inefficiency, they have proven to be a most efficient form of organization for industrial society. Max Weber, the German sociologist who first analyzed the nature of bureaucracy, wrote of its efficiency as an organizational form despite his deep concern about its negative consequences.

What are the major characteristics of bureaucracy? From a study of history and observation of events of his day, Weber was able to sketch the basic organizational principles that were being developed to promote rational and efficient administration. Weber identified an organization with the following characteristics as a bureaucracy.

- *The organization has a division of labor based on the principle of specialization.* In an organization, each status has certain functions or tasks attached to it for which the occupant is responsible. (See Figure 6.2 for an organizational chart outlining the division of labor in a modern university.) A specialized division of labor improves organizational performance because individuals can become experts in a limited area of organizational activity.

- *The organization has a hierarchy of authority.* To clarify the concept of authority, it is necessary first to define power. **Power** refers to the ability to control the behavior of others even against their will. **Authority** is power that produces compliance because those subjected to it believe that obedience is the proper way to behave. Authority, in short, involves the exercise of legitimate power. Bureaucratic organizations tend to have a pyramidal shape. The greatest authority is concentrated in the few positions at the top, with decreasing amounts of authority for the ever-increasing number of positions lower on the organizational chart. (See Figure 6.2.) Organizational effectiveness and efficiency are enhanced by a hierarchy that can coordinate the many statuses involved in a highly specialized organization.

- *Organizational affairs are based on a system of rules and procedures.* These rules and procedures guide the performance of work and provide a framework for decision making. They stabilize the organization because they are designed to cover most situations and to coordinate activities.

- *Members of the organization maintain written records of their organizational activities.* These records are known as the files. This organizational "memory" is essential to smooth functioning, stability, and continuity.

- *Statuses in the organization—especially managerial ones—are considered to be full-time jobs.* Bureaucratic statuses are not considered to be avocational. This increases the commitment of the organization's members and demands their full attention.

- *Relationships within the organization are impersonal; they are to be devoid of favoritism.* For example, personnel are supposed to be selected on the basis of technical and professional qualifications and promoted on the basis of merit. Similarly, relationships with clients are to be conducted without regard to personal considerations. The norms prohibiting the influence of favoritism and personal relationships in organizational activities are intended to insure equal treatment for those who work in the organization as well as for those who are served by it.

- *Employees of bureaucratic organizations do not own their positions.* Bureaucratic positions cannot be sold or inherited; this fact allows organizations to

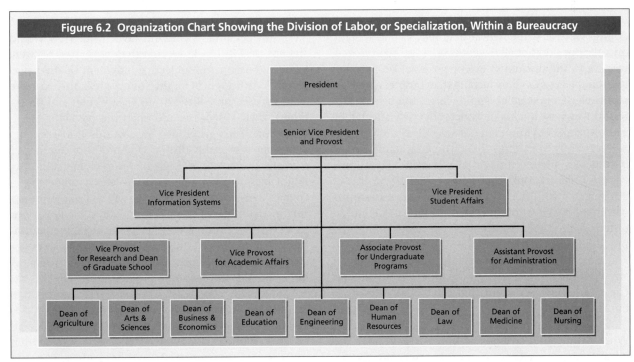

Figure 6.2 Organization Chart Showing the Division of Labor, or Specialization, Within a Bureaucracy

President

Senior Vice President and Provost

Vice President Information Systems

Vice President Student Affairs

Vice Provost for Research and Dean of Graduate School

Vice Provost for Academic Affairs

Associate Provost for Undergraduate Programs

Assistant Provost for Administration

Dean of Agriculture

Dean of Arts & Sciences

Dean of Business & Economics

Dean of Education

Dean of Engineering

Dean of Human Resources

Dean of Law

Dean of Medicine

Dean of Nursing

Each position has certain tasks associated with it. The connecting lines indicate who reports to whom and who has authority over whom.

fill positions with the most qualified people. Those in bureaucratic positions are held responsible for their use of equipment and facilities provided by the organization.

For Weber, then, bureaucratic organization involved a particular set of strategies for achieving the ordering of human relationships. The achievement of order in human relationships was seen by Weber as increasingly difficult in societies undergoing industrialization. In traditional societies, activities in one aspect of life had been closely related to activities in every other aspect of life. As societies modernized, people's activities became less closely related and traditional systems of norms and values began to weaken. Weber saw bureaucracy, with its hierarchy of authority, system of rules and procedures, and various safeguards, as a way to maintain control in an increasingly complex society.

The spread of the bureaucratic form of organization to all spheres was part of a general process of rationalization in modern society. The reason why it would inevitably spread was because its characteristics of precision, continuity, discipline, strictness and reliability make it technically the most satisfactory form of organization for those who sought to exercise organizational control (Thompson, 1980:10–11).

Do modern organizations have all the characteristics of bureaucracy? In his formulation of bureaucratic characteristics, Weber used what is known as the **ideal-type method.** This method involves isolating, even to the point of exaggeration, the most basic characteristics of some social entity, in this case bureaucracy. It involves the construction of a pure type. Because real life is seldom as pure as a theoretical description, bureaucratic organizations do not all have Weber's ideal-type characteristics to the maximum extent possible. Universities, for example, are less bureaucratic than business firms, in part because universities employ large numbers of professionals whose quality of work is adversely affected by a bureaucratic environment (Litwak, 1961).

Practically speaking, then, the characteristics of bureaucracy are best thought of as a series of continua; various organizations emphasize each characteristic of bureaucracy to different degrees. In fact, some units within an organization may be more bureaucratic than others (Hall, 1962). In a university, for example, the president's office is much more bureaucratic than an academic department such as philosophy or biology. At any rate, the more closely an organization is based on the characteristics identified by Weber, the more bureaucratic it is.

What are the advantages of bureaucracy? Of the advantage of bureaucracy over other forms of organization, Weber wrote:

The decisive reason for the advance of bureaucratic organization has always been its purely technical superiority over any other form of organization. The fully developed bureaucratic mechanism compares with other organizations exactly as does the machine with the nonmechanical modes of production (Gerth and Mills, 1958:214).

Weber's point of comparison was bureaucracy versus other existing forms of organization, which were very inefficient and inappropriate for changing society. For example, in monarchies, the administration usually lacked talent, structure, precision, and continuity. Leadership positions were filled on the basis of favoritism and politics. Administrators had little or no training for the work they were doing. Because they were wealthy, administration was only an avocation or hobby for them. And because high positions were politically based, power could be lost overnight. Moreover, when a king or queen was replaced, the entire leadership structure changed drastically.

As Weber correctly saw, earlier types of organizational structure were out of step with the rise of the capitalist market economy, which required steadiness, precision, continuity, speed, efficiency, and minimum cost—advantages that bureaucracy could offer. **Rationalism**—the solution of problems on the basis of logic, data, and planning rather than tradition and superstition—was on the rise, and the characteristics of bureaucracy were consistent with this sweeping trend.

The bureaucratic type of organization is also designed to protect individuals, as strange as that might sound. Without a system of rules and procedures and the norm of impersonal treatment, which everyone seems to hate, decision making would be arbitrary and based on personal whim and favoritism. How would students feel if they knew that their course grades were based purely on the subjective judgment of their professors? Although it might sound great not to have exams (based on college or university rules and procedures), their absence would open the possibility for a professor to assign grades on grounds that students would most probably dislike. One professor might give a greater proportion of high grades to women; another might want to prove that many athletes don't belong in college. This, of course, is not to say that most professors in a nonbureaucratic setting would be abusive or that favoritism does not occur in bureaucratic organizations. Nevertheless, the existence of formally established rules and procedures and the norm of impersonality insures some protection for equal treatment. At the very least, their presence permits a grounds for appeal against unfair treatment.

Another famous account of the benefits of specialization is found in *The Wealth of Nations,* a book written by Adam Smith, a moral philosopher and the father of economics. Published in 1776, this book attempted to answer the question, What is the source of the wealth in a society? For centuries, the mercantilists had found wealth to lie in the gold and silver in the royal treasury. Smith, expressing the new economic thought of the French physiocrats, gave a different answer—the wealth of nations resides in the productive power of labor, power that is best promoted by the division of labor. Smith goes on to offer a description of a pin factory as an illustration of the benefits of the division of labor that comes when "one man draws out the wire, another straightens it, a third cuts it, a fourth points it, a fifth grinds it at the top for receiving the head . . ." (Smith, 1965:4). By this process, Smith notes that ten men working in the factory can make forty-eight thousand pins daily, whereas a man working alone could make twenty or so pins. Although recognizing pin making as a "trifling" example, Smith emphasizes that the principles of the division of labor are readily applicable to any type of "art and manufacture."

What are the disadvantages of bureaucracy?
Although Weber praised bureaucracy as the most efficient type of organization, he was well aware of some of its undesirable side effects. He expressed a concern that the formal rationality of bureaucracy would not necessarily lead to the achievement of socially desirable outcomes. For one thing, Weber noted that bureaucracy represented "the concentration of the material means of management in the hands of the master" (Gerth and Mills, 1958:221). This leads to concerns about who will ultimately control the effects of bureaucratic organizations. In the words of Charles Perrow, one of the foremost contemporary scholars of complex organizations:

> *At present, without huge, disruptive, and perilous changes, we cannot survive without large organizations. Organizations mobilize social resources for ends that are often essential and even desirable. But organizations also concentrate those resources in the hands of a few who are prone to use them for ends we do not approve of, for ends we may not even be aware of, and, more frightening still, for ends we are led to accept because we are not in a position to conceive alternative ones. The investigation of these fearful possibilities has too long been left to writers, journalists, and radical political leaders. It is time that organizational theorists began to use their expertise to uncover the true nature of bureaucracy. This will require a better understanding both of the virtues of bureaucracy and its largely unexplored dangers (Perrow, 1986:6).*

Weber also raised the specter of what he called the *iron cage of rationality*. He feared the likelihood of rationality spreading to all aspects of social life, creating a dehumanizing social environment entrapping everyone. George Ritzer (1993) uses the *McDonaldization of society* as a metaphor to express Weber's iron cage of

rationality. McDonald's, Ritzer argues, is the model expression of rationalism in organizations. On the dehumanizing effects of extreme rationality, Ritzer writes that

> *. . . the fast-food restaurant is often a dehumanizing setting in which to eat or work. People lining up for a burger, or waiting in the drive-through line, often feel as if they are dining on an assembly line, and those who prepare the burgers often appear to be working on a burger assembly line. Assembly lines are hardly human settings in which to eat, and they have shown to be inhuman settings in which to work (Ritzer, 1993:12).*

It is the McDonaldization of everything from lube jobs to medicine that Ritzer believes threatens modern society with Weber's iron-cage nightmare come true. If the nightmare is still there when we awaken, then the rational has become irrational.

Although Weber and others in his tradition were deeply concerned about the societal implications of bureaucracy, others have focused on bureaucracy's alleged administrative inefficiencies (Hummel, 1987; Leibenstein, 1987). Prominent among these inefficiencies are goal displacement and trained incapacity.

Eating at McDonald's seems like a harmless activity (fat content aside). For many sociologists, however, the rationality and standardization undergirding fast-food restaurants promote dehumanization.

When organizational rules and regulations become more important than organizational goals, **goal displacement** occurs (Merton, 1968). Professors who emphasize grades more than learning are practicing goal displacement. Social workers who become more concerned with eligibility requirements than with delivering services to their low-income clients are engaging in goal displacement.

According to Thorstein Veblen (1933), **trained incapacity** exists when previous training prevents someone from learning to adapt to new situations. Trained incapacity involves tunnel vision, inflexibility, and inability to change based on past training and experience. Let me offer an illustration based on personal experience. At one point during my graduate training in sociology, the federal government was giving me financial support. With a child on the way, I felt it was necessary to earn some additional money. The chairperson of the sociology department permitted me to teach one course, providing the university administration agreed that the rules would permit receiving money from the federal government and the university at the same time. I posed this question to the appropriate university official and watched a perplexed bureaucrat fumble through the code of regulations for a pat answer, which simply would not present itself. After minutes of obvious discomfort—stemming from his inability to find a rule to support the refusal he seemed bent on issuing—he found what he was looking for. With the relief of a winner at Russian roulette, this man said excitedly, "Here it is!" Although the regulation actually could have been interpreted as permitting the earning of additional income by a recipient of a federal research grant, this administrator chose the interpretation that bolstered his original inclination to deny permission. I later learned that former policy had prohibited persons receiving federal money from obtaining additional financial support from the university. This bureaucrat had become so accustomed to this old way of doing things that he could not interpret the new ruling in any way other than as supporting the viewpoint that had become so indelibly his own. By past training and experience, he had developed an incapacity to meet the new organizational realities.

A similar thing happened to the crew of the USS *Vincennes,* when on July 3, 1988, it mistook an Iranian passenger airliner for an Iranian F-14 jet fighter bearing down on it. Instead of destroying a menacing military plane, the *Vincennes* crew killed all of the nearly three hundred civilian passengers on a regularly scheduled flight over the Persian Gulf. This regrettable accident occurred because the *Vincennes* was created to operate on the open sea, not on a small body of water with all sorts of air and water vehicles constantly coming into its radar sights. Instead of identifying the passenger airliner for what it was, which the crew could have done, the crew acted on the basis of its training and did what it thought was necessary for self-protection. The crew's training resulted in a tragedy because it acted in the way it had been taught to act without adapting to the new environment in which it found itself. The crew's incapacity stemmed from the nature of its training.

Do alternatives to bureaucracy exist?   Weber developed the theory of bureaucracy to explain the kinds of organizations that would be appropriate in the changing society in which he lived. During the 1960s, several social theorists contended that enough changes had occurred since Weber's time that bureaucracy was becoming an outmoded form of organization. Warren Bennis (1965), for example, contended that bureaucracy was no longer able to deal with the individual goals of employees and increasing rates of social change. He therefore predicted that bureaucracies would be replaced by **organic-adaptive systems**— organizations based on rapid response to change rather than on the continuing implementation of established administrative principles.

> *The key word will be "temporary;" there will be adaptive, rapidly changing temporary systems. These will be "task forces" organized around problems-to-be-solved. The problems will be solved by groups of relative strangers who represent a set of diverse professional skills. The group . . . will evolve in response to the problem rather than to programmed role expectations. The "executive" thus becomes coordinator or "linking pin" between various task forces. . . . People will be differentiated not vertically, according to rank and role, but flexibly and functionally according to skill and professional training (Bennis, 1965:34–35).*

Bureaucracy, of course, has not disappeared, and Bennis (1979) has qualified some of his earlier predictions. Increasing numbers of organization researchers would agree, however, that bureaucracy is best suited for relatively stable, predictable situations. In situations characterized by rapid change and uncertainty, other forms of organization are likely to emerge (Mintzberg, 1979). Such attempts to relate the structure of organizations to the situations they face have been labeled the **contingency approach** to organization theory (Burns and Stalker, 1966; Lawrence and Lorsch, 1967; Shepard and Hougland, 1978). Although the contingency approach has been criticized on some grounds (Pennings, 1975; Pfeffer, 1978), it is considered a useful corrective to the traditional assumption that there exists "one best way" to organize.

A strong advocate of alternatives to the bureaucratic form of organization, Joyce Rothschild (1979, 1986) has studied cooperative organizations, organizations intended to be on the other end of the spectrum from bureaucracies. Cooperative organizations, owned and managed by their employees, are characterized by full membership participation, minimization of rules, promotion of primary social relationships, hiring and promotion based on intimate ties and shared values, elimination of status differences, and a deemphasis on job specialization.

Although some analysts contend that nonbureaucratic forms of organizations are best suited to new business technologies and the future environmental context (Drucker, 1988; Peters, 1988), most scholars agree that bureaucracy will remain the dominant organizational form. Bureaucracies, it is contended, persist in part because they work best with the technologies and environments characteristic of the modern era. This does not mean, however, that some organizations cannot be nonbureaucratically structured. It is just that such organizations will be few in number. Nonbureaucratic principles of organizing are thought to be most useful as a supplement to bureaucratic organizations, as in the research laboratory of a computer software firm (Robbins, 1990).

What are some of the differences among formal organizations? Sociologists, such as Amitai Etzioni (1975), have attempted to take differences in formal organizations into account by classifying organizations into categories. Some organizations, such as churches, political parties, universities, and service clubs, are joined because people want to join; they are joined because of personal interest or emotional commitment. Because

members are free to join or not, these are *voluntary* organizations. Etzioni refers to them as normative organizations, because shared understandings provide an important basis for coordination of members' activities. Organizations that people are forced to join are *coercive* organizations. Examples include prisons, custodial mental hospitals, concentration camps, forced-labor camps, elementary schools, and military systems using the draft. Other organizations are *utilitarian;* people join them because of the benefits they derive from membership. Work organizations, including factories, offices, and professional firms, are examples of utilitarian organizations. Utilitarian organizations fall somewhere between the first two types. People are not forced to work (it is not a coercive organization), and yet they must work to be self-supportive (it is not a voluntary organization). Thus, utilitarian organizations are neither entirely voluntary nor entirely coercive; they have some elements of each.

Whereas Etzioni classified organizations according to the motivation for membership, Peter Blau and Richard Scott (1962) have classified organizations according to the recipients receiving the most benefit. In *mutual-benefit* organizations (political parties, professional associations, labor unions, social clubs), the prime beneficiary is the membership. These organizations exist to promote the interests of those who belong to them. *Business* organizations, on the other hand, are expected to serve the interests of their owners. Because these organizations are profit oriented, their emphasis

FEEDBACK

1. A formal organization is _____

   _____

2. List the major characteristics of bureaucracy according to Max Weber.

3. The ability to control the behavior of others even against their will is _____.

4. Which of the following has *not* been seen as an advantage of bureaucracy?

   ____ a. its avoidance of the use of inappropriate criteria in hiring employees

   ____ b. its use of rules to provide definite guidelines for behavior within the organization

   ____ c. its success in hiding the true nature of authority relationships

   ____ d. its encouragement of administrative competence in managers

5. The nature of bureaucracy raises questions about who will ultimately control its effects. *T or F?*

6. Social workers who become more concerned with eligibility requirements than with delivering services to their low-income clients are engaging in _____.

7. Which of the following is *not* a characteristic of organic-adaptive systems?

   ____ a. reliance on a set of rules

   ____ b. use of task forces

   ____ c. emphasis on coordination of executives

   ____ d. reliance on skill and professional training in determining work assignments

8. Attempts to relate the structure of organizations to the problems they face have been labeled the _____ approach to organization theory.

9. Formal organizations differ in some ways, but sociologists have not been able to develop meaningful classifications of these differences. *T or F?*

Answers: 1. a group that has been deliberately and consciously created for the achievement of one or more specific goals 2. division of labor, hierarchy of authority, system of rules and procedures, written records, full-time jobs, relationships devoid of favoritism, positions not owned by their occupants 3. power 4. c 5. T 6. goal displacement 7. a 8. contingency 9. F is on the achievement of maximum gain at the least cost. The prime beneficiary of a *service* organization is the organization's clients. The goal of social-work agencies, schools, and hospitals is to aid those who qualify for these organizations' professional services. *Commonwealth* organizations (the military, the Department of State, police and fire departments, the Environmental Protection Agency) are intended to serve the general public.

THE DYNAMICS OF FORMAL ORGANIZATIONS

The discussion of formal organization and its structure thus far has described the framework within which organizational actions occur. But organizations are also the settings for a number of dynamic processes. These processes both result from the structure of the formal organization and lead to the modification of that structure (Hall, 1996). This section is concerned with the informal structures that exist within all formal organizations, power and its effects on formal organizations, and the environments within which formal organizations exist.

Informal Organization

One does not have to observe any formal organization for long to realize that it contains patterns of interaction and relationships that are not anticipated by the formal organization chart. Bureaucratic organizations, for example, are designed to allow for secondary relationships only. Relationships among organization members are supposed to be governed by the hierarchy of authority, by rules and procedures, and by impersonality. As anyone who has worked in a bureaucratic organization knows, primary relationships in fact are woven into the formal structure. Primary relationships are found within what is called the **informal organization**—a group in which personal relationships are guided by rules, rituals, and sentiments not provided for by the formal organization. Based on common interests and personal relationships, informal groups are usually formed spontaneously, without any conscious effort on the part of the participants.

The existence of informal organization within bureaucracies was first documented in the mid-1920s, when a group of Harvard researchers were studying the Hawthorne plant of the Western Electric Company in Chicago. In a study of fourteen male operators in the Bank Wiring Observation Room, F. J. Roethlisberger and William Dickson (1964, originally published in 1939) observed that work activities and relationships on the job were based on informal norms and social sanctions.

Informal groups, which normally develop spontaneously, help to meet the social needs of their members. The importance of informal groups within formal organizations has long been documented by research.

Informal norms prohibited doing too much work ("rate busting"), doing too little work ("chiseling"), and telling group secrets to supervisors ("squealing"). Conformity to these norms was maintained through ridicule, sarcasm, criticism, hostility, and hitting deviants on the arm ("binging"). The group also had a sharp status structure, with soldermen at the bottom, wiremen in the middle, and inspectors at the top.

Why does informal organization exist?   Informal groups spontaneously develop in part because they meet the needs of those involved. Most modern organizations are deliberately designed to be impersonal, but informal groups permit the expression of personal feelings and behavior—love, hate, anger, favoritism, and humor. Informal organization thereby humanizes formal organization. Informal organization is also a means of control for the group. Within the group, informal rules, punishments, and rewards encourage conformity. Group solidarity also protects the informal group from maltreatment by organization members outside the group.

How does informal organization affect the formal organization?   Informal organization has both positive and negative effects. Early studies, starting with the Bank Wiring Observation Room, concluded that informal organization resulted in behavior counterproductive to the organization's goals. In the Hawthorne study, for example, the workers enforced the informal norm of producing an average of two production units a day. Workers who produced more than two units per day were subsequently forced by the group to meet the informal quota.

Later research has shown, however, that informal organization has consequences other than restriction of production, some of which promote worker satisfaction as well as organizational effectiveness. Raymond Van Zelst (1952) found that establishing work groups among carpenters and bricklayers on the basis of friendship patterns increased the workers' job satisfaction, decreased friction and anxiety among work partners, and created a friendly, cooperative atmosphere on the job—all of which resulted in lower turnover and higher productivity.

Joseph Bensman and Israel Gerver (1963) described the use of an illegal tool in an airplane factory. This tool, called a tap, is used to thread nuts so they will fit with wing bolts that are slightly out of alignment. The tap is outlawed—its use can lead to dismissal—because a tapped nut does not have proper holding power under flight conditions. However, use of the tap was tolerated by management as long as it was handled discreetly. In fact, without its use, the productivity of the plant would have suffered. Thus, the use of the tap was controlled informally by supervisors, inspectors, and workers themselves. In short, to avoid detection by Air Force inspectors, supervisors and workers evolved an informal organization complete with norms and sanctions for the systematic use of an illegal device. The formal organization considered itself to have benefited from this informal subterfuge. (Of course, the same cannot be said for the passengers who travel in the planes that are involved.)

The Role of Power in Formal Organizations

Power (defined earlier as the ability to control the behavior of others even against their will) affects all human relationships. Power plays a particularly important role in many organizations because of the diversity of interests and goals that exist among the participants. In the typical industrial organization, for example, the preferences of executives and wage laborers are likely to differ; the perspectives of sales managers and production managers may vary as these managers attempt to decide what is most beneficial for the organization. Because a "rational" choice between these conflicting views requires more information than most decision makers can assemble, the decision is likely to become a "political" one (Cyert and March, 1963; Simon, 1976; Pfeffer, 1981).

Social scientists have become increasingly aware that the exercise of power is a normal process in organizations and should be analyzed as such. Under many circumstances, the use of power may lead to benefits for the organization by facilitating organizational change and adaptation to changing conditions. However, power may also be used for the benefit of a small, unrepresentative group within the organization. The fear of this happening was most clearly articulated by German sociologist Robert Michels (1949, originally published in 1911) when he developed the iron law of oligarchy.

What is the iron law of oligarchy?   The **iron law of oligarchy** states that power tends to become concentrated in the hands of a few members of any social structure. Michels observed that even in organizations intended to be democratic, leaders eventually gain control and members become virtually powerless. In fact, he concluded, concentration of power at the top always occurs sometime after an organization is formed because those in power want to remain in power.

It is organization which gives birth to the dominion of the elected over the electors, of the mandatories over the mandators, of the delegates over the delegators. Who says organization, says oligarchy (Michels, 1949:401).

The government in communist China is a prime example of oligarchy. Not subject to popular election, the aging individuals at the top are able to consolidate power over a long period of time. Each of the leaders is able to build a loyal staff, control communications, and marshal resources (money, jobs, favors).

Why does organization lead to oligarchy?   According to Michels, three factors encourage organizations to become oligarchic: the need for a hierarchy of authority, the advantages held by those at the top of the hierarchy, and the characteristics of the rest of the members of the organization. Agreeing with Weber, Michels contended that the organizational problems of any large group of people desiring to achieve a goal requires the development of a bureaucracy. Even in highly democratic organizations, such as the socialist parties of Europe that Michels was deeply interested in, it is necessary for the masses to delegate much of the decision making to a few leaders. Delegation means creating a hierarchy of authority.

Leaders at the top of the hierarchy are able to monopolize power in part because of their own characteristics and in part because of the situation within the organization. Those who become the leaders are able to do so because of their social and political skills. Once in power, they use these skills to maintain their positions. Moreover, merely by doing their jobs, leaders of the organization gain special knowledge not available to the membership. They create a staff that is loyal to them; they control the channels of communication; they can legally use the resources of the organization to maintain or increase their power.

DOING RESEARCH

Rosabeth Moss Kanter— Men and Women of the Corporation

While working as a clinical sociologist, Rosabeth Moss Kanter (1993) became intrigued with the effects of modern bureaucracies on employees. In an intensive five-year case study of a large manufacturing firm (Indsco), she identified three job-related factors (*structural determinants*) that she contends fundamentally mold employee attitudes and behavior: opportunity, power, and relative numbers.

Kanter used the case study method, drawing on data she gathered over a five-year period as consultant, participant observer, and researcher at Indsco. Specifically, her sources of information included surveys (conducted by Kanter or the company); personal interviews; analysis of company documents; recorded group discussions; participant observation in meetings, training programs, and so forth; individual conversations; and key informants within Indsco.

Opportunity refers to employees' expectations regarding future promotion. People who perceived themselves to be on the "fast track"—variously referred to as "water walkers," "wonders," "high fliers," "one performers," or "superstars"—were more committed to work and were willing to take on extra tasks and responsi-

bility. Kanter observed the "Gerald Ford syndrome" (not wanting to be president until becoming president) in employees whose prospects for promotion unexpectedly increased. One twenty-year employee who had never wanted to be anything but a secretary was rapidly upgraded from executive secretary to assistant manager to manager. She changed her tune from "I never really thought about another career" to "Now I'm ambitious . . . I will probably work twenty more years, and I expect to move at least six grade levels, maybe to vice-president." Opposite behavior and attitudes appeared in individuals at a dead end. Known in the company as "mummies," "zombies," or "mystery men," these people entered the "writeoff" stage of their careers, withdrawing from as much

---

Oligarchy in organizations is, according to Michels, also encouraged by the characteristics of the followers. The masses tend to defer to the greater skills possessed by their leaders. Followers feel their leaders are more articulate and are doing a better job than they could do. In addition, the masses are passive, even indifferent; they do not want to spend a lot of time in organizational activities. The general membership, in fact, feels grateful to their leaders for the leadership they are providing.

What are the consequences of oligarchy? Democratic organizations—the type of organization in which oligarchy presents the greatest problem—suffer from the development of oligarchy, contended Michels, because those in power come to care more about perpetuating their power than about organizational goals. Oligarchic leaders become excessively careful and conservative as maintenance of their positions becomes more important to them than the interests of the membership. As a result, the desires of the membership may be sacrificed.

Must oligarchy always develop? Research done since Michels published his theory has tended to support the iron law of oligarchy, the operation of which has been documented in a variety of organizations, including political parties, legislatures, churches, the American

Legion, the American Sociological Association, the American Medical Association, and labor unions.

There is, however, opposition to oligarchy in organizations; otherwise, all organizations would be oligarchic. According to Alvin Gouldner, if there is an iron law of oligarchy, there must also be an iron law of democracy: "If oligarchical waves repeatedly wash away the bridges of democracy, this eternal recurrence can happen only because men doggedly rebuilt them after each inundation" (Gouldner, 1955:506). Sentiment from the masses is continuously bubbling up to depose oligarchic leaders in democratic organizations. The survival of democracy in the face of oligarchy can be seen in the democratic removal of entrenched labor leaders and politicians. Although the odds against turning out entrenched leaders are high, it happens often enough to remind us that oligarchy does not completely destroy democracy. Moreover, oligarchic leaders have to pay attention to the desires of their followers if they are to remain in power. Although mere responsiveness to the membership may not be the ideal form of democracy, its presence indicates that even oligarchic leaders often do not have a totally free hand to do as they please (Sayles and Strauss, 1967).

It is also possible for members of democratic organizations to erect barriers to the development of oligarchy. A responsibility as possible. Other responses to blocked opportunity included turning to junior people for recognition, criticizing those who had moved up, and resisting any proposed changes.

The second structural determinant—*power*—Kanter defined as the capability of marshaling organization resources to accomplish objectives. At Indsco, relatively powerless managers, insecure about their position in the company, tended to overcontrol their subordinates. They made all decisions, resisted delegating work, and tried to control all communication in and out of their units. Jealously guarding company perquisites and privileges was another way to maintain an illusion of power. One manager, approached by a young employee to have lunch, accused the man of

manipulating him to gain access to the executive dining room. Powerless managers often emphasized territoriality; they clutched their domains with viselike grips, emphasizing the importance of their functions, contributions, and expertise.

By *relative numbers,* the third structural determinant, Kanter referred to the ratio of women to men in the organization. Because men overwhelmingly outnumbered women in the upper ranks, women found themselves in a minority status. Successful women in male peer groups became symbols of what-women-can-do. On the one hand, this made them highly visible in an environment in which success depends on visibility. On the other hand, they sometimes felt the loneliness of being different from the other

group members and had to sacrifice their privacy and anonymity. Because of their visibility, these women also felt extra pressure to avoid making mistakes. Even women who had experienced little or no discrimination felt they had to work harder than the average man to be successful.

Critical feedback

1. Which of the major theoretical perspectives do you think guided Kanter's study? Explain and illustrate your choice.
2. Several examples were selected to illustrate the consequences of each of the structural determinants identified by Kanter. From your understanding of these three factors, speculate on some additional consequences.

---

classic example of this is the International Typographical Union (ITU), which is composed of typesetters. According to Seymour Lipset, Martin Trow, and James Coleman (1956), the ITU has been able to avoid oligarchic leadership because of its internal two-party system. Every two years, each party puts up a complete slate of candidates for all offices. Because the two parties are of equal strength, the turnover in office compares favorably with the turnover in national political elections.

Some other research concludes that concentration of power does not inevitably lead to an abandonment of the organization's goals for the selfish benefit of the leaders. Research on policies of churches toward civil rights and other controversial issues has led James Wood to note that "many organizations exist precisely to foster values that are precarious when left to individuals" (Wood, 1981:101). Wood found that such values—including those that would favor the enactment of controversial policies regarding civil rights—can most effectively be implemented when leaders have a strong organizational base of support. Contrary to Michels' predictions, leaders in such situations often enacted policies that resulted in controversy, a loss of membership, and a loss of professional positions within the organization. Given a choice between the leaders' personal benefits and the maintenance of the organiza-

tion's values, the leaders often chose to maintain the organization's values.

Joyce Rothschild and J. Allen Whitt (1987) also question the alleged inevitable connection between organization and oligarchy. They challenge the iron law of oligarchy through an investigation of *cooperatives*—organizations owned and operated by workers themselves. Although not all cooperatives prosper or even survive, Rothschild and Whitt maintain that the very existence and persistence of cooperatives contradict the common social-science assumption that where organization appears, the eventual concentration of power in the hands of a select few must follow.

Organizations and Their Environments

The **organizational environment** consists of all the forces outside the organization that exert an actual or a potential influence on the organization. An organization's environment includes the general technological, economic, political, cultural, and related conditions in which the organization operates (Hawley, 1968; Meyer and Scott, 1985; Hannan and Freeman, 1989; Scott, 1992). Interest in environments has increased in part because of the growing realization that organizations can survive only to the extent that they generate enough support among environmental actors to continue receiving needed resources (in the form of money, new members, or whatever) from the environment. It is equally true, however, that many organizations have sufficient resources to exert significant influence on their environments (Florida and Kenney, 1991). Both when they are influencing outside actors and when they are receiving influence, organizations are frequently involved in interorganizational relationships.

What is the nature of interorganizational relationships? An **interorganizational relationship** is a pattern of interaction among authorized representatives of two or more formally independent organizations. Interorganizational relationships may be entered for many reasons, but most researchers would agree that a need for resources possessed by other organizations and an awareness of the availability of those resources in the other organizations contribute to the formation of interorganizational relationships (Levine and White, 1961; Cook, 1977; Hougland and Sutton, 1978; Alter and Hage, 1992).

Although organizations become involved in interorganizational relationships because outside resources are needed, all parties are concerned that the interests of their own organizations be protected. Figure 6.3, which is based on a study conducted by Richard Hall and John Clark on a social-control system for problem youth, shows that though the police must interact with fourteen other organizations in the area of problem youth, only three of these relationships are regularly cooperative. Six of the relationships, including those with such important organizations as the juvenile court and the detention center, are regularly marked by conflict (Hall, 1996).

A COMPARATIVE VIEW OF FORMAL ORGANIZATION: JAPAN

The discussion of organizations thus far has assumed a broadly common cultural background, namely that of the West. An examination of some of the unique features of formal organizations in Japan demonstrates that culture can be a powerful player in shaping the structure and dynamics of formal organization (Lincoln and McBride, 1987; McCormack and Sugimoto, 1988; Lincoln and Kalleberg, 1990).

Formal Organization in Japan

Bureaucracies in both the East and the West share the fundamental characteristics identified by Weber. Culture, however, impinges to create some distinctive differences between the Western and the Japanese styles of bureaucracy (Pascale and Athos, 1981; Hasegawa, 1986; Reischauer, 1988; Ouchi, 1993). (See Table 6.2.)

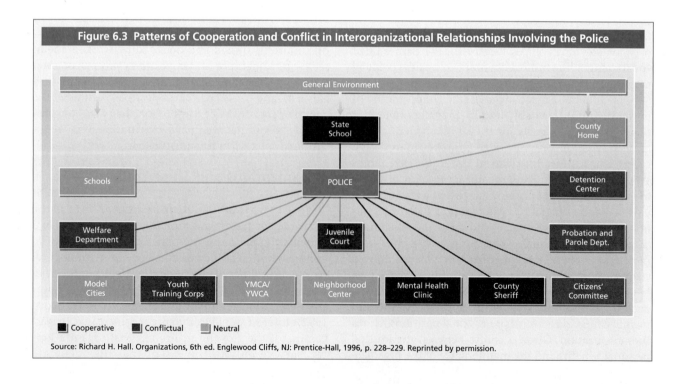

Figure 6.3 Patterns of Cooperation and Conflict in Interorganizational Relationships Involving the Police

Source: Richard H. Hall. Organizations, 6th ed. Englewood Cliffs, NJ: Prentice-Hall, 1996, p. 228–229. Reprinted by permission.

TABLE 6.2	Contrasting Features of Western and Japanese Organizations
Western	**Japanese**
Short-term employment	Lifetime employment
Rapid evaluation and promotion	Slow evaluation and promotion
Specialized career paths	Broad organization training
Individual decision making	Collective decision making
Lack of involvement in employees' nonwork lives	Blending of employees' work and nonwork lives

Source: William G. Ouchi, *Theory Z.*, pg. 58, © 1981 by Addison-Wesley Publishing Company, Inc. Reprinted by permission of the publisher.

The following discussion applies only to the largest Japanese organizations, which employ about one-third of the Japanese labor force. Like most newly industrialized nations, Japan has a dual structure in its economy (Wokutch, 1992; Schnitzer, 1994). At the bottom are small handicraft industries, many service industries, and most retailing companies, which have not changed much from premodern times. On top of the economy sit new large-scale industrial corporations. Thus, the system about to be described does not apply to most workers in the lower tier of Japan's dual economy or to most women.

Neither Western workers nor their companies view employment as a lifetime commitment. Individuals typically change organizations to suit their advantage; and as a matter of policy, most organizations fire, lay off, and permanently terminate their employees. In contrast, one of the prime characteristics of the larger Japanese organizations is lifetime employment. Once hired, an employee is not expected to leave until mandatory retirement (age 55) for anything short of a criminal act. In fact, those managers who are lured away by other companies are seen as deviants or as possessing serious character flaws. This policy of lifetime employment creates a strong sense of organizational loyalty and a workforce that is enthusiastic, proud of company products, and pleased to work overtime. Many employees do not take full vacation time, and workers are so conscientious that quality control inspectors are not needed.

Rapid employee turnover in Western firms requires frequent evaluation, financial reward, and promotion, partly to reduce the rate of departure. In large Japanese firms, employees who join an organization at the same time typically receive the same pay increases and promotions for about ten years. Only at the end of ten years is a formal evaluation of an employee made and differential promotions awarded. Consequently, much

competition and friction are eliminated, both within and across age groups. Individuals within the same age group do not try to outshine their peers, and superiors in the organization do not have to face ambitious subordinates who want to replace them.

Specialization is very high in Western firms, both on the factory floor and in the managerial and professional ranks. This is by design on the part of higher management. At the same time, because American workers cannot expect to spend their lives in one firm, they prepare themselves to be marketable. A prime way to accomplish this is to become competent in a set of specialized skills that any number of employers find valuable. In large Japanese organizations, training is much broader. At the peak of their careers, Japanese managers will have been prepared to be experts in every function, specialty, and office of their firms. Whereas Westerners are more likely to identify themselves as engineers, sales executives, electronics experts, or economists, the Japanese identify themselves as Toyota, Sony, or Nippon Steel employees.

By tradition, Westerners have tended to admire the image of the strong leader who takes charge and is responsible for tough decision making. In organizations, this has been expressed in "top-down" management—department heads, division managers, vice presidents, and presidents want to take full responsibility for decisions they make. Even the "participative" style of management that is beginning to be practiced in Western firms does not approach the "bottom-up" style of decision making typical of large Japanese organizations. Decision making in large Japanese organizations involves everyone likely to be affected. Sometimes this means that sixty or more people may be directly involved in making a decision. Although time-consuming, this approach increases the employees' commitment to a decision and enables employees to be more effective in the implementation of the decision.

The relationship between employees and their employers in Western organizations tends to be narrowly limited to work-related matters. Large Japanese firms, in contrast, are very involved in their employees' nonwork lives. Concern for the total welfare of their workers extends to organizational provision of such things as housing, loans, education, and recreational outlets. Imagine the intimate and holistic ties that begin to be forged even during the induction of a group of bank trainees (Rohlen, 1974). The bank president tells the parents that the firm will accept responsibility for their children's physical, moral, and intellectual development. The parent representative for the ceremony rises to admonish the new employees to show their new family the loyalty they have exhibited toward their biological families. Finally, a trainee pledges for the entire class a commitment to meet bank and parental expectations.

Origins of Japanese Organizational Structure

Obviously, differences between Japanese and Western organizations can be traced to environmental factors in one way or another. The process of linking environmental conditions to Japanese organizational characteristics, however, is not simple. As a broad generalization, the distinctive characteristics of large Japanese organizations reflect a culture that places the group above the individual: "The Japanese company is in effect a community of motivated people. In Japan a company sees itself as a cohesive group of people working toward a common goal" (Hasegawa, 1986:1).

Although the nature of the Japanese organization is consistent with a collectivistically oriented culture, it does not mean that the organizational structure emerged totally formed from premodern times. For example, the practice of lifetime employment for nonexecutives, although consistent with a cultural emphasis on group solidarity, has not always existed in Japanese industry. Lifetime employment for all employ-ees became a feature of large Japanese organizations only in response to the need for skilled workers to perform more technically complicated work. To overcome the shortage of skilled workers following World War II, large Japanese corporations adopted the practice of lifetime employment for all employees. Much-needed skilled labor, the Japanese found, could be retained through offering employees a variety of special privileges, of which job security was the mainstay.

Environmental change continues in Japan. Although the Japanese system will probably remain intact for some time, Edwin Reischauer (1988) points to some signs of erosion. Among them is the beginning of some disenchantment among younger workers who are less attracted to the idea of working hard in the present for future rewards. An increasing desire for more personal freedom makes company paternalism somewhat stifling. To the shock of the older generation, more younger people, including executives, are showing less company loyalty as they display a greater willingness to accept new job offers. The reward system itself is moving away from seniority toward skills and productivity.

FEEDBACK

1. A group in which personal relationships are guided by rules, rituals, sentiments, and traditions not provided for by the formal organization is known as a(n) _____ organization.
2. The effects of informal organization on the formal organization are almost always negative. *T or F?*
3. The theory known as the _____ emphasizes the danger of power becoming concentrated in the hands of a few members of any organized group.
4. Research conducted since Robert Michels first formulated the iron law of oligarchy has suggested that
   ____ a. oligarchy never develops in organizations.
   ____ b. oligarchic leaders are often overthrown by other organization members.
   ____ c. oligarchic leaders always ignore the desires of their followers.
   ____ d. concentration of power always leads to the abandonment of an organization's goals.
5. Forces outside an organization that exert an actual or a potential influence on the organization are labeled the organizational _____.
6. Regular patterns of interaction among authorized representatives of two or more formally independent organizations are referred to as _____.
7. A need for resources possessed by other organizations provides a major motivation for entering into interorganizational relationships. *T or F?*
8. Cooperative patterns tend to dominate in interorganizational analysis. *T or F?*
9. Japanese organizational structure has proven to be the exception to Weber's analysis. *T or F?*

Answers: 1. informal 2. F 3. iron law of oligarchy 4. b 5. environment 6. interorganizational relationships 7. T 8. F 9. F

SUMMARY

1. A fundamental characteristic of humans is their extensive participation in group life. Even though this is so, Americans are socialized to think in individualistic terms.

2. A primary group is composed of individuals who know one another well, interact informally, and like being together. A secondary group is goal oriented, and its members interact without personal or emotional involvement. Likewise, a primary relationship is intimate and personal, whereas a secondary relationship is impersonal.

3. Primary groups are more likely to develop when a relatively small number of people are involved, when interaction is face-to-face, when people are in contact over a long period of time, and when the social environment is appropriate. Primary groups are formed in part because of the emotional support and sense of belonging they provide members. This type of group serves society by aiding in the socialization process and by helping to maintain social control.

4. We use reference groups to evaluate ourselves and to acquire attitudes, values, beliefs, and norms. Reference groups may be groups we merely aspire to join or may be groups we use as negative role models.

5. Particularly intense feelings of identification and loyalty envelope in-groups. The feelings involve opposition, antagonism, and competition toward out-groups. Efforts must be made to maintain group boundaries, the consequences of which have significant social implications.

6. A social network (not a group per se) links a person with a wide variety of individuals and groups. The social relationships that are formed may be "strong"—parents, child—or "weak"—distant relatives, members of a college class. Social networks provide a sense of belonging, social support, and help in the job market. More women are now using networking, once the sole province of men, for career advancement.

7. Five types of social interaction are basic to group life. In cooperation, individuals or groups combine their efforts to reach some common goal. When individuals or groups work against one another to obtain a larger share of the valuables, they are in conflict. In conflict, defeating one's adversary is thought to be necessary. Social exchange is a form of interaction in which one person voluntarily does something for another person with the expectation that a reward will follow. Coercion exists when one person or group is forced to behave according to another person's or group's demands. When a person behaves as others expect, conformity occurs.

8. Most groups experience both cooperation and conflict. This is partly because groups satisfy some personal needs, such as social acceptance, whereas they thwart others, such as personal freedom. Social scientists have been very interested in the ways in which interaction within groups can achieve group goals while fulfilling the needs of individual group members.

9. Moving beyond the trait theory of leadership, social scientists have focused on appropriate leader behavior. More recently, studies of leadership suggest that appropriate leadership behavior depends on the situation a leader faces.

10. In some situations, group decision making has been shown to be superior to individual decision making. Intellectual tasks, for example, are better handled through groups. On the other hand, group pressure may damage the quality of decision making because of the excessive conformity of group members. Many group decisions are reached in situations in which group members are discouraged from expressing their reservations. At the extreme, this is called groupthink.

11. A formal organization—the epitome of a secondary group—is deliberately created to achieve some goal. There are many types of formal organizations in modern society, but most share one feature—they are bureaucratic. According to Max Weber, bureaucracy is the most efficient type of formal organization. Using the ideal-type method, Weber outlined the characteristics that give bureaucracy its technical superiority over other types of organizations.

12. Although Weber saw bureaucracy as an efficient organizational form, he was well aware of its negative side. When rules and regulations become more important than organizational goals—as they often do—those who are supposed to be served by the bureaucracy suffer. Adherence to rules and regulations may also make bureaucrats overly conservative; such people learn their jobs so well that they cannot adapt to changing times and circumstances. Such problems have led researchers to examine alternatives to bureaucracy. Although most social scientists agree that bureaucracy is appropriate for many situations, contingency theorists have suggested that other types of organizations may fit particular circumstances.

13. The existence of primary groups and primary relationships within formal organizations has received considerable attention during the twentieth century. These informal groups can either help or hinder the achievement of formal organizational goals, depending on the circumstances.

14. The exercise of power in organizations is receiving increasing attention. Power can benefit organizations by facilitating change and adaptation to changing conditions, but power can also be used for the benefit of a small, unrepresentative group within the organization. The applicability of the iron law of oligarchy to modern organizations is a continuing source of debate among sociologists.

15. Organizations exist within an environment. They depend on factors within the environment for necessary resources, and they also attempt to exert influence on portions of the environment. Both of these situations encourage involvement in interorganizational relationships, which allow organizations to adapt to environmental pressure but involve conflict as well as cooperation.

16. An examination of the unique characteristics of organizations in Japan demonstrates that culture can shape the structure and dynamics of formal organizations. It will be interesting to see the effects of future environmental change on the nature of formal organizations in Japan.

LEARNING OBJECTIVES REVIEW

After careful study of this chapter, you will be able to:

- Define the concept of the group and differentiate it from social categories and social aggregates.
- Outline the basic characteristics of primary and secondary groups.
- Describe the five major types of social interaction.
- State the nature of social processes in groups relative to task accomplishment, leadership styles, and decision making.
- Define the concept of formal organization and identify the major characteristics of bureaucracy.

- Discuss the advantages and disadvantages of bureaucracy.
- Distinguish between formal and informal organization.
- Describe the iron law of oligarchy and demonstrate its importance with some examples.
- Compare and contrast the major features of Western organizations and Japanese organizations.

CONCEPT REVIEW

Match the following concepts with the definitions listed below them:

____ a. social exchange
____ b. rationalism
____ c. ideal-type method
____ d. bureaucracy
____ e. out-group

____ f. iron law of oligarchy
____ g. organizational environment
____ h. group
____ i. conflict
____ j. secondary group

____ k. formal organization
____ l. primary relationship
____ m. social category
____ n. conformity
____ o. in-group

1. A type of formal organization based on rationality and efficiency.
2. Occurs when individuals or groups work against one another to obtain a larger share of the limited valuables in a society.
3. A type of social interaction in which an individual behaves toward others in ways expected by the group.
4. A social structure deliberately created for the achievement of one or more specific goals.
5. Several people who are in contact with one another; share some ways of thinking, feeling, and behaving; take one another's behavior into account; and have one or more interests or goals in common.
6. A method that involves isolating the most basic characteristics of some social entity.
7. A group toward which one feels intense identification and loyalty.
8. The principle stating that power tends to become concentrated in the hands of a few members of any social structure.

9. All the forces outside the organization that exert an actual or potential influence on the organization.
10. A group toward which one feels opposition, antagonism, and competition.
11. A relationship that is intimate, personal, based on a genuine concern for another's total personality, and fulfilling in itself.
12. The solution of problems on the basis of logic, data, and planning rather than tradition and superstition.
13. A group that is impersonal and task oriented and involves only a segment of the lives and personalities of its members.
14. A number of persons who share a social characteristic.
15. When one person voluntarily does something for another expecting a reward in return.

CRITICAL THINKING QUESTIONS

1. Think of a club you belong to on campus or a club you joined as a high school student. Was this a primary or secondary group? Support your answer with the use of sociological concepts and actual experiences.

   _____

   _____

2. Recall an in-group of which you are (or were) a member. Discuss the consequences this group had for others. Defend the right of this group to exist on sociological grounds.

   _____

   _____

3. Most Americans think of bureaucracies in a very negative way. Develop an argument you could use the next time someone endorses the prevailing view of bureaucracy in your presence.

   _____

   _____

4. Do you think that the Japanese type of organizational structure would work in the United States? Defend your position using the characteristics of Japanese organizations and your knowledge of American culture.

   _____

   _____

MULTIPLE CHOICE QUESTIONS

1. **Several people who are in contact with one another; share some ways of thinking, feeling, and behaving; take one another's behavior into account; and have one or more interests or goals in common, is known as a**
   a. commune.
   b. clique.
   c. group.
   d. social category.
   e. social aggregate.

2. **Which of the following are considered the two principal types of groups?**
   a. primary and small groups
   b. primary and reference groups
   c. secondary and small groups
   d. secondary and corporate groups
   e. primary and secondary groups

3. **A group is sometimes confused with a_____ , a number of people who happen to be in the same place at the same time.**
   a. social collectivity
   b. social category
   c. social class
   d. social aggregate
   e. mob

4. **A web of social relationships that joins a person directly to other people and groups and, through those individuals and groups, indirectly to additional parties is known as a**
   a. social network.
   b. relationship web.

   c. social structure.
   d. safety net.
   e. social complex.

5. **Robert Nisbet has described five types of social interaction basic to group life. Which of the following is *not* one of them?**
   a. cooperation
   b. conflict
   c. coercion
   d. consensus
   e. conformity

6. **Which of the following statements has been found to be *most* accurate concerning group decisions?**
   a. Individual decisions are always better than group decisions.
   b. Group pressure has little effect on group decisions.
   c. Group decisions can be superior to individual decisions.
   d. Mistakes made by groups are harder to correct.
   e. Group decisions are always better than individual decisions.

7. **The type of formal organization based on rationality and efficiency is known as a/an**
   a. commune.
   b. ecclesia.
   c. hierarchy.
   d. profession.
   e. bureaucracy.

8. According to the text, _____ refers to the ability to control the behavior of others even against their will.
   a. authority
   b. prestige
   c. privilege
   d. ostracism
   e. power

9. The solution of problems on the basis of logic, data, and planning rather than on the basis of tradition and superstition is known as
   a. groupthink.
   b. rationalism.
   c. mythology.
   d. existentialism.
   e. jurisprudence.

10. A judge who is more concerned with following legal procedures than reaching just decisions is engaging in
    a. contingency thinking.
    b. categorical thinking.
    c. goal displacement.
    d. coercion.
    e. trained incapacity.

11. The part of a formal organization in which personal relationships are guided by rules, rituals, and sentiments not provided for by the formal organization itself is known as
    a. the enforcement of mores zone.
    b. the application of legitimate authority network.
    c. informal organization.
    d. the Peter Principle unit.
    e. the contingency group.

12. _____ states that power tends to become concentrated in the hands of a few members of any social structure.
    a. The iron law of oligarchy
    b. The Peter Principle
    c. Parkinson's Law
    d. The law of demographic transition
    e. The iron law of responsibility

13. _____ consists of all the forces outside the organization that exert an actual or a potential influence on the organization.
    a. Organizational environment
    b. Bureaucratic environment
    c. Organic-adaptive environment
    d. Organizational context
    e. External organization

14. Which of the following is *not* characteristic of formal organization in Japan?
    a. collective decision making
    b. rapid evaluation and promotion
    c. lifetime employment
    d. broad organizational training
    e. blending of employees' work and nonwork lives

FEEDBACK REVIEW

TRUE-FALSE

1. Social scientists agree that leaders must have certain traits if they are to be effective. *T or F?*
2. The effects of informal organization on the formal organization are almost always negative. *T or F?*
3. A need for resources possessed by other organizations provides a major motivation for entering into interorganizational relationships. *T or F?*

FILL IN THE BLANK

4. State the three functions of primary groups. _____

   _____

   _____

5. List the major characteristics of bureaucracy according to Max Weber. _____

   _____

   _____

6. Attempts to relate the structure of organizations to the problems they face have been labeled the _____ approach to organization theory.

MULTIPLE CHOICE

7. Which of the following is *not* a condition that promotes the development of primary groups?

    a. small group size

    b. face-to-face contact

    c. continuous contact

    d. interaction on the basis of status or role

8. Which of the following is *not* a characteristic of organic-adaptive systems?

    a. reliance on a set of rules

    b. use of task forces

    c. emphasis on coordination by executives

    d. reliance on skill and professional training in determining work assignments

MATCHING

9. Listed below are some examples of primary and secondary relationships. Indicate which examples are most likely to be primary relationships (P) and which are most likely to be secondary relationships (S).

    ____ a. a marine recruit and his drill instructor at boot camp

    ____ b. a married couple

    ____ c. a manager and his professional baseball team

    ____ d. professors and students

    ____ e. car salespersons and their prospective customers

10. Match the following terms and statements:

    ____ a. group   (1) skinheads

    ____ b. in-group   (2) spectators of a Fourth of July fireworks display

    ____ c. social aggregate   (3) orchid growers

    ____ d. social category   (4) teenage subculture

    ____ e. reference group   (5) members of a regular Saturday night poker game

GRAPHIC REVIEW

Table 6.1 displays the sales, assets, and profits of corporations by size of the organization. Answering the following questions will check your understanding of this information as it relates to the material in this chapter.

1. Choose one characteristic of bureaucracy to show that this type of organization makes possible the development of extremely large corporations.

    _____

    _____

    _____

2. Relate the concentration of power in the top one hundred companies to the iron law of oligarchy.

    _____

    _____

    _____

REVIEW GUIDE

ANSWER KEY

Concept Review	Multiple Choice	Feedback Review
a. 15	1. c	1. F
b. 12	2. e	2. F
c. 6	3. d	3. T
d. 1	4. a	4. provide emotional support, contribute to socialization, promote conformity
e. 10	5. d	5. division of labor, hierarchy of authority, system of rules and procedures, written records, full-time jobs, relationships devoid of favoritism, positions not owned by their occupants
f. 8	6. c	
g. 9	7. e	
h. 5	8. e	
i. 2	9. b	6. contingency
j. 13	10. c	7. d.
k. 4	11. c	8. a.
l. 11	12. a	9. a. S
m. 14	13. a	b. P
n. 3	14. b	c. S
o. 7		d. S
		e. S
		10. a. 5
		b. 1
		c. 2
		d. 3
		e. 4

Chapter Seven

Deviance and Social Control

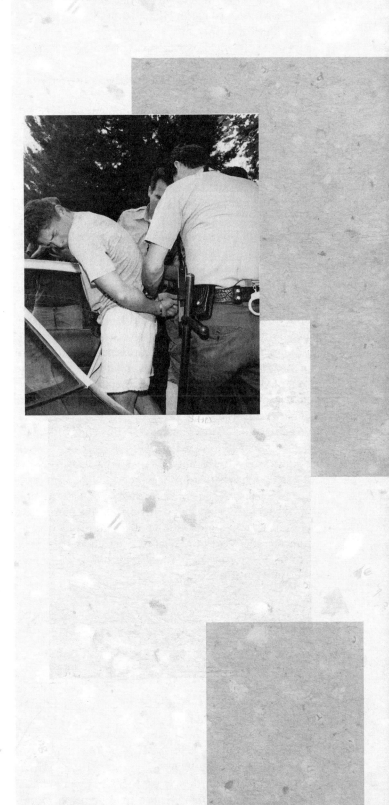

LEARNING OBJECTIVES

After careful study of this chapter, you will be able to:
- Define deviance and describe the major factors promoting the relativity of deviant behavior.
- Define social control and identify the major types of social control.
- Describe the biological and psychological explanations of deviance.
- Discuss the positive and negative consequences of deviance.
- Differentiate the major functional theories of deviance.
- Compare and contrast cultural transmission theory and labeling theory.
- Discuss the conflict theory view of deviance.
- Distinguish the four approaches to crime control.

SOCIOLOGICAL IMAGINATION

Is rape normally committed by a stranger? A woman concerned about rape is usually wary of either a surprise attack coming in a dark place outdoors or a stranger at her door posing as something other than a rapist. Although this is a warranted concern, over 80 percent of all rapes in the United States, in fact, are committed by individuals acquainted with their victims. Experts now use the terms *date rape* or *acquaintance rape* to describe this reality.

As date or acquaintance rape illustrates, deviant behavior may involve perversion. Deviance, however, is not limited to "nuts, sluts, and perverts" (Liazos, 1972). It exists in any situation in which group norms have been violated. Social control, which involves attempts to discourage deviance, is not the exclusive prerogative of official groups such as the police and prison guards. Informal social control exercised by group members is a central feature of social life. These points are amply illustrated in this chapter.

DEVIANCE AND SOCIAL CONTROL

The Nature of Deviance

Without conformity and predictability in human behavior, society could not exist. Order and stability are the cornerstones of social life. We know, however, that social life involves people violating norms. Some people violate norms by robbing banks, assaulting others, or committing murder. Some incidents of deviance have been high profile because they involved promi-

nent sports figures whose behavior was captured on national television. Dennis Rodman, a forward with the Chicago Bulls basketball team, cross-dresses, has tattoos all over his body, and regularly changes the color of his hair (yellow, red, blue, green, purple and orange, multicolored, and so forth). During games in 1996, Rodman head-butted a referee and deliberately kicked a photographer in the groin. More dramatically, Mike Tyson, former heavyweight boxing champion, twice bit the ears of current champion, Evander Holyfield. On the first bite, Tyson actually bit off the top of Holyfield's right ear and spat it onto the ring mat (Starr and Samuels, 1997). Each of these examples represents **deviance,** or violation of significant social norms.

The possibility of deviating from some norms is likely to exist in a society as complex as modern American society. This is reflected in a study by Jerry Simmons in which Simmons asked a sample of respondents to list categories or types of persons they considered deviant. Here is the summary of the results:

The sheer range of responses predictably included homosexuals, prostitutes, drug addicts, radicals, and criminals. But it also included liars, career women, Democrats, reckless drivers, atheists, Christians, suburbanites, the retired, young folks, card players, bearded men, artists, pacifists, priests, prudes, hippies, straights, girls who wear makeup, the President of the United States, conservatives, integrationists, executives, divorcees, perverts, motorcycle gangs, smart-alec students, know-it-all professors, modern people, and Americans (Simmons, 1969:3).

To this list, Leslie Lampert would add obese people. For a week she wore a "fat suit," adding 150 pounds to her 100-plus pound normal body weight, in order to experience firsthand what it feels like being an overweight

Sexual abuse is much more common on college campuses than the public thinks. Three-fourths of college women report being victims of sexual aggression; almost one-fourth report being victims of date rape.

woman in American society. According to Lampert, American "society not only hates fat people, it feels entitled to participate in a prejudice that at many levels parallels racism and religious bigotry" (Lampert, 1993:154).

In fact, to exist in American society, or in most *any* society for that matter, is to deviate from some social norms. This is why sociologists conceptualize deviance as the violation of *significant* social norms. Significant norms are those thought to be highly important either by most members of a society or by those in a society with the most power. **Deviants** are those who have violated one or more of society's most highly valued norms, especially those valued by the elite.

Identifying deviance is a challenge in large part because deviance is a relative matter. Insight into the relativity of deviance began early in the history of sociology. According to William Graham Sumner (1840–1910), norms determine whether behavior is deviant or normal (Sumner, 1906). Because norms vary from group to group, society to society, and time to time, the behavior considered to be deviant varies. Emile Durkheim contended that no behavior is deviant in itself, because deviance is a matter of social definition.

From Sumner, Durkheim, and many other sociologists, we have learned that what is considered deviant depends on the social status and power of the individuals involved, the social context in which the behavior occurs, and the historical period in which the behavior takes place.

How do social status and power influence the definition of deviance? No researcher has better demonstrated the importance of social status and power in defining deviance than William Chambliss, who, as a participant observer, watched two teenage gangs for two years at Hannibal High School. While at Hannibal High, Chambliss observed the school and nonschool activities of the Saints, a gang of eight upper middle-class white boys, and the Roughnecks, a gang of six lower-class white students. Although the Saints were "good students" headed for college, they were truant for part of nearly every school day and spent much of their weekends drinking, driving wildly, stealing, and vandalizing. The Saints' activities included "accidentally" spilling soft drinks in a favorite cafe, shouting obscenities at women and girls, deliberately running red lights, destroying empty houses, and removing warning lights and signs from road repair sites to trap unsuspecting motorists. Members of the community—including the police—believed that the Saints were merely good boys sowing wild oats. How could such well-dressed, well-mannered boys be anything but fun-loving kids playing pranks? Not a single Saint was arrested during the two years Chambliss observed the school.

The Roughnecks engaged in many of the same types of behavior as the Saints, but they did not drive cars, were not well-mannered, and did not dress well. The boys in this lower-class gang were looked down on by community members as worthless kids who were poor students, heavy drinkers, and violent individuals (they did fight a lot, especially among themselves). Moreover, the Roughnecks were constantly in trouble with the police. Over the two years that Chambliss was in the community, each Roughneck was arrested at least once, several were arrested many times, some spent at least one night in jail, and two were given six-month sentences in a reform school.

Why, Chambliss asked, "did the community, the school, and the police react to the Saints as though they were good, upstanding, nondelinquent youths with bright futures but to the Roughnecks as though they were tough, young criminals, who were headed for trouble?" (Chambliss, 1973:28). For one thing, the Roughnecks were more visible. Whereas the Saints could drive out of town in their own cars for delinquent activities, the Roughnecks, lacking transportation, spent most of their time in the center of the city. Second, the police knew that arresting a Saint meant dealing with parents angry at them for making a fuss over "harmless pranks"; they knew the parents of the Roughnecks would not put any pressure on them. Finally, the Saints acted apologetic and penitent when caught by the police. Roughnecks, on the other hand, could barely conceal their contempt for any authority figure. In short, the different treatment of these two gangs seemed to be traceable to their social-class backgrounds. Lower-class people are more likely to be labeled deviants than are middle- or upper-class individuals, even when they behave in similar ways.

How do definitions of deviance vary with social context? Behavior can be considered conforming or deviant only within a particular social context; behavior considered deviant in one society may be acceptable and even encouraged in others. Suppose that a group believes that gods live in rocks and trees and that these spirits are capable of granting them power over their enemies, the ability to heal the sick, and the foresight to predict the future. Among the traditional Alaskan Eskimo, such people would be awarded high social status and deference. But in a highly industrialized society, members of such a group would be considered strange.

Deviance is also relative from group to group within the same society. Conformity to the norms of one group may earn intense disapproval from another. Teenagers, for example, are usually torn between conforming to the demands of their peers and meeting the expectations of their parents. The possibility of being considered deviant from some group's point of view is especially likely in a society as complex as that of modern America.

How do definitions of deviance vary over time?   Ideas regarding deviant behavior change from time to time. Homosexual organizations, for example, have made considerable efforts to eradicate the image of homosexuality as a mental disorder. Their efforts were rewarded when the American Psychiatric Association removed homosexuality from its list of mental disorders (Spector and Kitsuse, 1977). If the American public comes to accept this new definition, homosexual pressure groups will have progressed in their drive to change the public perception of homosexuality. The gay struggle is far from over as can be seen in public attitudes toward homosexuality in general and AIDS in particular, and in the continuing public debate on the right of gays to serve in the armed forces (Murray, 1996). Another example is cigarette smoking. Gerald Markle and Ronald Troyer (1988), who studied the changing definition of cigarette smoking in American society, found that definitions of smoking have varied considerably. The definition of smoking was quite negative during the early 1900s when smoking was banned by fourteen states. It became positive during the 1950s when advertising began to imbue the habit with glamour and sophistication. Today, smoking is increasingly being prohibited in places as diverse as restaurants, government buildings, corporate offices, airplanes, elevators, physicians' offices, and university and college buildings. A final illustration of changing definitions of deviance is the current gambling craze in the United States. In response to this demand, many state governments are adopting lotteries to generate needed funds. Gambling casinos are appearing in places as diverse as riverboats and Native American lands (Popkin and Hetter, 1994).

Forms of Social Control

Human behavior is predictable most of the time. We generally know what to expect from others and what others expect of us. As you know by now, this is no accident. Just as the laws of gravity relieve us of the worry of our dishes rising off the table during dinner, social and cultural mechanisms exist to promote order, stability, and predictability in social life. We feel assured that drivers will remain on their side of the road, waiters will not pour wine over our heads, and strangers will not expect us to accept them as overnight guests. Without **social control**—means for promoting conformity to a group's or society's rules—social life would be capricious, even chaotic. There are two broad types of social control: internal controls, which lie within individuals, and external controls, which exist outside individuals.

What is the distinction between internal and external social control?   Internal social control is self-imposed and is acquired during the socialization process. For

Deviants are not just people who engage in obviously deviant acts such as prostitution. Any departure from social norms is considered deviance. This homeless woman is deviating from social expectations and makes most Americans feel uncomfortable.

example, most people most of the time refrain from stealing, not because they fear arrest or lack the opportunity to steal but because they consider theft to be wrong. The norm against stealing has become a part of the individual. This is known as the internalization of social norms.

The internalization of norms is not complete in all societies or groups. Otherwise, we would have no deviance at all. Because the process of socialization does not insure the conformity of all people all of the time, external social control must also be present.

External social control is based on social sanctions, or rewards and punishments, designed to encourage desired behavior. (Refer to Chapter 3, "Culture.") Positive sanctions such as awards, salary increases, and smiles of approval are used to encourage conformity. Negative sanctions such as fines, imprisonment, and ridicule are intended to stop people from repeating socially unacceptable behavior. Sanctions may be formal or informal. Ridicule and gossip are informal sanctions; imprisonment, capital punishment, exile, and honorary titles are formal sanctions.

BIOLOGICAL EXPLANATIONS OF DEVIANCE

As explained in Chapter 3 ("Culture"), sociologists, despite the rise of sociobiology, are still skeptical of biological explanations of human behavior. The search for some biological foundation for deviance, however, has a long history and is still occurring.

1. _____ is behavior that violates significant social norms.

2. A major problem sociologists have in defining deviance is its _____.

3. _____is the means for promoting conformity to a group's or society's rules.

Answers: 1. Deviance 2. relativity 3. Social control

Search for a Biological Basis for Deviance

What are some biological explanations of deviance? One of the earliest proponents of biological causes of deviance was Cesare Lombroso (1918), an Italian physician who saw a relationship between physical traits and crime. Lombroso believed that criminals were accidents of nature, throwbacks to an earlier stage of human evolutionary development. American psychologist William Sheldon (1949), a direct intellectual heir of Lombroso, identified three basic types of body shape: endomorphs (soft and round), mesomorphs (muscular and hard), and ectomorphs (lean and fragile). Sheldon concluded that criminals were more likely to be mesomorphs.

The theories of Lombroso and Sheldon are not generally considered valid today, but the search for a connection between physical characteristics and deviant behavior continues. Contemporary research in this area is much more sophisticated methodologically than the work of Lombroso and Sheldon (Fuller and Thompson, 1978; Rowe, 1983, 1986; Mednick, Gabrielli, and Hutchings, 1984; Brunner et al., 1993). The most influential contemporary attempt to establish a link between biology and criminal behavior appears in a book entitled *Crime and Human Nature* by James Q. Wilson and Richard Herrnstein (1985). The authors argue that Lombroso was thinking in the right direction. Although criminals are not born, the authors contend that many people are born with "constitutional factors"—personality, intelligence, anatomy—that predispose them to serious criminal behavior. Wilson and Herrnstein point to research on twins and adopted children. Criminality is said to be more common among identical than fraternal twins. This is taken to be important because identical twins are the same genetically, whereas fraternal twins are only as genetically similar as brothers or sisters. Sons of chronic offenders raised in noncriminal homes are reported to have a three times greater probability of following a life of crime than sons of nonoffenders who have been adopted by law-abiding parents. Wilson and Herrnstein are careful to point out that they are referring only to genetic predispositions toward criminal behavior. People with these biological predispositions do not necessarily become criminals; these predisposi-

tions may be offset by social learning in an emotionally supportive atmosphere such as the family.

How do sociologists evaluate biological explanations of deviance? Sociologists generally have not placed much stock in biological explanations of deviance. There are five main reasons for this. First, biological theories ignore the fact that deviance is more widely distributed throughout society than are the hereditary and other physical abnormalities that are supposedly the causes of deviance. Second, biological theories almost totally discount the influence of social, economic, and cultural factors in creating deviant behavior. These theories fail to explain how a person can have a genetic predisposition for something that is relative. Third, early biological theories of deviance were based on methodologically weak research. Fourth, there are ideological problems and controversial implications inherent in the biological approach. Forced sterilization of "defectives" and related eugenic strategies for reducing crime have created a storm of protests. Finally,

Definitions of deviance change from time to time. Although homosexuality is still viewed as a form of deviance by the majority of Americans, prohibitions have loosened to the point that many gays, such as Massachusetts Congressman Barney Frank, are willing to go public.

1. Recent research has established a definite casual relationship between certain human biological characteristics and deviant behavior. *T of F?*
2. According to James Q. Wilson and Richard Herrnstein,
   a. Studies of adopted children cannot help in understanding possible biological antecedents of criminal behavior.
   b. Intelligence is not correlated with criminality.
   c. Emotionally supportive families cannot help prevent criminal behavior.
   d. Constitutional factors predispose individuals to criminality.

Answers: 1. F 2. d.

biological factors are more often invoked to explain the deviance of armed robbers, murderers, and heroin addicts than, say, the crimes of corporate executives, government officials, and other high-status persons (Empey, 1982).

PSYCHOLOGICAL THEORIES OF DEVIANCE

Personality as the Source of Deviance

What common focus do psychological theories of deviance share? As varied as they are, psychological explanations of deviance all locate the origin of criminality in the individual personality (Hagan, 1994b). This is to be expected because psychology, as discussed in Chapter 1 ("The Sociological Perspective"), concentrates on the development and functioning of mental-emotional processes in human beings. More specifically, psychological theories of deviance take for granted the existence of a "criminal personality," a pathological personality with measurable characteristics that distinguish criminals from noncriminals. Pathological personality traits may be either inherited (and, therefore, related to biological explanations of deviance) or socially acquired (Raine, 1993).

Although Sigmund Freud did not attempt to explore the origins of deviance, his biologically based, instinctual approach to human personality lends itself to such exploration. Psychologists who have adapted Freud's theories to the study of deviance believe that deviance is due to unconscious personality conflicts created in infancy and early childhood. For example, criminality is said to be the result of biologically based desires beyond the control of a defective ego and superego. Other psychological explanations of crime are based on the identification of specific personality traits. Delinquents are said to be more aggressive, hostile, rebellious, or extroverted than nondelinquents (Glueck and Glueck, 1950; Eysenck, 1977). Some advocates of the psychological search for the criminal personality contend that criminals are born rather than made, an approach closely akin to biological explanations of deviance (Yochelson and Samenow, 1994). Other researchers attribute criminality to IQ differences. According to Robert Gordon

(1987), for example, differences in the delinquency rates of blacks and whites is due more to variations in intelligence than to social and economic characteristics.

What is the sociological critique of psychological theories of deviance? Sociologists are no more inclined to adopt psychological theories of deviance than they are biological explanations. Some of the objections are the same. First, psychological theories, like biological ones, ignore social, economic, and cultural factors shown by sociological research to affect the likelihood, frequency, and types of deviance. Second, psychological theories focus on deviance such as murder, rape, and drug addiction with relatively little to say about such deviance as white-collar crime. Third, psychological theories tend to view deviance as the result of physical or psychiatric defects rather than as actions considered deviant by social and legal definitions. Fourth, psychological theories of deviance tend to be theoretically and methodologically flawed. Fifth, psychological theories of deviance cannot explain why deviant behavior is engaged in by individuals not classifiable as pathological personalities. Finally, psychological theories, with their emphasis on pathology—deviants are "psychopaths" or "antisocial" personalities—suggest eugenic solutions to the crime problem that are unacceptable to some segments of society. For example, because more deviance occurs among certain racial and ethnic groups, birth control is a solution to the crime problem that is the most consistent with psychological theories. Many African Americans and Latinos view any such solution as an attempt to reduce their numbers and their power in society.

The topic of power brings us to sociological explanations of deviance. The specific theories of deviance will now be discussed within the frameworks of functionalism, symbolic interactionism, and conflict theory.

FUNCTIONALISM AND DEVIANCE

Functionalism has shed light on deviance, both in a general sense and in the strain theory of Robert Merton. Both of these contributions toward understanding deviance need to be examined.

1. Which of the following is *not* one of the psychological approaches to deviance?
    a. extroversion
    b. race
    c. unconscious conflicts
    d. IQ
2. Psychological theories of deviance still have not developed the breadth to be able to explain white-collar crime.
    T or F?

Answers: 1. b 2. T.

The Functions of Deviance

Functionalism attempts to explain how any social element affects a society. It is easy to think of deviance as having negative social consequences. It is harder to imagine any benefits of deviance to society, but functionalism tells us that there are some. (See *Doing Research.*)

What are the negative effects of deviance on society? Widespread violation of norms threatens the foundation of social life. Social life requires predictability created by members of society doing what is expected of them. What would happen if bus drivers decided to create their own routes each morning, if local television stations aired what they preferred rather than what the schedule called for, if parents took care of their children's needs only when it was convenient? The result of such nonconformity is social disorder, perhaps even chaos. Out of this disorder comes unpredictability, tension, and conflict.

Deviance erodes trust. We expect others to behave in certain ways. If bus drivers do not follow the planned route, if television stations constantly violate their schedules, if parents support their children sporadically, trust will be undermined. A society characterized by widespread suspicion and distrust cannot function smoothly. This fact is illustrated in the increasing rate of rape in the United States. As pointed out in the *Sociological Imagination* opening this chapter, rape overwhelmingly occurs between people who know each other. Almost one-fourth of college women report being victims of rape. Therefore, it is not surprising that their increasing distrust frequently creates a barrier to normal dating relationships.

Unpunished deviance encourages others to be nonconformists. If bus drivers regularly pass people waiting for the bus, those people might begin to throw rocks at the passing buses. If television stations offer random programming, they will be under siege. If parents stop caring for their children, delinquency will skyrocket. Deviance stimulates deviance.

Deviance is expensive because it diverts resources that can be used elsewhere to control nonconformists. Police who have to deal with bus drivers, irate passengers, and mobs of violent television viewers are taken

Deviant groups, such as this Los Angeles Latino street gang, often engage in random, unpredictable acts in violation of social norms. The reputation for such activities among gang members creates a sense of mistrust among those outside the group.

**Kai Erikson—
Wayward Puritans**

As noted in the discussion of the functions and dysfunctions of deviance, Emile Durkheim observed that deviance, generally thought to be totally negative, is actually integral to all healthy societies. According to Durkheim, deviant acts, much like other emergencies—wars, floods, famines, earthquakes—bring a group's sense of mutuality and togetherness to the fore. Solidarity is also increased when group members focus their disapproval on people who have violated highly valued rules. Although this thesis is a sociological classic, it has not been well researched.

Kai Erikson (1966) takes the Durkheimian (or functional) viewpoint of deviance in a study of the Puritan settlers who settled in the Massachusetts Bay area in the early seventeenth century. Erikson thought that the Puritans, because of their relative physical isolation, would provide an exceptional laboratory for social research. Erikson's aim was not to discover new things about the Puritan community in New England but to use the Puritan experience to enhance the general understanding of deviant behavior.

Because Erikson obviously could not conduct an experiment, do a survey, or be a participant observer, he had to rely on precollected data. He turned to original documents, including court records, treating them as the "living" voices of the Puritans themselves. Beyond these original documents, Erikson depended on historical accounts of Puritan life. Erikson concentrates on three

guiding hypotheses, each of which rests on Durkheim's view of the social benefits of deviance.

In the first hypothesis, Erikson predicts that as a community's definition of itself is threatened, the types of behavior considered deviant will change. Identifying and punishing deviants is a community's way of handling a fear that its "boundaries" (or sense of what it represents) are being blurred. Erikson associates each of three crime waves (one of which was the Salem witch trials) with an antecedent shift in the European political climate perceived by the Puritans to threaten their unique place in the world. Combating these self-generated crime waves served to bolster the Puritans' sense of being different from the European society they left.

According to Erikson's second hypothesis, the amount of deviance in a community is likely to remain fairly stable over time. Support for this prediction would

from their normal duties. If children turn to delinquency because of widespread parental neglect, the criminal justice system will be kept from handling other types of crime.

How does deviance benefit society?   Another contribution of Emile Durkheim was his recognition of the benefits of deviance. Deviance clarifies norms. When "creative" bus drivers and television station managers are brought into line, others are taught (or reminded of) the norms and the limits of tolerance for deviance from them. When parents are taken to court or lose their children because of neglect, society's expectations are demonstrated to other parents and children.

Deviance can be used as a temporary safety valve. Fathers and mothers who are upset about the demands of child rearing may overreact negatively to their children's unreasonable demands. Although not socially approved, such punitive behavior may help preserve the family, because these parents can then resume fulfilling their socially defined roles.

Deviance can increase unity within a society or group. If deviance reminds people of something they value, it may strengthen their commitment to it. The

exposure of spies selling government secrets to an enemy may intensify feelings of patriotism among citizens. Learning about parents who have abused their children may cause other parents to rededicate themselves to proper child care.

Deviance may promote needed social change. Suffragettes who took to the streets in the early 1900s scandalized the nation but helped bring women the right to vote. Prison riots may lead to the reform of inhuman conditions. As deplorable as it sounds, the discovery of sexual molestation of young children may lead to an improvement of day-care facilities. Looting and other deviant behavior in South Central Los Angeles in 1992 may lead to government efforts to face the problems of American inner cities (Morganthau, 1992).

Strain Theory

According to Durkheim (1964a, originally published in 1893), **anomie** is a social condition in which norms are weak, conflicting, or absent. Anomie causes societies to become disorganized and individuals to lose a sense of shared values and norms. In the absence of suggest that societies require a certain quota of deviance and function to insure meeting that quota. Using the records of the Essex County Court, which he assumes provide complete coverage of all criminal behavior in the northeast corner of Massachusetts, Erikson reports the existence of relative stability in deviance (as measured by crime rates) for the period 1651 to 1680.

Finally, Erikson explores the hypothesis that all societies have one or more "deployment patterns" by which they regulate the needed supply of deviant persons. The Puritans' unique deployment pattern was grounded in the theology of predestination. This deployment pattern involved the assumption that people do not change very much as a result of different experiences; a person's character is predestined. People who engage in certain types of deviant behavior—such as Ann Cole, one of the colony's first

"witches," who admitted that the Devil had had frequent carnal knowledge of her—were believed to be locked into permanent deviant roles. Unlike many deployment patterns, Puritan society made it difficult for deviants to return to normal life. This did not mean that everyone who committed an infraction was "lost"; deviants could be chastised or fined in order to be returned to their senses. But when a person was assumed to be of irredeemably bad character, a line of no return was drawn: ". . . when they branded him on the forehead or otherwise mutilated his face they were marking him with the permanent emblem of his station in life and making it extremely difficult for him to resume a normal social role in the community" (Erikson, 1966:197). Banishment and execution, of course, were punishments meant to convey the same sense of final judgment.

Erikson's study is a classic. Its high regard among sociologists is based on its inventive methodology and the support it provides for Durkheim's thesis that deviance can contribute to the stability and order of society.

Critical feedback

1. From your understanding of functionalism, explain why Erikson's findings were (or were not) supportive of Durkheim's assertion regarding the benefits of deviance to society.
2. Are you convinced that Durkheim's assertion is valid? Why or why not?
3. Does labeling or conflict theory have any applications to Puritan society as described by Erikson? Explain.
4. Do contemporary American attitudes regarding deviants share anything in common with those of the Puritans? Be specific if possible.

shared values and norms, individuals are uncertain about how they should think and act. Robert Merton (1968) has adapted the concept of anomie to deviant behavior in his strain theory.

What is strain theory?   According to Merton's **strain theory,** deviance is most likely to occur when there is a discrepancy between culturally prescribed goals and legitimate (socially approved) means of obtaining them. The resulting strain causes some people to engage in deviant behavior. Merton argued that culture determines the things people should want (goals) and the legitimate ways (means) of obtaining these things. In American society, an important goal is success and the acquisition of the material possessions that usually accompany it. Although everyone is taught to value material success, contends Merton, some are denied access to the legitimate means for achieving it.

Merton suggests five general modes of adaptation open to those denied access to the approved means for achieving success. (See Table 7.1.) One possible response is *conformity*—pursuing culturally approved goals through legitimate means. Poor minority members who continue to work hard in conventional jobs in

the hope of improving themselves illustrate this response. The four other responses are all considered deviant.

In *innovation,* the goal of success is accepted but illegitimate means for achieving it are adopted. Teenagers

TABLE 7.1	Modes of Adaptation to the Discrepancy Between Goals and Means (Strain Theory)	
Modes of Adaptation	**Cultural Goals (Success)**	**Legitimate Means**
Conformity	Accept	Accept
Innovation	Accept	Reject
Ritualism	Reject	Accept
Retreatism	Reject	Reject
Rebellion	Reject and substitute new goals	Reject and substitute new means

Source: Adapted with permission from Robert K. Merton. *Social Theory and Social Structure,* rev. ed. New York: Free Press, 1968, p. 194. Copyright, 1967 by Robert K. Merton.

may steal an expensive automobile. Others may use prostitution, drug dealing, or other lucrative criminal behavior to be successful. Innovation is the most widespread and obvious type of deviant response.

In *ritualism,* success is rejected but legitimate means are maintained. One example is the bureaucrat who continues to go about the daily routines of work while abandoning any idea of moving up the success ladder. Some college students begin to believe that their college careers are without meaning, that the whole experience has been a big disappointment, and that the dual goals of intellectual expansion and economic success are not going to be realized. Yet because of family expectations or lack of alternatives, they continue to go through the motions of getting a college education.

Retreatism is a deviant response in which the legitimate means and goals are rejected. Skidrow alcoholics, dope addicts, and bag ladies have dropped out. Some are considered by criminologists to be double failures, because they have failed to reach success by either legitimate or illegitimate means.

In *rebellion,* individuals reject both success and the legitimate means for achieving it but substitute a new set of goals and means. Revolutionaries of various types are good examples.

How has strain theory been applied?   Strain theory has been used most extensively in the study of juvenile delinquency. Albert Cohen (1977) uses it to explain why gang delinquency occurs most often among lower-class youth. According to Cohen, lower-class youth are poorly equipped to succeed in the middle-class world of school, which rewards verbal skills, academic success, neatness, and the ability to delay gratification. To protect their self-esteem, they reject these rules of success and self-worth and create their own status standards—standards they are able to achieve. Being tough, destructive, and daring are a vital part of this status system.

According to Richard Cloward and Lloyd Ohlin (1966), deviance does not always result from the strain created by a discrepancy between cultural goals and socially approved means for reaching these goals. Cloward and Ohlin refine strain theory by pointing out that deviant behavior is not merely an automatic response; like any other type of behavior, deviance must be learned. And learning must take place through interaction with and observation of others. Thus, deviance is more likely to occur as it increases.

The concept of opportunity to engage in deviance helps to explain why deviance occurs more in some parts of a society than in others. In middle- and upper-class homes, young people are encouraged to value education—the most important socially approved means for success—and are exposed constantly to people who have been successful by using legitimate means. Young people in poor ghettos, on the other hand, can more easily model their own behavior after pimps, drug dealers, and other types of criminals who have illegitimately acquired all the material symbols of success—money, clothes, cars, and women. And when juvenile delinquents are sent to reform schools or prisons, they have the opportunity to learn how to be criminals from well-practiced veterans. Cloward and Ohlin's use of strain theory overlaps with the symbolic interactionist approach to deviance, discussed in the next section.

How can strain theory be evaluated?   Merton's strain theory has had great staying power, in part because of its applicability to juvenile delinquency and crime. Another plus is its emphasis on social structure rather than on individuals.

Strain theory does have some shortcomings. Because it is in the functionalist tradition, it assumes a consensus in values—it assumes that everyone values success and defines success in economic terms. Also, strain theory is not very helpful in explaining why individuals choose one mode of adaptation over another. Why, for example, do some people choose to conform rather than to rebel? Finally, although strain theory helps explain crime and delinquency, it offers no help in explaining mental illness or drug abuse.

Control Theory

What is control theory?   Travis Hirschi's (1972) control theory is also based on Durkheim's view of deviance. According to **control theory,** conformity to social norms is based on the existence of a strong bond between individuals and society. If the strength of that bond is too weak, or is broken for some reason, deviance occurs. In control theory, it is assumed that people are inclined toward deviance by nature. Deviance requires no further explanation. What is not explained is why conformity occurs. This is where the social bond enters the picture. Individuals conform because of fears that deviance will harm the relationships they have with others. It is concern about "losing face" with family members, friends, or classmates, not fear of legal punishment, that encourages conformity. According to Hirschi, the social bond has four intertwined dimensions:

1. *Attachment.* The likelihood of conformity varies with the strength of the ties one has with parents, friends, and basic institutions such as schools and churches.
2. *Commitment.* The greater one's commitment to legitimate social paths such as educational attainment and occupational success, the more likely one is to conform. The commitment of adults exceeds that of teenagers, and the commitment

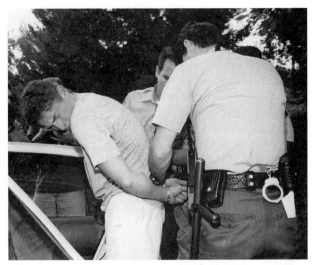

According to control theory, the likelihood of conformity to social norms rests on the presence of a strong bond between individuals and society. When social bonds are weak, the probability of deviance increases. When social bonds fail to control deviance, formal authorities such as the police attempt to supply the control.

of individuals who believe their hard work will be rewarded is greater than it is for those who think they cannot compete within the system.

3. *Involvement.* Participation in legitimate social activities increases the probability of conformity, both because it absorbs time and energy and because it places one in contact with others whose good opinion is valued.

4. *Belief.* Attachment to norms and values of society promotes conformity. Respect for the rules of social life strengthens one against the temptations of deviance.

In short, when social bonds are weak, the chances for deviance increase. Individuals who lack attachment, commitment, involvement, and belief have little reason to follow the rules of society. In the absence of a sufficiently strong social bond, the natural tendency toward deviance takes over.

How has control theory been received by other social scientists? Generally speaking, empirical support for Hirschi's control theory has been relatively strong. Several studies have shown the ability of control theory to explain nondelinquent and deviant behavior (Krohn and Massey, 1980; Wiatrowski, Griswold, and Roberts, 1981; Lyerly and Skipper, 1982; Bernard, 1987; Rosenbaum, 1987; Sampson and Laub, 1990).

It is when Hirschi attempted to broaden his control theory of delinquency into a general theory encompassing all types of crime (Hirschi and Gottfredson, 1990; Gottfredson and Hirschi, 1990) that scientific support weakened. Critics are skeptical that a specific theory of conformity and delinquency can explain crime in all its varieties, ranging from insider trading to murder (Gottfredson and Hirschi, 1990; Polk, 1991). The existence of weak social bonds may satisfactorily account for pimps, drug dealers, and carjackers. But it is much more difficult for control theory to account for socially well-integrated individuals who embezzle corporate funds, use public monies for private purposes, or dump toxic waste into rivers (Hagan, 1994b; McCaghy and Capron, 1997). In fairness, insufficient research has been conducted to judge adequately the expansion of control theory to cover all types of criminal behavior.

SYMBOLIC INTERACTIONISM AND DEVIANCE

According to theories based on symbolic interaction, deviance is learned through socialization. Concepts

---

FEEDBACK

1. Which of the following is *not* one of the benefits of deviance for society?
   a. decreases suspicion and mistrust among members of a society
   b. promotes social change
   c. increases social unity
   d. provides a safety valve
   e. promotes clarification of norms

2. According to _____ theory, deviance is most likely to occur when there is a discrepancy between culturally prescribed goals and the socially approved means of obtaining them.

3. A college professor who simply goes through the motions of teaching classes without any thought of success is an example of which of the following responses in strain theory?
   a. rebellion
   b. conformity
   c. ritualism
   d. innovation
   e. retreatism

4. The higher commitment to societal values and norms among adults than teenagers is consistent with the principles of _____ theory.

Answers: 1. a 2. strain 3. c 4. control

such as the looking-glass self, significant others, primary group, reference group, and the generalized other, discussed in Chapter 4 ("Socialization Over the Life Course"), underlie both cultural transmission theory and labeling theory.

Cultural Transmission Theory

What is cultural transmission theory? Cultural transmission theory emerged from the "Chicago School" during the 1920s and 1930s. According to **cultural transmission theory,** deviance is part of a subculture that is transmitted through socialization. Clifford Shaw and Henry McKay (1929) observed that delinquency rates stayed high in certain neighborhoods, even though the ethnic composition changed over the years. According to Shaw and McKay, delinquency in these neighborhoods persisted because it was transmitted through play groups and gangs. As new ethnic groups entered the neighborhoods, they learned delinquent behavior through interacting with the people already there.

Edwin Sutherland and Donald Cressey (1992), who were part of the Chicago School, developed the differential association theory to explain how this cultural transmission of deviance occurs from generation to generation and from one ethnic group to another. According to Sutherland, delinquent and criminal behavior are learned just like religion, politics, and basketball, principally in primary groups. Sutherland's **differential association theory** posits that crime and delinquency are more likely to occur among individuals who have been exposed more to unfavorable attitudes toward the law than to favorable ones. Those exposed to more people who advocate breaking the law are more likely to become criminals.

At least three other factors impinge on the likelihood of becoming a deviant through differential association. One is the ratio of deviants to nondeviants to which one has been exposed. Also, a person is more likely to become a deviant if his or her significant others are deviants. Finally, the earlier one is exposed to deviants, the more likely one is to become a deviant.

How has cultural transmission theory been evaluated? Cultural transmission theory reveals that conformity in one setting may be considered deviant in another. Cultural transmission theory also explains why individuals placed in prison may become even more committed to deviance; in prison one is exposed almost exclusively to criminals. **Recidivism**—a return to crime after prison—makes sense in the light of cultural transmission theory because ex-convicts normally return to the significant others, primary groups, and reference groups through which they originally learned to be deviants.

Cultural transmission theory leaves some blank spaces. It cannot account for the reason some individuals become deviant without exposure to great numbers of people who advocate breaking the law. Middle-class delinquents are an illustration. Conversely, cultural transmission theory cannot say why most people reared in crime-ridden environments do not become deviants. Cultural transmission theory is limited in scope because it deals only with crime. Finally, although cultural transmission theory explains how deviance is learned, it fails to explain why some behaviors are considered deviant whereas others are not. An explanation of this relativity of deviance is a major strength of labeling theory.

Labeling Theory

What is labeling theory? Although strain theory and cultural transmission theory help us understand deviance, neither explains the relativity of deviance. Labeling theory attempts to fill this gap. In this theory, no act is inherently deviant; deviance is always a matter of social definition. According to **labeling theory,** deviance exists when some members of a group or society label others as deviants. Howard Becker, a pioneer of labeling theory, states the heart of this perspective:

> Social groups create deviance by making the rules whose infraction constitutes deviance, and by applying these rules to particular people and labeling them as outsiders. From this point of view, deviance is not a quality of the act the person commits, but rather a consequence of the application by others of rules and sanctions to an "offender." The deviant is one to whom that label has successfully been applied; deviant behavior is behavior that people so label (Becker, 1991:9).

Labeling theory allows us to understand the relativity of deviance. It explains, for example, why unmarried pregnant girls are more severely ostracized than the biological fathers. An illicit pregnancy requires two people, but only one of the pair is labeled a deviant. This labeling occurs because of cultural ideas about women and their sexuality. Women are to guard their sexual access. When they become pregnant out of wedlock, they have violated this norm and are considered deviant. Men are not considered as deviant, not because they do not literally bear the child but because our ideas about their sexuality (and their responsibility) are different. And, of course, it is easier to stigmatize women because advanced pregnancy is so visible. Labeling theory also explains why a middle-class youth who steals a car may well go unpunished for "borrowing" the vehicle, whereas a lower-class youth goes to court for stealing. Lower-class youth are expected to be criminals; middle-class youth are not.

SOCIOLOGY IN THE NEWS

Teenage Pregnancy as Deviance?

This CNN report highlights the problem of unmarried American teenagers having children. While the rate of teenagers giving birth has decreased in recent years, it is estimated that there are now three thousand new teenage mothers each day in the United States. By tradition, having a child out of wedlock has been considered deviant behavior in America because it violates a significant social norm.

1. Are teenage mothers still considered to be deviants? Explain.

_____

_____

2. Is race a major factor in how most Americans view unwed teenage mothers? Why or why not?

_____

_____

Pregnancy among unwed teenage mothers is a significant problem among whites and blacks in America today. While unwed motherhood is still thought to constitute deviance, young women no longer have to leave their communities in shame to hide their situation.

What are the consequences of labeling? Edwin Lemert's distinction between primary and secondary deviance helps clarify the consequences of labeling. In cases of **primary deviance**—isolated acts of norm violations—there is frequently relabeling and thus no long-term consequences. For example, when college students are asked to respond to a checklist of unlawful activities, most admit to having violated one or more norms. Yet, the vast majority of college students have never been arrested, convicted, or labeled as criminals. Certainly, those who break the law for the first time do not consider themselves criminals. If their deviance stopped at this point, they would have engaged in primary deviance; deviance would not be a part of their lifestyle or self-concept. Juveniles, likewise, may commit a few delinquent acts without becoming committed to a delinquent career or regarding themselves as delinquents.

Secondary deviance, on the other hand, refers to deviance as a lifestyle and personal identity. A secondary deviant is a person whose life and identity are organized around the facts of deviance (Lemert, 1972). In this case, the deviant status engulfs individuals and becomes a status that overshadows all other statuses. Individuals identify themselves primarily as deviants and organize their behavior largely in terms of deviant roles. Other people respond to them chiefly as deviants. When this occurs, these individuals usually begin to spend most of their time committing acts of deviance; deviance becomes a way of life, a career (Kelly, 1996). Secondary deviance is reflected in the words of Carolyn Hamilton-Ballard, known as Bubbles to her fellow gang members in Los Angeles:

Because of my size, I was automatically labeled a bully-type person. . . . I mean, people saw that Bloods jacket and since everybody thought I was crazy, I started acting crazy. At first it was an act, but then it became me. After being the target for drive-bys and going through different things, that became my life-style. I started retaliating back and I got more involved (Johnson, 1994:209).

The labeling of people as deviants can cause pain and suffering. Erving Goffman revealed some of the negative effects of labeling in his writings on **stigma**—undesirable characteristics used by others to deny the deviant full social acceptance. Prostitutes, ex-convicts, and the physically challenged are not considered normal. Why? Because stigmatic labels—such as *whore, jailbird,* and *cripple*—are used to spoil an individual's entire identity. One attribute—a missing leg, a stutter, a prison record—is used to form a general conception of an individual, and a stigma is used to discredit the individual's entire worth. The words of a 43-year-old bricklayer unemployed during the Depression illustrate this point.

How hard and humiliating it is to bear the name of an unemployed man. When I go out, I cast down my eyes because I feel myself wholly inferior. When I go along the street, it seems to me that I can't be compared with an average citizen, that everybody is pointing at me with his finger. I instinctively avoid meeting anyone. Former acquaintances and friends of better times are no longer so cordial. They greet me indifferently when we meet. They no longer offer me a cigarette and their eyes seem to say, "You are not worth it, you don't work" (quoted in Goffman, 1963:17).

Does labeling cause deviance? Labeling theory suggests that labeling is all that is required for an act to be deviant. Deviance, however, cannot be defined without some reference to norms or rules. It cannot be defined solely in terms of the reactions of others. If an act is not deviant unless it is labeled, there can be no such thing as undetected deviance.

Labeling theory also seems to imply that those who react against deviants are the culprits. Lawmakers do not cause people to commit criminal acts. It seems inappropriate to say that a police officer who arrests a person who has violated the law is somehow at fault. If the criminal law were abolished, acts that result in injuries would not disappear. In other words a "mugging by any other name hurts just as much" (Nettler, 1984).

Deviance cannot be defined without some reference to norms, but this does not mean that labeling processes are unimportant. Future behavior may be affected by the way others respond to an individual's deviance. In this sense, labeling produces deviants. To assign a person the label of delinquent, drug addict, or prostitute is a form of social penalty that may lead to rejection and exclusion from conventional groups and, as a consequence, may evoke further deviant acts. That is, labeling does lead to deviance when it produces secondary deviance.

Recall the study of the Roughnecks and the Saints, two teenage high school gangs. According to Chambliss, labeling the Roughnecks as deviants promoted additional acts of deviance, whereas most of the Saints—labeled as bound for bright futures—did become successful.

What has been the evaluation of labeling theory?

Labeling theory has contributed several insights. First, it has established that deviance is a matter of social definition, that deviance is relative. Second, labeling theory demonstrates, through concepts such as secondary deviance and stigma, the acquisition of a deviant self-concept, which is likely to lead to a career of deviancy. Third, labeling theory applies to a greater variety of types of deviance than the other theories. It can be applied to mental illness and prostitution as easily as to drug dealing and homosexuality.

Labeling theory also has its limitations. It does not explain habitual deviants who have never been detected and labeled. Nor does labeling theory deal with isolated acts of deviance (primary deviance). It is also possible that labeling may halt rather than create deviance, as in the case of a man who frequents child pornography movies until he is suddenly discovered and labeled by his friends. Also, with its accent on relativity, labeling theory tends to overlook the fact that some acts, such as murder and incest, are banned in virtually all societies. Finally, labeling theory tends to generate sympathy for deviants who are depicted as innocent victims of labeling by others.

Mental Illness and Labeling Theory

It is especially interesting to examine mental illness from the labeling perspective, because mental disorders, being a part of the medical field, are assumed to have an objective basis like cancer or arteriosclerosis. Labeling theory offers a unique way of viewing mental illness.

How does labeling theory approach mental illness?

By virtue of its foundation in symbolic interactionism, labeling theory views mental illness as the result of the process of social interaction in which others respond to us and we imagine what those responses mean. Mental illness is considered to be a matter of social definition (Collier, 1993).

According to psychiatrist Thomas Szasz (1986), mental illness, as defined by the medical community, is a myth. Szasz sees the behaviors associated with mental disorder as adaptations to interaction-based stresses threatening to overwhelm an individual. In response to "problems of living," people (primarily unconsciously) impersonate a sick person in order to obtain help from others.

Thomas Scheff (1974, 1984) has formulated a theory of mental illness (based on the labeling perspective) in which most psychiatric symptoms of mental illness are seen as violations of social norms. Because of temporary pressures, contends Scheff, anyone might exhibit behaviors that violate norms—withdrawing socially, becoming overly suspicious, muttering to oneself. This primary deviance is usually considered transitory and is ignored by others. If this rule breaking is exaggerated and distorted by others, it can lead to a label—mentally ill. If the process of labeling goes far enough, mental illness may become a career path; it may reach the stage of secondary deviance in which it becomes part of a person's self-concept. Thus, a person may be successfully labeled mentally ill by a social process when, in fact, the original deviance may have been only temporary acts to handle temporary pressures.

Scheff's theory has been challenged by critics who present findings that the prime factor determining the negative reaction of others to mental patients is the patients' behavior, not the label such people have acquired (Phillips, 1963, 1964; Link and Cullen, 1983). Some more recent research concludes that labels do matter and that labeling theory cannot be dismissed as a framework for understanding responses to the mentally ill (Link, 1987; Link et al., 1987).

CONFLICT THEORY AND DEVIANCE

Deviance in a Capitalist Society

Marxist sociologists have carried out the most important line of research and theory development on

FEEDBACK

1. Which of the following did Edwin Sutherland mean by differential association?
   a. Crime is more likely to occur among individuals who have been exposed more to unfavorable attitudes toward the law than to favorable ones.
   b. People become criminals through association with criminals.
   c. Crime is not transmitted culturally.
   d. Crime comes from differential conflict.
2. _____ theory is the only sociological theory that takes into account the relativity of deviance.
3. _____ refers to deviance as a lifestyle and personal identity.
4. _____ is an undesirable characteristic used by others to deny the deviant full social acceptance.
5. According to labeling theory, mental illness is a matter of _____.

Answers: 1. a. 2. Labeling 3. Secondary deviance 4. Stigma 5. social definition

deviance from a conflict perspective (Pelfrey, 1980; Quinney, 1980; Chambliss and Seidman, 1982). Focusing on class conflict, Marxist criminologists see deviance as a product of the exploitative nature of the ruling class. Deviance is simply behavior that the rich and powerful see as threatening to their interests. Consequently, the rich and powerful determine which acts are deviant and to what extent deviants should be punished.

According to the Marxian interpretation, the rich and powerful use the law to maintain their position. They can do this, argue the Marxists, because a society's beliefs, values, and norms (including, of course, laws) exist to buttress the interests of the ruling class.

By what means do capitalism and deviance become linked? Steven Spitzer (1980) identifies some basic ways in which the culture of a capitalist society supports the nature of the society's economic system. For one thing, people such as socialists are likely to be considered deviants because their beliefs challenge the economic, political, and social basis of capitalism. Second, because capitalism requires a willing workforce, those who will not work are considered deviants. Third, those who threaten private property, especially that belonging to the rich, are prime targets for punishment. Fourth, as discussed in Chapter 6 ("Groups and Formal Organizations"), formal organizations are based on a hierarchy of authority. Because of society's need for respect of authority, people who show a lack of respect for authority—agitators on the job, people who stage nonviolent demonstrations against established practices—are treated as deviants. Fifth, certain activities and characteristics are encouraged or discouraged depending on how well they mesh with the requirements of the economic system. Athletics are approved because they foster competition, achievement, teamwork, and winning. Males who are not athletic may be ostracized as "sissies" or "wimps."

The Marxian perspective is not the only way to understand deviance from the viewpoint of conflict theory. The link between race, ethnicity, crime, and punish-

ment is another way (Hagan, 1994a). White-collar crime is yet another (Coleman, 1987, 1989; Green, 1990).

Race, Ethnicity, and Crime

What is the relationship between race, ethnicity, and crime? Advocates of the conflict perspective point to the differential treatment minorities receive in the American criminal justice system. African Americans and Latinos are dealt with more harshly than whites at all points in the criminal justice process—arrest, indictment, conviction, sentencing, and parole (Schaefer, 1993). African Americans and Latinos are more likely than whites to be convicted of the same crimes, and they serve more time in prison than whites for the same criminal offenses (Huizinga and Elliott, 1987; Bridges and Crutchfield, 1988; Klein, Turner, and Petersilia, 1988). This differential treatment within the legal system is reflected in the fact that African Americans constitute a significantly larger proportion of those in prison with death sentences. Although African Americans account for only 13 percent of the total population in the United States, more than 43 percent of inmates under the death penalty are African American. This situation cannot be explained by the fact that the murder rate is higher among African Americans than whites—51 percent versus 46 percent (U.S. Bureau of the Census, 1996a).

Consider some illustrations of the differentially negative treatment African Americans receive in the criminal justice system. During the four years following 1977, when the death penalty was legalized again in the United States, state attorneys were twice as likely to ask for the death penalty when an African American was found guilty of murdering a white person than when an African American murdered an African American or a white killed a white (McManus, 1985). Although more than one-half of all homicide victims in the United States are African American, the overwhelming majority of prisoners on death row are there through convictions of murdering whites (U.S. Bureau of the Census, 1996a). Prosecutors are less likely to ask for the death penalty when an African American has been killed, and juries and judges less prone to impose the death penalty in cases involving African American victims.

How can this differential treatment of minorities within the legal system be explained? First, to conflict theorists, minorities' interests appear to be less important than the interests of whites. Crimes committed against minority members are more likely to be reduced from felonies to misdemeanors. This process has been labeled *victim discounting*—reducing the seriousness of crimes based on the lower social value placed on the worth of the victims (Gibbons, 1985). The logic behind victim discounting states that if the victim is less valuable, the crime is less serious, and the penalty is less severe. Second, conflict theorists point to the fact that minorities generally do not have the economic resources to buy higher quality legal protection. Consequently, minorities have a higher probability of being arrested, convicted, and more severely punished.

White-Collar Crime

What is white-collar crime? According to Edwin Sutherland (1940, 1983), the originator of the term, **white-collar crime** is any crime committed by respectable and high-status people in the course of their occupations. Lower status people commit crimes of the streets; higher status people engage in "crimes of the suites" (Nader and Green, 1972). Though higher status individuals may commit murder or rape, the term *white-collar crime* is reserved for such acts as price fixing, insider trading, illegal rebates, embezzlement, bribery of a corporate customer, manufacture of hazardous products, toxic pollution, and tax evasion (Rowen, 1991; Schlegel and Weisbund, 1994; Geis et al., 1995).

How can white-collar crime be viewed from the conflict perspective? Advocates of the conflict perspective point out that white-collar crime is extremely harmful to society. Its perpetrators, however, are treated more leniently than other criminals because of their class position.

The costs of white-collar crime are higher than is popularly known. According to the U.S. Department of Justice, white-collar crime costs run up to $200 billion annually, an amount exceeding the costs of street crime by eighteenfold. Illegal working environments (for example, factories that expose workers to toxic chemicals) are judged to account for about one-third of all work-related deaths in the United States. Five times more Americans are killed each year from illegal conditions on their job than are murdered on the streets. None of these figures, of course, reflect the costs to society of a demoralized citizenry constantly bombarded by news of criminal acts by corporate, political, and religious leaders (Simon and Eitzen, 1982; Hagan and Palloni, 1986).

Despite this harm, much more tolerance is shown white-collar criminals than criminals in the lower classes. Penalties are both tougher and more likely to be imposed for crimes committed by lower-class people than for those committed by people in higher social classes. Detected drug-law violators (more likely to be from the lower class), for example, are treated more harshly than embezzlers. Although fewer than four hundred embezzlers are in prison, over eight thousand drug-law violators are incarcerated. In federal court, where most white-collar cases are tried, probation is granted to 40 percent of antitrust-law violators, 61 percent of fraud defendants, and 70 percent of embezzlers. In general, convicted white-collar criminals are less likely to be incarcerated and receive shorter average sentences if they are sentenced to prison terms. Moreover, when sentenced, white-collar criminals are more likely to be placed in a prison with extra amenities (for example, tennis courts or private rooms), which critics have dubbed "Club Fed" (Gest, 1985; U.S. Department of Justice, 1987; Reiman, 1990). Pablo Escobar Gaviria, head of the Colombian Medellin cocaine cartel who surrendered to his government in 1991, served time in a palatial compound that could only laughingly be called a prison.

What are the strengths and weaknesses of the social conflict approach to deviance? Of course, not all definitions of deviance can be attributed to the exploitation of the weak by the powerful. Laws against murder and rape are meant to protect both the rich and the poor. Moreover, some laws—consumer and environmental laws—exist despite the opposition of big business. Most proponents of the conflict perspective would not conclude that deviance exists only because of cultural and social arrangements supportive of the ruling class. Nevertheless, the conflict approach to deviance, like labeling theory, underscores the relativity of deviance. Perhaps the greatest strength of conflict theory is its ability to link deviance to social inequality and power differentials.

CRIME IN THE UNITED STATES

Crime, the most important type of deviance in any modern society, is a major problem in America. Many Americans think of crime—acts in violation of the law—as including a narrow range of behavior. On the contrary, there are more than 2,800 acts considered crimes by federal law alone, and many more acts violate state and local statutes. If a society is to prevent and control crime, it must be aware of the extent of its occurrence.

1. Marxist criminologists see deviance as a product of the exploitative nature of the _____.
2. Which of the following is *not* one of the basic ways the culture of a capitalist society supports the nature of the society's economic system?
   a. People whose beliefs clash with capitalism are labeled deviants.
   b. Capitalism requires a willing workforce.
   c. Innovation is rewarded.
   d. People who fail to show respect for authority are likely to be considered deviant.
3. The process of reducing the seriousness of crimes based on the lower social value placed on the worth of victims is called _____.
4. _____ is any crime committed by respectable and high-status people in the course of their occupations.

Answers: 1. ruling class 2. c 3. victim discounting 4. White-collar crime

The Extent of Crime

The major source of statistics on crime is the FBI's *Uniform Crime Reports* (U.S. Department of Justice, 1996). These official statistics are gathered by the FBI from police departments across the country. During 1992, reports were submitted voluntarily by law enforcement agencies serving 95 percent of the nation's population.

Data on arrests are gathered for twenty-nine crime categories, and information on crimes known to the police is collected for eight index crimes: murder, forcible rape, robbery, aggravated assault, burglary, larceny-theft, motor vehicle theft, and arson. In 1995, the number of index crimes known to the police totaled 13,867,143. Violent crime—murder, forcible rape, aggravated assault, and robbery—makes up about 13 percent

of the known crimes, and property crime—burglary, larceny-theft, motor vehicle theft—accounts for about 87 percent, as shown in Table 7.2. Figure 7.1 shows the *Uniform Crime Reports* estimate of the frequency at which crimes occur in the United States. Figure 7.2 displays the percent distribution of crime in America.

The Measurement of Crime

Are the *Uniform Crime Reports* accurate? Although they provide considerable information about crime, the FBI statistics have serious limitations. The official statistics are widely believed to be biased. The lower classes and minorities are overrepresented, whereas the middle and upper classes are undercounted. Thefts of inexpensive items and assaults that do not result in serious injury are not as likely to be reported to the police as

TABLE 7.2	Crimes in the United States			
	1995		**1991–1995**	
Types of Crime	**Number of Crimes**	**Crime Rate***	**Percent Change in Crime Rate**	**Percent Change in Number of Crimes**
Violent crime	1,798,785	684.6	−9.7	−5.9
Murder	21,597	8.2	−16.3	−12.6
Forcible rape	97,464	37.1	−12.3	−8.6
Robbery	580,545	220.9	−19.0	−15.6
Aggravated assault	1,099,179	418.3	+0.6	−3.5
Property crime	12,068,358	4593.0	−10.6	−6.9
Burglary	2,594,995	987.6	−21.1	−17.8
Larceny-theft	8,000,631	3044.9	−5.7	−1.7
Motor vehicle theft	1,472,732	560.5	−14.9	−11.4
Arson**				

*Per 100,000 inhabitants.
**No figures for arson are available for 1995 for the total United States.
Source: U.S. Department of Justice, Federal Bureau of Investigation, *Uniform Crime Reports, 1995* (Washington, DC: U.S. Government Printing Office, 1996).

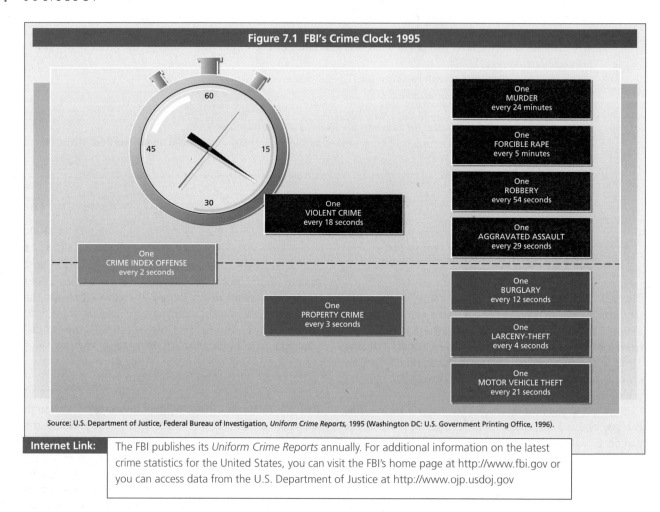

Figure 7.1 FBI's Crime Clock: 1995

One
MURDER
every 24 minutes

One
FORCIBLE RAPE
every 5 minutes

One
ROBBERY
every 54 seconds

One
AGGRAVATED ASSAULT
every 29 seconds

One
VIOLENT CRIME
every 18 seconds

One
CRIME INDEX OFFENSE
every 2 seconds

One
PROPERTY CRIME
every 3 seconds

One
BURGLARY
every 12 seconds

One
LARCENY-THEFT
every 4 seconds

One
MOTOR VEHICLE THEFT
every 21 seconds

Source: U.S. Department of Justice, Federal Bureau of Investigation, *Uniform Crime Reports,* 1995 (Washington DC: U.S. Government Printing Office, 1996).

Internet Link: The FBI publishes its *Uniform Crime Reports* annually. For additional information on the latest crime statistics for the United States, you can visit the FBI's home page at http://www.fbi.gov or you can access data from the U.S. Department of Justice at http://www.ojp.usdoj.gov

The Crime Clock should be viewed with care. Being the most aggregate representation of UCR (*Uniform Crime Reports*) data, it is designed to convey the annual reported crime experience by showing the relative frequency of occurrence of the Index Offenses. This mode of display should not be taken to imply a regularity in the commission of the Part I Offenses; rather, it represents the annual ratio of crime to fixed time intervals.

murders and auto thefts. Prostitutes and intoxicated persons are subject to arrest in public places but are relatively safe in private settings, where the police cannot enter without a warrant. Also, about two-thirds of crimes in the United States are not reported to the police. In addition, reporting of crime varies from jurisdiction to jurisdiction and from crime to crime. Few violations committed by government officials, corporate executives, and other occupational offenders are included in official crime statistics. Finally, in such areas as unfair advertising and job discrimination, the laws are enforced through special administrative boards, so the criminal process is rarely used in these areas (Box, 1981; Wright, K., 1987; Ermann and Lundman, 1996; Coleman, 1997).

In response to criticisms of the *Uniform Crime Reports,* the National Crime Survey was launched in the early 1970s. A semiannual crime survey is conducted

for the Bureau of Justice Statistics by the U.S. Census Bureau, which uses state-of-the-art survey research techniques. (U.S. Department of Justice, 1997). This annually published survey of crime victimization contradicts the *Uniform Crime Reports*. For example, the National Crime Survey indicates that the FBI statistics underestimate the total amount of crime in the United States. Although the FBI statistics reported just under fourteen million crimes in 1994, the National Crime Survey set criminal victimizations at 42.4 million.

Whom are we to believe? There is no completely accurate accounting of crime in the United States. Both the *Uniform Crime Reports* and the National Crime Survey have their unique strengths. Because most people know relatively little about the technical definitions of crime, a major strength of the *Uniform Crime Reports* is the publication's use of legally experienced police

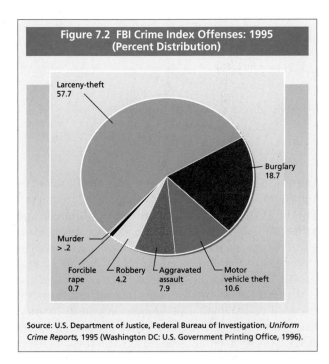

Figure 7.2 FBI Crime Index Offenses: 1995 (Percent Distribution)

Larceny-theft 57.7

Burglary 18.7

Murder > .2

Forcible rape 0.7

Robbery 4.2

Aggravated assault 7.9

Motor vehicle theft 10.6

Source: U.S. Department of Justice, Federal Bureau of Investigation, *Uniform Crime Reports*, 1995 (Washington DC: U.S. Government Printing Office, 1996).

is to compare crime rates in the United States with those of other countries.

Global Crime Comparisons

How do the levels of crime in other countries compare with crime in the United States?   Data on crime rates in the 1990s reveal some striking differences between the United States and other countries (Kalish, 1988; MacKellar and Yanoghishita, 1995). The following specific statements can be made:

- The U.S. homicide rate per 100,000 population is 10.0; the rate of homicide in Europe is less than 2 per 100,000.
- The U.S. rape rate was 36 per 100,000, roughly seven times higher than the average for Europe.
- U.S. rates for robbery at 205 per 100,000 compared to the average European rate of less than 50 per 100,000.
- For crimes of theft and auto theft, a comparison of U.S. rates to average European rates was roughly two to one.
- Burglary was the only crime examined for which the U.S. rate was less than double the average for European countries.
- U.S. crime rates were also higher than those of Canada, Australia, and New Zealand, but the differences were smaller compared to Europe. For burglary and auto theft the rates were quite similar (Hagan, 1994a:70).

Clearly, Americans watching the evening news do not have to fear being deceived by sensationalizing reporters telling them of the homicides, rapes, and robberies that seem to be occurring everywhere in the country. There is simply a lot of crime in the United States. Of the major industrialized countries, the United States has the highest murder, rape, and robbery rates, and keeps pace in burglaries and auto thefts. Despite the rise in crime rates in Europe during the 1990s, crime remains a significantly greater problem in America than in any country.

Are these cross-cultural data really to be trusted?   You read earlier in this chapter about the difficulties associated with gathering accurate information on crime

officers to decide whether an act has violated a statute. The National Crime Survey has two attractive advantages. First, it helps to compensate for the massive official underreporting of crime. Second, the semiannual surveys are very sound methodologically. At the very least, the National Crime Survey is an increasingly important supplement to the FBI's official statistics. Together they provide a more complete account of the extent and nature of crime in the United States than either alone could provide (Wright, 1987; U.S. Department of Justice, 1997).

GLOBAL DIFFERENCES IN CRIME

Crime rates and the incidence of crime vary within any society. Sociological factors such as gender, age, race, ethnicity, social class, geographic region, and neighborhood each contributes to variations in the extent of crime. Crime rates also vary from society to society. One way to examine cross-cultural differences in crime

FEEDBACK

1. According to the FBI's *Uniform Crime Reports,* crime in the United States has increased since 1991. *T or F?*

2. According to the National Crime Survey, the *Uniform Crime Reports* underestimates or overestimates the amount of crime in America? _____

Answers: 1. F 2. underestimates.

1. In which of the following categories of crime is the United States *not* the leader in the world?
   a. murder
   b. rape
   c. robbery
   d. burglary
2. Crime in Japan supports/refutes a predicted positive relationship between modernization and crime.

Answers: 1. d. 2. refutes

within the United States. It is not unreasonable to question the accuracy of data coming from so many different cultures. There are problems associated with data recording, data collection, and varying definitions of crime from one country to another. Despite some notable dissenters (Balrig, 1988; Gerber, 1991), experts generally agree that the differences in crime rates across societies are too great and too consistent to be denied credibility.

Crime and Modernization

According to one line of argument, increased crime is associated with economic development (Fenwick, 1982). If this is the case, then the world's economic leaders will predictably have the highest levels of crime. This prediction, however, is contradicted by the relatively low rates for most types of crimes in such highly industrialized countries as Japan, Germany, and Switzerland.

APPROACHES TO CRIME CONTROL

A variety of organizations have been created to control crime. The **criminal justice system** is comprised of police, courts, and correctional system (Inciardi, 1996). Criminal justice systems may draw on four approaches to the punishment of criminals for purposes of crime control—deterrence, retribution, incapacitation, and rehabilitation. These approaches may be used in varying combinations.

Deterrence

The **deterrence** approach to punishment emphasizes the intimidation of members of society into compliance with requirements of the legal system. There is a debate on the effectiveness of deterrence (DiIulio and Piehl, 1991).

Does punishment deter crime? Until the mid-1960s, social scientists generally rejected the idea that punishment deters crime. This rejection was based partly on earlier studies, which indicated that the death penalty

did not deter murder any more effectively than life imprisonment. More recent research has led to a reevaluation of the relationship between punishment and crime. Although the evidence to support this point is inconclusive, social scientists may have discounted the deterrence doctrine prematurely (Greenberg and Kessler, 1982; Logan, 1982; Hook, 1989).

Investigation of the complicated relationship between punishment and crime is only in its infancy, but recent research indicates that punishment does deter crime if potential lawbreakers feel they are likely to get caught and if the punishment for crime is severe. The punishment for crime, however, is usually not certain, swift, or severe. Consequently, punishment does not have the deterrent effect that it would have if criminals knew that they were likely to be caught and if they knew that the punishment for their crimes would be severe (Pontell, 1984).

Capital punishment is a special case. Murder is usually committed not after a rational consideration of the act or its consequences but during an outburst of emotion. Under such circumstances, one would not expect the fear of capital punishment to be a deterrent, and research shows that it is not. If the death penalty were a deterrent to murder, a declining use of it should be followed by an increase in the murder rate. Research in many countries, including the United States, however, indicates that the murder rate remains constant or declines even when the trend is away from the use of the death penalty. Other research indicates that the use of capital punishment as a sanction is neither swift nor certain (Fattah, 1981; Andersen, 1983; Bailey and Peterson, 1989; Sellin, 1991).

Despite those findings, about three-fourths of Americans believe that the death penalty acts as a deterrent to murder. Actually, attitudes regarding the deterrent ability of the death penalty do not seem to affect attitudes toward the death penalty itself. Of those Americans who favor the death penalty, over three-fourths indicate they would continue to favor it even if confronted with conclusive evidence that the death penalty does not act as a deterrent to murder and that it does not lower the murder rate (Longmire, 1996). Feelings of revenge and a desire for retribution, then, appear to contribute more to support of capital punishment than its deterrent effects. When asked to

There is a debate among social scientists regarding the effectiveness of imprisonment on the deterrence of crime. Whether from model prisons like this one or from more primitive prisons, between 30 and 60 percent of released inmates are reincarcerated within two to five years.

choose, a significantly higher proportion of the American population supports the death penalty for murder (71 percent) than opposes it (19 percent; Longmire, 1996).

As shown in Figure 7.3, 77 percent of white Americans favor the death penalty as the punishment for murder compared with 40 percent of African Americans and 52 percent of Latinos. This is not surprising given the earlier discussion of race, ethnicity, and crime within the context of conflict theory. Because African Americans and Latinos convicted of murder are more likely to receive the death penalty than whites, their greater opposition to capital punishment is to be expected (Spohn, 1995).

Retribution

What is the philosophy behind the retributive approach to criminal punishment? The kernel of **retribution** is the public demand from criminals of compensation equal to their offense against society. An eye taken requires that an eye be returned. If an individual whose husband is a victim of homicide exacts the price of wrongdoing by shooting her husband's killer, she is acting in revenge. Although legal systems support the idea of compensation, they do not easily tolerate acts of vengeance. The legal system removes responsibility for the punishment of a criminal from the hands of individuals and places it in the hands of

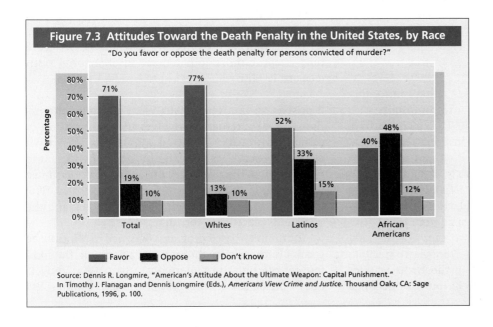

Figure 7.3 Attitudes Toward the Death Penalty in the United States, by Race

"Do you favor or oppose the death penalty for persons convicted of murder?"

Favor · Oppose · Don't know

Source: Dennis R. Longmire, "American's Attitude About the Ultimate Weapon: Capital Punishment." In Timothy J. Flanagan and Dennis Longmire (Eds.), *Americans View Crime and Justice*. Thousand Oaks, CA: Sage Publications, 1996, p. 100.

officially recognized representatives of society. A convicted criminal repays the state rather than the harmed individual(s). Retribution, in the name of justice, is a social rather than a personal matter.

Legal action carried out strictly under the retributive philosophy is concerned with past rather than future crime. Like deterrence, the incapacitation and rehabilitation approaches to punishment, to which we now turn, are centered on the prevention of crime.

Incapacitation

What is the incapacitation justification for punishment? In 1994, Michael Fay, an American teenager convicted of car vandalism, was lashed four times with a four-foot rattan cane in a Singapore prison (Elliott, 1994). The experience of being "caned" is said to be sufficiently bad that future crime is effectively discouraged. As Figure 7.4 reveals, Singapore is by no means alone in the use of corporal punishment on criminals. Practices such as caning are designed to protect society through deterrence. The incapacitation justification for punishment of criminals, also intended to protect society, is quite different. **Incapacitation** is the removal of criminals from society. Criminals who are not on the street cannot commit crimes against citizens on the outside.

Incapacitation can be an answer to crime control in itself. The prison door can be locked and the key figuratively thrown away. In societies during extremely repressive eras, such as the former Soviet Union under Communist rule, prisoners may spend their entire lives in prison camps for infractions ranging from murder to political opposition. The incapacitation philosophy, of course, may take less extreme forms. The United States in recent years, for example, has placed increased emphasis on incapacitation. An exploration of America's experience with rehabilitation sheds some light on the effectiveness of incapacitation in more moderate forms.

Rehabilitation

Rehabilitation is the approach to crime control that attempts to resocialize criminals into conformity with the legal code of a society. Rehabilitative efforts have traditionally been made in penal institutions.

Do prisons rehabilitate criminals? It is difficult to believe that past approaches to rehabilitation in the United States have been successful, considering that 30 percent to 60 percent of those released from penal institutions will be back in prison two to five years later (Allen et al., 1981; Greenfeld, 1985; Zawitz, 1988; Elikann, 1996). Violent crime in the United States increased between 1982 and 1992, but has declined

somewhat through 1995, and the prison population tripled between 1982 and 1995.

Such high rates of recidivism (past offenders returning to prison) can be attributed to characteristics of the offenders or the problems they face as a result of having come into contact with the criminal justice system, such as the stigma of being an ex-convict or the difficulty of securing a job. The tendency toward recidivism can also be partially attributed to the inadequacies of prisons. When the problem is viewed from the standpoint of differential association theory, a major inadequacy of prisons is the institutions' provision of an ideal setting for learning unfavorable attitudes toward the law and law-abiding behavior. Prisons are, in effect, schools for crime in the sense that ex-convicts often leave prison as more committed criminals than when they entered, with a higher likelihood of continued criminal involvement after release.

Social scientists also recognize that it is difficult to change attitudes and behavior in an isolated, artificial prison setting in which the prison subculture blocks efforts to rehabilitate. Although prison guards and officials hold power over inmates, the informal rules of the prison subculture have an even greater effect on prisoners' behavior. Conformity with the "inmate code," which stresses loyalty among inmates and opposition to correctional authorities, is enforced verbally and physically. There is evidence that among minority group inmates, defiance and involvement in criminal networks increase in prison, leading to a continued life of crime after release (Hagan, 1993; Sherman, 1993a).

The probability of continued criminal activity after release from prison is even higher because of the changing nature of employment in today's economy. Extended separation from a rapidly changing society leads to the obsolescence of employable job skills and to the deterioration of the network of contacts on the outside for finding jobs. Also, the toughness reinforced in prison life in the past transferred to manufacturing jobs. This transfer no longer works in the service economy because more jobs require interpersonal skills (Hagan, 1994b).

If prisons do not rehabilitate, what are some alternatives? One alternative involves the combination of prison and probation. A mixed or split sentence, known as *shock probation,* is designed to shock offenders into recognition of the realities of prison life. Prisoners must serve part of their sentence in an institution. The remainder of the sentence is then suspended, and the prisoner is placed on probation.

Community-based programs designed to reintroduce criminals into society through participation in the community offer another alternative. By getting convicts out of prison for at least part of the day, community-

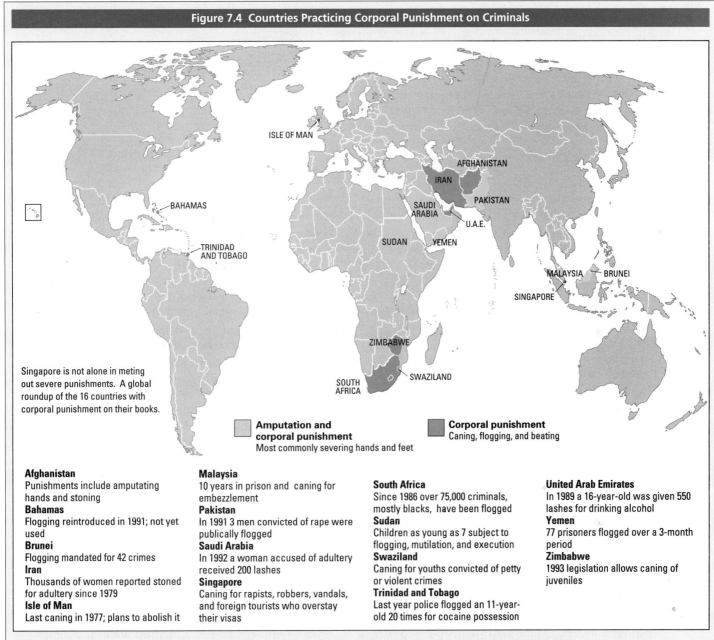

Figure 7.4 Countries Practicing Corporal Punishment on Criminals

Singapore is not alone in meting out severe punishments. A global roundup of the 16 countries with corporal punishment on their books.

Amputation and corporal punishment Most commonly severing hands and feet	**Corporal punishment** Caning, flogging, and beating

Afghanistan
Punishments include amputating hands and stoning
Bahamas
Flogging reintroduced in 1991; not yet used
Brunei
Flogging mandated for 42 crimes
Iran
Thousands of women reported stoned for adultery since 1979
Isle of Man
Last caning in 1977; plans to abolish it

Malaysia
10 years in prison and caning for embezzlement
Pakistan
In 1991 3 men convicted of rape were publically flogged
Saudi Arabia
In 1992 a woman accused of adultery received 200 lashes
Singapore
Caning for rapists, robbers, vandals, and foreign tourists who overstay their visas

South Africa
Since 1986 over 75,000 criminals, mostly blacks, have been flogged
Sudan
Children as young as 7 subject to flogging, mutilation, and execution
Swaziland
Caning for youths convicted of petty or violent crimes
Trinidad and Tobago
Last year police flogged an 11-year-old 20 times for cocaine possession

United Arab Emirates
In 1989 a 16-year-old was given 550 lashes for drinking alcohol
Yemen
77 prisoners flogged over a 3-month period
Zimbabwe
1993 legislation allows caning of juveniles

Source: Michael Elliott, Crime and Punishment, *Newsweek* (April 18, 1994):20. Data provided by Amnesty International and *Newsweek* reporting. Reprinted by permission.

based programs help break the inmate code. At the same time, prisoners have a chance to become part of society under professional guidance and supervision.

Another approach is known as *diversion*. This strategy is aimed at preventing or greatly minimizing the offender's penetration into the criminal justice system. Diversion involves a referral to a community-based treatment program rather than to a penal institution or a probationary program. By handling offenders outside the formal system of criminal law, authorities believe that offenders will not acquire stigmatizing labels and

other liabilities stemming from formal processing (Musheno, 1982; Rojek and Erickson, 1981–1982; Morris and Tonry, 1990).

Whether increased use of these alternatives will be successful in combating crime and delinquency remains to be seen. Most of the programs are relatively recent and have not yet been adequately evaluated. One of the major determinants of the continued or expanded use of these innovative programs will be the country's idea of the proper functions of prisons. These programs can exist only as long as rehabilitation has a

1. Match the approaches to punishment of criminals on the left side with the examples on the right.

    _____ a. rehabilitation   (1) imprisonment without parole

    _____ b. deterrence   (2) the practice of caning

    _____ c. retribution   (3) employment and educational programs in prison

    _____ d. incapacitation   (4) death penalty for murder

2. Contrary to the 1960s, prisons are now rehabilitating inmates. *T or F?*
3. It appears that social scientists have prematurely discounted the deterrence doctrine. *T or F?*
4. Research supports those who believe in capital punishment as a deterrent to murder. *T or F?*

Answers: 1. a. (3) b. (2) c. (4) d. (1) 2. F 3. T 4. F

high priority, support for which may be eroding because of the harsher view the American public is taking toward criminals (Stinchcombe et al., 1980; Gallup and Newport, 1990a).

CRIME CONTROL: DOMESTIC AND GLOBAL

Crime Control in the United States

A society could adopt various combinations of the four justifications for punishment of criminals. Singapore, for example, takes deterrence, retribution, and incapacitation very seriously. The upside is that Singapore has safe streets and clean subways. The downside is a reduction in individual liberty and increase in brutality toward prisoners, at least in the eyes of many Americans.

What is the approach to crime control reform in the United States? The United States does not have a consistent commitment to any of the major approaches to crime control (Bouza, 1993). Public opinion and public policy have been split in modern times between the less punitive approach of rehabilitation and the more punitive approaches of deterrence, retribution, and incapacitation. Currently, public pressure has politicians from the White House to the county courthouse publicly advocating more punitive steps associated with incapacitation such as "three strikes and you're out" (Booth, 1994; Shichor and Sechrest, 1996). These politicians, of course, are referring to life in prison without any possibility of parole for three-time criminal offenders (Morin, 1994). Public and political concern about violent crime has led to some new penal policies that reflect a turn from rehabilitation toward a more punitive approach to crime control. The approach taken involves more emphasis on incapacitation as a method of deterring crime and gaining retribution from criminals. During the 1980s, the prison population nearly tripled, standing at nearly 900,000 in 1992.

Even recent rehabilitative initiatives have been harsher. Boot camps, for example, have been created for young nonviolent first offenders. The tough, military-style environment is intended to resocialize these young people by instilling self-discipline, commitment to work, and respect for the law (Katel, 1994).

Although some of the $30 billion in the 1994 national crime bill is devoted to a variety of crime prevention programs such as the hiring of 100,000 more police, youth programs, and the banning of nineteen semiautomatic assault-type weapons, it also has a strong punitive accent. Funds will be made available to assist states in building new prisons. The legislation vastly increases the number of federal crimes subject to the death penalty and imposes mandatory life sentences for repeat violent offenders.

A cross-cultural comparison in approaches to crime control will be informative. It reveals some striking differences between the United States and other countries.

Global Crime Control

All nations of the world use incapacitation as an approach to crime control. However, significant variation exists in the extent of imprisonment across cultures (Christie, 1994). Even countries in Western Europe have varying imprisonment rates. England and Wales are near the top in inhabitants per 100,000 in prison. Germany, France, and Spain are near the middle, and Greece and the Netherlands are near the bottom. The imprisonment rate in the United States is more than four times that of any Western European country, standing at 504 inhabitants per 100,000.

Christie also gathered data on Eastern European countries. Currently, the United States stands out compared with these countries as well. For example, in Russia, the country's prosecutor-general estimates that two out of every three crimes are not even reported, and that a relatively small proportion of those reported are solved. Partly for these reasons, Russia's imprisonment rate is significantly lower than that of the United States ("Money Kills", 1996).

Not all industrialized countries have moved in the more punitive direction. In Sweden, emphasis remains on rehabilitation, treatment, and job training. Among

1. Toward which approach to crime control has the United States been moving?
   a. rehabilitation
   c. incapacitation
   b. deterrence
   d. retribution
2. Of these Western European countries, which has the highest rate of imprisonment?
   a. Spain
   c. Germany
   b. England and Wales
   d. Netherlands
3. The imprisonment rate in the United States is four times that of any Western European country. *T or F?*

Answers: 1. c 2. b 3. T

the Japanese, moral responsibility is placed on criminal offenders who are resocialized to recognize their moral shortcomings (Feeley and Simon, 1992). This stands in stark contrast to the current warehousing approach in the United States.

A FINAL NOTE ON CRIME CONTROL

Because the ineffectiveness of the incapacitation approach in the United States has already been docu-

mented in this chapter, the trend toward higher rates of imprisonment seems surprising and counterproductive. In fact, evidence suggests that the most effective approach in the juvenile justice system involves job-training programs outside penal institution walls (Lipsey, 1991). It will be interesting to watch American public opinion and political actions in the future as a skyrocketing amount of tax dollars are spent on an unpromising incapacitation approach to crime control.

SUMMARY

1. Although social life requires conformity and order, it also involves deviance, or the violation of significant social norms. Deviance is difficult to define because it is a relative matter. What is considered deviant depends on the characteristics of the individuals, the social context, and the historical era.

2. All societies need some means of social control to promote conformity to their rules. Internal social control, implanted through the socialization process, enables individuals to control their own behavior. External social control is applied by other people.

3. According to biological explanations of deviance, a link exists between physical characteristics and deviant behavior. Early on, Cesare Lombroso believed that criminals were individuals who were throwbacks to an earlier stage of human evolutionary development. William Sheldon attributed crime to body shape. Despite later research efforts, there is still no convincing proof that genetic characteristics cause people to be deviant.

4. Psychological theories of deviance attribute deviance to pathological personality traits. These traits include unconscious mental conflicts, hostility, extroversion, and IQ. Sociologists remain as skeptical of psychological theories of deviance as they are of those rooted in biology.

5. According to functionalists, deviance has both negative and positive consequences for society. On the negative side, deviance encourages social disorder, erodes trust, encourages further nonconformity in others, and diverts

resources from other social needs. On the positive side, deviance may help clarify norms, offer a safety valve, increase social unity, and bring about needed social change. Strain theory holds that deviance occurs because of a discrepancy between cultural goals and socially acceptable means of achieving those goals. Control theory attempts to explain conformity rather than deviance. Conformity, which, of course, excludes deviance, is based on the existence of a strong bond between individuals and society.

6. The symbolic interactionist perspective has yielded two theories of deviance. Cultural transmission theory contends that deviance is learned just like any other aspect of culture. According to labeling theory, an act is deviant only if other people respond to it as if it were deviant. Isolated norm violation, or primary deviance, may have no serious consequences for individuals. If individuals are labeled as habitual deviants, however, they may organize their lives and personal identities around deviance. Mental illness illustrates this theory.

7. The conflict perspective has also been applied to the study of deviant behavior. This perspective on deviance emphasizes social inequality and power differentials. That is, the most powerful members of a society determine the group norms and, consequently, who in the group will be regarded as deviant. The conflict perspective has been effectively used by criminologists who relate deviance to capitalism. Conflict theorists also point

to the relationship between race, ethnicity, and crime and to white-collar crime.

8. The official crime statistics—the *Uniform Crime Reports* published annually by the FBI—are useful in providing an estimate of the extent of crime, but they are underestimations. Because of limitations in the *Uniform Crime Reports,* the National Crime Survey has been launched. A problem exists in the inconsistency of the findings of these two sources.

9. The United States has the greatest crime problem in the world. The gap between the United States and other countries remains substantial, despite the rising crime rates in Europe during the 1990s.

10. The four approaches to crime control are deterrence, retribution, incapacitation, and rehabilitation. Deterrence is intended to intimidate people into legal conformity. Compensation equal to the criminal offense lies at the heart of the retributive approach. Incapacitation involves removal of criminals from society. Rehabilitation aims at

resocializing criminals into law-abiding citizens. The United States is torn between the less punitive approach of rehabilitation and the more punitive approaches of deterrence, retribution, and incapacitation.

11. It is increasingly recognized that prisons do not really rehabilitate criminals. In the more distant past, attempted reform took place within prisons. In recent years there has been an interest in rehabilitating criminals outside prisons. During the 1970s, Americans began to take a harsher view of criminals. Even social scientists are beginning to believe that punishment can deter crime under certain circumstances, although no convincing evidence exists that capital punishment affects the murder rate.

12. Of late, the United States has increasingly moved toward incapacitation as a solution to crime control. The imprisonment rate in the United States is currently over four times that of any Western European country and nearly double that of the former Soviet Union at the beginning of the 1990s.

LEARNING OBJECTIVES REVIEW

After careful study of this chapter, you will be able to:

- Define deviance and describe the major factors promoting the relativity of deviant behavior.
- Define social control and identify the major types of social control.
- Describe the biological and psychological explanations of deviance.
- Discuss the positive and negative consequences of deviance.

- Differentiate the major functional theories of deviance.
- Compare and contrast cultural transmission theory and labeling theory.
- Discuss the conflict theory view of deviance.
- Distinguish the four approaches to crime control.

CONCEPT REVIEW

Match the following concepts with the definitions listed below them:

____ a. labeling theory
____ b. differential association theory
____ c. retribution
____ d. control theory

____ e. social control
____ f. anomie
____ g. white-collar crime
____ h. cultural transmission theory

____ i. deviance
____ j. recidivism

1. Behavior that violates significant social norms.
2. A social condition in which norms are weak, conflicting, or absent.
3. The conception that conformity is based on the existence of a strong bond between individuals and society.
4. The theory that deviance is a part of a subculture transmitted through socialization.
5. The public demand of compensation from criminals equal to their offense against society.
6. The theory that crime and delinquency are more likely to occur among individuals who have been exposed to

more unfavorable attitudes toward the law than to favorable ones.
7. The theory that deviance exists when some members of a group or society label others as deviants.
8. Means for promoting conformity to a group's or society's rules.
9. A crime committed by respectable and high-status people in the course of their occupations.
10. A return to crime after imprisonment.

CRITICAL THINKING QUESTIONS

1. Identify a time in your life when you were considered a deviant by others. Discuss the attempts at social control you experienced. Be specific as to the types of social control and their concrete application to you.

   _____

   _____

2. Name some type of deviant behavior you would like yourself and/or others to engage in because you think society would benefit from it. State the sociological case for the social benefit of this type of deviance.

   _____

   _____

3. Describe someone you know who falls into one of four deviant responses identified by strain theory. Use specific characteristics of this person to show that he or she fits the response you selected and not the other three responses.

   _____

   _____

4. Everyone has observed someone who has been labeled deviant by some members of society. Discuss the consequences of this labeling for the person identified as a deviant.

   _____

   _____

5. The text outlined several distinct approaches to crime control. Choose one approach and explain why you believe it has been successful or unsuccessful in terms of functionalism, conflict theory, or symbolic interactionism.

   _____

   _____

MULTIPLE CHOICE QUESTIONS

1. **Behavior that violates significant social norms is known as**
   a. deviance.
   b. anomie.
   c. stigma.
   d. recidivism.
   e. social control.

2. **Emile Durkheim contended that no behavior is deviant in itself, because deviance is a matter of social**
   a. pressure.
   b. biological determination.
   c. definition.
   d. control.
   e. indifference.

3. **The means for promoting conformity to a group's or society's rules are known as**
   a. coercion.
   b. influence.
   c. social cohesion.
   d. social control.
   e. groupthink.

4. **Imprisonment, capital punishment, medals, and honorary titles are examples of**
   a. informal sanctions.
   b. internal social control.
   c. formal sanctions.
   d. stigmas.
   e. differential rewards.

5. **According to Emile Durkheim, _____ is a social condition in which norms are weak, conflicting, or absent.**
   a. chaos
   b. ennui
   c. enmity
   d. melancholia
   e. anomie

6. **The strain theory has been used most extensively in the study of**
   a. white-collar crime.
   b. scientific fraud.
   c. alcoholism.
   d. prison riots.
   e. juvenile delinquency.

7. **Theories based on symbolic interaction say that deviance is the product of**
   a. behavior modification.
   b. cerebral mutations.
   c. socialization.
   d. trial and error.
   e. psychological adaptation.

8. **Clifford Shaw and Henry McKay used _____ to explain why delinquency rates remained high in certain neighborhoods even though the ethnic composition changed over the years.**
   a. cultural transmission theory
   b. differential association theory
   c. functionalism
   d. labeling theory
   e. deviant zone theory

9. **According to _____, deviance exists when some members of a group or society label others as deviants.**
   a. labeling theory
   b. functionalism
   c. strain theory
   d. cultural transmission theory
   e. symbolic interactionism

10. **Isolated acts of norm violations are known as**
    a. secondary deviance.
    b. noncriminal activity.
    c. learned deviance.
    d. primary deviance.
    e. independent deviance.

11. **Undesirable characteristics used by others to deny the deviant full social acceptance are known as**
    a. social shields.
    b. social brands.
    c. social stigmas.
    d. informal sanctions.
    e. formal sanctions.

12. **Marxist criminologists view deviance as a product of**
    a. socialization.
    b. language acquisition.
    c. differential association.
    d. the harmful effects of religion.
    e. class conflict.

13. **The official statistics from the FBI *Uniform Crime Reports* are known to be biased mainly due to**
    a. undercounting of the lower class.
    b. overcounting of the middle and upper classes.
    c. overrepresentation of the lower classes and minorities.
    d. human error in compiling the statistics.
    e. the upper class not being included due to their high social status.

14. **The _____ is comprised of the police, the courts, and the correctional system.**
    a. justice complex
    b. social legal charter
    c. system of jurisprudence
    d. legal elite
    e. criminal justice system

15. **Which of the following is the most accurate statement regarding crime and punishment?**
    a. Crime and punishment are not related in industrial societies.
    b. Punishment deters crime if potential lawbreakers believe they are likely to be caught and severely punished.
    c. Punishment deters crime if potential lawbreakers believe that the criminal justice system is fair.
    d. Punishment deters crime only if children are taught that an eye for an eye is the best philosophy.
    e. Potential lawbreakers are less likely to commit crimes if they have friends or relatives who have been punished by the criminal justice system.

FEEDBACK REVIEW TEST

True-False

1. Recent research has established a definite casual relationship between certain human biological characteristics and deviant behavior. *T or F?*
2. It appears that social scientists have prematurely discounted the deterrence doctrine. *T or F?*
3. The imprisonment rate in the United States is four times that of any Western European country. *T or F?*

Fill in the Blank

4. A major problem sociologists have in defining deviance is its _____.
5. According to labeling theory, mental illness is a matter of _____.
6. The process of reducing the seriousness of crimes based on the lower social value placed on the worth of victims is called _____.

Multiple Choice

7. Which of the following is *not* one of the benefits of deviance for society?
   a. decreases suspicion and mistrust among members of a society
   b. promotes social change
   c. increases social unity
   d. provides a safety valve
   e. promotes clarification of norms

8. Which of the following did Edwin Sutherland mean by differential association?
   a. Crime is more likely to occur among individuals who have been exposed more to unfavorable attitudes toward the law than to favorable ones.
   b. People become criminals through association with criminals.
   c. Crime is not transmitted culturally.
   d. Crime comes from differential conflict.

9. In which of the following categories of crime is the United States *not* the leader in the world?
   a. murder   c. robbery
   b. rape   d. burglary

Matching

10. Match the approaches to punishment of criminals on the left side with the examples on the right.
    ____ a. rehabilitation   (1) imprisonment without parole
    ____ b. deterrence   (2) the practice of caning
    ____ c. retribution   (3) employment and educational programs in prison
    ____ d. incapacitation   (4) death penalty for murder

GRAPHIC REVIEW

Attitudes toward the death penalty for murder vary among Americans. The results of a Gallup Poll contained in Figure 7.3 clearly reflect a difference between blacks and whites. Answer these questions to explore the meaning of these results.

1. Do you think that the strong support for capital punishment for murder among white Americans is due more to its deterrent effects or to the desire for revenge? Explain.

   _____

   _____

2. Why do you think African Americans are less supportive of the death penalty for murder than white Americans?

   _____

   _____

ANSWER KEY

Concept Review	Multiple Choice		Feedback Review	
a. 7	1. a	11. c	1. F	9. d
b. 6	2. c	12. e	2. T	10. a. 3
c. 5	3. d	13. c	3. T	b. 2
d. 3	4. c	14. e	4. relativity	c. 4
e. 8	5. e	15. b	5. social definition	d. 1
f. 2	6. e		6. victim discounting	
g. 9	7. c		7. a	
h. 4	8. a		8. a	
i. 1	9. a			
j. 10	10. d			

Chapter Eight

Social Stratification

LEARNING OBJECTIVES

After careful study of this chapter, you will be able to:

- Explain the relationship between social stratification and social class.
- Compare and contrast the three dimensions of stratification.
- State the major differences among the functionalist, conflict, and symbolic interactionist approaches to social stratification.
- Identify the distinguishing characteristics of the major social classes in America.
- Discuss the measurement and perceptions of poverty in the United States.
- Evaluate the cultural and situational explanations of poverty.
- Describe the state of the effort to combat poverty in the United States.
- Outline some of the consequences of social stratification.
- Describe upward and downward social mobility in the United States.
- Discuss the major features of global stratification.

SOCIOLOGICAL IMAGINATION

Is federal spending on social welfare programs out of control? Critics of social welfare, backed by most of the American population, hold very negative beliefs about welfare and its recipients. They are unhappy with the amount of tax dollars being spent on the poor. Although total social welfare spending has increased significantly since 1960, less money in real terms was spent on Aid to Families with Dependent Children (AFDC) recipients in 1996 than in 1970. In 1996, when welfare reform was passed by the U.S. Congress, AFDC recipients accounted for only 3 percent of total federal social welfare expenditures, a reduction of two-thirds since 1970, and less than two-fifths of 1 percent of the total gross national product (GNP) of the United States.

Negative attitudes toward welfare recipients is part of the cultural heritage in America that stigmatizes the poor. Although all societies have social inequality based on the unequal distribution of economic resources, not all societies develop negative (and false) stereotypes of those at the bottom of the scale. This chapter begins with a discussion of the tendency for all societies to engage in *social stratification,* and proceeds immediately to describe the dimensions on which *social classes* are created. Poverty in the United States and the world is covered later in the chapter.

DIMENSIONS OF STRATIFICATION

As a result of the human tendency to rank one another, inequality emerges. Soon some people are at the top of the hierarchy and others are at the bottom. This tendency is a recurring theme in literature. Dr. Seuss writes of the Sneetches, birds who rank themselves according to the presence or absence of a large star on their stomachs. Star-bellied Sneetches have high status, plain-bellied Sneetches have low status. In *Animal Farm,* a novel by George Orwell, the pigs ultimately take over the classless animal society by altering one of their cardinal rules from "All animals are equal" to "All animals are equal—but some animals are more equal than others." These fictional creatures are engaged in **social stratification**—the creation of layers (or strata) of people possessing unequal shares of scarce desirables, the most important of which are income, wealth, prestige, and power.

Once people in some layers have greater amounts of the scarce desirables than those in other layers, social inequality exists. Each of these layers is a **social class**—a segment of a population whose members hold a relatively similar share of scarce desirables and who share attitudes, values, norms, and an identifiable lifestyle. In the United States, we often speak of the upper class, middle class, working class, and lower class. The number of social classes within a stratification structure varies. A stratification structure might include upper-upper, middle-upper, lower-upper, upper-middle, middle-middle, lower-middle, upper-lower, middle-lower, and lower-lower classes. Or, as in old coal mining towns in America or in underdeveloped countries, there might only be an upper class and a lower class.

Karl Marx and Max Weber are the two sociologists who have made the most significant early contributions to the study of social stratification. Marx is credited with demonstrating the importance of the economic foundations of social classes, and Weber is recognized for calling attention to the prestige and power aspects of any stratification structure. Together, the work of these two great minds allows us to see that stratification is a multidimensional affair (Gerth and Mills, 1958).

The Economic Dimension: Karl Marx

Although both Marx and Weber were concerned with inequality, their emphasis was quite different. For Marx, the economic factor was an independent variable explaining the existence of social classes. Weber, on the other hand, viewed the economic dimension as a dependent variable. That is, Weber was more concerned with the economic consequences of stratification.

The process of social stratification creates layers of people—social classes—possessing unequal shares of scarce desirables such as financial resources, prestige, and power. These members of India's ancient caste system are relegated by birth to the very bottom of the stratification structure.

Marx recognized the existence of several social classes in nineteenth-century industrial society—laborers, servants, factory workers, crafts people, proprietors of small businesses, moneyed capitalists—but predicted that capitalist societies would ultimately be reduced to two social classes. Those who owned capital—the bourgeoisie—would be the rulers; those without ownership of the means of production—the proletariat—would be the ruled. The propertied capitalist class could exploit the labor of those without property. Even though the workers performed all the labor, they would be kept at a subsistence wage level, and the capitalists would enjoy the economic surplus created by that labor. In short, capitalists would both rule and exploit the working class because they owned the means of production (factories, land, etc.). The working class would have nothing to sell but its labor.

Marx's analysis went much deeper: Marx contended that all aspects of capitalist society—work, religion, government, law, morality—were economically conditioned. Consequently, the capitalists controlled all social institutions, which they used to their advantage. They could structure the legal system, educational system, and government to suit their own interests. Marx thus saw all of capitalist society as a superstructure resting on an economic foundation; the economy was the society to him.

Although Marx predicted that there would ultimately be only two social classes, Weber foresaw the development of several social classes. Whereas Marx focused on the relationship to the means of production as the cause of social stratification, Weber examined the consequences of people's relationships to the economic institution. These consequences Weber termed **life chances**—the likelihood of securing the "good things of life," such as housing, education, health, and food. The probability of possessing these desirables was directly related to such economic resources as real estate, wages, inheritance, and profits from various investments.

By almost any measure—food, clothing, housing, or health—the standard of living has steadily improved in Europe and America. Still, there is considerable economic inequality even within Western countries, and the United States is no exception.

To understand the extent of economic inequality, it is necessary to distinguish between income and wealth. **Income** refers to the amount of money received by an individual or group. **Wealth** refers to all the economic resources possessed by an individual or group. Wealth is distributed more unequally than income, as can be shown in the case of the United States.

What is the extent of economic inequality in America?
In 1995, over thirty-six million Americans were living in poverty (U.S. Bureau of the Census, 1996b). There are over five million millionaire households and some fifty billionaires in the United States today (Phillips, 1991; Stanley and Danko, 1996; Sloan, 1997).

Paul Samuelson described the inequality of income in America in these words: "If we made an income pyramid out of a child's blocks, with each layer portraying $500 of income, the peak would be far higher than Mt. Everest, but most people would be within a few feet of the ground" (Samuelson and Nordhaus, 1995:355). The truth in Samuelson's statement is supported by government figures on the distribution of income. (See Figure 8.1.) In 1995, the richest 20 percent of American families received over 48 percent of the nation's income; the

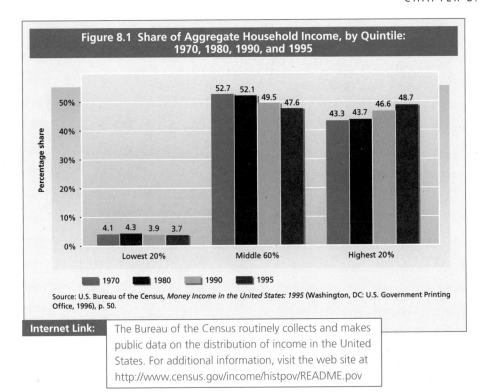

Figure 8.1 Share of Aggregate Household Income, by Quintile: 1970, 1980, 1990, and 1995

Source: U.S. Bureau of the Census, *Money Income in the United States: 1995* (Washington, DC: U.S. Government Printing Office, 1996), p. 50.

Internet Link: The Bureau of the Census routinely collects and makes public data on the distribution of income in the United States. For additional information, visit the web site at http://www.census.gov/income/histpov/README.pov

poorest 20 percent controlled under 4 percent. Income inequality decreased slightly between 1950 and the late 1960s, but it has increased significantly since (Phillips, 1991, 1993; Wolff, 1992; Lynch, 1996).

The1980s and 1990s saw the pre-tax income of the poorest fifth of the nation decline by 5 percent as the income of the richest fifth increased by 31 percent. Income among the top 5 percent rose by 58 percent. Clearly, income inequality between wealthy and other Americans has increased markedly:

> . . . *the richest 1 percent of all Americans now receive nearly as much income after taxes as the bottom 40 percent of Americans combined. Stated another way, the richest 2.5 million people now have nearly as much income as the 100 million Americans with the lowest incomes (Johnson, 1994:37).*

This gap is the largest since the U. S. Census first reported on the distribution of income nearly thirty years ago (Reich, 1997).

Although these income distribution figures reveal persistent economic inequality, they do not show the full nature of the concentration of wealth in America. Because data on national income are published regularly by the federal government, income comparisons are not difficult to make. Accurate data on wealth holding are much harder to obtain (Gilbert and Kahl, 1993).

Some relatively recent attempts have been made to ascertain the distribution of wealth in the United States. The results of a survey on net worth conducted by the Bureau of the Census are shown in Figure 8.2. It is obvious that wealth is highly concentrated in the United States. On the low side, over 25 percent of American households have a net worth (assets minus liabilities) below $5,000; one-third show a net worth below $10,000; and about 43 percent of American families have a net worth below $25,000. At the other extreme, just over 3 percent of American households have a net worth of $500,000 or more (U.S. Bureau of the Census, 1994a).

Another survey indicates an even higher concentration of wealth—almost three-fourths of the country's privately held assets are estimated to be in the hands of only 10 percent of American families (U.S. Congress, 1986). This survey also reported that the share of wealth held by the top one-half of 1 percent of American households (those with assets of $2.5 million or more) had increased from 25 percent to 35 percent from 1963 to 1983. Income distribution trends during the 1980s and 1990s suggest the existence of an even greater concentration of wealth today.

What are the consequences of growing income inequality?   The personal disadvantages to the bulk of the American population losing income are obvious. Inferior health care, restricted diets, low quality housing, and poor police protection are only a few of the personal disadvantages brought about by the growing disparity between the rich and poor in the United States. The widest rich-poor gap since the U.S. Bureau of the Census began keeping records in 1947 also has

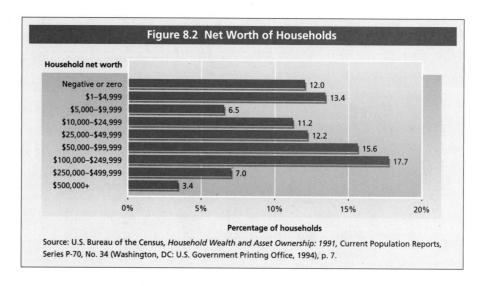

Figure 8.2 Net Worth of Households

Source: U.S. Bureau of the Census, *Household Wealth and Asset Ownership: 1991*, Current Population Reports, Series P-70, No. 34 (Washington, DC: U.S. Government Printing Office, 1994), p. 7.

important negative social consequences. The growing economic disparity between the rich and poor is lowering educational levels—as income decreases so do college enrollment rates and college graduate rates. This, in turn, decreases the proportion of the workforce with college backgrounds, thereby retarding the growth of the skilled workforce. The quality of the workforce is further damaged because the increase in low-income students leads to poorer overall academic performance at all grade levels. In addition, the rate of job growth is affected by the income gap (Bernstein, 1994).

The Power Dimension: Max Weber

You will recall that according to Weber, **power** is the ability to control the behavior of others even against their will. Sociologists have devoted considerable attention to power in society because individuals or groups who possess power are able to use it to enhance their own interests, often at the expense of others.

Why is power a separate dimension of stratification? According to Marx, an individual's power is a reflection of the individual's relationship to the means of production. Those who own and control capital have the power. Sociologists agree that economic success increases the chances of gaining power. The wealthy, for example, have the power to influence tax laws in their favor. However, many sociologists, beginning with Weber, have argued that economic success and power do not always overlap completely. This argument is based on several points. First, the fact that money can be used to exert power does not mean that it will be used in this way. Money is a resource that can be used to enhance power, but a decision must be made to use money for that purpose. For example, there are other families as wealthy as the Kennedys who have not used their economic resources to gain political power.

Second, money and ownership of the means of production are not the only types of resources that can be used as a basis for power. Expert knowledge can be used to enhance power, too. For example, many lawyers convert their special knowledge into substantial amounts of power in political affairs. Other possible resources include eloquence as a speaker and fame. In 1952, for example, Albert Einstein was offered the presidency of Israel, which he refused by saying, "I know a little about nature, and hardly anything about men."

Power is also attached to the social positions we hold. Elected officers in organizations have more power than rank-and-file members. People who control the mass media are powerful, even if they themselves do not have great wealth.

Finally, we can overcome a scarcity of resources if we have large numbers of people on our side or if we are skillful at organizing the resources we have. Hitler, for example, was able to transform limited resources into a mass political movement. He was able to exchange his promise of delivering Germans from their destitution for absolute power.

The Prestige Dimension: Max Weber

A third dimension of social stratification is **prestige**—social recognition, respect, and admiration that a society attaches to a particular social position. Prestige is always a cultural and social matter. In the first place, favorable social evaluation is based on the norms and values within a group. Honor, admiration, respect, and deference are extended to dons within the Mafia, but despite America's fascination with the underworld, Mafia chiefs do not have high prestige outside their own circles. Second, honor, deference, and the like must be given to one person by another person or group of persons. Scientists cannot proclaim themselves Nobel Prize winners, journalists cannot award themselves Pulitzer

Homes in most societies connote prestige, whether high, middle, or low. Large, contemporary houses such as this one suggest that its inhabitants are reasonably high on the prestige hierarchy.

from Jersey, Hereford, or Shorthorn cows (Wallechinsky and Wallace, 1981). There are many contemporary illustrations of status symbols as well. While a top executive at the Ford Motor Company, Lee Iacocca and his colleagues regularly enjoyed lunches flown in from all over the world at an estimated $104 a head. This was in 1960 dollars! (Iacocca and Novak, 1984). Millionaire Malcolm Forbes's son paid $156,450 for an almost 200-year-old bottle of wine believed to have belonged to Thomas Jefferson.

How is prestige distributed?   The basis for assigning prestige within any society coincides with that society's particular value system. Those statuses that are considered most important, or are valued most highly, acquire the most prestige. Because we value the acquisition of wealth and power in American society, we tend to assign the highest prestige to those persons who occupy the positions that are also rewarded with wealth and power. The most stable source of prestige in modern society is associated with occupations (Nakao, Hodge, and Treas, 1990; Nakao and Treas, 1990).

The prestige rankings of some selected occupations in the United States are presented in Table 8.1. These rankings are based on survey results from a random sample of Americans. Generally speaking, there are few surprises in this ranking of occupational prestige, because those in the sample were asked to rank each occupation according to its general standing in American society. Obviously, white-collar occupations (doctors, ministers, schoolteachers) tend to have higher prestige than blue-collar occupations (carpenters, plumbers, mechanics). As expected, physicians, lawyers, and dentists have high occupational prestige, whereas janitors, cab drivers, and garbage collectors have low prestige. You may find it somewhat surprising, however, that priests and college professors have more prestige than bankers—demonstrating that wealth and power are not always the determinants of prestige.

Prizes, and corporate executives cannot grant themselves honorary doctorates. Recognition must come from others, or it does not exist. Third, persons or families who are accorded a similar rank in the prestige hierarchy form social classes that share identifiable lifestyles. The offspring of upper-class homes are more likely to attend private universities and Episcopalian churches; children from lower-class homes are less likely to attend college at all and tend to belong to fundamentalist religious groups. In fact, some sociologists view social classes as subcultures.

Members of social classes with sufficient economic resources may use their wealth to attempt to enhance their prestige. They may consume goods and services to display their wealth to others. This phenomenon Thorstein Veblen (1995, originally published in 1899) called **conspicuous consumption,** examples of which are numerous and often astonishing. Of course, it is among the super-rich that status symbols are the most lavish. Guests at Cornelius Vanderbilt's seventy-room, ten-million-dollar "summer cottage" at Newport could choose between using salt water and fresh water in the bathrooms. Guests preferring milk at Baron Alfred de Rothschild's mansion had their choice of milk

| TABLE 8.1 | Prestige Rankings of Selected Occupations in the United States | | | | | | |

Occupation	Prestige Score	Occupation	Prestige Score	Occupation	Prestige Score	Occupation	Prestige Score
Surgeon	87	Banker	63	Cattle rancher	48	Sales clerk in a store	36
Astronaut	80	Druggist	63	Jazz musician	48	Piano tuner	35
Lawyer	75	Author	63	Locomotive engineer	48	Baker	35
College professor	74	Economist	63	Interior decorator	48	Tax collector	33
Airline pilot	73	Advertising		Mail carrier	47	Bus driver	32
Psychiatrist	72	executive	63	Insurance agent	46	Dry cleaner	32
Dentist	72	Veterinarian	62	Secretary	46	Logger	31
Priest	71	Ship's captain	62	Airline stewardess	46	Skycap	30
Engineer	71	Building contractor	61	Mechanic	46	Waitress	29
TV anchorwoman	70	Owner of an art		Disc jockey	45	Taxicab driver	28
Secret Service agent	70	gallery	61	Income-tax preparer	45	Garbage collector	28
Geologist	70	Police officer	61	Photographer	45	Used car salesperson	25
Cosmetic surgeon	70	Actor	60	Plumber	45	Bartender	25
School principal	69	Journalist	60	Crop-duster pilot	45	Bill collector	24
Medical technician	68	TV anchorman	60	Bank teller	43	Faith healer	24
Pharmacist	68	Businessperson	60	Automobile dealer	43	Nut roaster	23
Owner of manu-		Actress	59	Tool and die maker	43	Janitor	22
facturing plant	68	Motel owner	57	House carpenter	43	Bicycle messenger	22
Optometrist	67	Restaurant owner	55	Deep-sea diver	43	Filling station	
Registered nurse	66	Nursery school		Retiree	43	attendant	21
High school teacher	66	teacher	55	Tailor	42	Soda jerk	20
Army colonel	66	Fashion designer	55	Landlord	41	Grocery bagger	18
Accountant	65	Ballet dancer	53	Travel agent	41	Dishwasher	17
Podiatrist	65	Firefighter	53	Evangelist	41	Prostitute	14
Air traffic controller	65	Airplane mechanic	53	Prison guard	40	Street-corner	
Professional athlete	65	Commercial artist	52	Tree surgeon	40	drug dealer	13
Paramedic	64	Social worker	52	Auto mechanic	40	Fortune teller	13
Public grade school		Surveyor	51	Telephone operator	40	Panhandler	11
teacher	64	Electrician	51	Auctioneer	39		
Meteorologist	63	Housewife	51	TV repair person	39		
Radiation control		Vineyard owner	51	Dynamite blaster	38		
engineer in a		Funeral director	49	Roofer	37		
power plant	63	Ambulance driver	49	Barber	36		

Source: General Social Survey, National Opinion Research Center, 1992.

What is the basis for occupational prestige? According to Robert Hodge, Paul Siegel, and Peter Rossi (1964), all societies appear to rely on the same set of factors in determining occupational prestige. In fact, Robert Hodge, Donald Treiman, and Peter Rossi (1966) have shown that the ranking of occupations on the basis of prestige is similar in both advanced and underdeveloped societies. Occupational prestige scores have traditionally been found to vary according to compensation; education, skills, and ability required; power associated with the occupation; the importance of the occupation to the society; and the nature of the work (mental or white-collar work versus manual or blue-collar work). More recently, occupational prestige has been shown to be affected by gender.

How does gender affect occupational prestige? Brian Powell and Jerry Jacobs (1984a, 1984b) found significant differences between the occupational prestige of males and females. Higher prestige is attached to men than to women in occupations where men predominate (for example, firefighters) and to women in occupations held mostly by women (for example, dressmakers). Christine Bose and Peter Rossi (1983) had reported similar results in an earlier study. Moreover, according to their results, the average prestige score in predominantly female jobs is twelve points lower than in primarily male jobs, and the maximum prestige score that can be attained within the traditional women's labor market is over twenty points lower than that potentially attainable within traditional male jobs. Gender, it appears, must be considered in future studies of occupational prestige.

Why do different cultures evaluate occupations similarly?   Donald Treiman has attempted to answer this question through an examination of nearly a hundred studies of occupational prestige in sixty countries. From these studies, which included countries of Western and Eastern Europe, North America, Africa, and South America, Treiman concluded that all societies must accomplish certain objectives. For example, domestic peace must be maintained, and a variety of goods—including food, housing, clothing, and transportation—must be produced and distributed. Culturally dissimilar societies develop similar occupational positions because of the functions each must perform. This is clearly a functionalist explanation.

Not only do societies around the world develop similar occupational positions, but these positions tend to be arranged in occupational prestige hierarchies that are substantially similar throughout the world.

> In all societies, ranging from highly industrialized nations like the United States to peasant villages in up-country Thailand, the basic pattern of occupational evaluations is the same—professional and higher managerial positions are most highly regarded, lower white-collar and skilled blue-collar jobs fall in the middle of the hierarchy, and unskilled service and laboring jobs are the least respected (Treiman, 1977:103).

This is the case, according to Treiman, because certain occupations in all societies have greater control over scarce resources than do other occupations. Differential access to these resources—skill, authority, property—creates differential power. The power created by control over scarce resources is used to acquire special privilege. Power and privilege, in turn, are used to garner prestige.

In this way, a single, worldwide occupational prestige hierarchy is created. This is a conflict explanation.

Are the economic, power, and prestige dimensions related?   Generally, a close relationship exists among these three dimensions of stratification. Those who are high on one dimension tend to be high on the other two. Power, particularly political power, is a decided economic advantage to its holders. For example, the wealthy pay for political favors either directly (as with money) or indirectly (as through inside information on lucrative business deals). And George Bush is by no means the only American who has used his wealth to assist in achieving high political rank.

Prestige and wealth are also closely intertwined. High prestige is heavily dependent on economic resources. The lifestyle associated with high prestige cannot be maintained without the economic clout to purchase such necessities as owning a nice home in the proper neighborhood or the ability to send one's children to the right schools. But those with high prestige normally have no cause for concern, for with prestige come special economic rewards and unique business opportunities for generating wealth. Most of the wealth amassed by Arnold Palmer and Jack Nicklaus did not come directly from prize money in professional golf tournaments. Because of their prestige, people want to buy Arnold Palmer sweaters and Jack Nicklaus golf clubs as much as corporations want their endorsements on products. In fact, Jack Nicklaus now owns the MacGregor sporting goods company, the firm that first used his name on its golf equipment. In the light of a $40 million contract with Nike coming at age twenty-one, it is hard to imagine the wealth Tiger Woods will accumulate.

Prestige and power often mix quite well. Prestige is attached to social positions requiring important decision making. And prestige may be converted into power. Consider Ronald Reagan, the actor who became a California governor and President of the United States, or Jack Kemp, the professional quarterback who became a member of the House of Representatives, a presidential candidate, a member of George Bush's cabinet, and a vice-presidential candidate in the 1996 elections.

FEEDBACK

1. _____ involves the creation of layers of people possessing unequal shares of scarce desirables.
2. Match the concepts with the examples:

    ____ a. life chances   (1) the respect accorded doctors
    ____ b. power   (2) a politician conforms to the interests of a lobby
    ____ c. prestige   (3) living in a neighborhood with good police protection
    ____ d. conspicuous consumption (4) a lavish wedding

3. The top 0.5 percent of American households hold _____ percent of the total wealth:
    a. 2.5   b. 10   c. 20   d. 35
4. The most stable source of prestige in modern societies is associated with _____.

Answers: 1. Social stratification 2. a. (3) b. (2) c. (1) d. (4) 3. d. 4. occupations

EXPLANATIONS OF INEQUALITY

Obviously, social class exerts a powerful influence on social life. The social class into which people are born, and to which they rise or fall, affects the nature of their existence as much as any social factor. For this reason, if for no other, it is of benefit to understand why the process of stratification is a mainspring of human society.

Inequality—in wealth, power, and prestige—seems to be a basic fact of life in most societies. Even societies deliberately organized to eliminate inequality are not classless. The history of the former Soviet Union since the Communist revolution represented an attempt to create a classless or unstratified society. We know, however, that great differences in rank and rewards were attached to different statuses in the former Soviet Union. Leading members of the Communist Party, the dominant elite, were able to obtain privileges for themselves and their families. For example, according to Hedrick Smith (1976), Politburo members and national secretaries of the Communist Party drove black Zil limousines, hand-tooled cars worth over $100,000 each, while the majority of Russian people had no chance of ever owning any kind of car. The Chinese Communist Party has also apparently become an entrenched, self-interested ruling class (Wortzel, 1987).

Why does inequality appear to be a nearly universal characteristic of social life? There are several explanations. According to the functionalist explanation, inequality exists because it is beneficial and necessary to society. Conflict theory, on the other hand, contends that inequality exists because it benefits individuals and groups with the power to exploit others. According to the evolutionary theory, the existence of stratification depends on the stage of technological evolution. Each of these explanations is developed more fully in this section, starting with the functionalist theory.

Functionalist Theory of Stratification

How do the functionalists explain inequality? The functionalist theory of stratification, proposed by Kingsley Davis and Wilbert Moore (1945) and pursued by many later sociologists, states that inequality in the distribution of desirables exists as a device for ensuring that the most important positions are filled by the most qualified people and that people in these positions perform their tasks competently. The functionalist theory recognizes that certain jobs in society are more important than others and that these jobs often involve special talent and training. To encourage people to make the sacrifices necessary (years of education, and so forth) to fill these jobs, special monetary rewards and

According to the functional theory of stratification, higher monetary rewards are distributed to the most qualified people in the most important positions. If this is so, we can feel more comfortable knowing that airline pilots earn high salaries.

prestige are attached to these positions. For example, doctors make more money than electricians because a high level of skill is required in the medical profession and our society's need for highly qualified doctors is great. Some college teachers may not make more money than high school teachers, but they have greater prestige as a reward for the years of extra schooling they have had.

Getting the most qualified people into the most important positions is only one aspect of the use of rewards to motivate people in stratification structures. Once people have prepared themselves to assume positions, they must be motivated to do their jobs competently. Therefore, greater rewards go to those who do their work better than others. Thus, more competent doctors, secretaries, and salespersons should receive more of the desirables than those performing their tasks less well. (However, even incompetent doctors receive more desirables than highly competent secretaries.) According to the functionalist theory of stratification, then, social inequality is necessary or functional for a society.

The functionalist theory has dominated sociological thinking for a long time, but it has been a very difficult theory to test. Although some studies support at least some aspects of the functionalist theory (Grandjean, 1975; Cullen and Novick, 1979), several weaknesses should be noted.

What are the weaknesses of the functionalist theory?
Although the functionalist theory is elegant, concise, and consistent with traditional American beliefs about inequality, it has several weaknesses (Tumin, 1953). First, there are many people who have power, prestige, and wealth whose contributions to society do not seem very important. Many top athletes, film stars, and rock singers make many times as much money as the President of the United States. Why should scientists make almost no money from groundbreaking books reporting their research whereas screenwriters regularly receive $500,000 plus per script?

Second, the functionalist theory ignores the barriers to competition faced by some members of society—the poor, women, the aged, African Americans, and various ethnic groups. A wealth of talent exists among people in these categories, but it has never been tapped, because such people have been denied equal opportunity to compete. They have been denied the opportunity to become educated because they are too poor, not the right color, or the wrong sex. And even if they are technically qualified, they have often been denied higher level positions in favor of less qualified individuals who happen to possess the right social characteristics.

Third, the functionalist theory overlooks the inheritance of social class level. In some societies, one's social class is totally determined at birth. Even in modern societies, in which the rate of mobility is supposed to be high, children tend to remain in or near the social class of their parents. This means that talented daughters and sons from lower-class homes cannot compete on an equal basis with the offspring of middle- and upper-class people. Consequently, many talented individuals are truck drivers, plumbers, and short-order cooks who might have been surgeons, lawyers, or executives.

Finally, the functionalist theory of stratification has an ethnocentric bias. That is, it assumes that all people in all societies will be motivated to compete to acquire more of the scarce desirables than others have. In fact, in some societies, greater emphasis is placed on having a more or less equitable distribution of wealth, power, and prestige than on encouraging individual competition for larger shares. (See the discussion of the evolutionary theory of stratification in the section entitled "Synthesizing the Two Perspectives.")

Melvin Tumin (1953) concluded that stratification exists not because it is necessary for the benefit of society but because it helps people holding the most power and economic resources to maintain the status quo. From his perspective, the functionalist theory is used as a justification for keeping some people at the top of the class structure and some at the bottom. This viewpoint is consistent with the conflict theory of stratification.

Conflict Theory of Stratification

How does conflict theory account for inequality?
According to the conflict theory of stratification, inequality is a matter of some people having the power and the willingness to exploit others. According to this theory of stratification, those in the upper levels of a stratification structure hold a monopoly over the community's or society's desirables and use their monopoly to dominate others. Conflict theorists, then, see societies as run by force rather than by consent. Those with wealth, power, and prestige are able to maintain, perhaps even increase, their share of the desirable things in a society.

The conflict theory of stratification is largely based on Marx's ideas regarding class conflict. For Marx, all of history has been a class struggle between the powerful and the powerless, the exploiters and the exploited. Capitalist society is the final stage in the development of the class struggle. Although the capitalists are outnumbered, they are able to control the workers by creating a belief system that legitimates the status quo, with all of the advantages that holds for them. Those who own the means of production are able to spread their ideas throughout the land—through the schools, the media, the churches, the government, and so forth. In this way, the dominant ideas of the ruling elite become the dominant ideas of the ruled. Marx used the term *false consciousness* to refer to the situation in which the working class accepts the dominant ideology as true although this ideology contradicts its own class interests. As long as the working class remains unaware of its own power and the extent to which it is being exploited, capitalists are safe.

Because Marx believed that class conflict occurs in every historical epoch, he predicted that the working class would eventually shed its false consciousness—and its chains—for a revolution based on a true class consciousness. That is, it would realize its exploitation and reject the dominant belief system. After overthrowing the capitalists, the working class would create a socialist society in which the means of production and all property would be owned equally by the people. This would be a transitional stage between capitalist society and the classless society of communism. Ultimately, socialism would be replaced by communism, to put an end to human misery and exploitation.

Marx's work has been somewhat neglected by American sociologists since the turn of the century, although it has indirectly influenced the field through many sociologists' negative reactions to him. Why has Marx been neglected? For one thing, many of his ideas—exploitation, revolution, classless society, communism—do not fit well with the American emphasis on capitalism, achievement, and upward mobility. In addition, many of Marx's predictions appear to have been wrong. Socialist revolutions occurred in noncapitalistic societies; communism did not produce classless societies; the middle class rather than the working class has expanded in capitalist societies.

Marxists do not consider that any of the preceding factors disprove Marx's theory. First, the so-called socialist and communist revolutions in precapitalist societies were essentially the result of misinterpreting Marx, who insisted that capitalism was an essential precondition that must occur before a communist society is possible. Second, Marx's theory defines the middle class as part of the working class, because the middle class also does not own the means of production and its gradual expansion fulfills Marx's prediction of the increasing "proletarianization" of labor (that is, more and more people, including small business owners, doctors, and other professionals, lose ownership of their labor power and come to work for someone else). Finally, it is naive to assert that Marx was wrong simply because capitalism has not yet disappeared. Marx gave no time table for the socialist revolution, and Marxist theorists see evidence of increasing contradictions within capitalism (Therborn, 1982, 1984; Wright, 1985).

Recently, however, many sociologists have begun to look past the apparent miscalculations in Marx's theories to appreciate some of the assumptions on which they are based. There is currently substantial appreciation of Marx's insight into the ways the elite use society's institutions, including the state, to achieve their own selfish ends.

German sociologist Ralf Dahrendorf (1959) advocates shifting interest from conflict between two classes to conflicts between various groups within a society, for example, unions and corporations. His emphasis is less on ownership of property than on who controls whom. Some other sociologists have joined Dahrendorf in support of Marx (Collins, 1975, 1979; Wallerstein, 1979a, 1979b; Domhoff, 1986; Giddens, 1987, 1997). They point out that America's legal system is often used by the wealthy for their own benefit, that both factory and office work is repetitive and alienating, that education is used by the elite to perpetuate their power, and that economic wealth is concentrated in the hands of a few.

According to these sociologists, then, Marx was not wrong in his basic thrust. Classes are created by a struggle over scarce resources, and those who secure the most resources use their ideology in conjunction with the power of the state to protect and further their interests.

The realization that there is truth in both the functionalist and conflict theories has led to attempts to combine the theories (Collins, 1994). The most well-known attempt at synthesis is Gerhard Lenski's (1984) evolutionary theory of stratification.

Synthesizing the Two Perspectives

According to functionalists, social inequality exists because of the contribution it makes to the operation and survival of society. The mechanism for distributing rewards in the functionalist explanation lies in the increasing of rewards in line with increases in positional rank. This occurs, the functionalists argue, for two reasons. First, social positions with the highest rank are the most crucial for society and require the greatest talent or training. Second, rewards are used to motivate people to undergo the preparation necessary to qualify for the most important positions and to perform successfully the duties of those positions. In the conflict perspective, rewards are distributed not to serve society but to serve those with the most power. Scarce rewards are said to be captured by those who, in conflict with others, are able to work their will on others. Lenski believes that each of these theories is valid under certain conditions. Social stratification systems take different forms, depending on the society's stage of technological evolution.

Lenski places the pursuit of self-interest or group interest near the center of human nature. Lenski formulates two laws of distribution consistent with this view of human nature—one that applies to the distribution of necessities for survival, another that applies to the distribution of any existing surplus of goods and services. The first law of distribution, drawing from the functional theory of stratification, says that people will share the fruits of their labor to the extent that it is required to "insure the survival and continued productivity of those others whose actions are necessary or beneficial to themselves" (Lenski, 1984:44). In effect, then, the basic resources needed for survival are distributed according to functionalist principles—the highest rewards go to incumbents of the most important positions. This first law of distribution does not explain the distribution of any existing surplus of goods and services. According to Lenski's second law of distribution, power determines the distribution of almost all the surplus goods and services produced by a society.

> If we assume that in important decisions human action is motivated almost entirely by self-interest or partisan group interest, and if we assume that many of the things men most desire are in short supply, then . . . this surplus will inevitably give rise to conflicts and struggles aimed at its control (Lenski, 1984:44).

Those with the most power will, as the conflict theory states, acquire more of the surplus from these conflicts and struggles. According to the evolutionary theory of stratification, then, the degree of inequality varies directly with the size of the surplus of goods and

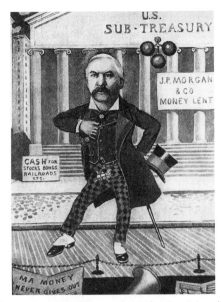

According to the conflict theory of stratification, inequality is a result of the use of power to create and maintain inequality. The enormous power of nineteenth-century industrialists such as J. P. Morgan is reflected in this caricature.

services in a society and the size of the surplus depends on the level of technological development of a society.

What is the relationship between technological development and inequality?   In hunting and gathering societies, the technological base is so primitive that no surplus can be created. Because no surplus exists, there can be no competition and no power struggles through which some can gain more wealth than others. Hence, stratification does not exist; all members of this type of society are relatively equal. Even simple horticultural societies, which rely on the digging stick, can produce some resources beyond what is needed for survival. Consequently, there is greater inequality than in hunting and gathering societies. As the surplus increases in advanced horticultural societies (based on the hoe) and agricultural societies (based on draft animals), the degree of inequality increases dramatically. In advanced agricultural societies, a definite and rigid social-class system emerges, including rulers, a governing class, a merchant class, a priestly class, a peasant class, an artisan class, and a variety of "outsider" classes at the very bottom of the stratification structure.

Industrial society, based on inanimate sources of energy, is capable of producing a surplus beyond comprehension in earlier types of societies. Based on past history, the degree of inequality should increase considerably. This is not the case. Inequality is less pronounced in industrial society than in agricultural society. In other words, "The appearance of mature industrial societies marks the first significant reversal in the age-old evolutionary trend toward increasing inequality" (Lenski, 1984:308).

What are the strengths and weaknesses of the evolutionary theory?   The evolutionary theory of stratification is broader in scope than either the functionalist or the conflict theory. The latter two theories focus on the creation of inequality within a society at a particular time, whereas the evolutionary theory deals with inequality in different historical eras and from one culture to another. Also, the functionalist and conflict theories of stratification tend to compete on moral grounds. The functionalist theory is benign—social inequality helps society. For conflict advocates, inequality is undesirable because it is the result of the powerful exploiting the powerless. Because the evolutionary theory is morally neutral, it avoids value-laden arguments regarding good and evil. On the down side, the evolutionary theory explains little about the distribution of scarce desirables beyond the idea that the controllers of the technology creating the surplus will rise to the top.

Symbolic Interactionism and Stratification

Why was the inequality between former Soviet political leaders and the average Russian tolerated by the masses? Why does the poorest one-fifth of Americans (possessing less than 1 percent of the wealth) not revolt against the richest one-fifth of the population (holding three-fourths of the wealth)? Social stratification systems can persist over the long haul only if people believe in their legitimacy. Symbolic interactionism helps us understand this legitimation process.

How is the legitimacy of a stratification structure incorporated into the minds of individuals?   If legitimacy is to be sustained, it must occur through socialization. Symbols, in the form of language, are used to explain to the young the existence of stratification structure and the reasons for people being located in particular strata. In American society, children are taught that one's place in the stratification structure is the product of talent and effort. Those on top have worked hard and used their abilities, whereas those on the bottom lack the talent and will to succeed. "Losers" must accept the blame for their lowly place in the stratification structure. Because it is their fault, it is not fair to challenge the system. (More is said about this in the next chapter.) Moreover, Americans are taught to prepare themselves and work hard to improve their position on the social-class ladder.

Thus, the legitimacy of the stratification structure is established and perpetuated through teaching beliefs to the young and continuously reinforcing them. In this way, people come to believe in the fairness of the existing system of inequality.

These ideas, as the symbolic interactionists have shown, penetrate even deeper into individuals. The deepest penetrations occur because beliefs about the legitimacy of a stratification structure are incorporated into the self-concepts of individuals.

How do ideas about inequality become part of the self-concept? Obviously, certain self-images are implied in the set of beliefs regarding the legitimation of stratification in America. "Losers" would be expected to have low self-esteem and "winners" to have a high self-regard. As will be seen in the chapter on racial and ethnic minorities, those at the bottom of the stratification structure do suffer from lower self-esteem. How could it be otherwise when messages from all sides tell them of their inferiority? Their lower self-esteem, as the symbolic interactionists have documented, comes from their perceptions of others' opinions of them.

Moreover, the level of an individual's sense of self-worth affects the amount of rewards the individual feels entitled to receive. Worthless people feel that they deserve little. Thus, people at the bottom of the heap tend not to challenge the stratification structure because of the deep-down conviction that they are undeserving. Others blame the victims; the victims blame themselves.

Emphasis has been placed on the importance of the self-concept among people at the bottom in maintaining a system of inequality. The same logic, however, applies to those in other nonelite classes. The reverse is also true for the higher classes. Those profiting most from the existing stratification structure tend to appear to have higher self-esteem. This, in turn, fuels their conviction of the justness of the present arrangement. In short, people's self-concepts help preserve the status quo.

SOCIAL CLASSES IN AMERICA

Americans have always been aware of inequality, but they have never developed a sharp sense of **class consciousness**—a sense of identification with the goals and interests of the members of one's social class (Centers, 1949; Hodge and Treiman, 1968; Jackman and Jackman, 1983). In part because the American public has historically shown relatively little interest in class differences, American scholars began to investigate inequality rather late. It was not until the 1920s that sociologists began systematically to identify social classes in American communities. Since that time, however, the material available on this subject has been voluminous. Early efforts to study stratification were intensive case studies of specific communities (Warner and Lunt, 1941; Hollingshead, 1949; Hollingshead and Redlich, 1958; Warner, 1960). Only in relatively recent times have attempts been made to describe the stratification structure of America as a whole.

Any attempt to describe the social-class structure of American society is hazardous. Social classes are fluid, replete with exceptions, and abstract. Nevertheless, sociologists have identified the major features of the American class structure (Rossides, 1980,1990a; Gilbert and Kahl, 1993). According to Daniel Rossides (1990a), the upper class comprises 1 to 3 percent of the population, the upper-middle class 10 to 15 percent, the

---

FEEDBACK

1. Match the theories of stratification with the examples:
   - _____ a. functionalist theory
   - _____ b. conflict theory
   - _____ c. evolutionary theory
   - _____ d. symbolic interactionism

   (1) Corporate executives make more money because they determine who gets what in their organizations.
   (2) Engineers make more money than butlers because of the education they possess.
   (3) A society is stratified in part because a surplus of goods and services exists.
   (4) Ghetto children tend to have low self-esteem.

2. Which of the following is *not* a criticism of the functionalist theory of stratification?
   a. There are people at the top whose jobs do not seem very important.
   b. It has a bias toward cultural relativism.
   c. It overlooks the barriers to competition.
   d. It glosses over the inheritance of social-class level.

3. The conflict theory of stratification is based on Marx's ideas on class _____.

4. According to the evolutionary theory, in which of the following types of societies is stratification *least* likely to occur?
   a. hunting and gathering   b. horticultural   c. agricultural   d. industrial   e. postindustrial

5. According to the symbolic interactionists, the deepest penetration of the beliefs supporting the stratification structure comes as a result of
   a. the "I"   b. evolution   c. conflict   d. the self-concept

Answers: 1. a. (2) b. (1) c. (3) d. (4) 2. b. 3. conflict 4. a 5. d.

TABLE 8.2	Model of the American Class Structure: Classes by Typical Situations			
Proportion of Households	Class	Education	Occupation of Family Head	Family Income 1990
1%	Capitalist	Prestige university	Investors, heirs, executives	Over $750,000, mostly from assets
14%	Upper middle	College, often with postgraduate study	Upper managers and professionals; medium business owners	$70,000 or more
60%	Middle	At least high school; often some college or apprenticeship	Lower managers; semiprofessionals; nonretail sales; crafts-people; foremen	About $40,000
	Working	High school	Operatives; low-paid craftspeople; clerical workers; retail sales workers	About $25,000
25%	Working poor	Some high school	Service workers; laborers; low-paid operatives and clericals	Below $20,000
	Underclass	Some high school	Unemployed or part-time; many welfare recipients	Below $13,000

Source: From *The American Class Structure*, 4th ed. by Dennis Gilbert and Joseph A. Kahl, © 1993 by Wadsworth, Inc. Reprinted by permission of Wadsworth, Inc.

lower-middle class 30 to 35 percent, the working class 40 to 45 percent, and the lower class 20 to 25 percent. Despite the hazards involved, a brief description of each of these social classes follows. (Table 8.2 presents a similar picture of the American class structure.)

The Upper Class

Despite America's emphasis on the openness and fluidity of its class structure, its upper class contains only a fraction of its population (1 to 3 percent). Moreover, there is even stratification within the upper class. At the pinnacle rests the upper-upper class, the 1 percent of the American population that we recognize as the aristocracy. Its members represent the "old" money amassed for generations by families whose names appear in the *Social Register*—Ford, Rockefeller, Vanderbilt, du Pont. Gaining entrance into the upper-upper class is a long-run proposition because the basis of membership is blood rather than sweat and tears. Money must age over generations to have any chance of entering the orbit of the blue bloods. Obviously, members of the upper-upper class pay enormous attention to lineage, which they protect by sending their children to the best private secondary schools and universities, maintaining rather exclusive interaction with one another, and promoting a high rate of intramarriage.

The lower-upper class, the "new rich," are in the upper class on the basis of achievement rather than birth; membership is based more on earned income than inherited wealth. The lower-upper class is composed of people with much more varied socioeconomic backgrounds. Many members of the lower-upper class inherited some wealth, although usually not fortunes. Others may have amassed fortunes by being shrewd investors on the stock market, creating a chain of hardware stores, becoming a Metropolitan opera star, or earning millions from athletic talent. Although they may actually be better off financially, people who have entered the lower-upper class in their own lifetimes are looked down on by those on the top and are generally barred from the exclusive social circles in which members of the upper-upper class travel.

The upper class as a whole wields tremendous power over domestic and international decisions and over events affecting vast numbers of people (Dye, 1990). If the United States has a power elite, this is it. (See Chapter 13, "Political and Economic Institutions.")

The Middle Class

The upper-middle class (10 percent to 15 percent of the population) is composed of those who have been successful in business, the professions, politics, and the military. Basically, this class is made up of individuals and families who benefited from the tremendous corporate and professional expansion following World War II. Members of this class earn enough to live pretty well and to save some money. They are typically college-educated and have high educational and occupational aspirations for their children. Although they do not exercise national or international power, their influence is felt in their communities in which they tend to be active in voluntary and political organizations.

The lower-middle class is large (30 percent to 35 percent of the population) and heterogeneous. Its members are small business owners; small farmers; small, independent professionals (small-town doctors and lawyers); semiprofessionals (clergy, teachers, nurses, firefighters, social workers, police officers); middle-level managers; and sales and clerical workers. Their income level, which is at about the national average, does not permit them to live as well as the upper-middle class. Typically, members of the lower-middle class have only a high school education, although many may have had some college or have completed college. Members of this class are interested in civic affairs. Their participation in political activities is less than in the classes above them but higher than either the working or the lower class.

The 1980s witnessed a dramatic decrease in the proportion of Americans that can be classified as middle class. During the 1980s, there was a 20 percent reduction in the proportion of middle-income Americans (Morin, 1991; Wallich and Corcoran, 1992).

The Working Class

The working class comprises about one-third of the population. Its members are typically either blue-collar workers who work with their hands, such as roofers, delivery truck drivers, and machine operators, or lower-level sales and clerical workers (Rubin, 1994). The economic resources of its members are, on the whole, fewer than the classes above, although some workers (plumbers, for example) may earn more than some middle-class people. Members of the working class experience considerable economic pressure because their incomes are below the national average, their employment is more unstable, and they are generally without the hospital insurance and retirement benefits enjoyed by those in the middle class. The specter of unemploy-

The American working class, the largest single segment of the population, is composed primarily of blue-collar workers. These construction workers taking a coffee break are representative of the working class.

ment or illness is both real and haunting. Outside of union activities, members of the working class have little opportunity to exercise power, and their participation in organizational life, including politics, is scant. Little evidence exists that the working class—even the higher-income members—is likely to be incorporated into the middle class. Members of the working class, in fact, take great pride in working with their hands.

The Lower Class

Lumped into the lower class (20 percent to 25 percent of the population) are the working poor, the permanently unemployed, the frequently unemployed, and the underpaid. The most common shared characteristics of the lower class is lack of value in the labor market. There are many routes into the lower class—birth, physical or mental disability, age, abandonment, occupational fail-

1. _____ refers to a sense of identification with the goals and interests of the members of one's social class.
2. The upper-upper class in the United States contains about _____ percent of the total population.
3. A major difference between the upper-upper class and the lower-upper class is that the latter has relied on _____ rather than birth.
4. The largest social class in the United States is _____.

a. the upper-middle class   c. the working class
b. the lower-middle class   d. the lower class

ure, alcoholism. There are, however, very few paths out. In fact, some refer to the poorer of the lower class as America's permanent "underclass" (Wilson, 1987; Wilson et al, 1988; Gilbert and Kahl, 1993; Katz, 1993). The underclass will be explored in the following chapter on racial and ethnic minorities. Poverty in the United States, another way to discuss the underclass, is the topic of the next section of this chapter.

POVERTY IN AMERICA

Contrary to popular opinion, poverty is widespread throughout the United States. Great poverty existed when Michael Harrington helped to make it a political issue with his book *The Other America* (1971). Over a decade later, Harrington (1984) continued to underscore the presence of poverty in affluent America, calling special attention to undocumented immigrant laborers, out-of-work blue-collar workers, and women raising children alone. In all, 36.4 million Americans were below the government's poverty line—$15,569 for a family of four—in 1995 (U.S. Bureau of the Census, 1996b).

Measuring Poverty

On an absolute scale, being poor involves a lack of housing, food, medical care, and other necessities for maintaining life. **Absolute poverty** is usually defined as the absence of enough money to secure life's necessities. On a relative scale, it is possible to have those things required to remain alive—and even to live in reasonably good physical health—and still be poor. **Relative poverty** is measured by comparing the economic condition of those at the bottom of a society with other segments of the population. Poverty is deter-

mined by the standards that exist within a society. Thus, poverty in India may not be the same as poverty in the United States.

How can absolute poverty be measured? Poverty in America has traditionally been measured in an absolute way. Absolute measures of poverty involve drawing a "poverty line," an annual income level below which people are considered poor. (See Figure 8.3.) The poor, as measured by the federal government standard, comprised 13.8 percent of the American population in 1995 (U.S. Bureau of the Census, 1996b).

How can relative poverty be measured? According to advocates of relative measures of poverty, the poverty threshold should be raised as the standard of living in a country rises. A rising amount of spendable income does not make people feel less poor if the income levels of those around them are increasing at the same or a faster rate. A common measure of relative poverty is a comparison of the lower fifth of the population in terms of income with the other four-fifths. (Refer to Figure 8.1.) Such a measure means, of course, that a segment of the population will always be poor unless the total income distribution approaches equality.

Who are America's poor? Poverty is particularly prevalent among minorities; within households without husbands/fathers; among children under eighteen years of age; and among the elderly, the disabled, and people who live alone or with nonrelatives. Although nearly three-fourths of the poor in America today are white, the poverty rate for African Americans and Latinos is higher than that for whites. (See Figure 8.4.) In 1995, just over 11 percent of the white population fell below the poverty threshold, compared with about 29 percent among African Americans and 30 percent

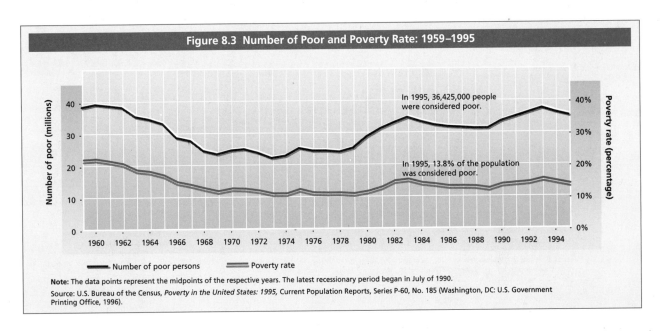

Figure 8.3 Number of Poor and Poverty Rate: 1959–1995

In 1995, 36,425,000 people were considered poor.

In 1995, 13.8% of the population was considered poor.

—— Number of poor persons   ═══ Poverty rate

Note: The data points represent the midpoints of the respective years. The latest recessionary period began in July of 1990.

Source: U.S. Bureau of the Census, *Poverty in the United States: 1995*, Current Population Reports, Series P-60, No. 185 (Washington, DC: U.S. Government Printing Office, 1996).

DOING RESEARCH

E. E. LeMasters—
Blue-Collar Aristocrats

During the late 1960s and early 1970s, E. E. LeMasters spent a lot of time in a working-class tavern that he called The Oasis. As a participant observer, LeMasters was investigating a group of "blue-collar aristocrats," 90 percent of whom were skilled construction workers. LeMasters's objective was to see whether these affluent blue-collar workers were adopting a middle-class lifestyle. In the process, LeMasters provides a case study of the powerful and enduring consequences of social class.

LeMasters's ploys to gain acceptance into this group are interesting. Initially, he pretended to want just to drink beer and play pool with the patrons. After a while, some people (probably including his wife and department chair) began to question the amount of time he was spending drinking and playing pool. After the patrons decided that he was "too nice a guy to be a cop" for the state liquor commission, they concluded that he was an alcoholic spending his time at a blue-collar tavern out of sight of his professional colleagues. Finally, LeMasters revealed himself as a college professor who wanted to know more about blue-collar people. LeMasters credits his full acceptance to his competitive ability at The Oasis pool table.

LeMasters offers many examples of the patrons' desire to main-

tain the working-class lifestyle. A few of the findings regarding sex, marriage, and child rearing support his interpretation.

Boys from blue-collar backgrounds tend to have heterosexual activities earlier than boys from middle-class homes. This pattern held among men at The Oasis, for whom the average age for the first sexual encounter was sixteen. LeMasters never heard chastity defended for either sex at The Oasis. Virginity, for both boys and girls, was viewed as a state of affairs to be corrected as soon as possible. According to one man, "if a girl is big enough, she's old enough, and if she's old enough then she's big enough" (LeMasters, 1975:93).

A happily married woman characterized her husband as a "real nice bastard." This statement seems to capture the marriage-as-a-struggle attitude expressed by

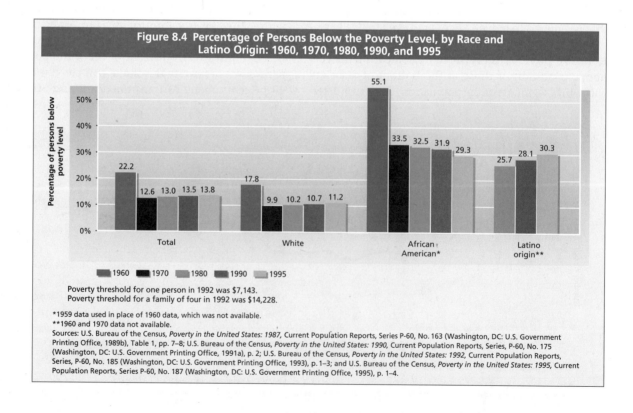

Figure 8.4 Percentage of Persons Below the Poverty Level, by Race and Latino Origin: 1960, 1970, 1980, 1990, and 1995

Poverty threshold for one person in 1992 was $7,143.
Poverty threshold for a family of four in 1992 was $14,228.

*1959 data used in place of 1960 data, which was not available.
**1960 and 1970 data not available.
Sources: U.S. Bureau of the Census, *Poverty in the United States: 1987*, Current Population Reports, Series P-60, No. 163 (Washington, DC: U.S. Government Printing Office, 1989b), Table 1, pp. 7–8; U.S. Bureau of the Census, *Poverty in the United States: 1990*, Current Population Reports, Series, P-60, No. 175 (Washington, DC: U.S. Government Printing Office, 1991a), p. 2; U.S. Bureau of the Census, *Poverty in the United States: 1992*, Current Population Reports, Series, P-60, No. 185 (Washington, DC: U.S. Government Printing Office, 1993), p. 1–3; and U.S. Bureau of the Census, *Poverty in the United States: 1995*, Current Population Reports, Series P-60, No. 187 (Washington, DC: U.S. Government Printing Office, 1995), p. 1–4.

both husbands and wives, a sharp contrast to the "togetherness" model of marriage typical of the white-collar middle class. This blue-collar type of marriage seems to be an extension of the general battle of the sexes, in which the basic ingredients are distrust, suspicion, and fear. Men commonly described women as sneaky; women regularly referred to men as dumb. Often heard in the tavern was the sentiment that "you can't live with 'em and you can't live without 'em."

That the men at the tavern wanted to preserve their way of life was clearly reflected in their child-rearing attitudes and behavior. They taught their sons to defend themselves physically just as their fathers had taught them. Their sons, like themselves, were also trained to control women, never be a fall guy, and never act spoiled. The men's view of girls

also remained traditional—girls need less education than boys; girls should not be spoiled; and girls should not trust men. These blue-collar fathers did not share the middle-class willingness to "do anything for my children." Finally, any conflict between the woman's roles as wife and mother was to be settled in favor of the man.

Mothers tended to be more oriented to the middle class in their child-rearing attitudes and behavior. For example, they shared some middle-class ideas such as emphasizing the unique worth of each of their children. Many of the mothers also preferred white-collar husbands for their daughters.

LeMasters concluded that affluence has not destroyed the cultural distinctions between the American working and middle classes; the consequences of social class have persisted among these people. Whether this will con-

tinue remains to be seen. That a break with the traditional blue-collar lifestyle may occur in the future is suggested by the partial white-collar orientation of many of the women.

Critical feedback

1. From reading about LeMasters's research, do you believe that the working class constitutes a subculture? Why or why not?
2. Do you believe that the lifestyle distinctions between the working and middle classes will continue to exist? Defend your answer.
3. Which of the major theoretical perspectives is the most helpful in understanding LeMasters's findings? Explain your choice.

among Latinos (U.S. Bureau of the Census, 1996b). Although African Americans account for just over 12 percent of the total population, they make up 27 percent of the poor population.

There is a heavy concentration of the poor in households without husbands/fathers. About half of the poor are members of female-headed families, whereas almost 95 percent of the nonpoor are in families with a husband/father present. The poverty rate for female-headed families is about 32 percent (compared to just 11 percent for all families). Of all those living in poverty, about 40 percent are children under eighteen years of age. About 15 million American children under eighteen live in impoverished households. The poverty rate for children under six years is about 24 percent, the highest poverty rate for any age group in the United States (U.S. Bureau of the Census, 1996b). All of this documents the **feminization of poverty,** the trend toward more of the poor in the United States being women and children (Thomas, 1994).

The elderly account for another large segment of the poor. About 9 percent of the poor are sixty-five or older and about 11 percent of those sixty-five or older

live in poverty (U.S. Bureau of the Census, 1996b). Another large segment of the poor are the disabled—those who are blind, deaf, crippled, or otherwise disabled—who account for some 12 percent of America's poor. Finally, in the neighborhood of 21 percent of the poor—one out of every five—either live alone or live with nonrelatives.

This description should not leave the impression that able-bodied poor in the United States do not work (Schwarz and Volgy, 1992). Just over 23 million of the poor in America are over sixteen years of age. Of these, nearly 9.5 million worked either full-time or part-time in 1995. In other words, 41 percent of poor Americans of working age were working (U.S. Bureau of the Census, 1996b). Fifty percent were ill or disabled, retired, at home or not working for family reasons, or looking for work. Almost 10 percent were in school or not working for some unspecified reason. According to a recent study, at any particular point in time, about one-third of welfare mothers are working (Harris, 1993). Nearly all of the poor in America are under-age children, working, or not working for legitimate reasons. This evidence contradicts the prevailing assumption that the poor are able-bodied individuals simply waiting for others to take care of them. (Schwarz and Volgy, 1992; O'Hare, 1996; Swartz and Weigert, 1996).

Perceptions of Poverty

How do Americans view the poor? The traditional American view of the poor has deep historical roots, roots that extend as far back as the ancient Greeks. In the seventh or eighth century B.C., the Greek poet Hesiod expressed for his contemporaries this view about work and the poor:

> *Work! Work, and then Hunger will not be your companion, while fair-wreathed and sublime Demeter will favor you and fill your barn with her blessings. Hunger and the idling man are bosom friends. Both gods and mortals resent the lazy man, a man no more ambitious than the stingless drones that feed on the bees' labor in wasteful sloth. Let there be order and measure in your own work until your barns are filled with the season's harvest. Riches and flocks of sheep go to those who work. If you work, you will be dearer to immortals and mortals; they both loathe the indolent (Hesiod, 1983:74–75).*

The American view of the poor also has historical roots in the adaptation of Charles Darwin's theory of biological evolution to the social world. According to Darwin, only the fittest plants and animals survive in a particular environment. Those plants and animals that survive carry with them physical characteristics that give them a competitive edge for continued survival. The process of natural selection ensures that the fittest survive. When applied to human society, this view is called social Darwinism. It contends that the fittest individuals survive as a result of social selection and that the inferior are eliminated by the same process.

Although social scientists have long rejected social Darwinism, this philosophy has contributed to that part of America's basic value system that is variously labeled the work ethic, the achievement and success ideology, and the ideology of individualism. Individualism in America involves several central beliefs:

1. That each individual should work hard and strive to succeed in competition with others.
2. That those who work hard should be rewarded with success (seen as wealth, property, prestige, and power).
3. That because of widespread and equal opportunity, those who work hard will in fact be rewarded with success.
4. That economic failure is an individual's own fault and reveals lack of effort and other character defects (Feagin, 1975:91–92).

The elderly constitute a significant portion of America's poor, particularly older members of minorities. Nearly one-tenth of the poor are sixty-five or over and about one-tenth of the elderly live in poverty.

According to the ideology of individualism, then, those at the bottom are where they belong because they lack the ability, energy, and motivation to survive in a competitive social world. From this vantage point, it is the poor who are to blame for their condition—those who fail deserve to do so (Ryan, 1976). In its most extreme form, the ideology of individualism contends that the enactment of legislation to protect the poor is harmful because protective legislation interferes with the process of natural selection, thus slowing social evolution (Lux, 1990).

To test whether the ideology of individualism still affects the American public's view of the poor, James Kluegel and Eliot Smith (1986) conducted a nationwide survey. Respondents were asked to judge the importance of a list of reasons some people give to explain why there are poor people in America. Table 8.3 shows that individualistic explanations of poverty were the most popular. Lack of thrift, lack of effort, lack of ability, and loose morals and drunkenness—all individualistic causes—are considered to be the most important reasons for poverty. Unlike advanced European nations, much less emphasis is placed on structural explanations that locate the causes of poverty in social and economic factors beyond the control of the poor. Although structural explanations for poverty have grown in importance in recent years, the individualistic interpretation remains dominant, particularly among affluent, higher-status, white, older Americans (Nilson, 1981; Kluegel, 1990; Wilson, 1997).

Is this negative image of the poor accurate? Several studies have challenged the belief that the poor are lazy and opposed to work and prefer to live off others. Studies of welfare mothers indicate that about three-fourths prefer employment to staying at home. Most say they were not presently working because of their

Reasons for Poverty	Percentage Replying		
	Very Important	Somewhat Important	Not Important
INDIVIDUALISTIC EXPLANATIONS			
Lack of thrift and proper money management by poor people	64%	30%	6%
Lack of effort by the poor themselves	53	39	8
Lack of ability and talent among poor people	53	35	11
Loose morals and drunkenness	44	30	27
STRUCTURAL EXPLANATIONS			
Low wages in some businesses and industries	40	47	14
Failure of society to provide good schools for many Americans	46	29	26
Prejudice and discrimination against blacks	31	44	25
Failure of private industry to provide enough jobs	35	39	28
Being taken advantage of by rich people	20	35	45

TABLE 8.3 Perceived Reasons for Poverty in the United States

Source: Adapted with permission from James R. Kluegel and Eliot R. Smith, *Beliefs About Inequality: Americans' Views of What Ought To Be*, New York: Aldine de Gruyter, 1986, p. 79. Copyright © 1986 by James R. Kluegel and Eliot R. Smith.

children, housework, bad health, or lack of skills (Poddel, 1968; Warren and Berkowitz, 1969). A national survey indicates a strong work ethic among the poor. A representative sample of poor Americans was asked, "Which do you think is more important in life: working hard and doing what is expected of you or doing the things that give you personal satisfaction and pleasure?" Seventy-eight percent subscribed to the idea of hard work, which was higher than for the nonpoor in the sample (65 percent). In addition, the poor favored the idea of workfare—requirement by the government that physically able poor people work before they can receive poverty benefits—to a somewhat greater degree than the nonpoor (59 percent versus 51 percent). This evidence suggests that the poor believe in the same work-related values and attitudes as other Americans (Lewis and Schneider, 1985). Others report no difference in desire to work between the poor and the nonpoor. Whether male or female, black or white, young or old, poor people appear to share the middle-class identification of self-esteem with work and self-support. The aspirations of the poor for achievement match those of the nonpoor; both groups desire an adequate education and a home in a good neighborhood (Goodwin, 1972; Schiller, 1973; Wilson, 1997).

In summary, Americans tend to blame the poor for their poverty. This belief persists despite evidence showing that the poor are not the lazy, shiftless, willingly dependent people they are popularly thought to be. Both the attitudes and the behavior of the poor contradict the negative label Americans generally apply to them.

Explanations of Poverty

Are the rich and the poor different from each other? Consider the following two passages from American literature.

Let me tell you about the rich. They are different from you and me. They possess and enjoy early, and it does something to them, makes them soft where we are hard, and cynical where we are trustful, in a way that, unless you were born rich, it is difficult to understand. They think, deep in their hearts, that they are better than we are because we had to discover the compensations and refuges of life for ourselves. Even when they enter deep into our world or sink below us, they still think that they are better than we are. They are different.

F. Scott Fitzgerald
"The Rich Boy," 1926

The rich were dull and they drank too much, or they played too much backgammon. They were dull and they were repetitious. He remembered poor Julian and his romantic awe for them and how he had started a story once that began, "The very rich are different from you and me." And how someone had said to Julian, "Yes, they have more money."

Ernest Hemingway
The Snows of Kilimanjaro, *1936*

Fitzgerald is saying that the rich are really quite unlike other kinds of people, but Hemingway contends that the only thing separating the rich and the nonrich is money. Something of a parallel to this literary disagreement exists in the social science debate between two explanations for poverty: one cultural, the other situational.

What is the cultural explanation of poverty? There are those who contend that the poor participate in a subculture of poverty. That is, the poor are said to have evolved certain ways of thinking, feeling, and behaving that are different from those of participants in the larger culture and that are passed on from generation to generation.

Oscar Lewis (1978), a noted anthropologist, sees the creation of a subculture among the poor as an adaptation, an attempt at self-defense on the part of people at the bottom of society. One notable aspect of the subculture of poverty is identified as the absence of a success orientation. Realizing that they cannot experience success according to the terms of the larger society, states Lewis, the poor attempt to deal with feelings of discouragement and despair by creating and subscribing to values, beliefs, and attitudes better suited to their deprived condition.

An important implication of the cultural explanation is that changing the objective situation of the poor does not ensure that they will make the effort needed to get out of poverty. Their cultural traditions, originally created to help them cope with their social and economic helplessness, are said to be passed from generation to generation and to persist even when the conditions that gave rise to them no longer exist. These subcultural ways of thinking, feeling, and behaving are thought to stand as a barrier to rising above poverty. Lewis is quite explicit on this point: "By school age, children of poverty are so thoroughly socialized in their subculture and so psychologically stunted that they are seldom able to capitalize on opportunities they may encounter in later life" (Lewis, 1978:6).

Richard Ball (1968) also sees subcultural traits among the poor as an adaptation to the reality of living a denying and frustrating existence. He contends that persons in poverty face such overwhelming problems that they build a way of life based on dependence, belligerence, and fatalistic resignation. Although the subculture that is developed does not permit the poor to deal effectively with their situation, it soothes the pain of day-to-day living and provides some relief from the discomforts of frustration. Like Lewis, Ball believes that the subculture of poverty becomes a way of life and that it endures even when the conditions that created it no longer exist.

Although he recognizes the influence of lower-class subculture, Edward Banfield (1970) argues that the primary cause of poverty is the poor's attitude toward providing for the future. According to Banfield, the poor have difficulty imagining their future and lack the capability to make sacrifices today in order to have a better tomorrow. Their view of the world is said to be situated exclusively in the present time.

The lower-class individual lives from moment to moment. If he has any awareness of a future, it is of something fixed, fated, beyond his control: Things happen to him, he does not make them happen. Impulse governs his behavior either because he cannot discipline himself to sacrifice a present for a future satisfaction or because he has no sense of the future. He is therefore radically improvident: whatever he cannot consume immediately he considers valueless. His bodily needs (especially for sex) and his taste for "action" take precedence over everything else—and certainly over any work routine. He works only as he must to stay alive, and drifts from one unskilled job to another, taking no interest in the work (Banfield, 1970:53).

What is the situational explanation of poverty? Those who favor a situational explanation of poverty contend that the nonpoor have social, educational, and occupational opportunities that are not available to the poor. Change this situation and the poor will also change is the message of those who support the situational explanation of poverty.

According to Lee Rainwater (1967), the lower class only appears to have rejected middle-class values. In fact, this apparent rejection serves as a camouflage for those whose life experiences have hammered home the assumption that failure is inevitable when one attempts to succeed without the necessary resources and opportunities. It follows, Rainwater asserts, that a radical change in available social and economic opportunity would reveal the true desire among lower-class persons to achieve occupational success and to live the "good life."

William Ryan (1976) views the subculture-of-poverty explanations as a new form of "blaming the victim," of finding the cause of poverty in the poor themselves. Whereas in the past, in the context of social Darwinism, the poor were thought to be biologically defective, the new version locates the defects of the poor in their social environment.

Which explanation of poverty is correct? Each of these two theories of poverty is incomplete when considered alone. The theories are complementary explanations that in combination increase our understanding of poverty.

An early attempt to include elements of both the cultural and situational theories of poverty has been made by Hyman Rodman (1963). Like the culturalists, Rodman believes that lower-class values, beliefs, and attitudes help the poor adjust to their deprived and frustrating situation. But he attempts to clarify the nature of this adaptation by way of the concept of the "lower-class value stretch." According to Rodman, lower-class persons who realize the impossibility of living up to general societal goals minimize potential pain and frustration by settling for less. They do not decide that success is unimportant, just that less of it will suffice. Middle-class values are not abandoned; they are stretched to fit the lower-class condition.

Lower class persons in close interaction with each other and faced with similar problems do not long remain in a state of mutual ignorance. They do not maintain a strong commitment to middle class values that they cannot attain, and they do not continue to respond to others in a rewarding or punishing way simply on the basis of whether these others are living up to the middle class values. A change takes place. They come to tolerate and eventually to evaluate favorably certain deviations from the middle class values. In this way they need not be continually frustrated by their failure to live up to unattainable values. The result is a stretched value system with a low degree of commitment to all the values within the range, including the dominant, middle-class values (Rodman, 1963:209).

Focusing on African Americans, William Julius Wilson (1991, 1997) has called for an end to the simplistic either/or distinction between the cultural and the situational explanations of poverty. Wilson wishes to synthesize the situational and the cultural explanations. He begins by documenting the "weak labor-force attachment" of the inner-city poor. Weak labor-force attachment does not refer to an unwillingness or lack of desire to work. Rather, some groups are weakly attached to the labor force because limited opportunities built into the structure of the economy render them more vulnerable to joblessness than others. On the cultural side, Wilson utilizes the social cognitive theory of Albert Bandura (1982, 1986). Specifically, Wilson focuses on "perceived self-efficacy"—belief in one's ability to do what is necessary to achieve the goals in a particular situation. When members of some groups, such as people living in inner-city ghettos, continuously experience failure to be strongly attached to the labor force, they begin to share a lower "collective efficacy" as their individual perceived self-efficacy declines. The transmission of low self- and collective efficacy is, of course, a cultural phenomenon. If the structural problem of weak labor-force attachment were solved for the ghetto poor, it follows, perceived self- and collective efficacy would increase. Put another way, economic, political, and social improvements will produce positive changes in the cultural and behavioral characteristics of those in poverty.

Those who advocate a cultural or a situational explanation of poverty do not agree on the likelihood of the poor's changing if their situation were to improve, but they do agree that poverty is a problem of structure. The philosophy of letting the poor help themselves ignores the existing barriers to continuous employment and self-support that are beyond the control of poor people. It glosses over the ways in which educational and economic structures are arranged to induce failure among the poor. Inferior education, dead-end and low-paying

jobs, and discrimination cannot easily be overcome by individual motivation and hard work. These are characteristics of social structures rather than of individuals. Oscar Lewis, a proponent of the cultural explanation, also sees poverty as a problem of structure.

There is nothing in the concept [of the subculture of poverty] that puts the onus of poverty on the character of the poor. Nor does the concept in any way play down the exploitation and neglect suffered by the poor. Indeed, the subculture of poverty is part of the larger culture of capitalism, whose social and economic system channels wealth into the hands of a relatively small group and thereby makes for the growth of sharp class distinctions (Lewis, 1978:20).

Lewis touches on the fundamental source of poverty. Whether or not participation in a subculture of poverty helps to keep the poor in poverty, the basic cause of poverty is the economic and political nature of American society.

Combating Poverty

Prior to the mid-1960s, poverty was not a prominent concern of the federal government. The New Deal policies and programs of the 1930s were generally conceived as short-range measures to handle the economic crisis of the Depression. The interest in poverty as a nationwide problem publicized by Presidents John Kennedy and Lyndon Johnson, however, led the federal government to launch the broadest attempt to combat poverty ever made in America. In August 1964, President Lyndon Johnson began the War on Poverty by signing into law the Economic Opportunity Act (Danzinger and Weinberg, 1986; Trattner, 1989; Levitan, 1990).

The philosophy behind the Economic Opportunity Act of 1964 was to help poor people help themselves (Patterson, 1986). The intent was to enhance the income-generating capacity of the poor. If the chains of poverty were to be broken, President Kennedy had said, it had to be through self-improvement rather than through temporary relief. The Economic Opportunity Act took as its primary target the young, those who were the most likely to rise out of poverty. The antipoverty programs concentrated on either providing employment or offering services that would lead to employment. Thus, almost 60 percent of the original budget for the Economic Opportunity Act was earmarked for youth opportunity programs and the work experience program (a program of work and job training designed primarily for welfare recipients and unemployed fathers). The bulk of the remaining funds was allocated to urban and rural community-action programs (33 percent).

Some changes in the social welfare policy of the federal government have occurred over the past several years. One important change has been a gradual move away from workforce development and education toward cash payments, in-kind transfers (provision of goods rather than money), and direct services (legal services and medical care). Although workforce development and educational improvement are still part of the government's programs, they account for a smaller percentage of the poverty budget (O'Hare, 1987).

What has been the progress against poverty in recent years? Figure 8.3 clearly shows that according to the government's fixed standard of poverty, both the number of poor people and the percentage of the total population that the poor represent declined between 1959 and 1973. In 1959, approximately 22 percent of the American population (39.5 million persons) lived in poverty. By 1973, that figure had dropped to about 11 percent (23 million persons). The decline in poverty was relatively steady during the years when the federal government's commitment to poverty policies and programs was at its peak. The trend toward a reduction in poverty, however, began to level off in 1973, and poverty increased in the 1980s. By 1995, 36.4 million Americans—13.8 percent of the population—remained below the official poverty line.

The plight of the poor is worse than it appears. There is reason to believe that the federal government's official poverty threshold is unrealistically low (Wilson, 1991). The Bureau of the Census now identifies those Americans with annual incomes at least 50 percent below the official poverty line. In 1975, 30 percent of those below the poverty line fell in this "poorest of the poor" category; the percentage rose to 38 percent by 1995.

Has America's effort to help the poor been a failure? According to Charles Murray, the war on poverty in America has clearly been lost. Murray's conclusion, presented in his book *Losing Ground* (1984), rests on what he calls the "poverty/spending paradox"—poverty has not declined since 1970 despite a dramatic increase in federal welfare spending. The reason poverty has not been reduced, Murray contends, lies in a welfare policy that breeds dependency among recipients of government welfare aid. In particular, Murray asserts that alterations between 1960 and 1970 in the policies underlying Aid to Families with Dependent Children (AFDC) encouraged recipients to abandon work and marriage and to have children out of wedlock in order to increase their benefits. George Gilder (1981) believes that welfare benefits (read as AFDC) provide such a good living that the poor avoid work. According to Gilder, the only proven path away from poverty is work.

> *The first principle is that in order to move up, the poor must not only work; they must work harder than the classes above them. Every previous generation of the lower class has made*

America's efforts to help the poor have been defended by its supporters and criticized by its detractors. Both groups, however, tend to agree that the Head Start program has been a success.

> *such efforts. But the current poor, white even more than black, are refusing to work hard. . . . The poor choose leisure not because of moral weaknesses, but because they are paid to do so (Gilder, 1981:68).*

In response to Murray and other "warriors of welfare," Theodore Marmor, Jerry Mashaw, and Philip Harvey wrote *America's Misunderstood Welfare State* (1992). Although these authors concede that total federal social welfare spending increased greatly between 1960 and 1996 (when welfare reform was enacted by the U.S. Congress), less money in real terms was spent on AFDC recipients in 1996 than in 1970. The AFDC program, which is what most people mean when they cite the failure of poverty programs, constituted only 3 percent of total federal social welfare expenditures prior to 1996. (See Figure 8.5.) By bashing AFDC, critics of welfare are said to commit the logical error of "extension"—they extend their opponent's position to an easy-to-attack position that actually doesn't exist (Browne and Keeley, 1990). They attribute to AFDC a fiscal role that does not remotely correspond to reality. Social welfare in America is, they argue, not reducible to AFDC. Benefiting primarily the elderly, social insurance programs—Social Security, public employee retirement, railroad retirement, unemployment insurance, workers' compensation, and a few other programs—account for over 70 percent of all welfare expenditures. Programs based on need, which constitute the "safety net," currently comprise less than 30 percent of all social welfare spending.

Programs of the American welfare state, including AFDC, cannot eliminate poverty, Marmor, Mashaw, and Harvey contend, because this was never their sole purpose. According to these authors, social welfare programs have been designed to encourage the poor to

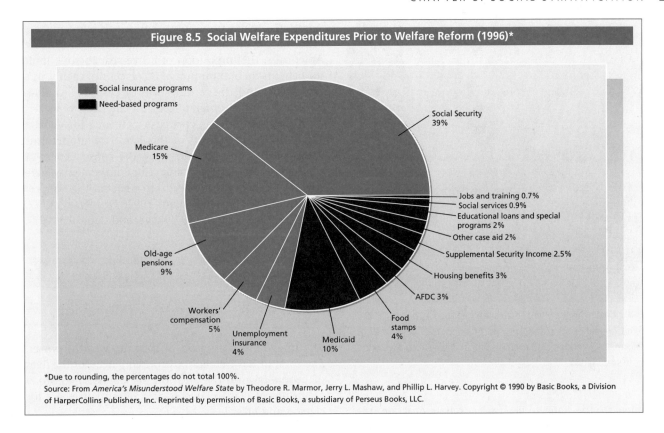

Figure 8.5 Social Welfare Expenditures Prior to Welfare Reform (1996)*

Social insurance programs

Need-based programs

Social Security
39%

Medicare
15%

Jobs and training 0.7%
Social services 0.9%
Educational loans and special
programs 2%
Other case aid 2%
Supplemental Security Income 2.5%
Housing benefits 3%
AFDC 3%

Old-age
pensions
9%

Food
stamps
4%

Workers'
compensation
5%

Medicaid
10%

Unemployment
insurance
4%

*Due to rounding, the percentages do not total 100%.

Source: From *America's Misunderstood Welfare State* by Theodore R. Marmor, Jerry L. Mashaw, and Phillip L. Harvey. Copyright © 1990 by Basic Books, a Division of HarperCollins Publishers, Inc. Reprinted by permission of Basic Books, a subsidiary of Perseus Books, LLC.

behave in more socially acceptable ways, such as working at any available jobs; to provide a "safety net" for the "deserving poor"; and to protect recipients from economic insecurities engendered by sickness, injury, retirement, involuntary unemployment, and widowhood.

This debate regarding the government's role in combating poverty is not over. Murray's view, however, is clearly imprinted on the contents of the Personal Responsibility and Work Opportunity Reconciliation Act, the welfare reform legislation passed by the U.S. Congress in the summer of 1996.

Welfare Reform

What is the target of welfare reform in 1996?   When the term *welfare reform* is used in the United States, the reference generally is to the AFDC program. As shown in Figure 8.5 and highlighted in the *Sociological Imagination* at the beginning of this chapter, this program accounted for 3 percent of all federal governmental social welfare expenditures prior to 1996. And of all federal, state, and local government means-tested welfare programs—where eligibility is determined on the basis of need—AFDC accounted for only about $21 billion of a total outlay of $210 billion prior to welfare reform. Still, the AFDC program remained the hot button of welfare reform (Vobejda, 1994).

How hot is the welfare reform button?   As you have seen earlier in this chapter, Americans feel very negatively about the poor in general and welfare recipients

in particular. However, prior to welfare reform, only 7 percent of the American public favored eliminating all welfare programs entirely, and only 25 percent agreed to cutting the amount of money given to all people on welfare. Still, 81 percent advocated fundamental welfare reform and 16 percent endorsed minor reform (Gibbs, 1994a).

Although the American public favored helping the poor, it overwhelmingly endorsed "workfare." Americans strongly indicated that able-bodied welfare recipients, primarily women with children, should work. Although the public did not wish to abolish welfare, it favored a social policy based on pro-family and pro-work principles (Califano, 1993; Gueron, 1993). Consistent with public opinion, the 1996 welfare reform legislation was based on the workfare philosophy.

What is the nature of welfare reform?   During his 1992 presidential campaign, Clinton consistently promised "to end welfare as we know it." Although major differences exist between a reelected President Clinton in 1996 and the Republican majority in Congress, the major point of agreement was the need for the welfare system to incorporate the concept of workfare with limits on the amount of time the able-bodied can receive welfare payments. The most recent welfare reform legislation reflects this agreement.

This legislation places the federal government's relationship with the poor on a fundamentally different basis (O'Hare, 1996). The bill's three major elements are reductions in welfare spending, the transfer of more

Public opinion in the United States reached such a peak in 1996 that Congress passed, and President Clinton signed, a welfare reform law. Workfare and eligibility limitations were cornerstones of this legislation.

power to the states to determine welfare program rules, and the creation of new restrictions on welfare eligibility (for example, the denial of benefits to children of unwed teenage mothers unless their mothers remain in school and reside with an adult as well as to many legal immigrants who are not U.S. citizens and a restriction in eligibility of disabled children for Social Security benefits).

Additional major provisions of this legislation include elimination of the guarantee of benefits to eligible Americans, the imposition of a five-year lifetime limit for benefits for a minimum of 80 percent of each state's recipients, and the termination of cash aid to able-bodied adults if they fail to get a job after two years.

Welfare reform has created a new area for debate: What will be the effects of this latest welfare reform? Although it is too early to assess the consequences of this legislation, defenders and critics of the new welfare reform law have already begun to argue for their opposing interpretations (Edelman, 1997; Grunwald, 1997; Harris and Havemann, 1997; Meyer and Cancian, 1997).

CONSEQUENCES OF STRATIFICATION

Social Class, Attitudes, and Behavior

The unequal distribution of income, wealth, power, and prestige has tremendous consequences for the lives of people beyond the creation of poverty. Two broad categories of class-related social consequences are *life chances* and *lifestyle*.

Life Chances

As noted earlier, life chances refers to the likelihood of possessing the good things in life, including health, happiness, wealth, legal protection, and life itself. The

FEEDBACK

1. _____ poverty is defined as the absence of enough money to secure life's necessities.
2. _____ poverty is measured by comparing the economic condition of those at the bottom of a society with other segments of the population.
3. Which of the following is *not* one of the major categories of the poor in the United States?
   a. children under eighteen
   b. able-bodied men who refuse to work
   c. the elderly
   d. the disabled
   e. people who live alone or with nonrelatives
4. According to the ideology of _____, those at the bottom of the stratification structure are there because of their own inadequacies.
5. America's poor have clearly different work-related values and behaviors than other Americans. *T or F?*
6. According to the _____ explanation of poverty, changing the objective situation of the poor does not ensure that the poor will make the effort needed to get out of poverty.
7. According to the _____ explanation of poverty, the difference between the nonpoor and the poor is that the former have social, educational, and occupational opportunities that are not available to the latter.
8. The philosophy behind the Economic Opportunity Act of 1964 was
   a. to provide temporary relief for the poor
   b. to make the poor dependent on the federal government
   c. to make it easy for the poor to receive help
   d. to help poor people help themselves
9. As of the late 1980s, poverty in America was
   a. increasing slightly
   b. decreasing dramatically
   c. increasing sharply
   d. at about the 1980 level
10. The American public is strongly opposed to the continued existence of a welfare system. *T or F?*

Answers: 1. Absolute 2. Relative 3. b. 4. individualism 5. F 6. cultural 7. situational 8. d. 9. d. 10. F.

TABLE 8.4	Life Chances by Social Class*		
	Lower Class	Middle Class	Upper Middle and Upper Class
In excellent health	28%	37%	53%
Victims of violent crime per 1,000 population	50	29	21
Psychologically "impaired" per 100 "well"	470	135	46
Feel lonely frequently or sometimes	46%	35%	27%
Obesity in native-born women	52%	43%	9%
Children, 18–24, in college	15%	38%	54%
Dissatisfied with personal life	22%	15%	5%

*Classes defined by income.

Source: Dennis Gilbert and Joseph A. Kahl, *The American Class Structure*, Belmont, CA: Wadsworth, 1993, p. 3. Reprinted by permission. Copyright 1993.

probability of acquiring and maintaining the material and nonmaterial rewards of life is significantly affected by social-class level. Power, prestige, and economic rewards increase with social-class level. The same is true for education, the single most important gateway to these rewards. But additional life chances exist, much more subtle ones, that vary among social classes. (See Table 8.4.) These less obvious life chances are in good part the product of inequality in the distribution of education, power, prestige, and economic rewards.

How does social class affect more subtle life chances? The probability of possessing life itself—the most precious life chance—declines with social-class level. Whether measured by the death rate or by life expectancy, the likelihood of a longer life is enhanced as people move up a stratification structure (U.S. Bureau of the Census, 1996a). This disparity in the most fundamental life chance is due to differences in the value placed on medical attention, concern with proper nutrition, attention to personal hygiene, and ability to afford these things.

In the light of differences in life expectancy, it is not surprising that physical health is affected by social-class level. Those lower in a stratification structure are more likely to be sick or disabled and to receive poorer medical treatment once they are ill (Califano, 1985; Wilkinson, 1986; Conrad and Kern, 1990). It is no different for mental health. Persons at lower-class levels have a greater probability of becoming mentally disturbed and are less likely to receive therapeutic help, adequate or otherwise (Hollingshead and Redlich, 1958; Goodman et al., 1983; Link, Dohrenwend, and Skodol, 1986).

Innumerable other inequalities exist in life chances, only a few of which can be mentioned. The poor often pay more for the same goods and services. They are more likely to get caught for committing a crime, and they stand a greater chance of being convicted and serving prison time for their alleged crimes. They are

less likely to have connections to help them beat the system, as in obtaining tickets to a football game or getting a favor from a political figure. And the public services they receive—garbage collection, police protection, street repair—are inferior (Caplovitz, 1963; Thompson and Zingraff, 1981).

Lifestyle

How does social class affect lifestyle? As research has shown, the rich and the poor are separated by much more than money. Social-class differences have been observed in many areas of American life, including education, marital and family relations, child rearing, political attitudes and behavior, and religious affiliation.

People in higher social classes tend to marry later, display greater family stability (for example, through lower divorce rates), and have better marital adjustment (Kitson and Raschke, 1981; Fergusson, Horwood, and Shannon, 1984). Although working-class and middle-class parents are more alike in child-rearing practices than in the past, some differences remain. Compared to the middle class, lower-class parents tend to be less attentive to their children's social and emotional needs; they are more inclined to use physical punishment rather than logic and reasoning in disciplining their children. Middle-class parents are more interested in helping their children develop such traits as concern for others, self-control, and curiosity. Finally, working-class parents emphasize conformity, orderliness, and neatness, whereas middle-class parents stress self-direction, freedom, initiative, and creativity (Kohn, 1963, 1977, 1990; Wright and Wright, 1976; Williamson, 1984).

The incidence of voting and involvement in politics increases with social-class level. Political attitudes are also associated with location in a stratification structure. The middle and upper classes are more likely to be Republicans, whereas the working and lower classes tend to register as Democrats. Those in the lower class tend to be more liberal than middle- and upper-class Americans on economic issues but more conservative on social issues. Thus, lower-class people are more in favor of labor unions, government control of business, and social welfare programs but tend to exhibit less tolerance and sympathy than do higher social classes for social issues involving civil rights and international affairs (Wolfinger and Rosenstone, 1980; Syzmanski, 1983).

The rate of church membership and attendance is lowest at the extremes of the stratification structure. Among those of all social classes who do attend church, there seems to be a discernible pattern of affiliation by social class. Episcopalian, Congregational, and Presbyterian churches are significantly less populated by members of the lower class. Lower-class Americans lean more toward Baptist and fundamentalist churches. Methodist and Lutheran denominations fall in between. (See Chapter 14, "Religion.")

SOCIAL MOBILITY

Social mobility refers to the movement of individuals or groups within a stratification structure. In the United States, the term *mobility* implies an elevation in social-class level. In fact, it is possible to move down in social class or to move with little or no change in social class.

Types of Social Mobility

Movement can be horizontal or vertical. Both types of mobility can be measured either within the career of an individual—**intragenerational mobility**—or from one generation to the next—**intergenerational mobility.**

What is horizontal mobility? A change from one occupation to another at the same general status level is called **horizontal mobility.** Examples of intragenerational horizontal mobility would be an Army captain who becomes a public school teacher, a minister who becomes a psychologist, and a restaurant waiter who becomes a taxi driver. The daughter of an attorney who becomes an engineer illustrates intergenerational horizontal mobility. Because horizontal mobility involves no real change in occupational status or social class,

sociologists have not been very interested in investigating it. On the other hand, vertical mobility has been investigated extensively.

What is vertical mobility? **Vertical mobility** occurs when occupational status or social class changes upward or downward. Vertical mobility can also be intragenerational or intergenerational. The simplest way to measure intragenerational mobility is to compare individuals' present occupations with their first ones. Someone who began as a dockworker and later became an insurance salesperson would have experienced upward intragenerational mobility. Intergenerational mobility involves the comparison of a parent's (or grandparent's) occupation with the child's occupation. If a lawyer's son becomes a carpenter, downward intergenerational mobility has occurred.

Caste and Open Class Stratification Systems

The extent of vertical mobility varies from society to society. Some societies have considerable mobility; others have little or none. This is the major difference between caste and open class systems.

What is a caste system? In a theoretically closed class or **caste system,** there is no social mobility, because social status is inherited and cannot be subsequently changed. In a caste system, statuses (including occupations) are ascribed or assigned at birth; individuals cannot change their statuses through any efforts of their own. By reason of religious, biological, magical, or legal justification, those in one caste are allowed to marry only within their own caste and must limit relationships of all types with those below and above them in the stratification structure. Because the caste system is an ideal type, there are few if any pure cases in real life. Ancient India and the pre–Civil War South are the best illustrations. But even within these caste systems, some small degree of social mobility did occur.

What is an open class system? An **open class system** is one in which an individual's social status is based on merit and individual effort. Rank in this type of stratification structure, then, is achieved. Individuals move up and down the stratification structure as their abilities and efforts permit. Inequality does exist in an open

FEEDBACK

1. Identify the following as examples of life chances or lifestyle.
   - ____ a. divorce rate
   - ____ b. quality of neighborhood street repair
   - ____ c. life expectancy
   - ____ d. child-rearing practices
   - ____ e. voting behavior
   - ____ f. church attendance

Answers: 1. a. lifestyle b. life chances c. life chances d. lifestyle e. lifestyle f. lifestyle class system, but it is supposed to be based on differences in personal worth and accomplishment.

Just as some social mobility occurs in an actual caste system, the opportunity for upward mobility is sometimes denied individuals or groups within an open class stratification structure. Prominent illustrations in American society today—which is relatively, but not purely, an open class system—are African Americans, Native Americans, and Latinos. (See Chapter 9, "Inequalities of Race, Ethnicity, and Age.")

Upward Social Mobility

The conventional wisdom in American society is that economic and occupational success are the products of talent and willingness to work. Sociologists, who have consistently challenged this belief, have attributed success or failure to factors associated with the social and economic background of individuals. Those most likely to succeed are, for example, those whose parents have been successful.

How much upward mobility exists in America? The Horatio Alger stories—in which a down-on-his-luck boy makes good through honesty, pluck, and diligence—embody the long-standing belief that America is the land of opportunity. The only thing standing between any American citizen and success, it is argued, is talent, willingness to work, and perseverance (Kluegel and Smith, 1986). The careers of Abraham Lincoln, Cornelius Vanderbilt, John D. Rockefeller, and Henry Ford are used to demonstrate the openness of American society. Actually, these men are exceptions to rather than examples of the rule. Countless Americans have failed to be upwardly mobile despite their talents and dedication to work.

The actual extent of upward mobility has never been so great as the rags-to-riches myth would have us believe. American presidents, contrary to popular belief, were not born into the lower strata of society (Pessen, 1984). This does not mean that upward mobility does not occur in American society. In fact, the considerable mobility that has occurred has been upward. It does mean, however, that great leaps in social-class level have always been rare. Studies have consistently shown that upward mobility, when it occurs, is usually a small improvement (Sorokin, 1927; Jackson and Crockett, 1964; Blau and Duncan, 1967; Featherman, 1971; Davis, 1982; Gilbert and Kahl, 1993).

Why is upward social mobility not dramatic intergenerationally? One reason that upward social mobility does not occur in large leaps, even in an open class system, is that one's occupational level is strongly influenced by the occupational level of one's parents. Sociologists have traditionally shown this through data on the influence of fathers' occupations on their sons'

occupations. More recently, mobility studies have begun to include women (Gilbert and Kahl, 1993; Baxter, 1994; Kalmijn, 1994).

In the United States, there is considerable occupational succession (sons at the same occupational level as their fathers). Despite a significant amount of occupational mobility up and down the occupational ladder, there is a barrier between white-collar and manual jobs. About two-thirds of sons born on either side of this manual-nonmanual line do not manage to permanently cross it in their work lives. Moreover, most upward social mobility in the United States occurs in small increments. For example, the son of a janitor becomes a bus driver or the son of a retail sales worker becomes a commercial artist.

The occupational level of females is also highly correlated with their fathers' occupational level. Just over half of the daughters of upper white-collar fathers are in similar positions; only one-fourth of the daughters of lower manual workers are in upper white-collar occupations. Similarly, nearly 40 percent of the daughters of fathers in lower manual jobs are in manual work, whereas a relatively small percentage of the daughters of white-collar men hold blue-collar jobs.

Why has upward mobility occurred in the United States? Although upward mobility in the United States has not typically involved dramatic leaps, it has occurred frequently. According to Peter Blau and Otis Dudley Duncan, there are three basic causes of upward mobility in contemporary society. One cause is the labor pool of less qualified workers from other countries or from rural areas. As the less qualified take lower level jobs, the more qualified urban dwellers fill higher level positions. A second cause lies in the differential fertility rate of the social classes. Because the higher social classes have fewer children, members of the lower social classes are needed to fill higher level positions. Another important reason for social mobility is that industrialization produces greater dependence on machines and the elimination of lower level jobs. At the same time, modern technology creates more jobs at higher levels. The shrinking of job opportunities at the bottom means that some of the children of lower level manual workers simply cannot follow in their parents' occupational footsteps. Some of them have to prepare themselves to move up to higher level occupations. This type of social mobility, which occurs because of changes in the distribution of occupational opportunities, is called **structural mobility** (Lipset and Zetterberg, 1964; Tyree, Semyonov, and Hodge, 1979; Hope, 1982).

Is American society more open than other industrialized societies? Evidence presented by Seymour Martin Lipset and Reinhard Bendix (1964) indicates a similarity in the amount of upward and downward mobility across the manual-nonmanual (working-class and middle-class) line within industrialized countries: United States (34 percent), Sweden (32 percent), Great Britain (31 percent), Denmark (30 percent), Norway (30 percent), France (29 percent), West Germany (25 percent), Japan (25 percent), and Italy (22 percent). In other words, about one-fourth to one-third of the non-farm male population moves across the manual-nonmanual line in all industrialized countries (Blau and Duncan, 1967; Connor, 1979). If these rates of social mobility across the manual-nonmanual line in these societies are not identical, they do seem to bear a reasonable resemblance to one another. The United States, in this respect, is not significantly more open to mobility than other industrialized countries (Hope, 1982; Grusky and Hauser, 1984; Kaelble, 1986; Erikson and Goldthrope, 1992).

Although there is a clear trend toward more equal rates of mobility in all industrial societies, some evidence suggests that social mobility is somewhat higher in the United States, particularly into the elite level (Wong, 1990). Blau and Duncan compared the United States, Japan, and five Western European countries in terms of movement from the bottom of the stratification structure to the top. This is their conclusion: "Upward mobility from the working class into the top occupational stratum of society is higher in the United States than in other countries. . . . There is a grain of truth in the Horatio Alger myth" (Blau and Duncan, 1967: 434–435). The proportion of Americans who move from the bottom to the top of the stratification structure is greater than in other industrialized societies primarily because higher education is more accessible in the United States than in any other country (Lipset, 1982).

Lack of Mobility and Downward Mobility

Why may social mobility decrease in the future? Prior to about 1970, structural mobility—mobility fueled by changes in the distribution of occupational opportunities—contributed to the relatively high rate of upward social mobility in the United States. Since 1970, an entire generation of Americans has witnessed sluggish economic expansion and a new trend in job creation. The rapid growth of high-paying white-collar and skilled blue-collar jobs, characteristic of the American economy since the end of World War II, is being eclipsed by a polarized pattern of growth in both relatively low-skill, low-paying jobs and high-skill, high-paying jobs. The most rapidly increasing jobs in the 1990s, however, are expected to be on the low end, especially in the service area—retail salespersons, home health-care aides, waiters and waitresses, hairdressers, taxi drivers (Howe and Strauss, 1991). The exportation of high-paying skilled jobs by American corporations, along with a loss of America's competitive economic

advantage over the rest of the world (Thurow, 1992; Albert, 1993; Hampden-Turner and Trompenaars, 1993) has led some to write of "deskilling" among a significant proportion of American workers (Reich, 1989, 1993; Reskin and Padavik, 1994).

America has always been viewed as the land of opportunity, as a place in the world where one's children (if not one's self) could experience upward mobility. Americans have come to expect that they and their children will have more in the future than they have today. This, as we have just seen, very likely will not be the case for future generations. Many will fail to be upwardly mobile; many others will be downwardly mobile. There will be negative social-psychological repercussions in both instances.

What are the social-psychological costs of a lack of social mobility? Because some degree of social mobility occurs among members of all classes in an open class stratification structure, those who fail to move up may pay a social-psychological price. Because mobility is permitted, people feel a responsibility to keep trying to improve their status. And because failure to achieve higher status cannot be blamed on a system in which upward mobility is supposed to occur, the blame is placed on the nonmobile themselves, whom others judge to be inferior or unwilling to put forth effort.

Americans who have failed to move up from the working class feel a heavy responsibility to help their children over the class barrier to dignity and responsibility through an education they themselves failed to obtain. Richard Sennett and Jonathan Cobb offer these examples:

> "If you don't have them degrees, they're gonna treat you like you was nothing," says a garbage collector to his children. "I say to Sheila," remarks an electrician, "you do that homework, or you'll wind up in the same boat like me . . . It's for your own good you got to study" (Sennett and Cobb, 1973:186–187).

Although not being upwardly mobile is difficult, moving down the stratification structure is probably worse. In fact, economic and social changes are forcing one employed American in five down the mobility ladder. These social and economic forces are displacing millions of comfortable middle-class people—divorced women, laid-off blue-collar workers, fired managers, indebted farmers, unemployed professionals (Levy, 1987; Jost, 1993; Johnson, 1994).

What are the social-psychological costs of downward mobility? In *Falling from Grace*, Katherine Newman (1988) contends that America's traditional belief in the fruits of hard work prevents recognition of the problem of downward mobility for many members of the middle class. And, she argues, the consequences are enormous for people in a society that measures self-worth by occupational status. The downwardly mobile experience a loss of a sense of honor, lowered self-esteem, despair, depression, and feelings of powerlessness. More importantly, Newman concludes, those Americans currently sliding down the stratification structure are disillusioned with the promise of the American Dream. Consider this cynical conclusion from a respondent in the research of another author:

> In Peoria, Illinois, a twenty-six-year-old fifth-grade public school teacher is called in by her principal and given a pink slip. She has been teaching one year, then becomes one of nine teachers laid off because of a budget crisis. All of those let go have the least seniority. With a $10,000 student loan to pay off, she is unable to meet rent, car, and health insurance payments after her school benefits run out. She moves back into her parents' home. To earn money to go back to school and start over, she answers an ad in the paper for blackjack operators on a new gambling riverboat, the Paradise, in the Illinois River off Peoria. "I feel I did everything the way I was supposed to," she says, explaining why she believes the American Dream is dead. "I worked hard in high school. I worked hard in college. I always made the honor roll. I did my job well, and the bottom just fell out. So here I am. I want to be a teacher and what am I doing? I'm going to blackjack school!" (Johnson, 1994:25)

Consequently, writes Newman, downwardly mobile members of the middle class are without cultural guidelines for regaining their lost niche or for advising their children regarding the achievement of success and prosperity.

Because self-dislike is a very painful attitude, those skidding down a stratification structure may shift the blame to forces outside themselves. Minorities make convenient scapegoats. Downwardly mobile individuals may accuse African Americans, Jews, and Latinos of taking job opportunities away from them. By choosing an external target, the downwardly mobile can excuse themselves and vent their hostility and frustration on others. This may be quite effective, because it plays on existing prejudices against minorities. Thus, people not only can attempt to rationalize away personal accountability for failure but also can do so with considerable support from those around them.

In a later book, *Declining Fortunes*, Newman (1993) examines what she terms "the withering of the American Dream." This book is discussed in the *Doing Research* feature in Chapter 18, "Social Change in Modern Society."

GLOBAL STRATIFICATION

Thus far, the focus has been on stratification within societies. Scarce desirables—income, wealth, prestige, power—are also differentially distributed among nations. (See Figure 8.6.) World stratification is generally depicted in three rank-ordered layers. The widely disparate conditions prevailing among these categories of nations leads to entirely different worlds of daily life. The gross domestic product (GDP) of a country—the total value of the goods and services it produces in one year—is a reasonably good indicator for classifying countries into economic categories (*The World Bank Atlas 1997*, 1997). GDP is useful both because it is a single, measurable economic indicator and because a nation's rank economically is highly correlated with the extent of its prestige and power.

FEEDBACK

1. _____ refers to the movement of individuals or groups within a stratification structure.
2. Match the major types of social mobility with the illustrations:
   ____ a. intergenerational mobility (1) a restaurant waiter becomes a taxi driver
   ____ b. vertical mobility (2) an auto worker becomes a manager
   ____ c. horizontal mobility (3) the daughter of a hairdresser becomes a college professor
3. In a _____ system, social status is inherited at birth from one's parents and cannot be changed.
4. In an _____ system, rank on the stratification structure is achieved.
5. Upward mobility in America usually involves only a small improvement in status. *T or F?*
6. The son of a steelworker finds that the steel industry has shrunk so much that a job in the mill is not available. Instead of following in his father's footsteps, he attends a community college and becomes a computer operator. This is an example of
   a. horizontal mobility c. caste mobility
   b. intragenerational mobility d. structural mobility
7. Compared to other industrial countries, the United States has _____ bottom to top mobility.
   a. more b. less c. the same
8. In America, working-class individuals who fail to experience upward mobility claim that it is the system rather than themselves. *T or F?*

Answers: 1. Social mobility 2. a. (3) b. (2) c. (1) 3. caste 4. open class 5. T 6. d 7. a 8. F

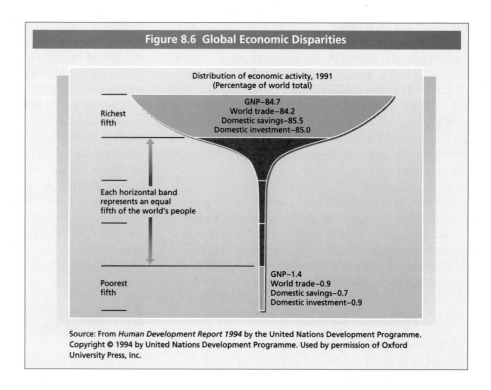

Figure 8.6 Global Economic Disparities

Distribution of economic activity, 1991
(Percentage of world total)

Richest fifth

GNP–84.7
World trade–84.2
Domestic savings–85.5
Domestic investment–85.0

Each horizontal band represents an equal fifth of the world's people

GNP–1.4
World trade–0.9
Domestic savings–0.7
Domestic investment–0.9

Poorest fifth

Source: From *Human Development Report 1994* by the United Nations Development Programme. Copyright © 1994 by United Nations Development Programme. Used by permission of Oxford University Press, Inc.

High-Income Economies

The richest countries include Western European countries such as England, Germany, France, Norway, Finland, Denmark, Sweden, and Switzerland; the United States and Canada in North America; Australia and New Zealand in Oceania; and Japan in Asia. A few oil-rich nations such as Kuwait and the United Arab Emirates also have high-income economies. (These are the countries in light blue in Figure 8.7.) Despite some variations, these countries are based on capitalist economies. In 1997, high-income economies, spanning approximately one-quarter of the earth's land surface in continents around the world, had just 15 percent of the world's population (approximately eight hundred million people). Yet, these countries are in control of most of the world's wealth. Because of this concentration of wealth, the standard of living of nearly all the persons living in these countries, even most of its poor, is higher than the average person in low-income economies.

Middle-Income Economies

Middle-income economies (those in light green in Figure 8.7) include, but are not limited to, nations historically founded on, or influenced by, socialist or communist economies. This encompasses members of the former Soviet Union, many renamed and now part of the new Commonwealth of Independent States created in 1991. Other countries that fell into the Soviet orbit but are not part of the Commonwealth of Independent States include Poland, Bulgaria, the Czech Republic, the

Slovak Republic, Hungary, Romania, Estonia, Latvia, Lithuania, and Cuba.

From the end of World War II until 1989, all these countries, under the political, economic, and military control of the former Soviet Union, constituted the Eastern bloc in the cold war against the West. These countries are no more densely populated than high-income economies—they occupy 15 percent of the earth's land with only 10 percent of the world's population (about five hundred million people). Although more prosperous than low-income countries, citizens of middle-income economies do not enjoy the standard of living characteristic of the high-income economies. The extent of industrialization in these countries is variable. Although most inhabitants of these countries live in urban areas, a higher percentage than the population in high-income economies live in rural areas and participate in agricultural economic activities. In 1989, Eastern European countries, freed of Soviet domination and its socialist economic structure controlled by government bureaucracy, gravitated to capitalism and its market mechanism. One of the major sources of political and economic turmoil now occurring in the former Soviet Union itself revolves around the struggle between the socialist and capitalist roads to economic development. (See Chapter 13, "Political and Economic Institutions," for further discussion.)

The middle-income category contains other countries that have never been part of the old Soviet bloc. These countries are spread from Mexico, Latin America, and South America to parts of Africa to Thailand, Malaysia, and New Guinea in the Far East.

Figure 8.7 Groups of Economies

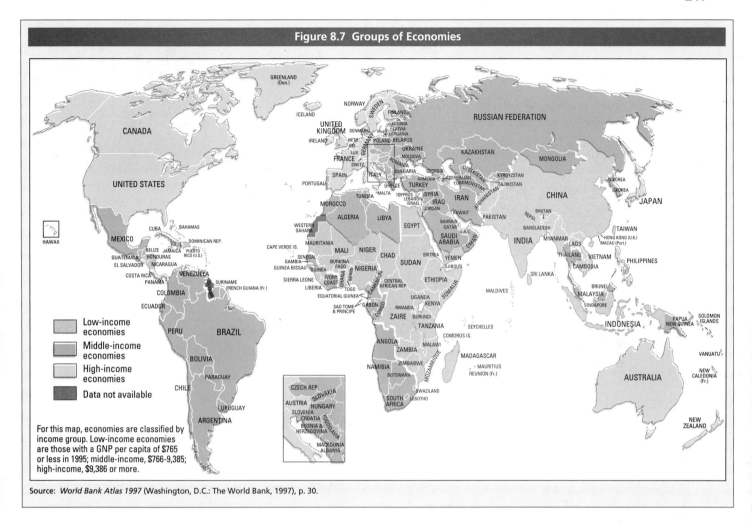

Low-income economies

Middle-income economies

High-income economies

Data not available

For this map, economies are classified by income group. Low-income economies are those with a GNP per capita of $765 or less in 1995; middle-income, $766-9,385; high-income, $9,386 or more.

Source: *World Bank Atlas 1997* (Washington, D.C.: The World Bank, 1997), p. 30.

Low-Income Economies

The economic base of low-income economies is primarily agricultural rather than industrial. These countries span the globe, from south of the border of the United States to Africa to China to Indonesia. (These are the countries in dark gray in Figure 8.7.) Population density within these countries as a whole is very high. In fact, just over two-thirds of the world's population (over five billion people) live on 60 percent of the earth's land.

No single economic system is typical of these economies. Capitalism, socialism, and various combinations of each are easily found. If these countries share no economic system, they do share one important economic characteristic—they are unimaginably poor. Problems of the poor in high-income countries, which have already been explored in this chapter, are real and significant. It does not make light of the poor in the affluent societies to point out that most people in low-income economies live annually on less than 7 percent of the amount of money used to determine the official poverty line in the United States (U.S. Bureau of the Census, 1996a). The plight of the poor in low-income

countries is particularly bleak because of a double bind in which they are trapped. These countries fall further behind the rest of the world each year because of their exploding populations created by a decline in their infant mortality rates due to improved medical care and the maintenance of their traditionally high birthrates. (See Chapter 16, "Population and Urbanization.")

Global Poverty

Poverty persists despite the development of a $425 trillion global economy. A recent international report details the extent and distribution of global poverty (*Human Development Report,* 1997).

What is the state of poverty at the end of the twentieth century? Despite progress, one-quarter of the earth's population remains in poverty. One-third of the people in developing countries—1.3 billion—have incomes of less than $1 daily. Poverty is most extensive in South Asia—over 500 million people are poor by this standard. South Asia, East Asia, Southeast Asia, and the Pacific have over 400 million of these 1.3 billion people. Sub-Saharan Africa now has 200 million of its

SOCIOLOGY IN THE NEWS

Stratification in Brazil

This CNN film clip reports on the place of blacks in the Brazilian stratification structure. Afro Brazilians, comprising about 50 percent of the population, are at the bottom of a stratification pyramid with a very wide base. Most of them live in poverty. According to their leaders, Afro Brazilians are worse off than blacks were in South Africa under apartheid, because the racism that keeps them down is denied by whites and ignored by most blacks.

1. What would symbolic interactionists point out about the situation of Afro Brazilians?

   _____

   _____

2. Does the cultural or situational explanation of poverty best describe Afro Brazilians? Why?

   _____

   _____

Despite worldwide economic progress, one-fourth of the world's population remains in poverty. Afro Brazilians in South America are among the most impoverished.

people in poverty and the rate is increasing. According to estimates, one-half of the people in this area will be in poverty by the turn of the century. In Latin America and the Caribbean, over 100 million people have less than $1 a day on which to live.

In the more economically developed world, the greatest increase in poverty in the last ten years is found in Eastern European countries and among members of the Commonwealth of Independent States, countries of the former Soviet Union. Whereas poverty in these areas was formerly not widespread, about a third of these people (120 million) now have incomes below $4 daily. Even in industrial nations, over 100 million people subsist below the established poverty line (one-half of individual median income within a country).

Has there been no progress in combating global poverty? Poverty has actually been significantly reduced in many areas of the world. Poverty has decreased in some ways in nearly all nations. In fact, the past fifty years has seen poverty decrease more than in the preceding five centuries in all regions of the world. In less than twenty years, China and fourteen other countries whose populations exceed 1.6 billion have cut in half the proportion of their people existing in poverty. An additional ten nations, who number almost one billion, have cut their poverty rate by one-fourth. Still, as just indicated, poverty is pervasive throughout the world and may increase as a result of mounting global forces such as slowed economic growth and persisting conflict within and between countries.

FEEDBACK

1. Classify the following characteristics as high-income (HI), middle-income (MI), or low-income (LI):
   - ____ a. Commonwealth of Independent States
   - ____ b. agricultural economic base
   - ____ c. movement toward capitalism
   - ____ d. eight hundred million people
   - ____ e. double population bind
   - ____ f. one-fourth of the earth's land surface
2. Do you think that classifying all the nations of the world into three broad categories leads to some distortion of reality? Explain.

   _____

   _____

3. What proportion of the earth's population is considered poor today?
   a. one-tenth   c. one-fourth
   b. one-fifth   d. one-third
4. Over the past fifty years, global poverty has actually decreased. *T or F?*

Answers: 1.a. (MI) b. (LI) c. (MI) d. (HI) e. (LI) F. (HI) 3. c. 4. T.

SUMMARY

1. There is an almost universal tendency to create inequality within human groups. When individuals are ranked by the amount of wealth, prestige, and power associated with them, social stratification exists. Any stratification structure is composed of either social classes or social castes—segments of a population whose members have a similar share of the desirable things and who share attitudes, values, norms, and an identifiable lifestyle.

2. A stratification structure in a community or society is based on economic, power, and prestige dimensions. Income inequality has always existed in American society, but it has increased dramatically since 1980. Prestige, especially in developed societies, is based primarily on occupation. Those occupations that generally have the greatest prestige are those that pay the most; require the greatest amount of training, skill, and ability; provide the most power; and are the most important.

3. Those individuals who are ranked high on one dimension of stratification are usually high on the other dimensions. Wealth, power, and prestige tend to go together.

4. Functionalists contend that stratification is necessary to motivate people to prepare themselves for difficult and important jobs and to motivate them to perform well once they are in those jobs. According to the conflict theory of stratification, inequality exists because some people have more power than others and are willing to use it to promote their self-interest. Because each of these theories has some validity, sociologists are becoming interested in combining them. Lenski's evolutionary theory of stratification, which looks at the relationship between technological development and inequality, attempts to synthesize the functionalist and conflict perspectives.

5. The functionalist, conflict, and evolutionary theories shed light on the reasons for the existence of stratification. Symbolic interactionism aids in understanding the manner in which stratification structures are perpetuated. This perspective emphasizes the use of symbols learned through the process of socialization and in the development of the self-concept.

6. Although most sociologists believe that class lines in America are difficult to draw, they agree that the following social classes have been identified: the upper class, the middle class, the working class, and the lower class. Sociologists agree that the poor exist as an underclass.

7. Poverty can be measured in an absolute or a relative way. Absolute measures establish an annual income level below which people are considered to be poor. Relative poverty is measured by comparing those at the bottom of the stratification structure with income groups above them. Because of the continuing influence of social Darwinism, Americans still tend to blame the poor for their own situation, despite evidence to the contrary.

8. There are two major explanations for poverty. According to the cultural explanation, the poor are different from the nonpoor because they are enmeshed in a subculture of poverty. Because the poor have distinctive ways of acting, thinking, and feeling, they probably could not take immediate advantage of any removal of barriers to their upward mobility. According to the situational explanation, on the other hand, the poor are poor only because social, educational, and occupational opportunities are not available to them. Change this situation by providing opportunities, and the poor would respond.

9. The federal government became serious about combating poverty during the mid-1960s. During the period of intense government effort, poverty in fact declined. This decline began to level off in 1973, and the poverty level began to rise during the early 1980s. In 1995, the percentage of Americans in poverty was higher than in 1973.

10. Americans do not wish to abolish all welfare benefits, but they do want a system that forces the able-bodied off the rolls. In 1996, the U.S. Congress passed a sweeping welfare reform act that changed dramatically the relationship between the poor and their government.

11. Social classes can be thought of as subcultures. For this reason, distinctive patterns of thinking, feeling, and behaving are associated with the various social classes. Both life chances and lifestyle characteristics are affected by one's social-class level.

12. Social mobility, the movement of individuals or groups within the stratification structure, is usually measured by changes in occupational status. Sometimes movement within a stratification structure is lateral, or horizontal. Sociologists, however, are much more interested in upward or downward (vertical) mobility, whether within an individual's lifetime or from one generation to the next. Closed class societies, such as India, permit little vertical mobility; open class societies, such as those in industrialized countries, allow considerable upward mobility. Even in open class societies, however, an upward move tends to be a small one. Although considerable mobility exists in the United States, it is not appreciably more open than in other major industrialized countries, except for movement from the bottom of the stratification structure to the top.

13. The lack of mobility and downward mobility, likely to increase in American society, have some negative social-psychological consequences. Those who cannot move from the lower levels of the stratification structure in an open class system tend to blame themselves. The downwardly mobile suffer lowered self-esteem, despair, depression, and other negative repercussions. Sometimes minorities are used as scapegoats by those sliding downward.

14. Nations are also stratified because they share unequally in available scarce desirables. The richer nations of the world, primarily in the West, are classified as high-income economies. Industrialized nations less economically advanced, including those historically associated with the former Soviet Union, can be classified as middle-income economies. The rest of the world, poor and resting primarily on an agricultural base, are low-income economies. In these societies, poverty is widespread.

LEARNING OBJECTIVES REVIEW

After careful study of this chapter, you will be able to:

- Explain the relationship between social stratification and social class.
- Compare and contrast the three dimensions of stratification.
- State the major differences among the functionalist, conflict, and symbolic interactionist approaches to social stratification.
- Identify the distinguishing characteristics of the major social classes in America.
- Discuss the measurement and perceptions of poverty in the United States.
- Evaluate the cultural and situational explanations of poverty.
- Describe the state of the effort to combat poverty in the United States.
- Outline some of the consequences of social stratification.
- Describe upward and downward social mobility in the United States.
- Discuss the major features of global stratification.

CONCEPT REVIEW

Match the following concepts with the definitions listed below them:

____ a. life chances
____ b. horizontal mobility
____ c. caste system
____ d. intergenerational mobility

____ e. relative poverty
____ f. power
____ g. social mobility
____ h. wealth

____ i. conspicuous consumption
____ j. structural mobility

1. All the economic resources possessed by an individual or a group.
2. The consumption of goods and services in order to display one's wealth to others.
3. Social mobility that takes place from one generation to the next.
4. The ability to control the behavior of others, even against their will.
5. Movement of individuals or groups within a stratification structure.
6. The type of stratification structure in which there is no social mobility because social status is inherited and cannot subsequently be changed.
7. The likelihood of securing the "good things in life," such as housing, education, good health, and food.
8. The measurement of poverty by a comparison of the economic condition of those at the bottom of a society with other segments of the population.
9. A change from one occupation to another at the same general status level.
10. Mobility that occurs because of changes in the distribution of occupational opportunities.

CRITICAL THINKING QUESTIONS

1. Stratification can occur in any social setting. Think about a group with which you are familiar that is stratified. Discuss the "unequal shares of scarce desirables" used to rank members.

2. Consider the existence of a class at the bottom of the American social stratification structure. Contrast and compare the explanations for the enduring presence of this class that would be given by functionalism and conflict theory.

3. More and more of the poor in the United States are women and children. Relate the feminization of poverty to the prevailing view of the poor held by most Americans. Is there an inconsistency? Why or why not?

4. Evaluate the statement that the rich and the poor are different from each other within the context of attempts to combat poverty.

5. You are a member of a particular social class. Discuss your life chances and lifestyle in sociological terms.

6. Analyze the social mobility that has occurred in your family across as many generations as you can. Use specific sociological concepts in your analysis.

MULTIPLE CHOICE QUESTIONS

1. **The creation of layers of people possessing unequal shares of scarce desirables is known as**
   a. social layering.
   b. social grading.
   c. social tracking.
   d. social stratification.
   e. resource accumulation.

2. **A _____ is a segment of a population whose members hold a relatively similar share of scarce desirables and who share attitudes, values, norms, and identifiable lifestyles.**
   a. social network
   b. social nexus
   c. social group
   d. social category
   e. social class

3. **Which of the following statements regarding Marx's analysis of class differences is *false*?**
   a. All aspects of capitalistic societies are economically based.
   b. The capitalists control all social institutions, which they use to their advantage.
   c. Capitalists will both rule and exploit the working class because they own the means of production.
   d. Prestige is the prime determinant of social stratification.
   e. Several social classes existed in nineteenth-century industrial society.

4. **In contrast to Karl Marx, Max Weber placed emphasis on**
   a. the consequences of people's relationship to the economic institution.
   b. the economic dimension as an independent variable.
   c. the development of a single social class.
   d. social stratification from an evolutionary perspective.
   e. the "culture of poverty."

5. Which of the following statements regarding economic inequality in the United States is *true*?
   a. There is surprisingly little economic inequality in the United States.
   b. The majority of wealth in the United States is held by the richest fifth of the population.
   c. The middle fifth of the American population holds approximately half of the wealth.
   d. Economic inequality is declining in the United States today.
   e. The majority of wealth in the United States is controlled by religious organizations.

6. Max Weber argued that economic success and power do not always overlap completely. Which of the following statements does *not* reflect his line of argument?
   a. Money and ownership of the means of production are not the only types of resources that can be used as a basis of power.
   b. The fact that money can be used to exert power does not mean that it will be used in this way.
   c. An individual's power is a reflection of his or her relationship to the means of production.
   d. Power is also attached to the social positions we hold.
   e. Individuals can overcome a scarcity of resources if they have large numbers of people on their side.

7. The most stable source of prestige in modern society is associated with
   a. occupation.
   b. education.
   c. social class.
   d. physical beauty.
   e. athletic performance.

8. According to the text, which of the following statements *best* describes the relationship among the economic, power, and prestige dimensions of stratification?
   a. Individuals who are low on the prestige dimension tend to be high on the other two dimensions.
   b. Individuals who are high on the economic dimension tend to be low on the other two dimensions.
   c. Individuals who are high on one dimension tend to be high on the other two dimensions.
   d. Individuals who are low on the power dimension tend to be high on the other two dimensions.
   e. Sociologists now know that it is impossible to generalize about the relationship among the economic, power, and prestige dimensions of stratification.

9. According to the functionalist theory of stratification, _____ is necessary to ensure that the most important positions are filled by the most qualified people.
   a. social equality
   b. comparable worth
   c. conflict
   d. cooperation
   e. social inequality

10. According to Gerhard Lenski's evolutionary theory, human social stratification emerged with the
    a. development of advanced hunting skills.
    b. invention of the plow.
    c. creation of surplus goods and services.
    d. establishment of organized religion.
    e. development of a large brain capacity.

11. _____ refers to a sense of identification with the goals and interests of the members of one's social class.
    a. Class consciousness
    b. Class diversion
    c. Goal placement
    d. Class preference
    e. Class knowledge

12. The absence of enough money to secure life's necessities is known as
    a. relative poverty.
    b. chronic poverty.
    c. undeserved poverty.
    d. lifestyle deprivation.
    e. absolute poverty.

13. _____ mobility is measured by an upward or downward change in occupational status over an individual's lifetime.
    a. Intergenerational
    b. Horizontal
    c. Lateral
    d. Intragenerational
    e. Caste

14. The world poverty rate currently stands at _____ of the world's population.
    a. one-eighth
    b. one-quarter
    c. one-third
    d. one-half
    e. three-fourths

FEEDBACK REVIEW

True-False

1. The American public is strongly opposed to the continued existence of a welfare system. *T or F?*
2. In America, working-class individuals who fail to experience upward mobility claim that it is the system rather than themselves. *T or F?*

Fill in the Blank

3. The most stable source of prestige in modern societies is associated with _____.
4. The conflict theory of stratification is based on Marx's ideas on class _____.
5. A major difference between the upper-upper class and the lower-upper class is that the latter has relied on _____ rather than birth.

Multiple Choice

6. According to the evolutionary theory, in which of the following types of societies is stratification *least* likely to occur?
    a. hunting and gathering
    b. horticultural
    c. agricultural
    d. industrial
    e. postindustrial
7. Compared to other industrial countries, the United States has _____ bottom to top mobility.
    a. more
    b. less
    c. the same

Matching

8. Match the concepts with the examples:
    ____ a. life chances
    ____ b. power
    ____ c. prestige
    ____ d. conspicuous consumption

    (1) the respect accorded doctors
    (2) a politician conforms to the interests of a lobby
    (3) living in a neighborhood with good police protection
    (4) a lavish wedding

9. Match the theories of stratification with the examples:
    ____ a. functionalist theory
    ____ b. conflict theory
    ____ c. evolutionary theory
    ____ d. symbolic interactionism

    (1) Corporate executives make more money because they determine who gets what in their organizations.
    (2) Engineers make more money than butlers because of the education they possess.
    (3) A society is stratified in part because a surplus of goods and services exist.
    (4) Ghetto children tend to have low self-esteem.

10. Match the major types of social mobility with the illustrations:
    ____ a. intergenerational mobility
    ____ b. vertical mobility
    ____ c. horizontal mobility

    (1) a restaurant waiter becomes a taxi driver
    (2) an auto worker becomes a manager
    (3) the daughter of a hairdresser becomes a college professor

REVIEW GUIDE

GRAPHIC REVIEW

Figure 8.5 displays social welfare expenditures by the federal government prior to the 1996 welfare reform legislation. Answer these questions to make sure you understand the information presented.

1. If you were a politician making a speech against Aid to Families with Dependent Children (AFDC), would you use Figure 8.5 to buttress your case? Why or why not?

2. What does this pie chart tell us about the likelihood of significantly reducing social welfare expenditures at the federal level?

ANSWER KEY

Concept Review	Multiple Choice	Feedback Review
a. 7	1. d	1. F
b. 9	2. e	2. F
c. 6	3. d	3. occupations
d. 3	4. a	4. conflict
e. 8	5. b	5. achievement
f. 4	6. c	6. a
g. 5	7. a	7. a
h. 1	8. c	8. a. 3
i. 2	9. e	b. 2
j. 10	10. c	c. 1
	11. a	d. 4
	12. e	9. a. 2
	13. d	b. 1
	14. b	c. 3
		d. 4
		10. a. 3
		b. 2
		c. 1

Chapter Nine

Inequalities of Race, Ethnicity, and Age

LEARNING OBJECTIVES

After careful study of this chapter, you will be able to:
- Distinguish among the concepts of minority, race, and ethnicity.
- Describe the patterns of racial and ethnic relations.
- Differentiate prejudice from discrimination.
- Illustrate the different views of prejudice and discrimination taken by functionalists, conflict theorists, and symbolic interactionists.
- Describe the condition of minorities in the United States relative to the white majority.
- Describe the increasing racial and ethnic diversity in America.
- Distinguish between age stratification and ageism.
- Compare and contrast the ways in which functionalism, conflict theory, and symbolic interactionism approach ageism.
- Outline the inequality experienced by America's elderly.

SOCIOLOGICAL IMAGINATION

Have African Americans reached near income parity with whites in the United States? Many people believe that as a result of civil rights legislation passed during the 1960s black and white family incomes are now approaching comparable levels. This belief is contradicted by the evidence. The average income of African American households is considerably less than that of white households. Moreover, at each level of education (the gateway to jobs) black males gain less of an increase in income than their white male peers. White high school graduates, for example, earn on the average nearly as much annually as African American men who hold college degrees.

Sociologists attribute this enduring economic inequality to prevailing and widespread negative racial attitudes and behavior among the majority of white Americans. Like all minorities, blacks suffer from the consequences of prejudice and discrimination. Latinos, Native Americans, Jewish Americans, white ethnics, and other minorities have felt the same sting of prejudice and unequal treatment. More recently, attention has been called to other categories of people who are subject to the fallout from prejudice and discrimination, particularly women and the elderly. Inequalities of age are explored later in this chapter. Gender inequality is covered in the subsequent chapter.

RACIAL AND ETHNIC MINORITIES

All Americans are either immigrants or the descendants of immigrants. Even Native Americans are thought to have migrated to this continent many centuries ago. One result of the migration of so many different peoples to the United States has been the occurrence of prejudice and discrimination directed at members of minorities thought to be racially and ethnically different. The emergence of minorities is a recurring theme in the American experience as immigrant and native groups alike have encountered barriers to their full integration into American society.

The Meaning of Minority

In everyday usage, minority refers to a relatively small number of people. But a sociological definition of a minority can apply to a people numerically larger than others in a society. Blacks in South Africa and in parts of the southern United States, for example, are minorities even though they outnumber whites. There is obviously something more than size that distinguishes a minority.

If small numbers do not necessarily make a minority, what does? Louis Wirth has noted the most important characteristics of a **minority.**

> We may define a minority as a group of people who, because of their physical or cultural characteristics, are singled out from the others in the society in which they live for differential and unequal treatment, and who therefore regard themselves as objects of collective discrimination. The existence of a minority in a society implies the existence of a corresponding dominant group with higher social status and greater privileges. Minority carries with it the exclusion from full participation in the life of the society (Wirth, 1945:347).

This definition holds several key ideas. First, *a minority must possess distinctive physical or cultural characteristics that can be used to distinguish its members from the majority.* Such characteristics are used both as a reminder to the majority that the minority is different and as a means for determining whether a given person is a member of the minority. This is more evident for physical characteristics than it is for cultural ones. Culturally, a minority member can pass as a member of the majority by a change in name, the loss of an accent, or the adoption of the majority's culture. Experience suggests, however, that where such differences are not sufficiently visible to allow easy identification, other means of identification may be imposed. The forced wearing of yellow stars by Jews during the Nazi era in Germany illustrates this point.

Second, *whatever its numerical size, the minority is dominated by the majority.* This dominance is reflected especially in a society's stratification structure. Almost any society has desired goods, services, and privileges. Because these "good things" are usually in limited supply, there is competition for them. The majority dominates largely because it has accumulated enough power to obtain an unequal share of the desired goods, services, and privileges. The majority can deny minority members access to desirable resources by limiting opportunities to compete on an equal basis. Minorities, for example, can be denied good jobs by being limited to inferior schooling or by not being hired even when qualified.

Third, *members of a minority are denied equal treatment.* The distinctive cultural or physical traits of a minority are usually judged by the majority to be inferior to their own. This presumed inferiority is then used as justification for the unequal treatment given the minority. In other words, the alleged inferiority of a minority becomes part of the majority's **ideology**—a set of ideas used to justify and defend the interests and actions of those in power in a society (Eagleton, 1994). Members of a majority can more easily discriminate in occupations if, for example, its ideology contains the belief that members of the minority are a shiftless and lazy lot.

Fourth, *because members of a minority regard themselves as objects of discrimination by the majority, they have a sense of common identity.* Within the minority there is a sense of "consciousness of kind." It is from this sense of common identity that a "we" and "they" vocabulary is accepted within the minority. This vocabulary reflects a strong sense of solidarity and loyalty.

Finally, *membership in a minority is ascribed.* People do not normally make an effort to become part of a minority; they simply become members by virtue of being born to existing members. Because membership in a minority is an ascribed status, it is not easily shed. One does nothing to achieve this status, and there is little one can do to avoid it. Although some members of ethnic minorities may, through great effort and a name change, leave this ascribed status behind, it is nearly impossible for members of racial minorities to do so.

Throughout history, various peoples have been treated as minorities. For this reason, it is important to understand the nature of race and ethnicity.

The Meaning and Significance of Race

What is race?   A **race** is a category of people who are alleged to share certain biologically inherited physical characteristics. Biologists have used such physical characteristics as skin color, hair color and texture, facial features, head form, eye color, and height to create broad racial classifications. The most commonly used system of racial classification has three major racial categories—Negroid, Mongoloid, and Caucasoid—along with some unclassified racial categories.

Are racial classifications valid?   Racial categories, first developed by nineteenth-century biologists, are now considered by sociologists to be arbitrary and misleading. Races are social rather than biological classification

A miniority is a people who possess some distinctive physical or cultural characteristics, are dominated by the majority, and are denied equal treatment. You can easily identify members of various American miniorities in this photograph.

categories that help to perpetuate inaccurate and damaging stereotypes. Even defenders of the usefulness of racial classification acknowledge that the number of identifiable categories exceeds thirty (Dobzhansky, 1962).

If racial classification categories are this numerous, sociologists conclude, they lose any meaningful capacity to differentiate human beings on the basis of their physical characteristics. When we continue to attribute validity to racial categories, we are actually imposing our social definitions on rather superficial physical differences.

For example, in the United States a person born of a black-white couple was traditionally considered to be "black." If "white" blood were merely a matter of biology, then there would have been no legal penalty attached to offspring of black-white couples who claimed to be Caucasian in the United States before the Civil War. However, attempting to "pass for white" during this period carried heavy legal sanctions. Some southern and border states legally defined a person as African if one of the great-grandparents was African ($^1/_8$) or if one of the great-great-grandparents was African ($^1/_{16}$). In practice though, it took far less than $^1/_{16}$ of the blood to be "black" to be considered African—anyone with any known African ancestry was usually deemed an African. This was known as the "one-drop" rule (Spickard, 1992). Absent the socially imposed restrictions attached to having "black" blood, these individuals could, if they preferred, have classified themselves as "white." In fact, because the legal system in the United States now recognizes the arbitrariness of these racial labels, parents are able to decide for themselves into which census category their children are to be classified. Also consider Nazi Germany, a prime example of a society attaching social significance to physical characteristics. Jesse Owens's five-gold-medal performance in Germany's 1936 Olympics drew Hitler's fury only because a "black" man consistently defeated the allegedly superior blond, blue-eyed athletes of the Third Reich.

Are physical differences significant? Racist thinkers, from Count de Gobineau in the mid-nineteenth century to Adolf Hitler in the twentieth century, have attempted to link physical differences among people to innate superiority or inferiority. Alleged differences in the intelligence and achievement of a people have been attributed by racists to inherited biological characteristics over which no one has any control. In fact, **racism** can be defined in part as an ideology that links the physical characteristics of a people with their psychological or intellectual superiority or inferiority (van den Berghe, 1978; Miles, 1989; Wievorka, 1995).

The weight of current scientific evidence does not support the racists' connection between genetically determined physical characteristics and innate superi-

ority or inferiority. Contemporary social scientists overwhelmingly believe that there are no significant innate (that is, biologically inherited) differences in intelligence among the various races. Most social scientists endorse the following statement from a UNESCO Statement on Race prepared by a distinguished group of social scientists:

> *According to present knowledge, there is no proof that the groups of mankind differ in their innate mental characteristics, whether in respect of intelligence or temperament. The scientific evidence indicates that the range of mental capacities in all ethnic [or racial] groups is much the same (Klineberg, 1950:466).*

A subsequent review of research supports UNESCO's earlier conclusion ("Statement on Race and Intelligence," 1969). Both of these reports attribute any existing differences in measured IQ among racial groups to differences in social environment, training, and education.

The consequences of practicing racial discrimination on the basis of such alleged differences are real. Alleged innate racial differences are used by the majority to make judgments regarding superiority and inferiority and to justify prejudice and discrimination against racial minorities. For this reason, many social scientists have become more interested in "social" races than in biological races (Farley, 1995; Feagin, 1996). African Americans, Native Americans, Latinos, and Asian Americans are examples of socially defined races in America that are still considered to be inferior. In whatever definitional form it takes, it does not seem unreasonable to conclude with Omi and Winant (1994) that race will always be at the heart of the American experience.

The Meaning and Significance of Ethnicity

A people can be considered a racial category at one time and not at another time. The Irish and Italians, for example, were once considered to be inferior racial groups by Anglo-Saxon Americans. Later, as it became clear that they had no distinctive physical characteristics, the Irish and Italians became ethnic minorities.

What is an ethnic minority? Ethnicity comes from the Greek word *ethnos*, originally meaning "people" or "nation." Thus, the original Greek word referred to cultural and national identity. Today an **ethnic minority** is socially identified and set apart by others and by itself on the basis of unique cultural or nationality characteristics.

Because of their differences from the host culture, ethnic minorities can be considered subcultures. They have a way of life that is based on their own language,

Racial classification categories are arbitrary and misleading. The futility of placing social definitions on physical differences can be seen in Jesse Owens's 1936 Olympic performance in which he showed a black man to be superior to Hitler's blond, blue-eyed athletes.

values, beliefs, norms, and customs. Like any subculture, they are part of the larger culture—they work in the larger economy, send their children to the dominant educational system, are subject to the laws of the land—but they are also separate from it. This separation persists either because an ethnic minority wishes to maintain its cultural and national origins or because the majority erects barriers to prevent it from blending in with the larger culture. Michael Novak (1996), himself a Slovak, has contended that members of white ethnic minorities from Southern and Eastern Europe—

Poles, Slavs, Italians, Greeks—have not been permitted to blend completely into American society because they are more culturally different from white Anglo-Saxon Protestants (WASPS) than are immigrants from Western and Northern Europe. Western Europeans, for example, have an alphabet similar to the one used in English, even if the language is foreign, and they have similar religions.

Are ethnic minorities considered inferior? Just as physical characteristics are used by the majority to belittle racial minorities, cultural differences are used to demonstrate the alleged inferiority of ethnic minorities. Negative attitudes toward ethnic minorities exist in part because of the development of ethnocentrism (the judgment of others in terms of one's own cultural standards). The majority, out of loyalty to and preference for its own values, beliefs, and norms, uses the cultural differences of ethnic minorities as a sign of inferiority. Because members of ethnic minorities do not measure up to the majority's conception of appropriate ways to think, feel, and act, it is assumed that something is wrong with these minorities. These ethnocentric judgments are often expressed in prejudice and discrimination against ethnic minorities. Jews, for example, have had to contend with prejudice and discrimination based on their religious beliefs.

A society's attitudes toward ethnic (and racial) minorities vary from minority to minority and from time to time (Gallup, 1994a). Figure 9.1 illustrates the opinions of Americans toward a variety of immigrant minorities. Clearly, older immigrant minorities (Irish and Polish) are viewed more favorably than later arrivals (Cubans, Haitians, Iranians). You may want to

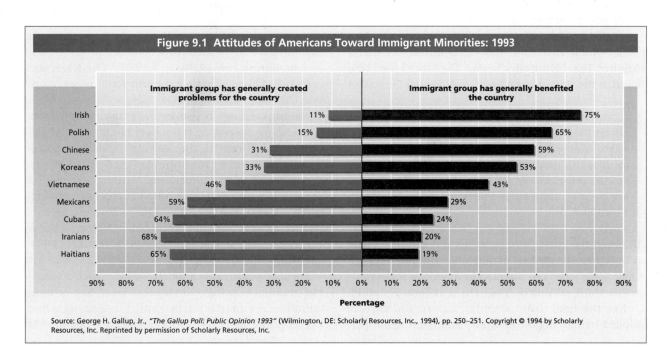

Figure 9.1 Attitudes of Americans Toward Immigrant Minorities: 1993

	Immigrant group has generally created problems for the country	Immigrant group has generally benefited the country
Irish	11%	75%
Polish	15%	65%
Chinese	31%	59%
Koreans	33%	53%
Vietnamese	46%	43%
Mexicans	59%	29%
Cubans	64%	24%
Iranians	68%	20%
Haitians	65%	19%

Percentage

Source: George H. Gallup, Jr., *"The Gallup Poll: Public Opinion 1993"* (Wilmington, DE: Scholarly Resources, Inc., 1994), pp. 250–251. Copyright © 1994 by Scholarly Resources, Inc. Reprinted by permission of Scholarly Resources, Inc.

compare your attitude toward various American ethnic and racial minorities with those shown in Figure 9.1.

Patterns of Racial and Ethnic Relations

When people of various racial and ethnic backgrounds come into contact with one another, a wide range of outcomes is possible. It is helpful, however, to divide these many outcomes into two major types: patterns of assimilation and patterns of conflict.

What are the patterns of assimilation? **Assimilation** can be defined as "those processes whereby groups with distinctive identities become culturally and socially fused together" (Vander Zanden, 1990:274). The emphasis in assimilation is on the basically egalitarian outcome wherein the racial or ethnic minority becomes integrated into the surrounding society and is accepted into full participation in all aspects of that society.

The classic works in the study of assimilation are those of Milton Gordon, who has defined three basic assimilation patterns in American society: Anglo-conformity, melting pot, and cultural pluralism (Gordon, 1964, 1978).

Anglo-conformity is a general term referring to a type of assimilation in which Anglo or American institutions are maintained. Basic to this pattern of assimilation is the lack of opposition to immigrants as long as immigrant minorities attempt to conform to the generally accepted standards of the society. Anglo-conformity has been the most prevalent pattern of assimilation in America. It is the least egalitarian of assimilation patterns in that the immigrant minority is required to conform to general social values and, by implication, to either give up or suppress its own values.

A second pattern of assimilation is that of the *melting pot* in which all ethnic and racial minorities blend together. Israel Zangwill's drama *The Melting Pot,* produced in the United States in 1908, expressed the goal of

> . . . the completion of a vast American symphony which will express [the immigrant's] deeply felt conception of [the] adopted country as a divinely appointed crucible in which all the ethnic divisions of mankind will divest themselves of their ancient animosities and differences and become fused into one group, signifying the brotherhood of man (Gordon, 1978:193).

Although the ethnic and racial pot did boil, especially in the cities, the melting of difference and fusion into a single society has not occurred in American society.

The third assimilation pattern is *cultural pluralism.* Unlike the image of a singular society and culture implied by the Anglo-conformity and melting-pot pat-

terns of assimilation, cultural pluralism recognizes the durability of the traditions immigrants bring with them and the immigrants' desire to maintain at least some of their "old" ways. Ethnic enclaves or settlements characterized American cities of the past century, and many survive today. Such enclaves allowed immigrants to maintain their traditional ways while learning American values and norms.

The extreme of cultural pluralism is *accommodation,* which occurs when a minority, despite its familiarity with the beliefs, norms, and values of the dominant culture, continues totally with its own culturally unique way of life. An accommodated minority learns how to deal with the dominant culture where necessary—education, economy, politics—but remains independent in language and culture. Cubans in Miami are an example of a distinct community within the larger community (Wilson and Martin, 1982).

Cultural pluralism emphasizes the ethnic and racial diversity that still characterizes American society and is a more democratic and egalitarian pattern of assimilation than either Anglo-conformity or the melting pot. Like the other two patterns, however, cultural pluralism also overlooks the continuing inequities that characterize American racial and ethnic relations today. Each of these assimilation patterns stresses a kind of egalitarian outcome and thus omits patterns of conflict.

What are the patterns of conflict? Conflict patterns stress patterns of dominance among the diverse racial and ethnic minorities of a society. Three basic patterns of conflict are genocide, population transfer, and subjugation (Mason, 1970).

Genocide involves systematic efforts to kill or destroy an entire population. Genocide is a tragic outcome of efforts by a society to establish dominance over one or more minorities (Fein, 1993). One of the best-known examples is Hitler's attempt to murder all the Jews in Europe during the 1930s and 1940s. The Nazis succeeded in killing some six million Jews before World War II came to an end in 1945. Less well known is the "Rape of Nanking" begun in 1937, during which the Japanese massacred an estimated 260,000 to 350,000 Chinese men, women, and children. Actually, genocide is more common in world history than might be supposed. The Serbian campaign of "ethnic cleansing" of Bosnian Muslims and the ethnic slaughter of anywhere from 200,000 to 500,000 people in Rwanda are two recent expressions of genocide (Nelan, 1993a; Whitaker, 1993; Hammer, 1994; Michaels, 1994; Sullivan, 1996).

Population transfer is another way in which a society can attempt to achieve dominance over one or more minorities. In this pattern of conflict, a minority is banished or otherwise forced to move to a remote location

Many sociologists contend that the majority in a society uses prejudice and discrimination as weapons of power to satisfy its own interests. The infamous Trail of Tears symbolizes the destruction of Native American culture by the white majority that systematically and relentlessly confiscated valuable territory.

or even to emigrate from the society. The experience of Native Americans is an excellent illustration of this pattern. Population transfer was the most common policy toward Native Americans, especially during the late 1800s. Tribes that once lived in widespread regions of what eventually became the United States can now be found on reservations in Oklahoma and other parts of the western and plains states. For example, the Cherokees were once a single tribe or nation in the Southeast. Reflecting the Trail of Tears experience (a forced march by the U.S. government of seventy thousand Native Americans from the southeastern United States to Oklahoma reservations during which approximately twenty thousand died) and resistance to it, Cherokee reservations can today be found in North Carolina as well as Oklahoma. Native Americans lost their lands, were forced to move to reservations that were often far from their home areas, and were placed in a dependency status as wards of the U.S. government. Such a pattern of conflict survives today and characterizes the relations between many Native Americans and the federal government. As a partial function of the reservation system, many Native Americans have migrated to cities to find work, a move that threatens the survival of some tribes.

Subjugation is the most common pattern of conflict. This pattern most clearly reflects the characteristics of a minority relationship in that the majority enjoys greater access to the culture and lifestyle of the larger society than the minority. Inequities consistently appear in such areas as power, economics, and education as well as in other important indicators of the quality of life, such as health and longevity.

In cases of subjugation, the majority and the minorities may occupy the same general area and may participate together in at least some aspects of social life, such as work. However, members of the minority are clearly subordinate to members of the majority. Such subjugation may be based on the law—**de jure subjugation**—as, for example, in the case of the era of segregation of African Americans that followed the Civil War and lasted until the mid-1960s. Subjugation may also be based in common, everyday practice—**de facto subjugation**—rather than in the law. In this case, subjugation is visible mainly in objective indicators of inequality, such as political representation and occupational distribution in the labor market. The continuing practice of housing discrimination in the United States is an illustration of subjugation. This practice has existed over many decades, even in the face of direct efforts, including open housing laws, designed to overturn it. According to a 1989 survey, African Americans and Latinos experience discrimination more than half the time they attempt to rent or purchase a home (Turner, Struyk, and Yinger, 1991).

The patterns of race and ethnic relations are many and diverse. In any society, we can find examples of any of these patterns coexisting. Also, in the history of the relations between two peoples, we can find wide variation through time in the ways the people have interacted. Many factors such as the nature of their first contact, the reasons for contact and interaction, their visibility, the views held by their respective members, and general social conditions can influence the pattern of racial and ethnic relations adopted.

1. Which of the following is *not* always a characteristic of a minority?
   a. distinctive physical or cultural characteristics
   b. smaller in number than the majority
   c. dominated by the majority
   d. denied equal treatment
   e. a sense of collective identity
2. _____ is an ideology that links a people's physical characteristics with their alleged intellectual superiority or inferiority.
3. A(n) _____ is a people within a larger culture that is treated as a minority because of its distinctive cultural characteristics.
4. _____ refers to those processes whereby groups with distinctive identities become culturally and socially fused.
5. The pattern of conflict in which one people is forced to move to a remote location is called _____.

Answers: 1. b 2. Racism 3. ethnic group 4. Assimilation 5. population transfer

THEORIES OF PREJUDICE AND DISCRIMINATION

A variety of theoretical perspectives have been used to explain the existence of prejudice and discrimination. Before introducing these theories, however, the nature of prejudice and discrimination needs to be discussed.

The Nature of Prejudice and Discrimination

What is prejudice? Prejudice is a general characteristic of human relationships. Students can be prejudiced against professors, professors against university administrators, Catholics against Protestants, and employees against employers. In sociology, however, the concept of **prejudice** has a narrower meaning and refers to negative attitudes toward some minority or its individual members. Such attitudes may be related to strong emotional feelings, or they may simply be rooted in unchallenged false ideas. In any case, prejudiced attitudes are often difficult to change, even in the face of overwhelming evidence that should call such attitudes into question.

Prejudice involves an either/or type of logic: a minority is either good or bad, and it is assumed that every member of the minority possesses the characteristics attributed to the group itself. Prejudice involves overgeneralization based on biased or insufficient information. Such overgeneralizations are usually defended by citing limited personal experiences with minority members or by reciting stories told by others about their experiences. New information that contradicts one's prejudices tends to be denied because of selective perception. New evidence is viewed emotionally in the light of what is already believed. To many Americans, for example, an African American who fails to fit the stereotype is simply an exception. Consequently, prejudiced attitudes generally are not altered either by new personal experiences or by accounts from others.

What is discrimination? Prejudice refers to attitudes. **Discrimination** refers to the unequal treatment of individuals based on their membership in some minority. Prejudice does not always result in discrimination, but it often does. Discrimination takes various forms, including avoiding social contact with members of a minority, excluding minority members from certain places of employment and neighborhoods, and physically attacking or killing minority members (Allport, 1958).

What is the relationship between prejudice and discrimination? Prejudice is usually considered the cause of discrimination. Although this is often the case, discrimination may also occur without prejudice. Discrimination in some situations may also be the cause of prejudice. For example, unskilled workers may believe that their jobs are in jeopardy because of the massive immigration of a new ethnic or racial group. This fear of economic threat may lead to the unequal treatment of members of that minority. To justify this discrimination, threatened workers may attempt to show why a minority deserves unequal treatment. This occurs in part through the creation of stereotypes. This is exactly what happened to Chinese Americans during the nineteenth and early twentieth centuries (Lyman, 1974).

What is a stereotype? Like prejudice, stereotyping appears throughout a society. Athletes are thought not to be very smart, and politicians are accused of being corrupt. In stereotyping, it is assumed that all members of a category share the same characteristics. As used in sociology, a **stereotype** is a set of ideas based on distortion, exaggeration, and oversimplification that is applied to all members of a social category. Stereotypes may be used as a justification for prejudiced attitudes and discrimination. An example is the colonists' treatment of Native Americans. Early relationships between the colonists and Native Americans were relatively peaceful and cooperative, but as the population of the colonies grew, conflicts became more frequent and intense. Population growth led to the deeper and deeper penetration of whites into the wilderness. To rationalize this expansion, the colonists developed a stereotype of the Native Americans. It was easier to develop a picture of the lying, thieving, murdering savage who was pagan in religion and racially stupid except for a kind of animal cunning. Such a person had no rights; the only good Indian was a dead Indian. This picture was frequently not held by the trappers and traders who moved individually among the Indians and, because of friendly contact, tended to judge the Indian differently. But the prejudice of the farmer-settler prevailed, leading to the continued seizure of Native American lands with a minimum of compensation and the dramatic reduction of the Native American population from as many as five to seven million to as few as a quarter-million in 1890 (Snipp, 1996).

Americans are still willing and able to rate group members purely on the basis of race and ethnicity. Tom Smith (1990) found that with the exception of Jews, minority groups are judged more negatively than whites in general. No group other than Jews scored above whites (that is, showed a positive mean) on the characteristics of wealth, industriousness, nonviolence, intelligence, self-support, and patriotism. Asian Americans and white Southerners are ranked behind Jews (second or third) on nearly every dimension. African Americans and Latinos are ranked last or next to last on almost all characteristics.

Prejudice and discrimination against certain minorities cannot be accounted for by any single factor. The causes are many, complex, and interrelated. Explanations for the existence of prejudice and discrimination are both psychological and sociological.

The Psychological Perspective

Psychological explanations of prejudice and discrimination focus on the prejudiced person's personality, how it developed, and how it functions in the present. Such questions as the following have been asked about prejudiced persons: What was their relationship with their parents or with their significant others? What are their values, attitudes, and beliefs? How high is their self-esteem? Two prominent psychological explanations of prejudice and discrimination are the frustration-aggression explanation and the authoritarian personality.

According to the frustration-aggression explanation, prejudice and discrimination are the products of the deep-seated hostility and aggression that stem from frustration. Aggression is most likely when hostility that cannot be directed at the actual source of frustration builds up. Pent-up hostility and frustration may subsequently be redirected toward some substitute object that is less threatening than the one causing the

frustration. These substitute objects, known as scapegoats, serve as convenient and less feared targets on which to place the blame for one's own troubles, frustrations, failures, or sense of guilt.

John Dollard (1939), originator of the frustration-aggression explanation, contended that the frustrations experienced by the Germans after World War I help account for their acceptance of anti-Semitism. The loss of the war, the disappearance of international prestige, the forced acceptance of the Treaty of Versailles, and a ruined economy all contributed to tremendous frustration among the German people. Because direct aggression against the Allies was not an alternative, the Germans channeled their aggression toward other targets, one of which was the Jews. The best scapegoats are those who have already been singled out by the majority for unequal treatment and who therefore have the least chance to defend themselves.

Another psychological explanation contends that there is a personality type—the authoritarian personality—that tends to be more prejudiced than other types. The **authoritarian personality** is characterized by excessive conformity; submissiveness to authority figures; inflexibility; repression of impulses, desires, and ideas; fearfulness; and arrogance toward persons or groups thought to be inferior. T. W. Adorno and his colleagues, the creators of this theory, summarized it this way:

The most crucial result of the [study of the authoritarian personality] is the demonstration of the close correspondence in the type of approach and outlook a subject is likely to have in a great variety of areas, ranging from the most intimate features of family and sex adjustment through relationships to other people in general, to religion, and to social and political philosophy. Thus a basically hierarchical, authoritarian, exploitative parent-child relationship is apt to carry over into a power-oriented, exploitatively dependent attitude toward one's sex partner and one's God and may well culminate in a political philosophy and social outlook which has no room for anything but a desperate clinging to what appears to be strong and a disdainful rejection of whatever is relegated to the bottom (Adorno et al., 1950:971).

The Functionalist Perspective

Functionalism generally looks for the positive contributions that aspects of a society make to the society's stability and continuity. As noted in Chapter 7 ("Deviance and Social Control"), Emile Durkheim identified the ways in which deviance contributes to the cohesiveness of society. The functional theory of stratification, as shown in Chapter 8 ("Social Stratification"), contends that social inequality helps society to channel the most qualified people into the most important positions and to ensure that people in these positions are motivated to perform their tasks competently.

Although functionalists have emphasized the positive aspects of social stratification, they have also focused on the dysfunctions of prejudice and discrimination. Despite the emphasis of functionalism on the negative consequences of prejudice and discrimination, the perspective can identify some potential benefits. But, as will be seen, for every benefit of prejudice and discrimination, there is a down side.

What can functionalism say about prejudice and discrimination? Functionalist Robert Merton (1968) demonstrates that any aspect of a society can have both positive and negative consequences. What is functional for one segment of society may be dysfunctional for other segments. Possible functionalist arguments for why it is desirable for women to stay at home with the children would be familiar. Males can concentrate on earning a living, knowing that the home-related work is being done simultaneously. The psychological and social development of children is promoted by having a mother at home full-time. Family stability is enhanced by the traditional division of labor. On the other hand, functionalists point out, women are damaged by a system that promotes low self-esteem and limits their opportunities. Society also loses something important, because the skills and abilities of half the population are not given free expression.

Prejudice and discrimination, a la Durkheim's argument regarding deviance, may contribute to social solidarity of the majority. This happens because the attempted exclusion of minorities rests in part on **ethnocentrism**—the tendency to judge others in terms of one's own cultural standards. It is by judging one's own kind to be superior that others can be subjected to stereotypes and treated as scapegoats. Once a people is convinced of its superiority, it can unite around its own way of life and exclude entrance to others. This strengthens the boundaries of the majority.

As the latter part of this chapter demonstrates, the dysfunctions of prejudice and discrimination for minorities are both wide and deep. The social, political, educational, and economic costs of exploitation and oppression to minorities are extremely high. Moreover, the safety and stability of the larger society are at risk because minorities may resort to violence in attempts to throw off the shackles of long-term prejudice and discrimination.

The Conflict Perspective

According to the **differential power** explanation, a majority uses prejudice and discrimination as weapons of power in the domination of one or more subordinate minorities. This theory, then, traces the existence of prejudice and discrimination to majority interests rather than to personality needs. Domination by the majority may be motivated by the majority's desire to gain or increase its control over scarce goods and services. It has been argued that a majority uses prejudice and discrimination in such a struggle. Eliminating or neutralizing a minority as a competitor is especially effective when those with power in a society are able to persuade the majority that a minority should be subjugated (Wilson, 1973).

Members of a majority tend to think of minorities as people unlike them but similar to one another. Members of racial and ethnic minorities know this is not true. Despite being a common target for the majority's prejudice and discrimination, minorities tend to view one another as competitors rather than as allies in their struggle for scarce resources (Olzak and Nagel, 1986). Conflict among minorities, particularly African Americans and Latinos, is increasing in the United States as whites leave cities and African Americans assume political power. To many urban African Americans, Latinos are the beneficiary of the civil rights movement to which they contributed nothing. Many Latinos believe that African Americans are using their power to favor themselves. In 1988, for example, a Latino member of Chicago's Board of Education charged the five African American members on the board with instituting "apartheid" on Latino children by voting to send more African American children to integrated schools while overlooking the overcrowding suffered by Latino students. It remains to be seen if urban African Americans and Latinos will become allies for their mutual welfare or will engage in fierce conflict over the scarce resources available to them (Salholz, 1988).

From a Marxian viewpoint, the ruling class (capitalists) benefits from a continuation of this sort of conflict among minority groups. If various minorities fight with each other and attribute their job losses to other minority groups, then the ruling elite can downsize, move jobs offshore, and replace workers with technology with little or no resistance. In other words, so long as the conflict is not too extreme, capitalists are the beneficiary of a divided working class.

Prejudice and discrimination cannot be accounted for solely by psychological, functional, and conflict explanations. They do not explain why particular groups are selected as targets for prejudice and discrimination. And they have nothing to say about why prejudice and discrimination are directed at some specific minority even when no self-interest can be satisfied by doing so. In this regard, symbolic interactionism adds to our understanding of the creation and persistence of prejudice and discrimination.

The Symbolic Interactionist Perspective

Prejudice and discrimination, just like other aspects of culture, are learned through socialization. Members of a society learn to be prejudiced in much the same way that they learn the countless other aspects of their culture. Gordon Allport (1958) has described two stages in the learning of prejudice. In the first stage—the *pregeneralized* learning period—children have not yet learned to categorize people into general groupings. The idea that African Americans, Jews, and Latinos each form a distinct grouping is beyond their understanding. A child who hears a parent object to people of some specific minority moving into the neighborhood is introduced to the second stage of learning prejudice—*total rejection*. By this stage, the child has learned the name of the minority he or she is supposed to dislike and can identify individuals who belong to it. When a child reaches the second stage, all members of the minority are rejected, on all counts, in all situations.

James Jones (1972) points out that even our language provides a context supportive of prejudice and discrimination. For example, the terms *black* and *white* as used in Anglo culture help create such a climate. Although it is good to be "in the black" financially, most references using black are negative. Such terms as *blackball, blacklist, black mark,* and *black eye* illustrate the generally negative connotations given to the term *black*. By comparison, it is difficult to think of similar instances in which *white* is used negatively. Such cultural referents, learned as a part of everyday culture, create a context in which it is easy to assume that the well-dressed African American male standing outside a fine restaurant is the doorman rather than a customer. These responses support the continuation of prejudice and discrimination.

The interactionist perspective also underscores the labeling process that occurs between the majority and various minorities. Robert Blake and Wayne Dennis (1943) graphically illustrated the attempt to label African Americans in their early study, which included fourth and fifth graders in an all-white Southern school.

Faced with a list of questions about the virtues of African Americans and whites (for example, Who is more cheerful? Who is more cruel? Who is more emotional?), the children thought that nearly all virtues were exclusive property of their own kind. They thought, for example, that African Americans were less cheerful and easygoing and had less of a sense of humor than whites. Although the prevalence of labeling African Americans by whites seems to be declining, it is still occurring. In 1996, Texaco Inc. agreed to pay more than $115 million as reparation for the economic effects of racism within the company. This settlement was fueled in part by a tape recording made by a high corporate official on which one executive appeared to call blacks "niggers" and another joked that "black jelly beans were stuck to the bottom of the bag" (Solomon, 1996).

The use of symbolic interactionism in understanding prejudice and discrimination is reflected in the **self-fulfilling prophecy**—when an expectation leads to behavior that causes the expectation to become a reality (Merton, 1968). For example, if two nations are convinced they are going to war, they may engage in hostile interaction that actually leads to war. Similarly, if members of a minority are constantly treated as if they are less intelligent than the majority, they may eventually accept this definition of themselves. This, in turn, may lead them to place less emphasis on education and subsequently appear to themselves and others to be less intelligent. Or, members of a minority may be socialized to believe that they are not good enough to hold important positions in a society and at the same time are locked out of such positions. Given this negative interaction and the lack of opportunities to show their abilities to themselves and others, these minority members may become locked in lower level jobs.

INSTITUTIONALIZED DISCRIMINATION

Although it is popular to think of the United States as a free society in which all individuals have an equal

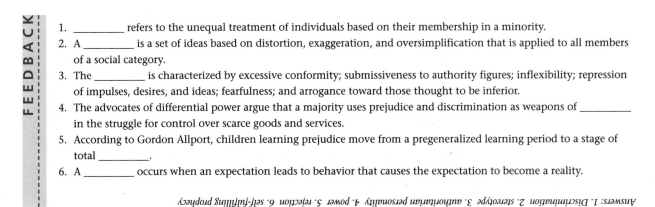

FEEDBACK

1. _____ refers to the unequal treatment of individuals based on their membership in a minority.
2. A _____ is a set of ideas based on distortion, exaggeration, and oversimplification that is applied to all members of a social category.
3. The _____ is characterized by excessive conformity; submissiveness to authority figures; inflexibility; repression of impulses, desires, and ideas; fearfulness; and arrogance toward those thought to be inferior.
4. The advocates of differential power argue that a majority uses prejudice and discrimination as weapons of _____ in the struggle for control over scarce goods and services.
5. According to Gordon Allport, children learning prejudice move from a pregeneralized learning period to a stage of total _____.
6. A _____ occurs when an expectation leads to behavior that causes the expectation to become a reality.

Answers: 1. Discrimination 2. stereotype 3. authoritarian personality 4. power 5. rejection 6. self-fulfilling prophecy chance to achieve their desired lifestyles and to practice their cultural values, such freedoms have always been limited by the structure of American institutions. This is the *American dilemma* of which Gunnar Myrdal wrote with regard to African Americans (Myrdal, 1944). At various times in American history, such practices as slavery and the internment of Japanese Americans during World War II have reflected the open and legal practice of discrimination against members of various minorities. Virtually all minorities have encountered such practices to a greater or lesser degree, with some representing targets of long standing (Feagin and Feagin, 1984; Schaefer, 1997).

With the passage during the 1960s of a series of civil rights laws, many Americans felt that at long last the surviving remnants of a tawdry past had been overcome. A careful look at events since the passage of these acts and the current lifestyles of a number of racial and ethnic minorities, however, suggests that the legacy of two to three centuries of discrimination is not easily erased from American life, even if the apparent decline in racial and ethnic prejudice does turn out to be real (Kinder, 1986; Ladd, 1987; Lipset, 1987; Firebaugh and Davis, 1988; Gallup and Hugick, 1990; "Black/White Relations in the United States," 1997).

To help clarify discriminatory practices, Joe and Clairese Feagin (1984) have distinguished between direct and indirect institutionalized discrimination. **Direct institutionalized discrimination** refers to organizational or community actions intended to deprive a racial or ethnic minority of its rights. **Indirect institutionalized discrimination** refers to unintentional organizational or community actions that negatively affect a racial or ethnic minority. Examples of direct institutionalized discrimination include laws that segregated African Americans or that denied Mexican American children the right to speak their native language in public schools. Examples of indirect institutionalized discrimination are seniority systems that discriminate against the promotion of newly hired workers. Because members of minorities are just now beginning to enter jobs affected by seniority systems, their hopes for promotion remain slim even though no mention of race or ethnicity is made in the rules governing the seniority systems.

Institutionalized discrimination in the United States can be best illustrated by looking at selected minorities and their experiences. In each case, we will look at the history of the minority and analyze the minority's current situation. The minorities to be included are African Americans, Latinos, Native Americans, Asian Americans, white ethnics, and Jewish Americans. For all minorities, the social and economic costs of past discrimination have been enormous, and the costs have continued through the 1990s (Andrews, 1996; Spencer, 1997).

African Americans

African Americans constitute the largest of America's minorities, numbering almost 34 million, about 13 percent of the total population. (See Figure 9.2.) They are also one of the oldest minorities, having first been brought to America as indentured servants in 1619.

Why have African Americans remained a target for so long? Because African Americans are physically identifiable, it has been much more difficult for them to be absorbed into the larger society than it has been for white minorities.

A second reason for African Americans being targeted for such a long time lies in the nature of slavery in America. According to Frank Tannenbaum (1947), the attitudes and practices associated with *manumission*—the transition from slave to free person—are the most important aspects of any slave system. Under the law in the United States, slavery was generally a permanent condition. Manumission existed, but support for it was weak, and there were numerous barriers to it. Freed slaves constantly ran the risk of being returned to slavery. For instance, in Maryland in 1717, any freed

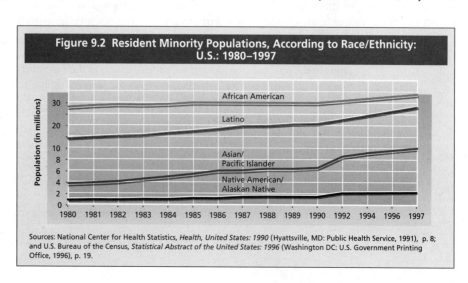

Figure 9.2 Resident Minority Populations, According to Race/Ethnicity: U.S.: 1980–1997

Sources: National Center for Health Statistics, *Health, United States: 1990* (Hyattsville, MD: Public Health Service, 1991), p. 8; and U.S. Bureau of the Census, *Statistical Abstract of the United States: 1996* (Washington DC: U.S. Government Printing Office, 1996), p. 19.

slave who married a white person was returned to slavery for life. In Virginia, an emancipated slave could be sold into slavery again if he or she remained in the state after one year. Those slaves who did manage to gain and keep their freedom were barred from most jobs, holding public office, and voting.

As a consequence of the nature of the American form of slavery, slaves freed before the Civil War were denied upward social mobility. According to Richard Hofstadter:

> The Anglo-Americans of the North American mainland quickly became committed to sharp race separation, took a forbidding view of manumission, defined mulattoes simply as Negroes, and made outcasts of free Negroes. Hence there was as little upward mobility from slavery as possible, especially in the Southern colonies and states, and even where masters chose to manumit slaves (Hofstadter, 1973:114).

The denial of opportunity for upward mobility did not end with the Civil War. Although slavery was legally abolished in the 1860s, the legacy of prejudice and discrimination produced in part by American slavery affects African Americans to this day. By the late 1800s, de jure segregation (the separation of blacks and whites based on law) was institutionalized, especially in the South. Such practices continued until the late 1960s, when they were finally made illegal by the passage of civil rights legislation and by court decisions. In a very real sense, then, African Americans have been legally free in much of the United States for barely

thirty years. The gap between African Americans and whites in areas such as education, income, and employment represents both the legacy of centuries of prejudice and discrimination against African Americans and the continuation of such practices (Harris and Wilkens, 1988; Lemann, 1991; Feagin and Vera, 1995).

What is the situation for African Americans today?
African Americans lag behind whites on all three dimensions of stratification—economic, prestige, and power—discussed in Chapter 8 ("Social Stratification"). The median income of African American households ($22,393) is substantially below that of white households ($35,766). (See Figure 9.3.) The poverty rate for African Americans (29 percent) is almost triple that of whites (U.S. Bureau of the Census, 1996b). There is some good news on the income front for blacks. Whereas the family median income of whites decreased by about 1 percent between 1990 and 1995, black family income increased by 4 percent (U.S. Bureau of the Census, 1996a). Still, African American income is approximately 61 percent that of whites. This means that in terms of averages, for every $100 a white family earns, an African American family earns $61 (U.S. Bureau of the Census, 1996a).

Not surprisingly, there is a vast gulf in wealth (home and car equity, net business assets, net liquid assets) between African Americans and whites. The average African American family holds less than one-quarter of the wealth of the average young white family (U.S. Bureau of the Census, 1994a).

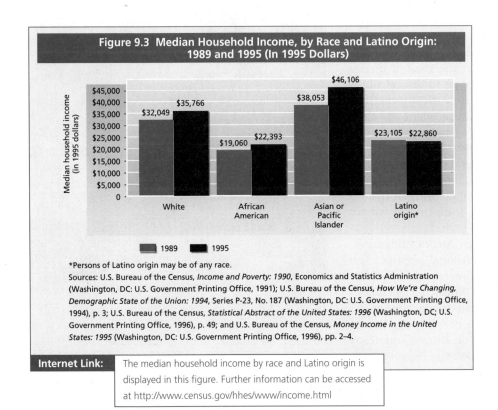

Figure 9.3 Median Household Income, by Race and Latino Origin: 1989 and 1995 (In 1995 Dollars)

*Persons of Latino origin may be of any race.
Sources: U.S. Bureau of the Census, *Income and Poverty: 1990*, Economics and Statistics Administration (Washington, DC: U.S. Government Printing Office, 1991); U.S. Bureau of the Census, *How We're Changing, Demographic State of the Union: 1994*, Series P-23, No. 187 (Washington, DC: U.S. Government Printing Office, 1994), p. 3; U.S. Bureau of the Census, *Statistical Abstract of the United States: 1996* (Washington, DC: U.S. Government Printing Office, 1996), p. 49; and U.S. Bureau of the Census, *Money Income in the United States: 1995* (Washington, DC: U.S. Government Printing Office, 1996), pp. 2–4.

Internet Link: The median household income by race and Latino origin is displayed in this figure. Further information can be accessed at http://www.census.gov/hhes/www/income.html

Part of this economic disadvantage stems from long-standing overrepresentation in low-prestige, low-paying jobs and the underrepresentation in high-prestige, better-paying jobs of African American men and women. Thirty-five percent of African American men are employed in the highest occupational categories: professional, managerial, technical, and administrative. Forty-nine percent of white men have jobs in these categories. Similarly, about 61 percent of African American women are employed in these occupational categories, compared to 74 percent of white women. At the same time, African Americans are almost twice as likely to work in low-level service jobs than are whites (U.S. Department of Labor, 1997).

New, long-term economic trends threaten to make matters worse. These trends include the shift to lower-paying service jobs from higher-paying manufacturing jobs; technological replacement of workers; downsizing due to global competition; and global economic integration, which transfers jobs from high-wage countries like the United States to low-wage countries. Although these long-term economic trends create new jobs and industries, they contribute to an accelerated decline of America's urban economies. Reduced job opportunities for African Americans and other minorities are an important negative consequence (Jacob, 1994).

The disadvantage of African Americans compared to whites can also be seen in unemployment. Joblessness rates among African Americans is over double that of whites (U.S. Department of Labor, 1997). But when *hidden unemployment*—discouraged workers and people with part-time jobs who would prefer full-time jobs—is considered, the actual rate of African American unemployment approaches the Depression figure of one out of every four workers (Swinton, 1989). It is among African American teenagers that the greatest unemployment problem exists. According to official statistics, about one out of every three African American teenagers is unemployed. (U.S. Department of Labor, 1997). With hidden unemployment taken into account, it is estimated that over 40 percent of all African American teenagers are unsuccessfully looking for full-time work. Consequently, thousands of African American youths are becoming adults without the job experience vital to securing jobs in the future (Pearlstein and Brown, 1994).

Education is the traditional way Americans increase their economic gain and occupational prestige. The educational story for African Americans is mixed. By 1995, 83 percent of whites had finished high school compared to 74 percent of African Americans. On the other hand, whereas 24 percent of whites had completed college, only 13 percent of African Americans had done so. As noted in the *Sociological Imagination* at the outset of this chapter, an additional problem is that higher educational attainment doesn't have as high a payoff for African Americans as it does for whites. Although income tends to rise with educational level for both African Americans and whites, it increases much less for African American men (and for women of both races) than for white men. At each level of schooling, black men tend to gain less than their white peers. White male high school dropouts have an average income almost equal to African American men with high school diplomas. White high school graduates, on the average, earn nearly as much each year as African American men with college degrees (U.S. Bureau of the Census, 1993a).

African Americans have enjoyed considerable political success since 1970. This progress is due to legal changes that occurred in the 1950s and 1960s, including the increase in African American voter registration and the fact that African Americans constitute a growing political majority in many large cities. More than 4,800 African Americans are serving as city and county officials, up from 715 in 1970. There are nearly 8,000 African American elected officials in the United States, a fivefold increase since 1970 (U.S. Bureau of the Census, 1996a).

It is in the state houses (there has been only one African American governor elected in this century—

THE SLAVE DECK OF THE BARK "WILDFIRE," BROUGHT INTO KEY WEST ON APRIL 30, 1860.—[FROM A DAGUERREOTYPE.]

Damage to the blacks brought to America on slave ships did not stop even after the abolition of slavery. The particular nature of slavery in the United States limited the upward mobility of freed slaves both before and after the Civil War.

In 1996, California became the first state to repeal its affirmative action laws. It may not be long before residents of other states mount protests against such actions as in the case of these California residents.

L. Douglas Wilder of Virginia) and at the national level that African Americans have experienced the least gain in power. Despite the visibility and influence generated by Jesse Jackson's two presidential bids and the naming in 1989 of an African American as chairman of the national Democratic Party for the first time, less than 1 percent of national political offices are held by African Americans. This is true even though there were more African Americans (40) in the U.S. House of Representatives sworn in on January 3, 1995, than at any time in American history. This is more than double the number of House members in 1981. And for the first time since 1978, an African American, Carol Moseley Braun, was elected to the U.S. Senate in 1992 (*Black Elected Officials,* 1994; Ornstein, Mann, and Malbin, 1994). Although African Americans are an emerging political force, they must accomplish much before they can claim proportionate political representation (O'Hare, 1991; *Black Elected Officials,* 1994; Dentler, 1995).

In fact, a 1995 U.S. Supreme Court decision appears to represent a road block in the path to political power for African Americans. The Court ruled that race could no longer be the "predominant factor" in determining congressional districts—or, it seems to follow, *any* jurisdiction, from school boards to state legislatures (Fineman, 1995). More broadly, the Court created at about the same time a legal test that will make it hard, if not impossible, to keep in place government programs providing an advantage to minorities and women. In fact, "antipreference" forces are making significant inroads in repealing affirmative action laws. The Universities of California and Texas have aban-

doned all racially based preferences, the State of California has repealed affirmative action laws, and Arizona, Ohio, New Jersey, and Washington are seriously considering similar legislation (Ponnuru, 1997).

Has no progress been made in recent years?   During the 1960s and 1970s, some progress was made (Farley, 1984, 1985; Farley and Allen, 1989). The number of African Americans in professional and technical occupations—doctors, engineers, lawyers, teachers, writers—increased by 128 percent. The number of African Americans who are managers or officials or are self-employed is more than twice as high as it was in 1960. As a result of the recent upward mobility of educated African Americans, some scholars see the emergence of two black Americas—the black underclass composed of the permanently poor trapped in inner-city ghettos and a growing black middle class (Glasgow, 1981; Wilson, 1984; Landry, 1988). In Richard Freeman's (1977) terms, a black elite is emerging in America. David Swinton offers this summary statement on the mixed state of African Americans:

> *The empirical evidence does show that some individual blacks have made impressive economic gains. In fact, there has been an increase in the proportion of blacks who can be classified as upper middle class. This limited upward mobility for the few cannot offset the stagnation and decline experienced by the larger numbers of black Americans whose economic status has deteriorated. The central tendency for the group as a whole is revealed by the trends in the averages. And these trends tell a consistent story of stagnation or decline (Swinton, 1989:130).*

U.S. Census data confirm the existence of two black Americas—the rich and the poor (O'Hare, 1991). The number of African American households earning a minimum of $50,000 annually has more than tripled since 1970, rising from 5.2 percent to 17 percent (two million). About 28 percent of African American households earn between $25,000 and $50,000 each year. The percentage of poor African Americans, however, continues to dwarf that of whites. Although less than 6 percent of white families have annual incomes under $10,000, almost one-fifth of African Americans are at that low income level. (See Figure 9.4.)

Is the significance of race declining?   The argument for the declining significance of race is applied to both successful African Americans and the black underclass. Let's examine each in turn.

Although African Americans as a whole remain behind whites on all dimensions of stratification, gains made since the 1960s have led some to conclude that race is of declining importance in America. According to some analysts, race is now less important than economic resources in determining the life chances of young, well-educated African Americans. In other words, it is argued, well-educated African Americans can now compete on equal ground with their white peers. Those African Americans with money and education are said to be no longer shut out because of their color.

William Julius Wilson concurs that race is now less important than economic class for successful African Americans. Wilson goes further by saying that race is declining in significance even for the African American urban poor who are part of the **underclass**—those in poverty who are either continuously unemployed or underemployed because of the absence of job opportunities and/or required job skills. In some of his early work on this topic, Wilson stated it this way:

The recent mobility patterns of blacks lend strong support to the view that economic class is clearly more important than race in predetermining job placement and occupational mobility. In the economic realm, then, the black experience has moved historically from economic racial oppression experienced by virtually all blacks to economic subordination for the black underclass (Wilson, 1980:152).

In his book *The Truly Disadvantaged,* Wilson (1987) elaborates on the thesis of the declining significance of race in the United States. It is true, Wilson acknowledges, that many African Americans (and other minority group members) are in the underclass as a result of historic patterns of discrimination. This, however, is in the past. Currently, African Americans are in the underclass because of features of the American economy. For one thing, inner-city African Americans concentrated in the Midwest and Northeast are being adversely affected by the deindustrialization of the American economy. In the past, African Americans were drawn to Detroit, Chicago, and other large cities by the availability of high wages and stable employment in factories. These manufacturing jobs, which traditionally required little education, have gone to low-income countries where labor is cheap and labor unions are nonexistent. Consequently, today's inner-city poor are denied the type of jobs used for upward mobility by all American minority groups. A second contributing economic factor is the internal movement of business and industry from central cities to the suburbs. A third economic factor contributing to the perpetuation of the underclass is the movement of upwardly mobile African Americans to the suburbs. The poor remaining in the inner cities, lacking the positive role models available in the past, now witness the use of drug dealing, prostitution, and other illegal economic activities as primary avenues to success (Vobejda, 1991; Clayton, 1996; Massey and Denton, 1996).

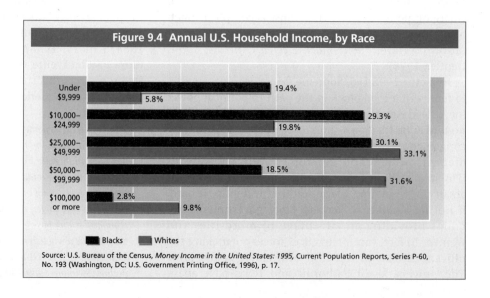

Figure 9.4 Annual U.S. Household Income, by Race

Income	Blacks	Whites
Under $9,999	19.4%	5.8%
$10,000–$24,999	29.3%	19.8%
$25,000–$49,999	30.1%	33.1%
$50,000–$99,999	18.5%	31.6%
$100,000 or more	2.8%	9.8%

■ Blacks   ■ Whites

Source: U.S. Bureau of the Census, *Money Income in the United States: 1995,* Current Population Reports, Series P-60, No. 193 (Washington, DC: U.S. Government Printing Office, 1996), p. 17.

If economic factors are perpetuating the underclass, Wilson argues, solutions are to be found in the economic sphere. Wilson calls for a federal economic policy designed to create a higher employment rate and better jobs for all Americans. Wilson looks to such things as the creation of higher-paying jobs, upgrading of education, job training, relocation support to encourage members of the underclass to secure and keep decent jobs, publicly supported day care for the low-income employed, a markedly higher minimum wage, and medical insurance for the employed in low-paying jobs (Wilson, 1993, 1997).

Use of the term *underclass* has been questioned (Katz, 1993). Wilson, himself, suggests substituting the term *ghetto poor* because the concept of underclass has become a code word for inner-city African Americans and is used by journalists to highlight unflattering behavior in the ghetto (Wilson, 1991). Herbert Gans (1990) sees the term underclass as a pejorative, value-laden term now being used to describe the "undeserving" poor. Joining a growing number of social scientists, Gans believes that the term underclass is hopelessly mired in ideological connotations of undeservingness and blameworthiness. These social scientists, including Wilson, do not wish to obscure the harsh reality faced by the urban poor by continuing to use the term underclass if such use contributes to the blaming of the victim. It remains to be seen whether the term underclass will lose its currency in sociological terminology. At any rate, some term, perhaps Wilson's ghetto poor, will be utilized to capture the problems associated with the inner-city poor.

Some evidence exists that young African American college graduates are reaching an economic par with their white peers and that many African Americans in general have made significant gains in the past twenty years (Farley, 1984). It is not certain, however, that this situation is a permanent one. The strides since the 1960s may simply be the result of tremendous changes in the past two decades that will not be maintained in the future.

Many critics disagree with the idea that race is a declining force in the situation of African Americans, whether or not some progress has been made (Thomas and Hughes, 1986; Carnoy, 1994; Hacker, 1995; Cancio, Evans, Maume, 1996). Doris Wilkinson (1995) believes that discrimination still determines the life chances of African Americans. The current racial dominance, she argues, is simply more sophisticated and subtle. Joe Feagin (1991) has recently added to this chorus of dissenting voices. (See *Doing Research.*) Others who disagree with Wilson's position on the declining significance of race point to inferior education in America's inner cities, a situation tied in part to racial discrimination. (See Chapter 12, "Education.") Still other critics point to housing discrimination and segregation as an enduring barrier to large-scale African

American migration to the suburbs where many jobs are being moved (Massey and Denton, 1987; Massey, 1990; Massey and Eggers, 1990). Finally, some argue that the concept of a black underclass applies to rural areas. According to this viewpoint, the black underclass is more densely concentrated in rural parts of the South than in the inner cities of the North (O'Hare, 1992a, 1992b; O'Hare and Curry-White, 1992).

Latinos

Who are Latinos? Latinos in the United States number almost 29 million and are composed of many diverse ethnic minorities, including Mexican Americans, Puerto Ricans, Cubans, and increasing numbers of people from Central and South American countries. High birthrates and immigration combine to make Latinos one of the most rapidly growing minorities in the United States (del Pinal and Singer, 1997). In fact, early in the twenty-first century, Latinos will become America's largest ethnic minority. By 2050, the Latino population is projected to reach 100 million, constituting 24 percent of the U.S. population. (See Figure 9.5.)

Over one-half of Latinos today are of Mexican descent, whereas Puerto Ricans make up a little less than one-tenth of the total Latino population. Most Puerto Ricans are concentrated in or near New York City, although greater geographic dispersion is occurring. Cubans make up the third largest category of Latinos, with about 1.2 million people, most of whom are located in the Miami, Florida area (U.S. Bureau of the Census, 1996a).

Latino peoples are quite diverse in many important ways. Each came to the United States under different

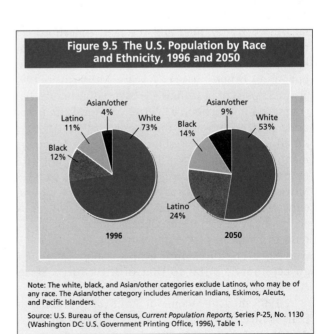

Figure 9.5 The U.S. Population by Race and Ethnicity, 1996 and 2050

Note: The white, black, and Asian/other categories exclude Latinos, who may be of any race. The Asian/other category includes American Indians, Eskimos, Aleuts, and Pacific Islanders.

Source: U.S. Bureau of the Census, *Current Population Reports,* Series P-25, No. 1130 (Washington DC: U.S. Government Printing Office, 1996), Table 1.

Latinos, including Mexican Americans, Puerto Ricans, Cuban Americans, and a variety of peoples from Central and South America, comprise a diverse minority in American society. Their socio-economic condition ranges from destitution to some measure of success. The Mexican Americans involved in this parade are among the small proportion of Latinos who have done relatively well.

circumstances and retains a sharp sense of its own identity and separateness. In addition, it is not difficult to find significant internal disagreements within the larger Latino minorities. For example, there are obvious differences between the first Cuban immigrants who came to the United States, most of whom were either middle or upper class on arrival, and some later Cuban immigrants, many of whom were either convicted criminals or relatively uneducated (Cafferty and McCready, 1985; Gann and Duignan, 1986; Portes and Truelove, 1987; Stavans, 1996; Darder and Torres, 1997).

What is the socioeconomic condition of Mexican Americans, Puerto Ricans, and Cubans? Latinos are far behind white Americans in formal education. The median educational level for the total American population is about 12.7 years, or high school graduation, but for all Latinos, it is 11.7 years. Although in 1995 about 53 percent of Latinos age twenty-five or over had completed high school, 83 percent of non-Latinos had done so (U.S. Bureau of the Census, 1996a). Mexican Americans have the lowest levels of educational attainment, Cubans the highest. This is accounted for by the migration pattern and the history of these peoples. Mexican migrants came to the United States with little education, unable to speak English, and willing to accept the lowest level jobs. Cubans, on the other hand, tend to be middle- and upper-middle class migrants who had been educated and economically successful in an economy similar to that of the United States (Valdivieso and Davis, 1988).

A 1995 median income of $24,313, is lower than that of African Americans, and is significantly lower than that of whites ($40,884). Cubans typically are the most affluent Latinos, but even their median income is only about 75 percent that of whites. The poorest among the large Latino groups are the Puerto Ricans, whose income is about 51 percent that of whites. Approximately 33 percent of Latino families were living below the poverty level in 1995 compared to about 9 percent among whites of non-Latino origin. Puerto Rican families had the highest poverty rate, Cubans the lowest (U.S. Bureau of the Census, 1996b).

As one might expect from such income figures, Latinos, with some exceptions, tend to be concentrated in lower paying and lower status jobs as semiskilled workers and unskilled laborers. Farm work is also still common, especially among Mexican Americans. Cuban men belong to the only Latino minority that approaches the occupational profile of Anglos (Moore and Pachon, 1985).

No Latino minority is very close to the national average of 66 percent home ownership. Even the relatively well-off Cubans show a figure of only 52 percent.

The increase in Latino-owned businesses must be noted. In 1969, there were only 100,000 Latino-owned businesses with less than $5 billion in revenue. By 2000, there will be about 800,000 Latino-owned firms with a total revenue of nearly $50 billion.

Politically, Latinos are now becoming a nationally visible—though comparatively small—force shaping American politics. Although there are currently no Latino senators, seventeen seats in the House are held by Latinos, a tripling from 1981. Of these, thirteen are Mexican Americans, three are of Cuban descent, and one is of Puerto Rican ancestry (*The World Almanac of U.S. Politics,* 1995). Below the national level, there are almost six thousand elected Latino public officials (U.S. Bureau of the Census, 1996a).

Issues such as the need for increased education and concern for illegal immigration as well as the basic issues of income and the quality of life promise to keep Latinos politically active. Only time will tell whether Latino political activism will counter negative stereotypes or whether old patterns of prejudice and discrimination will persist.

Native Americans

If a single word can be used to describe Native Americans today it is diversity. About five hundred tribes and bands have been identified (U.S. Bureau of the Census, 1988). This diversity is generally unrecognized because of old stereotypes and the larger society's lack of knowledge of Native Americans. Actually, the Navajo and Sioux are entirely different nations with different cultures, social structures, and problems. Tribal groups are as different from one another as they are from the dominant Anglo culture and as different from one another as white Americans are from Italians or Brazilians. Today, Native Americans number just under two million, three-fourths of whom live off reservations (Snipp, 1992, 1996).

What has been the U.S. policy toward Native Americans?   Over the years, the federal government has vacillated between a policy of complete paternalism (that is, total domination and care) and a policy of almost total neglect. Impoverishment, oppression, and deceit have typically, though not always, characterized relations between the government of the United States and various tribal governments. Jonathan Turner summed up both past and present government policy toward Native Americans as follows:

The legacy of economic exploitation, especially the great "land grabs" by whites in this century, has forced Indians into urban areas because they can no longer support themselves on their depleted reservations. Yet the legacy of isolation on the reservation prevents many Indians from making the cultural and psychological transition to urban, industrial life. And the burden of change has been placed on the indi-vidual Indian, for white institutions—from factories to labor unions and welfare agencies—display little flexibility in adjusting to Indian patterns. The contemporary Indian is therefore faced with impoverishment no matter what course of action he chooses: to stay on the reservations results in poverty, but to leave the reservation and encounter the hostile white economic system in urban areas also results in poverty (Turner, 1976:185).

Both the paternalistic and neglectful approaches have left Native Americans outside the mainstream of social and economic opportunity.

What is the current situation of Native Americans? Poverty remains a major fact of life among Native Americans, especially for many who remain on reservations. Approximately 27 percent of the Native American population is below the poverty line. The median income is $21,619 per year and 14 percent of the population has an income of over $50,000 per year, as compared with about 32 percent of the white population. A gap in education exists. Of the Native American population twenty-five years of age or older, only 66 percent were high

Historically, the U.S. government's policy toward native Americans, alternating as it has between paternalism and neglect, has been harmful. The greatest impoverishment is experienced by those remaining on reservations.

DOING RESEARCH

Joe R. Feagin—The Enduring Significance of Race

According to many scholars—most notably William Julius Wilson (1980, 1984, 1987, 1988, 1991)—it is economic class rather than racism and discrimination that now detrimentally affects African Americans. Through in-depth interviews with middle-class African Americans in several cities, Joe Feagin (1991) questions this line of argument.

Feagin starts with this stipulation from Title II of the 1964 Civil Rights Act: "All persons shall be entitled to the full and equal enjoyment of the goods, services, facilities, privileges, advantages, and accommodations of any place of public accommodation . . . without discrimination or segregation on the ground of race, color, religion, or national origin." Public accommodations under-scored in the Act are restaurants, hotels, and motels, although public places of all types are covered.

Feagin wished to study African Americans in the middle class because they would have the economic resources needed to take advantage of these public accommodations. His research was guided by these major questions: Do middle-class African Americans still experience racism in public accommodations and other public places? If they do, how is discrimination manifested? What means are used by middle-class African Americans to handle this form of discrimination? What are the effects of discrimination on its middle-class victims?

To answer these research questions, Feagin used thirty-seven in-depth interviews drawn from a large study of 135 middle-class African Americans in Boston, Buffalo, Baltimore, Washington, D.C., Detroit, Houston, Dallas, Austin, San Antonio, Las Vegas, and Los Angeles. The data came from a survey done by African American interviewers in 1988–90. Initially, respondents were identified as members of the African American middle class by knowledgeable consultants in the cities. Additional respondents were identified by eliciting names from the first wave of respondents. (This is known as "snowball" sampling.)

Middle class was defined as "those holding a white-collar job (including those in professional, managerial, and clerical jobs), college students preparing for white-collar jobs, and owners of successful businesses" (Feagin, 1991:103). The subsample of thirty-seven middle-class African Americans indicating public discrimination to the interviewers is reported to be fairly representative of the larger sample on a series of demographic characteristics—occupation, age, income, education, sex, and geographic area of residence.

From his analysis, Feagin concludes that middle-class African Americans do still experience discrimination based on color. Although several types of discriminatory actions were reported by the respondents, rejection or poor service was the most common in public accommodations, and verbal or physical threat by white citizens or police officers was most likely to occur in the street. (See Table 9.1.) The most frequent responses by the African American targets of racism in public accommodations were verbal retorts or resigned acceptance. If discrimination occurred in the street, the most common responses were verbal retorts or withdrawal. (See Table 9.2.)

According to Feagin, the most tragic cost of this enduring discrimination for middle-class African Americans is the physical and psychological drain created by the need for eternal vigilance. Although isolated discriminatory acts may appear insignificant to whites, years of being the victim of discriminatory actions—blatant avoidance, verbal attack, physical

school graduates versus 81.4 percent of the total U.S. population. Only 9.4 percent have completed four years or more of college, as contrasted with 24 percent of the white population (U.S. Bureau of the Census, 1993i).

Native Americans have made only a scant penetration into the upper levels of the occupational structure. Although some gains were made during the 1960s and 1970s, only about 20 percent of all employed Native American men and women are in professional, managerial, or administrative positions. About one-third are in blue-collar jobs (craftsworkers, supervisors, operatives, and nonfarm laborers; U.S. Bureau of the Census, 1993e).There are now no Native American members of the House or Senate (U.S. Bureau of the Census, 1996a).

The situation for Native Americans living on reservations is even worse. Although the poverty rate for all native Americans is 27 percent, half of those on reservations are below the poverty line. Native Americans have the lowest annual income of all minority groups in the United States ($21,619); those on reservations earn only $16,000. Only 54 percent of Native Americans on reservations complete high school compared to 66 percent overall. The Native American rate of college education or better among those living off reservations is nearly double that of those on reserva-

TABLE 9.1	Percentage Distribution of Discriminatory Actions by Type and Site: Middle-class African Americans in Selected Cities, 1988–1990

	Site of Discriminatory Action	
Type of Discriminatory Action	Public Accommodations	Street
Avoidance	3	7
Rejection/poor service	79	4
Verbal epithets	12	25
Police threats/harassment	3	46
Other threats/harassment	3	18
Total	100	100
Number of actions	34	28

Source: Joe R. Feagin, "The Continuing Significance of Race: Antiblack Discrimination in Public Places," *American Sociological Review,* 56(February 1991):104. Reprinted by permission of the American Sociological Association and the author.

TABLE 9.2	Percentage Distribution of Primary Responses to Discriminatory Incidents by Type and Site: Middle-class African Americans in Selected Cities, 1988–1990

	Site of Discriminatory Incident	
Response to Discriminatory Incident	Public Accommodations	Street
Withdrawal/exit	4	22
Resigned acceptance	23	7
Verbal response	69	59
Physical counterattack	4	7
Response unclear	—	4
Total	100	99
Number of responses	26	27

Source: Joe R. Feagin, "The Continuing Significance of Race: Antiblack Discrimination in Public Places," *American Sociological Review,* 56(February 1991):104. Reprinted by permission of the American Sociological Association and the author.

abuse, subtle and covert slights—have a cumulative effect. African Americans adopt the use of a "second eye" to analyze interracial situations. As one respondent said:

I think that it causes you to have to look at things from two different perspectives. You have to decide whether things that are done or slights that are made are made because you are black or they are made because the person is just rude, or unconcerned and uncaring. So it's kind of a situation where you're always kind of looking to see with a second eye or a second antenna just what's going on (Feagin, 1991:115).

What may appear to whites as "black paranoia," then, is instead a realistic sensitivity to interracial encounters generated and reinforced by the types of cumulative discrimination just described.

Despite decades of legal protection, Feagin concludes, African Americans have not attained the full promise of the American dream. Although middle-class African Americans have worked hard for their success, their color too often overshadows the symbols of their achieved social-class level.

Critical feedback

1. Do you think that it is economic class, rather than racism and discrimination, that currently hurts African Americans? Why or why not?
2. Did Feagin adequately test his hypothesis? Explain your conclusion.
3. Which of the three major theoretical perspectives best fits Feagin's research study? Defend your choice.

tions—9.3 percent versus 5 percent (U.S. Bureau of the Census, 1993e, 1993i).

Is there any hope for change?   The relationship between the government and various tribal governments has been changing recently. According to Vine Deloria and Clifford Lytle, a new government policy grants tribes special political status. As a result, "many tribes have created signs proclaiming their nationhood, that traditional Indians believe themselves to be sovereign entities endowed with almost mystical powers, and that groups of Indians have recently appeared on the world scene demanding some form of represen-

tation in the United Nations" (Deloria and Lytle, 1984:7).

Deloria and Lytle also cite several possible directions for change. There is, first of all, the need for reform in tribal governments to allow continuity with the past as well as forward movement. There is also the need to establish a place for Native American cultures in the modern world. Another need is the creation of relationships with federal, state, and local governments based on mutual respect and parity. Such changes will not come easily, given the tradition of inequality and discrimination that has characterized Native American life for so long.

Asian Americans

There are over nine million Asians in the United States, or 3.6 percent of the total population. Asian Americans, the fastest growing minority in the United States, increased by 80 percent between 1980 and 1989 (O'Hare and Felt, 1991; Palen, 1992). In 1984, more Asian immigrants entered the United States than came between 1930 and 1960. In 1994, approximately 300,000 Asians were permitted to immigrate to the United States. The Asian American population is predicted to increase by 65 percent by the year 2010 and nearly quadruple by 2050 (U.S. Bureau of the Census, 1996a). If there is any racial minority success story in America, it is among Chinese and Japanese Americans, particularly the latter. Even for Chinese and Japanese Americans, however, the road has not been smooth (Kasindorf and Shapiro, 1982; Doerner, 1985; Kitano and Daniels, 1988).

How have Chinese Americans fared over the years? Chinese Americans first began immigrating to the United States in the mid-1800s. Racial prejudice against the Chinese existed in America prior to their immigration. Consequently, they were greeted with prejudice and discrimination. When heavy Chinese immigration to California began in 1848, a great furor arose. Americans feared that the presence of yet another race would encourage the institution of slavery, cause internal discord, and lead to the introduction of incurable diseases. Although Chinese labor was accepted, even exploited, in the mining and railroad industries in the West, there was a fear that it would spread across the country. This fear, combined with long-standing prejudice and the belief that the Chinese would deprive white men of work in the declining mining and railroad industries, resulted in an anti-Chinese movement that lasted into the first part of the twentieth century. During this period of more than fifty years, there was much violence against the Chinese, including riots, lynchings, and massacres. When a group of Chinese in Tonapah, Nevada, during the late 1880s asked for time to collect their belongings before being expelled from the town, they were robbed and beaten, and one of them died as a result (Lyman, 1974).

Finally, the "Chinese question" became a political matter of high priority. Many state laws were created to restrict Chinese from holding certain jobs that could deprive whites of employment, from attending certain schools, and from living in certain neighborhoods. The Chinese Exclusion Act passed by the U.S. Congress in 1882 prohibited the entry of any skilled or unskilled Chinese laborers or miners into the United States for a ten-year period. Strict federal legislation against Chinese immigration continued to be passed from then until after 1940. During this period, Chinese Americans were driven into large urban ghettos known as Chinatowns, where they are still mainly concentrated.

Although Chinese Americans in many ways remain isolated from American life to this day, their situation began to improve after 1940. American-born college graduates began to enter professional occupations, and Chinese American scholars and scientists began to make publicly recognized contributions to science and the arts. Their dedication to hard work and their contributions to American society have been widely recognized.

What has been the history of Japanese Americans? Unlike the Chinese, the earliest relations between Americans and Japanese were positive. Early diplomatic relations, thanks in part to Admiral Perry's opening of the door to Japan, were warm and cordial. Beginning in 1885, large numbers of Japanese men immigrated to the West Coast of the United States, but the timing for this massive immigration was wrong. The entry of the Japanese came on the heels of America's attempt to exclude, harass, and harm the Chinese. Although the Japanese suffered prejudice and discrimination during these early years, they moved from being laborers in certain industries (railroads, canning, logging, mining, meat packing) to being successful farmers (Kitano, 1993).

When the Japanese began to compete with white farmers, however, anti-Japanese legislation was passed. The California Alien Land Bill of 1913, for example, permitted Japanese to lease farmland for a maximum of three years and did not allow land they already owned to be inherited by their families. In 1924, the U.S. Congress halted all Japanese immigration, and the 126,000 Japanese already in the United States became targets for prejudice, discrimination, stereotyping, and scapegoating. In 1942, President Roosevelt signed an executive order that sent more than 110,000 of the 126,000 Japanese in America—two-thirds of whom were American citizens—to concentration camps that were euphemistically referred to as "relocation centers" (Nagata, 1993). The argument that the Japanese posed a security threat during World War II can hardly be supported, because German and Italian Americans were not relocated even though their countries were also at war with the United States (U.S. Commission on Wartime Relocation and Internment of Civilians, 1983).

In spite of past and present prejudice and discrimination, Japanese Americans have been one of the most successful racial minorities in the United States. Despite their internment, Japanese Americans have not had to overcome the centuries of prejudice and discrimination endured by African Americans and Native Americans. Through a cultural emphasis on education and hard work, Japanese Americans represent an amazing story of economic and occupational success (Montero, 1981).

Japanese Americans faced prejudice and discrimination in the United States from the time they first arrived. Prejudice and discrimination reached its horrific peak in 1942 when President Roosevelt signed an executive order placing more than 100,000 of the 126,000 Japanese in America in concentration camps. Two-thirds of them were American citizens whose property, as well as their liberty, was taken from them. German Americans and Italian Americans did not suffer the same maltreatment despite Germany's and Italy's involvement in the war.

Why have Asian Americans been so successful? Asian Americans have been particularly successful at using the educational system for upward mobility (Gibson and Ogbu, 1991). This fact is reflected in the academic achievement of schoolage Asian Americans—over 38 percent of Asian Americans have completed four years of college compared to about 24 percent of whites and 9 percent of Latinos (U.S. Bureau of the Census, 1996a).

Some claim that Asian Americans are academically successful because they are innately more intelligent than other Americans. Most scholars believe that Asian academic excellence is due to culture, socialization, and influence of the family (Gibson and Ogbu, 1991; Caplan, Choy, and Whitmore, 1992). Most Asians see education as the key to success. Moreover, in the Confucian ethic, which is so much a part of the Chinese, Japanese, Vietnamese, and Korean cultures, academic excellence is the only way of repaying the debt owed to parents. According to school principal Norman Silber, "Our Asian kids have terrific motivation. They feel it is a disgrace to themselves and their families if they don't succeed" (McGrath, 1983:52). In addition, Asian immigrants are accustomed to a tougher academic regimen. For example, the average number of school days in America is 180, but it is 225 days in Japan. Whatever the reasons, Asian Americans have been particularly successful in part because of their willingness to work hard inside and outside of school (Barrett, 1990).

The struggle, of course, is not over. Although Asian Americans have been more successful than other American racial minorities, they still feel the effects of prejudice and discrimination. In fact, the relative success of Asian Americans has hurt them in some ways. The stereotype of success has led the public to conclude erroneously that Asian Americans are all well educated, overrepresented in higher level occupations, and making as much money as or more money than white Americans. The facts speak otherwise. Vast socioeconomic differences exist among groups within Asian American communities. Popular emphasis on success has led to the overlooking of those Asian Americans who have not done very well (Buckley, 1991). Moreover, the accent on success covers over the fact that even "successful" Asian Americans are worse off than similarly educated and employed white Americans. As a consequence of the stereotype of success, many experts believe that Asian Americans have been neglected and ignored by governmental agencies, educational institutions, private corporations, and other sectors of society (U.S. Commission on Civil Rights, 1980).

It is true, however, that a record number (seven) of Asian Americans hold seats in the House of Representatives, including the first ever Korean American to serve in Congress (*The World Almanac of U.S. Politics,* 1995). In 1996, Gary Locke became the first Asian American governor on the U.S. mainland (Puente, 1996).

SOCIOLOGY IN THE NEWS

Institutionalized Discrimination—Asian Americans

Asian Americans constitute the fastest growing minority in the United States, numbering over nine million. This minority will more than double by 2010 and quadruple by the middle of the twenty-first century. Although Asian Americans are generally thought to be uniformly successful after entering the United States, they actually suffer the same social and economic consequences of institutionalized prejudice and discrimination as other minorities. This CNN report highlights this reality.

1. What do you think is the most important stereotype Americans hold of Asian Americans?

   _____

   _____

2. Are Asian Americans more the victims of direct or indirect institutionalized discrimination? Explain.

   _____

   _____

About 3.6 percent of the American population is Asian (nine million persons). Despite their general success educationally, many Asian Americans do not enjoy a middle-class lifestyle.

White Ethnics

Who are white ethnics?   White ethnics are the descendants of immigrants from Eastern and Southern European nations, particularly Italy and Poland, but they also include Greek, Irish, and Slavic peoples (Rubin, 1994). The majority of them are blue-collar workers living in small communities surrounding large cities in the eastern half of the United States (Palen, 1992). During the 1960s, white ethnics gained the false reputation of being conservative, racist, pro-war, and "hardhats." Peter Schrag wrote early of "the forgotten American:"

> There is hardly a language to describe him, or even a set of social statistics. Just names: racist-bigot-redneck-ethnic-Irish-Italian-Pole-Hunkie-Yahoo. The lower middle class. A blank. The man under whose hat lies the great American desert. Who watches the tube, plays the horses, and keeps the niggers out of his union and his neighborhood. Who might vote for [George] Wallace (but didn't). Who cheers when the cops beat up on demonstrators. Who is free, white and twenty-one, has a job, a home, a family, and is up to his eyeballs in credit (Schrag, 1977:129).

What are white ethnics really like?   This portrait of white ethnics is not an accurate one. In fact, the evidence is just the contrary. Surveys conducted during the 1960s showed white ethnics to be more against the Vietnam War than white Anglo-Saxon Protestants. Catholic blue-collar workers were found to be more liberal than either Protestant blue-collar workers or the country as a whole—they were more likely to favor a guaranteed annual wage, more likely to indicate they would vote for an African American presidential candidate, and more likely to express concern about the environment. Finally, white ethnics were more likely to be sympathetic to government help for the poor and were more in favor of integration than WASPS were (Greeley, 1974).

According to David Featherman (1971), white ethnics have not traditionally been the victim of occupational or income discrimination on purely ethnic grounds. In fact, Andrew Greeley (1976) has contended that white ethnics have been so successful that it is unfair to label them working class. Their success has not generally been recognized, according to Greeley, because America's elite is not willing to abandon the myth of the blue-collar ethnic.

Despite their relative success, many white ethnics have in recent years become very conscious of their cultural and national origins (Schaefer, 1997). There is, in fact, a white ethnic "roots" movement. The new trend toward white ethnic identity began with the black power movement of the 1960s. Just as many African Americans decided that they wanted to preserve their cultural and racial identity, many white ethnics now believe that "white ethnicity is beautiful." Many think that the price of abandoning one's cultural and national roots is simply too high.

Lillian Rubin (1994) links the continuing accent on white ethnicity to the economic decline of the last twenty years and the rising demands and presence of racial and ethnic minorities. White ethnics, she contends, are attempting to establish a public identity to use in self-defense against the claims of race. At the same time, this identity enables white ethnics to take a seat at the multicultural table.

A few sociologists contend that white ethnicity is actually fading in the United States. According to Richard Alba (1985, 1990), the remaining relatively small number of white ethnics may not survive for very long. Ethnic identity, he argues, cannot be maintained in the face of disappearing ethnic families, neighborhoods, and communities. Mary Waters (1990) agrees with Alba's description of the "twilight" of white ethnics, reporting that ethnicity was not very important to the white ethnics she studied. Although her respondents felt that being Italian, Polish, or Irish might make one distinctive in some way, ethnicity had little effect on where they lived, who they married, or what they did for a living.

Jewish Americans

The existence of Jewish Americans as a white ethnic group also argues against the stereotype of white ethnics. The United States has the largest number of Jews of any country in the world. America's six million Jews outnumber even Israel's three and a half million. The majority of Jewish Americans are concentrated in several northeastern states, including New York, New Jersey, Massachusetts, and Connecticut. A sizable number live in California, Florida, and Pennsylvania. The most recent immigrants—primarily well-educated and highly skilled—have come from Israel and the former Soviet Union. It is estimated that about 10 percent of all Jewish Americans have been in the United States less than ten years (Kass and Lipset, 1980).

The first Jewish immigrants to the United States landed in New Amsterdam in 1654. They were discouraged from immigrating by all of the colonies, who legally prohibited them from holding political office or voting. Still, by the 1840s, Jews began to arrive in large numbers. Anti-Semitism reached its peak in the 1920s and 1930s, particularly among America's upper and upper-middle classes. New York was often called "Jew York," and Jewish Americans were subjected to occupational and social segregation (Davis, 1978; Parrillo, 1997).

Throughout the first half of the twentieth century, Jewish Americans were almost totally excluded from top positions in most major industries, denied membership in social and recreational organizations, and subjected to quotas in institutions of higher learning. Although such exclusion has abated since World War II, a study in the late 1960s showed that four-fifths of elite men's clubs in America still had no Jewish members (Carlson, 1969). Lack of acceptance into these pres-

The existence of Jewish Americans as a white ethnic group argues against the stereotype of white ethnics. Despite prejudice and discrimination, Jewish Americans are one of the most successful ethnic groups ever to migrate to the United States.

tigious social clubs appears to persist (Zweigenhaft and Domhoff, 1982).

Although Jewish Americans have been limited in their opportunities for movement into the top circles of the most economically powerful American corporate entities (Zweigenhaft and Domhoff, 1982), they nevertheless are one of the most successful ethnic groups ever to migrate to the United States. This is partly because Jewish men, particularly later immigrants, entered the country as skilled workers in a proportion far above the norm for immigrant groups. In addition to bringing skills necessary to fit into industrial society, Jews valued hard work, sacrifice, perseverance, family responsibility, and education. Although working at skilled blue-collar jobs, Jewish parents urged their children to pursue higher education for entrance into the professions such as teaching, law, and medicine. Consequently, Jewish Americans represent an above-average proportion of college graduates; they comprise 2 percent of the population, but they account for 5 percent of college graduates (Gallup, 1993). Among Jewish Americans, the mean level of educational attainment is just over 15 years compared to under 13.6 for the non-Jewish population (Chriswick, 1993).

Because of the exclusion they experienced, Jewish Americans have tended to remain socially isolated from the rest of American society. This is currently undergoing some change. Whereas less than 6 percent of Jewish

Americans married non-Jews in the 1950s, approximately 47 percent now do so (Goldman, 1993). Although the social isolation of Jewish Americans has diminished somewhat, some separation will probably continue for two basic reasons. First, for whatever reason—prejudice on the part of some African Americans, the Middle East conflict, the upsurge of fundamentalist religious teachings—anti-Semitic sentiment appears to be increasing once again in the United States. Second, many Jewish Americans do not want to abandon their cultural roots (Sanoff, 1983).

Beyond Direct Institutionalized Discrimination

The socioeconomic gap between many minorities and the general white population is doubtlessly due largely to past and present prejudice and discrimination. The situation is now even more complex. Minorities are concentrated at the bottom of the stratification structure in the United States, in part because of the structure of the economy, a factor that operates beyond the influence of direct or consciously intended institutional discrimination. This factor has been explored through the context of the **dual labor market**—the existence of a split between core and peripheral segments of the economy and the division of the labor force into preferred and marginalized workers (Hodson and Sullivan, 1990). This perspective, which attributes persisting racial inequality to economic factors, originated in an effort to understand the continuing survival of large-scale racial (and gender) inequalities in the face of programs designed to overcome such disparities (Hodson and Kaufman, 1982).

What is the dual labor market theory?   Traditionally, it has been assumed that only one labor market existed in which all workers competed for jobs. It was also assumed that occupational advancement was based on hard work, education, and training. These assumptions have been successfully challenged by advocates of conflict theory. The existence of a dual labor market means that hard work, education, and training pay off differently in different segments of the labor market. Workers involved in the *core* sector of the labor market—durable manufacturing and petroleum industries, for example—enjoy high wages, good opportunities for advancement, and job security. Those involved in the *peripheral* sector, including such industries as textile manufacturing and retail trades, are employed in low-paying jobs with little hope for advancement. African Americans and other minority peoples tend to be trapped in these secondary labor markets and lack the resources to alter their situation. Thus, it is concluded, historical practices of racial and ethnic discrimination now interact with contemporary economic processes to lock a significant and growing number of minority peoples out of the core economy, thereby reducing the likelihood of any improvement in their life chances and lifestyles (Bonacich, 1976; Szymanski, 1976; Cummings, 1980; Lord and Falk, 1982).

How does indirect institutionalized discrimination operate in the dual labor market?   Minority groups, who are overrepresented in the peripheral sector of the American economy and, consequently, tend to hold low-skill, low-paying jobs, are subjected to indirect institutionalized discrimination. A long-standing division of labor automatically reduces minority access to jobs in the core sector despite their education and training. In addition, minority members are disproportionately shut out of the more desirable jobs and out of the education and training needed to move into core sector jobs on the grounds that they actually prefer, and are better suited to, the marginal jobs they hold (Hodson and Sullivan, 1990).

FEEDBACK

1. _____ discrimination refers to unintentional organizational or community actions that negatively affect a racial or ethnic minority.
2. The attitudes and practices associated with _____ in the American slave system help account for the long-term prejudice and discrimination experienced by African Americans.
3. The evidence clearly shows that race is declining in importance in America. *T or F?*
4. Which of the following Latino ethnic minorities is in the best socioeconomic condition?
   a. Puerto Ricans   b. Cubans   c. Mexican Americans
5. The federal government's policy toward Native Americans has vacillated between _____ and almost total neglect.
6. Contrary to popular opinion, white ethnics in the United States are politically more liberal than Protestant blue-collar workers. *T or F?*
7. Anti-Semitism in America during the 1920s and 1930s was most prevalent among the _____ and upper-middle classes.
8. Within the dual labor market, minorities tend to be disproportionately trapped in the _____ sector.

Answers: 1. Indirect institutionalized 2. manumission 3. F 4. b 5. paternalism 6. T 7. upper 8. secondary

INCREASING RACIAL AND ETHNIC DIVERSITY
IN THE UNITED STATES

INCREASING RACIAL AND ETHNIC DIVERSITY IN THE UNITED STATES

The United States is currently moving from a predominately white population of Western origin and culture to a society comprising larger and larger numbers of diverse racial and ethnic minorities. Almost one-half of all Americans will be members of a minority group by the middle of the twenty-first century (O'Hare, 1992b; Parrillo, 1996). This rapid growth and increasing diversity of the minority population of the United States is one of the most important changes now occurring. Although concern for prejudice and discrimination aimed at minorities is inherently a problem within the value system of America, this trend to diversity increases its urgency.

Minorities' Share of the Total Population

In 1996, the combined number of African Americans, Latinos, Asian Americans, and Native Americans exceeded 71 million, up from just under ten million in 1900 and 21 million in 1960. If these minority groups resided in a separate country, they would be the thirteenth most populous nation in the world. Great Britain, France, Italy, and Spain would all be smaller.

Figure 9.6 shows the share of the American population minorities represented in 1900 and is expected to represent in 2050. For most of the twentieth century, America's minority population constituted a relatively constant percentage of the total population—the share of the population represented by minorities only increased from 13.1 to 14.9 percent between 1900 and 1960. Since then, however, the minority proportion of the population has increased to about 30 percent and, as noted earlier, is expected to constitute about 50 percent by 2050.

Relative Growth of Minority Groups

Whereas the total population of the United States is expected to increase by 45 percent between 1995 and 2020, the minority population is projected to grow 129 percent. The predominant minority in the United States for most of the twentieth century has been African American. A rather dramatic change in the relative growth rate of minorities is expected between 1995 and 2050. Although the African American population is expected to expand by about 83 percent, Latinos are projected to increase by almost 260 percent, Native Americans are expected to grow by almost 95 percent, and Asian Americans will have mushroomed over 265 percent. (See Figure 9.7.)

These changes in the relative size and diversity of the minority population promise to alter the nature of American society. Reconsideration of the status of racial and ethnic minorities seems probable.

AGE STRATIFICATION AND AGEISM

Thus far in our exploration of social inequality, we have examined race and ethnicity as major factors in the differential distribution of the scarce desirables in a society. Because chronological age is another basis for social ranking, sociologists are interested in **age stratification**—when the unequal distribution of scarce desirables (power, wealth, privileges) in a society are based on chronological age. Although age can operate to the advantage or disadvantage of any age group in a society,

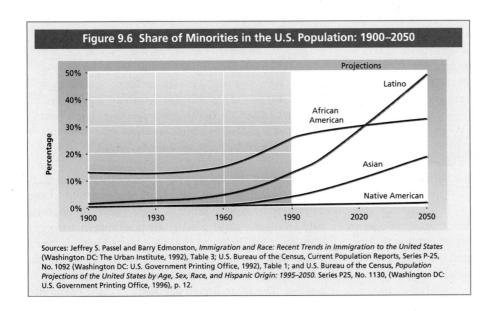

Figure 9.6 Share of Minorities in the U.S. Population: 1900–2050

Sources: Jeffrey S. Passel and Barry Edmonston, *Immigration and Race: Recent Trends in Immigration to the United States* (Washington DC: The Urban Institute, 1992), Table 3; U.S. Bureau of the Census, Current Population Reports, Series P-25, No. 1092 (Washington DC: U.S. Government Printing Office, 1992), Table 1; and U.S. Bureau of the Census, *Population Projections of the United States by Age, Sex, Race, and Hispanic Origin: 1995–2050*. Series P25, No. 1130, (Washington DC: U.S. Government Printing Office, 1996), p. 12.

Figure 9.7 Projected Population Increases, by Ethnicity and Race: 1995–2050

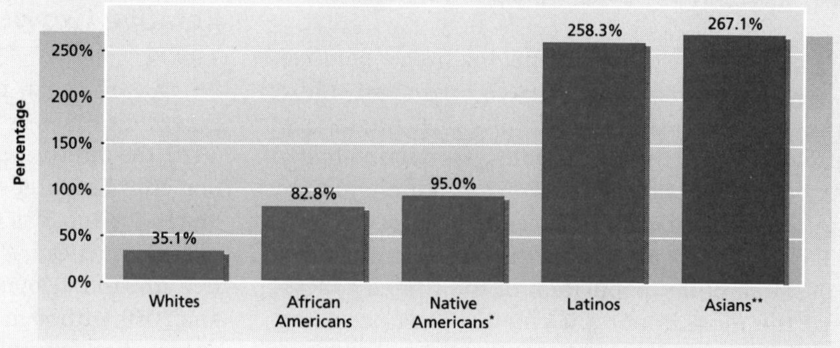

(middle series)
*Includes Alaska natives
**Includes Pacific Islanders
Source: U.S. Bureau of the Census, *Population Projections of the United States by Age, Sex, Race, and Hispanic Origin: 1995–2050* (Washington DC: U.S. Government Printing Office, 1996), p.14.

1. Almost _____ percent of all Americans will be members of minority groups in 2050.
2. Rank order (1–4) the following minority groups in terms of their relative proportion of the total American minority population projected for 2050.
   _____ a. Asian Americans
   _____ b. Native Americans
   _____ c. Latinos
   _____ d. African Americans

Answers: 1. 50 2. a. (3) b. (4) c. (1) d. (2)

the remainder of this chapter focuses on inequality among the elderly.

Age stratification, like most aspects of a society, must exist within a social and cultural context. Age stratification must make sense to those involved; age stratification must be justified. The justification for negative age stratification among the elderly in a society comes in the form of **ageism**—a set of beliefs, attitudes, norms, and values, used to justify age-based prejudice and discrimination. An understanding of ageism—in this context directed at the elderly—can be further enhanced by examining it within the context of the three major theoretical perspectives.

THEORETICAL PERSPECTIVES ON AGEISM

Functionalism and Ageism

Functionalism focuses on the contributions of ageism to the order and stability of society. According to functionalists, the elderly are not treated the same in preindustrial and industrial societies because of the different functions they serve in each.

Aging is not stigmatized in all societies. In agricultural society, elderly males usually, though not always, play important roles, such as priests or elders. Donald Cowgill and Lowell Holmes point out why in such societies the aged are usually respected and highly valued.

In all of the African societies, growing old is equated with rising status and increased respect. Among the Igbo, the older person is assumed to be wise: this not only brings him respect, since he is consulted for his wisdom, it also provides him with a valued role in his society. The Bantu elder is "the Father of His People" and revered as such. In Samoa, too, old age is "the best time of life" and older persons are accorded great respect. Likewise, in Thailand, older persons are honored and deferred to and Adams reports respect and affection for older people in rural Mexico (Cowgill and Holmes, 1972: 310).

Because of the stability contributed by the elderly in early colonial America, no stigma was attached to age. In fact, to be elderly brought respect and the opportunity to fill the most prestigious positions in the community. It was believed that God looked with favor on those who reached old age: The longer one lived, the more likely he or she was to have been elected to go to heaven. The Bible linked age with living a moral life:

"Keep my commandments, for length of days and long life shall they add unto thee." During the 1600s and 1700s, Americans even tried to appear older than they actually were. Men and women wore clothing that made them appear older and covered their hair with white, powdered wigs. During the 1700s, people often inflated their age when making reports to census takers (Fischer, 1977).

Conflict Theory and Ageism

Competition over scarce resources lies at the heart of the conflict perspective. According to this perspective, the elderly constitute a social stratum that competes with other age strata for economic resources, power, and prestige. In preindustrial society, the elderly are often accorded a relatively large share of the scarce resources. The elderly are more likely to be deferred to by the young and middle-aged. Younger age strata in industrial society, in contrast, do not permit the elderly to compete as freely for scarce resources (Palmore, 1981).

Why are the elderly not permitted to compete? According to the conflict perspective, the elderly are treated relatively well in preindustrial society because their labor is needed by the other age strata. Because work in preindustrial society is labor intensive, all available hands must be utilized. In preindustrial society a major problem is the shortage of labor, whereas industrial society has the opposite problem. Industrial society has more workers than it needs, so the elderly are removed from the competition for jobs. In addition, industrial societies save scarce resources by replacing high-priced older workers with less costly younger ones. Forced retirement is one of the weapons of the more powerful age strata in removing the elderly as a competitor for scarce resources.

Conflict theory, then, traces ageism—and the prejudice and discrimination that accompany it—to group interests and competition for scarce resources. According to the theory of differential power, prejudice and discrimination are used by the majority as weapons in the domination of subordinate groups. In this way, the minority cannot effectively compete with members of the majority for scarce goods and services. If the elderly can be depicted as intellectually dull, closed-minded, inflexible, and unproductive, forcing their retirement from the labor market becomes relatively easy. This leaves more jobs available for younger workers.

Symbolic Interactionism and Ageism

Prejudice and discrimination, as noted earlier, are justified in part through the creation and use of stereotypes. As with racism, ageism involves negative stereotyping of the elderly. Symbolic interactionists say that children learn negative images of older people just as they learn other aspects of culture. Through the process of socialization, stereotypes of the elderly are often firmly implanted into a child's view of the world. Third-grade children have been shown to possess unfavorable images of the elderly. In fact, negative images of the elderly have been observed in three-year-old children (Hickey, Hickey, and Kalish, 1968; Seefeldt et al., 1977).

At the same time, stereotypes harm older people in part because they are too often used to justify prejudice and discrimination against the elderly. They also harm the elderly because the stigma attached to aging promotes the development of a negative self-concept (Ward, 1977; Friedan, 1993).

What are the stereotypes of the elderly?   Robert Butler has succinctly summarized the many negative stereotypes of the elderly:

An older person thinks and moves slowly. He does not think as he used to or as creatively. He is bound to himself and can no longer change or grow. He can learn neither well nor swiftly and, even if he could, he would not wish to. Tied to his personal traditions and growing conservatism, he dislikes innovations and is not disposed to new ideas. Not only can he not move forward, he often moves backward. He enters a second childhood, caught up in increasing egocentricity and demanding more from his environment than he is willing to give to it. Sometimes he becomes an intensification of

Societies have very different views of the elderly. In traditional Eskimo society, it was socially acceptable to kill the elderly. Whereas the practice of gerontocide has, in fact, been documented in several hunting and gathering societies, the elderly in America are encouraged to take steps (such as exercising in fitness centers) to prolong their lives.

himself, a caricature of a life-long personality. He becomes irritable and cantankerous, yet shallow and enfeebled. He lives in his past; he is behind times. He is aimless and wandering of mind, reminiscing and garrulous. Indeed, he is a study in decline, the picture of mental and physical failure. He has lost and cannot replace friends, spouse, job, status, power, influence, income. He is often stricken by diseases which, in turn, restrict his movement, his enjoyment of food, the pleasures of well-being. He has lost his desire and capacity for sex. His body shrinks, and so too does the flow of blood to his brain. His mind does not utilize oxygen and sugar at the same rate as formerly. Feeble, uninteresting, he awaits his death, a burden to society, to his family, and to himself (Butler, 1975: 6–7).

Are these stereotypes accurate? By definition stereotypes are usually inaccurate because they do not apply to all members of a group. Stereotypes of the elderly are no exception, as much research has shown. The majority of the elderly are not senile. Old age is not a sexless period for the majority of those over sixty-five. There are few age differences on job-related factors. The majority of the elderly are able to learn new things and adapt to change (Palmore, 1980, 1981; Hendricks and Hendricks, 1987; Atchley, 1997).

In summary, there is enough evidence to challenge the truth of popular stereotypes of the elderly. It is, of course, true that some older people fit one or more of these stereotypes (as do some younger people) and that more individuals are likely to fit one or more of them as they advance into their seventies and beyond. This, however, is no justification for applying these stereotypes to *all* older people at any age or for mindlessly applying them to individuals in their fifties and sixties.

INEQUALITY AMONG AMERICA'S ELDERLY

Racial, ethnic, and religious groups have long been recognized as minorities. Some sociologists say that

women (Hacker, 1951) and the elderly are also minority groups. With respect to the elderly, Jack Levin and William Levin (1980) contend that the field of **social gerontology**—the scientific study of the social dimensions of aging—has in the past blamed the elderly for their situation in much the same way that Americans in general have blamed the poor for their plight. In fact, the early gerontological literature tended to attribute the problems of the elderly to their own deterioration and decline. For example, this literature, based largely on the study of older people in institutions, suggested that the elderly should be denied full participation in society because of their diminished mental and physical capacities. This perspective coincided with and reinforced the American public's negative view of the elderly. Many scholars join Levin and Levin in the belief that the best way to avoid blaming the elderly for their situation is to view them as a minority (Barron, 1953; Comfort, 1976; Palmore, 1978; Barrow, 1996).

Other sociologists do not believe that either the elderly or women qualify as minorities. Gordon Strieb (1968) emphasizes the heterogeneity of the elderly. The elderly, he notes, includes both sexes as well as members of all races, ethnic groups, and social classes who have reached old age. As a result of this diversity in background and characteristics, the elderly are more likely to identify with their lifelong statuses than with their newly acquired one. The elderly not only do not have a common identity, Strieb is saying, but also do not share any distinct cultural traits (although they do possess distinct physical traits). According to this line of argument, the elderly are more a social category or statistical aggregate than a genuine group of any kind, minority or otherwise.

It is not necessary to consider the elderly (or women) as a minority to recognize them as victims of prejudice and discrimination. With this recognition has come the creation of the concept of ageism, defined earlier as a set of beliefs, attitudes, norms, and values used to justify age-based inequality. It seems undeniable that there is age-based inequality in America (Posner, 1996).

FEEDBACK

1. _____ exists when the unequal distribution of scarce desirables in a society is based on chronological age.
2. _____ refers to a set of beliefs, attitudes, norms, and values used to justify age-based prejudice and discrimination.
3. Below are several statements about the elderly. Identify each statement with one major theoretical perspective: functionalism (F), conflict theory (C), symbolic interactionism (S).
   ____ a. Ageism results in part from an oversupply of labor.
   ____ b. The young may wish to distance themselves from the elderly to avoid facing their own mortality.
   ____ c. The stigma attached to aging promotes a low self-concept among the elderly.
   ____ d. Ageism is associated with industrialization.
   ____ e. Older people are stereotyped.
   ____ f. Ageism exists in part because older workers are more expensive than younger ones.

Answers: 1. Age stratification 2. Ageism 3. a. (C) b. (S) c. (S) d. (F) e. (S) f. (C)

Economic Situation Among the Elderly

The economic situation among America's elderly has improved since 1960. As a result, the public has begun to think of the elderly not as people in need but as an age group that is well off and actually taking economic resources from the young (Duff, 1995). It is true that the poverty rate for Americans over sixty-five years of age has declined since 1960. (See Figure 9.8.) This generalization, however, fails to capture the complexity of their economic condition. Several factors mask the true economic situation of the elderly. Factors include, but are not limited to, the way poverty is measured, race, ethnicity, and gender.

What is being left out of this picture? Before concluding that the elderly are very financially secure, let us consider some complicating factors. First, the poverty line is drawn at a higher dollar amount for the elderly than for younger Americans. Despite the fact that the elderly spend proportionately more on health care and housing than younger people, the federal government assumes a lower cost of living for the elderly. If the standard used for younger age categories were applied to the elderly, their poverty rate would exceed 15 percent (Villers Foundation, 1987). Second, gross poverty rates fail to take into account the just over 7 percent of the elderly who are officially considered to be "near poor." Counting these at-risk elderly, about one in five of those over sixty-five is poor (U.S. Bureau of the Census, 1996b). Third, official statistics do not include the "hidden poor" among the elderly who either live in institutions or with relatives because they cannot afford to live independently. Inclusion of these people would substantially raise the poverty rate among the elderly.

Not all of America's elderly are at the bottom of the stratification structure. Consider this affluent retired couple with their grandchildren.

Diversity is the fourth factor complicating analysis of the economic situation of the elderly. In the first place, a wide financial disparity exists within the ranks of the elderly in general. We have been concentrating on the financial difficulties of the elderly as an undifferentiated age category. Some of the elderly, however, have moderate to high incomes based not on Social Security benefits but on dividends from assets, cash savings, and private retirement programs. Almost 6 percent of households headed by an elderly individual have annual incomes in excess of $50,000 (Davis and Rowland, 1991). Diversity is also reflected in race, ethnicity, and gender.

How do race and ethnicity compound age inequality? The elderly who are members of racial or ethnic groups are, of course, in much worse condition than white Americans. Whereas the median income of whites over sixty-five was approximately $18,700 in 1994, the median income for elderly African Americans was approximately $12,500 and for elderly Latinos was about $13,100 (U.S. Bureau of the Census, 1996b). The poverty rate among African American elderly is nearly triple that of whites. For older Latinos, the poverty rate is two and one-half times that of white Americans. Unquestionably, this disparity is intimately linked to the lifetime of prejudice and discrimination experienced by these racial and ethnic groups that becomes magnified in old age.

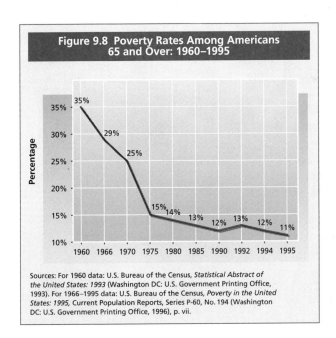

Figure 9.8 Poverty Rates Among Americans 65 and Over: 1960–1995

Sources: For 1960 data: U.S. Bureau of the Census, *Statistical Abstract of the United States: 1993* (Washington DC: U.S. Government Printing Office, 1993). For 1966–1995 data: U.S. Bureau of the Census, *Poverty in the United States: 1995*, Current Population Reports, Series P-60, No. 194 (Washington DC: U.S. Government Printing Office, 1996), p. vii.

How does gender compound age inequality?
American women are clearly at a disadvantage on all dimensions of stratification. Nowhere is this disadvantage as obvious or as poignant as it is for women in poverty. A dramatic change has taken place in the face of poverty in the United States, a change known as the **feminization of poverty**—the trend toward more of the poor in America being women and children (Thomas, 1994). The percentage of poor families headed by females has increased sharply over the years—23.7 percent in 1960 to 47.9 percent in 1995. Over one-third of all families headed by females are poor (Gelpi et al., 1986; U.S. Bureau of the Census, 1996b; Sidel, 1996).

The consequences of gender inequality are compounded by the inequalities of age. Elderly women constitute one of the poorest segments of American society. Women over sixty-five are twice as likely to live in poverty than their male counterparts (U.S. Bureau of the Census, 1996b). Elderly women most likely to be poor are single women who either have never married or are divorced, separated, or widowed. About 56 percent of women over sixty-five have annual incomes below $10,000 compared to 24 percent among older males. Although the median income of elderly women is $9,355, it is $16,484 for males over sixty-five. Thus, the average income of elderly women is about 57 percent that of males, lower than the amount that American women under sixty-five earn relative to working males (U.S. Bureau of the Census, 1996b). This is not surprising, because the roots of poverty among older women lie in the work-related experiences women have had throughout their life cycle (Sidel, 1996).

The situation, of course, is even worse when the cumulative effects of being elderly, female, and a member of a racial or ethnic minority are felt. Among elderly African American women, 31.1 percent are in poverty, for elderly Latino women the poverty rate is 28.9 percent. This compares very unfavorably to elderly white males whose poverty rate is only 5.4 percent. Poverty rates are at their peak among minority women living alone. For example, nearly 50 percent of elderly African American women living alone are poor. The income picture is expectedly consistent. Just to cite one case, white males sixty-five to seventy-four years of age earn over $16,576 annually compared to $8,586 among white females, $6,749 among female African Americans, and $6,013 among female Latinos. Finally, compared to elderly white males, elderly female members of racial and ethnic minorities experience longer periods of unemployment, hold lower-paying jobs (both of which lower Social Security benefits), and hold (or are in) jobs without pension benefits.

Undeniably, then, the elderly in the United States are economically better off than they were three and a half decades ago. Despite this improvement, large segments of Americans over sixty-five years of age are either in poverty or near poverty. This is especially true for elderly members of racial and ethnic minorities and for elderly women.

Political Power and the Elderly

Clearly, any power that the elderly have cannot be based on wealth or occupational prestige. In fact, in avoiding the consequences of ageism, their major

The elderly who are members of minority racial or ethnic groups experience the highest rates of poverty. This African American man living on a farm is existing close to the level of subsistence.

source of influence must be through the political process, primarily via voting in elections, holding political office, and joining political interest groups.

What is the voting turnout among the elderly?
Voting turnout in the United States increases with age. Since the mid-1980s, Americans sixty-five and over have been the most active voters in presidential and congressional elections. In 1996, for example, 67 percent of the elderly voted in the presidential election compared to just over 30 percent of the eighteen- to twenty-four-year-olds and just over 49 percent of the twenty-five- to forty-four-year-olds (U.S. Bureau of the Census, 1997a).

The high political participation through voting among the elderly is important for two reasons. First, a relatively large proportion of the elderly express their political preferences via the ballot. Second, the elderly represent a large segment of the total voting population, somewhere around 15 percent to 20 percent. And, of course, with the graying of America, the proportion of the voting population that is elderly will rise further.

The potential political power of the elderly is not fully realized because of the diversity among the older population. Because the elderly cut across many important divisions in American society—social class, ethnicity, race, geographic area, religion—they do not speak with a unified political voice. In fact, the elderly do not vote as a bloc on any political question, even on issues related directly to the interests of older people.

Do the elderly use interest group participation to their advantage?
Because large numbers do not necessarily translate potential political power into reality, the elderly also need to attempt to exercise their power through political interest groups. The assault on ageism is being waged by millions of Americans who belong to such "gray lobby" organizations as the American Association for Retired Persons (AARP), the Gray Panthers, the National Council of Senior Citizens (NCSC), the National Association of Retired Federal Employees (NARFE), and the National Caucus and Center on Black Aged (NCCBA). These national organizations are political rather than recreational or social in nature.

Do the elderly actually exercise significant political power in the United States?
The debate on whether the elderly are an effective political force in the United States has not been settled. On the one hand, some analysts, who write of "senior power," see the elderly as a strong political force whose political issues and activities influence both the election and decision making of public officials. Some analysts even fear excessive gray power. Others believe that the elderly are not a significant political force in the United States. The elderly, this line of argument goes, have been successful only in a defensive sense—they have successfully shielded their benefits programs from budget cuts and prevented certain legal changes detrimental to their interests. On the other hand, the elderly have not been very effective in modifying present policies that harm them or securing the enactment of new policies that would help them (Jacobs, 1990). In addition, some researchers assert, the elderly are too heterogeneous to be truly politically effective.

Like most significant debates on social issues, the truth seems to lie between these two polar positions. Although the actual political clout of the aged today may be overestimated, their voting power and national political organizations make them a political force that cannot be ignored. Heterogeneity, though it certainly dilutes the influence of any political bloc, does not rule out age as an important focal point for the exercise of political power.

One of the most interesting questions revolves around whether the political power of the elderly will increase dramatically as America continues to become more gray. Because of the stigma associated with aging, many older Americans cling to an identification with the "middle aged." If the elderly are to become a truly formidable political force in the future, they must develop a sense of identification with their chronological age. Such a step, leading to greater political power, would constitute a significant move against the negative effects of age stratification and ageism experienced by elderly Americans (Powell, Williamson, and Branco, 1996).

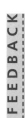

FEEDBACK

1. The scientific study of the social dimensions of aging is known as _____.
2. Of the following, which is an accurate statement?
   ____ a. Since 1960, the economic situation for the elderly in the United States has deteriorated.
   ____ b. The poverty rate for Americans over sixty-five is lower than the official count indicates.
   ____ c. The elderly who are members of racial or ethnic minorities are in much worse condition than other older people.
   ____ d. Only about one-fourth of American families in poverty have female heads of household.
3. America's elderly for the first time now speak in a unified political voice. *T or F?*

Answers: 1. social gerontology 2. c 3. F

SUMMARY

1. A minority consists of more than merely small numbers. A minority possesses some distinctive physical or cultural characteristics, is dominated by the majority, and is denied equal treatment. Minorities also have a sense of collective identity, tend to marry within their own kind, and inherit their minority status at birth.

2. A race is a category of people who are alleged to share certain biologically inherited physical features. Racists use these physical characteristics as an index of a race's superiority or inferiority. Despite the lack of scientific support for this viewpoint, prejudice and discrimination are often justified by alleged differences in intelligence and ability among races.

3. Ethnic minorities can be thought of as minorities with distinct subcultures. Unique cultural characteristics of ethnic minorities are used by the majority as a basis for prejudice and discrimination against them.

4. Patterns of racial and ethnic relations assume two general forms: patterns of assimilation and patterns of conflict. Patterns of assimilation include Anglo-conformity, melting pot, and cultural pluralism. Genocide, population transfer, and subjugation are the major patterns of conflict.

5. Prejudice—negative attitudes toward some minority—is very difficult to change because prejudiced individuals tend to reject information that contradicts their existing attitudes. Whereas prejudice refers to attitudes, discrimination refers to behavior. When members of a minority are denied equal treatment, they are being discriminated against. Although prejudice usually leads to discrimination, there are exceptions. In fact, in some instances discrimination against a minority may lead to the creation of prejudiced attitudes, usually embodied in racial or ethnic stereotypes.

6. Many attempts have been made to explain the existence of prejudice and discrimination. Some psychologists attribute prejudice and discrimination to deep-seated frustration. Hostility and frustration are said to be released on scapegoats who had nothing to do with causing their development in the first place. Other psychologists point to the existence of a personality type—the authoritarian personality—that tends to be very prejudiced against those thought to be inferior. Although functionalism emphasizes the negative consequences of prejudice and discrimination, it does point to some benefits. According to the differential power explanation, prejudice and discrimination are used by the majority as weapons of power to dominate subordinate minorities. Other social scientists believe that prejudice and discrimination are simply learned through the process of socialization, like any other aspect of culture.

7. Virtually all minorities have felt the consequences of institutionalized discrimination. Institutionalized discrimination may be direct, as when segregation laws are enacted; it may be indirect, as when seniority systems prevent upward mobility of newly hired minorities.

8. Institutionalized discrimination has been harmful to America's minorities. As a result, African Americans, Native Americans, Latinos, some Asian Americans, and white ethnics lag behind the white majority occupationally, economically, and educationally. Successful cases do exist, but the gains of all minorities remain fragile.

9. Some scholars contend that the socioeconomic disadvantage minorities are currently experiencing is due to economic as well as social factors. Because of the dual labor market, minorities are trapped in the peripheral labor market, which offers low-paying, dead-end jobs.

10. Racial and ethnic diversity is increasing dramatically in the United States. By the middle of the twenty-first century, one-half of the entire American population will be composed of minority group members. There are differential growth rates among minorities. African Americans, traditionally the predominant minority in the United States, will be surpassed in size by Latinos in the early part of the twenty-first century.

11. Age stratification—the unequal distribution of scarce desirables in society based on chronological age—can affect any age group. The social inequality reflected in age stratification is always justified socially and culturally. The justification for the relatively low social standing among the elderly in a society is based on ageism—a set of beliefs, attitudes, norms, and values that permit age-based prejudice and discrimination.

12. Each of the three major theoretical perspectives has a unique slant on ageism. According to functionalism, ageism increases as industrialization proceeds because the contributions to society by the elderly change. Conflict theorists attribute ageism to a battle for scarce resources. Whereas the elderly are needed to work in preindustrial society, they become competitors with younger workers in industrial society, where a surplus of labor exists. Symbolic interactionism traces ageism to the process of socialization in which stereotypes of the elderly are learned. Although erroneous, these stereotypes of the elderly are used to justify prejudice and discrimination that can harm the elderly's self-concept.

13. Although the economic situation of America's elderly has improved over the last several decades, their poverty rate still stands at over 10 percent. Moreover, the gross poverty rate masks the true economic situation of the bulk of elderly Americans. The elderly who are members of racial or ethnic minorities, or are female, are in the worst economic condition.

14. Given the situation of the elderly, any power the elderly hold cannot be financially or occupationally based. The major avenue for the exercise of power among the elderly must be through the political process. The elderly exert political influence through their high voting rate and their support of interest groups advocating issues important to them. The political influence of the elderly, however, is diluted significantly because of their racial, ethnic, social-class, religious, and geographic diversity.

LEARNING OBJECTIVES REVIEW

After careful study of this chapter, you will be able to:

- Distinguish among the concepts of minority, race, and ethnicity.
- Describe the patterns of racial and ethnic relations.
- Differentiate prejudice from discrimination.
- Illustrate the different views of prejudice and discrimination taken by functionalists, conflict theorists, and symbolic interactionists.
- Describe the condition of minorities in the United States relative to the white majority.
- Describe the increasing racial and ethnic diversity in America.
- Distinguish between age stratification and ageism.
- Compare and contrast the ways in which functionalism, conflict theory, and symbolic interactionism approach ageism.
- Outline the inequality experienced by America's elderly.

CONCEPT REVIEW

Match the following concepts with the definitions listed below them:

____ a. ethnic minority
____ b. underclass
____ c. race
____ d. minority
____ e. de facto subjugation
____ f. assimilation
____ g. self-fulfilling prophecy
____ h. ageism
____ i. feminization of poverty
____ j. differential power
____ k. discrimination
____ l. ideology
____ m. age stratification

1. Those processes whereby groups with distinctive identities become culturally and socially fused.
2. Subjugation based on common, everyday social practices.
3. A theory according to which a majority uses prejudice and discrimination as weapons of power in the domination of subordinate minorities.
4. Unequal treatment of individuals based on their minority membership.
5. A people that is socially identified and set apart by others and by itself on the basis of its unique cultural or nationality characteristics.
6. A set of ideas used to justify and defend the interests and actions of those in power in a society.
7. A people who possess some distinctive physical or cultural characteristics, are dominated by the majority, and are denied equal treatment.
8. A distinct category of people who are alleged to share certain biologically inherited physical characteristics.
9. When an expectation leads to behavior that causes the expectation to become a reality.
10. Those in poverty who are continuously unemployed or underemployed because of the absence of job opportunities and/or required job skills.
11. The unequal distribution of scarce desirables (power, wealth, privileges) in a society based on chronological age.
12. A set of beliefs, attitudes, norms, and values used to justify age-based prejudice and discrimination.
13. The trend toward more of the poor in the United States being women and children.

CRITICAL THINKING QUESTIONS

1. Do you think that both the pattern of assimilation and the pattern of conflict accurately describe the state of racial relations between blacks and whites in America today? State your position using terms and evidence in the text.

2. Compare and contrast the functionalist and conflict perspectives on prejudice and discrimination. Is one more explanatory than the other? Are they complementary? Defend your conclusion.

3. Taking into account the past and present situation of African Americans in the United States, construct a line of argument for or against affirmative action. Be specific.

4. Does ageism exist in America today? Substantiate your position by drawing on functionalism, conflict theory, and symbolic interactionism.

5. Considering the occupational and economic situation of America's elderly, do you think ageism is at work? Provide the detailed information you need to convince someone you know will not agree with your stance.

MULTIPLE CHOICE QUESTIONS

1. People who possess some distinctive physical or cultural characteristics, are dominated by the majority, and are denied equal treatment are known as a/an
   a. minority.
   b. race.
   c. ethnic group.
   d. subculture.
   e. caste.

2. A distinct category of people who are alleged to share certain biologically inherited physical characteristics is known as a/an
   a. minority.
   b. race.
   c. ethnic group.
   d. subculture.
   e. caste.

3. Which of the following statements *best* represents contemporary social scientists' beliefs regarding differences in intelligence among various races?
   a. The weight of current scientific evidence supports a connection between genetically determined physical characteristics and innate superiority or inferiority of certain races.
   b. There is no scientific proof that races differ in their innate mental characteristics.

c. Recent reports attribute existing differences in measured IQ to racial differences rather than to the social environment.
d. Social scientists are now becoming less interested in biological than in social races.
e. Social scientists have stopped doing research on what they consider a closed issue.

4. _____ is a pattern of assimilation in which immigrants are allowed to maintain their traditional cultural ways while at the same time learning the values and norms of the host culture.
   a. The melting pot
   b. Anglo-conformity
   c. Cultural pluralism
   d. Accommodation
   e. Population transfer

5. The unequal treatment of individuals based on their membership in some minority is known as
   a. stereotyping.
   b. prejudice.
   c. racism.
   d. the ideal-type method.
   e. discrimination.

6. According to the _____ explanation, the majority uses prejudice and discrimination as weapons of power in the domination of one or more subordinate minorities.
   a. authoritarian personality
   b. differential association
   c. differential power
   d. ethnocentric
   e. functionalist

7. According to dual labor market theory, minority workers tend to be trapped in the _____ labor market, with low-paying jobs and little hope for advancement.
   a. core
   b. primary
   c. inferior
   d. malfunctioning
   e. secondary

8. Age stratification refers to
   a. the age distribution within the various cohorts in a society.
   b. the unequal distribution of scarce desirables based on age.
   c. changes in the function of the elderly.
   d. differential treatment of the elderly based on race and gender.
   e. the stigma attached to the loss of the ability of the elderly to contribute society.

9. According to the functionalist theory, industrialization is associated with
   a. a loss of power and prestige among the elderly.
   b. an increase in prestige among the elderly because there are more jobs that do not require great physical strength or stamina.
   c. an increase in economic resources among the elderly.
   d. a change in the role of the elderly with no real change in prestige.
   e. a destigmatization of aging.

10. In the study of aging, functionalism focuses on
    a. the role of the elderly in the class structure.
    b. the socialization of the elderly to accept a subordinate position in society.
    c. the contributions of ageism to the stability and order of a society.
    d. the struggles between the elderly and other interest groups.
    e. the ways in which the elderly learn to adjust to old age.

11. According to the conflict perspective, the elderly
    a. constitute a stratum that competes with other strata for scarce resources.
    b. add to the orderly function of society through the stabilizing influence they bring.
    c. keep investment returns high because of the capital accumulated in retirement accounts.
    d. lose status because they are competing with the majority of people who are trying to make enough money to support their families.
    e. have failed to use their political power to protect their interests.

12. A topic of primary importance to symbolic interactionists who study aging is
    a. the contribution the elderly make to the socialization of children.
    b. how the process of socialization perpetuates age stereotypes.
    c. the ways in which the elderly were socialized in comparison to the modern methods used today.
    d. the extent to which the aged now live with their children.
    e. the political reasons for the elderly's loss of status.

13. The major reason for the upcoming increase in the number of elderly in the United States is
    a. the decreasing stigma of aging in postindustrial societies.
    b. the number of children born to pre–World War I immigrants.
    c. the need for more older workers in service and fast-food jobs.
    d. the aging of the baby boom generation.
    e. the reverberations of the "echo" generation.

14. The primary reason many elderly are returning to work is
    a. the satisfaction associated with following a routine schedule.
    b. to improve their health and happiness.
    c. to conform to the American value of staying active.
    d. the economic necessity of doing so.
    e. to help their grandchildren have the extra things in life.

15. The major source of political influence of the elderly must be based on their
    a. prestige.
    b. children's voting patterns.
    c. ability to contribute financially to PACs.
    d. association with nonvoluntary groups.
    e. power.

FEEDBACK REVIEW

True-False

1. The evidence clearly shows that race is declining in importance in America. *T or F?*
2. Contrary to popular opinion, white ethnics in the United States are politically more liberal than Protestant blue-collar workers. *T or F?*

Fill in the Blank

3. The advocates of differential power argue that a majority uses prejudice and discrimination as weapons of _____ in the struggle for control over scarce goods and services.
4. According to Gordon Allport, children learning prejudice move from a pregeneralized learning period to a stage of total _____.
5. The attitudes and practices associated with _____ in the American slave system help account for the long-term prejudice and discrimination experienced by African Americans.
6. Almost _____ percent of all Americans will be members of minority groups in 2050.

Multiple Choice

7. Which of the following is *not* always a characteristic of a minority?
   a. distinctive physical or cultural characteristics
   b. smaller in number than the majority
   c. dominated by the majority
   d. denied equal treatment
   e. a sense of collective identity
8. Of the following, which is an accurate statement?
   a. Since 1960, the economic situation for the elderly in the United States has deteriorated.
   b. The poverty rate for Americans over sixty-five is lower than the official count indicates.
   c. The elderly who are members of racial or ethnic minorities are in much worse condition than other older people.
   d. Only about one-fourth of American families in poverty have female heads of household.

Matching

9. Rank order (1–4) the following minority groups in terms of their relative proportion of the total American minority population projected for 2050.
   ____ a. Asian Americans
   ____ b. Native Americans
   ____ c. Latinos
   ____ d. African Americans
10. Below are several statements about the elderly. Identify each statement with one major theoretical perspective: functionalism (F), conflict theory (C), symbolic interactionism (S).
   ____ a. Ageism results in part from an oversupply of labor.
   ____ b. The young may wish to distance themselves from the elderly to avoid facing their own mortality.
   ____ c. The stigma attached to aging promotes a low self-concept among the elderly.
   ____ d. Ageism is associated with industrialization.
   ____ e. Older people are stereotyped.
   ____ f. Ageism exists in part because older workers are more expensive than younger ones.

GRAPHIC REVIEW

Some attitudes of Americans toward various immigrant minorities are summarized in Figure 9.1. Answering these questions will test your understanding of these data.

1. If you had to write a one-sentence summary of Figure 9.1 for the CNN evening news, what would you say?

_____

_____

2. What do the data in Figure 9.1 suggest regarding the future of ethnic relations in the United States?

_____

_____

ANSWER KEY

Concept Review	Multiple Choice	Feedback Review
a. 5	1. a	1. F
b. 10	2. b	2. T
c. 8	3. b	3. power
d. 7	4. c	4. rejection
e. 2	5. e	5. manumission
f. 1	6. c	6. 50
g. 9	7. e	7. b
h. 12	8. b	8. c
i. 13	9. a	9. a. 3
j. 3	10. c	b. 4
k. 4	11. a	c. 1
l. 6	12. b	d. 2
m. 11	13. d	10. a. C
	14. d	b. F
	15. e	c. S
		d. F
		e. S
		f. C

Chapter Ten

Inequalities of Gender

After careful study of this chapter, you will be able to:

- Distinguish the concepts of sex, gender, and gender identity.
- Demonstrate the relative contributions of biology and culture to gender formation.
- Outline the diverse perspectives of gender taken by functionalists, conflict theorists, and symbolic interactionists.
- Differentiate among liberal, radical, and socialist feminism.
- State the major consequences of sex stereotypes and gender definitions.
- Describe the inequality of women in the United States with respect to work, law, politics, and sport.
- Discuss the factors promoting resistance to changing traditional gender roles as well as the factors promoting change in gender roles.
- Describe the state of knowledge regarding the future of gender roles.

SOCIOLOGICAL IMAGINATION

How well has the United States done in achieving income equality for women in the same jobs as men? Even if they recognize that women are not paid on a par with men, most Americans believe that at least the United States is making more progress than other developed countries. America, in fact, is near the bottom in male-female income parity among modern countries. Only Luxembourg and Japan rank lower than the United States. Whereas women in manufacturing jobs, for example, earn 89 percent of the wages paid men in Sweden (the top ranked country), females in these same jobs in the United States earn only 61 percent of the wages paid men.

Throughout history, the predominant pattern has been for men to perform social, political, and economic functions outside the home, whereas the responsibility for child care and household tasks has been assigned to women. Although this division of labor does not necessarily imply any status difference between the sexes, in practice it generally has. Women—thought to be dependent, passive, and deferring—have usually been considered subordinate to independent, aggressive, and strong men. Thus, the division of labor based on sex has almost always led to sexual inequality. This chapter will show many of the disadvantages attached to the female gender role. But first comes an exploration of the biological, cultural, and social foundations of gender roles.

THE BIOLOGICAL AND CULTURAL BACKGROUND OF GENDER

Why do sexual differences in behavior and sex-based inequality exist? The popular explanation is biological: Anatomy, people believe, is destiny. Behavioral differences between the sexes and sex-based inequality are attributed to **sex**—the biological distinction between male and female. Males are naturally supposed to be providers and protectors, whereas females are designed for domestic work. If this popular conception were true, men and women would behave in their unique ways in all societies because of innate biological forces they cannot control.

Actually, this is not so. Women in other cultures have performed roles popularly associated with men only. Thus, **gender**—the expectations and behaviors associated with a sex category within a society—is acquired through socialization.

From the moment of birth, males and females are treated differently on the basis of their obvious external biological characteristics. Few parents in American society point with pride to the muscular legs and broad shoulders of their baby girl or to the long eyelashes, rosebud mouth, and delicate curly hair of their baby boy. Rather, parents stress the characteristics and behavior that fit the society's image of the ideal male or female, including modes of dress, ways of walking and talking, play activities, and aspirations for work and life. In response, children gradually conform to these expectations and learn to behave as boys and girls are supposed to behave within their society. From this process comes **gender identity**—an awareness of being masculine or feminine. Margaret Andersen succinctly captures the difference between sex and gender:

The terms sex and gender have particular definitions in sociological work. Sex refers to the biological identity of the person and is meant to signify the fact that one is either male or female. . . . Gender refers to the socially learned behaviors and expectations that are associated with the two sexes. Thus, whereas "maleness" and "femaleness" are biological facts, becoming a woman or becoming a man is a cultural process. Like race and class, gender is a social category that establishes, in large measure, our life chances and directs our social relations with others. Sociologists distinguish sex and gender to emphasize that gender is a cultural, not a biological, phenomenon (Andersen, 1997:31).

The debate over explanations for gender differences persists, and the nature (biology) versus nurture (socialization) distinction remains at the heart of the controversy (Wilson, 1993). A notable difference in this debate today, however, is that research is beginning to replace bias and opinion. Explanations for definitions of masculinity and femininity are now based on research rather than on tradition and false common knowledge. This research has been biological and cultural (Moir, 1991; Andersen, 1997).

Biological Evidence

In what ways are men and women biologically different? The obvious biological differences between the sexes include distinctive muscle and bone structure and fatty tissue composition. The differences in reproductive organs, however, are much more important, because they result in certain facts of life: Only men can impregnate; only women are able to produce an ovum, carry and nurture the developing fetus, and give birth; and only women can secrete milk for nursing infants. In addition, the genetic composition of the body cells of men and women is different. All human beings have twenty-three pairs of chromosomes (the components of cells that determine heredity), but one of those pairs is different in men and women. Males have an X and a Y chromosome in the pair that determines sex, and females have two X chromosomes. At about the eighth week of prenatal development, the information coded into the chromosomes determines whether the embryo—which is still sexually neutral—will begin to develop testicles or ovaries. These reproductive glands discharge the hormones that are characteristic of either males (testosterone and androgens) or females (estrogen and progesterone). These hormone combinations influence development throughout life.

More recently, some research indicates that the brains of men and women are slightly different in structure (Gur et al., 1995). For example, men showed more activity in an older region of the brain thought to be more tied to adaptive, evolutionary responses such as fighting. Women had more activity in a newer region of the brain, which is more highly developed and thought to be linked to emotional expression.

Do biological differences produce different behavior? In general, researchers attempting to establish behavioral differences between the sexes due to biological differences have not been able to do so consistently (Keller, 1985; Sapiro, 1985; Fausto-Sterling, 1987; Andersen, 1997). One researcher's findings tend to contradict another's. Furthermore, many studies seek to find differences but ignore the similarities between males and females. To compound the problem, researchers often fail to document the variation in characteristics that exist within sex categories. Some men, for example, tend to be submissive, weak, and noncompetitive, and some women are aggressive, strong, and competitive.

Studies have been conducted of infants intentionally raised in the wrong gender by parents. According to the

These two young children at play are engaging in the stereotypical behavior associated with their respective genders. The small girl, with a bow in her hair, is taking care of her "baby." The small boy, with a baseball cap on his head, is playing with a truck.

results of such studies, individuals can easily be socialized into the gender of the opposite sex. What's more, after a few years, these children resist switching to the gender that their biological characteristics would seem to dictate. Although biology may predispose the sexes in certain directions, such tendencies are sufficiently weak that they are subject to cultural and social influences (Schwartz, 1987; Shapiro, 1990; Ridley, 1996; Sapolsky, 1997).

Cultural Evidence

What is the cross-cultural evidence? In her classic study of three primitive New Guinean peoples, Margaret Mead (1950) demonstrated the influence of culture and socialization on gender role behavior. Among the Arapesh, Mead found that both males and females were conditioned to be cooperative, unaggressive, and concerned about the welfare of others. Both men and women in this tribe behaved in a way that is consistent with our conception of the female gender role. Among the Mundugumor, both men and women were trained to be "masculine"—they were aggressive, ruthless, and unresponsive to the needs of others. In the Tchambuli tribe, the gender roles were the opposite of those found in Western society. Women were dominant, impersonal, and aggressive, and men were dependent and submissive. On the basis of this evidence, Mead concluded that human nature is sufficiently malleable to rule out biological determination of gender roles.

Cross-cultural research since Mead's landmark work has clearly demonstrated that gender roles are not fixed at birth (Janssen-Jurreit, 1982). Societies such as the Arapesh, Mundugumor, and Tchambuli are exceptions to the rule. The general pattern in cross-cultural studies of gender roles in preliterate societies is one of male dominance and female nurturance (Zelditch, 1955; Barry, Bacon, and Child, 1957; Collins, 1971; Reiss, 1976; Lee, 1977; Sinclair, 1979). Although both men and women are involved in economic tasks that provide for family welfare, women's tasks usually involve domestic chores (such as caring for the home and producing and preparing food), raising children, and promoting family emotional harmony. Men, on the other hand, are more likely to provide for the family and to represent the family in activities outside the home (Tilly and Scott, 1978). This is probably because, since early hunting societies, men have had the advantage of greater size and physical strength, and women have had the disadvantage of pregnancy and lactation. The physical advantage of men has led to social, economic, and political advantages. Once gained, these advantages have been retained by men in almost all known societies.

Even in societies that have attempted to eliminate traditional gender differences, males still tend to dominate (Epstein and Coser, 1981). Communist and socialist societies encourage women to enter the labor force, and they provide child-care and household assistance to free the women from housework. Consequently, fewer differences in gender roles would be expected in socialist countries than in capitalist ones. Comparing the United States to Sweden and Denmark, Jeanne Block (1973) did find less gender role differentiation in socialist countries. The situation of women in the kibbutzim of Israel, however, is quite similar to that in capitalist countries. Dorit Padan-Eisenstark (1973) described the kibbutz occupational structure as being even more traditional than in Israeli society in general; kibbutz women are concentrated in feminine occupations. A similar status fell to Soviet women, as described by Caroline Lund:

> *Fifty percent of the wage earners in the Soviet Union are women, but they generally work at lower-paying jobs and are not proportionately represented in supervisory positions. For instance, women are 32 percent of all engineers but only 12 percent of plant directors; they are 73 percent of primary and secondary schoolteachers but only 23 percent of school directors; they are 42 percent of scientists but only two of the 204 members of the Soviet Academy of Scientists are women. Women are 79 percent of the doctors in the Soviet Union, but doctors receive two-thirds the wage of a skilled worker, so men tend not to want to be doctors. In the political arena, only three of the 195 members of the Communist Party Central Committee are women (Lund, 1970:11–12).*

Even in what was formerly the Soviet Union, where sexual equality was an explicit goal, women were relegated to lower occupations and men dominated the society.

Do gender definitions change? Cultural definition and redefinition of gender roles can be observed in the earlier history of the United States. In his book *American Manhood*, E. Anthony Rotundo (1993) identifies three distinct, culturally created conceptions of "manhood" first developed in the Northern middle class prior to the twentieth century. The first phase, *communal manhood,* was developed in the socially integrated society of colonial New England. During this period, the definition of manhood was embedded in one's obligations to community. A man's social class of birth gave him his place in the community. He fulfilled himself through public usefulness more than through economic success. As head of the household, a man expressed his value to his community and bestowed a social identity on his wife and children. During this era before 1800, men were considered superior to women. In particular, men were held to be more virtuous. Thought to possess greater powers of reason, men were considered better able than women to control passions such as ambition and envy. *Self-made manhood* became the dominant male gender definition in the first decades of the nineteenth century. Social status of birth was replaced with a social identity based on personal achievement. A man's work role replaced his role as head of the household as the source of male identity. "Male" passions were then allowed fuller expression. Because a man was supposed to prove his superiority, the drive for dominance was viewed as virtuous. As males turned from the tradition of public usefulness to the pursuit of self-interest, the nature of women was redefined. Women were attributed with a stronger moral sense than men and assigned the role of protecting the common good in an age of rampant male individualism. Despite her moral superiority, a woman's social identity still depended on her husband's place in society, and she was denied participation in the pursuit of individualism. Her primary purposes were to control men's passions, to make others happy, and to take care of her husband and children. Late in the nineteenth century came another shift. The period of *passionate manhood* emphasized not just achievement, but ambition, combativeness, and aggression. Male toughness was admired, tenderness scorned. The gender definition of men and women changed over this period. What remained constant, however, was the exclusion of women from economic, political, and social power. In these arenas, women were consistently relegated to a subservient position throughout each phase of the changing social conception of manhood.

Where does the nature/nurture debate on gender stand? The nature/nurture controversy has unfortunately been posed as an either/or situation by both the mass media and some researchers. It is more reasonable to grant a place for both biological and social factors in gender formulation. Biological characteristics can be altered considerably through social influences. Also,

1. _____ is the biological distinction between male and female.
2. _____ refers to the expectations and behaviors associated with a sex category within a society.
3. Gender is acquired through the process of _____.
4. _____ is an awareness of being masculine or feminine.
5. Societies that do not follow traditional gender definitions are exceptions to the rule. *T or F?*
6. Societies attempting to eliminate gender differences have been uniformly successful. *T or F?*

Answers: 1. Sex 2. Gender 3. socialization 4. Gender identity 5. T 6. F

this is a good time to remember that human behavior is the result of multiple causes. (Refer to Chapter 2, "Sociologists Doing Research.") For this reason, it seems fruitless to search for a single determinant of gender formation (Lowe, 1983; Fausto-Sterling, 1987; Andersen, 1997).

THEORETICAL PERSPECTIVES ON GENDER

As you would expect, functionalism, conflict theory, and symbolic interactionism view gender differently. The functionalist perspective focuses on the original development of gender, whereas conflict theory concentrates on the reason gender continues to exist. Symbolic interactionism attempts to explain the ways in which gender is acquired.

Functionalism and Gender

If some pattern of behavior is not beneficial, functionalists argue, it will ultimately cease to be important. According to functionalism, then, the division of labor based on sex has survived because it is beneficial and efficient for human living. Early humans found a division of labor based on sex to be efficient. In part because of their size and muscular strength, men were assigned the hunting and defense tasks. In addition, men were assigned these tasks because they were more expendable than women. Because these duties were more dangerous and because a single male could insure sufficient procreation to sustain the survival of a group if there were several women (the same is not true of a single woman with several men), it hurt the group less to lose a man. Women were assigned domestic and child-care tasks. Because this sexual division of labor promoted the survival of the species, assert the functionalists, it has been retained.

Although giving credit to the functionalist perspective for its insights into the original emergence of gender roles, some critics contend that it is too conservative and inadequate to explain recent changes in gender (Peplau, 1983). Gerald Maxwell (1975), like most sociologists, concedes that the traditional division of labor

may have been appropriate at one time. Today, however, argues Maxwell, rapid social change has undermined the value of functional theory in understanding gender. Maxwell believes that it is to the advantage of modern society to treat people on the basis of their ability rather than on an outdated conception of instrumental and expressive roles. According to Maxwell and other critics, then, the contribution of the functionalist view of gender is restricted to an explanation of the development of gender. Functional theory is said to be of little help in analyzing gender today.

Conflict Theory and Gender

Conflict advocates are concerned with the early origins of the traditional conception of gender and with gender's persistence. The focus here will be on the latter. According to conflict theory, men and women have differential access to the resources needed to succeed outside the home. Moreover, men have an interest in keeping it that way. It is to the advantage of men (higher status members of the society) to prevent women (lower status members of the society) from gaining access to political, economic, and social resources. If men can prevent women from developing their potential, they can maintain the status quo. They can keep the traditional division of labor intact, which will continue to preserve the privileges they now enjoy.

Conflict theorists see traditional gender roles as outdated. Although conventional gender roles may have been appropriate in hunting and gathering, pastoral, and agricultural societies, they are not appropriate for the industrial and postindustrial era. Male physical strength may have been important when hunting was the major means of subsistence, but work in modern society does not place men at an advantage over women. In addition, advocates of the conflict perspective argue, the demographic characteristics of women in modern society make them increasingly available for work outside the home: Women are marrying later, are having fewer children, are younger when their last child leaves home, are remaining single in greater numbers, and are increasingly choosing to be single parents. Moreover, say the conflict theorists, those women who do not want to stay at home, who prefer careers in fields formerly reserved for men only, have every right to do so, whether or not it is "functional" for society. For example, any potential harm to the family must now be weighed against the damage done to women who are not permitted to develop their potential outside the home. As a matter of fact, recent research suggests that there are benefits to the family associated with working wives and mothers (Hoffman and Nye, 1975; Moen, 1978). Finally, the barriers blocking women from entry into higher level jobs waste talents and skills that modern societies need.

One increasingly emphasized version of conflict theory is found in Marxist and socialist feminist theories. These theories stress that the position of women in capitalist society is the result of a combination of patriarchal (or male-dominated) institutions and the historical development of industrial capitalism. That is, women are relegated to subordinate roles because their unpaid labor in the home is necessary to produce the future labor force and to maintain the status quo. Home work provides for the care and feeding of workers, the creation of new workers and consumers (through childbearing and child rearing), and the creation of people who can devote their time to the consumption activities necessary to keep factories running and workers working. In addition, women provide a reserve army of cheap labor for times of crisis (for example, war) and for expansion of the labor force when more workers are needed. Thus, the subordination of women is seen as a key component in the maintenance of the political and economic institutions in capitalist society. Simultaneously, maintaining the traditional division of labor is one means of maintaining sexual inequality (Sokoloff, 1980; Chafetz, 1984, 1990; Jaggar and Rothenberg, 1993).

The conflict perspective is built into the concept of **sexual harassment**—the use of one's superior power in making unwelcome sexual advances. Superior physical power may be used in sexual harassment. This was clearly the case in the 1992 Tailhook scandal in which many women Navy personnel, including officers, were sexually harassed by naval aviators at a Las Vegas convention. The subsequent investigation, however, revealed a pattern of sexual harassment in the Navy that went beyond the physical power of an individual or the collective physical strength of a group.

Sexual harassment in all branches of the military has subsequently been revealed to be quite high (Thomas and Vistica, 1997). According to a 1995 Army survey, 4 percent of all female soldiers reported that they had been the victim of an actual or attempted rape or a sexual assault within the previous year. That is almost ten times greater than the rate of rape and sexual assault in the general population. The highly publicized court cases involving the sexual abuse of military recruits by their superiors during the mid- and late-1990s are apparently only the tip of an ugly and frightening iceberg.

Work-related sexual harassment, which involves the use of superior power in making unwelcome sexual advances, is usually committed by men. Since sexual harassment is often based on social and economic power, men may be harassed by women.

Sexual harassment, best understood from the conflict perspective, is a matter of social and economic power rather than mere physical superiority. Two prominent cases in the United States should be mentioned, both of which illustrate the difficulties involved in establishing legally that one is a victim of sexual harassment. Anita Hill, a law professor, brought the issue of sexual harassment to public attention in 1991 while testifying in televised Senate Judiciary Committee confirmation hearings of now Supreme Court Justice Clarence Thomas. Professor Hill accused Justice Thomas of sexually harassing her during the time she worked on his staff. The Senate committee, then totally comprised of men, initially discounted the sexual harassment charge and subsequently confirmed Justice Thomas. The truth was not determined. The second case involves Senator Robert Packwood of Oregon, who was accused in 1993 of sexual harassment by several former employees. A legal struggle involving the senator and his diaries, the women involved, the Senate Ethics Committee, and the Justice Department followed. Packwood was eventually forced to resign.

Symbolic Interactionism and Gender

Symbolic interactionism sheds light on gender formation through its emphasis on socialization. According to this theoretical perspective, gender is acquired in large part from interaction with parents, teachers, and peers. In addition to occurring through face-to-face interaction with others, gender socialization occurs through the mass media. The socialization process is very powerful because gender is incorporated into the self-concept through role taking and the looking-glass self.

How do parents contribute to gender socialization? Parents are especially important in the process of learning gender. Usually intentionally, although frequently unconsciously, parents transfer their values and attitudes regarding the ways in which boys and girls should behave as they respond to their children's behavior. Parents also evaluate children according to their level of conformity to gender definition.

The learning of gender begins at birth and is well established by the time the child is two and a half years old (Mussen, 1969; Frieze et al., 1978; Davies, 1990). Actually, the groundwork for gender socialization is begun by parents before the child is born. A name is often painstakingly selected by the parents to represent the gender-related characteristics the parents expect their child to possess. Immediately after birth, friends and relatives give gifts to reflect the child's gender—blue or pink blankets, frilly dresses or baseball playsuits. In fact, a study of first-time parents showed that even one-day-old infants are viewed in terms of gender (Rubin, Provenzano, and Luria, 1974). Fathers in this study described their newborn sons as being firmer, more coordinated, more alert, bigger, stronger, and hardier; the other fathers judged their newborn daughters to be softer, more awkward, less attentive, weaker, and more delicate.

More important, numerous studies report that these gender preconceptions lead to differential treatment of boys and girls. Studies of infant care have found that girls are cuddled more, talked to more, and handled more gently. Parents expect boys to be more assertive, and they discourage them from clinging (Lewis, 1972). In one study, the same six-month-old infant dressed as a boy in blue pants and then as a girl in a pink dress was presented to eleven mothers (Will, Self, and Datan, 1976). The five mothers who played with "Adam" and the six mothers who played with "Beth" exhibited very different behavior. For play, the mothers gave Adam "masculine" toys such as a train and Beth a "feminine" toy such as a doll. The mothers with Beth smiled more and held her closer than did the mothers placed with Adam. Adam and Beth were, of course, the same baby dressed differently.

Differential treatment of the young by parents does not stop at infancy. The bedrooms of preschoolers reflect masculine and feminine themes—girls' rooms contain dolls and household items, whereas the rooms of boys are more likely to contain guns, footballs, and trucks (Rheingold and Cook, 1975). Gender is even taught and reinforced in the assignment of family chores. In an investigation of almost seven hundred children between the ages of two and seventeen, Lynn White and David Brinkerhoff (1981) found that boys were often given "masculine" jobs such as cutting grass and shoveling snow. Girls were more often assigned "feminine" chores such as washing dishes and cleaning up the house.

In what ways do schools reinforce gender socialization? Although the most critical period of gender socialization occurs within the family during early childhood, gender socialization also occurs through interaction in schools. Although teachers deny giving differential treatment to or having differential expectations of boys and girls (Guttentag and Bray, 1977), classroom observation of preschool teachers has revealed a clear sex-based difference in the teacher-student relationship. Myra and David Sadker, in an extensive study of fourth-, sixth-, and eighth-grade students, found boys to be more assertive in class, whereas girls sit patiently with their arms raised. Boys were eight times more likely than girls to call out answers. Sadker and Sadker linked this classroom behavior to the differential treatment given boys and girls by teachers. Teachers, report these researchers, are more likely to accept the answers given by boys who call out answers. Girls who call out in class are given such messages as, "In this class we don't shout out answers, we raise our hands." According to Sadker and Sadker, the message is subtle and powerful—"Boys should be academically assertive and grab teacher attention; girls should act like ladies and keep quiet" (Sadker and Sadker, 1985:56).

In their later book *Failing at Fairness,* Myra and David Sadker (1994), examine the evidence on sexism in schools from elementary school through college. Their conclusion is that America's schools shortchange females through differential treatment. The gender bias in America's schools results in an inferior education for girls. Although girls typically outperform boys academically in the early years of school, their often well-intentioned teachers cripple their potential through the transmission of sexist values. Girls, they conclude, are systematically taught deference to the alleged superior abilities of boys, a dislike of math and science, and passivity. Females carry the damage done to them in school into adult life and the working world where its disadvantageous effects take their full toll.

How do peers contribute to gender socialization? Even children in day-care centers and nursery schools appear to be aware of gender definitions and encourage conformity to them (Lamb, Easterbrooks, and Holden, 1980). When three-year-olds behave as their gender dictates, other children their age encourage, praise, and imitate them. Children in the study who acted like the opposite sex met considerable opposition within the group. This opposition included criticism and various efforts to change behavior to conform to traditional gender definitions.

A great deal of the interaction among adolescent peers also affects gender conceptions. This is because adolescence is the time when an individual's identity is being firmly established (Erikson, 1964, 1982), and identity is closely linked with definitions of masculinity and femininity. Adolescents want to be approved of and accepted by their peers. Acceptance or rejection by peers has a significant effect on the adolescent's self-concept. Teenagers who most closely fulfill gender definitions—such as football players and cheerleaders—are given the greatest respect, whereas feminine boys and masculine girls are accorded low status. This social pressure encourages teenagers to conform to society's gender definitions. To do otherwise is to risk rejection and a significant loss of self-esteem.

What is the role of the media in gender socialization?
All members of society are subject to the continual bombardment of images from television, books, magazines, radio, movies, and advertising. Many studies document the distorted view of women's roles prevalent in the media and the ways in which this distorted view affects people. In general, the media present the most stereotyped version of men's and women's gender definition, thus reflecting and reinforcing the limits on the options available to both sexes (Tuchman, Daniels, and Benét, 1978). Some illustrations from children's books, television, and advertising demonstrate the influence of the media on gender conformity.

Illustrated children's books present a conservative, traditional image of the sexes, with women underrepresented in titles and as major characters. One study examining over 2,700 children's stories found five stories about boys for every two stories about girls, 119 biographies of men compared to twenty-seven biographies of women, and sixty-five stories degrading girls compared to two degrading boys (American Association of School Administrators, 1974). In a study of a library display of children's books, Deane Gersoni-Stavn (1974) found less than half included pictures of women. Of these books, 80 percent included a picture of a woman wearing an apron. In the few books that pictured women working outside the home, the jobs were traditionally feminine occupations such as teaching and nursing or unusual female roles such as queen or trapeze artist. In a content analysis of several prizewinning children's books—the Caldecott Medal winners for the years 1967–1971—Leonore Weitzman and Deborah Eifler (1972) found a distinct gender bias. Human male characters outnumbered human female characters by eleven to one; the ratio of male to female nonhuman animals was ninety-five to one. Males were the central characters who dominated the action, whereas females were cast in passive roles.

A later study of Caldecott winners published between 1972 and 1979 showed some improvement—

Gender is learned through various agencies of socialization, including family, teachers, and peers. Strong social pressures still exist to channel American females into gender-related activities like cheerleading.

human males appeared more often than females by a ratio of 1.8 to 1; nonhuman male animals outweighed females by 2.66 to 1 (Kolbe and LaVoie, 1981). A gender bias reminiscent of the Weitzman and Eifler study was still apparent in these books, however. The characters in seventeen of the nineteen books portrayed the traditional gender definitions, and not one female character was shown to be employed outside the home.

When researchers later replicated Weitzman and Eifler's study, they also found only slight improvement. Boys were still depicted as competitive, active, and creative; girls were cast as submissive, dependent, and passive (Williams et al., 1987; McDonald, 1989; Bigler and Liber, 1990). Although male and female characters were about the same numerically and one-third of the principal characters were female, only one female, a waitress, was shown working for pay. Despite the increased appearance of females, they still were unequally represented in active roles. The power of children's literature to influence young minds is now being recognized. Although this heightened sensitivity is currently reflected in the publication of more books with tempered depictions of the sexes, all traces of gender bias have not disappeared (Purcell and Stewart, 1990). Myra and David Sadker (1994) examined the content of grade-school textbooks. The ratio of boys/men to girls/women in the language arts textbooks they studied, for example, ranged from two-to-one to three-to-one in favor of males. Out of a 631-page textbook on world history, to cite only one more illustration, seven pages related to women.

Television also reflects traditional gender roles. Most of those women on television who are portrayed as working outside the home are in female-dominated jobs such as nursing and secretarial work (Kalisch and Kalisch, 1984). Male characters are either heroes or villains, whereas females tend to be adulterers or victims (Goff, Goff, and Lehrer, 1980). In general, television consistently depicts men as aggressive, independent, and in charge of events. Women are portrayed as dependent and passive. Characters such as the females in television's "Murphy Brown" exist only as dramatic exceptions.

The gender images seen on television do appear to affect the values and attitudes of viewers. Terry Fruch and Paul McGhee (1975) found children who viewed television in excess of twenty-five hours per week to have more traditional gender conceptions than children viewing television ten or fewer hours a week. Similar findings have been reported among preschoolers and young adults (Gross and Jeffries-Fox, 1978; Eisenstock, 1984).

Feminist Theory

Although there are numerous variations to contemporary feminism, **feminism** can be generally defined as a social movement aimed at the achievement of sexual equality, socially, legally, politically, and economically. **Feminist theory** links the lives of women and men to the structure of gender relationships within society. Margaret Andersen (1997) has distinguished three frameworks within the broad umbrella of feminist theory—liberal feminism, radical feminism, and socialist feminism. Each of these theoretical frameworks attempts to explain the relationships among gender, social class, and minority status.

What is liberal feminism? Advocates of *liberal feminism* focus on equal opportunity for women and heightened public awareness of women's rights. Operating within the existing structure of society, liberal feminists look to legal and social change to rectify the abridgment of women's rights. If women are to have treatment already accorded to men, long-standing laws, social policies, and social structures must be altered. Equal pay for equal work, abortion rights, and civil rights are among the issues uniting the liberal feminist perspective.

What is the difference between liberal and radical feminism? Liberal feminism attributes the subordination of women to unequal rights and to the ways in which males and females are taught by society to view themselves. *Radical feminists* trace the oppression of women to the fact that societies are dominated by men. A **patriarchal society** is controlled by males who use their power, prestige, and economic advantage to dominate females, including their sexuality. Male-controlled institutions—family, religion, economy,

Radical feminists find the roots of sexism in the long-standing tradition of patriarchy—the domination of society and women by men. Men are said to use their institutionalized power to perpetuate female subordination. Radical feminists see no humor in this situation which persists around the world.

education, government—define women as subordinate to men and are used to insure the perpetuation of female subordination. Radical feminists are especially interested in the ways men gain and maintain the power to control all social institutions. Much of the theorizing of radical feminists centers on various aspects of the central fact of male supremacy.

What does socialist feminism add to radical feminism? *Socialist feminism,* actually more radical than radical feminism, adds capitalism to patriarchy as the source of female oppression. Their perspective on capitalism, however, goes beyond the Marxist interpretation of capitalism. Socialist feminists contend that the problem of female oppression extends beyond the fact that under capitalist societies women are the property of men. Because female oppression exists in societies without a capitalist economic base, socialist feminists reason that more than capitalism is at work in female oppression. That something else is patriarchy. Patriarchy is the means for male domination of females in preindustrial and socialist societies. In capitalist society, the power relations of the class structure combine with the force of patriarchy to create and maintain male oppression of women.

Despite some significant differences in theoretical perspectives, feminists share at least two common

FEEDBACK

1. According to the _____ perspective, the division of labor based on sex has survived because it is beneficial and efficient for human living.
2. According to the _____ perspective, traditional gender definitions exist because they provide greater rewards and privileges to men than to women.
3. _____ is the theoretical perspective that attempts to explain the ways in which gender is acquired.
4. Which of the following is *not* a true statement?
    a. Boys are less assertive in class than girls.
    b. Illustrated children's books present very traditional gender definitions.
    c. Parents' perceptions of the physical characteristics of their children are heavily influenced by their children's gender.
    d. Nursery school children who fail to conform to their "appropriate" gender are met with resistance from their peers.
5. Match the words or phrases on the right with the three feminist theoretical frameworks on the left.
    ____ a. liberal feminism   (1) patriarchy
    ____ b. radical feminism   (2) capitalism and patriarchy
    ____ c. socialist feminism   (3) equality of opportunity
                                 (4) social and legal reform
6. One of the major points of agreement among the three feminist theoretical perspectives is that gender and gender relationships are psychological in origin. *T or F?*

Answers: 1. functionalist 2. conflict 3. Symbolic interactionism 4. a 5. a. (3), (4) b. (1) c. (2) 6. F

themes. First, they believe that sociology carries the weight of bias from all these years of theoretical and research work done from one composite viewpoint shared by white, middle-class males from Western Europe and North America. Second, the nature of gender and gender relationships are sociological rather than psychological—they are embedded in social structures.

GENDER AND SEX STEREOTYPES

Sex Stereotypes

Chapter 9 ("Inequalities of Race, Ethnicity, and Age") defined *stereotype* as a set of ideas based on distortion, exaggeration, and oversimplification that is applied to all members of a social category. A stereotype does not take into account individual differences, so every member of the group is assumed to have the same traits, strengths, and weaknesses. The stereotype is the standard used to judge individuals. Men in American society are expected to be virile, brave, sexually aggressive, unemotional, logical, rational, mechanical, practical, dominating, independent, aggressive, confident, and competitive. Women are expected to be the opposite: weak, fearful, sexually passive, emotional, insecure, sentimental, "arty," dependent, submissive, modest, shy, and noncompetitive (Williams and Best, 1990). In American society, these stereotypes reflect a belief in the superiority of men and the inferiority of women.

How do sex stereotypes and gender differ? Gender involves expectations and behaviors associated with each sex. It represents an ideal. In reality, we know that people do not always behave as expected. Although mothers are expected to be patient and loving with their children, many are not. Consequently, individuals can decide to emphasize different aspects of their gender according to their preferences and abilities. Thus, there is a wide range of actual feminine and masculine behavior. Not all men try to be strong, fearless, and aggressive; not all women attempt to be sweet, submissive, and deferring.

Stereotypes, on the other hand, do not allow for a range of alternatives. One label is applied to all members of the group. Moreover, stereotypes tend to be exaggerations or extremes infrequently found in real life. For example, consider the stereotype of men offered by Marc Fasteau:

The male machine is a special kind of being, different from women, children, and men who don't measure up. He is functional, designed mainly for work. He is programmed to tackle jobs, override obstacles, attack problems, overcome difficulties, and always seize the offensive. He will take on any task that can be presented to him in a competitive framework. His most positive reinforcement is victory.

He has armor plating that is virtually impregnable. His circuits are never scrambled or overrun by irrelevant personal signals. He dominates and outperforms his fellows, although without excessive flashing of lights or clashing of gears. His relationship with other male machines is one of respect but not intimacy; it is difficult for him to connect his internal circuits to those of others. In fact, his internal circuitry is something of a mystery to him and is maintained primarily by humans of the opposite sex (Fasteau, 1975:60).

Some Consequences of Sex Stereotypes and Gender Definitions

What is wrong with sex stereotypes? There are at least three good reasons for dispelling sex stereotypes. First, traditional sex stereotypes do not fit the actual behavior of modern men and women. They promote a myth of women and men that is not consistent with the facts. Because American women are having fewer children, marrying later, and entering the labor force in greater numbers, motherhood is no longer a lifelong, full-time job for many of them. At the same time, many American men wish to find an alternative to the traditional "macho" ideal. Second, sex stereotypes are used to justify prejudice and discrimination. Women, for example, are blocked from certain jobs because they are thought to be emotionally fragile. Men are ridiculed if they enter traditionally female occupations such as nursing, hairdressing, or elementary school teaching. Finally, evidence suggests that sex stereotypes are psychologically damaging to both males and females who do not meet the ideal-typical characteristics. At the extreme, some individuals who are confused or ambivalent about their past experiences as boys or girls and who are unwilling to conform to gender roles assigned to them at birth may become so distraught that they willingly turn to sex-change operations in which they "are carved up on the operating table to gain acceptance to the opposite sex role" (Billings and Urban, 1982:278).

Sandra Bem (1975) compared men and women with masculine, feminine, and androgynous (synthesis of the strengths of masculine and feminine characteristics) gender identities. She found that men with the traditional male identity lacked the ability to express warmth, playfulness, and concern and that women with the traditional female identity were not independent in judgment or assertive in stating their preferences. However, individuals with an androgynous gender identity performed equally well on masculine and feminine tasks and were able to be both assertive and affectionate. In brief, sex stereotypes and gender roles place limits on the potential range of behavior for both sexes.

What are the negative consequences of sex stereotypes and gender roles? Sex stereotypes carry many costs for women, including a damaged sense of self-worth, self-hatred, and dislike of their gender identity. These negative attitudes are a result of women's tendency, at least in the past, to accept their alleged inferiority. Although boys show a strong preference for male

interests and activities at an early age, young girls tend to prefer male games and male friends, even though most of their actual activities are basically feminine. Girls, it seems, become aware at an early age of the advantages of being male. Boys seldom wish to be mothers, but many girls express the desire to be fathers (Mussen, 1969). Many studies have shown that both sexes in America place a higher value on men than on women (Broverman et al., 1970).

To test the idea that women are prejudiced against themselves, Philip Goldberg (1968) asked a group of college women to read and evaluate a series of six articles in a booklet. In one booklet, three of the six articles were said to be authored by women, and three carried the name of a male author. Another booklet contained the same articles but with the female and male authors reversed. The female college students tended to rate articles as more valuable and their authors as more competent when the apparent authors were men, even though the articles were exactly the same as those bearing female names. It may well be that "women have long suffered from an image of the self that paralyzes the will and short-circuits the brain, that makes them deny the evidence of their senses and internalize self doubt to a fearful degree" (Gornick and Moran, 1974:xx).

Studies of mental health find women are two to five times more likely to be diagnosed as suffering from depression than men are (Dohrenwend and Dohrenwend, 1977). Although these statistics may result in part from the greater likelihood of women to seek treatment for depression, the fact remains that women are more often depressed than men. According to Maggie Scarf, sex stereotypes account for this difference because "being female means (frequently) never being encouraged to become a self-sufficient person" (Scarf, 1979:51). Women are socialized to be the victims of learned helplessness, a condition closely related to depression.

Sex stereotypes and gender roles are also damaging to men. American men have a life expectancy eight years shorter than the life expectancy for women. Men have a greater incidence of cardiovascular disease, hypertension, ulcers, and migraine headaches. Men commit suicide three times more often than women do and are much more likely to become alcoholics (Goldberg, 1987). These characteristics, of course, are merely symptoms or consequences of deeper problems. One of these problems is dealing with the heavy and conflicting demands placed on men by sex stereotypes and gender roles.

First, men are expected to be strong, aggressive, confident, and in control of all situations at all times. Because many men find it difficult to live up to this masculine ideal, they may feel a loss of self-esteem, overcompensate with "machismo" or supermasculinity (Aramoni, 1972), or constantly pretend to be something they are not. Second, males are permitted less lee-way in gender role modeling than females. Girls can be tomboys, but it is not acceptable for boys to act like sissies. Consequently, boys must learn not to cry when they hurt and are often pushed into sports regardless of their talents or preferences. Third, men are forced to prove—to themselves and to others—that they are masculine, as these statements by two men suggest:

I became certain of my masculinity by becoming an accomplished football player after repeatedly being told by both coaches and friends that I was too small.

Learning to box (prize fight) as a youngster and winning a lot of medals helped me feel masculine, but making hard decisions in life...made me convinced of my masculinity (Tavris, 1977:39).

Men, in other words, are made to feel more masculine when they are competing, winning, and achieving. Although all men are expected to strive toward higher and higher goals, most of them must face the reality that only a few can really make it. For many men, the mirage of success becomes the reality of failure (Rubin, 1994). Finally, the rigidly defined masculine gender role inhibits men from developing close, emotional, personal relationships with others, including their children, male peers, and women (Wetcher, 1991). Mirra Komarovsky (1973), for example, found that half of the college men she interviewed felt insecure on dates with bright women. Because men are accustomed to competing, they probably feel threatened by women who are intellectually their equals or superiors.

A major purpose of gender socialization is to promote a sense of masculinity or femininity. Males in American society are made to feel more masculine when engaged in competitive activites.

What role conflicts and strains are created by gender roles and sex stereotypes? Women suffer from conflicts and strains built into the female role. They must balance the requirement of working outside the home with the demands of housework and child care (Burros, 1993; Hochschild, 1997). Women must juggle the expectation that they will move from one place to another as their husbands climb occupationally against their concern for the effects of rootlessness on their

---

SOCIOLOGY IN THE NEWS

Sex Stereotypes and Gender Inequality

Women are still victims of gender stereotypes that are applied to them by others and too often accepted by themselves. These stereotypes, which generally portray women as inferior to men, are used to socialize females and males into their culturally defined gender identities. For example, this CNN report points out that American females are taught that they should place emphasis on beauty and popularity and leave occupational achievement to males. As a result, women are steered away from the highly rewarded jobs in science, engineering, and technology.

1. State the view conflict theorists would take of sex stereotypes.

_____

_____

2. Identify some of the aspects of this situation that interest symbolic interactionists.

_____

_____

One aim of the women's movement is to help young American women move from the traditional female emphasis on beauty and popularity toward preparation for occupations in male-oriented fields such as science and engineering.

children. Finally, they are supposed to be inwardly strong but outwardly dependent.

A changing society is creating even greater conflicts. Women are increasingly working outside the home. (See Figure 10.1.) At the same time, women generally retain responsibility for domestic affairs.

Today, most married women share responsibility for the breadwinner role; but men, by and large, have been slower to share domestic responsibilities. Furthermore, no other institution in society, neither public nor private, has stepped in to help working women. As a result, women have shouldered the responsibilities of work and family largely on their own (Reskin and Padavic, 1994:147).

According to past research, housewives report higher rates of illness than women in the labor force (Nathanson, 1980). This apparent positive effect on women's health of working outside the home, however, may now be offset by the increasing stress that accompanies the combined demands of work and family (Rosenfield, 1989). In addition, greater stress and depression occur among women whose husbands do not support their working than among women with supportive

husbands (Ulbrich, 1988). Poorer health is also more likely among minority working women with husbands opposed to their employment (Andersen, 1997).

Like women, men experience conflicts and strains within the traditional masculine role. Many working-class men, whose wives must work just to support their families, are bothered by a feeling of having failed at their task of provider:

A guy should be able to support his wife and kids. But that's not the way it is these days, is it? I don't know anybody who can support a family anymore, do you?...Well, I guess those rich guys can, but not some ordinary Joe like me (Rubin, 1994:78).

Men are expected to be occupationally and economically successful while simultaneously devoting time to their wives and children. Men are encouraged to admire other women sexually, but they are expected to remain faithful to their wives. The "macho man" ideal conflicts with women's preferences for warm, gentle, romantic lovers. Men also have to handle the inconsistency between being cool and unemotional at work and loving at home (Doyle, 1995).

Role conflict and strain for men are also intensifying in today's environment of changing gender roles. Although many women want to be liberated from male domination, they sometimes still expect men to be protective of them. This is partly because appropriate gender role behavior is currently conditioned by the situation. For example, some women want to be treated as equal to men while at work but treated in more traditional ways off the job. Other women resent any attempt by men to treat them differently from the way they would treat other men. Consequently, men are often confused about what women expect from them. Moreover, although many men wish to shed the John Wayne image, they are less certain than women about the role model that should replace it (Wetcher, 1991).

Strains associated with changing gender roles tend to be greater within working-class families. In *Worlds of Pain*, Lillian Rubin (1992) found that working-class Americans advocate more traditional gender roles, yet were more likely to be required for economic reasons to work outside the home. She also reported that working-class husbands with wives in the labor force do not participate very much in domestic tasks. This assertion has since been bolstered by other research reporting that less educated, lower income men are less likely to share household labor when their wives work outside the home (Miller and Garrison, 1982; Sweet, Bumpass, and Call, 1988). In addition, Rubin concluded, working-class husbands of working women see their wives' employment as a challenge to their self-esteem and masculinity. The contradiction between traditional

Figure 10.1 Marital Status of Women in the U.S. Workforce: 1890, 1940, and 1994

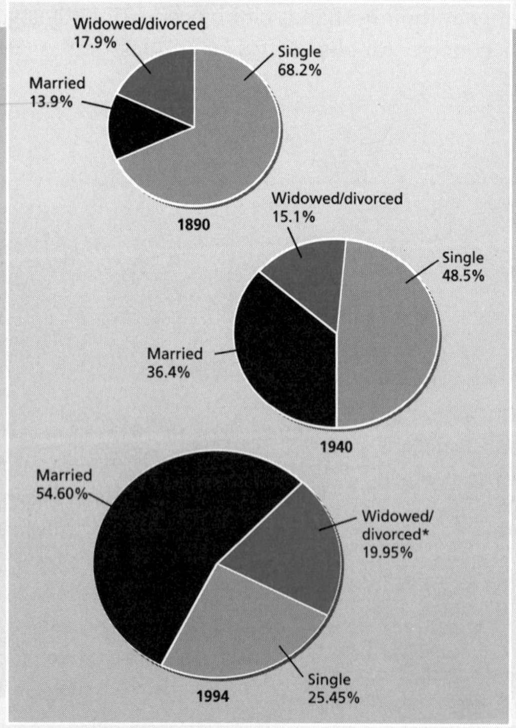

*Includes women who are separated.

Source: U.S. Bureau of the Census, *Statistical Abstract of the United States: 1995* (Washington DC: U.S. Government Printing Office, 1995), Table 637.

gender role attitudes and the sharing of the provider, household, and child-care roles among working-class couples, Rubin found, led to personal strain. Rubin wrote that, among other things, this contradiction led working-class men to demand more submission and obedience from their wives and children in order to bolster their own egos and sense of masculinity.

In her latest book, *Families on the Faultline,* Rubin (1994) found that although more contemporary working-class men now believe in a woman's right to full independence, they still want women with sufficient dependency and need to make them feel manly. Nearly all the working-class men Rubin studied did some work inside the family—washing dishes, vacuuming, grocery shopping—but only 16 percent of the men shared the family work equally with their wives. Almost all of these men, however, either worked different shifts than their wives or were unemployed. The high price paid for the retention of traditional sex stereotypes and gender roles remains much higher for working-class women.

GENDER INEQUALITY

Although women numerically constitute a slight majority of the population in the United States, they are nevertheless treated in some ways as a minority group (Hacker, 1951). In 1944, Swedish economist Gunnar Myrdal drew a parallel between the social and economic situations of blacks and women. He contended that both groups are victims of a paternalistic society. If blacks are subjected to racism, women are hurt by **sexism**—a set of beliefs, norms, and values used to justify sexual inequality. A central sexist belief is that men are naturally superior to women. Sexist ideology is used to justify men's right to leadership positions and power in the economic, social, and political spheres of society and to explain women's dependence on men.

Although there is no factual basis to support the maintenance of sexism, its influence remains. Women can easily be identified on the basis of certain biological and cultural characteristics and are subject to many forms of discrimination.

But isn't this all changing? The answer is yes and no. Some segments of American society now have more

positive attitudes about the role of women in society. Still, a careful examination of sexual inequality in the United States reveals many gaps between the sexes in social rights, privileges, and rewards. These gaps, although closing somewhat in recent years, are reflected in the continuing occupational, economic, legal, and political inequality experienced by American women (Bianchi and Spain, 1996; Riley, 1997).

Occupational and Economic Inequality

By far the most important labor market development over the last thirty years has been the dramatic increase in the number and proportion of working women (U.S. Department of Labor, 1997a). From 1972 through 1989, more than a million additional women have entered the labor force each year. In 1960, 23 million women worked outside the home for pay; by 1996 the figure had reached almost 62 million. Fifty-nine percent of women now work outside the home compared to 75 percent of men. (See Figure 10.2.) Women represent approximately 46 percent of the U.S. labor force. (See Figure 10.3.)

The increase in the labor-force participation rate among women in the central age category (twenty-five to fifty-four years) has been particularly steep. This, of course, includes women in the childbearing and child-rearing years. For example, the labor-force participation rate of married women with children under six years of age stood at 19 percent in 1960. By 1975, this percentage had risen to 37 percent and stood at 63.5 percent in 1995 (U.S. Bureau of the Census, 1996a).

Hasn't increased labor-force participation eliminated occupational inequality? Despite this increased labor-force participation, women are still concentrated in different occupations than men. This is known as **occupational sex segregation** (Charles, 1992; Tomaskovic-Devey, 1993; Reskin, 1993). Women are underrepresented in high-status occupations and overrepresented in low-status occupations. For example, 11 percent of engineers and about 27 percent of lawyers are females, whereas nearly all *pink-collar* jobs—secretaries, clerks, stenographers—whose purpose is to support the work of those higher up the occupational ladder are held by women (U.S. Department of Labor, 1997a). Moreover, when women are in high-status occupational

FEEDBACK

1. The idea that victory is the most positive reinforcement for all men is an example of
   a. sex.   c. a gender definition.
   b. a sex stereotype.   d. a value uniquely held by men.
2. Gender and sex stereotypes are essentially the same thing. *T or F?*
3. Because gender definitions are changing, men and women are experiencing less role conflict and strain. *T or F?*

Answers: 1. b 2. F 3. F

Harriet Bradley— Men's Work, Women's Work

Bradley's study is an excellent example of the combination of sociology and history. Focusing on work-related gender job segregation and job sex-typing since the Industrial Revolution in Britain around 1750, Bradley presents case studies of various industries and occupations—food and raw material production, agriculture, fishing, mining, manufacturing, pottery, hosiery, shoemaking, shopwork, medicine, teaching—to document the development of the division of labor

along gender lines. Her database consisted of a combination of secondary sources in the form of previously published research, which she synthesized, and new material she collected from primary sources.

As noted, this research focuses on job sex-typing. Bradley begins with the observation that the pervasive sex-typing of jobs—the allocation of certain work-related activities to men and to women—results in men and women rarely being allocated the same types of jobs. Even when men and women are working side by side, Bradley points out, they are doing different things. This allocation of certain work to men and certain work to women, she argues, is based on a simple principle—men occupy jobs that place them in control and women do work that requires obedience. In the fields

men cut and women gather; in the factory men stamp out parts, women sew them together; in the office men handle accounts, women do the typing and filing.

The social definition of what is "women's work" and "men's work" may vary from time to time and place to place. Jobs thought of as the province of women may have historically been assigned to men or vice versa. Prior to the industrialization of the cotton industry, for example, men were weavers, whereas women did the spinning. These roles were flip-flopped when power-driven machinery replaced hand labor. What does not vary, Bradley asserts, is the assignment of jobs based on the principle of men as superordinate and women as subordinate. Weavers had higher status prior to industrialization and lower status afterwards.

Figure 10.2 U.S. Labor Force Participation Rates, by Sex: 1890–1996

Source: U.S. Department of Labor, Bureau of Labor Statistics, *Employment in Perspective: Women in the Labor Force,* No. 865 (Washington DC: U.S. Government Printing Office, 1993), p. 1; and U.S. Department of Labor, Bureau of Labor Statistics, *Employment and Earnings* (Washington, DC: U.S. Government Printing Office, 1997), p.6.

Internet Link: For additional information on employment and earnings of American women and men in the United States, you can visit the website of the Department of Labor, Bureau of Labor Statistics at http://stats.bls.gov/

Although gender job segregation and job sex-typing existed in preindustrial society, Bradley contends that capitalism served to increase segregation and sex-typing and either obliterated or marginalized the traditional skills of women. From the case studies, she concludes that the tremendous expansion of capitalist production in the 1880s and 1890s created the foundation for the current patterns of gender job segregation and job sex-typing.

According to Bradley, hiring women in jobs labeled as "women's work" served both the interests of capitalists and male workers. Capitalists could increase their profits by hiring women for lower wages because of the traditionally low social standing of female labor. In addition, capitalists used women as a threat in their battle to shatter the control

male workers held as a result of their expertise in preindustrial craft production techniques. As men fought back by demanding that certain jobs remain "men's work," job segregation deepened and women were forced into low-paying "female" jobs.

Gender-based ideologies, Bradley concludes, continue to steer males and females into socially appropriate lines of work by playing on their desire to assert and solidify their unique sexual identities. Gender-based ideologies emphasize respective differences from the opposite sex. She uses this quote from Cockburn (1986) to accent the point:

What a man "is," what a woman "is," what is right and proper, what is possible and impossible, what should be hoped and what should be feared. The hegemonic ideology of masculinism involves a

definition of men and women as different, contrasted, complementary and unequal. It is powerful and it deforms both men and women (Bradley, 1989:229).

Critical feedback
1. With which framework of feminist theory—liberal, radical, socialist—do you think Bradley would most likely identify? Why?
2. Relate the concept of patriarchy to the results of Bradley's study.
3. Do you think that the current changes in gender relations in the United States will alter the state of gender segregation described by Bradley? Explain.

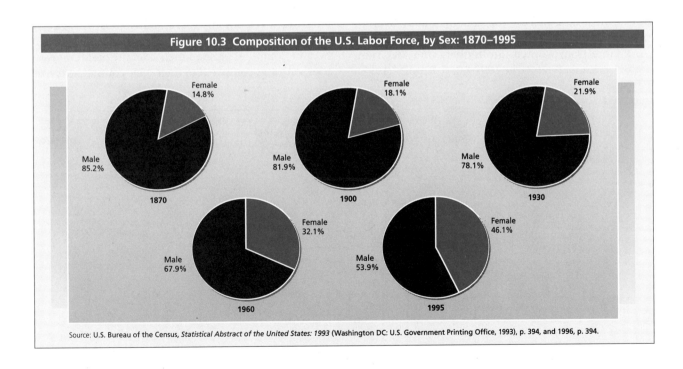

Figure 10.3 Composition of the U.S. Labor Force, by Sex: 1870–1995

Female 14.8%
Male 85.2%
1870

Female 18.1%
Male 81.9%
1900

Female 21.9%
Male 78.1%
1930

Female 32.1%
Male 67.9%
1960

Female 46.1%
Male 53.9%
1995

Source: U.S. Bureau of the Census, *Statistical Abstract of the United States: 1993* (Washington DC: U.S. Government Printing Office, 1993), p. 394, and 1996, p. 394.

groups, they are more concentrated in low-prestige, low-paid jobs. Even within female-dominated occupations, a disproportionate share of higher positions are filled by men. Occupational sex segregation, then, produces a **dual labor market**—a split between core and peripheral segments of the economy and the division of the labor force into preferred and marginalized workers. (See Chapter 9, "Inequalities of Race, Ethnicity, and Age.")

What is the financial situation of today's women?
Earnings inequality has followed a pattern since the 1960s. After increasing in the late 1960s, earnings inequality stabilized for most of the 1970s, decreased during the 1980s, and increased somewhat at the beginning of the 1990s (Grubb and Wilson, 1992). Thus, there remains a wide discrepancy between the earnings of American women and men. By 1995 women who worked full-time earned only seventy-four cents for every dollar earned by men. To put it another way, women now work about seven days to earn as much as men do in five days. The good news is that this salary gap has decreased since 1980, when women were earning 60 percent of men's earnings. The bad news is that the overall salary gap in 1995 is too close to what it was in 1955. (See Figure 10.4.)

In virtually every occupational category, men outstrip the earning power of women (U.S. Bureau of the Census, 1996a). The earnings gap between women and men persists, regardless of educational attainment. Women in the same professional occupations as men earn less than their male counterparts. (See Figure 10.5.) This is true even for women who have pursued careers on a full-time basis for all of their adult lives. In fact, males in female-dominated occupations typically earn more than women (Stoltenberg and McCrum, 1986; Williams, 1989; "The 1997 Salary Report," 1997).

What is comparable worth?   According to the advocates of **comparable worth,** similar wages should be paid for jobs requiring equivalent skills, responsibility, and effort, even if the specific duties performed are dissimilar (Acker, 1988; England, 1989; Evans and Nelson, 1989; Sorensen, 1994). Supporters of comparable worth point out that federal law prohibits paying people on the same job differently because of sex and that sex discrimination is built into the current pay structure because male-dominated occupations are accorded higher prestige and pay than female-dominated occupations (Blum, 1991; Bellas, 1994).

Opposition to comparable worth remains strong. According to many critics, valid job evaluations cannot be made. Many government agencies and corporations are opposed to comparable worth because of the billions of dollars it would cost them if enacted. Still others fear that establishing the doctrine of comparable worth could promote socialism.

Supreme Court decisions in 1989 added to the barriers to comparable worth by shifting the burden of proof of pay inequities from employers to employees. Still, some cities have already put comparable worth programs in place, and a large number of state legislatures are considering comparable worth as a partial solution to pay inequities among government workers.

Is there no hope of closing the salary gap?   Although the salary gap will remain indefinitely, there are a few signs of hope. Women now earn 74 percent of what men earn. (See Figure 10.4.) Probably the most encouraging word is that two-thirds of the women currently entering the labor force are taking jobs historically dominated by men. For example, as more women became computer programmers over the past few years, their earnings compared to their male counterparts rose

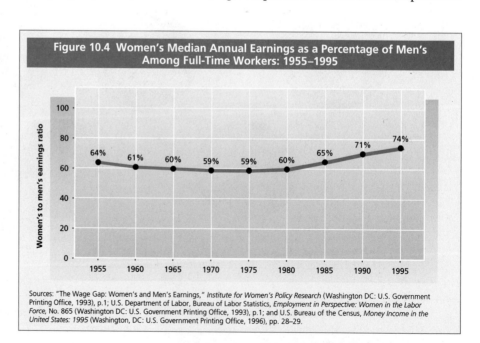

Figure 10.4 Women's Median Annual Earnings as a Percentage of Men's Among Full-Time Workers: 1955–1995

Sources: "The Wage Gap: Women's and Men's Earnings," *Institute for Women's Policy Research* (Washington DC: U.S. Government Printing Office, 1993), p.1; U.S. Department of Labor, Bureau of Labor Statistics, *Employment in Perspective: Women in the Labor Force*, No. 865 (Washington DC: U.S. Government Printing Office, 1993), p.1; and U.S. Bureau of the Census, *Money Income in the United States: 1995* (Washington, DC: U.S. Government Printing Office, 1996), pp. 28–29.

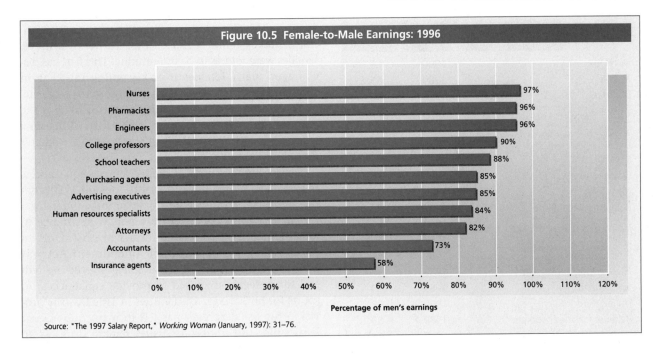

Figure 10.5 Female-to-Male Earnings: 1996

Occupation	Percentage
Nurses	97%
Pharmacists	96%
Engineers	96%
College professors	90%
School teachers	88%
Purchasing agents	85%
Advertising executives	85%
Human resources specialists	84%
Attorneys	82%
Accountants	73%
Insurance agents	58%

Percentage of men's earnings

Source: "The 1997 Salary Report," *Working Woman* (January, 1997): 31–76.

from 70 percent to almost 85 percent ("The 1997 Salary Report," 1997).

More generally, as the number of females as a percentage of full-time workers in many professional, managerial, and technical occupations has increased, the female-to-male income ratio has improved in a number of these occupations. Also, increasing numbers of women are leaving the wage system behind as they become entrepreneurs. Excluding large corporations, women now own about one-fourth of all businesses. However, females are still more likely than males to hold jobs that pay relatively low wages, and, as you have seen, the differential in female-to-male earnings persists within occupational categories.

Why does the salary gap persist? Research on wage differences has shown that even taking into account all possible factors that might reasonably account for wage differences between women and men—for example, age, experience, amount of absenteeism, and educational background—a large difference remains. With the elimination of all other explanations, the conclusion reached from these studies is that sex discrimination is the major source of economic differences.

One of the most serious indicators of economic inequality is the disproportionate number of women living in poverty. More than half of poor families have a female head of household. Individuals living in female-headed households are about three times as likely to live below the poverty level; female-headed households among minority groups are about four times as likely to live in poverty. This is what is meant by the **feminization of poverty** (Zopf, 1989; Goldberg and Kremen, 1990; Thomas, 1994; Sidel, 1996; U.S. Bureau of the Census, 1996b).

How do American women fare globally? There is a surprise in Table 10.1. As noted in the *Sociological Imagination* opening this chapter, women in the United States do not fare very well economically relative to women in other developed countries. In 1989, American women in manufacturing earned only 61 percent of that earned by men in the same sector of the economy. Although women in the United States are not at the bottom (that distinction goes to Japanese women, who live in one of the most paternalistic cultures in the economically developed world), they are fourth from the bottom. This is in dramatic contrast to Scandinavian countries where the male-female relative wage is in the eighties. (See Table 10.1.)

American working women have an income about three-fourths that of men. A gap in earnings between men and women is not obliterated even when education and occupation are comparable. This female teacher in all probability is paid less than her male counterparts.

TABLE 10.1	Male/Female Relative Wages in Manufacturing in Developed Societies	
Sweden	1990	89
Norway	1989	85
Denmark	1990	85
Australia	1989	80
France	1987	79
Greece (wage earners)	1989	78
Finland	1989	77
Netherlands	1989	75
New Zealand	1990	75
Belgium (wage earners)	1989	74
Gibraltar	1989	71
Ireland	1989	69
Czechoslovakia	1989	68
Greece (salaried workers)	1989	68
Switzerland	1989	68
United Kingdom	1990	68
Belgium (salaried workers)	1989	63
United States	1989	61
Luxembourg (wage earners)	1989	60
Luxembourg (salaried workers)	1989	55
Japan	1990	41

Source: U.S. Bureau of the Census, Center for International Research, *Gender and Generation in the World's Labor Force* (Washington, DC: U.S. Government Printing Office, 1993).

Sexism in Sport

Women have felt the consequences of sexism in athletics just as they have in other areas. The source of this sexism has roots in culture dating back at least as far as the ancient Greeks, whose gods had masculine characteristics. Greek gods were depicted as athletic, strong, powerful, competitive, rational, physical, and intellectual. Greek goddesses, with few exceptions, were passive, beautiful, physically weak, supportive, unathletic, and sexually attractive. The few active, strong goddesses were usually neither attractive to nor attracted by men. To Greek males, women who were physically or intellectually superior to them were unfeminine. It is from the Greeks, then, that Western culture inherited the definition of women as nonaggressive, weak, inferior, and dependent. These gender definitions have survived the past 2,500 years of human history. Their influence is exhibited in sport as much as in other aspects of social life, thanks to child-rearing practices, schools, churches, and the media.

What are some of the consequences of sexism for females in sport? One aspect of sexism is the existence of negative stereotypes that have traditionally discouraged females from participating in athletics. For

centuries the idea has existed that involvement in athletics makes females masculine. To be an athlete, females were told, is to be unfeminine. This fear has discouraged many females from becoming involved in sport and has tyrannized many of those who did participate. Another deterrent to female sport activity is the long-standing argument that such activity will harm a woman's health, particularly as it relates to a woman's reproductive system.

Sexism has traditionally denied females equal access to organized sports. It was not until the mid-1970s that the national Little League organization, under legal threat, rescinded its males-only policy. Resistance to females at the local level continues to exist. Only in 1972, when the Educational Amendment Act was passed, were high schools and colleges receiving federal funds legally required to provide equal access to sports for both sexes. Ambiguities in Title IX—the provision for equal access—has led to many subsequent legal suits, and important issues remain unresolved. Despite some significant gains, equal access of females in sport is not a reality (Nakamura, 1997). As late as 1984, the U.S. Supreme Court ruled that Title IX applies only to "specific" programs or departments receiving federal funds, a departure from the original interpretation that an institution receiving federal funds has to provide equal opportunity to females in "all" of its sport programs. This opens the door to administrative decisions that could undermine the progress made since 1972.

Women are still denied equal access to the power structure of sport. In fact, although Title IX increased equality for female athletes, it led to a decrease in the number of coaching and administrative positions held by women. In the early 1970s, women's intercollegiate teams were headed almost entirely by women; now more than one-half of women's teams at the top of the NCAA's structure are coached by men. The same trend is appearing among female athletic directors.

Currently there are two women's professional basketball leagues, a volleyball league, golf tour, and tennis circuit. As with both white and minority men, relatively few women athletes make it to the professional ranks. Those women who do overcome the odds to become professionals make on the average significantly less money than their male counterparts. In golf, for example, one of the few professional sports offering significant opportunities for women, the leading money winner on the men's tour typically earns more than twice as much as the leading money winner on the women's tour (*Golf World,* 1997).

Legal and Political Inequality

Sexism is also built into America's legal and political institutions. National, state, and local legal codes reflect sexual bias, and important differences exist between women and men in terms of political power.

How is sexism built into the law? According to the U.S. Civil Rights Commission, more than eight hundred sections of the U.S. legal code are sexually biased. Legal inequities and complexities are compounded by overlapping state and local jurisdictions. Because state and local statutes apply as long as they do not conflict with federal law, the legal situation of women varies greatly from one place to another. For example, some states have enacted state versions of the Equal Rights Amendment, which, if enacted, you may recall from Chapter 2 ("Sociologists Doing Research"), would enter this language into the United States Constitution: "Equality of rights under the law shall not be denied or abridged by the United States or by any state on the basis of sex" (Andersen, 1997:292). Other states, however, have retained archaic nineteenth-century laws based on the idea that women are the property of either a husband or a father. Georgia, for example, has an 1863 law that "defines a woman's legal existence as 'merged in the husband'."

The dependent legal status of women shows up in all areas of the law. A U.S. Supreme Court decision refused to grant women health insurance benefits for pregnancy-related medical costs despite the fact that medical coverage for medical conditions unique to men—such as prostate problems and vasectomies—was routinely provided. Some states refuse women the right to their own surnames after marriage. Others have so-called protective legislation restricting women's rights. Protective legislation restricts the number of hours women can work, the conditions under which they can work (such as requiring lounges for women workers), and the kinds of work they can do (such as limiting the amount of weight they can lift). Consequently, women have been barred from certain jobs, including some of the more lucrative ones. Furthermore, these laws assume that women need protection from the exploitation of employers but that men can take care of themselves (Andersen, 1997).

There are differences by sex in criminal law as well. Certain crimes are typically associated with one sex or the other. For example, crimes such as prostitution are generally only enforced against the women; male customers go free. On the other hand, sexual relations with a minor female, even though she consents, may result in prosecution of the man for statutory rape.

Rape has been one of the most controversial legal areas. Nowhere is sexism more explicitly built into the language and application of the law. For example, there is no other crime besides rape requiring independent corroboration that a crime has been committed. There are two assumptions of many rape laws: women will lie about whether a rape has taken place, and unless evi-

dence of rape is produced, no prosecution can occur. For women raped without witnesses (presumably the majority of cases) or for women who submit to rape because of the threat of violence, there is often no legal protection. Furthermore, when cases actually do come to trial, the victim is often treated more harshly than the defendant. Many observers have described the conduct of rape cases as a trial of the victim, with evidence of the victim's past sexual activity used to undermine her testimony and acquit the accused. Even convicted rapists have been allowed to go free on the grounds of provocation. A notorious case in the late 1970s involved a Wisconsin judge who refused to send a convicted rapist to prison on the grounds that the way women dress today causes men to lose control.

How does sexism relate to political power? One possible explanation for the bias built into the law is the underrepresentation of women in the institutions responsible for making laws. Between World War II and the mid-1970s, only 3 percent of the members of the U.S. Congress and 8 percent of state legislatures were women. By 1997, 11 percent of the members of Congress were female and 25 percent of state legislators were women. (See Figure 10.6.) Of all mayors in cities with over thirty thousand people, 21 percent were female in 1997. Thus, the participation of women in local and state elective politics appears to be increasing. Recently, there has been an increase in female governors, lieutenant governors, and attorneys general. This is also true at the national level. In 1988, Geraldine Ferraro became the first female vice-presidential candidate in the history of the United States. Madeleine Albright was named the first female Secretary of State in 1996.

Women are notoriously underrepresented in positions of political power in the United States. Madeleine Albright, who became the first female Secretary of State in 1996, is one of the growing number of women in high political office.

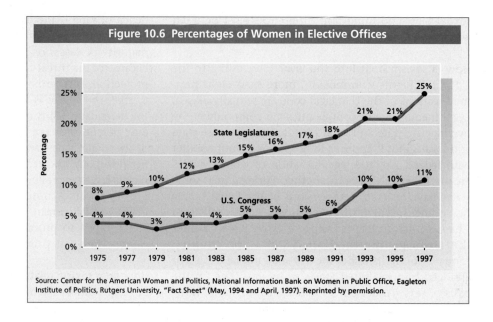

Figure 10.6 Percentages of Women in Elective Offices

Source: Center for the American Woman and Politics, National Information Bank on Women in Public Office, Eagleton Institute of Politics, Rutgers University, "Fact Sheet" (May, 1994 and April, 1997). Reprinted by permission.

Still, although women constitute more than half the population, they hold a small proportion of important political positions. (See Table 10.2.) Women occupied only 11.7 percent of the seats in the U.S. House of Representatives in 1997. It is significant, however, that the number of female House members increased to 51 from only 19 a decade earlier. Female U.S. Senators increased from two to nine over this same time period, still only representing 9 percent of the Senate (Center for the American Woman and Politics, 1997). Women in Congress have risen to few positions of power. There are only three female chairs of House or Senate standing committees, and no women are ranking members of these committees.

Female exclusion from political power is not restricted to elective offices. The record for women in appointed offices is also poor. Although there have been recent increases in the number of appointments, the total number is extremely small. When President Jimmy Carter appointed two women to his cabinet, he doubled the number ever to have held such positions at one time. President Clinton appointed three women to cabinet posts. Still, the total number of women who have ever served as cabinet officers is minute. President Reagan appointed the first woman Supreme Court justice, Sandra Day O'Connor, in 1981, and Clinton elevated Ruth Bader Ginsburg to the high court in 1993. Only a small percentage of federal judges are women. The "glass ceiling" extends beyond the highest elected and appointed political offices to the federal bureaucy as a whole (McAllister, 1992). Although women are in nearly one-half of the federal government's white-collar jobs, they have failed to gain access in large numbers to top executive and supervisory jobs (U.S. Merit Systems Protection Board, 1992). Women are in nearly two-thirds of lower grade jobs (GS-1 to GS-8) and fill 86 percent of the 300,000 clerical jobs. Only one-quarter of all federal supervisors are female and only 10 percent of senior federal executives are women.

The political position of women in the United States is among the worst in the Western world. With some notable exceptions, Western European nations have much greater female political participation. In the Scandinavian countries, for example, up to 20 percent

FEEDBACK

1. _____ is a set of beliefs, norms, and values used to justify sexual inequality.

2. On the average, women earn about _____ of what men earn.
   a. three-fourths   c. two-thirds
   b. one-half   d. 95 percent

3. _____ is the idea that similar wages should be paid for jobs requiring equivalent skills, responsibility, and effort, even if the duties performed are dissimilar.

4. Despite Title IX, females are still not beginning to be represented more equally in the collegiate sport power structure. *T or F?*

5. Although women are discriminated against in civil law, they are not at a disadvantage in criminal law. *T or F?*

6. The exclusion of women from political power applies mainly to elective offices. *T or F?*

Answers: 1. *Sexism* 2. *a* 3. *Comparable worth* 4. *T* 5. *F* 6. *F*

TABLE 10.2	Women in National, State, and Local Political Positions, 1995–1996	
Position	**Number of Women**	**Percentage Female**
Federal Legislative Branch		
U.S. Representative	51	12
U.S. Senator	9	9
House Leadership Post	3	1
Senate Leadership Post	1	1
House Standing Committee Chair	2	0
Senate Standing Committee Chair	1	1
Federal Judicial Branch		
Supreme Court Justice	2	22
U.S. Court of Appeal, Chief Judge	2	15
U.S. District Court Judge	7	8
Federal Executive Branch		
Cabinet Member	3	21
Executive Agency Head	2	22
Principal Advisor, Office of the President	4	36
State Executive Branch		
Governor	2	4
Lt. Governor	18	36
Attorney General	8	16
Secretary of State	12	24
State Treasurer	10	20
State Legislative Branch		
State Representative	1,233	23
State Senator	363	18
Local Executive Branch		
Mayors of 100 Largest Cities	12	12
Mayors of Cities over 30,000	202	21
All Mayors and Municipal Council Members in Cities over 10,000	4,513	21

Source: Center for the American Woman and Politics, Rutgers University, "Fact Sheet," April, 1997.

of the members of parliament are women. The 1979 election of Margaret Thatcher to the position of prime minister of England marked a significant breakdown in that country's barrier keeping women from the highest public office. The election of a woman president in the United States, on the other hand, seems highly unlikely in the near future.

CHANGING GENDER ROLES

Gender Roles and Social Change

How important will gender roles be in the future? What are the possibilities for change in gender roles? What are the consequences of gender role changes? One way to begin to answer these questions is to look at the factors discouraging and encouraging gender role changes.

Why is there resistance to changing the traditional gender roles? Several factors contribute to the resistance to changing gender roles. First, although men may not be aware that their suppression of women is to their own advantage, like all dominant groups, they can see little reason to change the way things have always been. Second, even if men were willing to relinquish their dominant position, it would be costly to them to do so. Third, some women resist giving up some advantages they see for themselves in the traditional gender role arrangement. Some fear losing the option to be full-time homemakers and do not want to relinquish their husbands' legal obligation to provide for them and their children. Others do not wish to lose conventional male courtesies and protection. Finally, an ideology still exists to support members of both sexes who wish to maintain the traditional gender roles. This belief system provides those favoring the status quo with reasons for its existence.

Resistance to gender role change can hardly be surprising to anyone who understands the durability of deeply entrenched aspects of culture. Traditional gender roles are embedded in the long-standing cultural myth of male superiority. How long-standing? The idea of male superiority, revealed as a myth today, actually began as a myth in the first writings of the ancient Greeks sometime in the seventh or eighth century B.C. In his account of the birth of the Greek gods and the social structure of daily life, the poet Hesiod, who was writing before the Greeks used prose, chronicled the creation of the myth of male superiority (Hesiod, 1983). Zeus, the king of gods, decided to act harshly after another god, his son Prometheus, went too far in his constant deviousness by giving fire to man. In addition to making subsistence require hard work, the myth goes, Zeus created woman as the carrier of all the evils that she eventually unloaded on men. Even in modern rural Greece, peasants only reluctantly acknowledge that they have a female child, and it is not uncommon for a man to say that he has three children (sons) and one girl (Hesiod, 1983:97–98).

It was, of course, from this myth, and similar ideas from most of the world's major religions (O'Brian and Palmer, 1993), that male supremacy was founded. This supremacy underlies the establishment of **patriarchy**—a hierarchical social structure in which women are dominated by men. (See Chapter 11, "Family," for further discussion of patriarchy.)

What are some of the factors promoting change in gender roles? There are a number of sources promoting change in gender roles. Demographic trends—longer life expectancy, smaller families, higher education, increased numbers of jobs requiring skilled labor—have had a profound impact on everyone's lives. For women, these trends have led to their increased participation in activities outside the home—in school, work, politics, and voluntary organizations. In fact, it is expected that by 2000, three-fourths of all married women will hold jobs outside the home. Another source of change in gender roles is the feminist movement. Although the goals and objectives of the feminist movement appear radical to many people, its central ideas are not new.

The Feminist Movement

What is the feminist movement? Recent feminism is best viewed as the reemergence of a long historical struggle that began as far back as 1792 with the publication in England of *A Vindication of the Rights of Women* by Mary Wollestonecraft. The movement picked up tempo in the United States in 1848, when a group of women abolitionists drew up the Declaration of Sentiments, which demanded equal rights for

A patriarchy is a hierarchical social structure in which women are dominated by men. These Arab women live in a patriarchal society.

women. By the end of the nineteenth century, feminism was a mass movement (Kraditor, 1971; Evans, 1989; West and Blumberg, 1990; Andersen, 1997; Tobias, 1997).

The first wave of the women's movement almost disappeared with the passage of the Nineteenth Amendment in 1920, which gave women the vote. However, many of the problems that had prompted feminism in the first place—the economic dependence of women and the restriction of their full participation in work and politics—remained. These remaining problems, coupled with other developments, such as effective birth control technology and the growth of certain sectors of the labor market (clerical, sales, technical), kept the women's movement from perishing (Taylor, 1989).

During the 1960s, what appeared to be a new social movement called the women's movement gained national attention through its organized efforts for gender role and societal changes. Beginning with a few isolated voices, such as the publication of Betty Friedan's (1963) enormously influential book *The Feminine Mystique*, the women's movement quickly gained support. By the 1970s, it was a mass movement with widespread support (Evans, 1980).

One basic argument of feminism is that male-oriented social, legal, political, and economic structures have inhibited women in their development as complete human beings. The results of early socialization, as well as some blatant forms of discrimination, have restricted women's options and personal growth. (More recently, it has been recognized that men are also the victims of limited options based on gender roles, although with different outcomes.) Also central to the feminist movement is the realization that problems created by sexism are not created by individuals and cannot be solved by individuals. Like racism, the problem of sexism is embedded in the entire structure of society, and therefore its eradication requires changes in social structures as well as in individuals. To obtain such changes, group action is required (Ferree and Hess, 1994; Tobias, 1997).

Gender Roles in the Future

Can we predict the future of gender roles? Making predictions about the future of gender roles is risky. For one thing, an economic recession or depression could seriously curtail the occupational progress of women, who could easily fall victim once again to the "last hired, first fired" policy of the past. Second, people's attitudes toward gender roles are inconsistent and unstable. For example, although polls show that most Americans favor equal pay for equal work, there is far less agreement on whether or not sex bias actually exists in American society. Consequently, there is disagreement over whether special steps to insure sexual equality, such as affirmative action programs and quotas, should be taken. Because some Americans support traditional gender roles and others favor a "unisexual" society, it is difficult to know in which direction the society will go.

The ups and downs of the Equal Rights Amendment (ERA) are a good indicator of the uncertainties involved. First introduced in the U.S. Congress in 1923, it was almost fifty years before it passed both the Senate and the House of Representatives and was sent to the individual states for ratification. As of 1979, thirty-five of the necessary thirty-eight states had ratified it. Opponents from a wide variety of right-wing political and religious groups mobilized opposition to the ERA on grounds ranging from fear that women would be drafted to the specter of unisex toilets. The strong opposition of President Reagan and many of the conservative politicians elected in 1980 signaled the defeat of the amendment in 1982. Efforts were immediately under way, however, to reintroduce it.

Apart from the ERA, the 1980s and early 1990s have been a time of both right-wing backlash against the aims of the feminist movement and increased mobilization of supporters (Faludi, 1992). Efforts to ban abortion, to restrict access to contraception, and to weaken Title IX regulations to insure equality in education are all part of attempts to reassert traditional relations between the

sexes. At the same time, feminists have mobilized, with increasing militancy, around such issues as spouse abuse, rape, sexual harassment, and pension discrimination (Richardson and Taylor, 1983; Taylor, 1983; Andersen, 1997). Such inconsistencies illustrate the problem of forecasting the future of gender roles.

Despite these problems, some speculation about the future can be made. The changes that have already taken place provide a foundation for discussing the future of gender roles and the consequences of gender role changes.

What will gender roles be like in the future? Women are living longer, bearing fewer children, becoming better educated, and entering the labor force in greater numbers than at any time in history (Reskin and Padavic, 1994). If these trends continue—and there is every reason to believe that they will—certain gender role changes should occur. For example, more and more women can be expected to plan for and pursue careers. No longer will female gender roles define work and career as unfeminine pursuits, nor will certain occupations be considered off limits to women. The increasing enrollment of women in medical, law, and business schools indicates that this development is already under way.

Women may be placing less emphasis on children and marriage as the standards for femininity. Men may become less preoccupied with work and success. Many observers have speculated on the nature of gender roles in future generations. Robert Hefner and Rebecca Meda (1979) have summarized the various alternatives in this way:

- *Emergent pluralism.* There would be no restrictions on the options available to either sex. Individuals following the traditional gender roles would exist alongside those who are pursuing a different course.
- *Conservative pluralism.* Men's and women's roles would be different but valued equally. In other words, men might still have major economic responsibility for the family, and women would perform housework and child care. Or men might be doctors and lawyers, and women would work as nurses and secretaries. However, all roles, whether male or female, would be considered equally important and rewarded accordingly.
- *Melting pot.* There would be no important differences in gender roles for men and women. Male and female roles would be combined to create androgynous individuals who combined the strengths of both traditional roles without their weaknesses. Both men and women would be strong and self-sufficient and yet capable of expressing emotion and sensitivity.

- *Assimilation to the male model of success.* Women would be encouraged to follow the traditional masculine gender role if they wanted to participate in traditionally male activities. Women could become doctors, lawyers, and businesspersons as long as they behaved just like men. This alternative would assume that feminine traits are only a hindrance and can make no contribution to the performance of societal roles.
- *Female exclusion.* A continuation of traditional gender roles. On the assumption that sex differences are biological, men and women would be excluded from exchanging tasks. In particular, women would be discouraged from assuming

male roles, and their own tasks would still be underrewarded.

It is difficult to predict which of these gender role patterns will prevail in the future, because unforeseen events could result in the reversal of present trends. However, the changes that have already occurred suggest a future in which more alternatives will be available to both men and women. Therefore, the most restrictive pattern (female exclusion) should continue to break down, and a variety of factors and efforts will probably be directed toward creating an atmosphere that permits individuals to choose from a number of options (emergent pluralism).

FEEDBACK

1. The women's movement can be traced as far back as the _____ century.
2. _____ is a social movement aimed at the achievement of sexual equality.
3. A central assumption of the feminist movement is that the problems created by sexism can be solved by individuals. *T or F?*
4. Match the gender role alternatives for the future that are listed below with the situations beside them:

____ a. emergent pluralism
____ b. conservative pluralism
____ c. melting pot
____ d. assimilation to the male model of success
____ e. female exclusion

(1) Male and female gender roles are basically identical.
(2) Traditional gender roles exist along with alternatives.
(3) Traditional gender roles continue.
(4) Women are socialized into the masculine role.
(5) Men's and women's roles are different but equally valued.

Answers: 1. *eighteenth* 2. *Feminism* 3. F 4. *a.* (2) *b.* (5) *c.* (1) *d.* (4) *e.* (3)

SUMMARY

1. All societies have definitions of gender—expectations and behaviors associated with a sex category. Through socialization, members of a society acquire an awareness of themselves as masculine or feminine.
2. Aside from obvious biological differences, such as genital characteristics and muscle structure, there are no established, genetically based behavior differences between the sexes. Behavioral differences between men and women are culturally and socially created.
3. Research by anthropologists in various cultures reveals the potential diversity in gender characteristics. Although gender definitions quite different from those we are familiar with exist in other societies, the traditional gender definitions remain the predominant pattern. Even societies attempting to abolish sexual inequality have not been uniformly successful.
4. Functionalists and conflict theorists have different explanations for traditional gender definitions. According to functionalists, a division of labor based on sex was a convenient and obvious solution in preindustrial societies. Conflict theorists, who are more concerned with the continued existence of traditional roles than with their origins, contend that men want to maintain the status quo

to protect their dominance of women and society. According to conflict advocates, traditional gender definitions are not appropriate in the modern world.
5. Symbolic interactionism's focus on socialization promotes our understanding of gender along different lines. According to symbolic interactionism, gender definitions are imparted through the socialization process; they are learned and reinforced through interaction with parents, teachers, and peers. This interaction is especially important because gender becomes intertwined with the self-concept of those affected by it. Gender socialization, which begins before birth and continues throughout life, occurs through elements of the mass media such as books, television, and advertising.
6. Feminist theory can be divided into three frameworks—liberal feminism, radical feminism, and socialist feminism. Although important differences exist among these theoretical frameworks, they each link the lives of women and men to the structure of gender relationships within society.
7. Gender definitions permit some deviation from the ideal, but sex stereotypes are labels applied to all members of each sex. Sex stereotypes are employed to encourage all

men to be masculine and all women to be feminine and allow for no variation from the generalized picture.

8. Sex stereotypes are damaging to both men and women. Changing gender definitions are intensifying conflict and strains for both men and women.

9. Although both men and women are damaged by sex stereotypes, it is women who feel the full effect of sexism—an ideology used to justify sexual inequality. Women, like other minorities, are subject to prejudice and discrimination occupationally, economically, politically, legally, and socially.

10. There are several trends promoting gender role changes in the future, but there are also barriers. Consequently, it is risky to make concrete predictions. Most likely, the exclusion of women from traditionally male domains will continue to diminish, and both sexes will increasingly be able to choose from a variety of gender role alternatives.

LEARNING OBJECTIVES REVIEW

After careful study of this chapter, you will be able to:

- Distinguish the concepts of sex, gender, and gender identity.
- Demonstrate the relative contributions of biology and culture to gender formation.
- Outline the diverse perspectives of gender taken by functionalists, conflict theorists, and symbolic interactionists.
- Differentiate among liberal, radical, and socialist feminism.
- State the major consequences of sex stereotypes and gender definitions.

- Describe the inequality of women in the United States with respect to work, law, politics, and sport.
- Discuss the factors promoting resistance to changing traditional gender roles as well as the factors promoting change in gender roles.
- Describe the state of knowledge regarding the future of gender roles.

CONCEPT REVIEW

Match the following concepts with the definitions listed below them.

____ a. sex
____ b. gender
____ c. patriarchal society
____ d. feminism

____ e. occupational sex segregation
____ f. sexual harassment
____ g. dual labor market

1. A split between core and peripheral segments of the economy and the division of the labor force into preferred and marginalized workers.
2. The type of society that is controlled by men who use their power, prestige, and economic advantage to dominate women.
3. A social movement aimed at the achievement of sexual equality, socially, legally, politically, and economically.

4. The expectations and behaviors associated with a sex category within a society.
5. The concentration of women in different occupations than men.
6. The biological distinction between male and female.
7. The use of one's superior power in making unwanted sexual advances.

CRITICAL THINKING QUESTIONS

1. Suppose that one of your high school teachers invited you back to speak to her class on the nature versus nurture gender debate. Write a brief essay you could use to summarize the state of scientific knowledge and initiate a class discussion.

2. Describe the major influences on the development of your gender identity. Use the categories of influence in the text in laying out the specifics of your experiences.

_____

_____

3. Some sociologists draw a parallel between racism and sexism. Discuss the consequences of sex stereotypes to substantiate or refute this assertion.

_____

_____

4. Draw on the information in your text on gender inequality to discuss the need for affirmative action programs. Substantiate a case either for or against continuation of these programs.

_____

_____

5. Use information throughout this chapter to predict the future of gender roles. Choose one of the five alternatives for the future of gender roles in the text as the framework for your analysis.

_____

_____

MULTIPLE CHOICE QUESTIONS

1. Whereas _____ is a biological term, _____ is a psychological and cultural term.
   a. gender; sex
   b. identity; role
   c. gender; gender identity
   d. sex; gender
   e. phenotype; genotype

2. The debate over explanations for sex differences persists, with the _____ distinction still at the heart of the controversy.
   a. nature versus nurture
   b. culture versus socialization
   c. innate forces versus genetic forces
   d. gender roles versus gender identity
   e. testosterone versus estrogen

3. Although given credit for its insights into the original emergence of gender roles, functionalism is often criticized on the grounds that
   a. it tends to support a male bias.
   b. it is too conservative and inadequate to explain recent changes in gender definitions.
   c. its proposition that gender roles are functional is detrimental to women.
   d. it does not address the issue of gender-based inequality.
   e. it fails to take the biological basis of gender roles into account.

4. According to conflict theory, gender roles persist because
   a. men and women have differential access to the resources needed to succeed outside the home.
   b. they are beneficial and efficient for human living.
   c. gender roles are incorporated into the self-concept through role-taking and the looking-glass self.
   d. they are essential to human survival.
   e. women are naturally more well-suited to perform certain tasks.

5. Symbolic interactionists believe that traditional gender roles persist because
   a. men and women have differential access to the resources needed to succeed outside the home.
   b. they are beneficial and efficient for human living.
   c. gender is incorporated into self-concept through role-taking and the looking-glass self.
   d. they are essential for human survival.
   e. of innate biological differences.

6. Which of the following statements regarding the contribution of peers to gender socialization is *true*?
   a. Nursery schools now attempt to ignore gender roles and do not encourage conformity to them.
   b. Teenagers who most closely fulfill gender role expectations are given the greatest respect.
   c. Acceptance or rejection by peers is having less and less influence on the adolescent's self-concept.
   d. Boys are more concerned than girls with gender role socialization.
   e. In schools today, children are often encouraged to act out the roles appropriate to the opposite sex.

7. Women are hurt by the set of beliefs, norms, and values used to justify sexual inequality known as
   a. sexism.
   b. prejudice.
   c. gender identity.
   d. discrimination.
   e. gender roles.

8. The concept of patriarchy is *most* closely associated with
   a. liberal feminism.
   b. socialist feminism.
   c. conservative feminism.
   d. Marxist feminism.
   e. radical feminism.
9. On the average, females in the United States are now earning _____ cents for every dollar earned by men.
   a. forty-five
   b. fifty-two
   c. sixty
   d. seventy-four
   e. eighty-three
10. According to advocates of _____, similar wages should be paid men and women for jobs requiring equivalent skills, responsibility, and effort, even if the specific duties performed are dissimilar.
    a. pay equity
    b. comparable worth
    c. equal rights at work
    d. workfare
    e. job parity
11. According to the text, the political position of women in the United States is
    a. among the worst in the Western world.
    b. more favorable than that of women in Scandinavian countries.
    c. harmed by "women's lib."
    d. actually superior to that of men because women vote in greater numbers.
    e. measured by the dependency ratio.
12. The women's movement was an identifiable mass movement by the end of the _____ century.
    a. sixteenth
    b. seventeenth
    c. eighteenth
    d. nineteenth
    e. twentieth
13. The social movement aimed at the achievement of sexual equality is generally known as
    a. sexism.
    b. feminism.
    c. genderism.
    d. communism.
    e. comparable worth.
14. According to the text, the most likely form of gender roles in the future is
    a. conservative pluralism.
    b. the melting pot.
    c. assimilation to the male model of success.
    d. female inclusion.
    e. emergent pluralism.

FEEDBACK REVIEW

True-False
1. Societies that do not follow traditional gender definitions are exceptions to the rule. *T or F?*
2. One of the major points of agreement among the three feminist theoretical perspectives is that gender and gender relationships are psychological in origin. *T or F?*

Fill in the Blank
3. Gender is acquired through the process of _____.
4. According to the _____ perspective, the division of labor based on sex has survived because it is beneficial and efficient for human living.
5. According to the _____ perspective, traditional gender definitions exist because they provide greater rewards and privileges to men than to women.
6. _____ is the theoretical perspective that attempts to explain the ways in which gender is acquired.

Multiple Choice
7. Which of the following is *not* a true statement?
   a. Boys are less assertive in class than girls.
   b. Illustrated children's books present very traditional gender definitions.
   c. Parents' perceptions of the physical characteristics of their children are heavily influenced by their children's gender.
   d. Nursery school children who fail to conform to their "appropriate" gender are met with resistance from their peers.
8. The idea that victory is the most positive reinforcement for all men is an example of
   a. sex.
   b. a sex stereotype.
   c. a gender definition.
   d. a value uniquely held by men.

Matching

9. Match the words or phrases on the right with the three feminist theoretical frameworks on the left.

_____ a. liberal feminism

_____ b. radical feminism

_____ c. socialist feminism

(1) patriarchy

(2) capitalism and patriarchy

(3) equality of opportunity

(4) social and legal reform

10. Match the gender role alternatives for the future that are listed below with the situations beside them.

_____ a. emergent pluralism

_____ b. conservative pluralism

_____ c. melting pot

_____ d. assimilation to the male model of success

_____ e. female exclusion

(1) Male and female gender roles are basically identical.

(2) Traditional gender roles exist along with alternatives.

(3) Traditional gender roles continue.

(4) Women are socialized into the masculine role.

(5) Men's and women's roles are different but equally valued.

GRAPHIC REVIEW

Figure 10.4 displays American women's median annual earnings as a percentage of men's from 1955 to 1995. Attempting to reply to the questions below will test your understanding of this information.

1. How would you explain this trend line to your sixteen-year-old daughter (if you had one)?

_____

_____

2. Assuming that the upward trend since 1980 continues, which of the five gender role patterns outlined in the text would be most descriptive? Why?

_____

_____

ANSWER KEY

Concept Review

a. 6
b. 4
c. 2
d. 3
e. 5
f. 7
g. 1

Multiple Choice

1. d
2. a
3. b
4. a
5. c
6. b
7. a
8. e
9. d
10. b
11. a
12. d
13. b
14. e

Feedback Review

1. T
2. F
3. socialization
4. functionalist
5. conflict
6. Symbolic interactionism
7. a
8. b
9. a. 3,4
   b. 1
   c. 2
10. a. 2
    b. 5
    c. 1
    d. 4
    e. 3

Chapter Eleven

Family

After careful study of this chapter, you will be able to:
- Describe the types of family structure, dimensions of family structure, types of mate-selection norms, and types of marriage.
- Compare and contrast the view of the family taken by functionalists, conflict theorists, and symbolic interactionists.
- Discuss the extent and cause of divorce in America.
- Describe the extent and nature of family violence in the United States.
- Outline the contemporary major alternatives to the traditional nuclear family structure.
- Discuss the future of the family in the United States.

SOCIOLOGICAL IMAGINATION

Is the rising divorce rate an indication that marriage is in jeopardy? Popular depictions of family life by films, television, music, print media, and the Internet certainly suggest that family life in the United States is on the ropes. Actually, data on divorce provide some grounds for optimism regarding the future of the American family. Although a dramatic rise in the divorce rate did begin in the early 1960s, the divorce rate began to decline in 1985.

The facts regarding divorce in America, then, are not as clear-cut as commonly believed. As you will see, the family is not as simple an institution as it appears on the surface. It is a complex social unit with many dimensions and facets. In addition, there is great diversity in the structure of the family among various societies—even within one society.

FAMILY AND MARRIAGE DEFINED

In the simplest terms, a **family** is a group of people related by marriage, blood, or adoption. The **family of orientation** is the family in which you were born. It provides you with a name, an identity, and a heritage. Your family of birth "orients" you to the neighborhood, community, and society you live in. The forms you filled out to join such groups as the Boy or Girl Scouts or to register for college commonly ask, What is your father's name? What is your mother's name? Where do you live? Where do your parents work? The answers you give to these questions indicate who you are and to whom you belong. The family of orientation locates you in the world. The **family of procreation** is the one you establish when you marry. The key word is *procreation*. The marriage ceremony signifies that it is legal for a couple to have children and to give them a family name. **Marriage,** in fact, can be defined as a legal union, in which public approval is given to sexual activity, having children, and assuming mutual rights and obligations (Eshleman, 1997). The family of procreation becomes the family of orientation for the children created from this marriage.

All known societies have families and marriage; however, the permutations are staggering. Exposure to these variations requires somewhat of an anthropological excursion.

CROSS-CULTURAL ANALYSIS OF FAMILY AND MARRIAGE

An examination of variations in the structure of family and marriage involves going across many types of cultures. This cross-cultural examination provides a firm fix on the predominant family and marriage forms in modern society as well as an appreciation for alternative lifestyles. Table 11.1 compares the major characteristics of the traditional family with those of the modern family.

Types of Family Structure

The **nuclear family,** the smallest group of individuals that can legitimately be called a family, is generally composed of a mother, a father, and children. Because the nuclear family is usually based on marriage, it is sometimes called the *conjugal* family. However, the two are not always synonymous. A nuclear family can be composed of a single parent and children or of a brother and sister. The **extended family** consists of two or more adult generations of the same family sharing a common household and economic resources. Extended families contain grandparents, children, grandchildren, aunts, uncles, and so forth. Because it is based on blood ties, the extended family is often identified as the *consanguine* family.

In general, extended families are most characteristic of preindustrial societies and rural parts of industrial societies. This is not to say, however, that nuclear families were not common in preliterate societies or in preindustrial Europe (Nydeggar, 1985). Moreover, it would be inaccurate to conclude that the nuclear family—most characteristic of modern society—exists in isolation from a larger kinship network. Both of these issues—the association of the nuclear family with industrialization and the question of the isolation of the modern nuclear family—require elaboration.

TABLE 11.1	Major Characteristics of Traditional and Modern Families	
Characteristics	**Traditional Family**	**Modern Family**
Family structure	Extended (also some nuclear)	Nuclear (also some extended)
Basis of family bond	Blood (consanguine)	Marriage (conjugal)
Line of descent and inheritance	Patrilineal (male lineage) Matrilineal (female lineage)	Bilateral (dual parental lineage)
Locus of control	Patriarchal (male dominance) Matriarchal (female dominance)	Democratic (sexes share power)
Place of residence	Patrilocal (husband's parents) Matrilocal (wife's parents)	Neolocal (independent)
Marriage structure	Monogamy (one spouse) Polygyny (several wives) Polyandry (several husbands)	Monogamy (one spouse)

How are industrialization and the nuclear family related?   As Figure 11.1 indicates, there is a curvilinear relationship between types of family structure and industrialization. Before humans began to domesticate animals and cultivate crops, most economies were based on hunting and gathering. Small bands of nuclear families followed herds of animals and the changing seasons, moving around constantly, never staying long in any one place. When humans learned to domesticate animals to help with tilling the soil and cultivating crops, they no longer needed to be mobile to maintain a food supply. Families began to farm, settle down, and establish roots. The emergence of family farms and stable permanent residences permitted the development of the extended family. When agriculture became the dominant mode of subsistence, the extended family became the major type of family. As societies move from agricultural economies to industrialized ones, the extended family tends to be replaced by the nuclear family (Goode, 1970).

What accounts for the relationship between industrialization and nuclear family structure?   One reason is increasing geographic mobility. People in rural areas must leave their relatives to secure industrial employment in cities. Furthermore, those already living in urban areas come to accept as a way of life frequent changes in residence for occupational reasons. In any event, geographic mobility makes it difficult to maintain close ties with large numbers of relatives; usually only the most immediate kin can be taken along. Also, before industrialization, family members were expected to take care of the sick and the elderly. (Actually, they scarcely had a choice, because there were few outside sources to rely on.) With the growth of special purpose organizations and government programs, however, the sick can be cared for in hospitals—often paid for by insurance companies—and the old can be placed in homes for the elderly to live on their Social Security payments. In short, because outsiders have taken over many of the services formerly performed by the family, relatives do not have to depend on one another as much as they once did. Thus, certain conditions associated with industrialization are more compatible with the nuclear family than with the extended family.

How isolated is the modern nuclear family?   Although families in industrialized societies tend to have a nuclear structure and maintain separate residences, there is evidence that they are not as isolated from their relatives as was once believed. Early research depicting the American nuclear family as isolated was conducted prior to some recent changes—changes that are now occurring in other industrialized societies as well. For one thing, modern transportation and communication make it easier for relatives to maintain family ties even though separated geographically. Second, with metropolitan growth, children no longer need to leave their parents and other relatives to go to college or to pursue their occupations.

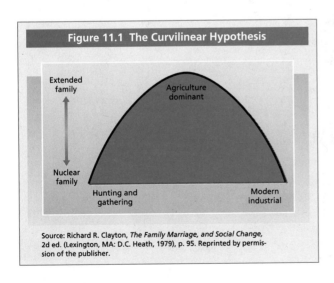

Figure 11.1 The Curvilinear Hypothesis

Source: Richard R. Clayton, *The Family Marriage, and Social Change,* 2d ed. (Lexington, MA: D.C. Heath, 1979), p. 95. Reprinted by permission of the publisher.

DOING RESEARCH

Theodore Caplow—
Families in Middletown, USA

From the spring of 1976 until the fall of 1978, Theodore Caplow and his colleagues conducted an intensive study of a community called Middletown (Muncie, Indiana). Actually, their effort was the third study of this community. In the mid-1920s, Robert and Helen Lynd first studied the total culture of this community, the findings of which they published in *Middletown* (1929). They replicated their earlier study eight years later

to document the speed and direction of social change, publishing their conclusions in *Middletown in Transition* (1937). Together these two studies form a landmark in sociological research—the scientific study of a total community; the followup was the first attempt to document social change within a total community.

Caplow and his research team took turns living in Middletown. In the process, they employed a wide variety of data-gathering techniques. They read local publications and attended public meetings. Interviews were conducted with community leaders, and representative surveys of the population were carried out. In total, thirteen major and several minor surveys took place, three of which used questionnaires originally

designed by the Lynds for their studies.

Several books and numerous articles have been published from the data gathered in the "Middletown III" project. One of the books, *Middletown Families* (1982), is the subject of this "Doing Research." Because the book covers the gamut of family life—from family roles to religion and the family—it is impossible to summarize all its findings. Thus, we will focus on the future of the American family. Caplow's conclusions are based on the assumption that "Middletown families are reasonably similar to American families in general. What we have discovered about marriage, divorce, child raising, housework, employment of women, home ownership, family festivals, kin-

Early on, Marvin Sussman (1953) observed considerable cooperative effort and mutual aid between adults in urban nuclear families and their parents in the United States. According to Sussman, families of orientation and procreation are linked together through a kin network. He found, for example, parents offering a variety of services to their married children, including financial aid, home repair and maintenance, babysitting, and various kinds of help during illnesses.

Subsequent researchers have also found this network of kin relationships (Litwak, 1960; Adams, 1968, 1970; Troll, 1971). A study of Muncie, Indiana, by Theodore Caplow and his colleagues (1982) showed that a high proportion of the close kin of the residents surveyed lived within fifty miles of the city. Most of the residents with parents still living saw their parents at least once a month, and nearly half saw them either weekly or more often. (See *Doing Research* for more on this study.)

The dominance of the nuclear family is a fairly recent phenomenon. At the beginning of the twentieth century, the extended family—in which several adult generations live together—was the dominant family type in the United States.

ship networks, and erotic behavior in Middletown does not diverge greatly from what is known for the country as a whole" (1982:323).

According to Caplow, the Middletown (and American) family is in extremely fine shape. In fact, Caplow states that the condition of the family in Middletown has improved since the 1920s: ". . . we discovered increased family solidarity, a smaller generation gap, closer marital communication, more religion, and less [geographic] mobility" (1982:323).

Caplow raises this interesting question: Why do Americans keep talking about the deterioration (if not destruction) of the family? He points to the existence of the "sociological myth" of the declining family. This myth, argues Caplow, was created by some early sociologists and has been perpetuated by scholars, politicians, poets, and the public in general. According to Caplow, the myth of the downward slide of the family survives in part because of the comfort people derive from its existence: "When Middletown people compare their own families and their own relationships with spouses, children, and parents with the 'average' or the 'typical' family's, nearly all of them discover with pleasure that their own families are better than other people's" (1982:328). The perception that the family in general is in terrible shape permits the people of Middletown (and America) to feel superior in their performance as husbands, wives, children, parents, housekeepers, and providers.

Critical feedback
1. Caplow and his colleagues offer Middletown, a midwestern city of 80,000, as typical of America as a whole. Agree or disagree with this assumption and defend your position.
2. Which of the major theoretical perspectives is used to explain why the "myth of the declining family" continues to survive? Explain your choice.
3. According to this study, the American family is healthy. Agree or disagree with this conclusion, drawing on any of the material in this chapter.

Apparently, the overwhelming majority of Americans maintain frequent contact with a substantial number of relatives. At the very least, it must be concluded that the modern family has either a modified nuclear or an extended structure, family types that fall somewhere between the isolated nuclear family and the classic extended family.

Whatever the family structure—nuclear or extended—several dimensions of family life must be recognized. They relate to descent and inheritance, authority within the family, and place of residence.

Dimensions of Family Structure

Who owns what? Different arrangements exist for determining descent and inheritance. If the arrangement is **patrilineal,** descent and inheritance are passed from the father to his male descendants. Should descent and inheritance be transmitted from the mother to her female descendants, the arrangement is a **matrilineal** one. In some societies, descent and inheritance are **bilateral**—they are passed equally through both parents. Thus, both the father's and mother's relatives are accepted equally as part of the kinship structure.

Who controls whom? In **patriarchal control,** the oldest man living in the household controls the rest of the family members. In its purest form, the father is the absolute ruler. Usually, patriarchal families are extended families structured along blood lines. So rare is **matriarchal control,** in which the oldest woman living in the household holds the authority, that controversy exists over whether any society has ever had a genuinely matriarchal family structure. In **democratic control,** authority is split evenly between husband and wife.

Who lives where? With a nuclear family, a married couple lives with neither set of parents and establishes a new residence of its own. Such a residential arrangement is a **neolocal residence.** Extended families, of course, have different residence norms. A **patrilocal residence** calls for living with or near the husband's parents. Residing with or near the wife's parents is expected under a **matrilocal residence.**

Mate Selection

In the United States, most people assume that they will have complete freedom of choice in choosing a mate. It is certainly true that families generally are no longer expected to arrange marriages for their children. American women are no longer expected to fill hope chests with quilts, china, silverware, and other household goods in anticipation of marriage. An American man is not expected to negotiate with a future father-in-law a price for the woman he wishes to marry. Therefore, we assume that freedom of choice prevails in the selection of a marriage partner. Actually, this is a false assumption. All societies, including the United States, have norms regarding who may marry whom.

Exogamy refers to mate-selection norms requiring individuals to marry someone outside their kind. Exogamous norms are usually referred to as incest taboos. In the classic Chinese family, for example, two people with identical family names could not marry unless their families were known to have diverged at least five generations previously (Queen, 1985). In the United States, you are not legally permitted to marry a son or daughter, a brother or sister, a mother or a father, or a first cousin. In some states, marriage to a second cousin is prohibited, and it is considered taboo to marry a former mother-in-law or father-in-law. Incest is almost universally prohibited. In fact, exogamous mate-selection norms are so strongly enforced that incest has seldom existed as an established pattern of mate selection; royalty in ancient Europe, Hawaii, Egypt, and Peru are prominent exceptions. Even in these instances, most members of the royal families chose partners to whom they were not related by blood.

Endogamy involves mate-selection norms encouraging individuals to marry within their kind. In the United States, an endogamous norm requires that marriage partners be of the same race. The degree to which racial and religious norms of endogamy are enforced is reflected in the small percentage of interracial and interfaith marriages.

Homogamy refers to mate-selection norms requiring individuals to marry someone with similar social characteristics. Included in these characteristics are age, social class, education, previous marital status, and geographic location of one's parental home. Most marriages in the United States occur between individuals who are about the same age. It is rare for the son or daughter of a multimillionaire to marry someone from the lower class. Most people who have never before married become married to someone who has also always been single. Divorced people tend to marry those who have been previously married. Finally, people tend to choose marriage partners from their own communities or neighborhoods.

As you can see, who may marry whom in a society is not completely a matter of choice. The field of eligible marital partners is limited by exogamous, endogamous, and homogamous norms of mate selection (Kephart and Jedlicka, 1991).

Types of Marriages

When marriage ceremonies are mentioned, Americans commonly think of a church, a bride in a long, white gown walking down the aisle on the arm of her father, the husband-to-be waiting for his bride at the front of the church with the minister, the repetition of the

The marriage of one man to only one woman at a time (monogamy) is currently the only legal form of marriage in the United States. Whatever form the marriage ceremony takes—the traditional form, such as this couple, or a couple taking the vows while skydiving—the ceremony's ritualistic purpose is to announce to others that a new family has been created.

ancient vows ("to love and to honor, from this day forward, in sickness and in health, for richer, for poorer, 'til death do us part"), an exchange of rings, and a kiss. In fact, modern couples get married on mountaintops, while diving out of planes, while swimming underwater, and in the nude. Whatever form it takes, the marriage ceremony is an important ritual whose purpose is to announce to the world that a man and woman have become husband and wife, that a new family has been formed, and that any children born to the couple can legitimately inherit the family name and property. The traditional marital arrangement found in the United States is merely one of several possibilities.

Monogamy—the marriage of one man to only one woman at a time—is the most widely practiced form of marriage. In fact, it is the only form of marriage that is legally acceptable in the United States and in most other societies. **Polygyny** is the marriage of one man to two or more women at the same time. The most obvious example of polygyny can be found in the Old Testament, in which King Solomon is reported to have had seven hundred wives and three hundred concubines. George Murdock (1957) found that polygyny was practiced in 75 percent of the 575 preliterate societies he studied. However, polygyny is not practiced widely in any existing society. Even in some Moslem countries, where the Koran allows a man to have as many as four wives, polygyny has been outlawed. In its statement on

1. A _____ is a group of people related by marriage, blood, or adoption.
2. A family of _____ is the family into which a person is born.
3. A family of _____ is the family established upon marriage.
4. The _____ family consists of two or more generations of the same family sharing a common household and economic resources.
5. There is a _____ relationship between types of family structure and industrialization.
6. Industrialization promotes the shift from the extended to the _____ family.
7. In a _____ family, the oldest man living in the household controls the rest of the family members.
8. Indicate which type of mate-selection norm is reflected in each of the following situations:

    ____ a. Jews are supposed to marry Jews.   (1) exogamy
    ____ b. A father is not permitted to marry his daughter.   (2) endogamy
    ____ c. Members of the same social class marry.   (3) homogamy

9. _____ is the marriage of a woman to two or more men at the same time.

Answers: 1. family 2. orientation 3. procreation 4. extended 5. curvilinear 6. nuclear 7. patriarchal 8. a. (2) b. (1) c. (3) 9. Polyandry

human rights, the United Nations explicitly frowned on the practice of polygyny. **Polyandry**—the marriage of one woman to two or more men at the same time—is an even rarer form of marriage. It is known to have been common in only two societies: in Tibet and among the Todas of India (Queen et al., 1985). Where polyandry has existed, it usually has consisted of several brothers sharing a wife. Instances of a father sharing a wife with his sons have been reported, but not with any degree of regularity. A *group marriage* consists of two or more men married to two or more women at the same time. This form of marriage is also rare. In fact, scientists cannot identify even one society in which group marriage has been the normal form. There have, however, been communes that have practiced group marriage (Berger, Hackett, and Millar, 1972; Kanter, 1972; Ferrar, 1977).

THEORETICAL PERSPECTIVES AND THE FAMILY

The family is as central to society as it is complex. It is therefore not surprising that each of the major theoretical perspectives shed some light on this institution.

Functionalism

The family is often referred to as the basic institution of society. You were born into a family with whom you lived for almost twenty years. You are likely to marry and, despite the high rate of divorce, the overwhelming majority of your adult life will be spent within a family.

The family performs many vital functions. First, it provides the initial learning experiences that make people human. Second, it is supposed to provide a warm and loving atmosphere that fulfills basic human social and emotional needs. Third, it is the only legitimate source of reproduction for a society. Fourth, it regulates sexual activity. Fifth, it places people in a social class at birth. Finally, the family serves an important economic function. In short, the family is the basic institution of society because of the vital functions it performs. These functions deserve some elaboration.

How does the family contribute to the socialization process? One of the major functions of the family begins immediately after a baby has been born. In addition to caring for the child's physical needs, new parents begin the vital process of teaching the child the things he or she needs to know to develop a personality and to fit into society. This begins in the first year through intimate contact—holding, touching, stroking, and talking to the baby. Around the first birthday, the infant begins to walk and to mimic the words and sentences others have used. During the second and third years of the child's life, the parents become more serious about socialization. They begin to teach the child customary ways of thinking, feeling, and behaving. The process of family socialization continues as the child enters new stages of development. The family enables the child to develop socially and psychologically.

What is the socioemotional function of the family? Another major function of the family is to provide what Meyer Nimkoff (1965) has called *socioemotional maintenance*. The family is supposed to be the one place in society where an individual is unconditionally accepted and loved, with no strings attached. The family is supposed to provide the kind of supportive environment in which people do not have to put up a front or "play games." Family members can accept one another as they are; every member is special and unique. Without the care and affection that is supposed to be provided by parents or other adults, children do not develop normally. They have low self-esteem, fear rejection, feel insecure, and eventually find it difficult to adjust to marriage or express affection to their own children. In fact, some young children who have not been given love and attention have become retarded or have died. (Refer to Chapter 4, "Socialization Over the Life Course.")

What is the reproductive function of the family?
Society cannot survive without new members. Childbirth signifies the continuation of the family line. The family's performance of the reproductive function provides an orderly mechanism for producing generation after generation. In virtually all societies, children are welcomed as new members of the group. It is customary in the United States to honor the mother-to-be with a baby shower. After birth, babies are displayed in hospital nurseries so that friends and relatives can marvel at them. The parents send out birth announcements that report their baby's weight, height, and name. Among the Sirono Indians of eastern Bolivia, there are three days of rituals celebrating the birth of a child (Holmberg, 1969). These rituals are designed to protect the life of the infant and to insure the child's good health.

In what ways does the family regulate sexual activity?
In no known society are people given total sexual freedom. Even in sexually permissive societies, like that of the Hopi Indians, there are rules about mating and marrying (Queen, Habenstein, and Quadagno, 1985). The function of sexual regulation is usually assigned to the family.

Norms regarding sexual regulation vary from place to place. In the Trobriand Islands, off the coast of New Guinea, all children are encouraged to engage in premarital sex (Malinowski, 1929). A boy's brother will take him aside and say, "Watch me and I'll show you how it's done." Thus, much of childhood is spent going from partner to partner, as a bee goes from flower to flower. In Moslem countries, in contrast, the family of the bride must guarantee that the bride is a virgin (Peters, 1965). In fact, the honor of the family rests on that guarantee. The United States has traditionally fallen somewhere between these two extremes.

Sexual norms also change within a society. In colonial America, sex outside of marriage was considered a sin. We have, indeed, come a long way since then. Most experts agree that a sexual revolution has occurred since the 1950s, when Alfred Kinsey's (1948, 1953) studies of sexual behavior hit the American public like a bombshell.

How does the family transmit social status? As noted in Chapter 8 ("Social Stratification"), an individual's place in the stratification structure is largely assigned at birth. If people move up or down the stratification structure, it is usually only a very slight move. It is primarily through the family that the transmission of social status occurs. This is so partly because our families provide us with a given level of economic resources that opens and closes certain occupational doors. The sons and daughters of high-income professionals, for example, are more likely to attend colleges and graduate schools than are the children of blue-collar workers. Consequently, the children of professionals are more likely to enter high-level professional occupations as adults. The family also transmits values that later affect social status. The children of professionals, to cite only one illustration, tend to have a greater need to be successful than their counterparts from blue-collar families. In these and many other ways, the family affects the placement of children in the stratification structure.

What is the economic function of the family?
Traditionally, one of the most important functions of the family was the economic support of the family unit. At one time, families were self-sufficient economic units whose members all contributed to the production of needed goods. Malls, corporations, and supermarkets were not part of the socioeconomic scene. Although the family still has an economic function in modern society, that function is one of consumption rather than production. Family members go outside the home to earn money to buy the things they want. Television advertising is designed to appeal demographically to specific family members for certain kinds of products. The Saturday morning television cartoons are likely to be sponsored by toy manufacturers and cereal companies, whereas "Frazier" is more likely to be sponsored by drug companies and health-care providers.

Conflict Theory

Functionalism emphasizes the ways in which the family attempts to contribute to the order and stability of society. The conflict perspective draws attention to matters such as sexual inequality.

How was the conflict perspective first applied to the family? As applied to the family, the conflict perspective can be traced to Friedrich Engels, Karl Marx's collaborator. Engels saw woman's oppression as the result, first, of woman's loss of a productive role, and second, of man's interest—with the development of private property and surplus wealth—in utilizing monogamy as a means of confirming paternity (Engels, 1942, originally published in 1884). To Engels, then, the family has historically been one means of maintaining male domination of females.

Do conflict theories still endorse this viewpoint? This theme, in fact, is still being advanced by conflict theorists, who point to at least three areas reflecting male dominance through the mechanism of the family: the traditional ownership of women by men, family rules of power and inheritance, and the male-dominated economic division of labor.

Women have traditionally been the property of men, first their fathers, then their husbands. It was only in 1920 that American women were permitted to vote in federal elections (after a tremendous struggle) and only lately that they have gained the right to make contracts and obtain credit independently of their husbands or fathers. If the traditional marriage ceremony is followed, women still must promise to honor and obey their husbands and have their fathers "give them away." The connotation of women as the property of men continues to survive, despite the trend toward equality between the spouses in advanced industrial societies (Collins, 1971).

As stated earlier in the chapter, most family structures throughout history have been patriarchal and patrilineal. That is, the control of family members and property has typically been passed through the males' bloodline. This has created built-in sexual inequality in most family systems; male dominance is accepted as "natural" and "legitimate." Males are dominant and in control, whereas females are expected to be their submissive helpers.

The predominant patrilineal and patriarchal family patterns are, of course, related to the traditional economic division of labor, which also perpetuates sexual inequality. In the traditional division of labor, males work outside the home for pay to support the family, whereas women remain at home to prepare meals, keep house, and care for the children. Thus, women are unpaid laborers who make it possible for men to earn wages. This economic power on the market is, in turn, used by males to maintain their dominance over the dependent, economically weak wife and mother.

The traditional patterns just described have been altered as industrialization has proceeded. Despite some changes, however, the insights of the conflict perspective still apply. The persistent social, economic, occupational, political, and legal inequality of women

continues to provide evidence for the conflict interpretation of the family, as we saw in Chapter 10 ("Inequalities of Gender").

Symbolic Interactionism

What is the relevance of symbolic interactionism to the study of the family? Symbolic interactionism is frequently used in the study of marriage and the family (Waller and Hill, 1951; Leslie and Korman, 1989). According to this perspective, the key to understanding behavior within the family is found in the interaction among family members and the meanings that members assign to the countless episodes of interaction that family life comprises. Within the family, all the major concepts of symbolic interactionism can be observed—socialization, looking-glass self, role taking, primary group, reference group, significant others, symbolic interpretation of events, symbolic communication.

Socialization begins within the primary group of the family. The family normally is characterized by intimate and personal relationships; family members are supposed to care about each other as total human beings. It is within this environment that infants interact with others to learn beliefs, norms, values, and symbols. As shared meanings are developed, children gradually develop self-concepts and learn to put themselves mentally in the place of others. As shown in Chapter 4, interaction with adults permits children to acquire the personality and social characteristics associated with human beings. With the repertoire of personality and social capabilities learned within the family, children are able to develop further as they interact with people outside the home.

According to symbolic interactionists, relationships within the family are constantly undergoing redefinition; consequently, family relationships are usually changing. A newly married couple will spend many months (perhaps years) testing their new relationship and roles. As time passes, the initial relationship will change, along with some aspects of the partners' personalities, including their self-concepts. These changes occur as the couple grapples with such issues as power struggles, personality clashes, sexual preferences, and in-law conflicts. With the arrival of children comes an entirely new set of adjustments. Parental views may differ as to child-rearing practices, number of children desired, and education. The possibilities for personality changes and redefinition of the situation are endless. The situation is made even more complex by the new member(s) of the family who must also become part of the interaction patterns.

This young mother saying good-bye before going to work represents a significant departure from the past. According to conflict theory, male dominance has been maintained in good part by controlling the economic division of labor. As industrialization advances, however, women increasingly work outside the home, and male dominance is challenged.

FAMILY AND MARRIAGE IN THE UNITED STATES

The Nature of the American Family

It would be easy to claim that the typical American family cannot be described. After all, the United States is a large, diverse society in which various groups—reflecting vastly different cultural heritages—have blended together. The early white settlers came primarily from Holland and England. Not long after they arrived, many blacks were brought in as slaves. Subsequently came waves of immigrants from Northern Europe, Ireland, Italy, the Slavic and Baltic countries, China, Japan, Southeast Asia, and, more recently, countries in Latin America. With such diversity, it might be more appropriate to ask whether there is really a typical American family type or rather many American family types.

The answer is that there are more similarities among American families than there are differences. As the various ethnic groups have been assimilated into the mainstream of American life, they have followed the traditional American family pattern. The typical American family contains only the parents and children in the same household (nuclear), traces lineage and passes inheritance equally through both parents (bilateral), divides family decision making equally (democratic), entails the establishment of a residence separate from other relatives (neolocal), and involves the marriage of one woman to one man at any particular time (monogamous).

A large family was once the norm in American society. In 1971, almost three-fourths of American adults thought that three or more children would be the most desirable number (Roper, 1985). Currently, the average number of children born to American women is about two (*World Population Data Sheet*, 1997).

Why have large families declined in popularity? Many factors contribute to the decline in the number of children desired by Americans. Both men and women are marrying later and staying in school longer. Increasing numbers of women are in the labor force for reasons of

both economic benefit and personal fulfillment (McLaughlin and Associates, 1988). Legalized contraception and abortion have also reduced family size. Two additional factors promoting this lower fertility rate are the high cost of supporting a child to adulthood and the removal of the stigma on childless marriages. (More is said about childless marriages later in the chapter.)

Only slightly more than one-fourth of American households consist of a married couple with children under age 18; clearly, Americans are increasingly choosing alternatives to the traditional family pattern of mother (as homemaker), father (as breadwinner), and children. (See Figure 11.2.) These new family and marriage arrangements—including single-parent families, child-free marriages, dual-employed marriages, single life, cohabitation, and gay and lesbian couples—are discussed later in the chapter.

Love as the Basis for Marriage

Love is one of the most widely abused and ill-defined words in the English language. It is used to describe feelings of affection for dogs, cats, horses, homes, motorcycles, cars, parents, children, wives, and mistresses. The word *love* can be molded, modified, and stretched to mean just about anything we want. Yet, when someone says, "I'm in love!" practically everyone else is able to empathize and say, "Yes, isn't it great!"

Love is widely considered a prerequisite of marriage in the United States; most people express feelings of love for the person they are about to marry. In a recent poll of the American public, 83 percent of both men and women rated "being in love" as the most important reason to marry. In fact, love is such an important

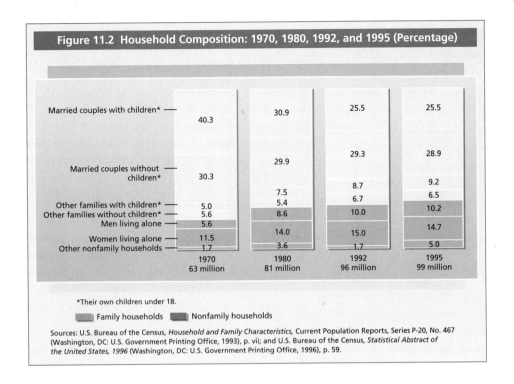

Figure 11.2 Household Composition: 1970, 1980, 1992, and 1995 (Percentage)

	1970	1980	1992	1995
Married couples with children*	40.3	30.9	25.5	25.5
Married couples without children*	30.3	29.9	29.3	28.9
Other families with children*	5.0	7.5	8.7	9.2
Other families without children*	5.6	5.4	6.7	6.5
Men living alone	5.6	8.6	10.0	10.2
Women living alone	11.5	14.0	15.0	14.7
Other nonfamily households	1.7	3.6	1.7	5.0
	63 million	81 million	96 million	99 million

*Their own children under 18.

■ Family households   ■ Nonfamily households

Sources: U.S. Bureau of the Census, *Household and Family Characteristics,* Current Population Reports, Series P-20, No. 467 (Washington, DC: U.S. Government Printing Office, 1993), p. vii; and U.S. Bureau of the Census, *Statistical Abstract of the United States, 1996* (Washington, DC: U.S. Government Printing Office, 1996), p. 59.

ingredient in courtship and mate selection that sexual intercourse is referred to as "making love."

Love has not always been this important. In fact, for most of human history, a marriage was arranged by the families of the bride and groom. The primary basis for marriage was economic. The bride's father would indicate the contents of his daughter's dowry—blankets, rugs, clothes, jewelry, animals, land, money. The groom's father would offer the bride's father a certain number of animals or amount of money—known as the bride price—to compensate for the loss of a hard worker. After the negotiations, a date for the marriage was set. Love between couples could develop after marriage, but it was not part of the mate-selection process. In fact, using love as a crucial factor in choosing a mate did not become common practice until the twentieth century. Moreover, arranged marriage is still the norm in many parts of the world, such as India, China, and Africa.

Are modern marriages based on love alone?   Love is almost always stated as *the* basis for marriage in modern societies, but it is seldom the only one. People marry for many reasons, and love may or may not be one of them. They marry to obtain regular sex partners, to legitimate living together, to enter rich and powerful families, or to advance their careers. But surely one of the strongest motivations for marriage is conformity. Throughout our formative years we are told what to do: "Feed yourself," "Join the Girl Scouts," "Learn to drive," "Go to college," "Get a job." Similarly, parents expect their children to marry after a certain age and worry about them—perhaps even pressure them—if they

remain single too long. Peers are another source of pressure to marry. Well over 90 percent of all adults in the United States marry, so conformity must certainly be a strong motivating factor. In fact, most Americans are married by their mid-twenties.

Divorce

What is the current divorce situation?   The **divorce rate** is reached by determining the number of divorces in a particular year for every 1,000 members of the population. Except for a peak and decline after World War II, the divorce rate in the United States increased only gradually and steadily between 1860 and the early 1960s. The dramatic increase in the divorce rate occurred over the next twenty years, during which time the divorce rate more than doubled (from 2.2 in 1960 to 5.3 in 1981). Since then, as indicated in the *Sociological Imagination* at the beginning of this chapter, the rate has leveled off, and it even declined slightly since 1985. The divorce rate, which peaked at 5.3 in 1981, stood at 4.6 in 1995. (See Figure 11.3.)

In 1940, the marriage rate was 12.1 per 1,000 people, whereas the divorce rate was only 2.0 per 1,000 persons. Now the rate of divorce (4.6) is half the rate of marriage. (See Figure 11.3.) Based on these data, it is correct to say only that the divorce rate of the United States is now half its marriage rate. It is wrong to conclude that half of all marriages end in divorce, because the divorce and marriage rates are snapshots of a single year and do not take into account the total number of divorces and intact marriages from all years past.

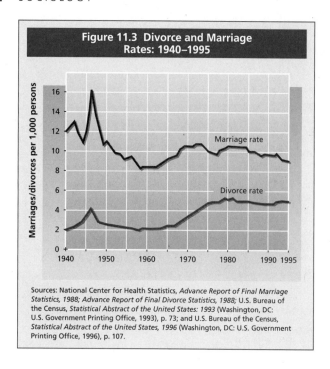

Figure 11.3 Divorce and Marriage Rates: 1940–1995

Sources: National Center for Health Statistics, *Advance Report of Final Marriage Statistics, 1988; Advance Report of Final Divorce Statistics, 1988;* U.S. Bureau of the Census, *Statistical Abstract of the United States: 1993* (Washington, DC: U.S. Government Printing Office, 1993), p. 73; and U.S. Bureau of the Census, *Statistical Abstract of the United States, 1996* (Washington, DC: U.S. Government Printing Office, 1996), p. 107.

This is why sociologists also employ the **divorce ratio**—the number of divorced persons per 1,000 persons in the population divided by the number of persons who are married and living with their spouses. In the United States in 1995, there were 17,653,000 divorced persons and 116,734,000 persons married and living with their spouses. Dividing the number of divorced persons by the number of married persons yields a divorce ratio of 151. This divorce ratio of 151 in 1995 is more than four times the 1960 ratio of 35. Women generally have a higher divorce ratio than men (174 versus 128 in 1995), reflecting the greater tendency of men to get remarried after divorce and to do so more quickly than women (U.S. Bureau of the Census, 1995b).

These figures provide some basis for optimism for the future of the American family, just as the divorce rates of recent years do. The divorce ratio in the United States increased by an annual average of 1.2 between 1960 and 1970 (from 35 in 1960 to 47 in 1970). During the 1970s the average annual increase rose sharply to 5.2 (from 47 in 1970 to 100 in 1980). Since 1980, the average annual increase has actually declined, to an average annual increase of 3.4 (from 100 in 1980 to 151 in 1995). In other words, although the divorce ratio is still rising each year, its rate of increase has declined since 1980 (U.S. Bureau of the Census, 1995b).

Is the rising divorce rate unique to the United States? Rates of divorce are also increasing in other industrialized countries. The divorce rate in Canada and the United Kingdom has risen more than fivefold since 1960, almost tripled in France, and doubled in Germany and Sweden. The divorce rate has gone up in Japan, but it has not been a significant increase. The doubling of the divorce rate in the United States since 1960 would be less dramatic if its divorce rate had not been significantly higher than any other industrialized society at that time (Ahlburg and De Vita, 1992).

What are the causes of divorce? There are no easy answers to this question; like the family and marriage, divorce is a complex matter. Those who study marriage and divorce have sought answers on both individual and societal levels (Knox and Schacht, 1997).

At the individual level, the following factors seem to be associated with divorce. First, the earlier one marries, the greater the likelihood of divorce. Second, the longer a couple has been married, the lower the probability that their marriage will end in divorce. The average length of first marriages ending in divorce in the United States is about six years. However, the largest number of first marriages ending in divorce occur between the second and third anniversaries. This suggests that poor decisions in mate selection may be as important as marital conflict in the termination of marriages. Third, as might be expected, divorce is related to the nature and quality of the marital relationship.

At the societal level, divorce rates seem to increase during times of economic prosperity and to decrease during times of economic recession or depression. Economic prosperity allows people to concentrate on issues beyond survival and to consider options for their personal happiness other than marriage. Second, the recent rise in the divorce rate reflects the passage of the baby-boom generation into the marriageable ages. Third, the increased participation of women in the labor force means increased economic independence, so that women are probably less hesitant about dissolving a bad marriage than they were in the past. Finally, values and attitudes about marriage and divorce are changing; the stigma of failure once associated with divorce, for example, is much weaker today.

What will the future divorce rate be like? Divorce is not a phenomenon that will disappear, but there are reasons that the divorce rate has declined slightly in recent years. First, we know that the later people marry, the less likely they are to divorce. (This happens in part because more mature individuals have more realistic expectations about their relationships and because they have fewer economic and career problems that create emotional strain.) The average age at first marriage is, in fact, increasing in the United States. In 1970, the average age at first marriage was 23.2 for men and 20.8 for women. By 1995, the average age had increased to 26.9 for men and 24.5 for women. (See Figure 11.4.) This trend is likely to continue into the twenty-first century. Second, the average age of the population of the United

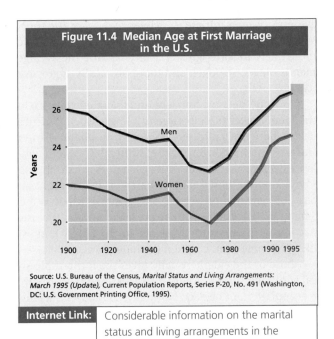

Figure 11.4 Median Age at First Marriage in the U.S.

Source: U.S. Bureau of the Census, *Marital Status and Living Arrangements: March 1995 (Update)*, Current Population Reports, Series P-20, No. 491 (Washington, DC: U.S. Government Printing Office, 1995).

Internet Link: Considerable information on the marital status and living arrangements in the United States can be accessed from the website of the U.S. Bureau of the Census at: http://www.census.gov/population/socdemo/ms-la/95his06.txt

Although the family does, as the functionalists contend, provide a safe and warm emotional haven, it can also be a hostile environment. Assault and murder are more likely to be the acts of someone living with you than someone outside the home. Over one-fifth of all cases of aggravated assault reported to the police involve domestic violence. The situation is actually much worse because episodes of domestic violence go largely unreported. Perhaps as little as 10 percent of such cases reach the authorities (Gelles and Loseke, 1993; Hampton et al., 1993; Gelles and Cornell, 1997).

Violence may rival love as the dominant theme of American family life. Domestic violence is not reserved for any particular members of the family; it involves children, spouses, siblings, and the elderly (Barnett, Miller-Perrin, and Perrin, 1997).

Parental abuse of children may begin in the womb. Approximately 40 percent of the assaults of wives by their husbands take place during pregnancy; birth defects and miscarriage are common results. In fact, such abuse may lead to more birth defects than all other childhood diseases combined (Gibbs, 1993b). Nor do all children grow up in a loving home, nurtured by parents who care. According to a national survey, almost one-quarter of adults in the United States reported being physically abused as children (Patterson

States is increasing as the baby boomers born after World War II grow older. It was this exceptionally large generation that set records for divorce in the late 1960s and 1970s. Baby boomers now range in age from the early to late 50s, which removes them from the age bracket that produces the highest divorce rates. Third, American couples are having fewer children spaced farther apart. This reduces the pressure on marriages.

The recent decline in the divorce rate is encouraging. At the very least, it is the first break in a long-term increase. Still, many experts are cautious about predicting the future, pointing to the fact that the divorce rate goes in cycles and that the recent decline may be only a temporary respite. Whatever happens to the divorce rate, Americans are currently participating in alternatives to the traditional nuclear family structure. More will be said about this in the next major section.

Family Violence

The 1990s brought heightened public attention to the problem of domestic violence (Ingrassia and Beck, 1994; Smolowe, 1994). Extensive media coverage made every effort to give Americans the details, some untrue, of several dramatic legal cases. Erik and Lyle Menendez admitted planning and carrying out the murder of parents they accused of abusing them. Lorena Bobbitt severed her husband's penis, while he slept, on the basis of past abuse, humiliation, and forced sex (Kaplan, 1994).

Although the family can provide a safe and warm emotional haven, it does not always do so. This woman has been the victim of a hostile family environment, as the expression on her face and the bruises on her face attest.

and Kim, 1991). Each year, then, thousands of children are mistreated in some way. Some suffer deliberate and willful injury; others are passively neglected. Neglect is thought to occur at double the rate of physical abuse, involving as many as nine million children each year. In most cases, physical violence involves a slap, a shove, or a severe spanking. However, kicking, biting, punching, beating, and threatening with a weapon are part of abusive violence on the part of many parents.

Spouses also use violence against each other. As many as one-half of all married women in the United States will be victims of spousal violence (Gibbs, 1993b). As many as four million women are battered by their spouses annually, no doubt an underestimation. More than four thousand women each year are killed through beating. Rape and sexual assault are not uncommon. As many as 14 percent of married women may be sexually attacked by their spouses each year. The extent of physical abuse may be underestimated, in part because three-fourths of spousal violence occurs during separation or after divorce; and most research is conducted among married couples (Knox and Schacht, 1997). Domestic violence, of course, is not limited to physical abuse. Verbal and psychological abuse appears to be a pattern in an overwhelming majority of marriages.

Frequently overlooked in the area of family violence are husband abuse, sibling abuse, and abuse of the elderly (Steinmetz, 1988; Wiehe, 1990). Although marital relationships in the United States are generally male dominated, there seems to be spousal equality in the use of physical violence. Straus, Gelles, and Steinmetz (1980) found that almost one-third of the husbands in their survey had acted violently against their wives and that wives were almost as likely to use physical violence against their husbands. According to later studies, husbands and wives seem to assault each other at about the same rate (Hotaling et al., 1988; Flynn, 1990). However, much of the violence on the part of women occurs in self-protection or in retaliation rather than as proactive aggression; men initiate the violence most of the time. Whoever initiates the violence, however, women are more likely to suffer greater injury because the average man is bigger, stronger, and more physically aggressive (Gelles, 1997).

Probably the most frequent and tolerated violence in the family takes place between siblings. Much violence occurs between children too young to channel their frustrations in nonphysical channels. Although violence declines somewhat with age, it does not disappear. Abuse among siblings may be based on rivalry and jealousy, disagreements over personal possessions, or incest. Sibling violence appears to be prevalent and on the rise. According to one study, violence in the family is most likely to come from children (Hotaling et al., 1988).

Abuse of the elderly usually takes the form of physical violence, psychological mistreatment, economic manipulation, or neglect. Normally, the abuser is a family member (Pillemer and Finkelhor, 1988). Less is known about abuse of the elderly because less research has been done. Estimates of elder abuse range from 500,000 to 2.5 million cases annually (Gelles, 1997). There is fear of an increase in abuse of the elderly as the population bulge represented by the baby boomers enters old age. For one thing, there will be fewer working adults to support a growing aging segment of the population (Kart, 1990). The most likely to be abused are white Protestant women over 75 years of age who are mentally, physically, or financially disadvantaged (Garbarino, 1989). Older women represent a higher proportion of the abused, in part, because they live longer than men. There are simply more elderly women with greater dependency needs. Abuse of the elderly is partly due to the psychological, physical, and economic strain of caring for elderly persons. These strains of caretaking may lead to hostility and depression, followed by violence. Elder abuse may involve a dependent adult child who has moved back into his or her parents' home following a divorce or due to financial problems (Gelles, 1997).

FEEDBACK

1. Which half of each of the following pairs best describes the typical American family?
   a. nuclear/extended
   b. patrilineal/bilateral
   c. authoritarian/democratic
   d. neolocal/matrilocal
   e. polygynous/monogamous

2. The use of love as a crucial factor in the mate-selection process did not become common practice until which century?
   a. seventeenth   b. eighteenth   c. nineteenth   d. twentieth

3. The _____ is the number of divorces in a particular year per 1,000 population.

4. Which of the following is *not* stated in the text as one of the factors associated with divorce?
   a. decline of religious influence
   b. age at first marriage
   c. length of marriage
   d. economic conditions

5. Contrary to what one would expect from the traditional male dominance in American family life, women are almost as frequent spousal abusers as men. *T or F?*

Answers: 1. a. nuclear b. bilateral c. democratic d. neolocal e. monogamous 2. d. 3. divorce rate 4. a. 5. T

LIFESTYLE VARIATIONS

American marital and familial arrangements are undergoing dramatic change. Although the future of the American family is difficult to predict, this chapter concludes with some projections. For the present, however, an examination of some current alternatives to the traditional nuclear family structure will provide some hints. Social approval of most of these alternative lifestyles is relatively recent; some are still considered illegitimate by many Americans.

Blended Families

The increase in the divorce rate has led to the creation of the **blended family**—a family formed by at least one remarried man or woman with a minimum of one child from a previous marriage. This type of family can become extremely complicated (Ganong and Coleman, 1994). Consider the number and complexity of relationships in the following blended family:

Former husband (with two children in the custody of their natural mother) marries new wife with two children in her custody. They have two children. Former wife also remarries man with two children, one in his custody and one in the custody of his former wife, who has also remarried and had a child with her second husband who also has custody of one child from his previous marriage. The former husband's parents are also divorced and both have remarried. Thus, when he remarries, his children have two complete sets of grandparents on his side, plus one set on the mother's side, plus perhaps two sets on the stepfather's side (Cox, 1993:648–649).

Part of this complexity is due to the fact that more marriages are ending through divorce rather than death. For children, a parent is being added, not replaced. Along with the third parent come his or her relatives. Blended families, then, create a new type of extended family, a family that is not based strictly on blood. As the example above shows, it is possible for a child in a blended family to have eight grandparents if each of their biological parents remarry. Although not all blended families will be this complicated, it is significant that about 39 percent of households in the United States contain biologically unrelated individuals.

Single-Parent Families

How widespread are single-parent families? Single-parent families are on the rise. Between 1982 and 1995, the percentage of unmarried American women with at least one child more than doubled, from 15 percent to nearly 37 percent (U.S. Bureau of the Census, 1995b). Approximately 27 percent of America's children under

eighteen years of age live with only one parent. One-parent families formed by never-married women are more prevalent among African Americans and Latinos than whites. In 1994, about 60 percent of all never-married African American women, 18 to 44, were single parents, compared with 31 percent among Latinos, and 21 percent among whites. While the proportion of one-parent families formed by never-married women has increased for all races since 1983, the gap between them has increased significantly. (See Figure 11.5.)

Over 84 percent of the heads of one-parent families are women. A father is the single parent in about 16 percent of the families with children under 18. Thus, the female single-parent household is a prominent type of family situation. This arrangement, accounting for nearly one-fourth of all American households with minor children, doubles the number in 1970 and triples the number in 1960. African American and Latino children are more likely than white children to live alone with their mothers because of the high divorce and illegitimacy rates combined with lower rates of marriage and remarriage (U.S. Bureau of the Census, 1995b). The current situation means that an increasing proportion of American children will live at least part of their childhood in a fatherless home.

What are the effects of single-parent families? Females heading single-parent households face considerable time and money problems (Kissman and Allen, 1993). Single parents have less time for parenting, although research is inconsistent on the effects of this fact. One study concluded that children in female-headed families are as socially, intellectually, and psychologically well adjusted as children from two-parent

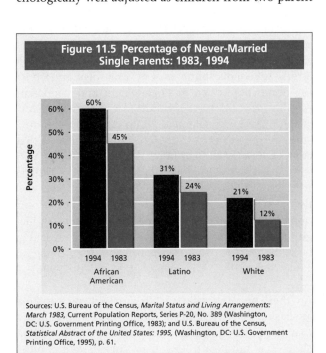

Figure 11.5 Percentage of Never-Married Single Parents: 1983, 1994

Sources: U.S. Bureau of the Census, *Marital Status and Living Arrangements: March 1983*, Current Population Reports, Series P-20, No. 389 (Washington, DC: U.S. Government Printing Office, 1983); and U.S. Bureau of the Census, *Statistical Abstract of the United States: 1995*, (Washington, DC: U.S. Government Printing Office, 1995), p. 61.

Divorce and Single-Parent Families

The high divorce rate in America today has led to an increase in the number of single-parent families. As this CNN report suggests, children usually remain with their mothers. In fact, almost 90 percent of these families are headed by women. Female-headed families face some unique problems, including difficulties related to a shortage of time and money.

1. Identify what you think are the most important problems faced by single-parent families._____

   _____

2. If you headed a single-parent family, how would you attempt to cope with these problems? _____

   _____

Children from single-parent homes often exhibit achievement and behavioral problems. According to some researchers, these problems may be due to expereiences before separation or divorce rather than to living with only one parent.

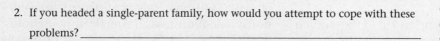

families (Cashion, 1982). In contrast, most of the research indicates negative effects on children. Sheppard Kellam, Margaret Ensminger, and R. Jay Turner (1977) found that children from one-parent families consistently score lower on tests of psychological well-being and are less able to adapt in social settings such as school than their counterparts from two-parent or two-adult families. They concluded that the problem lies not so much in the absence of the father, but in the aloneness of the mother. Children living in two-adult families fared almost as well as children from two-parent families. According to Sanford Dornbush and his colleagues (1985), adolescents who live with one parent or with a stepparent have much higher rates of deviant behavior than adolescents living with both natural parents. This study from a national sample of twelve- to seventeen-year-olds indicates that arrests, school discipline, truancy, runaways, and smoking occur more often in single-parent and stepparent families, regardless of income, race, or ethnic background. Daniel Mueller and Philip Cooper (1986), in a study of nineteen- to thirty-four-year-olds, found that compared to individuals from two-parent families, persons raised by single parents (especially single mothers) tend to have lower educational, occupational, and economic success. Adults who grew up in single-parent homes are also more likely to be divorced, have an illegitimate child, be receiving psychological counseling, be school truants, and use alcohol and drugs (Bianchi, 1990; Taylor, 1991; McLanahan and Sandefur, 1996).

Are these negative effects really due to divorce and living in single-parent homes? Some researchers contend that the achievement and behavioral problems noted above may not be attributable to single-parent families

alone. Such factors as poor parent-child relationships, failure to communicate, lack of economic resources, and family disharmony may be more important than the structure of the family (Demo and Acock, 1991). In longitudinal studies of children, one comparing children in the United States and Great Britain, some researchers conclude that the adjustment problems we associate with divorce and single-parent families may actually be reflections of conditions existing prior to separation or divorce (Block, Block, and Gjerde, 1986; Cherlin et al., 1991; Furstenberg and Cherlin, 1991).

This line of research will be carefully followed by other sociologists. Only later will we have surer knowledge of the effects of single-parent families in general and the causal relationship between divorce and the adjustment of children to divorce in particular.

On one point all researchers agree. Female single parents and their children suffer from insufficient economic resources. Although single mothers comprise only about one-fifth of all family households, they account for about half of households with minor children with income below the poverty line. Because only about 16 percent of the single-parent families headed by men are below the poverty line, experts are now writing about the feminization of poverty. (See Chapter 8, "Social Stratification" and Chapter 10, "Inequalities of Gender.") Earning a living is a problem for these single-parent mothers, both because of the difficulty of arranging for child care (which all working mothers must face) and because these women do not have the educational credentials to do well in the job market. Many were housewives or part-time workers before a divorce; others are young, uneducated women who never married. Thus, if they do not marry, these mothers are unlikely to work their way out of poverty.

Their children, of course, suffer the consequences along with them.

Child-Free Marriages

In part because of the assumption of the existence of a maternal instinct, there has been a stigma attached to marriages without children. Married women without children were considered to be deviants who were not fulfilling their biological destiny. Therefore, something must be wrong with them. As this traditional negative societal response to marriages without children diminishes, more American couples are choosing not to have children (Gallup and Newport, 1990).

To what extent are married couples choosing to be child-free? Around 20 percent of ever-married American women are child-free. This is up from about 15 percent in 1970 (U.S. Bureau of the Census, 1993f). Experts predict that married couples choosing never to have children is likely to increase (Benokraitis, 1996).

Why are married women choosing to be child-free? The reasons married women give for choosing not to have children are diverse. Some are so involved in their careers that children are seen either as an impediment or as too important to short-change if brought into the world. Others want to avoid children to keep their independence. According to still others, today's world is no place into which to bring a child. Finally, some women simply do not like children very much.

Aren't child-free couples missing an important part of life? Among child-free couples who want children, marital happiness may be lower than for married couples with children (Singh and Williams, 1981). However, this is not the case among women who prefer not to have children. Voluntarily child-free couples appear to be happier and more satisfied with their marriages and lives than couples with children (Renne, 1970, 1976; Polonko, Scanzoni, and Teachman, 1982; Callan, 1983).

Dual-Employed Marriages

In a dual-employed marriage, both husband and wife are in the labor market. This is quite different from the marriage in which the wife works at a job merely to supplement the family income, a pattern long observed in the lower and working classes. Wives in a dual-employed marriage are employed as much for the personal meaning it provides them as for the money the family derives (Bielby and Bielby, 1989). Although a relatively new trend, the dual-employed marriage is now the norm, as the discussion in Chapter 10 ("Inequalities of Gender") of the increased labor-force participation of American women clearly suggests.

Despite the happy expressions on the faces of this dual-employed couple, they face special strains. If the husband does not assume a significant role in household duties, the burden falls especially hard on his spouse.

Do dual-employed marriages create special strains? Women in dual-employed marriages are apparently expected to handle the bulk of the household and child care responsibilities in addition to the pursuit of their full-time jobs (Goldscheider and Waite, 1991; Thompson and Walker, 1991). Combining employment with child-care and domestic tasks, married working women work about fifteen hours more a week than men. This additional month of twenty-four-hour days a year Arlie Hochschild (1997) calls the *second shift*. Although men spend an average of between four and six hours per week in household and child-care duties, women clearly bear the larger burden (Berardo, Shehan, and Leslie, 1987).

In addition to this greater workload, women in dual employed marriages often must cope with conflicting role demands. They are torn between the time requirements of their jobs and their desire to spend more time with their children and husbands (Rubenstein, 1991).

Despite their general unwillingness to assume domestic responsibilities equal to their wives, men in dual-employed marriages also may feel the negative effects of conflicting roles and excessive demands on their time (Moen, 1992). In addition, having an employed wife, particularly if she earns more, may make a man feel like a failure as a provider (Menaghan and Parcel, 1991; Spitze, 1991).

The special strains experienced by dual-employed couples will probably not greatly decrease the number of people adopting this alternative. A national survey of 1991 high school seniors found that only 2 percent of the females questioned indicated they expected to be homemakers at age thirty, down 10 percentage points since 1976 (Morin, 1991).

The challenge in the future will be to find ways of reducing the strains placed on members of dual-employed families (Gilbert, 1993). Some pressure on working women could be relieved if men became more participative in domestic duties. Many experts believe that the combined pressures of job and family will lead to accommodative changes in the workplace. Alterations in company policies and practices will likely involve more flexible working schedules, family leaves for both males and females, work-at-home situations, and on-site child-care centers (Deutschman, 1991; Labich, 1991).

Some evidence suggests that men are assuming a more active role in child care. By 1991, 20 percent of all children under age five were being cared for by their fathers, up from 15 percent only three years earlier (O'Connell, 1993). This may be temporary or the start of a new trend. If it is the latter, it could signal movement toward greater willingness on the part of working fathers to offer more help around the home to their working wives.

How do women and men feel about their dual employment? Thus far, emphasis has been placed on the strains experienced by dual-employed couples. There is a positive side. On balance, the effects of employment on the psychological well-being of women have been beneficial (Moen, 1992; Crosby, 1993). Working outside the home provides a wider set of social relationships and feelings of control, independence, and heightened self-esteem. Employment also appears to provide a socioemotional cushion for women when their children leave home. Compared to women who do not

work outside the home, employed women have more alternative channels for meaningfully expressing themselves (Adelmann et al., 1989).

For men, benefits of a dual-employed marriage include freedom from sole responsibility of providing economic resources and increased opportunity for job changes and continuing their education (Spitze, 1991; Crosby, 1993). Men with employed wives can share the triumphs and defeats of the day with someone who understands. If their wives are happier working outside the home, husbands also enjoy a better marital relationship. Those husbands who take advantage of the opportunity can form a closer relationship with their children by being more active parents.

Whatever happens between partners, there seems to be a role for government and business in relieving the stress among dual-employed couples (Kamarck, 1992; *Starting Points,* 1994). As Table 11.2 shows, other industrialized countries have been more progressive in attempting to relieve strains associated with dual-employed arrangements by requiring employers to take certain steps. Sweden has been the most progressive nation, requiring employers to provide thirty-eight weeks of work furlough at 90 percent of pay following the birth of a child for either the father or mother and up to one year at a further reduced compensation rate.

Until the Family and Medical Leave Act of 1993 was passed under the Clinton administration, no employers have been required by federal law to offer any form of leave for parents having a child or facing a medical emergency. This act requires that employers with fifty or more employees provide up to twelve weeks of unpaid leave for either the father or mother of a newborn child after birth, the adoption of a child, the placement of a foster child, or for some other family member's medical emergency. A serious health condition is broadly defined to include illness, injury, and mental problems. The employer must permit the employee to return to the same or an equivalent job and continue group

TABLE 11.2	Parental and Maternity Leave Policies, United States vs. Selected Countries		
Country	Leave Duration (Weeks)	Percentage of Pay/Weeks	Recipient
Canada	17–41	60%/15 weeks	Mother
Italy	22–48	80%/22 weeks	Mother
West Germany	52	100%/14–18 weeks	Mother or father
Sweden	12–52	90%/38 weeks	Mother or father
Austria	16–52	100%/20 weeks	Mother
Chile	18	100%/18 weeks	Mother
United States	12	0	Mother or father

Source: John J. Sweeney and Karen Nussbaum. *Solutions for the New Work Force.* Cabin John, MD: Seven Locks Press, 1989, p. 108. Reprinted by permission.

health insurance for the employee during the leave period (Snarr, 1993; Stranger et al., 1993).

There are several shortcomings to this step, which is relatively modest in comparison to government policies in other industrialized societies. First, because most American workers are employed in organizations with fewer than fifty employees, this benefit is unavailable to them. Second, less than 40 percent of dual-employed couples can afford to take an unpaid six-week leave (Reskin and Padavic, 1994). Third, employees must have been employed by their company for at least twelve months to be eligible (Shaller and Qualiana, 1993).

Single Life

To what extent are Americans remaining single? The increased age at first marriage for both sexes and the high divorce rate have combined to create an increase in the percentage of single adults. In 1970, about 11 percent of women and 19 percent of men between the ages of 25 and 29 had never been married. By 1995, these percentages had increased to just over 35 percent for women, an increase of 24 percentage points, and 51 percent for men, an increase of about 32 percentage points. (See Figure 11.6.) Over twenty-four million Americans over the age of fifteen now live alone, a 130 percent increase since 1970. Although many of these people will eventually marry, an increasing percentage will remain single all their lives (U.S. Bureau of the Census, 1996a).

Why are more Americans choosing to live alone? Remaining single has always been an alternative, but it has also carried a stigma in the United States. During colonial days, bachelors were taxed more heavily than married men; spinsters were viewed as millstones around the necks of their families. Adults were expected to marry and to begin having children as soon as morally and physically possible. Failure to marry was evidence of inadequacy and deviance.

The stigma attached to remaining single has faded since colonial times, especially in the past two decades, and more Americans are either choosing this alternative or at least marrying later than previous generations. There are, however, other factors contributing to the popularity of singlehood in addition to the lifting of the stigma. More Americans are choosing to forgo having children, to obtain sexual gratification outside of marriage, to pursue careers, to have strictly homosexual relationships, to rear illegitimate children, to rear children from a former marriage alone, and to adopt children.

Will the current trend toward remaining single continue? It is too early to predict confidently whether the increase in singlehood among the young will eventually lead to a decline in marriage at all ages. But age groups that postpone having children for an extended period of time usually do not make up for it later. If this is true for singles, the proportion of unmarried Americans will increase as the younger generation ages. At any rate, it is surely safe to say that singlehood

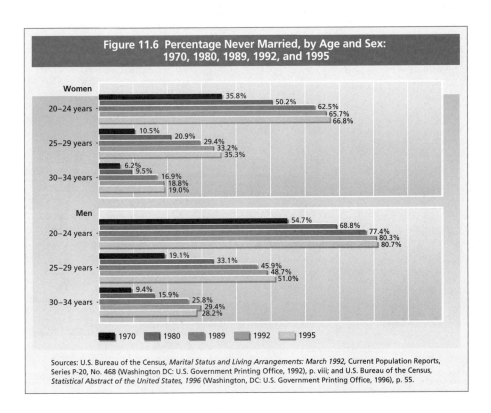

Figure 11.6 Percentage Never Married, by Age and Sex: 1970, 1980, 1989, 1992, and 1995

Women

20–24 years
- 35.8%
- 50.2%
- 62.5%
- 65.7%
- 66.8%

25–29 years
- 10.5%
- 20.9%
- 29.4%
- 33.2%
- 35.3%

30–34 years
- 6.2%
- 9.5%
- 16.9%
- 18.8%
- 19.0%

Men

20–24 years
- 54.7%
- 68.8%
- 77.4%
- 80.3%
- 80.7%

25–29 years
- 19.1%
- 33.1%
- 45.9%
- 48.7%
- 51.0%

30–34 years
- 9.4%
- 15.9%
- 25.8%
- 29.4%
- 28.2%

■ 1970 ■ 1980 ■ 1989 ■ 1992 ■ 1995

Sources: U.S. Bureau of the Census, *Marital Status and Living Arrangements: March 1992*, Current Population Reports, Series P-20, No. 468 (Washington DC: U.S. Government Printing Office, 1992), p. viii; and U.S. Bureau of the Census, *Statistical Abstract of the United States, 1996* (Washington, DC: U.S. Government Printing Office, 1996), p. 55.

is an increasingly popular alternative to traditional marriage. This does not necessarily mean a rejection of marriage, but it does imply a desire to expand the period of "freedom" after leaving home and an unwillingness to rush into the responsibilities of early marriage and parenthood.

Cohabitation

What is cohabitation and how widespread is it? **Cohabitation**—living with someone in a marriagelike arrangement without the legal obligations and responsibilities of formal marriage—has been a widely discussed alternative to traditional monogamy for some time (Clayton and Voss, 1977). It is still not known if cohabitation has increased because more Americans are delaying marriage or because cohabitation is being substituted for marriage.

It is presently impossible to predict accurately whether cohabitation will ever become a commonly accepted state in the mate-selection process, something experienced by virtually everyone. Certainly, as the average age at first marriage goes up, this is a possible outcome. Given the high divorce rate, it may be that cohabitation will become a more common step in selecting another marital partner. In fact, the number of American adults cohabiting increased over 400 percent between 1970 and 1992, from 523,000 to over three million. (See Figure 11.7.) According to a nationwide survey, 28 percent of adults in the United States have cohabited (Thompson and Colella, 1992). One-third of American women age fifteen to forty-four report that they have cohabited at some time in their lives. For women age twenty-five to twenty-nine, the figure is 45 percent (Vobejda, 1991).

Cohabitation has risen among people of all ages and marital statuses but particularly among the young and the divorced. By 1995, about 80 percent of all unmarried-couple households were maintained by someone under 45 years of age and about one-third involved at least one child under fifteen (U.S. Bureau of the Census, 1996a).

Gay and Lesbian Couples

How prevalent is homosexual cohabitation? Because of the stigma that still surrounds homosexuality, it is impossible to know what proportion of the American population is homosexual. The Institute of Sex Research, founded by Alfred Kinsey, estimates that homosexuals constitute 10 percent of the U.S. population (13 percent of the males, 5 percent of the females). Although estimating the number of cohabiting homosexual couples is hazardous, it is known that it is increasing, both on college campuses and in the general public. The number of homosexuals cohabiting is increasing despite the fact that homosexual "marriages" are not recognized by law in any state in the union (Curley, 1996).

Homosexual families—homosexuals living together with children—are also increasing in number, although their number is small compared with heterosexual marriages. Homosexual families are created in several ways.

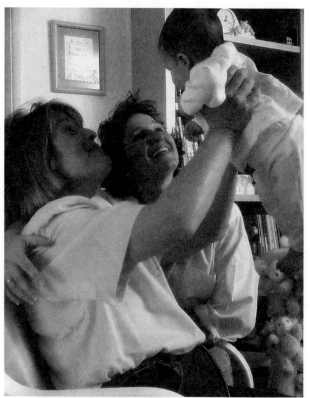

The number of homosexual families is small compared with heterosexual marriages in the United States, but it is on the rise. Whether by divorce, adoption, or artificial insemination, this lesbian couple has formed a family.

Figure 11.7 Cohabitation in the U.S.

Millions

1970	0.5
1992	3.3
1995	3.6

■ Without children ■ With children

Source: U.S. Bureau of the Census, *Statistical Abstract of the United States: 1996* (Washington, DC: U.S. Government Printing Office, 1996), p. 56.

A divorced person may take his or her children into a new homosexual relationship. In fact, about 15 percent of lesbian mothers are currently given custody of their children, and homosexual fathers are beginning to seek custody as well. Homosexual couples are also adopting children, a movement encouraged by a 1979 New York case in which the court gave permanent custody of a thirteen-year-old boy to an openly homosexual minister who had been caring for the boy (Maddox, 1982). Even before this case, welfare agencies in large cities were placing orphaned children with homosexual couples. Although it is as yet uncommon, lesbian couples can purchase frozen semen from sperm banks and have their families by artificial insemination, and homosexual male couples can hire surrogate mothers who are inseminated with the couple's mixed semen.

Adult Children Returning Home

In nineteenth-century America, the transition from childhood to adulthood (for males) varied in length and was very loose in its boundaries. As historian E. Anthony Rotundo writes:

> During this transitional phase, young males lived in settings that ranged from boardinghouses to college dormitories to their own family homes, and they often shuttled back and forth between these settings. A leading historian has aptly called this a period of "semidependence," since a youth's relation to his family was ambiguous and—in some cases—frequently shifting (Rotundo, 1993:56).

For different reasons, young American adults (eighteen to thirty-four year olds) of both genders have a much higher probability of living in their parents' home than they did thirty years ago. For example, the percentage of adults aged eighteen to twenty-four living at home increased from 43 percent in 1960 to 53 percent today. Moreover, 27 percent of adults eighteen to thirty-four years old now live with their parents (U.S. Bureau of the Census, 1996a). The label *boomerang kids* has been applied to these offspring because they are returning to their point of origin (Quinn, 1993a; Goldscheider and Goldscheider, 1994).

Why are more adult children returning home? Several factors combine to produce a higher proportion of young adults living with their parents (Ward, Logan, and Spitze, 1992; Cox, 1993). Because young adults are marrying later, more stay at home longer. More of them are continuing their education and find living at home the best solution to the costs of supporting themselves and paying schooling expenses. Many young adults return home even after completing their education because the higher cost of living outstrips their earning capacity. Parents often accept their children back after a

failed marriage, so the high divorce rate also increases the proportion of young adults living at home.

What are some negative consequences of this boomerang effect? An added financial burden on the parents can create significant strain in homes with boomerang adults. Costs associated with education, day-to-day living, and perhaps even a daughter or son starting a family contribute to the strain. Many parents complain that their adult children do not share in expenses, fail to help around the house, rob them of their privacy, and prevent them from developing relationships with spouses and friends. It is not surprising that higher marital dissatisfaction among middle-aged parents is associated with adult children living at home (Glick and Lin, 1986).

Adult children who find themselves in this situation suffer as well. The label boomerang kid is an uncomplimentary one. Adult children who have returned home normally do so from necessity rather than choice. They are likely to be having difficulties associated with balancing school and work, failing to make their way in the world economically, forming a family, or surviving the aftermath of a divorce. They know the burden they represent. In addition, returning home usually means forfeiting some of one's freedom and being subject to unwanted parental control.

The Sandwich Generation

The current middle-aged generation faces problems associated with new family forms. Due to prolonged life expectancy, refusal to place parents in nursing homes, and fewer siblings to share the burden, more middle-aged adults are finding mothers or fathers (their own or their spouse's) living with them (Knox and Schacht, 1997). *Sandwich generation* is the term applied to those adults caught between caring for their parents and the family they formed after leaving home. Sandwiching, of course, can occur whether or not the elderly parent(s) live with the younger couple.

What are the repercussions? On the positive side, elderly parents unable to take care of themselves usually receive better care from those who love them and feel responsible for their well-being. Older children may enjoy taking care of those who reared them. The caregiving is usually not one-way. Aging parents can offer emotional support and financial resources (Walker, Pratt, and Oppy, 1992).

There are also negative repercussions. Taking care of an elderly parent is often not easy. A parent with severe arthritis or Alzheimer's disease demands close and constant attention. Younger adults in this situation often experience strong feelings. They may resent the social and emotional intrusion in their own family life that aging parents may represent. Guilt feelings may arise because of this resentment and anger. Individuals who feel personal resistance to caring for their parents may suffer a loss of self-esteem because they perceive themselves to be selfish. Parents, too, may feel guilt and anger about the burden they are placing on others they love. They also experience stress and depression from the problems they see their adult children, their spouses, and their grandchildren undergoing (Pillemer and Suiter, 1991).

One additional negative consequence deserves special mention. The burden of caring for an aging parent falls much more heavily on women. According to one study, the primary caregiver is a female in 90 percent of the cases (Dychtwald, 1990). It is typically daughters, rather than sons, who take charge of the care for an aging parent (Brubaker, 1990; Spitze and Logan, 1990). The average woman in the United States is likely to devote more years to caring for her aging parents than she did to her own children (Clabes, 1989). According to one study, over one-fourth of nonworking women were out of the labor force because of their parents' needs (Hull, 1985).

THE FUTURE OF THE FAMILY

If the frequency of marriage and remarriage is any indication, the nuclear family is not disappearing. Over 90 percent of men and women in the United States marry sometime during their lives, and about three-fourths of men and women between the ages of twenty-five and sixty-five are married at any given time. It is estimated that two-thirds of all divorced persons remarry. In fact,

45 percent of all weddings each year are remarriages. Sixty percent of second marriages end in divorce, but three-fourths of people twice divorced marry a third or even a fourth time. This is known as *serial monogamy,* an alternative that is on the rise.

Despite all the current experimentation with alternatives, the nuclear family remains the most popular choice among Americans (see *Doing Research*). In 1995, 54 percent of all households were composed of married couples. Of all families, 78 percent were married couples (4.7 percent were families headed by males and 18 percent were families headed by females) (U.S. Bureau of the Census, 1996a).

The nuclear family, then, is not being abandoned. Contrary to a long-standing fear, most Americans are not avoiding marriage permanently; they are simply postponing it or sampling it often.

This is not to say that change in the American family is not occurring. The alternatives to the traditional model of the family currently being practiced testify to this fact. The proportion of all households with the traditional husband-wage earner, wife-homemaker, and two children is expected to continue to account for only about one-fourth of all American households. Continued increases are expected for the other family lifestyle alternatives described above, including the dual-employed family and single-parent family. (See Figure 11.8.) As Judith Stacey notes, there is no longer "a single culturally dominant family pattern to which the majority of Americans conform and most of the rest aspire" (1990:17). It is not a question of whether the family will survive. It is a question of the various forms the family will take in place of the traditional arrangement of homemaker-wife with a sole breadwinner-husband and dependent children.

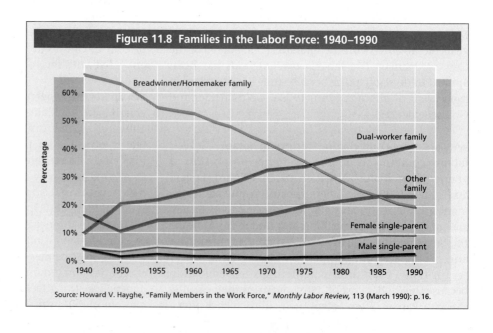

Figure 11.8 Families in the Labor Force: 1940–1990

Source: Howard V. Hayghe, "Family Members in the Work Force," *Monthly Labor Review,* 113 (March 1990): p. 16.

Americans, in fact, are spending less and less of their lives married. The marriage rate (the number of marriages per 1,000 population) in the United States has fallen to 9.1, the lowest rate since the late 1960s. (See Figure 11.3.) Although the average man in the mid-1970s had spent 55 percent of his years married, the percentage has decreased to 50 percent. For women, the figure has fallen from 48 percent to 44 percent (Vobejda, 1991).

Change, however, is not the equivalent of obliteration. Although variations will become more prevalent and socially acceptable, the family system is not about to disappear from American society. In the United States, as well as in other modern societies, the nuclear family is projected to remain the bedrock of the family system, despite the growing diversity in family structure (Kain, 1990; Ahlburg and DeVita, 1992).

The continued existence of the nuclear family does not ensure that all is well with the American family. Stresses associated with dual-employed marriages and family violence, for example, have already been documented. A final word needs to be said about the problems associated with the youngest Americans, those up to three years of age. As you saw in Chapter 4, the early years of child development are crucial. The current nature of American families appears to be placing the well-being of these children in jeopardy (Popenoe, 1996; Popenoe, Elshtain, and Blankenhorn, 1996).

In 1994, the Carnegie Corporation published a report with the expressed purpose of elevating the plight of families rearing children from birth to three years of age to the top of the list of America's challenges (*Starting Points,* 1994). The nation's future, the report states, is at stake. The report focuses on what is possible in attempting to reverse some alarming trends, given the relatively low level of resources and the perhaps questionable national will to mobilize whatever resources are available. This issue has already been addressed to some extent in the earlier section on dual-employed marriages, where you saw the relatively low commitment of American society to measures such as paid parental and maternity leave from work that would be beneficial to parents and their young children. (Refer back to Table 11.2.).

FEEDBACK

1. A family composed of at least one remarried man or woman with a minimum of one child from a previous marriage is a _____ family.
2. Although the proportion of single-parent families has increased for both blacks and whites since 1970, the gap between them has
   a. decreased somewhat.
   c. remained the same.
   b. increased significantly.
   d. decreased significantly.
3. Voluntarily child-free couples are generally
   a. less well educated than couples with children.
   b. less emotional than couples with children.
   c. less satisfied with their marriages and lives than couples with children.
   d. more satisfied with their marriages and lives than couples with children.
4. In a _____ marriage, both husband and wife are in the labor market.
5. In 1995, approximately _____ percent of women and _____ percent of men in America between the ages of twenty-five and twenty-nine had never married.
6. _____ involves living with someone of the opposite sex in a marriagelike arrangement without the legal obligations and responsibilities of formal marriage.
7. The percentage of adults aged eighteen to twenty-four who live with their parents now stands at over _____ percent.
8. Although the nuclear family is not disappearing , Americans are spending less and less of their lives married. T or F?

Answers: 1. blended 2. b 3. d 4. dual-employed 5. 35, 51 6. Cohabitation 7. 50 8. T

SUMMARY

1. A family is a group of people related by marriage, blood, or adoption. Each of us may belong to two families—the family into which we are born and the family we may create upon marrying.

2. The extended family and the nuclear family are two basic types of family structure. A curvilinear relationship exists between these two types of families and industrialization. In hunting and gathering societies, the nuclear family was the most prevalent type. With the rise of agriculture as the means of subsistence, the extended family came to prevail. The nuclear family regained popularity in modern industrial society.

3. Whether the family is nuclear or extended, there are several important dimensions of family structure. They pertain to descent and inheritance, family authority, and residence pattern.

4. Mate selection is never completely a matter of individual choice. To limit the field from which a man or woman can be chosen, exogamous, endogamous, and homogamous mate-selection norms exist in all societies.

5. There are four basic types of marriage: one man and one woman may be married; one man and several women may be married; one woman and several men may be married; or two or more men may be married to two or more women.

6. The family has always been the most important institution in all societies because of the functions it performs. It is the institution that produces new generations, socializes the young, provides care and affection, regulates sexual behavior, transmits social status, and provides an economic function.

7. Although functionalism emphasizes the benefits of the family for society, the conflict perspective depicts the traditional family structure as the instrument of male domination of women. Evidence of this domination is reflected in the traditional ownership of women by men, family rules of power and inheritance, and male-dominated economic division of labor.

8. Symbolic interactionism is widely used in the study of the family. It is within the family that the socialization of children begins and in which children develop self-concepts. Most all of the interactions within families can be analyzed within this theoretical perspective.

9. Modern marriages are mostly based on love between partners. This is a relatively new development in the formation of marriages. For most of history, marriages were made in the realm of economics—not heaven.

10. The divorce rate in the United States has risen dramatically in recent times. Factors promoting divorce are both individual and social and include the quality of the marital relationship, the increasing economic independence of women, and the social acceptability of divorce. The dissolution of marriage, however, is not expected to continue to escalate in the future. In fact, the divorce rate has declined slightly.

11. Although the American family still provides considerable social and emotional support for its members, violence in this intimate setting is not uncommon. Spouses use violence against each other, parents abuse children, siblings are physically violent with each other, and the elderly are abused.

12. There is considerable experimentation with new patterns of marriage and family. Despite the rise in blended families, single-parent families, child-free marriages, people remaining single, cohabitation, gay and lesbian couples, adult children returning home, and the sandwich generation, the nuclear family is not going to be replaced on a broad scale. The American family, however, will be affected by the rise of the dual-employed marital arrangement.

LEARNING OBJECTIVES REVIEW

After careful study of this chapter, you will be able to:

- Describe the types of family structure, dimensions of family structure, types of mate-selection norms, and types of marriages.
- Compare and contrast the view of the family taken by functionalists, conflict theorists, and symbolic interactionists.

- Discuss the extent and cause of divorce in America.
- Describe the extent and nature of family violence in the United States.
- Outline the contemporary major alternatives to the traditional nuclear family structure.
- Discuss the future of the family in the United States.

CONCEPT REVIEW

Match the following concepts with the definitions listed below them:

____ a. matrilinear descent   ____ e. polyandry   ____ i. patrilineal descent
____ b. exogamy   ____ f. blended family   ____ j. family of procreation
____ c. nuclear family   ____ g. marriage   ____ k. divorce ratio
____ d. family   ____ h. democratic control   ____ l. monogamy

1. Composed of a remarried man or woman and at least one child from a previous marriage.
2. The form of control in which authority is split evenly between husband and wife.
3. The number of divorced persons per 1,000 persons who are married and living with their spouses.
4. Mate-selection norms requiring individuals to marry someone outside their kind.
5. A group of people related by marriage, blood, or adoption.
6. The family group established upon marriage.
7. A heterosexual union in which public approval is given to sexual activity, having children, and assuming mutual rights and obligations.
8. The familial arrangement in which descent and inheritance are passed from the mother to her female descendants.
9. The form of marriage in which one man is married to only one woman at a time.
10. The smallest group of individuals (mother, father, and children) that can be called a family.
11. The familial arrangement in which descent and inheritance are passed from the father to his male descendants.
12. The form of marriage in which one woman is married to two or more men at the same time.

CRITICAL THINKING QUESTIONS

1. The text outlines six functions of the family. Describe how you have experienced these functions within your family of orientation. Be specific.

_____

_____

2. Does conflict theory explain any of the social relationships you have seen in your family or the families of your friends? Explain.

_____

_____

3. Focusing on the looking-glass self, show how symbolic interactionism can be used to explain your experiences in your family of orientation. Provide examples that apply to you or other members of your family.

_____

_____

4. What do you think the divorce rate trend line in the United States will be in the twenty-first century? Develop your answer within the context of information in this chapter.

_____

_____

5. Given all the current marital and familial lifestyle variations in the United States today, many observers argue that the nuclear family is going to become a choice of the minority of Americans. Do you agree or disagree? Use data and social trends to support your position.

_____

_____

MULTIPLE CHOICE QUESTIONS

1. The family that provides one with a name, an identity, and a heritage is known as the
   a. matrilineal family.
   b. family of orientation.
   c. patriarchal family.
   d. nuclear family.
   e. neolocality family.

2. The _____ family consists of two or more adult generations of the same family sharing a common household and economic resources.
   a. nuclear
   b. extended
   c. adult-child
   d. patriarchal
   e. matriarchal

3. As societies move from an agricultural to an industrial base, the _____ family tends to be replaced by the _____ family.
   a. nuclear; extended
   b. matriarchal; patriarchal
   c. matrilocal; patrilocal
   d. extended; nuclear
   e. democratic; patriarchal

4. The family arrangement in which descent and inheritance are transmitted from the mother to her female descendants is called a
   a. female-centric family.
   b. matrilineal family.
   c. neolineal family.
   d. matrilocal family.
   e. bilateral family.

5. A residential arrangement in which a married couple establishes a new residence of their own is known as a/an _____ arrangement.
   a. bilateral
   b. democratic
   c. matrilocal
   d. neolocal
   e. independent

6. A mate-selection arrangement in which norms encourage individuals to marry within their own kind is known as
   a. endogamy.
   b. exogamy.
   c. monogamy.
   d. polyandry.
   e. incest.

7. Which of the following is the *most* widely practiced form of marriage around the world today?
   a. polygyny
   b. polygamy
   c. polyandry
   d. monogamy
   e. group marriage

8. Which of the following is *not* considered to be a vital function of the family?
   a. The provision of initial learning experiences that make people human.
   b. The provision of a warm and loving atmosphere that fulfills basic human social and emotional needs.
   c. The legitimization of reproduction in a society.
   d. The regulation of sexual activity.
   e. The promotion of social mobility.

9. According to conflict theorists,
   a. the family has historically been an important means of maintaining male dominance over females.
   b. the family in modern society is helping to shatter male dominance.
   c. relationships within the family are failing to undergo the proper kinds of changes.
   d. male dominance over females is due to the dynamics of intimate and personal relationships within the family.
   e. the family remains the one place in society where an individual is unconditionally accepted.

10. The number of divorced persons per 1,000 persons in the population who are married and living with their spouses is known as the
    a. divorce rate.
    b. divorce ratio.
    c. marital failure rate.
    d. marriage-divorce ratio.
    e. marital success ratio.

11. According to research cited in the text, voluntarily child-free couples appear to be
    a. pathologically self-centered.
    b. more committed to religion than couples with children.
    c. declining in number compared to couples with children.
    d. lacking in self-esteem.
    e. more satisfied with their marriage and their lives than couples with children.

12. _____ involves living with someone in a marriagelike arrangement without the legal obligations and responsibilities of formal marriage.
    a. Dual-consent marriage
    b. Blended marriage
    c. Duolocal residence
    d. Homogamy
    e. Cohabitation

13. It is estimated that about _____ percent of the U.S. population is gay or lesbian.
    a. 2
    b. 5
    c. 10
    d. 12
    e. 20

14. Approximately _____ percent of American adults between the ages of eighteen and thirty-four live at home with their parents.
    a. 12
    b. 19
    c. 27
    d. 31
    e. 35

15. Which of the following is the *most* accurate statement regarding the future of the American family?
    a. The nuclear family is no longer the most popular choice.
    b. Childless couples are now so prevalent that the growth of the population is being seriously affected.
    c. Most Americans are not avoiding marriage permanently.
    d. Homosexual marriage is undermining the institution of the family.
    e. Americans are spending more and more of their lives married.

FEEDBACK REVIEW

True-False

1. Contrary to what one would expect from the traditional male dominance in American family life, women are almost as frequent spousal abusers as men. *T or F?*

2. Although the nuclear family is not disappearing, Americans are spending less and less of their lives married. *T or F?*

Fill in the Blank

3. Industrialization promotes the shift from the extended to the _____ family.

4. In a _____ family, the oldest man living in the household controls the rest of the family members.

5. In 1995, approximately _____ percent of women and _____ percent of men in America between the ages of twenty-five and twenty-nine had never married.

Multiple Choice

6. Which of the following functions of the family is *not* shared with any other institution?
    a. socialization
    b. reproduction
    c. socioemotional maintenance
    d. sexual regulation

7. Which of the following is *not* stated in the text as one of the factors associated with divorce?
    a. decline of religious influence
    b. age at first marriage
    c. length of marriage
    d. economic conditions

8. Although the proportion of single-parent families has increased for both blacks and whites since 1970, the gap between them has
    a. decreased somewhat.
    b. increased significantly.
    c. remained the same.
    d. decreased significantly.

Matching

9. Indicate which type of mate-selection norm is reflected in each of the following situations:
    ____ a. Jews are supposed to marry Jews.
    ____ b. A father is not permitted to marry his daughter.
    ____ c. Members of the same social class marry.

    (1) exogamy
    (2) endogamy
    (3) homogamy

10. Match the following examples with the major theoretical perspectives:
    ____ a. fathers "give away" brides
    ____ b. sexual regulation
    ____ c. development of self-concept
    ____ d. newly married couple adjusting to each other
    ____ e. child abuse
    ____ f. social class is passed from one generation to another

    (1) functionalism
    (2) conflict theory
    (3) symbolic interactionism

REVIEW GUIDE

GRAPHIC REVIEW

U.S. trends in age at first marriage for males and females are graphically depicted in Figure 11.4. Answer these questions to test your comprehension of the meaning of these trends.

1. If this upward trend in age at first marriage for both sexes continues, what do you predict would happen to the divorce rate?

_____

_____

2. Explain the reasons for the relationship between age at first marriage and the divorce rate you assumed in answering the first question.

_____

_____

ANSWER KEY

Concept Review	Multiple Choice	Feedback Review
a. 8	1. b	1. T
b. 4	2. b	2. T
c. 10	3. d	3. nuclear
d. 5	4. b	4. patriarchal
e. 12	5. d	5. 35, 51
f. 1	6. a	6. b
g. 7	7. d	7. a
h. 2	8. e	8. b
i. 11	9. a	9. a. 2
j. 6	10. b	b. 1
k. 3	11. e	c. 3
l. 9	12. e	10. a. 2
	13. c	b. 1
	14. c	c. 3
	15. c	d. 3
		e. 2
		f. 1

Chapter Twelve

Education

After careful study of this chapter, you will be able to:
- Describe the relationship between industrialization and education.
- Discuss various aspects of the bureaucratization and attempts to debureaucratize education in the United States.
- Outline the basic functions (manifest and latent) of the institution of education.
- Define meritocracy and explain the flaws in the meritocratic model.
- Define educational inequality and discuss the major issues surrounding it.
- Discuss the ways in which schools socialize students.
- Identify and describe the dominant issues in higher education.

SOCIOLOGICAL IMAGINATION

Is Head Start, a federally funded compensatory education program for disadvantaged children, another example of federal government waste of taxpayer dollars? Americans currently have a strong tendency to believe that the federal "bureaucracy," including its efforts in Head Start, is doomed to failure when it "interferes" with our lives. Research, to the contrary, has consistently shown that the Head Start program produces positive benefits (for example, improvements in IQ, general ability tests, and learning readiness) for poor children.

Education serves many important functions in modern society. Among these purposes is the provision of a medium for promoting a common identity among all members of society. Due to educational inequality among minority groups, the achievement of this integrative function has been retarded in the United States. Although Head Start has helped in this regard, barriers in the educational system remain. Before turning to the issue of educational inequality, however, some other topics require coverage. The chapter begins with the development and structure of the American educational institution.

DEVELOPMENT AND STRUCTURE OF EDUCATION

The basic purpose of education is the transmission of knowledge. Education was originally a family responsibility, but industrialization changed that dramatically.

Industrialization and Education

Why did schools emerge as the primary educational agent? Before industrialization, the family, along with the rural community and the church, was the major socializing group in a child's life. The family taught children the values, norms, and farming skills needed to survive in an agricultural society. However, this knowledge increasingly lost its relevance as societies industrialized. As knowledge increased in volume and complexity, family members gradually lost the ability to educate their children adequately. Very few parents today, for example, are qualified to teach their children biology, Shakespearean literature, and computer programming. The educational institution developed to fulfill society's need to transmit knowledge to the young.

What was the objective of early schools? There were conflicting ideas about the form and content of education in nineteenth-century America. The founding fathers thought schools would be an investment in democracy because they would create a literate, active, and informed public. Schools, they believed, would create some minimal "democracy of knowledge," thus preventing the privileged few from having a monopoly on knowledge. Business leaders favored simple skill training to prepare children to enter their factories. The teaching of democratic values and marketable skills seemed especially popular in those urban areas inundated by the many ethnic groups who immigrated to the United States between 1850 and 1910. Labor leaders, fearing the loss of adult jobs to child laborers, urged city governments to keep children in school. Many questions about education were raised. Should children attend school with other children like themselves? Should neighborhoods run schools? Should "professional" teachers be paid for their work? Should education be viewed as charity for children from the poorest families? On the whole, there was no consensus regarding the education of the young.

Eventually, some agreement about the mission of education was reached. Most late nineteenth-century proponents of public education believed the objective of the early school to be

> . . . not so much . . . intellectual culture, as the regulation of the feelings and dispositions, the extirpation of vicious propensities, the preoccupation of the wilderness of the young heart with seeds and germs of moral beauty, and the formation of a lovely and virtuous character by the habitual practice of cleanliness, delicacy, refinement, good temper, gentleness, kindness, justice, and truth (Katz, 1975:31).

Thus, early public schooling in America was not primarily designed to give children sophisticated intellectual

Government spending on the poor is under heavy attack, an attack based in part on the claim that these government programs do not work. Head Start, a federally funded program for preschoolers from disadvantaged homes, has proven to be a success.

skills; it was a mechanism of social control. Otherwise, it was feared, children would grow up "vicious" or preoccupied with the "wilderness of the young heart."

How did education develop after the turn of the century? Just as farming slowly gave way to manual factory labor in the latter part of the nineteenth century, relatively simple manual skills began to be less and less important after the beginning of the twentieth century. The opening up of new middle-management and

clerical jobs was seen as highly desirable by many Americans seeking to better their own lives and, more important, the future lives of their children. Business leaders and parents began to demand more advanced training in schools. School administrators responded by advocating and developing secondary schools emphasizing advanced learning. Figure 12.1 shows the percentage of the total American population over the age of twenty-five (as well as major racial and ethnic groups) that has at least completed high school.

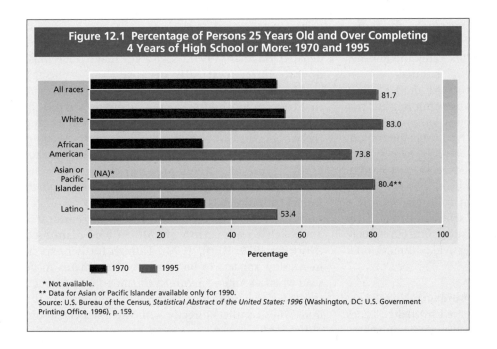

Figure 12.1 Percentage of Persons 25 Years Old and Over Completing 4 Years of High School or More: 1970 and 1995

- All races: 81.7
- White: 83.0
- African American: 73.8
- Asian or Pacific Islander: (NA)* 80.4**
- Latino: 53.4

Percentage (0, 20, 40, 60, 80, 100)

■ 1970   ■ 1995

* Not available.
** Data for Asian or Pacific Islander available only for 1990.
Source: U.S. Bureau of the Census, *Statistical Abstract of the United States: 1996* (Washington, DC: U.S. Government Printing Office, 1996), p. 159.

Education and Bureaucracy

Early school administration was based on the factory model of education. Children were to be educated in much the same way cars were mass-produced.

> *Schooling came to be seen as work or the preparation for work; schools were pictured as factories, educators as industrial managers, and students as the raw materials to be inducted into the production process. The ideology of school management was recast in the mold of the business corporation, and the character of education was shaped after the image of industrial production (Cohen and Lazerson, 1972:47).*

The concept of bureaucratized public education is, of course, still prevalent.

In what ways are schools bureaucratic? In keeping with the bureaucratic model, the process of educating children has been divided more and more into specializations. There are specialists who purchase materials, specialists who test and counsel students, and specialists who run libraries. There are even specialists who drive buses, serve lunches, and type memos.

Education can be accomplished more efficiently, according to the bureaucratic model, when students are more homogeneous in ability. In this way, one teacher can handle large numbers of students in the same way rather than having to adapt learning materials to the learning readiness of individuals. Age-graded classrooms, in which all students receive the same instruction, reflect the bureaucratic nature of schools.

It is also thought to be more efficient for schools to have all teachers of a given subject or a given grade teaching the same, or at least similar, material. When parents want to know what is being taught, those running the school can readily tell them. This practice also allows students to transfer from one school to another and still study approximately the same things.

Critics claim that the bureaucratic model is not appropriate for schooling, because children are not inorganic materials to be processed on an assembly line. Children are human beings who come into school with previous knowledge and interact socially and emotionally with other students. Moreover, it is hard to determine whether children have been "processed" correctly. How is one to know what takes place inside a child's head? Critics of formal schooling frequently charge that the school's bureaucratic nature fails to take into account the expressive, creative, and emotional needs of individual children.

Are there alternatives to the bureaucratic school? Interest in educational reform in the United States goes in cycles. The American progressive education move-

ment of the 1920s and 1930s can be seen as a reaction to the Victorian authoritarianism of early schools. The progressive movement, with its child-centered focus, lost out in the 1950s but had a resurgence in the 1960s and 1970s in the humanistic education movement. The humanistic movement advocated such steps as the elimination of restrictive rules and codes and the involvement of students in forming educational goals. The humanistic movement spawned a number of educational philosophies formulated to loosen up classroom structure and the learning environment (Ballantine, 1993).

Let's consider a representative philosophy, the *open classroom*, beginning with this summarizing statement by an early proponent:

> *In the 1960s, advocates of the open classroom proposed alternatives to the bureaucratically organized school. They agreed on an underlying philosophy: Children are naturally curious and motivated to learn by their own interests and desires. The most important condition for nurturing this natural interest is freedom, supported by adults who enrich the environment and offer help. In contrast, coercion and regimentation only inhibit emotional and intellectual development (Graubard, 1972:352).*

The guiding principles of the **open-classroom** philosophy advocate elimination of the sharp line of authority between teachers and students, the predetermined curriculum for all children of a given age, the constant comparison of students' performance, the use of competition as a motivator, and the grouping of children according to performance and ability. In short, schools in the open-classroom approach wished to avoid the bureaucratic features of traditional public schools (Holt, 1967; Kozol, 1968; Silberman, 1971; Kohl, 1984).

Approaches to school structure continue to be debated (Hallinan, 1995). Some argue that schools must operate bureaucratically because limited money requires less expensive mass instruction. Others argue that concern for children should supersede administrative needs and that individualized programs developing the whole personality of the child should be encouraged. According to yet another viewpoint, the bureaucratic nature of schools prepares children for the bureaucratic society in which they must live after their formal education has ended.

Evidence indicates that a less bureaucratic approach to education should not be dismissed (Horwitz, 1979; Hurn, 1993). Though further experimentation appears to be warranted, it remains to be seen whether Americans are in the mood for it. In fact, the "back to basics" movement which emerged in the mid-1980s conflicts directly with the open-classroom philosophy.

The American education system was founded on the model of the industrial factory. Although this mass production, bureaucratic conception of education is still prevalent in schools in the United States, classrooms are more informal than in the past.

What is the back to basics movement? In 1983, the National Commission on Excellence in Education issued a report, dramatically entitled *A Nation at Risk,* warning of a "rising tide of mediocrity" in America's schools. Because of deficiencies in its educational system, the report claimed, America is at risk of being overtaken by some of its economic competitors around the world (Gardner, 1983). As evidence, the report pointed to several indicators:

- Twenty-three million American adults, 40 percent of minority young people and 13 percent of all seventeen-year-olds, are functionally illiterate.
- Scholastic Aptitude Test scores dropped dramatically between the early 1960s and 1980.
- Only one-third of all seventeen-year-olds can solve a mathematics problem requiring several steps.
- American children are being outperformed by children of other nations on achievement tests.

Most of the solutions offered by the Commission tended toward the bureaucratic. High school graduation requirements should be strengthened, including four years of English, three years of mathematics, three years of science, three years of social studies, and a half year of computer science. School days, the school year, or both should be lengthened. Standardized achievement tests should be administered as students move from one level of schooling to another. High school students should be given significantly more homework than they are assigned at present. Discipline should be

tightened through the development and enforcement of codes of student conduct.

The Commission's report stirred intellectual debate (Adelson, 1984; Bunzel, 1985; Gross and Gross, 1985) and aroused the public. Americans generally favor the Commission's recommendations, especially its call for more homework and greater discipline (Barrett, 1990).

In a series of books, English professor E. D. Hirsch contends that American schoolchildren are not being taught the background knowledge they need to function in today's world (Hirsch, 1987, 1996). Hirsch has labeled American schoolchildren "culturally illiterate." Surprising numbers of schoolchildren don't know when the American Revolution occurred, who George Washington Carver is, or where Washington, D.C. is located. Although Hirsch is concerned for all Americans, he is particularly concerned that the failure of schools, and the accompanying cultural illiteracy, is dooming children from poor and illiterate homes to perpetual poverty. Keeping poor and illiterate children in the same condition as their parents is for Hirsch an unacceptable failure of American schools.

Late in 1987, U.S. Secretary of Education William Bennett under President Ronald Reagan proposed a back to basics curriculum for American high schools. Much emphasis was placed on a minimum core curriculum that all students must take (Bennett, 1988). In 1991, President George Bush's Secretary of Education, Lamar Alexander, formulated a national educational reform plan around six goals (*America 2000,* 1991):

1. Every child must begin school ready to learn.

"Charter" schools are part of the back to basics movement in education. Legal in only a few states, these publicly-supported schools are established under a charter specifying certain performance requirements. This charter school in Michigan features mandatory Spanish and French, daily homework, and uniforms.

2. The national high school graduation rate must be 90 percent.
3. Competence in core subjects must be shown after grades 4, 8, and 12.
4. American students should be the best educated in the world in math and science.
5. All adults must be literate and possess the skills necessary for citizenship and competition in a global economy.
6. Schools should be free of drugs and violence.

Implementation aspects of the plan included a longer school year, national standards and exams, and merit pay for teachers.

To date there is no evidence that the back to basics movement can live up to its advance billing. Implementation of some of the solutions could actually inflict greater harm. The dropout rate could increase as time spent in school and on homework assignments increases. The low self-esteem of many low-performing students could be reduced further. Stricter grading and increased academic requirements could discourage low-income and minority students even more. Bureaucratically oriented solutions to educational problems may not necessarily be the cure-all they seem to be on the surface. Despite all of the concern, it appears that the nation remains at risk (Kantrowitz, 1993a; Hancock and Wingert, 1995; Wingert, 1996).

Is there another educational reform cycle emerging?
The open-classroom philosophy, which, as you just read, became prominent in the 1960s and 1970s, is cur-

rently being implemented again, this time in the form of what is called **cooperative learning**—a non-bureaucratic classroom structure in which students study in groups with teachers as guides rather than as the controlling agents determining all activities:

> *Cooperative learning approaches are based in part on the assumption that students are likely to learn best when they are actively working with others in small heterogeneous groups. . . . It occurs when teachers have students work together in small groups on a task toward a group goal—a single product (a set of answers, a project, a report, a piece or collection of creative writing, etc.) or achieving as high a group average as possible on a test—then reward the entire group on the basis of the quality or quantity of its product according to a set standard of success (Oakes, 1985:208,210).*

According to the cooperative learning method, students learn more if they are actively involved with others in the classroom (Sizer, 1996). The traditional teacher-centered approach rewards students for being passive recipients of information and forces them to compete with others for grades and teacher recognition. Cooperative learning, in its focus on teamwork rather than individual performance, encourages kids to concentrate more on their work than on their performance compared to other students. Cooperation replaces learning as a form of competition involving classmates as rivals. Under this format, students typically work within small groups with a specific task. Credit for completion of a task is given only if all group members do their part.

Some benefits of the cooperative learning approach have been documented. Uncooperativeness and stress among students is reduced; academic performance is elevated; students have more positive attitudes toward school; racial and ethnic antagonism decreases; and self-esteem increases (Oakes, 1985; Aronson and Gonzalez, 1988; Children's Defense Fund, 1991).

Although Japanese schools have been using group learning in the early grades for years, resistance remains strong among many American teachers (Kantrowitz, 1993b). They fear loss of control and disciplinary problems among older kids, especially teenagers. Many American teachers and principals believe mapping a student's progress is more difficult in a cooperative learning setting. This approach also requires more preparatory work for teachers who must plan lessons so that students can work without frequent teacher intrusion. Grading is also more complex if an individual and group evaluation are combined for an overall grade.

FUNCTIONALIST PERSPECTIVE

Functions of Education

According to the functionalists, social institutions develop because they meet one or more of society's basic needs. The educational institution performs several vital functions in modern society: cultural transmission; social integration; selection and screening of talent; promotion of personal growth and development; the dissemination, preservation, and creation of knowledge.

How does the educational institution transmit culture? Cultural transmission must occur if a society is to endure. The educational institution contributes to the perpetuation of culture by formally passing on a society's cultural heritage from one generation to the next. When we see evidence of the teaching of Marxist-Leninist doctrine in communist societies, we call it indoctrination. Americans are less aware of the process of indoctrination when they learn about the merits of democracy through the study of Washington, Jefferson, and Lincoln. Thus, although schools teach basic academic skills such as reading, writing, and mathematics, they also pass on the basic values, norms, beliefs, and attitudes of the society. (More is said about this function of the educational institution in a later section on socialization and schools.)

How does education contribute to social integration? Although television is now a strong competitor in the United States, the educational system remains the major agent of social integration. Formal education provides a means of transforming a population composed of many ethnic groups and cultural backgrounds into a community of individuals who share to some extent a common identity. Learning an official language, sharing in national history and patriotic themes, and being exposed to similar educational sequences facilitate the development of a common identity. The result is a society with relatively homogeneous values, norms, beliefs, and attitudes. Schools in the United States have attempted to contribute to the melting-pot process. Newly arrived immigrant children without the ability to speak and write English soon learn to participate in the American way of life.

How do schools select and screen students? Scores on intelligence and achievement tests have been used for over fifty years as a means of grouping children by ability. The stated purpose of testing is to identify an individual's talents and aptitudes. Test scores have also been used for **tracking**—placing students in educational programs consistent with the school's expectations for the students' eventual occupations (Oakes, 1985; Children's Defense Fund, 1991; Hurn, 1993). Counselors use test scores and early performance records to predict careers for which individuals may be best suited. Counselors usually feel a moral obligation to ensure that students pursue careers consistent with their abilities. (The concept of tracking plays an important role in the conflict perspective of education, as discussed later.)

How do schools promote personal growth and development? Students are exposed in schools to a wide variety of perspectives and experiences that encourage

One of the functions of schools is to select and screen students. Scores on intelligence and achievement tests are used for purposes of "tracking"—placing students in educational programs consistent with the school's expectations for the types of occupations for which students are best suited. Lower achieving high school males are often channeled into a "shop" or manual work curriculum.

intellectual development, creativity, and development of verbal and artistic means of personal expression. In this way, education provides an environment in which individuals may improve the quality of their lives through intellectual, artistic, and emotional experiences. Schools also promote personal growth and development by preparing students to assume a place in the occupational world.

How does education encourage the dissemination, preservation, and creation of knowledge?   Because of our own learning experiences in elementary and high school, we are most familiar with the school's efforts at the dissemination of knowledge that most commonly takes the form of classroom teaching. Dissemination of knowledge also includes activities such as the development of written or audiovisual materials that can reach hundreds of thousands of people. Preservation of knowledge is perpetuated by the transmission of knowledge from generation to generation. The task of preserving knowledge is also achieved by such activities as deciphering ancient manuscripts; protecting artifacts; and recording knowledge in written, video, or audio form. Innovation—the creation or discovery of new knowledge through research or creative thinking—can take place at any level within the educational system, but it has traditionally received more emphasis at the university level.

Latent Functions of Education

As noted in Chapter 1 ("The Sociological Perspective"), functionalists distinguish between **manifest functions** (intended and recognized) and **latent functions** (unintended and unrecognized). The functions of schools discussed to this point have been recognized and intended. The educational institution has other functions that are unrecognized and unintended.

What are the latent functions of education?   We do not usually count among the functions of schools the provision of day-care facilities for dual-employed couples or single parents. Nor do we think that our tax money goes to schools so that people can find marriage partners. Also, schools are not consciously designed to prevent delinquency by holding juveniles indoors during the daytime. Finally, schools are not generally recognized for inculcating the discipline needed to follow orders in a bureaucratized society (DeYoung, 1989; also refer to *Doing Research* in Chapter 4, "Socialization Over the Life Course").

Each of these latent functions, although unrecognized and unintended, is considered a positive contribution to society. Recall from Chapter 1, however, that some consequences can be negative, or dysfunctional. Recall also that determining what is dysfunctional is relative to where one is standing when making a judgment. Tracking, for example, has just been presented as a positive function of the educational institution. Most of us would agree with this assessment. Most of us see this function as a manifest, positive function of schools, because it seems to fulfill a societal need. According to the conflict perspective, however, tracking serves a latent dysfunction—it is a means for those benefiting from the stratification structure to continue to benefit. In other words, tracking is one mechanism for perpetuating the social-class structure from generation to generation. Moreover, some evidence suggests that tracking is harmful to those placed on "slower" tracks (Hurn, 1993). The conflict perspective on education is presented in the following two sections: the first section covers the relationship between education and what is known as *meritocracy;* the second section covers *educational inequality.* These two topics are highly interrelated and are central to the relationship between education and social inequality, the central research question in the sociology of education since World War II (Hallinan, 1988).

FEEDBACK

1. The _____ institution developed to fulfill society's need to transmit knowledge to the young.
2. Contemporary public education is important primarily because of
    a. the quality of teachers in modern schools.
    b. efficient operation of school bureaucracies.
    c. historical changes in the economic and social structures of industrial societies.
    d. lack of qualified individual tutors.
3. Early public schooling in America was viewed as a mechanism of _____.
4. Schools in America today are based on the _____ model of organization.
5. _____ attempt to avoid most of the bureaucratic features of traditional public schools.
6. _____ is a nonbureaucratic classroom structure in which students study in groups with teachers as guides rather than as the controlling agents determining all activities.
7. The process of placing students in educational programs consistent with expectations of their eventual occupations is called _____.
8. When schools keep children for their working parents, they perform a _____ function.

Answers: 1. educational 2. c. 3. social control 4. bureaucratic 5. Open classrooms 6. Cooperative learning 7. tracking 8. latent

A **meritocracy** is a society in which social status is based on ability and achievement rather than background or parental status (Young, 1967; Bell, 1976). According to the meritocratic model, all individuals have an equal chance to develop their abilities for the benefit of themselves and their society. A meritocratic society, it is claimed, gives everyone an equal chance to succeed. It is supposed to be free of the barriers that prevent individuals from developing their talents.

Many Americans believe that the United States is a meritocratic society. However, a number of recurring problems suggest that the model does not always work the way we think it does (Olneck and Crouse, 1979; "Schooling and Success," 1982; Fallows, 1985; Krause and Slomczynski, 1985; Lieberman, 1993).

Flaws in the Meritocratic Model

Is America a meritocratic society? According to the meritocratic model, as a society becomes more efficient and modern, the nonmeritocratic barriers to equal opportunity for all people disappear. Although America purports to be a meritocratic society, researchers recognize the existence of barriers to achievement based on merit. For example, Marxist critics challenge the existence of meritocracy in American society, arguing that public education in America serves primarily the economic elite. Because schools are not equal for children of all social classes, students do not all have an equal starting point (Bowles and Gintis, 1976; Giroux, 1983). By continually responding to the employment needs of capitalist industries, such critics believe, schools guarantee an oversupply of technical workers and thus keep wages down. By channeling lower-class children into unskilled occupations, schools insure stiff competition for even the least desirable jobs. Schools also convince dropouts that their failure to get higher-status jobs is their own fault rather than the system's. Similarly, by emphasizing material pleasures and consumption of goods as supreme values, schools are said to guarantee the continuation of the system.

Other sociologists, such as Randall Collins (1971), have argued that America is not a meritocratic society at all. Collins contends that American society is composed of various "status groups" competing with one another for wealth, power, and prestige. Status groups are made up of individuals who share a sense of status equality based on their participation in a common culture, with similar manners, hobbies, opinions, and values. In their struggle for advantage in society, individuals frequently turn to others like themselves as a key resource in the struggle against members of other status groups. In America, for example, white Anglo-Saxon Protestant men dominate most of the positions of wealth, power, and prestige. Collins argues that this elite has structured public schools to reinforce the shared culture of their group. Entry into higher-status positions requires acceptance of the values and culture of the dominant elite. People from higher-status homes who have been educated in "better" schools, for example, have an easier time getting into the higher-level jobs. Those who are from lower social classes, who are of a different color or sex, or who do not possess the appropriate values, manners, language, and dress find it more difficult to obtain higher-level jobs.

In a later book, Collins (1979) takes the conflict interpretation even further with the concept of **credentialism**—the idea that credentials (educational degrees) are unnecessarily required for many jobs. Why, he asks, must American physicians be trained almost into middle age when Communist countries can produce qualified doctors in a relatively short time? Credentials are required, according to Collins, so that those who can afford them can maintain a strong hold on the elite occupations. Otherwise, why would people spend so much time in school learning things that are useless to them on the job?

Social class is, in fact, a major factor impeding achievement on the basis of ability and effort. This is because social class and educational level are related.

What is the relationship between social class and educational level? Daniel Levine and Robert Havighurst (1992) report that the vast majority of children from upper-class and upper-middle-class families attend college, unlike most children from lower- and working-class families. This difference cannot be attributed merely to differences in ability among social classes. Summarizing ten years of his own research on the relationship between social class and achievement, William Sewell concludes the following:

> We estimate that a high SES (socioeconomic status) student has almost 2.5 times as much chance as a low SES student of continuing in some kind of post-high school education. He has an almost 4 to 1 advantage in access to college, a 6 to 1 advantage in college graduation, and a 9 to 1 advantage in graduate or professional education. . . . Even when we control for academic ability by dividing our sample into youths according to the students' scores on standardized tests, we find that higher SES students have substantially greater post-high school educational attainment than lower SES students (Sewell, 1971:795).

Part of the difference in the educational levels of persons in the upper and lower social classes is attributable to differences in economic resources. But the effect of social class on educational attainment is more complex than the influence of economic resources alone,

1. _____ refers to a social system in which social status is a function of ability and achievement rather than background or parental status.
2. Lack of economic resources helps to account for the lower educational attainment among the lower social classes. Some of the class-related educational differential, however, is due to _____.
3. The idea that some educational requirements are unnecessary for many jobs today is called
   a. status grouping.   b. credentialism.   c. meritocracy.   d. conflict.
4. A significantly larger proportion of higher social class children attend college than lower-class children. This is largely due to differences in ability among social classes. *T or F?*

Answers: 1. Meritocracy 2. flaws in the meritocratic model 3. b. 4. F

because college is increasingly accessible to the members of all social classes. Some of the differential impact of social class on educational attainment is due to enduring educational inequality. An examination of the problem of educational inequality further illustrates the conflict interpretation of the educational institution.

CONFLICT PERSPECTIVE— EDUCATIONAL INEQUALITY

From the beginning of public education in America, stress was placed on educational equality. That originally meant that all children must be provided a free public education; that all children, regardless of their social and economic backgrounds, must be exposed to the same curriculum; that all children within a given locality must attend the same school; and that all schools within a locality should be equally financed by local tax dollars (Coleman, 1969; Gordon, 1972).

Despite this rhetoric, "all" children have not been thought deserving of a formal education. Native Americans and African Americans were specifically excluded. During the 1850s, it was believed that slaves could not and should not be educated. It was not until slavery was abolished and former slaves began to assume industrial jobs that African Americans eventually were included among those Americans thought to be educable. Still, African Americans were not given equal educational treatment. In the *Plessy v. Ferguson* decision of 1896, the Supreme Court made "separate but equal" part of America's legal framework. The doctrine held that equal treatment was being given when African Americans and whites had substantially equal facilities, even when these facilities were separate.

It was not until 1954 that the separate but equal doctrine was seriously challenged by the judicial system. In the *Brown v. Board of Education of Topeka* decision of 1954, Chief Justice Earl Warren wrote, as part of the Court's opinion, "Does segregation of children in public schools solely on the basis of race, even though the

physical facilities and other 'tangible' factors may be equal, deprive the children of the minority group of equal educational opportunities? We believe that it does." Warren went on to say that the separate but equal doctrine had no place in America's public educational structure and that separate educational facilities are inherently unequal. Still, it took the civil rights movement of the 1950s and 1960s to elevate the long-existing educational inequality for African Americans and other minorities to the status of a social problem of national importance. One result was the Civil Rights Act of 1964, which prohibits discrimination on the grounds of race, color, religion, or national origin in any educational program or activity receiving financial support from the federal government.

What is meant today by educational equality? The most recent concept of educational equality involves equality in the "effects" of schooling. Schools are equal when the students in them have the same general levels of academic performance. If African American and white students attend schools with equal resources but either race performs academically below the other, educational inequality exists. The most significant shift, then, has been from equal educational "opportunity" to equality in "results." This shift can be traced to a nationwide study conducted by James Coleman (1966).

In 1964, James Coleman was commissioned by Congress to conduct a survey on the impact of discrimination on educational inequality. From information on approximately 600,000 students and 60,000 teachers from some 4,000 schools, Coleman concluded that some consistent differences in black and white schools did exist. On the average, African American students had fewer school facilities—laboratories for physics, chemistry, and languages were less plentiful; textbooks were in shorter supply; and fewer books per pupil were available in school libraries. Coleman reported that minorities also suffered in relation to the majority when it came to curricular and extracurricular matters. African Americans in secondary schools were less likely to attend accredited schools, college-preparatory and accelerated curricula were less accessible to minorities, and whites generally enjoyed better access to such academically related extracurricular activities as debating teams and student newspapers. An important source of these differences was the inequality of school financing. Using the survey data collected by Coleman, Christopher Jencks (1972) reported that about 15 to 20 percent more funds were spent for the average white schoolchild per year than for the average African American student.

It was not surprising that Coleman found some differences in resources allocated to majority and minority schools. It was surprising, however, that the differences in resources accounted for only a small part of the differences in majority and minority student performance on standardized achievement tests.

According to Coleman, school quality affects student performance through student background outside of school. If educational equality is to exist, schools must overcome the educational inadequacies children bring with them from home. It is not enough, therefore, to provide minority children with equal school resources. **Educational equality** exists when schooling produces the same results (achievement and attitudes) for lower-class and minority children as it does for other children. Results rather than resources are the test of educational equality (Coleman, 1968).

Such logic supported the mandatory busing programs instituted in urban areas from the late 1960s to the present. As interpreted by the courts, public schools must provide not only equal resources such as texts, teachers, and buildings but also equal "social climates," which are the result of students' personality and motivational factors. Because students influence one another's learning capabilities, school districts must provide similar types of students within each school in terms of racial, social class, and ethnic characteristics (Coleman, 1990).

Can schools provide educational equality? The answer to this question appears to be no. Ray Rist (1977) has contended that teachers often evaluate the potential of students on the basis of their social-class, racial, and ethnic characteristics rather than on their objective abilities. Evidence exists to support this allegation. For example, in his classic study of "Elmstown's" youth, August Hollingshead (1949) found that when middle- and upper-class children had difficulty in class, their parents were usually called in to discuss learning problems. Similar learning difficulties, however, were interpreted as behavioral problems when they occurred among lower-class children. Aaron Cicourel and John Kitsuse (1963) later found the same phenomenon in their study of high school counseling procedures. When test scores were consistent with the counselor's expectations, there was no problem. The counselor, however, was much more willing to find

excuses for poor performance among high-status students than among low-status ones. Rist (1977) reported a similar pattern among classroom teachers. When asked questions in class, the students perceived as more capable by teachers were given more time to answer than "less able" students were given. Also, lower-class students were more likely to be considered by teachers to be less academically capable.

This tendency to evaluate students on nonacademic criteria is especially apparent in the practice of tracking (Kilgore, 1991; Gamoran, 1992). Most high schools have a series of curricula: college preparatory, commercial, vocational, and general (for those who are undecided about their occupational futures). As in the lower grades, social class and race are associated with placement. Researchers have found that social class and race heavily influence the placement of students in tracks regardless of their past academic achievement or IQ (Dreeban and Gamoran, 1986; Paul, 1987; Salamans, 1988). Subsequent levels of academic performance are influenced more by students' track than by their intelligence or past scholastic performance. Regardless of earlier school performance or IQ, the subsequent academic performance of those bound for college increases, whereas the performance of those on a noncollege track decreases.

The effects of tracking extend beyond academic matters. Placement in a noncollege track produces lower self-esteem, encourages students to drop out of school, lessens students' involvement in extracurricular school activities, and makes students more likely candidates for behavioral problems in and out of school.

According to this evidence, then, school personnel frequently respond to, advise, and track students based on social class rather than measured abilities or potential. In this way, schools deny educational equality to many students, especially minority children, who typically come from lower-class families (Oakes, 1985; Gibson and Ogbu, 1991).

Critics of American education see a threat to educational equality coming from yet another direction. The *school choice* movement, which began in the 1980s, advocates parental and student freedom to select the schools they prefer. **Vouchers**—tax allowances granted to cover part of the costs for an academic year at the school selected—are being supported as a means of implementing school choice. Vouchers would cover private schools if chosen. Proponents point to the freedom vouchers give parents, the increased quality that would presumably come from schools competing for students, and the ability to place students in educational environments best suited to their particular needs. Critics fear that free choice of school would make urban schools even more of a dumping ground for the poor and minorities because poor minority parents in inner cities cannot make up the difference

SOCIOLOGY IN THE NEWS

Gender Bias in the Classroom

One of the functions of schools is to place students on educational tracks in order to channel them into occupations for which they are judged to be best suited. As noted in this CNN report, schools steer females away from math and science on the assumption of alleged male superiority in these areas. This gender bias tends to produce a self-fulfilling prophecy—teacher expectations lead females to behave in ways that make the expectations come true.

1. Relate gender bias in the classroom to the "hidden curriculum" in schools.

   _____

   _____

2. Is the practice of tracking in schools functional or dysfunctional? Explain.

   _____

   _____

The "hidden curriculum" channels American boys and girls into different occupations. Such educational tracking is beginning to be challenged.

between the amount of the voucher and the cost of sending their children to higher quality schools:

> "Choice" plans of the kind the [Bush] White House has proposed threaten to compound the present fact of racial segregation with the added injury of caste discrimination, further isolating those who . . . have been consigned to places nobody would choose if he had any choice at all (Kozol, 1991:63).

In addition, the critics contend, open competition could lead to flight from public schools to religious schools and to private, for-profit schools, leaving the public school system in worse shape than it is now (Henig, 1994). Consequently, the existing disparity in the quality of education received by rich and poor youth would be perpetuated (Manski, 1992).

Educational Inequality and Intelligence Testing

Because intelligence testing is used as a means of sorting and tracking students, it contributes to educational inequality. Whenever the topic is discussed, the question of inherited intelligence always arises.

Is intelligence inherited?   Historically, it had been assumed that individual and group differences in measured intellectual ability were due to genetic differences. This assumption, of course, underlays social Darwinism.

Some researchers also take this viewpoint. Almost twenty years ago Arthur Jensen (1969), an educational psychologist, contended that the lower average IQ score among African American children may be due to

heredity. A recent book by Richard Herrnstein and Charles Murray (1994), entitled *The Bell Curve,* is also in the tradition of linking intelligence to heredity. According to these authors, humans inherit 60 to 70 percent of their intelligence level. Herrnstein and Murray further contend that the fact of inherited intelligence makes largely futile efforts to help the disadvantaged in general, and blacks in particular, through programs such as Head Start and affirmative action.

Social scientists have generally criticized advocates of genetic differences in intelligence for failing to consider adequately the effects of the social, psychological, and economic climate experienced by children of various racial and ethnic minorities. Even those social scientists who grant genetics an important role in intelligence criticize both the interpretations of the evidence and the public policy conclusions contained in *The Bell Curve.* They point to the body of research that runs counter to Herrnstein's and Murray's thesis. More specifically, they see intelligence not as an issue of nature versus nurture but as a matter of the effects of both genetics and environment (Morganthau, 1994; Wright, 1995). We know, for example, that city dwellers usually score higher on intelligence tests than do people in rural areas, that higher-status African Americans score higher than lower-status African Americans, and that middle-class African American children score about as high as middle-class white children. We also have discovered that as people get older, they usually score higher on intelligence tests. Findings like these have led researchers to conclude that environmental factors affect test performance more than genetic factors (Samuda, 1975; Schiff and Lewontin, 1987).

Robert Williams (1974) is one of the many earlier social scientists who argue that intelligence tests are designed for middle-class children and are so culturally biased that the results measure learning and environment rather than intellectual ability. Consider this IQ test item cited by Daniel Levine and Robert Havighurst (1992:67):

A symphony is to a composer as a book is to what?
( ) paper ( ) sculptor ( ) author
( ) musician ( ) man

Higher-class children find this question easier to answer correctly than lower-class children because they are more likely to have been exposed to information about classical music. Several studies have indicated that because most IQ tests assume fluency in English, minorities cannot do as well on intelligence tests. Some researchers have even suggested that many urban African American students are superior to their white classmates in several dimensions of verbal capacity, but this ability is not recognized, because intelligence tests do not measure these areas (Gould, 1981; Goleman, 1988; Hurn, 1993).

Some researchers have shown that the testing situation itself affects performance. Low-income and minority students, for example, score higher on intelligence tests when tested by adult members of their own race or income group. Apparently they feel threatened when tested in a strange environment by someone dissimilar to them. Middle-class children are frequently eager to take the tests because they have been taught the importance of tests and to be academically competitive. Because low-income children do not recognize the importance of tests and have not been taught to be academically competitive, they ignore some of the questions or look for something more interesting to do. Other researchers report that nutrition seems to play a role in test performance. Low-income children with poor diets may do less than their best when they are hungry or when they lack particular types of food over long periods of time.

Promotion of Educational Equality

Several avenues have been considered for creating educational equality. Four of them—school desegregation, compensatory education, community control, and private schooling—are discussed here.

Does school desegregation promote educational equality? The Coleman Report has been widely used to support the idea that desegregation will improve the school achievement of minority children. Specifically, James Coleman (1966) reported that African American children in classrooms that are more than one-half

white scored higher on achievement tests. The finding that minority children perform better academically when placed in desegregated, middle-class schools has been verified by several other studies. In a broad review of the research on school desegregation, Meyer Weinberg (1970) concluded that African American children consistently experience higher academic achievement in desegregated schools. Also, he states, research among African Americans indicates that desegregation tends to enhance African American aspirations, self-esteem, and self-acceptance.

Several studies have shown that racially balanced classrooms can have either positive or negative effects on the academic achievement of minority children. Mere physical desegregation may have detrimental impacts on both white and African American children. In contrast, desegregated classrooms in which an atmosphere of respect and acceptance prevails promote improved academic performance among African American children (U.S. Commission on Civil Rights, 1967; Katz, 1968; Longshore, 1982a, 1982b; Patchen, 1982; Orfield et al., 1992).

Some postschooling benefits of desegregation for minorities who attend desegregated public schools include higher occupational positions and higher incomes than minorities in segregated schools. These differences cannot be attributed solely to differences in educational attainment. Rather, increased social contact with middle-class whites contributes to better higher education and employment opportunities and provides information about the availability of jobs (Crain, 1970; Crain and Weisman, 1972; Hawley, 1985).

Some evidence suggests that school desegregation may also lead to better racial and ethnic relations because of exposure to people of differing backgrounds (Hawley, 1985; Hawley and Smylie, 1988). It is on this evidence that the promise of **multiculturalism**—an approach to education in which the curriculum accents the viewpoints, experiences, and contributions of women and diverse ethnic and racial minorities—rests. Multiculturalism attempts to make the traditions of minorities (and women) valuable assets to the broader culture and to dispel stereotypes. According to one study, school attendance and academic performance has increased in some schools with multicultural curricula (Marriott, 1991). Research in this area, however, is in the early stages.

Unfortunately, public schools in the United States are drifting back toward segregation. Reversing the trend toward integration from the mid-1960s until 1988, African American and Latino youngsters are becoming racially and ethnically segregated (Orfield, 1993). Seventy percent of African American and Latino students are now in classrooms with a predominately minority enrollment. In fact, according to an earlier assessment, nearly one-third of youngsters in these two minority

Some evidence suggests that ethnic and racial integration in schools improves attendance and academic achievement. Despite these potential advantages, over two-thirds of African American and Latino students are in largely segregated classrooms.

groups attend schools composed of 90 percent or more minority students (Children's Defense Fund, 1991).

Does compensatory education work?   Central to compensatory education is the idea that special preschool programs can benefit deprived children and overcome an inferior intellectual environment (Edelman, 1992). Contrary to the position of Herrnstein and Murray (1994), there is evidence that remedial education during early childhood successfully increases the school achievement of disadvantaged children (Entwisle and Alexander, 1989, 1993; Ramey and Campbell, 1991; Zigler and Styfco, 1993; Campbell and Ramey, 1994).

The best-known attempt at compensatory education is the Head Start program, a federally supported program for preschoolers from disadvantaged homes that originated as part of the Office of Economic Opportunity in 1965. Head Start was conceived in part as a means of preparing disadvantaged preschoolers for entering the public school system. Its goal is to provide disadvantaged children with an equal opportunity to develop their potential (Waldman, 1990-1991).

A majority of early studies indicated that children's scores on IQ tests, general ability tests, and learning readiness assessments significantly improve after exposure to Head Start programs. As asserted in the *Sociological Imagination* at the start of this chapter, many studies reported that in full-year programs, the scores of Head Start participants, who usually enter the program at a distinct disadvantage, eventually reach the national average on tests of general ability and learning readiness (*Report on Evaluation Studies of Project Head Start,* 1970). Attitudinal, motivational, and social progress have also been observed among children after some time in Head Start programs.

One early study concluded that although the Head Start participants are ahead of comparable non-Head Start children when both groups enter the first year of formal schooling, the advantage is either reduced or eliminated by the end of the year (*The Impact of Head Start,* 1969). This same study, however, concedes some worthwhile results. Head Start programs were judged to be more effective among African American children, children in central-city programs, and children in the southeastern portion of the United States. It can be argued that these are the groups having the greatest need for Head Start. Also, the parents of Head Start children were pleased with the program's impact on their children and with their own extensive involvement in it. Finally, Head Start participants were superior to non-Head Start children on certain tests of cognitive development in the area of learning readiness.

Other research that followed up on the latter progress of low-income children who had been in compensatory education programs reports positive long-term results. Low-income youngsters between the ages of nine and nineteen who were in preschool compensatory programs performed better in school, had higher achievement test scores, and were more motivated academically than low-income youth who had not been in compensatory education programs when they were younger (Bruner, 1982; Etzioni, 1982). Later studies have also been supportive of the benefits of Head Start (Schweinhart, 1992; Zigler and Muenchow, 1992).

Politicians, both Republican and Democrat, currently recognize the benefits of Head Start. After an overwhelming vote in both the House and Senate, new Head Start authorization legislation was signed by President Clinton on May 19, 1994. Under this legislation, funding was significantly increased and coverage extended to approximately 800,000 children (Dewar and Vobejda, 1994).

What is the relationship between community control and educational equality?   Traditionally, the legal authority to finance and administer elementary and high schools lies within the states rather than with the federal government. The states in turn delegate most of this authority to local school systems. Ultimately, then, power over schools rests with school administrators who carry out policies established by the local school boards.

The local school board is thought to represent the interests of the community residents, and it is assumed that administrators represent the professional needs of teachers and the educational needs of students. These assumptions have been challenged. It is argued that the current school governance arrangement has become so centralized and bureaucratic that accountability to parents and others in the local neighborhood is rare. It is also argued that political decentralization in urban areas, which would place direct control of each school in the hands of neighborhood residents, is desirable because the present remote, bureaucratic structure fails to meet the needs of children, needs that vary from one neighborhood to another. This is held to be a particular disadvantage by minorities, who charge that schools are oriented toward white students' needs and are not responsive to the problems of minority students.

Although many school administrators have agreed to share major educational decisions with parents, it remains to be seen whether they are willing to share their power with parents. There is also the question of whether neighborhood schools can raise enough money to insure quality education. Finally, there is the possibility that community control will increase school segregation.

Do private schools teach the disadvantaged better than public schools?   Amidst all the controversy surrounding equality of educational opportunity in public schools has come some interesting research by James Coleman and his colleagues (1981, 1982a, 1982b, 1982c) on the value of a private school education. Hypothesizing that most private schools are more immune than public schools to pressures for equalizing school policies, Coleman and his colleagues attempted to investigate whether actual achievement among students attending private schools was better than that among public school students. Much to the dismay of many public school educators, this research suggested that the private schools studied produced higher student achievement levels in mathematics and vocabulary even when the background characteristics of students in both types of schools were similar. Of particular interest to the researchers was the observation that minority children in these private schools, particularly Catholic schools, performed better than their counterparts in public schools. The authors argued that higher academic standards, stricter discipline, more homework, and student perceptions of academic emphasis within the private schools accounted for the differences in achievement levels. These findings were used by Coleman and his associates to argue for changes in public school policies that could enhance achievement levels for all students and to suggest that the growing private school movement might have beneficial consequences for many of America's children.

The results of the Coleman study have been criticized. Arthur Goldberger and Glen Cain (1981, 1982) accuse the researchers of being biased toward private schools. Others contend that Coleman has overstated his conclusions. The government agency sponsoring the Coleman research, for example, contends that the performance comparisons are inaccurate because a greater percentage of private school students (70 percent) pursue college-bound academic programs than that of public school pupils (about one-third). After a reanalysis of his data, Coleman agreed that precollegiate public school seniors perform as well on standardized tests as Catholic school seniors.

These criticisms should not be allowed to overshadow the results of the Coleman study. Diane Ravitch places the debate into proper focus:

> What it [the Coleman study] does show is that good schools have an orderly climate, disciplinary policies that students and administrators believe to be fair and effective, high enrollment in academic courses, regular assignment of homework, and lower incidence of student absenteeism, classcutting, and other misbehavior. These conditions are found more often in private than in public schools (though they are not present in all private schools, and they are not absent in all public schools) . . . (Ravitch, 1982:10).

FEEDBACK

1. Educational equality exists when schooling produces the same _____ for lower-class children and minorities as it does for other children.
2. Sociological research has concluded that schools can provide educational equality. *T or F?*
3. Most sociologists agree that IQ test scores do not measure inherited learning ability. *T or F?*
4. A _____ is a tax allowance granted to cover part of the costs for an academic year at the parental school of choice.
5. There is evidence that school desegregation reduces educational inequality. *T or F?*
6. _____ is an approach to education in which the curriculum accents the viewpoints, experiences, and contributions of women and diverse ethnic and racial minorities.
7. Compensatory education appears to benefit deprived children. *T or F?*
8. Public schools appear to teach the disadvantaged better than private schools. *T or F?*

Answers: 1. results 2. F 3. F 4. voucher 5. T 6. Multiculturalism 7. T 8. F

DOING RESEARCH

Robert Rosenthal and Lenore Jacobson—Pygmalion in the Classroom

Anecdotal evidence supports the self-fulfilling prophecy—one person's expectation of another person results in the expected behavior. Robert Merton (1968, originally published in 1948) applied this concept from his observation of many banks in the 1930s that eventually closed because depositors withdrew their money on the *false* belief that the banks were failing. Their expectation, or prophecy, led to its fulfillment. Psychologist Gordon Allport (1958) showed that nations expecting to wage war usually do so.

Robert Rosenthal and Lenore Jacobson (1989) wished to provide greater scientific legitimacy for the self-fulfilling prophecy. They chose Oak School, a public elementary school located in a predominantly lower-class community, as the setting for a natural experiment. They hypothesized that children whose teachers expected them to increase in intelligence would do so at a greater rate than comparable children whose teachers expected no particular IQ gains. At the beginning of the experiment, a test was given to all of the Oak School students. Although it was falsely purported to be a predictor of academic "blooming" or "spurting," it was actually a nonverbal intelligence test. Rosenthal and Jacobson subsequently identified for the teachers 20 percent of the children who allegedly were ready for a dramatic increase in intellectual growth. In fact, the names of these students had been selected by using a table of random numbers. The difference in potential for academic growth between the children in the experimental group (students said to be on the verge of blooming) and the control group (the rest of the students) existed only in the minds of the teachers.

Intellectual growth was measured by the difference between a

SYMBOLIC INTERACTIONISM

According to one of the basic assumptions of symbolic interactionism, the way we view things is based on the symbols and meanings learned from others. Thus, symbolic interactionists are very interested in the process of socialization, and schools are very much in the business of transmitting culture.

The Hidden Curriculum

Modern society places considerable emphasis on the verbal, mathematical, and writing skills an adult needs to obtain a job, read the newspaper, balance a checkbook, and compute income taxes. Consequently, children in the early grades spend the majority of their day being drilled in spelling lessons, multiplication tables, sentence construction, and reading comprehension. Anyone who walks into a classroom full of poems, graphs, books, and alphabet cards can see how important the written word is in contemporary education.

However, schools teach much more than basic academic skills. They also transmit to children a variety of values, norms, beliefs, and attitudes (Trimble, 1988; American Association of University Women, 1992; Sadker and Sadker, 1995). There is a **hidden curriculum** that goes far beyond the learning of grammar, mathematics, reading, and other academic skills (Jackson, 1968; Giroux and Purpel, 1983). The hidden curriculum teaches children such things as discipline, order, cooperativeness, and conformity—skills thought to be necessary for success in modern bureaucratic society, whether one becomes a doctor, college president, salesperson, or assembly-line worker. Schools, for example, enable children to make the transition from their closely knit, cooperative family to the loosely knit, competitive adult occupational world. According to Robert Dreeben (1968), the school provides systematic practice for acquiring adult norms. Schools teach children to think of themselves as students and to operate independently in the pursuit of academic achievement. The values of independence and achievement are emphasized through individual testing and grading. Children also learn that their teachers evaluate them as students, not as relatives, friends, or equals.

Harry Gracey (1977) has described kindergarten as an academic boot camp in which children are taught to follow rules and routines without question, even when the rules and routines make no sense whatever to them. Gracey was not advocating bureaucratic schools, but he did observe that this training in kindergarten (and later grades) ultimately makes the transition from school to employment in a bureaucratic society relatively easy. (Refer to *Doing Research* in Chapter 4, "Socialization Over the Life Course.")

child's IQ score at the end of the previous school year (pretest) and that same child's IQ score eight months after the next school year began (posttest). As Rosenthal and Jacobson expected, the children in the experimental group gained more IQ points than the control-group children (a 12-point gain versus an 8-point gain). The IQ gain of the experimental-group children over those in the control group was the most pronounced among first and second graders. First graders in the high expectancy group gained over 27 IQ points compared to 12 points in the control group. Among second graders, the advantage was 16.5 IQ points to 7.

The experiment also involved another IQ posttest two years after the study began to see if the observed changes persisted. A surprising twist occurred. The younger children lost their expectancy advantage, and the children in the upper grades increased theirs. Again, no definitive explanation for these results was given.

Despite the inconsistency of these results, the self-fulfilling prophecy appears to be an operating force in schools. Moreover, its presence has been observed in other social settings. Research subjects behave as they think researchers expect, and a client's progress in therapy is influenced by the therapist's expectations (Rosenthal and Rubin, 1978). People who are expected by others to be hostile will exhibit more hostile behavior (Snyder and Swann, 1978).

Critical feedback

1. By what means does the self-fulfilling prophecy work? That is, how are expectations transmitted from one person to another, and how do these expectations produce behavior?
2. What are the implications of the self-fulfilling prophecy for students? For teachers?
3. Design a study of the self-fulfilling prophecy that would improve on the one conducted by Rosenthal and Jacobson.

Textbooks and Teachers

How do textbooks socialize children? The school curriculum contains numerous political and social science courses. Potentially neutral courses such as history and government are slanted in favor of a particular society's view of history. Accounts of the American Revolution, for example, are not the same in British and American textbooks. The resistance of schools to presenting critical accounts of their history is especially apparent when a teacher attempts to introduce a controversial book, such as Alice Walker's *The Color Purple*. There are very few societies willing to admit that their way of life is less than perfect, and this viewpoint is reinforced in public schools.

School textbooks contain many implicit values and beliefs. Sociologists interested in gender stereotypes point out that elementary textbooks tend to show men engaged in challenging and aggressive activities but portray women as homemakers, mothers, nurses, and secretaries. Women are not only portrayed in traditional roles, but also appear far less frequently than men. Similarly, the little white house with the picket fence may be part of the worldview of middle-class Americans, but parents of low-income or inner-city children have complained that such pictures of middle-class life harm their children. Poor children who compare their homes with middle-class homes feel out of place in school (Trimble, 1988; Gibson and Ogbu, 1991).

How do teachers socialize children? Classroom teachers have a unique and important role in socializing children. Teachers are the first authority figures many children encounter on a daily basis outside the family. In addition, most parents urge their children to do as their teachers ask, because their children's futures may well be affected by their success or failure in school.

Teachers often intentionally socialize students by requesting them to perform academic tasks in prescribed ways. At the same time, teachers affect children unintentionally. In fact, the power of teachers in socializing students has been of great interest to social scientists. For example, in their book entitled *Pygmalion in the Classroom,* Robert Rosenthal and Lenore Jacobson (1989), demonstrated the **self-fulfilling prophecy**— when an expectation leads to behavior that causes the expectation to become a reality (Merton, 1968). In their study, elementary schoolteachers were given a list of children in their classrooms who, according to the researchers, were soon to blossom intellectually. Actually, these children were picked at random from the school roster and were no different from the student body in general. At the end of the year, the authors found that this randomly selected group of children had scored significantly higher on the same intelligence test they had taken at the beginning of the year. Their classmates as a group did not. According to Rosenthal and Jacobson, the teachers expected these "late bloomers" to spurt academically. Consequently, the teachers treated these students as if they were special. This behavior on the part of the teachers caused the students to become high academic achievers. (See *Doing Research*.) Although not all researchers have confirmed the operation of the self-fulfilling prophecy in the classroom, some have. Eleanor Leacock (1969), for example, also found the self-fulfilling prophecy at work in a study of second and fifth graders in black and white low- and middle-income schools. Both studies also demonstrated that negative self-impressions are transmitted by teachers through the self-fulfilling prophecy.

Subsequent research generally supports dramatic effects of the expectations of teachers (and parents) on the ability of students to learn (Gamoran, 1986). To quote the Children's Defense Fund: "Educational research shows that children adjust their performance and self-image to meet adult expectations, and children who are placed in lower ability classes soon come to believe that they cannot excel in school" (Children's Defense Fund, 1991:80).

Sexism is also being identified as a part of the hidden curriculum. Following a long line of earlier researchers, Myra Sadker and David Sadker (1995) contend that America's schools are failing to be fair to girls because teachers cripple them through differential treatment in the classroom. Gender bias robs girls of the quality of education equal to that received by boys. Well-meaning teachers unconsciously transmit sexist assumptions. Girls, for example, learn to talk softly, to avoid certain subjects (especially math and science), to defer to the alleged intellectual superiority of boys, and to emphasize beauty over brains. This not only makes young women second-class citizens at maturity, these researchers conclude, but also deprives the nation of the important contributions women could make to the nation and the world. There is additional evidence that girls are guided in school toward traditional female jobs and away from high-paying, powerful, and prestigious jobs in science, technology, and engineering (Millicent, 1992).

Postscript

The material in this chapter to this point provides an opportunity to demonstrate ways in which the major theoretical perspectives are sometimes intertwined. Functionalism, for example, focuses on ways in which the educational system socializes students into dominant values and selects and screens students through the process of tracking. For conflict theorists, these functions constitute means by which people are placed in "appropriate" class slots and the existing stratification structure is perpetuated. Both symbolic interactionism and conflict theory are used to analyze the hidden curriculum in schools. For symbolic interactionism, the hidden curriculum socializes students to function as members of society. Conflict theory, on the other hand, sees in the hidden curriculum a way of getting people on the bottom of the stratification structure to accept the status quo.

HIGHER EDUCATION

College Attendance

Over fourteen million students are now enrolled in one of over 3,600 American colleges and universities. Immediately following World War II, as a result of the G.I. Bill and the influx of veterans, enrollment in institutions of higher education began to grow. This increase in the proportion of high school graduates entering and completing college has continued to grow (U.S. Bureau of the Census, 1996a; also see Figure 12.2). Since 1980, the percentage of high school graduates attending college has increased from 50 percent to 62 percent. This increase has occurred despite predictions of a downturn in college enrollment, thanks mostly to the increased enrollment of persons twenty-five years and over (U.S. Bureau of the Census, 1996a).

FEEDBACK

1. According to symbolic interactionists, schools are very much in the business of transmitting _____.
2. The _____ teaches children such things as discipline, order, cooperativeness, and conformity—skills thought to be necessary for success in a modern bureaucratic society.
3. _____ are the first authority figures children encounter on a daily basis outside the family.
4. The _____ occurs when an expectation leads to behavior that causes the expectation to become a reality.
5. Match the following research topics with the three theoretical perspectives:
   a. functionalism
   b. conflict
   c. symbolic interactionism
   (1) requirement of unnecessary credentials by schools
   (2) benefit of schools for society
   (3) effects of teacher reactions on the self-esteem of students

Answers: 1. *culture* 2. *hidden curriculum* 3. *Teachers* 4. *self-fulfilling prophecy* 5. a. (2) b. (1) c. (3)

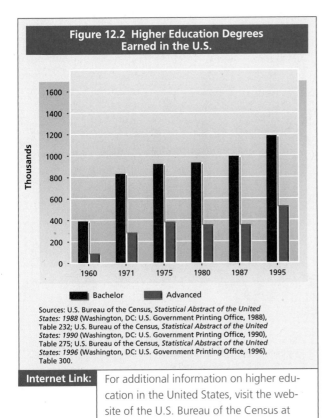

Figure 12.2 Higher Education Degrees Earned in the U.S.

Sources: U.S. Bureau of the Census, *Statistical Abstract of the United States: 1988* (Washington, DC: U.S. Government Printing Office, 1988), Table 232; U.S. Bureau of the Census, *Statistical Abstract of the United States: 1990* (Washington, DC: U.S. Government Printing Office, 1990), Table 275; U.S. Bureau of the Census, *Statistical Abstract of the United States: 1996* (Washington, DC: U.S. Government Printing Office, 1996), Table 300.

Internet Link: For additional information on higher education in the United States, visit the website of the U.S. Bureau of the Census at http://www.census.gov/

Why was a decline in college enrollment predicted? College enrollment was expected to drop in the 1980s because of a projected decline in the number of Americans eighteen to twenty-four years old, the traditional college-age category. This population category has in fact declined 11 percent since 1980, although it is now on the upswing. (See Figure 12.3.)

Why didn't the drop in college enrollment occur? Several trends account for the increase in college enrollment in the face of fewer eighteen- to twenty-four-year-olds. A substantially larger proportion of this age category is attending college. In addition, the number of students over age twenty-four attending college, particularly women, has continued to grow, and the number of students remaining in college after their first year has increased (U.S. Bureau of the Census, 1996a).

At least three factors account for these trends. Fearful of declining enrollments, colleges and universities mounted strong student recruitment campaigns to enlarge their applicant pools. This stimulated student demand. Also, low-paying service jobs have increasingly replaced high-paying manufacturing jobs, resulting in fewer attractive noneducational options for those with high school diplomas. With the loss of high-wage, unionized jobs in blue-collar industries such as automobiles and steel, the incentive to attend college has increased. Finally, the increase in community colleges and branch campuses offers an opportunity for many who could not have otherwise afforded to continue their education after high school.

Education, Income, and Occupation

Advanced industrialization has made education the ticket to better jobs and higher incomes. Although a college degree may not be worth what it used to be, it still has a significant payoff financially and occupationally (Guinzburg, 1983; Katchadourian and Boli, 1994).

What is the economic payoff of college? As Figure 12.4 shows, college education results in higher annual income regardless of race or gender. Among African American and white males who have a college education or more, annual average income is well more than

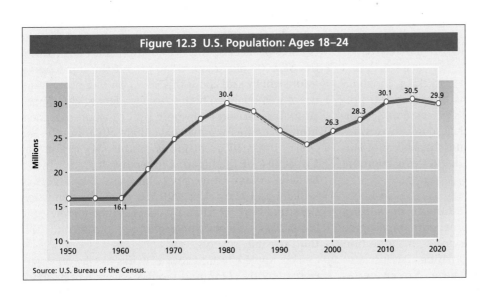

Figure 12.3 U.S. Population: Ages 18–24

Source: U.S. Bureau of the Census.

double that of high school dropouts. White male high school graduates earn annually about 58 percent of that earned by white college graduates. African American males with high school diplomas earn about 54 percent of that earned by African American male college graduates. The results are even more dramatic among females. White college graduates earn about three times the amount of white high school female dropouts; for African American females, the increase is also threefold. White female high school graduates earn each year only about 53 percent of that of white college graduates. African American females with high school diplomas earn only 51 percent of that of female college graduates.

The down side of the data in Figure 12.4 is African American, white, male-female comparisons. Although African Americans and women benefit from a college education, they clearly suffer in comparison to white males with identical educational credentials. (Refer to Chapters 9, "Inequalities of Race, Ethnicity, and Age," and 10, "Inequalities of Gender," for more detailed analyses of the socioeconomic situation of African Americans and women in the United States.)

To what extent are occupational status and education related? If anything, the benefits of college are even more apparent occupationally. For African Americans, whites, males, and females, occupational status (measured by blue-collar versus white-collar status) increases dramatically with higher education. The percentage for all race and gender combinations employed in white-collar occupations rises dramatically as the extent of college education increases, whereas the percentages in blue-collar jobs decline sharply. Although African Americans and females are more concentrated in lower-level jobs within the higher-status occupations, the significant relationship between occupational status and higher education cannot be dismissed (U.S. Department of Labor, 1997a).

Social Stratification in Higher Education

Earlier in the chapter we described tracking, in which secondary schools place students in educational programs consistent with the school's expectations for students' future employment. Colleges and universities also participate in the sorting and channeling of students. The two major tracks in higher education are two-year and four-year colleges.

How do four-year colleges contribute to the sorting and channeling process? The benefits of a degree from a four-year college were discussed in the section on education, income, and occupation. Because students must attend a four-year college to receive valuable undergraduate degrees, this type of college

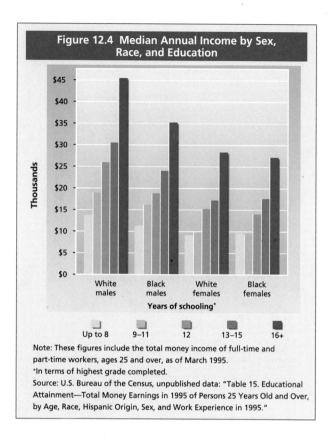

Figure 12.4 Median Annual Income by Sex, Race, and Education

Years of schooling*

Up to 8 | 9–11 | 12 | 13–15 | 16+

Note: These figures include the total money income of full-time and part-time workers, ages 25 and over, as of March 1995.
*In terms of highest grade completed.
Source: U.S. Bureau of the Census, unpublished data: "Table 15. Educational Attainment—Total Money Earnings in 1995 of Persons 25 Years Old and Over, by Age, Race, Hispanic Origin, Sex, and Work Experience in 1995."

represents the advanced track (Monk-Turner, 1990). Students from higher social class homes are more likely to be able to afford four-year colleges and have a better chance for occupational and financial success after receiving their degrees. Thus, four-year colleges tend to preserve the existing stratification structure.

The 1990s have brought a new, important wrinkle to the contribution of four-year colleges to the preservation of America's stratification structure. The spiraling cost of higher education is currently making it more difficult for bright middle-class and lower middle-class students to attend the nation's more prestigious schools, the colleges that traditionally increase the chances of success. This deescalation effect begins with the $20,000-plus required to attend the top schools. (More will be said about this later.) An increasing number of well-qualified high school graduates who cannot afford the price tag and whose family income bars them from governmental financial aid are attending the less costly but still prestigious "public ivys" such as University of Michigan, University of California at Berkeley, Miami University (Ohio), and the College of William and Mary. This shift has taken the admission slots traditionally filled by other bright middle- and lower-class students who are forced into second- and third-tier state schools.

This ripple effect, of course, works its way on down. Consequently, community college enrollment has risen from 4.7 million in 1985 to 6 million in 1991 (U.S.

Bureau of the Census 1996a). Almost 40 percent of the nation's college students, almost one-third of whose families earn less than $25,000 annually, now attend community colleges. This pushing of less affluent students down the educational hierarchy can only promote the perpetuation of the existing stratification structure, because different learning and earning opportunities have traditionally been associated with different levels of schools. Equal opportunity in education, a long-standing American ideal, is in some jeopardy (Toch and Slafsky, 1991).

How do two-year colleges sort and channel students?
American universities and colleges were faced with a dilemma following World War II. Because there was not enough room to accommodate all the high school students who wanted to go to college, many public universities had to limit the number of students by admitting only a few or by admitting too many students and identifying the best after such students' studies began. Needless to say, many students who wanted to go to college could not. A solution to this dilemma was the junior college. In his classic analysis, Burton Clark (1960) argues that the fantastic growth in junior colleges allowed higher education the appearance of providing a college education for everyone who wanted it—as the public wished—but actually channeled low-performing students into terminal programs, as high schools had traditionally done. Junior colleges were designed to prepare their higher-performing students for transfer to four-year colleges and to channel other students into the job market.

The sorting and channeling process in two-year schools, now more commonly referred to as community colleges, has changed (Grubb, 1991; Vaughan, 1991). During the 1970s, fierce competition for the relatively few jobs that required a college education discouraged students who had been led to believe that a college degree would bring economic and occupational security. The Carnegie Commission on Higher Education (1973) was concerned that public resources were being ill spent on students who didn't need a college degree. It feared the country might soon experience a political crisis from unhappy, underemployed, and unemployed college graduates and recommended the encouragement of students to enroll in terminal vocational programs within community colleges. Such training, it was argued, would permit students to enter middle-level occupations that would provide greater job satisfaction and security than they would have with college degrees. In a crowded labor market, a two-year vocational program was thought to serve many students better than a bachelor's degree (Pincus, 1989). Thus, the primary focus of most community colleges is on *career education*. Such schools tend to concentrate on training students for a variety of vocational and semi-

professional occupations in such areas as mechanics, business, computers, health care, and engineering.

This career emphasis has brought the charge that two-year colleges constitute a new form of tracking within higher education. By creating training programs to serve corporate needs, some critics believe that the long-standing liberal arts orientation of community colleges is being undermined. Career education, it is feared, reduces the chances of community college students being prepared to transfer to four-year schools (Pincus, 1989). In fact, of those who attend two-year colleges, only 12 percent graduate from four-year schools (Lieberman, 1989). Critics have also revived Burton Clark's thesis, contending that contemporary community colleges, like the earlier junior colleges, "cool out" students whose high aspirations outstrip their realistic chances in the job market by channeling them into vocational programs. As a result, lower socioeconomic and minority students are disproportionately channeled into two-year colleges (Clark, 1980; Eaton, 1990; Lee and Frank, 1990; Grubb, 1991). Fred Pincus, an early critic, stated that the stratification structure within higher education and within community colleges perpetuates the class and racial inequalities present in the larger society. Different schools and different tracks, he argued, would produce differential economic outcomes (Pincus, 1980). Pincus' conclusion has received some support by more recent researchers who also believe that community colleges contribute to the maintenance of the current stratification structure:

On average, community college entrants achieve a lower occupational status than four-year college entrants. . . . Community colleges do not provide two years of a college education; they provide a different kind of education. Community college entrants may take courses that are not readily transferable to four-year institutions. Tracking within community colleges may influence the deleterious impact of community college entrance on adult occupational attainment (Monk-Turner, 1990:725).

What are the benefits of the community college system? Supporters of the community college system disagree with the critics. They see corporations getting the types of employees they need, four-year colleges not wasting resources on probable dropouts, students finding appropriate employment, and minorities being better served (Brint and Karabel, 1989). In the fall of 1990, one-half of the minority students in higher education attended a two-year school (Eaton, 1991).

Whatever the merits and outcomes of this controversy, certain benefits of community colleges should not be overlooked. In addition to being relatively inexpensive, community colleges are accessible and convenient for students who wish to attend part-time or must take evening and weekend classes. Consequently, many

Community colleges, such as this one in Boston, perform an important function in the sorting and channeling of college students. Community colleges are designed for career education, training students for jobs in such areas as business, computers, and health care.

people attend college who would not otherwise. Moreover, because of the skyrocketing costs of attending four-year schools, increasing numbers of students are taking their first two years of college closer to home before entering a four-year college. This enables qualified students to obtain college degrees who otherwise probably would not. Moreover, many community colleges are working on programs to promote a higher rate of transfer to four-year colleges ("Transfer: Making It Work," 1987; Eaton 1990). In 1993, 39 percent of all college students in the United States were enrolled in two-year colleges (U.S. Bureau of the Census, 1996a).

Change in Higher Education

What are some of the ways universities are changing? David Riesman (1980) argues that college studies and college life have changed dramatically during the past twenty years. Riesman contends that college faculties and programs reached their peak in America during the ten to fifteen years before the war in Vietnam. During this period, institutions of higher learning were very selective, academic programs were extremely rigorous, and college faculties had extensive control over the direction of student learning and development. This academic revolution did not last long, however.

A shift from academic merit to student consumerism began in the mid-1960s. According to this trend, increased competition for declining numbers of students has forced most American undergraduate and graduate programs to attempt to package their curricula more attractively in an effort to recruit students. Thus, according to Riesman, universities during the past two

decades have increasingly lost their ability to shape development along academic and intellectual avenues. Instead, students increasingly demand courses and programs that are more instrumental in gaining employment. Riesman documents this contention by discussing the rise of vocational programs in community colleges, noting the shifts of students' majors into business and economics from the social sciences and the humanities, and pointing out recent court cases in which students have sued their former colleges for giving them low grades or poor instruction.

Stanley Aronowitz and Henry Giroux (1985) are also concerned about the influence of labor market considerations on higher education. They point to the rise of business enrollments and the concomitant decline of liberal arts majors. They see the social sciences and the humanities departments becoming service departments for nonmajors. They witness the diminution of graduate programs and faculty in liberal arts fields. As a result, they contend, a generation of liberal arts scholars has been lost. In the first place, pay in academia pales in comparison to the lure of business opportunities. Second, only those graduating with Ph.D.s from elite universities have any hope of finding teaching jobs beyond the high school level.

An explosion was set off in 1987 when Allan Bloom, a University of Chicago professor, published *The Closing of the American Mind*. This best-selling book became the touchstone for those who believe that higher education is on its way to intellectual chaos. According to these critics of higher education, academia is now under the control of administrators and faculty who were part of the radical left in the 1960s. Universities, they believe, are using a radical egalitarian philosophy to elevate works of popular culture to the level of Platonic studies. The most frequently cited illustration is a Stanford University faculty decision in 1988 to drop some traditionally classic works by white male authors from a required Western culture course in favor of some women and minority authors. A basic fear of these academic critics is that the gender, racial, and ethnic background of authors is supplementing merit as a basis for designing programs, courses, and reading lists. This is a debate that promises to continue for some time.

Another problem in higher education—declining public financial support—threatens the traditional high quality, low cost of public higher education for the foreseeable future. The Association of Governing Boards of Universities and Colleges describes the situation as "the new depression in higher education."

Since World War II, the public has supported higher education through its tax dollars. That commitment is currently diminishing. Between 1987 and 1992, state appropriations per full-time student in American state colleges and universities declined by 13 percent in real terms (Quinn, 1993b). Even the most prestigious public universities have been cut, causing them to become semipublic institutions at best. The University of California's system, consistently the world's finest, has seen the proportion of its state-funded financial base shrink from a high of 70 percent thirty years ago to less than 25 percent today. The University of Michigan now receives less than 20 percent of its funding from the state.

State colleges and universities have responded to reduced state funding in two basic ways. Private funding and tuition charges have increased in an attempt to make up for the state financial shortfall. As a result, the cost of a college education is skyrocketing (Morganthau and Nayyar, 1996). The low-cost, high-quality system of higher education provided the veterans after World War II and their children, the baby boomers (born between 1946 and 1964), is being denied the post-boomer generation. From 1990 to 1993 alone, tuition increases at state colleges and universities increased from 22 to 49 percent, depending on the type of school.

(See Table 12.1.) In recent years, due to the public outcry over rising tuition costs, these increases have leveled off, closely matching the inflation rate, averaging 4 to 6 percent per year. If the costs of higher education increase at 8 percent annually, students entering kindergarten when President Clinton was inaugurated in January 1993 can expect to pay about $300,000 for a degree from premier institutions such as Yale and Harvard, and as much as $75,000 for a degree at a public college or university (Johnson, 1995).

Parents and the general public are in shock. Tuition increased 135 percent between 1980 and 1991, whereas family income rose only 67 percent during the same period (de Witt, 1991). With college tuition rising at twice the rate of inflation, and with the current no-new-taxes mood of the country, the call is for universities to restructure. If companies such as IBM and Xerox can downsize and do more with less, the argument goes, so can public colleges and universities. "Time to Prune the Ivy" was the clever title of an article in *Business Week* (Farrell, 1993). Higher education and the public are obviously in conflict; the taxpayers are generally opposed to tax increases, and university administrators and professors fear a decline in quality due to restructuring and downsizing. Is the future of a country that has built its prosperity on higher education at risk?

A final issue on the horizon for higher education, if it emerges as many suspect, is the need for colleges and universities to redesign themselves. The technologically driven information age, contend these observers, will lead colleges and universities to move from a geographically bound campus to a virtual campus. Although the traditional residence-based design of higher education will continue, many believe that higher education will increasingly be characterized by geographically dispersed faculty and students. Distance learning will be necessary both because technology makes it possible and because higher education is being redefined from a set number of semesters to a life-long process of learning necessitated by a rapidly changing global economy (Brody, 1997).

TABLE 12.1	Average Annual Cost at a Public or Private College or University*		
	2-Year Public	**4-Year Public**	**4-Year Private**
Student living on campus	NA**	$9,649	$20,361
Student who commutes	$4,721	$7,747	$17,351
Tuition increase, 1995–96 to 1996–97	5%	6%	5%

*Includes tuition, fees, room, board, and books.
**Not applicable.
Source: College Board, Press Release, "1996–97 Increase in College Costs Averages Five Percent; Student Financial Aid at Record High," September 25, 1996.

1. As expected, college enrollment in the United States declined in the 1980s because of a decrease in the number of eighteen- to twenty-four-year-olds. *T or F?*
2. Although a college education pays off financially for white males, it does not benefit African American men and women. *T or F?*
3. Occupational status gains produced by a college education are less apparent than the relationship between income and higher education. *T or F?*
4. Four-year colleges appear to perpetuate the existing social stratification structure. *T or F?*
5. The primary focus of most community colleges is _____ education.
6. According to David Riesman, most colleges and universities have shifted from academic merit to _____.
7. Public commitment to higher education is as strong for the postboomer generation as it was for their parents. *T or F?*

Answers: 1. F 2. F 3. F 4. T 5. career 6. student consumerism 7. F

SUMMARY

1. As industrialization and urbanization have proceeded, the ability of the family to prepare its young for the future has declined. As a result, formal schooling has become necessary as preparation for living. The early emphasis in American schools was on "civilizing" the young, but after the turn of the century, the emphasis shifted to educating the young for higher-level jobs.

2. Early schools were modeled after businesses and have increasingly become more bureaucratic. Advocates of open classrooms and cooperative learning contend that bureaucratically run schools fail to take into account the emotional and creative needs of individual children. The evidence suggests that some aspects of these alternatives make some contributions that should not be overlooked. The back to basics movement, however, seems to clash with the open-classroom philosophy.

3. Functionalists see the emergence of the educational institution as a response to society's needs. The manifest functions of education include the transmission of culture; social integration; selection and screening of talent; promotion of personal growth and development; the dissemination, preservation, and creation of knowledge. Schools also serve latent functions.

4. America is believed by many to be a meritocracy in which social status is achieved rather than determined by family background or parental status. Advocates of the conflict perspective point to flaws in the meritocratic model, contending that elites use the educational institution to preserve their privileged positions.

5. Many of the flaws in the meritocratic model appear within the context of educational inequality, which exists when schooling produces different results for lower-class and minority children than it does for other children. Schools themselves promote educational inequality by reacting to students' social-class backgrounds and racial or ethnic characteristics rather than to their abilities and potential. Schools use intelligence testing, which ignores the environmental deficiencies of lower-class students, to channel students into noncollege tracks.

6. Although there is disagreement, some evidence suggests that school desegregation reduces educational inequality. Three other approaches to promoting educational equality are compensatory education, increased community control of schools, and private schooling.

7. Symbolic interactionists emphasize the nonacademic socialization that occurs within schools. Through the hidden curriculum, children are taught a variety of values, norms, beliefs, and attitudes thought to be desirable by a society. This nonacademic socialization is accomplished in large part through textbooks and teachers. Much of this socialization is designed to help young people make the transition from home to the larger society.

8. Contrary to expectation, college enrollment increased during the 1980s. This has occurred in part because of active college recruiting and because of the decline of high-paying unionized manufacturing jobs. College education still pays off handsomely in the United States. Both income and occupational prestige increase with higher education.

9. One important aspect of social stratification in higher education relates to the sorting and channeling of students in two-year and four-year colleges. Both types of colleges tend to perpetuate the existing stratification structure in American society. Four-year schools tend to draw students from more affluent homes and increase the chances of the financial and occupational success of such students. According to critics, two-year, or community, colleges perpetuate the stratification structure of American society by channeling young people from less affluent and minority backgrounds into lower-paying and less prestigious occupations.

10. Defenders of community colleges point to the benefits of two-year schools for students, including affordability, convenience, and the opportunity for many people to

attend college who would not otherwise. In addition, the rising cost of higher education is causing more students to use the community college as a stepping stone to a four-year college, and many community colleges are creating programs to increase the rate of transfer to four-year colleges.

11. Critics are disturbed about the change in higher education that began to occur in the mid-1960s. Some see a deleterious shift from an emphasis on academic merit to student consumerism. Others fear that programs, courses, and reading lists are being influenced too much by considerations of gender, race, and ethnicity. The financial crisis in higher education looms as a genuine threat to the long-standing reliance of American society on an affordable, high-quality system of public higher education. Some observers contend that colleges and universites will have to redesign themselves to adapt to the information age.

LEARNING OBJECTIVES REVIEW

After careful study of this chapter, you will be able to:

- Describe the relationship between industrialization and education.
- Discuss various aspects of the bureaucratization and attempts to debureaucratize education in the United States.
- Outline the basic functions (manifest and latent) of the institution of education.
- Define meritocracy and explain the flaws in the meritocratic model.

- Define educational inequality and discuss the major issues surrounding it.
- Discuss the ways in which schools socialize students.
- Identify and describe the dominant issues in higher education.

CONCEPT REVIEW

Match the following concepts with the definitions listed below them:

____ a. multiculturalism
____ b. hidden curriculum
____ c. vouchers

____ d. cooperative learning
____ e. self-fulfilling prophecy
____ f. dysfunction

____ g. manifest function

1. A negative consequence of some element of a society.
2. The nonacademic agenda in schools that teaches children skills thought to be necessary for success in modern bureaucratic society.
3. An intended and recognized consequence of some element of a society.
4. An approach to education in which the curriculum accents the viewpoints, experiences, and contributions of women and diverse ethnic and racial minorities.

5. When an expectation leads to behavior that causes the expectation to become a reality.
6. A nonbureaucratic classroom structure in which students study in groups with teachers as guides rather than as the controlling agents determining all activities.
7. Tax allowances granted to cover part of the costs for an academic year at the school selected.

CRITICAL THINKING QUESTIONS

1. The open-classroom philosophy represents an alternative to the bureaucratic model of education. Use the symbolic interactionist perspective to convince a local school board of the desirability of adopting the open-class approach.

_____

_____

2. Assume for the moment that schools cannot provide educational equality. Analyze the consequences of the persistence of educational inequality within the context of the functions of education discussed in the text.

_____

_____

3. Think back to your early school years with the "hidden curriculum" in mind. Provide specific examples from your own experience to support the existence of the hidden curriculum.

_____

_____

4. Identify what you believe are the two biggest issues facing higher education in the twenty-first century. Develop recommendations for solving these problems.

_____

_____

MULTIPLE CHOICE QUESTIONS

1. **Schools emerged as the primary educational agent in industrializing societies because**
   a. the skills needed for survival in an agricultural society became increasingly important.
   b. knowledge increased in volume and complexity and family members gradually lost the ability to educate their children adequately.
   c. too many parents were qualified to teach their children, and not enough parents were willing to work in agriculture.
   d. parents no longer received the income they needed to educate their children at home.
   e. the birthrate increased dramatically in the early 1800s.

2. **Critics of the bureaucratic model of education claim that**
   a. bureaucracies are notoriously inefficient.
   b. the United States should have more "kyoiku mamas."
   c. teachers should treat all students the same.
   d. administrators are useless and should therefore be eliminated.
   e. it is hard to determine whether children have been "processed" correctly.

3. **Education contributes to social integration by**
   a. allowing individuals to learn several languages.
   b. exposing individuals to similar educational traditions.
   c. exposing individuals to different ethnic and cultural traditions.
   d. presenting various religious doctrines.
   e. promoting personal growth and individual development.

4. **Schools promote personal growth and development by**
   a. preparing students to assume a place in the occupational world.
   b. discouraging emotional experiences.
   c. tracking lower-class students into vocational programs.
   d. discouraging group activities.
   e. teaching students the importance of following orders in a bureaucratic society.

5. **The school's efforts at the dissemination of knowledge most commonly take the form of**
   a. field trips outside the classroom.
   b. using television as a teaching tool.
   c. classroom teaching.
   d. personal experience.
   e. homework assignments.

6. **According to Randall Collins, credentials are required for many jobs because**
   a. they allow those who can afford to get them to maintain a strong hold on elite occupations.
   b. they perpetuate a meritocratic society.
   c. they fulfill a vital service for society by verifying educational attainment.
   d. they prevent the best talent from being lost.
   e. they provide worthwhile goals for those who lack advanced degrees.

7. **A significantly larger proportion of upper-class children attend college than do lower-class children. According to the text, this is due in part to differences in**
   a. academic ability.
   b. economic resources.
   c. desire for intellectual growth.
   d. motivation to learn.
   e. study habits.

8. **The most recent concept of educational equality involves**
   a. equality of resources.
   b. equality of opportunity.
   c. equality in busing.
   d. equality in effects of schooling.
   e. equal education for equal ability.

9. **Which of the following is a *false* statement?**
   a. City dwellers usually score higher on intelligence tests than people in rural areas.
   b. Performance on IQ tests increases with age.
   c. Middle-class African American children score about as high on intelligence tests as middle-class white children.
   d. Intelligence tests are culturally biased.
   e. Recent evidence has caused most sociologists to agree that about half of human intelligence is genetically based.

10. **Central to _____ is the idea that special preschool programs can help deprived children to overcome an inferior intellectual environment.**
    a. forced busing
    b. tracking based on standardized testing
    c. expanding the scope of the Special Olympics
    d. postsecondary school remedial reading
    e. compensatory education

11. **In his study on the value of a private school education, James Coleman found that**
    a. public schools produce higher student achievement levels.
    b. private schools produce higher student achievement levels.
    c. students in public and private schools produce similar levels of achievement.
    d. minority children in public schools perform better than minority children in private schools.
    e. minority students perform equally well in private and public schools.

12. **Harry Gracey has described kindergarten as an academic boot camp in which children are taught to follow rules and routines without question, even when the rules and routines make no sense to them. This is an example of**
    a. the self-fulfilling prophecy.
    b. a meritocracy.
    c. a bureaucracy.
    d. the hidden curriculum.
    e. the looking-glass self.

13. **When girls are successfully taught in school to see their self-worth in terms of their physical appearance rather than in terms of their abilities to contribute to society, the _____ has been at work.**
    a. functionalist perspective
    b. self-fulfilling prophecy
    c. "blooming" effect
    d. multicultural perspective
    e. policy of tracking

14. **Which of the following is a *false* statement?**
    a. A college education results in higher annual income regardless of race and gender.
    b. College-educated African American males and white males make more than double the annual income of high school dropouts.
    c. African American females earn annually three times as much as white high school dropouts.
    d. African American male high school graduates earn annually over one-half of that earned by African American college graduates.
    e. A college education pays off economically for white Americans, but not for African Americans.

15. **In his analysis of higher education, David Riesman contends that**
    a. university administrators are not paying enough attention to keeping educational organizations in business.
    b. universities should devote more attention to intercollegiate athletics in order to retain alumni interest and financial contributions.
    c. universities have increasingly lost their ability to shape development along academic and intellectual avenues.
    d. universities are now under the control of administrators who were part of the radical left of the 1960s.
    e. universities should do more to get students involved in activist causes and social movements.

FEEDBACK REVIEW

True-False

1. Sociological research has concluded that schools can provide educational equality. *T or F?*
2. Four-year colleges appear to perpetuate the existing social stratification structure. *T or F?*

Fill in the Blank

3. Early public schooling in America was viewed as a mechanism of _____.
4. When schools keep children for working parents, they perform a _____ function.
5. Educational equality exists when schooling produces the same _____ for lower-class children and minorities as it does for other children.

6. According to symbolic interactionists, schools are very much in the business of transmitting _____.

7. According to David Riesman, most colleges and universities have shifted from academic merit to _____.

Multiple Choice

8. Contemporary public education is important primarily because of
   a. the quality of teachers in modern schools.
   b. efficient operation of school bureaucracies.
   c. historical changes in the economic and social structures of industrial societies.
   d. lack of qualified individual tutors.

9. The idea that some educational requirements are unnecessary for many jobs today is called
   a. status grouping.
   b. credentialism.
   c. meritocracy.
   d. conflict.

Matching

10. Match the following research topics with the three theoretical perspectives:

   ____ a. functionalism (1) requirement of unnecessary credentials by schools
   ____ b. conflict (2) benefits of schools for society
   ____ c. symbolic interactionism (3) effects of teacher reactions on the self-esteem of students

GRAPHIC REVIEW

The trend in educational level among whites and major minority groups since 1970 is displayed in Figure 12.1. Answer these questions referring to the data in this figure.

1. Relate the educational levels in the various categories to the process of industrialization. _____

_____

_____

2. Would functionalists and conflict theorists interpret the gap between the educational level of whites and the educational attainment of African Americans and Latinos differently or the same? Explain._____

_____

_____

ANSWER KEY

Concept Review	Multiple Choice		Feedback Review	
a. 4	1. b	9. e	1. F	7. student consumerism
b. 2	2. e	10. e	2. T	8. c
c. 7	3. b	11. b	3. social control	9. b
d. 6	4. a	12. d	4. latent	10. a. 2
e. 5	5. c	13. b	5. results	b. 1
f. 1	6. a	14. e	6. culture	c. 3
g. 3	7. b	15. c		
	8. d			

Chapter Thirteen

Political and Economic Institutions

LEARNING OBJECTIVES

After careful study of this chapter, you will be able to:
- Distinguish power and authority and identify three basic forms of authority.
- Compare and contrast the perspectives of functionalism and conflict theory on the nation-state.
- Identify the major differences among democracy, totalitarianism, and authoritarianism.
- Differentiate the view taken of the distribution of power in America by functionalism and conflict theory.
- Describe the major characteristics of capitalism, socialism, and mixed economic systems.
- Discuss the effects of the modern corporation on American society.
- Describe America's changing workforce composition and occupational structure.
- Discuss corporate downsizing and its consequences.

SOCIOLOGICAL IMAGINATION

Is corporate downsizing necessary to insure the health of large American corporations? Corporate executives who engage in massive reductions of their workforces point to the need to cut costs and increase profits in order to survive in an increasingly competitive global economy. However, growing evidence suggests that this policy may backfire. Many companies that downsized are now less competitive than before.

The 1997 strike by the Teamsters Union against United Parcel Service was based on a strong objection to the corporate policy of changing permanent jobs into temporary ones, an important aspect of downsizing. UPS, which early in the dispute asked President Clinton to intervene, demonstrated the intimate connection between the political and economic institutions in modern society. The interrelationships between these two institutions will become apparent in this chapter.

Members of all societies must deal with the problems of economics and politics. Economic problems exist because of the need to produce goods and services. The institution for the production and distribution of goods and services is the **economy**. According to many thinkers, production is the most fundamental of all human activities, but decisions about the distribution of goods and services after they have been produced can lead to severe conflict. For this reason, every society must develop some means of deciding who gets what, when, and how. Decisions regarding the distribution of goods, services, and social privileges are made by those with the most power. The institution within which

power is obtained and exercised—the **political institution**—is covered next, followed by coverage of the economic institution.

POWER AND AUTHORITY

Definitions of Power and Authority

Max Weber (1946), one of the founders of political sociology, defined **power** as the ability to control the behavior of others even against their will. Floyd Hunter (1953), who conducted a classic study of power and decision making in Atlanta, Georgia, has written of James Treat, a man whose extensive power was not evident from his simple, nonthreatening appearance. Treat's power was based not on his visible personal characteristics but on his ability to influence the actions of others—including Georgia's governor. His power existed in his relationships with others.

Weber recognized that power can be exercised in the form of *coercion*—power through force. A male police officer might have sexual relations with a female pris-

The most stable form of power is authority or power viewed as legitimate by those subjected to it. This photograph of two lawyers conversing beside the massive columns of a federal building in Washington, D.C., reflects powerful symbols of authority.

oner despite the woman's opposition. A government might confiscate the property of some minority without the group's permission and without proper recompense. In neither case do the victims believe that this use of power is right. Moreover, because victims of coercive power see it as illegitimate, they normally are resentful and wish to fight back. Weber recognized the inherent instability of a political system based on coercive power. Because force only produces compliance in the short run and when those in power can exercise surveillance, Weber believed that the political institution tends to rest on a more stable form of power. This more stable form of power is **authority**—power accepted as legitimate by those subjected to it. Students take exams and accept the evaluations they receive because they believe their teachers have the right to determine grades. Citizens pay taxes because they believe their government—whether it rests on "divine right" or popular vote—can legitimately make them do so. Authority tends to produce compliance because people believe that obedience is the appropriate action. The idea that authority always rests with those to whom it applies means that people comply with directives from above because they believe it is right and proper for them to do so. Because authority is based on others' acceptance, it is the most stable form of power. Its existence reduces the likelihood that coercion will be needed to maintain order.

Weber identified three forms of authority—charismatic, traditional, and rational-legal. Those invested with these forms of authority can expect compliance from those who recognize them as legitimate holders of power.

Forms of Authority

What is charismatic authority? **Charismatic authority** exists when the control of others is based on an individual's personal characteristics. Charismatic leaders are obeyed because of their magnetic personalities or the feelings of trust they inspire in others. Hitler, Gandhi, Napoleon, Julius Caesar, and Mao Tse-tung are typical examples of leaders whose legitimacy rested on charismatic authority. Although the power of a charismatic figure may be virtually unlimited while it lasts, charismatic authority is too unstable to provide a permanent basis of power for modern nation-states because it is linked to an individual and is difficult to transfer to someone else. Charismatic leaders ultimately die or relinquish their power, leaving no rules or traditions to insure the continuation of their power. Adolf Hitler, a Christlike figure for many Germans, made a futile attempt at the end of World War II to name his successor and other high government officials. But as historian John Toland has noted:

Hitler's death brought an abrupt, absolute end to National Socialism. Without its only true [charismatic] leader, it burst like a bubble. . . . What had appeared to be the most powerful and fearsome political force of the twentieth century vanished overnight. No other leader's death since Napoleon had so completely obliterated a regime (Toland, 1976:892).

Therefore, even those governments controlled by charismatic leaders eventually come to rely on other types of authority. The two alternatives identified by Weber are traditional authority and rational-legal authority.

What is traditional authority? In the past, most states relied on **traditional authority,** in which the legitimacy of a leader is rooted in custom. Early kings, for example, often claimed that they ruled by a divine right based on the will of God. A key element of traditional authority was its perpetuation in an orderly way. An orderly transfer of power was possible because only a certain pool of individuals was eligible to become the ruler. Until fairly recently, custom was sufficiently strong to insure the legitimacy of those who inherited the right to rule. Kings in eighteenth-century Europe could actually count on the loyalty of their subjects. For this reason, traditional authority provided a more stable basis for states and governments than did charismatic authority.

What is rational-legal authority? Most modern governments are based on **rational-legal authority.** The power of government officials is based on the offices such leaders hold. Those who hold government office are expected to operate on the basis of specific rules and procedures that define and limit their rights and responsibilities. In other words, rational-legal authority is invested in statuses rather than in individuals; power is lost when a status with rational-legal authority is lost. Citizens are expected to follow the directives of the political leaders, but these leaders are expected to stay within their legal authority. Even presidents, as Americans saw with Richard Nixon, can lose their power if their excessive abuse of power is made public. The increasing reliance on legal authority has thus limited the power of most government officials.

Aren't these forms of authority sometimes mixed? In the real world, people may hold more than one form of authority. Weber knew this. He distinguished these three forms of authority through use of the **ideal-type method** defined in Chapter 6 ("Groups and Formal Organizations") as a method that involves isolating the most basic characteristics of some social entity. The purpose of this method is to crystallize the meaning of some social entity by removing the entity from the context in which it is embedded with other entities.

1. The institution within which power is obtained and exercised is the _____ institution.
2. _____ is the ability to control the behavior of others even against their will.
3. Power that is considered legitimate by those subjected to it is called _____.
4. On which of the following types of authority are kingships based?
   a. charismatic   b. traditional   c. rational-legal
5. Which of the following types of authority places the strongest limits on government officials' freedom of action?
   a. charismatic   b. traditional   c. rational-legal

Answers: 1. political 2. Power 3. authority 4. b 5. c

Stripped of the surrounding complexity, the essential characteristics of a social entity can be identified.

The power of any American president is fundamentally grounded in rational-legal authority. The concept of the "imperial presidency" also suggests that some element of traditional authority is attached to the office. Particular presidents (Theodore Roosevelt, John Kennedy, Ronald Reagan) may add charismatic authority, whereas other presidents (Gerald Ford, Jimmy Carter) do not.

At any rate, because authority is based on others' acceptance, it is the most stable form of power. This is particularly true of modern nation-states.

THE NATION-STATE

The Nature of the Nation-State

Nation-states are usually established when political leaders wish to consolidate power over a large geographic area. This involves combining several independent political units, as in the creation of Spain through the marriage of Ferdinand and Isabella. Similarly, the United States became a nation through its Constitutional Convention of 1787 in which some of the powers belonging to individual states were turned over to the federal government.

The nation-state is a fairly recent development in human history. Although 99 percent of human history was lived in small hunting and gathering bands, many forerunners of the nation-state existed thousands of years before the birth of Christ. Legal systems of control involving significant amounts of land remained quite rare. Even the Roman empire consisted, with a few insignificant exceptions, of self-governing cities (Jones, 1966; Service, 1975; Bottomore, 1993; James, 1996).

Nation-states began to emerge in Europe during the fifteenth century, but even one hundred years ago the earth contained territory claimed by no political unit, and many people lived without formal political rule. No more. The earth is presently divided into over one hundred political entities known as nation-states, and nearly everyone is a member of one of them. One now is born, lives, and dies only by the official recognition of a nation-state (Alford and Friedland, 1985; Skocpol, 1985; Skocpol and Amenta, 1986; James, 1996).

What is a nation-state?   Although the nature of nation-states may vary (as discussed subsequently), the **nation-state** is always the political entity that holds authority over a specified territory (Tilly, 1990). One of the key characteristics of a nation-state is the absolute sovereignty it has over its citizens. Citizens can appeal no higher than to the laws of their state, and the state has supreme authority over its citizens. Representatives of the state are the only ones who have a legal right to act in the name of the state. They are, for example, the only ones who can impose taxes, imprison criminals, or declare war. A second key characteristic of a nation-state is its devotion to *nationalism*—a people's commitment to a common destiny based on recognition of a common past and a vision of a shared future (Stoessinger, 1981).

It is important to distinguish between a state, which has ultimate authority over a certain territory, and a **government**—the political structure that rules a nation. All societies have government because some arrangement for settling disputes is necessary. A state exists, however, only if governmental functions are separated from those of such other institutions as the family and religion. As societies have become modernized, states have become increasingly common as governmental functions have been handled by groups of officials whose roles in society have been distinctly governmental.

The emergence of the state has been associated with advances in agriculture and industry. In the largely independent group typical of hunting and gathering societies, most decisions could be reached by consensus. Agricultural and industrial developments, however, led to surpluses that could be converted into wealth and power for at least some segments of society. It appears reasonable to suggest that when surpluses lead to conflict over who will control them, a clearly defined state—with control over sanctions as well as recognized bases of legitimacy—is needed to maintain order (Lenski, Nolan, and Lenski, 1995). As discussed later, such a suggestion has played an important role in the development of specific theories on why the nation-state emerged.

A Functionalist Perspective of the Nation-State

According to the functionalist perspective, nation-states were formed because they were needed by society. English philosopher Thomas Hobbes (1958, originally published in 1651) argued that unless a strong state exists, people will act only according to their own selfish interests. Without a state to control these selfish impulses, there would be a "war of every one against every one" and life would be "solitary, poor, nasty, brutish, and short." To escape such chaos, people agreed to create nation-states on the basis of a "social contract" of cooperation.

A somewhat similar argument was developed by Emile Durkheim (1964a, originally published in 1893), who feared that the increasingly rapid and extensive social changes being experienced within modern societies would lead to social disorder. Unless sufficient controls could be developed, modern societies would experience disruptive conflicts among individuals. Although all people would be attempting to act in their own self-interests, they would actually be creating a situation in which the disruption would prevent the achievement of meaningful personal goals. Durkheim, therefore, interpreted the role of the state as one of providing centralized regulation of economic life while encouraging the acceptance of norms that would define just social relationships. Like Hobbes, Durkheim believed that the absence of a state would create chaos.

Talcott Parsons (1959) also agrees with Hobbes that humans must have external constraints to live together. Social constraints are necessary to control unruly human passions. Out of these social constraints comes social order. For Parsons, the state rests on authority, on the consent of the governed. This authority exists because the state is fulfilling desirable functions. Consider a few. Through the development and enforcement of a system of laws, order and security are promoted by the state. The state mobilizes the society's resources to achieve commonly shared goals. Those in charge of the government assume responsibility for internal leadership, including planning and coordination functions. Government officials formulate policy decisions that are supposed to reflect the society's highest values. Relationships with other nation-states, whether political, economic, or military, are to be han-

dled by representatives of the state. Finally, the political institution defines the goals of a society.

A Conflict Perspective of the Nation-State

According to the conflict perspective, nation-states exist primarily to serve the interests of a society's elite. One of Hobbes's earliest critics, the French philosopher Jean-Jacques Rousseau, argued that in the original state of nature, people were noble savages living in harmony with their environment. They were probably indifferent rather than at war with one another. People had no reason to start fighting until the creation of private property led to the rise of social inequality. At this point, Rousseau said, a social contract may indeed have restored order by creating a state, but the social contract had advantages primarily for those who were already privileged. Rousseau described the terms of the social contract in this fashion:

> You need me, for I am rich and you are poor. Let us therefore make a contract with one another. I will do you the honor to permit you to serve me under the condition that you give me what little you have left for the trouble I shall take in commanding you (quoted in Cassirer, 1951:260).

Rousseau argued that such arrangements should be replaced by a new social contract in which obedience to the privileged class would be replaced by obedience to the "common will" for the welfare of all, rich and poor alike.

It is to Karl Marx, another of the founders of political sociology, that we look for the most influential expression of the conflict perspective on the nation-state. Marx contended that the nature of a society is determined by the relations between those who own property and those who do not. Industrial societies, in Marx's view, are dominated by the relatively small number of people who own factories, machines, and other productive resources. As was noted in Chapter 1 ("The Sociological Perspective"), those who owned productive property were the ruling class, and those who did not own property were the workers. The **bourgeoisie** (ruling class) and the **proletariat** (workers) were said to form the major social classes in capitalist societies which, according to Marx's analysis, would be characterized by a decrease in the proportion of capitalists at the top and an increase in the proportion of unpropertied wage earners at the bottom.

Marx believed the political life to be essentially a reflection of relations between social classes. Because of the ruling class's control over important resources, Marx said that the government would serve its needs. In his words, "The executive of the modern State is but a committee for managing the affairs of the whole bourgeoisie. . . ." (Tucker, 1972:337). The state provides a means for controlling workers through the power of force. In most cases, however, the use of force is unnecessary, because the state draws on common values and experiences to promote an ideology that encourages acceptance of the system. Marx used the term **false consciousness** to describe acceptance of a system that works against one's own interest.

According to Marx, then, those in charge of the mechanisms of the state in capitalist societies—branches of the government, the military, the police, the bureaucracy—make decisions that serve the interests of the ruling class. It is only through the revolution of the proletariat against the bourgeoisie that the state, which protects and promotes the class interests of the elites, will come to serve the interests of everyone alike. A "short-term dictatorship of the proletariat" was to replace the capitalist state as work proceeded to establish communism, the final stage of historical development. Because Marx spent his time analyzing the problems of capitalist society, he did not refine a conception of the communist society. It does seem, however, that the state under communism was to be nonauthoritarian and nonbureaucratic and was to exist for purely administrative reasons. Ownership of the means of production, including land, was to be in public hands, not in the hands of a few. To ensure that state functionaries did not form a new elite class, their pay was to be kept commensurate with those outside government.

Marx must be credited with identifying the fact that one of the primary goals of any capitalist state is to protect the interests of the ruling class. He failed, however, to foresee the difficulty of reducing the power of state bureaucracy and of persuading government officials to relinquish their positions and power (Orum, 1983). At any rate, modern conflict theorists have widened Marx's concept of class conflict to encompass social conflict of all types, whether between classes or between interest groups attempting to influence government policies and programs.

POLITICAL SYSTEMS

Two polar types of political systems exist in nation-states—*democracy* and *totalitarianism*. Because few societies are ever pure enough in their form of political rule to maintain all the characteristics defining each polar type, the real world contains variations in between, a point to which we shall return later (Orum, 1983).

Democracy

What is democracy, and what are its essential characteristics? There are two current conceptions of democracy. One conception, tied to the ancient Greeks, assumes a form of self-government in which all individual citizens participate actively in the political process. This classical conception of democracy brings to mind old New England town meetings and Athens in the fifth century B.C. A second conception, based on empirical realities of modern times, defines **democracy** as the type of political system in which elected officials are held responsible for fulfilling the wishes of the majority of the electorate.

The second conception of democracy is based on two important assumptions. First, a democratic political system is maintained through the mechanism of elections—political candidates who fail to satisfy the wishes of the majority of the people will not be elected. Second, citizens are not expected to be involved deeply in politics; they merely need to vote. This second assumption recognizes that not everyone in modern society can be actively involved in all political decision making. Given the minute division of labor in contemporary society, it would be hopelessly inefficient for all qualified citizens to be preoccupied with political activ-

---

FEEDBACK

1. A _____ is the political entity that holds authority over a specified territory.
2. A _____ is the political structure that rules a nation.
3. According to Thomas Hobbes and Emile Durkheim, nation-states exist
   a. to serve the interests of powerful, privileged social classes.
   b. because of custom.
   c. to control the selfish impulses that would otherwise make social order impossible.
4. According to Jean-Jacques Rousseau, it is primarily the privileged classes who benefit from the existence of the state. *T or F?*
5. According to Karl Marx, those in charge of the state in capitalist societies make decisions that serve the interests of the owners of the means of production. He used the term _____ to describe acceptance of a system that works against one's own interests.

Answers: 1. nation-state 2. government 3. c 4. T 5. false consciousness ities. Who would manage companies, teach school, design buildings, prepare taxes, rear children? The classical practice of democracy in Greece could exist only because the masses (women and slaves), excluded from the political process, were available to perform nonpolitical functions. Consequently, modern democratic societies are known as *representative democracies*.

Full-fledged democracies are highly prized because they are so relatively rare. Although the number of free countries is on the rise, nearly 80 percent of the world's people live in "partly free" and "not free" countries. (See Figure 13.1.) The collapse of Soviet communism and the end of the cold war have created new opportunities for more "free" societies. Still, there is little evidence of a sustained movement worldwide toward societies that protect basic political rights and civil liberties (Karatnycky, 1995). "Free" political systems are primarily associated with advanced economic development and are now found primarily in a few capitalist nations in Western Europe and their former colonies, notably Canada, Australia, and the United States, along with some Latin American countries, Japan, and some African nations.

Totalitarianism

What is the nature of totalitarianism? **Totalitarianism,** which lies at the other end of the political spectrum from democracy, is the type of political system in which a ruler with absolute power attempts to control all phases of social life. Characteristics of totalitarian states include a detailed ideology designed to encompass all phases of individuals' lives; a single political party typically controlled by one person; a well-coordinated system of terror; total control of all means of communication; a monopoly over military resources; and a planned economy directed by a state bureaucracy (Friedrich and Brzezinski, 1965). Classic illustrations of totalitarian states are Nazi Germany and the former Soviet Union prior to its recent turn toward democratization. A brief description of Nazi Germany in terms of the characteristics of totalitarianism should be enlightening.

Hitler laid the foundation for the ideology of the National Socialist Party in his autobiography, *Mein Kampf*. At the center of this ideology was the myth of Aryan racial superiority and its destiny in saving the

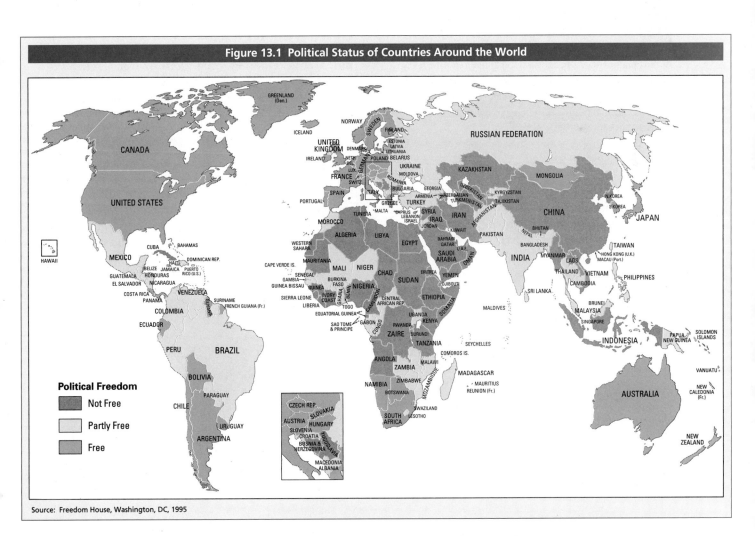

Figure 13.1 Political Status of Countries Around the World

Political Freedom

Not Free

Partly Free

Free

Source: Freedom House, Washington, DC, 1995

At the opposite end of the political spectrum from democracy is totalitarianism, the type of political system in which all aspects of society are controlled by an absolute ruler. This photo of Adolf Hitler and his staff touring Nuremberg at the zenith of Nazi power chillingly illustrates a totalitarian state.

bolstered by the Gestapo (secret police) and S.S. troops, whose terroristic practices functioned to put down dissenters and eliminate state enemies. With few exceptions, all news media were either seized by the Party or eliminated. Hitler dominated the armed forces, often devising military plans over the opposition of his generals. Nazi Germany's four-year plans, designed to run the economy, involved budgets, production, the organization of plants, and forced labor.

Authoritarianism

As noted at the beginning of this description of the types of political systems, many variations exist between democracy and totalitarianism. These political systems are usually labeled *authoritarian.*

Because authoritarianism is a residual (leftover) category between two polar types of political systems, it is more difficult to define than either democracy or totalitarianism. As a type of political system, it is, of course, closer to totalitarianism than to democracy. **Authoritarianism** refers to the type of political system controlled by nonelected rulers who generally permit some degree of individual freedom (Linz, 1964). Countless regimes have leaned toward totalitarianism but have fallen short of having all of its defining characteristics. These authoritarian governments include monarchies (the deceased Shah of Iran), military seizures of power or juntas (Chile's General Augusto Pinochet), and dictatorships (Saddam Hussein of Iraq).

world, in part by destroying Jews. Despite presenting a facade of democracy and the sharing of power with the state bureaucracy, Hitler and the National Socialist Party held the real power. In fact, Hitler helped to solidify his own power by setting up identical offices in the state bureaucracy and the Party. The ensuing competition allowed Hitler to divide and conquer any opposition to his control. Hitler's absolute control was also

POWER IN AMERICAN SOCIETY

Elitism is inevitable in all societies, even democratic ones, because democracy in the classical sense of self-government through the political participation of individual citizens is not feasible in modern mass society. Because the public cannot govern directly, it must delegate governing responsibilities to representatives. By this act of delegation, an elite group that makes political decisions is created (Dye, 1990). Still, citizens in

FEEDBACK

1. _____ is the type of political system in which elected officials are held responsible, through the mechanism of voting, for fulfilling the wishes of the majority of the electorate.

2. The United States is characterized by a
   a. classical democracy   c. representative democracy
   b. lineage democracy   d. democracy of hegemony

3. The type of political system in which a ruler with absolute power attempts to control all phases of social life is called _____.

4. _____ refers to the type of political system controlled by nonelected rulers who generally permit some degree of individual freedom.

Answers: 1. Democracy 2. c 3. totalitarianism 4. Authoritarianism democratic societies are supposed to be able to influence the actions of their representatives through voting (usually via political parties) and joining interest groups. Voting is covered next, and interest groups are discussed in the following section within the context of the pluralist model.

Influence of the Vote

Like all contemporary democracies, the United States emphasizes political participation through voting. Voting is supposed to be an important source of power for citizens, and it does, in fact, enable people to remove incompetent, corrupt, or insensitive officials from office and to influence issues at the local, state, and national levels.

To what extent do Americans exercise power through the ballot box? Voting does, however, have severe limitations as a means of exercising power. In the first place, the range of candidates that voters can choose from is limited because of the importance of **political parties**—organizations designed to gain control of the government through the election of candidates to public office. In part because American political parties are more concerned with winning elections than with maintaining a clear-cut ideological position, a politics of "blandness" exists. Usually, only candidates endorsed by the major parties have any chance of winning a major political office. And because the United States has only two major political parties, the candidates for any particular office are usually very much alike. Consequently, many groups are not effectively represented. The limited choice of political candidates is reinforced by the high cost of political campaigns.

Only people who are wealthy enough to help finance their own campaigns or who are able to attract large contributions from others can mount effective campaigns. Such people are probably not representative of the public, and powerful contributors expect favors once their candidates are in office.

In part because of the problems associated with increasingly weak political parties and the public's relatively low level of confidence in government leaders (see Figure 13.2), the public's interest in voting is declining. During the 1990s, an average of 44 percent of the voting-age population voted in national elections, the lowest voter turnout rate in the industrialized world. In fact, this placed the United States in the bottom 25 places of the 171 countries surveyed ("Voter Turnout from 1945 to 1997," 1997).

The lower class and the working class in the United States tend to vote in smaller proportions than the middle and upper classes. Members of minorities, people with less education, and people with less income are less likely to vote in both congressional and presidential elections. Whatever political influence Americans derive from voting, it is enjoyed more by those who are already receiving a larger share of society's rewards (U.S. Bureau of the Census 1997a).

The two major models of the distribution and exercise of political power in a democratic society are pluralism and elitism. According to **pluralism,** decision making is the result of competition, bargaining, and compromise among diverse special interest groups. In this view, power is widely distributed throughout a society or community. The pluralist model is consistent with the functionalist perspective. On the other hand, according to **elitism,** a community or society is controlled from the top by a few individuals or organizations. Power is said to

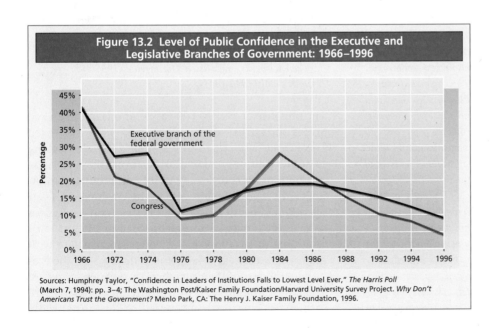

Figure 13.2 Level of Public Confidence in the Executive and Legislative Branches of Government: 1966–1996

Sources: Humphrey Taylor, "Confidence in Leaders of Institutions Falls to Lowest Level Ever," *The Harris Poll* (March 7, 1994): pp. 3–4; The Washington Post/Kaiser Family Foundation/Harvard University Survey Project. *Why Don't Americans Trust the Government?* Menlo Park, CA: The Henry J. Kaiser Family Foundation, 1996.

be concentrated in the hands of an elite group with common interests and backgrounds. The elitist model is based on the conflict perspective.

Functionalism: The Pluralist Model

How do pluralists view the distribution of power? According to David Riesman (1961), important political decisions are made in the United States after a variety of special interest groups (each of which has its own stake in the issue) have attempted to get their own way. In addition to attempting to attain their own special ends, interest groups also try to protect themselves from other interest groups that may wish to do them harm.

Interest groups are not new to American politics. They were active in promoting the abolition of slavery in the nineteenth century, and a history of twentieth-century America could not be written without describing a wide range of interest groups such as the Women's Christian Temperance Union and the early labor union movement. The 1960s—with controversies surrounding civil rights, the Vietnam War, the environment, the women's movement, and corporate power—strengthened many interest groups and led to the creation of a number of new ones.

What is an interest group? An **interest group** is a group organized to achieve some specific, shared goals by influencing political decision making. The goals such groups work to achieve may be those of their own members (as in the case of the National Association of Manufacturers) or those of a larger segment of society (as in the case of ecology-oriented groups such as the Sierra Club).

Gary Wasserman illustrates the process of bargaining and compromise that pluralists contend dominates American politics:

When Ralph Nader, the consumer advocate, proposes that automobile makers be required to install more safety devices such as air bags in their cars, numerous interests get involved in turning that proposal into law. The car manufacturers worry about the increased costs resulting in fewer sales and reduced profits, and they may try to limit the safety proposals. Labor unions may want to make sure that the higher costs do not result in lower wages. Insurance companies may be interested in how greater safety will affect the claims they have to pay out. Oil companies may worry about the effect on gasoline consumption if fewer cars are sold because of their increased cost. Citizens' groups like Common Cause may try to influence the legislation so that it provides the greatest protection for the consumer. The appropriate committees of the House and Senate and the relevant parts of the bureaucracy will weigh the competing arguments and pressures as they consider bills covering automobile safety. The resulting legislation will reflect the relative power of the competing groups

While the presence of interest groups is not new to American politics, their influence is increasing in part because of the skyrocketing costs of running for political office. Senator Fred Thompson headed a Senate investigation into possible illegal campaign contributions taking place during the 1996 national elections.

as well as the compromises they have reached among themselves (Wasserman, 1988:254).

In all of this, government leaders are seen as having the responsibility for balancing the public welfare with the self-centered desires of special interest groups.

How prevalent are interest groups? Alexis de Tocqueville depicted Americans as a nation of joiners:

As soon as several Americans have conceived a sentiment or an idea that they want to produce before the world, they seek each other out, and when found, they unite. Thenceforth they are no longer isolated individuals, but a power conspicuous from the distance whose actions serve as an example; when it speaks, men listen (Tocqueville, 1955:488, originally published in 1835).

The presence of over 100,000 associations in the United States today, however, might even have surprised Tocqueville. About two-thirds of all Americans belong to some voluntary formal association. Most of these associations, of course, are not interest groups attempting to affect government policy. The Washington, D.C., telephone directory, however, does contain listings for over 2,500 organizations whose names begin with "National Association of . . ." and which, along with many other interest groups, energetically push the economic interests of their members on a variety of issues. The number of lobbyists in Washington whose job is to influence Congress is estimated to be substantially

TABLE 13.1	Buried Treasure—A Partial Listing of the Beneficiaries of Special Breaks in the 1997 Tax-cut Bill

- Tax-exempt-bond issuers
- Computer companies
- Computer software exporters
- Drug manufacturers
- Churches
- Makers of vaccines against pertussis, measles, mumps and rubella
- D.C. developers
- Gasoline retailers
- Electric-car makers
- Natural gas companies
- Hard-cider makers
- State employee workers' compensation plans

- Virgin Islands bonds
- Farmers' cooperatives
- Mentally disabled people
- Cooperative hospital service companies
- Amtrak
- Federal criminal investigators
- Real estate investment trusts
- Distilled-spirits plants
- Wine importers
- Beer and wine producers
- Employee-owned companies
- Pension plans

Source: Patrice Hill, "Clinton May Trim Some Budget Pork," *The Washington Times* (August 5, 1997): A1, A14.

above the nearly six thousand formally registered to perform that function.

In a study of New Haven, Connecticut, Robert Dahl (1961), one of the leading proponents of the pluralist model, concluded that a number of quite separate groups controlled various areas of political decision making. The groups trying to influence political decisions on public education were not the same groups competing, bargaining, and compromising on the issue of highway construction. In short, power in New Haven was not concentrated in the hands of one elite group. Power was dispersed such that few segments of the community were devoid of power.

Pluralists would point to the beneficiaries of the 1997 tax-cut bill. Tax breaks come to both the wealthy, such as Microsoft, and to the interest groups with more modest resources, such as churches and the "mentally disabled" (Hill, 1997). Table 13.1 contains a sample of some of the beneficiaries. We can further examine the pluralist model by focusing on political action committees in the American political system today.

Political Action Committees

American political parties have been influential in the past because they controlled nominations for major offices and campaign financing and because many citizens tended to vote the straight party ticket. These bases of political party power appear, however, to be eroding. Less than one-fourth of Americans indicate that party affiliation makes a great deal of difference in their decision to vote for a candidate for Congress, whereas over one-third say that party affiliation would have little or no influence on their vote. Over one-third of Americans consider themselves as "Independents" (U.S. Bureau of the Census, 1996a), and 13 percent reg-

istered as Independents in 1992 (Brace, 1993). Moreover, 8 percent voted for Ross Perot, an Independent candidate for the presidency in 1996. This means that enough Americans are willing to vote for an Independent candidate to swing a presidential election. With the decline of political parties, political interest groups are currently most often represented by political action committees (Greider, 1992).

What are political action committees? Political action committees (PACs) are organizations established by interest groups to raise and distribute funds to selected political candidates. Although corporations and labor unions are prohibited from contributing directly to political candidates, legislation passed during the 1970s allows such organizations—as well as other interest groups—to use PACs to channel up to $10,000 into any candidate's election campaign. However, there is no legal restriction on the number of PACs that may support a candidate.

The number of PACs has mushroomed tremendously since PACs were authorized by the 1974 Federal Election Campaign Act and its 1976 amendments. Although only about six hundred political action committees existed in 1974, there are now over four thousand. (See Figure 13.3.) Moreover, contributions from PACs are beginning to dwarf the role of political parties in financing political campaigns. PACs now contribute more than nine times as much support to congressional candidates as do political parties. Dollar amounts of PAC contributions to federal candidates (so-called "hard" money, limited by legal restrictions) increased from just under $12 million in 1974 to over $217 million in 1995-96. Table 13.2 presents the top PAC contributors during the 1996 presidential and congressional elections (Federal Election Commission, 1997).

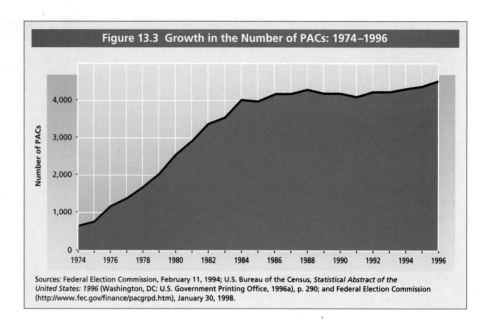

Figure 13.3 Growth in the Number of PACs: 1974–1996

Sources: Federal Election Commission, February 11, 1994; U.S. Bureau of the Census, *Statistical Abstract of the United States: 1996* (Washington, DC: U.S. Government Printing Office, 1996a), p. 290; and Federal Election Commission (http://www.fec.gov/finance/pacgrpd.htm), January 30, 1998.

For a fuller picture of campaign contributions, one has to add the $260 million in "soft" money (given to political parties without legal limitations) given by all contributors, including PACs (Common Cause, 1997a, 1997b).

Do PACs contribute to pluralism? PACs were originally designed to provide a mechanism for small contributors to become involved in politics, to allow candidates without personal wealth to run for office, and to encourage independence from the viewpoints of political parties. Critics are concerned, however, that PACs may be undermining pluralism, because a majority of PACs are connected with business-related interests (Eismeier and Pollock, 1988; Watzman, Youngclaus, and Shecter, 1997)

Philip Stern (1988) makes a strong case for the influence of business-related PACs on members of Congress. His book *The Best Congress Money Can Buy* is replete with purported connections between business-related PAC money and congressional votes. Rather than read a litany of names, consider the possibilities of political influence by business-related PACs associated with the Tax Reform Act of 1986. This tax bill brought forth considerable new PAC money. For example, as the congressional tax-writing panels were considering tax reform in 1985, PACs contributed $6.7 million to fifty-six members of the House Ways and Means and Senate Finance committees, $119,643 per member. This was two and a half times more money than the members of those same committees received from PACs two years before.

Does this mean that the pluralist model is invalid? Despite the mixed result of PACs, pluralism has not disappeared from the American political landscape.

Individuals in American society are represented only to the extent that they belong to groups—whether political parties or interest groups—that are capable of influencing political decision making. Therefore, interest groups are one of the most important avenues for political representation. Groups representing civil rights, environmental, and consumer concerns have won some concessions that otherwise would not have been obtained. They have accomplished this through lobbying, making contributions to political campaigns, gathering information, generating publicity to express their viewpoint, and filing lawsuits. The National Rifle Association, for example, launches major letter-writing campaigns whenever gun-control legislation is introduced. Many environmental organizations have taken legal action against companies that violate pollution laws and have forced some regulatory agencies to investigate illegal environmental practices that were usually ignored in the past.

Doesn't the public at large have any influence over its government? Political pressure is exercised by the general public. As noted earlier, the public's influence is routinely channeled through voting, despite the limitations of the ballot box. Two recent cases, moreover, demonstrate that an aroused public can have a political voice hearable by politicians above the din of special interest groups. The popular will was credited with the passage of the Brady handgun bill in 1993 and the ban on the sale of assault weapons in 1994 despite strong opposition by the National Rifle Association (Lacayo, 1993; Alter, 1994). The tobacco lobby is also on the wrong end of public opinion. For example, 96 percent of the American public believes that smoking should either be banned from workplaces or restricted to special areas

TABLE 13.2	Serious Money: The Top Fifty PACs, 1995–1996		
Rank	PAC Name	Contributions to Candidates	Primary Industry/ Interest
1	Democratic Republican Independent Voter Education Committee	2,611,140	Political Education
2	American Federation of State, County, & Municipal Employees	2,505,021	Public Sector Unions
3	UAW Voluntary Community Action Program	2,467,319	Labor Unions
4	Association of Trial Lawyers of America	2,362,938	Lawyers and Lobbyists
5	Dealers Election Action Committee of the National Automobile Dealers Association	2,351,925	Automobile Dealers
6	National Education Association	2,326,830	Public Sector Unions
7	American Medical Association	2,319,197	Health Professionals
8	Realtors Political Action Committee	2,099.683	Real Estate
9	International Brotherhood of Electrical Workers Committee on Political Education	2,080,587	Labor Unions
10	Active Ballot Club, a Department of United Food & Commercial Workers International Union	2,030,795	Labor Unions
11	Machinists Non-Partisan Political League	1,999,675	Labor Unions
12	Laborers' Political League	1,933,300	Labor Unions
13	United Parcel Service of America, Inc.	1,788,147	Transportation
14	Committee on Letter Carriers Political Education	1,715,064	Public Sector Unions
15	American Institute of Certified Public Accountants Effective Legislation Committee	1,690,925	Accountants
16	American Federation of Teachers Committee on Political Education	1,614,833	Public Sector Unions
17	NRA Political Victory Fund	1,565,821	Gun Policy
18	United Steelworkers of American Political Action Fund	1,524,650	Labor Unions
19	Build Political Action Committee of the National Association of Home Builders	1,472,849	Real Estate
20	Carpenters Legislative Improvement Committee, United Brotherhood of Carpenters & Joiners of America	1,470,106	Labor Unions
21	National Association of Life Underwriters	1,426,750	Life Insurance
22	National Beer Wholesalers' Association	1,324,992	Beer, Wine, & Liquor
23	American Bankers Association	1,298,850	Commercial Banking
24	CWA-Cope Political Contributions Committee	1,297,286	Labor Unions
25	American Dental Political Action Committee	1,286,775	Health Professionals
26	Transportation Political Education League	1,260,600	Transportation
27	American Telephone & Telegraph Company	1,244,134	Telecommunications
28	National Association of Retired Federal Employees	1,243,350	Public Sector Unions
29	AFL-CIO Committee on Political Education/Political Contributions Committee	1,190,514	Labor Unions
30	National Committee for an Effective Congress	1,108,110	Political Education
31	National Federation of Independent Business/Save America's Free Enterprise Trust	1,070,543	Small Business
32	Sheet Metal Workers International Association Political Action League	1,066,000	Labor Unions
33	Lockheed Martin Employees	1,026,250	Defense
34	American Maritime Officers, AFL-CIO Voluntary Political Action Fund	975,960	Labor Unions
35	United Association Political Education Committee	961,500	Political Education
36	Federal Express Corporation	948,000	Transportation
37	Americans for Free International Trade	936,800	Economic Trade
38	American Hospital Association	881,176	Health Professionals
39	Ernst & Young	880,615	Accountants
40	National Restaurant Association	880,119	Restaurants
41	Philip Morris Companies, Inc.	850,119	Tobacco/Beer/Food
42	Ironworkers Political Action League	828,615	Labor Unions
43	Air Line Pilots Association	827,000	Labor Unions
44	Service Employees International Union Political Campaign Committee	820,950	Labor Unions
45	Campaign America	819,681	Political Education
46	Union Pacific Fund for Effective Government	797,357	Commercial Banking
47	Associated General Contractors	791,050	Real Estate
48	American Nurses' Association	784,891	Health Professionals
49	Monday Morning Political Action Committee	771,500	Political Education
50	Committee for Thorough Agricultural Political Education of Associated Milk Producers, Inc.	750,000	Dairy

Source: Federal Election Commission (http://www.fec.gov/finance/paccnye.htm), July 1, 1997.

SOCIOLOGY IN THE NEWS

The Politics of Smoking

The tobacco industry in the United States constitutes an interest group. It is organized to reach the specific and shared objective of influencing political decision making regarding cigarette smoking. As noted in this CNN report, the tobacco industry is charged with buying favors from state law makers across the country. During 1995-96, three of the top ten contributors to Republican national party committees were tobacco companies. Philip Morris gave over twice as much as the second highest contributor ($2.5 million). Philip Morris also donated nearly one-half million dollars to Democrats during this period. Their aim, critics allege, is to get passed watered-down laws that appear to be effective but that do not discourage smoking among children and teenagers.

The American tobacco industry is currently experiencing intense scrutiny from medical, legal, and political quarters. According to one charge, cigarette makers are targeting minorities with advertising campaigns designed to promote smoking.

1. State the view of this situation through the eyes of a functionalist.

   _____

   _____

2. How would a conflict theorist interpret this situation?

   _____

   _____

(Farley, 1994). At least in good part as a response to public opinion, many steps are being taken to control smoking. Consider the national level alone. In 1994, President Clinton signed a bill prohibiting smoking in all public and some private schools. The U.S. Department of Defense, in that same year, banned smoking in all military work spaces, whether they be on military bases or inside battlefield tanks. The Smoke-Free Environment Act bans smoking from all buildings entered by ten or more people each day. All structures—bars, restaurants, workplaces—except homes would be smoke-free zones. Fines of up to $5,000 a day would be faced by violators. In 1994, a panel of the Food and Drug Administration concluded that nicotine is an addictive drug, setting the stage for possible regulation of nicotine products (Infante, 1994). Regulation of nicotine has been publicly supported by the American Medical Association. State and local governments are also becoming more aggressive, as are many private companies such as McDonald's and many owners of professional baseball franchises (Farley, 1994; Solomon, 1994). During 1997, a group of state attorneys general negotiated a settlement with the tobacco companies that called for the payment of $368 billion and measures designed to reduce smoking by children. For their part, the tobacco companies would be granted immunity from future civil lawsuits claiming damages. The U.S. Congress must approve the settlement before it becomes binding, an action not yet accomplished (Hall, 1997).

These two illustrations, of course, are not meant to pronounce dead the gun and tobacco lobbies.

According to functionalism and the pluralist model, interest groups such as these will continue to influence the exercise of political and economic power. They are simply part of the fabric of the American political system. At the same time, it cannot be assumed that the general American public is left out of all important political decisions. The public's will may be felt through strong collective expression of general opinion.

There is, however, a competing viewpoint. We turn now to the conflict perspective and the power elite model (Wilson, 1995).

The Conflict Perspective: The Power Elite Model

The pluralist model has been dominant in American social science since the mid-twentieth century. But C. Wright Mills (1956) and others (Hunter, 1953, 1980) have developed an alternative view that has become increasingly influential. Mills contends that the United States no longer has separate economic, political, and military spheres; rather, the top people in each sphere overlap to form a unified elite. Whereas Riesman and other pluralists see two levels of power—special interest groups and the public—Mills outlined three levels. (See Figure 13.4) Overshadowing special interest groups and the public is the **power elite**—a unified coalition of top military, corporate, and government leaders (the executive branch in particular). Mills placed U.S. senators and representatives at the middle levels of power rather than at the higher levels,

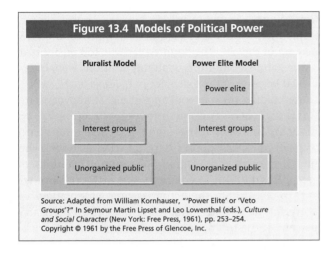

Figure 13.4 Models of Political Power

Pluralist Model   Power Elite Model

Power elite

Interest groups   Interest groups

Unorganized public   Unorganized public

Source: Adapted from William Kornhauser, "'Power Elite' or 'Veto Groups'?" In Seymour Martin Lipset and Leo Lowenthal (eds.), *Culture and Social Character* (New York: Free Press, 1961), pp. 253–254. Copyright © 1961 by the Free Press of Glencoe, Inc.

where decisions are made regarding national policy, war, and domestic affairs.

The bases of power within the power elite, according to Mills, are the common interests and social and economic backgrounds its members share. Elites tend to be educated in select boarding schools, military academies, and Ivy League schools; belong to Episcopalian and Presbyterian churches; and come from upper-class families. Members of the power elite have known one another for a long time, have mutual acquaintances of long standing, share many values and attitudes, and intermarry. All this makes it easier for them to coordinate their actions to obtain what they want.

Theorists who contend that society is dominated by an elite do not always use Marxian analysis. However, their explanations have often been influenced by Marx's thinking. Like Marx, elitists tend to see political leaders as captives of business and corporate interests. Pluralists, who are more likely to have been influenced by classical theorists of democracy than by Marx, view political leaders as referees among competing interest groups. (Table 13.3 compares the basic features of the pluralist and elitist models.)

In a democratic society, in which power is supposed to rest with the people, the powerful prefer that their activities not be widely advertised. Power is exercised behind the scenes and through informal channels. Although the invisibility of power is an obstacle, it has not stopped sociologists from studying it at both the community and national levels.

What is required to prove the existence of the power elite? We know from Chapter 8 ("Social Stratification") that an upper-upper class comprising only about 1 percent of the American population exists. G. William Domhoff (1986, 1990, 1996) has spent much of his career investigating a ruling class as described by Mills. Using such standards as appearing in the *Social Register,* attending exclusive private schools, and holding memberships in prestigious big-city clubs, Domhoff concluded that the very top of the upper class actually consists of less than 1 percent of the population (perhaps as little as 0.5 percent). Thomas Dye corroborates this conclusion from a review of the many studies of the socioeconomic characteristics of those in top political, military, and corporate positions. These top institutional leaders are not typical Americans: "They are recruited from the well-educated, prestigiously employed, older, affluent, urban, white, Anglo-Saxon, upper- and upper-middle-class male populations of the nation" (Dye, 1990:189). To offer some concrete illustrations, about half the elites are graduates of only twelve prestigious private universities, only 5 percent are female, and only one-fourth graduated from public universities.

Establishment of the existence of high socioeconomic homogeneity among military, political, and corporate institutional leaders does not, however, prove the existence of the power elite. To make a convincing case for a power elite, it is necessary to document the presence of an elite group that is unified and that cooperates for the promotion of the shared interests of the upper class.

TABLE 13.3	Characteristics of Two Models of Political Power	
	Functionalist Perspective	**Conflict Perspective**
Characteristics	**Pluralist Model**	**Power Elite Model**
Focus of analysis	Interest groups	Position holders in major institutions
Power base	Resources of interest groups	Leadership positions in major institutions
Distribution of power	Diffused among interest groups	Concentrated in hands of elites
Stability of power	Shifting constantly	Firm control by elites
Unity of interests among power centers	Disunity	High unity (disruptive internal conflict unlikely)
Degree of nonelite influence	Considerable; varies with issues	Very little influence on major issues
Type of political system	Democratic (in practice)	Democratic (in theory)

Is there a national power elite? It is impossible to assert a definitive answer to this question, although the evidence does tend to support the existence of such an entity. Thomas Dye (1990) identified 7,314 top institutional positions in thirteen different areas—industrial corporations, banking, investments, insurance, mass media, law, government, utilities, transportation and communications, foundations, the military, civic and cultural organizations, and education. The "elites" are those individuals who occupy the top positions in these institutional sectors of American society. As a whole, the occupants of these positions of leadership control one-half of America's industrial and financial assets, almost one-half of the assets of private foundations, and over two-thirds of the assets of private universities. In addition, they control the mass media (television networks, news services, leading newspapers), the most prestigious civic and cultural organizations, and all three branches of the federal government.

At the time of Dye's study, these 7,314 top positions were occupied by 5,778 individuals. The existence of fewer occupants than positions means there was *interlocking*—individuals holding more than one top position. A high degree of interlocking, of course, would imply high cohesion. Although only approximately 15 percent of the elites held more than one top position at a time, almost one-third of all top positions were interlocked because of the existence of *multiple interlockers*—those holding three or more positions. Moreover, there was a top set of multiple interlockers holding six or more top positions at the same time. These results lend some impressive support for the power elite model. The members of this inner group of multiple interlockers are in the position to communicate with one another and to coordinate the activities of a wide variety of powerful institutions. Still, it must be remembered, 85 percent of the top position holders were not interlockers. This leaves considerable room for disunity and conflict among the top institutional leaders, although this is offset somewhat by *sequential interlocking*, which occurs when individuals hold top leadership positions in a number of institutional areas over their careers. We hear constantly of prominent people moving from business to government, government to universities, universities to foundations, and so on (Useem, 1984; Mintz and Schwartz, 1985).

FEEDBACK

1. The model of political power known as _____ depicts political decision making as the outcome of competition, bargaining, and compromise among diverse special interest groups.
2. The model of political power known as _____ contends that political control is held by a unified and enduring few.
3. The majority of political action committees (PACs) are now associated with
   a. labor unions.   c. political parties.
   b. citizen groups.   d. corporations and other businesses.
4. Interest groups are an important means by which nonelites can influence political decision making. *T or F?*
5. The operation of PACs in America today invalidates the pluralist model. *T or F?*
6. According to C. Wright Mills, which of the following is *not* part of the power elite?
   a. military organizations   c. large corporations
   b. U.S. senators and representatives d. executive branch of the government

Answers: 1. pluralism 2. elitism 3. d 4. T 5. F 6. b

ECONOMIC SYSTEMS

The preceding description of interest groups showed that many groups are tied to corporate and other business interests. Discussions of elitist and pluralist models revealed that business interests play an important role in the distribution of power. These two points illustrate a very important fact. In any society, the polity and the economy are closely linked. Societies differ, however, regarding the kinds of political and economic systems they develop. Differences in economies have played particularly important roles in relationships between nations because fundamental differences in world views are likely to underlie differing solutions to problems associated with the production and distribution of goods and services. Because preindustrial societies were covered in Chapter 5 ("Social Structure and Society"), industrial societies will be featured in this chapter.

Some economies are organized on the assumption that individuals who have experienced success deserve to own and to control land, factories, raw materials, and the tools of production; others are organized on the assumption that such property belongs to the people of the society as a whole. Moreover, some societies are organized on the belief that every aspect of the

Risk taking and profit seeking are at the center of capitalism. Nowhere are these defining characteristics more apparent than on the floor of a stock market such as the Chicago Mercantile exchange.

economy must be subjected to careful planning by a centralized government, whereas others believe that economic success is most likely to be promoted by free competition in which government does not interfere. These contrasting worldviews have been involved in the development of capitalist and socialist economies. **Capitalism** refers to a system founded on belief in private property and the pursuit of profit. **Socialism** refers to a system founded on the belief that the means of production should be owned by the people as a whole and that government should plan and control the economy (Gregory and Stuart, 1989).

When we speak of capitalism and socialism, we are referring to ideals rather than actual operating systems. As we saw in Chapter 6 ("Groups and Formal Organizations"), however, much can be learned by considering ideal types that involve an isolation of the most basic characteristics of some social entity. No society is likely to embody all the principles described in the ideal type, but a contrast of ideal types will help us see how societies can become very different by placing greater emphasis on one set of ideals than on another (Schnitzer, 1997).

Capitalism

Capitalist societies are founded on the belief that private individuals and organizations have the right to pursue their own private gain and that society will benefit from their activities. Although economic activity under capitalism is motivated by the pursuit of profit, the individuals and organizations who are pursuing profit also run the risk of losing money.

In a typical year [in the United States] about four out of ten corporations report net losses. Of ten business firms started

in an average year, five close down within two years, and eight within ten years—lack of success being the main reason (Ebenstein and Fogelman, 1985:132).

This combination of profit seeking and risk taking is possible only under certain economic ownership arrangements.

Who owns the property in a capitalist system? Under capitalism, most property belongs to private individuals and organizations rather than to the state or the community as a whole. This private ownership is based on a belief that people have an inalienable right to hold and to control their own property. Those favoring capitalism recognize that private ownership of property leads to control over the lives of other people, but they believe that spreading this control over a large number of small owners is preferable to concentrating it in the hands of the state. Moreover, it is because of private ownership that the pursuit of profit is possible. Defenders of capitalism argue that the private ownership of property benefits society.

How is capitalism thought to benefit society? Adam Smith, the eighteenth-century English social philosopher, argued that a combination of private property and the pursuit of profit would bring important benefits to society. Because of competition from other capitalists trying to increase their own individual profits, Smith argued, individual capitalists would always be motivated to provide the goods and services desired by the public at prices the public was willing and able to pay. Capitalists who produced inferior goods or who charged too much would soon be out of business because the public would turn to their competitors. Smith reasoned that the public would benefit through the "invisible hand" of market forces. Not only would the public receive quality goods and services at reasonable prices, but capitalists would always be searching for new products and new technologies to reduce their costs. As a result, capitalist societies would use resources as efficiently as possible. Efforts by government to control any aspect of this would merely distort the economy's ability to regulate itself for the benefit of everyone.

Do contemporary capitalist economies work this perfectly?   We cannot say how perfectly or imperfectly a purely capitalist economy operates, because none exists. Because the United States is generally considered the purest existing example of a capitalist system, it is the most likely case to examine. Most Americans view the relative abundance of goods and services available to them as evidence that capitalism works as its staunchest advocates say that it should. (See Figure 13.5.) There are, however, some factors at work in the American economy that cause it to deviate from the ideal type.

Capitalism stifles competition through the creation of **monopolies**—separate companies controlling separate markets—and **oligopolies**—combinations of companies controlling a market. This stifling of competition is reflected in several ways. When capitalist organizations experience success, they tend to grow until they become giants within their particular industries. When this happens, new organizations and entrepreneurs simply cannot enter the industry with any hope of competing on an equal basis. Moreover, in the absence of fresh new competitors, established businesses may change drastically, losing their appetite for risk taking. In addition, the creation of monopolies and oligopolies permits the fixing of prices, leaving consumers with only the choice of buying or not buying a good or service (Ebenstein and Fogelman, 1985).

The United States does not have a complete hands-off policy with respect to private business. Government, for example, owns the Tennessee Valley Authority, a large public utility. The federal government is also heavily involved in the defense industry. The agricultural industry feels the influence of government through price controls, price supports, and grain embargos on exports to other countries. Antitrust legislation exists to control the growth of organizations. The federal government aids private industry through loans such as the one to the Chrysler Corporation (Reich and Donahue, 1986). The government is involved in the economy through such mechanisms as subsidies for many businesses and shared risk taking (such as insurance plans for financial institutions and guaranteed loans for some corporations).

The rise of widespread stock ownership can affect the profit motive. When a firm is owned and operated

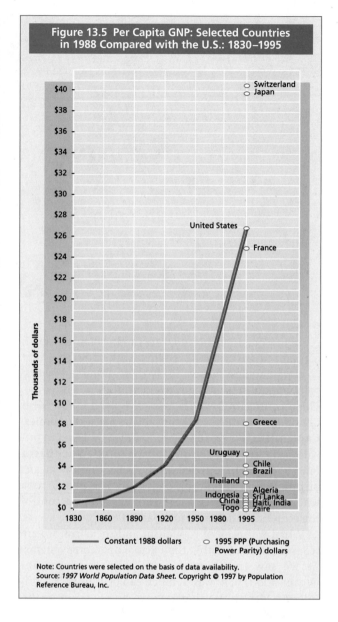

Figure 13.5 Per Capita GNP: Selected Countries in 1988 Compared with the U.S.: 1830–1995

Note: Countries were selected on the basis of data availability.
Source: *1997 World Population Data Sheet.* Copyright © 1997 by Population Reference Bureau, Inc.

by one person or a few individuals, the motivation to make a profit is usually very high. When a company is owned by thousands of shareholders and the managers are not the owners, the profit motive may have to compete with other motives, such as company growth or self-aggrandizement on the part of the managers.

Socialism

Private property and the pursuit of profit, according to many critics of capitalism, lead to serious inequities in society because workers are paid less than the value of what they produce. Such criticisms of capitalism led Karl Marx to suggest that contradictions within the capitalist system would lead to the system's overthrow by the proletariat, who would replace it first with socialism and later with communism. (It should be noted that no communist society as envisioned by Marx has yet been created.) Marx, however, was not the first to propose a socialist system. Plato's *Republic* and some portions of the Bible also contain elements of socialist thought (Ebenstein and Fogelman, 1985).

Who owns property in a socialist system? According to socialist thinkers, the means of production and distribution must be owned and controlled by the people as a whole rather than by a handful of wealthy capitalists. The state, as the people's representative, therefore owns and controls property. Through its control, the state is believed to be in a position to ensure that all members of a socialist society can share in the benefits created by the economy.

How is socialism thought to benefit society? Capitalism, according to Marx, was seriously deficient because its lack of regulation of production led to waste and other problems. According to socialist thought, adequate control over the economy can be achieved only if the state acts on behalf of the people to regulate economic activity on the basis of carefully designed plans. Capitalist thinkers, of course, have seen some role for the state in a capitalist economy. Adam Smith believed that the state should provide military protection, a system of criminal justice, and a limited number of public works and institutions (highways and public schools, for example) that benefit society but could never be profitable for private business. Socialist thinkers, however, have developed a far more ambitious role for the state. They see state control of the economy as a means for the people to regain control over the production and distribution of goods.

Socialist theory also points to important benefits for workers in a socialist economy. Although workers under capitalism are forced to receive wages that are less than the value of their labor and have little control over their work, workers under socialism benefit because both the state and the workplace exist for their benefit. As a result, Marx believed, workers would be able to exert significant control over both their work organizations and the policy directions of the society as a whole (Elliott, 1985).

Do contemporary socialist economies work this perfectly? Pure cases of socialism are as rare as pure cases of capitalism. In the former Soviet Union, for example, even before the dramatic breakup in 1991, some agricultural and professional work was performed privately by individuals who worked for a profit, and significant portions of housing were privately owned. Managers received salaries that were considerably higher than those received by workers, and they were eligible for such bonuses as automobiles and housing. Some socialist societies have even introduced economic incentives for higher productivity, a step that would seem to encourage the desire for profit on the part of workers. Thus, socialist systems have not been fully successful in eliminating income inequalities, nor have they been able to develop overall economic plans that will guarantee sustained economic growth.

Mixed Economic Systems

Because capitalist and socialist systems are based on drastically different assumptions, relationships between nations whose economies and political systems are based on one set of assumptions or the other are often strained. The frequent tensions between the United States and Russia are a case in point. In reality, however, every nation violates some of the assumptions of the economic system it has adopted, and most

Mainland China is attempting to develop a mixed economy or what has been termed a "socialist market economy." One test of mainland China's commitment to capitalism lies in its keeping the promise it made after the 1997 takeover not to interfere with Hong Kong's capitalistic economy.

nations fall between the extremes of capitalism and socialism. Most nations in Western Europe, for example, have developed capitalist economic systems in which both public and private ownership play important roles. In such societies, highly strategic industries (banks, transportation, communications, and some others) are owned and operated by the state. Other industries are privately owned but are regulated more closely than in the United States (Esping-Andersen, 1996).

Movement Toward Mixed Systems in Socialist Societies

Because the former Soviet Union lost control over its republics as well as over Eastern Europe, many communist countries have not been slow to move to some sort of market system. East Germany, now reunited with West Germany, is on the road to capitalism. Poland and Hungary are other prominent examples of this rapid adoption of the principles of a market economy (Sachs, 1990; Rezvin, 1990). If present trends continue, the future will likely see very few communist countries that do not follow the mixed economic model, including the republics of the former Soviet empire. In fact, in 1991, Cuba's Communist Party adopted some degree of capitalism by permitting handymen, plumbers, carpenters, and other tradespeople to work for their own profit.

Economic change has not been limited to Soviet-bloc countries (Nee and Stark, 1989). After the death of Mao Zedong, the long-time communist leader of the People's Republic of China, the Chinese began market-related economic reforms. The changes initiated in 1979—restoration of financial incentives, free-market prices, education, and economic relations with the West—picked up speed in the early 1980s. Market forces increasingly replaced central economic planning as farmers were permitted to work the land freely (rather than pay the government a predetermined amount of produce), markets and local entrepreneurship were allowed to develop, and foreign businesses were encouraged to open facilities on the mainland.

High inflation and public demand for greater political freedoms to match the newly granted economic liberation in the late 1980s led to political turmoil. Communist hardliner Li Peng gained control after a popular revolt in Tiananmen Square in June 1989 was suppressed.

Nevertheless, mainland China did not revert to central economic planning and continued its rapid move toward what its now deceased leader, Deng Xiaoping, called the creation of a "socialist market economy" (Nelan, 1993a, 1993b; Schell, 1997). When ownership of the port city of Hong Kong reverted to China in 1997, the Chinese government pledged not to interfere with its capitalistic economy, and so far, has lived up to that promise (Kraar, 1997). In fact, it is instructive to compare the economic progress of China and Russia. As Figure 13.6 reveals, the economic transition in China dwarfs that in Russia. Some evidence suggests that the downward economic spiral in Russia is slowing (Shapiro, 1994).

According to some scholars, the demise of communism insures the triumph of capitalism. The twenty-first century, it is argued, will be characterized by a global economy dominated by capitalism (Fukuyama, 1991; Thurow, 1992; Reich, 1993, 1997). Whether or not socialism passes from the economic and political scene, capitalism is clearly spreading rapidly.

The Global Cultures of Capitalism

What are the cultures of capitalism? Although several variants of capitalism can be identified (Hampden-Turner and Trompenaars, 1993), they can be collapsed into two basic types. **Individualistic capitalism** is founded on the principles of self-interest, the free mar-

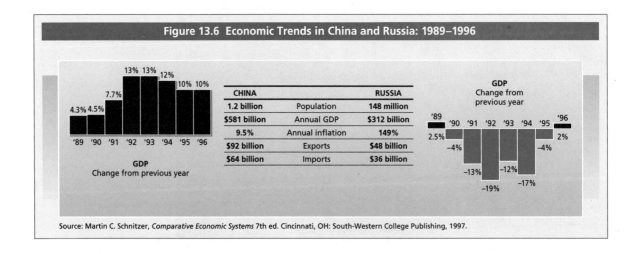

Figure 13.6 Economic Trends in China and Russia: 1989–1996

China GDP change from previous year: '89 4.3%, '90 4.5%, '91 7.7%, '92 13%, '93 13%, '94 12%, '95 10%, '96 10%

	CHINA		RUSSIA
1.2 billion	Population		148 million
$581 billion	Annual GDP		$312 billion
9.5%	Annual inflation		149%
$92 billion	Exports		$48 billion
$64 billion	Imports		$36 billion

Russia GDP change from previous year: '89 2.5%, '90 -4%, '91 -13%, '92 -19%, '93 -12%, '94 -17%, '95 -4%, '96 2%

Source: Martin C. Schnitzer, *Comparative Economic Systems* 7th ed. Cincinnati, OH: South-Western College Publishing, 1997.

ket, profit maximization, and the highest return possible on stockholder investment. This variant of capitalism dominated the world in the nineteenth century (the United Kingdom) and twentieth century (the United States). European and Japanese capitalism of the late twentieth century is known as **communitarian capitalism**—the type of capitalism that emphasizes the interests of employees, customers, and society. As will be seen, these two types of capitalism deserve the label "cultures of capitalism" because each rests on a distinctive value base.

What is culturally distinctive about individualistic capitalism? Individualistic capitalism is based on the ideas of John Locke and Adam Smith. Locke, a seventeenth-century philosopher who strongly influenced the thinking of the leaders of the American war for independence and the nature of the U.S. Constitution, placed individual rights, including the right to own property, at the heart of his *Second Treatise of Government* (1980, originally published in 1690). Although government is needed to protect the property of individuals from others, Locke argued, it must have very limited powers. Although Smith, an eighteenth-century philosopher-economist-sociologist, has been widely misinterpreted (Shepard, Shepard, and Wokutch, 1991; Werhane, 1991; Muller, 1993), some of his ideas were used to shape individualistic capitalism and continue to be used to maintain it. The concept of the "invisible hand," mentioned only once in the *Wealth of Nations* (1965, originally published in 1776), has been particularly influential in individualistic capitalism. If individuals are released to pursue their own self-interest, their separate activities will, through the free-market mechanism, automatically benefit everyone else by increasing the supply of consumable goods and services.

These cultural underpinnings are easy to observe in individualistic capitalism (Thurow, 1992). Both individuals and separate firms have their own economic strategies for success. The individual wants to prosper and firms wish to satisfy their stockholders. Firms seek to maximize profits because shareholders invest to maximize profits. Customers and employees are the firm's means to reach the goal of higher shareholder return. Because higher wages cut into profits, firms try to keep them as low as possible and lay off employees whenever necessary. Short-term thinking in the name of profit maximization is the norm. Employees are expected to leave one job for another if they can get higher wages. Employees and employers have no mutual obligations outside the legal employment contact.

Within individualistic capitalism, government is not to interfere with the workings of the free market. Government is supposed to make minimal rules and act as an umpire to settle disputes (Friedman, 1982). There

is no place for government in investment funding or economic planning. So long as government protects private-property rights, the pursuit of profit maximization will insure the greatest prosperity.

What is the value base of communitarian capitalism? Communitarianism, not to be confused with communism, is also rooted in the ideas of a philosopher. Jean-Jacques Rousseau, writing in the eighteenth century, was, like Locke, a social contract theorist—both believed that individuals forfeit some of their freedom to obtain protection on the part of a central government. Locke saw the social contract as an individualistic device, but Rousseau thought it a communitarian one. Locke's notion of individualism appeared in the seventeenth century and influenced the American Revolution. Rousseau's idea of the "General Will" emerged in the eighteenth century and helped to fuel the French Revolution. The General Will always puts the community first: "The French state thus became an instrument of intervention in the interest of the General Will, to plan and guide the nation's institutions" (Lodge, 1975:13).

Japan shares a communitarian viewpoint with France and many other western European countries. They each emphasize cooperation, interpersonal harmony, and the subordination of the individual to the community. Roots of Japanese communitarianism, however, lie in its own philosophic and religious ideas and in its unique demographic and geographic situation.

In communitarian capitalism, individuals and firms also have strategies (Thurow, 1992). Given the cultural foundations outlined above, those strategies are very different from strategies in individualistic capitalism. Individuals in communitarian capitalism attempt to select the best firm, and then work hard to be part of the company team; personal success or failure is identified with the fate of the company. Job switching is not widespread. In fact, in many Japanese firms, the person who voluntarily leaves a company is a "traitor" ("Graduates Take Rites of Passage into Japanese Corporate Life," 1991). At the same time, both Japanese and German workers are less often laid off.

Communitarian firms place the interests of shareholders far behind their employees and customers. Because employees are of first importance, high wages are a top priority. Maintenance of high wages and job security supersede profits. Long-range planning is emphasized over profit maximization, and shareholders, in fact, earn relatively low returns.

In both Europe and Japan, the government is expected to play a significant role in economic funding, planning, and growth. There is cooperation between business and government as well as between business and labor. Labor officials sit on the boards of directors of German firms, and German banks are major stockholders in German companies. Job training is the responsibility of both business and government. In short, there is no line of demarcation between private and public interests.

The contrast between the two cultures is conceptually clear, though the picture is more complicated in actual practice. As Figure 13.7 illustrates, communitarian and individualistic capitalism pass like ships in the night. Individualistic capitalists in Britain, Holland, Sweden, and the United States believe that by concentrating on individual self-interest they will automatically better serve customers and society. Communitarian capitalists, in contrast, assume that by

concentrating on serving customers and society they will automatically serve their own interests.

Which culture of capitalism is superior?   Some scholars believe that communitarian capitalism outperforms individualistic capitalism in the long run (Lodge and Vogel, 1987; Thurow, 1992; Albert, 1993). In fact, according to George Lodge, the American economy has already moved, kicking and screaming in resistance, more toward communitarian capitalism (Lodge, 1986). The question of which culture of capitalism will be most successful in the long run, however, can only be answered by unfolding events.

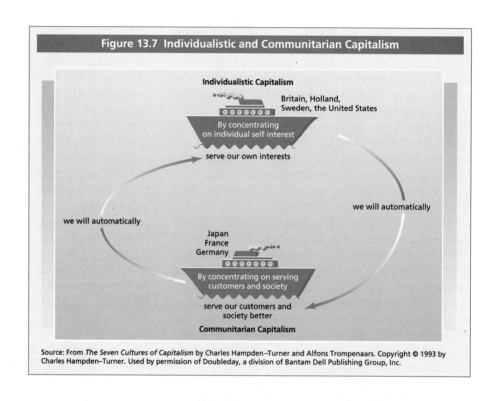

Figure 13.7 Individualistic and Communitarian Capitalism

Source: From *The Seven Cultures of Capitalism* by Charles Hampden–Turner and Alfons Trompenaars. Copyright © 1993 by Charles Hampden–Turner. Used by permission of Doubleday, a division of Bantam Dell Publishing Group, Inc.

FEEDBACK

1. Which of the following is not characteristic of capitalist thought?
   a. Private individuals have a right to pursue private gain.
   b. Society will benefit from attempts to make a profit.
   c. The state must make an active attempt to control the economy.
   d. Individuals pursuing a profit must be willing to risk losing money.
2. Under _____, most property belongs to private individuals and organizations rather than to the state or the community as a whole.
3. As capitalist systems mature,
   a. the economy is increasingly dominated by small businesses.
   b. the willingness to take risks increases.
   c. consumer goods become very scarce.
   d. many industries are dominated by a few large firms.
4. According to economic theorists, which of the following systems should have the greatest degree of control by the state?
   a. capitalism   c. democratic socialism
   b. socialism   d. welfare capitalism

continued on next page

FEEDBACK

5. Proponents of socialism contend that it will
   a. prevent workers from exerting significant control over social policy.
   b. lead to increased profits for owners.
   c. allow workers to exert significant control over work organizations.
   d. result in very inactive governments.

6. Currently, there seems to be a movement toward mixed economic systems in socialist societies. T or F?

7. Several characteristics of capitalist economic systems are listed below. Indicate in the space beside each characteristic whether it best describes individualistic capitalism (IC) or communitarian capitalism (CC).

   ____ a. government and business cooperation
   ____ b. high job mobility
   ____ c. profit maximization
   ____ d. more value placed on the interests of employees and shareholders
   ____ e. self-interest paramount
   ____ f. long-range economic planning

Answers: 1. c 2. capitalism 3. d 4. b 5. c 6. T 7. a. (CC) b. (IC) c. (IC) d. (CC) e. (IC) f. (CC)

THE CORPORATION

Within capitalist and most mixed economic systems, corporations have become extremely important actors. In their provocative book, *America Inc.,* Morton Mintz and Jerry Cohen contend that large corporations now own and operate the United States. In the introduction to the book, Ralph Nader sums up the authors' proposition when he states that government power is now a derivative of corporate power. Mintz and Cohen make this summary statement:

> *The danger that this superconcentration poses to our economic, political and social structure cannot be overestimated. Concentration of this magnitude is likely to eliminate existing and future competition. It increases the possibility for reciprocity and other forms of unfair buyer-seller leverage. It creates nation-wide marketing, managerial and financing structures whose enormous physical and psychological resources pose substantial barriers to smaller firms wishing to participate in a competitive market. . . . [This] superconcentration creates a "community of interest" which discourages competition among large firms and establishes a tone in the marketplace for more and more mergers. This leaves us with the unacceptable probability that the nation's manufacturing and financial assets will continue to be concentrated in the hands of fewer and fewer people (Mintz and Cohen, 1972:39).*

The political and economic effects of modern corporations made possible by this massive concentration of power are experienced at the local, national, and international levels, a point discussed later. For now, let's examine the nature of modern corporations.

The Nature of Modern Corporations

The economy of the United States has experienced substantial and continual growth. The gross domestic product (GDP)—the domestic output of goods and services—has steadily increased since 1933. This increase in the GDP has been accompanied by an increasing domination of the economy by large business corporations. The emergence of giant business organizations has contributed to economic growth but has also created a number of problems that can best be understood by examining the nature of modern corporations.

What is a corporation? A **corporation** is an organization owned by shareholders who have limited liability and limited control over organizational affairs. A key feature of the corporation is the separation of ownership from control (Berle and Means, 1968, originally published in 1932). Although shareholders receive reports of gains and losses from the corporation's transactions, they are not legally liable for them. And although shareholders are formally entitled to elect a board of directors, in reality they routinely approve the candidates recommended by management. The real control of a corporation rests with the corporation's managers. Some critics contend, however, that this widely assumed separation between ownership and management of large corporations is a myth.

How important are corporations today? At the beginning of the twentieth century, only a few American industries, such as railroading, shipping, steel, oil, and mining, were organized as corporations. Today, corporations dominate the economy, providing everything from diapers for babies to rest homes for the elderly.

Michael Useem—Insights into the American Business Elite

The influence business executives exert in America extends far beyond the boundaries of the corporations they manage. The governing boards of most major nonbusiness organizations, such as museums, high-level government bureaus, and many advisory panels, consist primarily of members of the business community. Of the many segments of the American labor force, only the professions approach business in terms of the number of members serving on governing boards. Some social scientists view this presence as primarily symbolic, but others view it as a fact of major sociological significance that insures the domination of both private and public life by business.

Michael Useem has attempted to enhance our understanding of the business elite through an innovative use of existing sources of information. He studied the directors of the 797 largest (as determined by *Fortune* magazine) corporations in the United States in 1969. Using published registers, he obtained the names of the 8,263 people who served as directors of these corporations. He then drew two samples for further analysis. One consisted of 1,570 people who held two or more directorships. The other comprised 433 directors who were affiliated with only one firm. For both samples, he collected information from *Who's Who in America,* official listings of individuals holding positions on federal government advisory committees, and lists of participants in exclusive social clubs and business policy groups. Thus, without asking corporate directors to complete any questions or interviews, Useem was able to obtain detailed information about their participation in noncorporate activities.

Useem found that directors for more than one corporation were more likely to participate in the control of several nonbusiness organizations. This finding was consistent for three distinct areas

---

More precisely, the United States is dominated by *large* corporations.

The Effects of Modern Corporations

The effects of corporations warrant an examination. Two specific issues—political decision making and the existence of multinational corporations—illustrate the type of effects large corporations are having.

How do corporations affect political decision making? Corporations represent massive concentrations of economic resources, which they know how to use. Many corporations, such as Lockheed, Citicorp, and Travelers have had little difficulty making their voices heard by government decision makers. Government policies regarding consumer safety, tax laws, and relationships with other nations often reflect corporate influence.

The tremendous influence of top corporate officials on government decisions exists for several reasons. First, corporate managers and directors have friendship, family, and organizational ties that lead them to be consulted informally about government policy matters. James Treat, the businessman whom Floyd Hunter found to be an influential adviser to Georgia's politicians, described his status in these terms: "I am asked occasionally to appear before state legislative committees, but I am not registered as a lobbyist. I just happen to know a lot of people, and I get called in on a lot of things. Sometimes, I think, on too many!" (Hunter, 1953:164). Mr. Treat, of course, happened to know a lot of people—and to exert tremendous influence—primarily because of his business position and the business activities of his relatives. On the federal level, many influential, wealthy individuals alternate between high corporate and political positions. For example, Robert McNamara was president of the Ford Motor Company before becoming Secretary of Defense in the Kennedy administration; Michael Blumenthal was chairman of Bendix Corporation before becoming Jimmy Carter's Secretary of the Treasury; and Donald Regan was chairman of the board and chief executive officer of Merrill Lynch before entering Ronald Reagan's cabinet as Secretary of the Treasury. The movement between government and corporate positions occurs on lower levels as well, as indicated by the National Association of Home Builders' tendency to hire former Federal Housing Authority (FHA) officials. The "revolving door" turns the other way as well. Consider the Department of Defense's proclivity for hiring people from defense contracting firms.

Second, corporate officials—because of their personal wealth and organizational connections—are able to reward and punish elected government officials through investment decisions. It is unlikely, for example, that Detroit's 500-million-dollar Renaissance Center would have been constructed without the personal and financial support of Henry Ford II. On a more of nonbusiness activity—governance of nonprofit private organizations; consulting for local, state, and national organizations; and participation in major business-policy associations. Those holding multiple directorships evidenced rates of participation in nonbusiness organizations two times greater than directors of only one corporation. Furthermore, those holding multiple directorships were more likely than those serving as directors of only one firm to have connections with relatively large firms. People who are involved in major social decisions, then, can be characterized as having "a greater capacity to mobilize corporate resources and . . . involvement in a national, transcorporate network of corporate directors" (Useem, 1979:567).

Because Useem had to rely on data from existing and accessible sources, he was unable to answer all his questions, and he has been careful to add qualifications to his conclusions. His overall pattern of findings, however, provides support for the idea that business elites constitute a special group likely "to express and represent its interests in places where other institutions are making decisions that can vitally affect business" (Useem, 1979:569). In other words, the business elite may exercise an important political leadership role that furthers the more general interests of the business community.

Critical feedback

1. Is it important to understand the influence business has on the decisions of nonbusiness organizations? Why or why not?
2. Describe the use Useem made of existing sources of information for this study. Did these sources give him all the information he needed to conduct an adequate study? Explain.
3. Which major theoretical perspective do you think underlies Useem's research? Support your choice.

ominous note, financially troubled International Harvester was able to extract considerable concessions from both Springfield, Ohio, and Fort Wayne, Indiana, by announcing that it would close its plant in one of the two cities. The decision to allow the Springfield plant to remain open was reached only after the city found $30 million in local and state funds to buy the plant and lease it back to the company.

Third, in addition to affecting political processes through investment decisions, wealthy members of society, including corporate officials, affect a political candidate's chance of being elected in the first place. State and national politicians who wish to be elected (or reelected) are hesitant to jeopardize their chances by offending corporate officials or wealthy potential contributors, who are frequently the same people.

Fourth, the political clout of large corporations is multiplied because of the existence of *interlocking directorates*—members of corporations sitting on one another's boards of directors. Although interlocking directorates are illegal if the corporations are competitors, this does not prevent noncompeting corporations with common interests from forming interlocking directorates. For example, members of the General Motors board of directors are also on the boards of twenty-nine other corporations, including Gulf Oil, Chase Manhattan Corporation, and Continental Oil. Moreover, these interlocks are only the beginning. The twenty-nine corporations having interlocking directors

with General Motors have interlocks with over seven hundred other corporations. Gulf Oil has eleven, Chase Manhattan Corporation has fifty-four, and Continental Oil has twenty-two. In addition, there are indirect interlocks that permit two board members of competing firms to be members of a third corporation. It is not difficult to imagine the political power emanating from this spider's web of interlocks among America's most powerful corporations (Mintz and Schwartz, 1981a, 1981b, 1985; Useem, 1984; Dye, 1990). In *Doing Research,* Michael Useem, one of the early researchers in this area, shows the extent to which business elites are in position to exercise political power that promotes their interests.

Fifth, the political power of corporations is further enhanced by existence of the *conglomerate*—a network of corporations in various, unrelated lines of business operated under a single corporate umbrella. The parent corporation does not actually produce a product or provide a service; it simply has controlling interest in a set of diverse enterprises. RJR Nabisco, Inc., for example, holds companies in such different areas as tobacco, pet foods, research and technology, candy, cigarettes, food products, and bubble gum. RJR Nabisco requires over one page in *Who Owns Whom* (1993) for a listing of its North American subsidiaries.

Finally, assets are concentrated in the hands of a relatively few giant corporations, which translates into political power. The top hundred corporations, which account for less than 0.1 percent of all corporations in the United States, control over 9 percent of total corporate assets; the top two hundred corporations control over 11 percent. This has changed little since 1960. There is also a very high concentration of assets "within" the top corporations. The top one hundred have three-fourths of the assets held by the top five hundred corporations, and the top two hundred corporations possess 88 percent. Although only 0.2 percent of American corporations have assets of $250 million or more, these corporations control over 81 percent of the corporate assets in the United States (U.S. Bureau of the Census, 1996a). In sharp contrast, the 98.5 percent of corporations with assets of less than $10 million control only about 7 percent of the nation's corporate assets.

The informal contacts and organized power of corporations, then, give corporations important advantages in the political decision-making process. And although corporate officials may disagree with one another and may not have the same goals, their ability to make effective use of their organizational resources when they see their interests being threatened cannot be ignored.

The dominance of large corporations is not confined to those corporations' countries of origin. The world is increasingly being influenced by **multinational corporations**—firms in highly industrialized societies with operating facilities throughout the world. Multinationals are concentrated in a few industries, each controlled by a small number of giant companies. The oil and automobile industries alone account for about half of all multinational activity.

Although multinational corporations have existed for centuries, they have become more powerful since World War II. Improvements in communication and transportation technology have allowed them to exert more control over their operations around the world.

How economically powerful are multinational corporations?

Fifty-one of the world's largest one hundred political and economic entities are multinational corporations rather than nation-states. Seven corporations based in the United States—Exxon, IBM, General Motors, Ford Motors, AT&T, Wal-Mart Stores, and General Electric—have sales volumes on a par with the gross national products of the world's forty most productive nations. Nations ranking below both General Motors and Ford Motors include Argentina, South Africa, Venezuela, Greece, Portugal, Israel, and Ireland (Steedley, 1994). If the three hundred largest multinational corporations were to band together, their combined resources would be exceeded only by those of the United States and Japan (*World Development Report 1994*).

What are the negative effects of multinational corporations?

Multinational corporations have many defenders who argue that they provide developing countries with access to technology, capital, foreign markets, and products that would otherwise be unavailable to them. Critics argue, however, that multinationals actually harm the economies of the nations in which they operate by exploiting their natural resources, disrupting local economies, introducing inappropriate technologies and products, and increasing the amount of income inequality. Critics note that multinationals often rely on inexpensive labor or abundant raw materials in developing nations while returning their profits to corporate headquarters and shareholders in rich nations. Multinationals' domination of their industries has made it difficult for the low economy nations to establish new companies that can effectively compete with the multinationals. As a result, multinationals may retard rather than promote the economic development of some regions of the world.

Although evidence on multinationals' effects is not always clear, these firms possess huge concentrations of power whose abuses cannot always be checked, even by national governments. Indeed, evidence suggests that executives of Internation Telephone and Telegraph Corporation (ITT) worked with elements of the U.S. government in an effort to overthrow the government of Chile, headed by Salvador Allende (Anderson, 1974; Wise, 1976). Although such actions may be atypical of multinationals, they illustrate the difficulties faced by a government wishing to control the activities of multinationals within its borders.

The available evidence suggests that some problems introduced by multinationals are not effectively controlled. For example, Haitian sugar cane cutters employed by Gulf & Western in the Dominican Republic are said by some observers to

> . . . labor under the most inhuman conditions known in Latin America in the twentieth century. They are transported to the Gulf & Western plantation in huge trucks, receiving only cane or a mixture of cane juice and brown sugar for nourishment. Their drinking water is the same water used to wash animals, and only 57 percent of the shacks have latrines (Kowalewski, 1982:144).

Environmental impacts are similarly problematic. For example, lakes in Jamaica are commonly polluted with "red mud," a by-product of aluminum-producing activities by such companies as Alcoa and Revere; and refineries operated in the Caribbean by Union Carbide and Sun Oil have been linked to increased incidences of respiratory disease (Kowalewski, 1982). Products marketed by multinationals may sometimes create problems because they are inappropriate for local conditions.

FEEDBACK

1. A _____ is an organization owned by shareholders who have limited liability and limited control over organizational affairs.
2. Which of the following is *not* one of the reasons behind corporate political influence?
   a. Corporate managers and directors have informal and organizational ties that lead them to be consulted about matters of political policy.
   b. Many corporate officials have reward and coercive power over elected government officials.
   c. Many elected officials are also top corporate officials.
   d. Interlocking directorates exist.
   e. Conglomerates are present.
3. Indirect interlocks permit two board members of different corporations to sit on each other's boards of directors. *T or F?*
4. Firms with operating facilities in several different countries are called _____.
5. Multinational corporations are concentrated in a few industries, each dominated by a small number of giant companies. *T or F?*
6. Multinational corporations
   a. are easily manipulated by the governments of low economy nations.
   b. sometimes create problems involving unsafe working conditions and inappropriate products.
   c. are known to have overthrown large numbers of low economy governments.
   d. carefully avoid environmental damage in their areas of operation.

Answers: 1. corporation 2. c 3. F 4. multinational corporations 5. T 6. b

Because of problems with local water supplies, infant formulas marketed by Nestlé have led to significant illness and death in several low economy nations.

Finally, the ability of multinationals to move operations to areas offering cheap labor and weak regulatory controls may have an increasingly severe impact on the economies of more developed nations, such as the United States. Multinational corporations, then, provide a clear illustration of the close relationship between economic and political power.

WORK IN THE CONTEMPORARY ORGANIZATION

Changing Workforce Composition and Occupational Structure

The contemporary economy in the United States is dynamic, if nothing else. Radical new alterations in the workforce composition are projected for the next century. Past trends in the occupational structure are expected to accelerate, with some increasingly serious consequences (Rifkin, 1995; Milkman, 1997). We will examine new workforce composition trends first.

What are the major workforce composition trends for the twenty-first century? There are five major trends that will revolutionize America's future workforce (Johnson and Packer, 1987; U.S. Department of Labor, 1988). First, because of the slowdown in population growth, America's workforce will be expanding by only 1 percent annually in the 1990s, compared to almost 3 percent per year in the 1970s. Effects of this slow labor force growth include retarded economic expansion, tightening of the labor markets for employers, and shift by employers to more capital-intensive (rather than human-intensive) production systems.

Second, the average age of the workforce (and the general population) will increase as the pool of young workers entering the labor market shrinks. The average age of the workforce will climb from 34 in 1985 to 39 in 2000 and 41 in 2020 as the baby boomers are followed by the baby-bust generation.

For a variety of reasons, the American economy is experiencing a decline in blue-collar manufacturing similar to the decline experienced by agriculture after World War II. This fundamental shift is creating a service economy in which more and more people will be employed in white-collar jobs.

Third, between 1990 and 2005 about one-half of new entrants into the U.S. labor force will be female, compared to 45 percent in 1988 (U.S. Department of Labor, 1992b). Whereas males accounted for 55 percent of new labor entrants in 1988, they will constitute only 50.5 percent by 2005. The proportion of the labor force that is female will have increased from 30 percent in 1950 to over 45 percent by 2005. (See Figure 13.8.) Although women will remain concentrated in lower-paying jobs compared to men, they will be rapidly entering many higher-paying professional and technical occupations (Maynard, 1994). Occupational inroads made by females will promote the growth of convenience industries and increase the demand for day care.

Fourth, just over one-third of the new entrants into the workforce between 1990 and 2005 will be nonwhites, twice the proportion of new entrants in 1988. Although this expanding share of a more slowly growing workforce might be expected to enhance the employment opportunities for America's minorities, there are reasons for pessimism. Minorities are concen-

trated in economically declining central cities, overrepresented in industries that are losing jobs, and underrepresented in the most rapidly growing occupations. Minority workers continue to be blocked from new jobs because of disadvantages in education and job training. Moreover, minority youth are still disproportionately born in poverty.

Finally, immigrants will constitute the largest share of the increase in the workforce (and the population) since World War I. During the 1970s, the foreign-born population of the United States rose by about 4.5 million, and almost half a million more immigrants a year are expected through the end of the century. Given the current rate of illegal immigration—approximately 750,000 annually—total immigration since 1970 would add over 16 million to the population and almost 7 million to the labor force. Although these immigrants come from a wide range of social and educational backgrounds, they share a strong resolve to advance themselves. Immigrants are more likely to hold two or more jobs as starting points. Concentrated

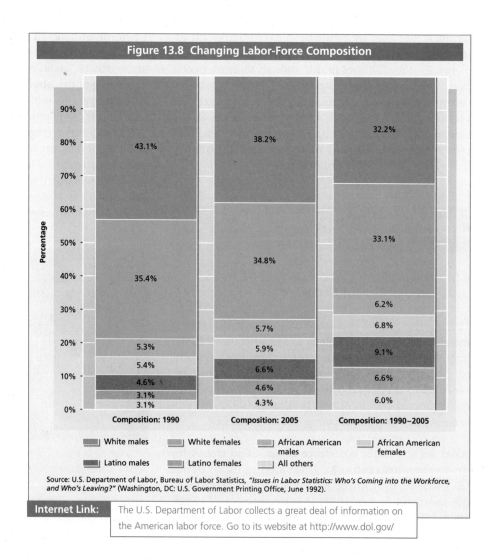

Figure 13.8 Changing Labor-Force Composition

Source: U.S. Department of Labor, Bureau of Labor Statistics, *"Issues in Labor Statistics: Who's Coming into the Workforce, and Who's Leaving?"* (Washington, DC: U.S. Government Printing Office, June 1992).

Internet Link: The U.S. Department of Labor collects a great deal of information on the American labor force. Go to its website at http://www.dol.gov/

in the South and West, these immigrants are expected to promote faster economic growth and create labor surpluses in these regions.

Taken together, these trends promise a very different looking pool of new labor market entrants in 2005 than exists today (Towers Perrin, 1992). The key characteristic of the workforce of the twenty-first century will be gender, racial, and ethnic diversity. (See Figure 13.8.)

Most economies contain agricultural, manufacturing, and service sectors, but the proportion of the labor force employed in each sector varies considerably. Preindustrial society was built on an agricultural base. With the coming of industrial society, manufacturing employed most of the labor force in producing goods. Postindustrial society now employs most of the labor force in service and white-collar occupations, including government, research, education, health, transportation, trade, and finance. One indication that the United States is in the postindustrial stage is the country's changing occupational structure (Milkman, 1997).

How has the occupational structure been changing? Since World War I, the fastest-growing occupations have been white-collar; in 1956, white-collar workers for the first time accounted for a larger proportion of the labor force than blue-collar workers. As Figure 13.9 shows, the percentage of American workers employed in manual jobs has been declining for more than two decades, and the percentage of workers employed in farming has declined throughout the century. Both declines are expected to continue. Changes in technology, increasing demand for services, and declines in traditional "smokestack industries" such as automobiles and steel are likely to mean that future growth will be in the service and white-collar sector rather than in the manual sector of the labor force (Ritzer and Walczak, 1986; Johnson and

Packer, 1987; Milkman, 1997). Consequently, the labor market is becoming more complex.

We saw in Chapter 9 ("Inequalities of Race, Ethnicity, and Age") that a dual labor market has developed in American society. The core industrial sector contains the large, profitable firms holding dominant positions within major industries. The automobile, aerospace, and petroleum industries are prime examples. In the peripheral industrial sector are small, less profitable firms operating in more competitive industries. Agriculture, textile, and retail trade are in this sector.

What are the effects of the changing occupational structure on employment and income? Although the core sector allows for the development of careers, the peripheral sector offers jobs characterized by high turnover, little or no job training, and few opportunities to develop skills or knowledge (Weakliem, 1990). Unfortunately, many of the industries in the core sector currently are laying off experienced workers rather than hiring new ones. As early as 1983, for example, a steel mill in Hibbing, Minnesota, that once employed 4,400 people had a payroll of only 650. Rockford, Illinois, a major tool-making center, lost 17,000 manufacturing jobs in a four-year period. In 1981, the steel industry laid off 86,000 workers in Pennsylvania's Monongahela Valley ("Left Out," 1983). As you will see shortly, this trend has accelerated.

Obviously, the industries that traditionally supplied choice industrial jobs in the United States are scaling down their operations. Some, such as steel, have lost much of their business to more technologically sophisticated foreign competitors. Others are moving their operations to less developed nations in which they can escape strong government regulation and effective labor unions. Currently several million workers in the United

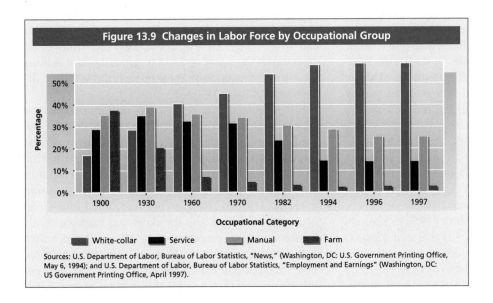

Figure 13.9 Changes in Labor Force by Occupational Group

Occupational Category

■ White-collar ■ Service ■ Manual ■ Farm

Sources: U.S. Department of Labor, Bureau of Labor Statistics, "News," (Washington, DC: U.S. Government Printing Office, May 6, 1994); and U.S. Department of Labor, Bureau of Labor Statistics, "Employment and Earnings" (Washington, DC: US Government Printing Office, April 1997).

Reemployment in the current American economy is becoming a significant problem. The white-collar worker (on the left) is engaging in the job-search process while eating lunch. Other Americans (on the right) are waiting in long lines outside personnel offices in hopes of landing a scarce job.

States, many of whom were employed in basic industries at one time, have been displaced by new technology or foreign competition. Between 1987 and 1992 the number of workers who lost jobs due to company failures, plant closings, plant relocations, or other curtailments in employment totaled 5.6 million. These job losses occurred disproportionately in goods-producing industries. Workers in service industries had the lowest probability of losing their jobs (Gardner, 1993).

The problem of reemployment is a significant one. Although the overwhelming majority of these displaced workers had been in full-time jobs, earning a wage or salary, only slightly more than one-half reported being reemployed in new full-time wage and salary jobs. Another one-third were either unemployed or were no longer in the labor force; the rest were in new jobs working either part time, in their own businesses, or as unpaid family workers. Finally, of those displaced workers who found full-time wage and salary jobs, about one-half were making less money than on their previous jobs.

The U.S. Bureau of Labor Statistics also forecasts slow job growth in the future—some 1.7 million new jobs each year to 2005. Of these, only one-third will come from large corporations, which customarily pay the most. As a result, the real average compensation per employee, which rose from $14,000 in 1950 to $24,000 in 1979, has fluctuated between $21,000 and $23,000 in the 1980s and 1990s (Fierman, 1993).

There is also the growing problem of what might be termed *downwaging* in the American economy. Just as

the job loss trend documented above does not seem to be temporary, a reduction in pay appears also to be a fixture in this new economic world. The continuation of this trend into the twenty-first century is portrayed graphically in projected job growth and decline. Of the ten jobs projected to expand between 1994 and 2005, five pay below the poverty level; only two of the top ten shrinking jobs fail to meet the poverty threshold. (See Figure 13.10.) The low pay scale associated with service jobs is rapidly replacing the high pay scale of traditional manufacturing jobs. These trends of job loss, slow job growth, and downwaging threaten the American Dream (Newman, 1993; Barlett and Steele, 1996; see *Doing Research* in Chapter 18, "Social Change in Modern Society"). One response has been that an apparent record number of Americans—as many as 25,000 annually—are leaving the United States permanently to build a better life elsewhere (Belsky, 1994).

It is estimated that at least one million people, and probably more, will be displaced each year in the future. Management experts suggest that blue-collar manufacturing is undergoing a decline similar to that experienced by agriculture since World War II. That is, employment will decline dramatically even as output increases. The sad truth is that traditional blue-collar jobs are diminishing and the skills that made currently displaced workers employable in the past are ill-suited to the future workplace. This problem can be solved only through a massive retraining thrust, an event unlikely to occur in the near future because of a lack of

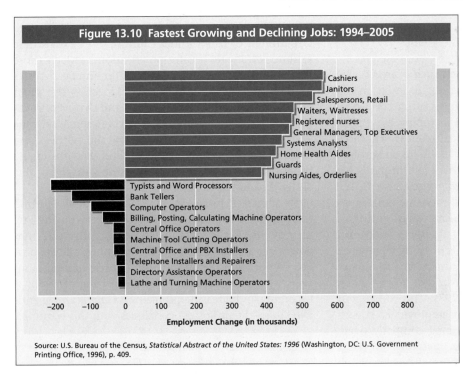

Figure 13.10 Fastest Growing and Declining Jobs: 1994–2005

Source: U.S. Bureau of the Census, *Statistical Abstract of the United States: 1996* (Washington, DC: U.S. Government Printing Office, 1996), p. 409.

financial commitment from the federal government and taxpayers (Reich, 1997).

Newer industries, such as those based on microchip technology, are beginning to replace the traditional ones in a few areas, including Massachusetts; Ann Arbor, Michigan; and California's Silicon Valley. Such industries, however, appear to be providing few jobs for those who were formerly employed in the traditional industries. Moreover, the majority of the jobs in the high-tech industries involve low skill levels and offer minimal wages with few chances for advancement. Responsible positions with high pay are held by a very small proportion of high-tech employees. Finally, many American high-technology firms are moving jobs to areas overseas with low labor costs, just as textile mills, auto and steel manufacturers, and producers of televisions did before them.

The dual labor market perspective has been criticized because some of its assumptions have not been clearly spelled out and because it does not always provide powerful predictors of the experiences that individuals will have in the labor market. Studies of industries and occupations as diverse as high technology electronics components, intercollegiate athletics, and longshore crane operations suggest that the actual experiences of individuals within a sector may vary considerably. Despite a need for more refinement, however, dual labor market theory points to a need to be alert for major shifts in the kinds of industries and firms that will be offering jobs in the future (Hodson and Kaufman, 1982; Kelley, 1990; Sakamoto and Chen, 1991).

Clearly, the labor force landscape in the United States has changed dramatically since the early 1980s, in part due to steps taken by top management in American corporations. Two of these interrelated steps, *downsizing* and *contingent employment,* deserve special attention.

Downsizing and the Emerging Contingent Employee

Downsizing is the decision on the part of top management to reduce its workforce in order to cut costs, increase profits, and enhance stock values. To justify downsizing, business leaders point to rising health-care costs, the need to become leaner and less hierarchically structured in an increasingly competitive environment, and the necessity of eliminating many levels of middle management because of computers and information technology (Gleckman et al., 1993; Huber and Korn, 1997).

As you shall see later, the negative consequences of downsizing for both organizations and employees may outweigh the apparent advantages. As William Snizek puts it, "company downsizing or right sizing is neither a necessary nor a sufficient condition for improved organizational functioning or profitability" (Snizek, 1994:4).

To what extent is downsizing occurring? The decision to downsize became a popular one in American corporations during the 1990s (Boroughs, 1996). Since 1985, it is estimated that over 4 million people have been trimmed due to downsizing alone, half of which held white-collar jobs Moreover, corporate leaders indicate this trend will continue (Sloan, 1996).

Why is downsizing taking place on such a large scale?
Part of the motivation for downsizing is based on top management's belief that it employs a surplus of people and that work can be done by fewer employees without reductions in efficiency and effectiveness. There is, however, more involved than the desire for an absolute reduction in the workforce. Part of the impetus for downsizing rests on the belief that work now in the hands and minds of permanent, full-time employees can be done in other, less expensive ways. Profitable companies are increasingly using the services of **contingent employees**—individuals hired by companies on a part-time, short-term, or contract basis to replace full-time employees ("The New World of Work," 1994; Moore 1996; Cappelli, et al., 1997). The contingent label is appropriate because their employment is contingent on daily, weekly, or seasonal company need. They may also be referred to as "marginal" workers, because they are out of the traditional system of job and economic security. Some even refer to them as "disposable" workers (Kirkpatrick, 1988; Castro, 1993).

Over one-third of the U.S. labor force (over forty million people) are now contingent employees, and it is expected that one-half of the labor force will be contingent workers by 2000. Temporary employment has increased over 250 percent since 1982. Temporary employees earn about 60 percent of the average full-time worker and are not entitled to health, retirement, or other benefits. The contingent part of the labor force has traditionally included groups from the bottom of the occupational structure—minorities, women, youth, the elderly. Most contingent workers have been in low-level, low-paying clerical, retail, and service jobs. Increasingly, however, this trend is encompassing higher level white-collar people—engineers, accountants, nurses, managers, physicians (Heckscher, 1995; Rifkin, 1995).

What will be the effects of downsizing and contingent employment? These new trends may portend a fundamental change in the U.S. labor market (Harrison, 1994; Heckscher, 1995). Some maintain that it marks the dawn of a brave new workplace of individual entrepreneurialism (Kiechel, 1993; Sherman, 1993b), whereas others insist it constitutes a fundamental change in the implied social contract between businesses and employees. Daniel Bell (1993), representative of the entrepreneurial view, cites the downfall of IBM, General Motors, and the American steel industry as evidence of a need for firms to become lean, adaptable, and flexible. Robert Kuttner (1993a, 1993b), on the other hand, points out that the age of job security and its underlying social contract is being fractured by these actions on the part of corporations. Similarly,

Robert Reich (1993, 1997), former Secretary of the U.S. Department of Labor under Clinton, asserts that these actions will create greater polarization between those that control capital and those that do not. Such critics believe that the disposable workforce is the most important trend in business today, and that it is fundamentally changing the relationship between Americans and their employers. Downsizing and the increased use of contingent workers, they contend, will foster the belief that a valued future relationship with the firm is either threatened or nonexistent.

As suggested in the *Sociological Imagination* opening this chapter, the disadvantages of extensive downsizing may outweigh the advantages. Many companies that downsized are not increasing their earnings and are now less competitive (Baumohl, 1993). Stress among managers who make and carry out downsizing decisions is becoming an increasing problem. A *Fortune* article entitled "Burned-Out Bosses" concludes with this warning:

> Constant restructuring has become a fact of business life in this era of change. Well and good, but companies that don't acknowledge the stress that survivors undergo and support those who are in danger of burning out may find that their glistening, reengineered enterprises end up being run by charred wrecks (Smith, 1994:5).

According to a survey of 2,500 employees across the United States, there is a growing split in employees' attitudes toward their work and their employers (*Towers Perrin Workplace Index*, 1997). Although employees express higher job satisfaction, there are indications of eroding trust in management's commitment to the partnership employers claim they want to create with their employees. For example, only 41 percent of those surveyed believe that their company considers their interests in decisions affecting them, down from a not very high 50 percent in 1995. Figure 13.11 shows evidence of employees' decreasing sense of reciprocity between themselves and their companies.

Negative reaction to the corporate policy of turning permanent jobs into temporary jobs is also being expressed in labor relations. The 1997 Teamsters Union strike against United Parcel Service, the first in the company's ninety-year history, was based squarely on union opposition to UPS's policy of increasing contingency employment (60 percent of its workers are part-time). John Sweeny, president of the AFL-CIO, publicly called the policy of contingency work "corporate greed" and cites it as an issue for the labor movement in general to pursue. The strike did not last very long, in part because the American public generally sided with the Teamsters (Greenhouse, 1997).

Research in this area is just beginning, so we do not know the full consequences downsizing and contingent employment will have on companies or employ-

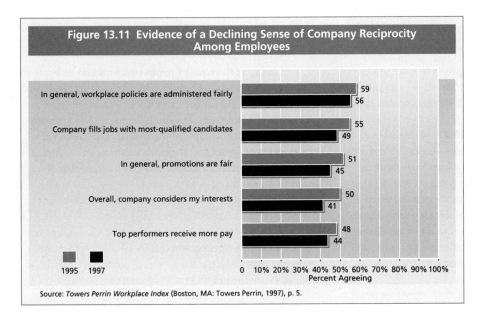

Figure 13.11 Evidence of a Declining Sense of Company Reciprocity Among Employees

Source: *Towers Perrin Workplace Index* (Boston, MA: Towers Perrin, 1997), p. 5.

ees. What will happen to employee loyalty, performance, and product quality? Will trust deteriorate between managers and employees? Will the survivors of downsizing suffer more illnesses due to greater workloads and higher stress? Will there be a deluge of lawsuits based on charges of wrongful discharge? Will employees, blue- and white-collar, turn more to unions (Watson and Shepard, 1997)? Characterizing survivors of downsizing as "victims," William Snizek, in a comprehensive review of existing research, concludes that:

"A growing body of evidence is beginning to accumulate which shows that the deleterious effects of corporate layoffs are as great, if not greater, on those who survive, as they are on those initially laid off" (Snizek, 1994:3).

One conclusion seems clear—management will have to become more aware of the problems associated with downsizing and contingency hiring for both people and organizations and take appropriate actions if expected economic benefits are to be realized (Sloan, 1996). Among these steps are maintaining honest and constant communication between top management and survivors of layoffs, setting goals with survivors, further empowering survivors on the job, exercising accessible and sensitive leadership, and creating a positive organizational culture for survivors (Snizek, 1994; O'Neill and Lenn, 1995).

A major fear associated with the development of a postindustrial economy, then, is that many workers will be unable to find employment or to find employment at the occupational level at which they are qualified and with a degree of compensation to which they are accustomed. A nationally representative sample of Americans were asked this question in a 1994 Harris poll: "And what is your greatest *personal fear* for something bad that might happen to you in 1994?" Job loss was the most prominent fear, loss of financial security the third (Taylor, 1994a).

FEEDBACK

1. Which of the following is *not* one of the major labor-force trends discussed in the text?
   a. slowdown in labor force growth
   b. aging of the labor force
   c. increased employment of females
   d. increase in blue-collar jobs
   e. increase in nonwhite proportion of the labor force

2. Most of the jobs that were added to the American economy over the past twenty years were in which of the following sectors?
   a. service and white-collar
   b. manufacturing
   c. mining
   d. skilled blue-collar

3. The _____ industrial sector contains large, profitable firms with dominant positions within major industries.

4. The _____ industrial sector contains small, less profitable firms operating in more competitive industries.

5. _____ is the decision on the part of top management to reduce its workforce in order to cut costs, increase profits, and enhance stock values.

Answers: 1. d 2. a 3. core 4. peripheral 5. Downsizing

REVIEW GUIDE

SUMMARY

1. The economy—the institution designed for the production and distribution of goods and services—and the political institution, within which power is obtained and exercised, are closely related.

2. Power—the ability to impose one's will on others whether or not they wish to comply—is central to social life. Authority is power accepted as legitimate by those subject to it. Nation-states can be based on one of three types of authority. Charismatic authority is too unstable, and traditional authority is too outdated to serve as the primary basis of power for modern nation-states. Today, most nation-states are based on rational-legal authority. Consequently, the power of government officials is limited by legal rules and regulations.

3. Although a nation-state is the political entity that holds authority over a specified territory, a government is the political structure that rules a nation. The nation-state is relatively new in human history and is associated with agriculture and industrial production. According to functionalism, the nation-state emerged to maintain social order. From the conflict perspective, nation-states exist primarily to promote the interests of societal elites.

4. Democracy and totalitarianism are the two polar types of political systems. In modern societies, democracies are representative with minimal citizen involvement in political affairs. Totalitarian political systems have absolute rulers who try to control all aspects of social life. Between these two polar types of political systems lies authoritarianism. This third type of political system involves nonelected rulers in absolute control, but some individual freedom is permitted.

5. Voting does not seem to be a highly effective means for nonelites to influence political decision making in the United States. This is partly because America does not have a high voter participation rate and partly because the most disadvantaged tend to be nonvoters.

6. The two major models of power distribution are pluralism and elitism. Pluralists, whose view is associated with functionalism, depict power in the United States as widely distributed among diverse special interest groups. Interest groups appear to be on the rise as the public become more politically sophisticated and as the influence of political parties declines. The growing importance of political action committees (PACs) reflects the same social conditions. Some powerful interest groups, notably those PACs associated with business interests, thwart public opinion by using their power for the attainment of narrow, selfish goals. Still, interest groups provide one of the few mechanisms for the representation of nonelite segments of a society aside from voting, and some of these interest groups have successfully challenged the political and economic establishment.

7. Advocates of the conflict perspective contend that American society is controlled by a unified and enduring elite. Proponents of the existence of a power elite see a unified coalition of top military, corporate, and government leaders in control of the United States. Although evidence tends to document the existence of a national power elite, it cannot be definitively asserted that one actually exists.

8. Economies differ in their organization and underlying assumptions. Capitalist economies are based on beliefs in private property and the pursuit of profit without government interference, whereas socialist economies are based on beliefs that the means of production should be owned by the people as a whole and that government should have an active role in planning and controlling the economy. There appears to be some movement toward capitalism in socialist countries. Capitalist and socialist ideologies appear to be becoming less of a major barrier in the relationships among nations.

9. Individualistic capitalism and communitarian capitalism are the two basic variants of capitalism. Self-interest, the free market, profit maximization, and highest return possible on stockholder investment are foundational principles for individualistic capitalism. In communitarian capitalism, the emphasis is on the interests of employees, customers, and society.

10. The rise of the corporate economy has intensified the link between political and economic institutions. Some critics contend that the political institution in the United States is now controlled by vast corporations whose economic assets are owned by a relatively small segment of the population.

11. Corporations are rapidly increasing in size, number, and power. Their influence is felt domestically and internationally. Corporations affect domestic political decision making and are influencing the political and economic institutions of countries around the world as well.

12. Workers in the contemporary economy face a changing labor-force composition. The labor force is growing more slowly, aging, and becoming more culturally diverse as more women, minorities, and immigrants enter the labor market. The occupational structure is also changing as more traditional industrial jobs are replaced with white-collar and service jobs. More American corporations are downsizing and replacing full-time employees with contingent or temporary workers. The full implications of these changes for workers and for the future of American society are not yet known. Evidence is beginning to appear showing that this corporate restructuring is having some negative consequences.

LEARNING OBJECTIVES REVIEW

After careful study of this chapter, you will be able to:

- Distinguish power and authority and identify three basic forms of authority.
- Compare and contrast the perspectives of functionalism and conflict theory on the nation-state.
- Identify the major differences among democracy, totalitarianism, and authoritarianism.
- Differentiate the view taken of the distribution of power in America by funtionalism and conflict theory.

- Describe the major characteristics of capitalism, socialism, and mixed economic systems.
- Discuss the effects of the modern corporation on American society.
- Describe America's changing workforce composition and occupational structure.
- Discuss corporate downsizing and its consequences.

CONCEPT REVIEW

Match the following concepts with the definitions listed below them:

_____ a. individualistic capitalism

_____ b. political action committees

_____ c. oligopoly

_____ d. government

_____ e. nation-state

_____ f. democracy

_____ g. monopoly

_____ h. downsizing

_____ i. communitarian capitalism

_____ j. authoritarianism

_____ k. traditional authority

_____ l. power elite

_____ m. rational-legal authority

_____ n. political party

_____ o. bourgeoisie

_____ p. elitism

_____ q. interest group

_____ r. corporation

1. The type of political system controlled by nonelected rulers who generally permit some degree of individual freedom.
2. Members of a society who own productive property.
3. The type of political system in which elected officials are held responsible for fulfilling the goals of the majority of the electorate.
4. The theory of power distribution that sees society in the control of a few individuals and organizations.
5. A group organized to achieve some specific, shared goals by influencing political decision making.
6. A situation in which a single company controls a market.
7. The political entity that holds authority over a specified territory.
8. Organizations established by interest groups for the purpose of raising and distributing funds to selected political candidates.
9. An organization designed to gain control of government through the election of candidates to public office.

10. A unified coalition of top military, corporate, and government leaders.
11. Authority based on rules and procedures associated with political offices.
12. The political structure that rules a nation.
13. Legitimate power rooted in custom.
14. The type of capitalism that emphasizes the interests of employees, customers, and society.
15. The decision on the part of top management to reduce its workforce in order to cut costs, increase profits, and enhance stock values.
16. The type of capitalism founded on the principles of self-interest, the free market, profit maximization, and the highest return possible on stockholder investment.
17. A situation in which a combination of companies controls a market.
18. An organization owned by shareholders who have limited liability and limited control over organizational affairs.

CRITICAL THINKING QUESTIONS

1. Like all organizations, universities are based on some form of authority. Discuss each of the three types of authority covered in the text to show which type of authority is most basic to the university structure. Be careful to show why each type of authority is applicable or inapplicable to the university's organizational structure._____

_____

_____

2. Functionalism and conflict theory each has a unique perspective on the nature and purpose of the nation-state. Do you think one of these perspectives more accurately describes American society than the other? Why or why not? _____

_____

_____

3. Define the concept of the power elite. Is American society best characterized as a pluralist society or a society controlled by a power elite? Defend your position. _____

_____

_____

4. Distinguish between the two global cultures of capitalism. Do you envision a convergence of these two approaches to capitalism within the rapidly developing global marketplace? Support your viewpoint. _____

_____

_____

5. American corporations have in recent years turned to the practices of downsizing and contingent employment. State the reasons for this managerial strategy and discuss whether or not it will promote the organizational goals it is designed to reach.

_____

_____

MULTIPLE CHOICE QUESTIONS

1. **Max Weber defined _____ as the ability to control the behavior of others even against their will.**
   a. authority
   b. charisma
   c. power
   d. elitism
   e. influence

2. **_____ is power accepted as legitimate by those subjected to it.**
   a. Pluralism
   b. Authority
   c. Elitism
   d. Influence
   e. Coercion

3. **The political structure that rules people in a specific territory is known as a**
   a. Leviathan.
   b. nation-state.
   c. democracy.
   d. political party.
   e. government.

4. **According to the conflict perspective, nation-states exist primarily to**
   a. serve the interests of a society's elite.
   b. eliminate tensions between social classes.
   c. solve the Hobbesian problem of order.
   d. perpetuate competition for new economic markets.
   e. establish the legitimacy of charismatic authority.

5. **What two types of political systems exist in nation-states?**
   a. democracy and authoritarianism
   b. totalitarianism and authoritarianism
   c. republican and democratic
   d. communism and republicanism
   e. democracy and totalitarianism

6. **According to _____, a community or society is controlled from the top by a few individuals or organizations.**
   a. symbolic interactionism
   b. elitism
   c. communism
   d. functionalism
   e. pluralism

7. **Decision making resulting from competition, bargaining, and compromise among diverse special interest groups is based on the _____ model.**
   a. elitist
   b. laissez faire
   c. conflict
   d. pluralist
   e. functional

8. **According to C. Wright Mills, power in the United States is controlled by**
   a. the people through a democratic process.
   b. labor unions who are able to exert covert influence on management and the general public.
   c. top military, corporate, and government leaders.
   d. Fortune 100 companies.
   e. competing interest groups.

9. A system founded on the belief that the means of production should be owned by the people as a whole and that government should have an active role in planning and controlling the economy is known as
   a. socialism.
   b. capitalism.
   c. communism.
   d. industrialism.
   e. pluralism.

10. As capitalist systems mature,
    a. consumer goods come under the influence of the law of diminishing returns.
    b. the willingness to take economic risks increases.
    c. the desire to turn to a mixed economic system increases.
    d. many industries are dominated by a few large firms.
    e. many small businesses are able to compete with larger firms.

11. An organizational form in which shareholders have limited liability and limited control over organizational affairs is known as a/an
    a. corporation.
    b. cooperative.
    c. unilateral organization.
    d. voluntary organization.
    e. employees' organization.

12. Which of the following statements is *false*?
    a. Communist countries have not been slow to move to some sort of market system after the fall of the former Soviet Union.
    b. Economic change in recent years has not been limited to Soviet-bloc economies.
    c. Cuba has now adopted some degree of capitalism.
    d. Capitalism now threatens to triumph over communism.
    e. China, with the repossession of Hong Kong, is now reverting to a planned economy.

13. According to the text, the key characteristic of the workforce in the twenty-first century will be
    a. planned obsolescence.
    b. cultural neutrality.
    c. cultural homogeneity.
    d. cultural diversity.
    e. an increasing work ethic.

14. Which of the following generalizations regarding corporate downsizing in the United States now appears to be *true*?
    a. Extensive downsizing is a fading corporate strategy.
    b. Contingent employment has passed its peak.
    c. The disadvantages of extensive downsizing may outweigh the advantages.
    d. Management no longer has to be concerned about problems associated with downsizing and contingent employment because employees have become accustomed to these practices.
    e. Since 1990, job loss has been receding from the pinnacle of employee fears.

FEEDBACK REVIEW

True-False

1. According to Jean-Jacques Rousseau, it is primarily the privileged classes who benefit from the existence of the state. *T or F?*
2. The operation of PACs in America today invalidates the pluralist model. *T or F?*

Fill in the Blank

3. According to Karl Marx, those in charge of the state in capitalist societies make decisions that serve the interests of the owners of the means of production. He used the term _____ to describe acceptance of a system that works against one's own interests.
4. _____ refers to the type of political system controlled by nonelected rulers who generally permit some degree of individual freedom.
5. Firms with operating facilities in several different countries are called _____.
6. The _____ industrial sector contains large, profitable firms with dominant positions within major industries.

Multiple Choice

7. The majority of political action committees (PACs) are now associated with
   a. labor unions.
   b. citizen groups.
   c. political parties.
   d. corporations and other businesses.

8. Which of the following is *not* one of the reasons behind corporate political influence?
   a. Corporate managers and directors have informal and organizational ties that lead them to be consulted about matters of political policy.
   b. Many corporate officials have reward and coercive power over elected government officials.
   c. Many elected officials are also top corporate officials.
   d. Interlocking directorates exist.
   e. Conglomerates are present.

9. Which of the following is *not* one of the major labor-force trends discussed in the text?
   a. slowdown in labor force growth
   b. aging of the labor force
   c. increased employment of females
   d. increase in blue-collar jobs
   e. increase in nonwhite proportion of the labor force

Matching

10. Several characteristics of capitalist economic systems are listed below. Indicate in the space beside each characteristic whether it best describes individualistic capitalism (IC) or communitarian capitalism (CC).
    ____ a. government and business cooperation
    ____ b. high job mobility
    ____ c. profit maximization
    ____ d. more value placed on the interests of employees and shareholders
    ____ e. self-interest paramount
    ____ f. long-range economic planning

GRAPHIC REVIEW

Figure 13.11 displays current U.S. data on employee sense of company reciprocity. Demonstrate your understanding of this figure by answering these questions.

1. In the light of recent corporate downsizing and contingent employment, do you think that this is a significant shift in employee attitudes? Why or why not? Defend your answer with research findings presented in the text.

   _____

   _____

2. If you were a corporate executive who took the results in this figure seriously, which two of the five factors would you attempt to address first? Justify your choice in the context of research results contained in the text.

   _____

   _____

ANSWER KEY

Concept Review		Multiple Choice		Feedback Review	
a. 16	j. 1	1. c	10. d	1. T	8. c
b. 8	k. 13	2. b	11. a	2. F	9. d
c. 17	l. 10	3. e	12. e	3. false consciousness	10. a. CC
d. 12	m. 11	4. a	13. d	4. Authoritarianism	b. IC
e. 7	n. 9	5. e	14. c	5. multinational	c. IC
f. 3	o. 2	6. b		corporations	d. CC
g. 6	p. 4	7. d		6. core	e. IC
h. 15	q. 5	8. c		7. d	f. CC
i. 14	r. 18	9. a			

Chapter Fourteen

Religion

After careful study of this chapter, you will be able to:
- State the sociological meaning of religion.
- Demonstrate the different views of religion taken by functionalists and conflict theorists.
- Distinguish the basic types of religious organization.
- Discuss the meaning and nature of religiosity.
- Define secularization and describe its relationship to religiosity in the United States.
- Differentiate civil and invisible religion in America.
- Describe the current resurgence of religious fundamentalism in the United States.
- Identify a wide variety of new religious movements in America.
- Outline the relationship of baby boomers with religion.

SOCIOLOGICAL IMAGINATION

Is the process of secularization destroying the expression of religious interests in the United States? A widely held popular belief is that secularization—the process through which religion loses influence over society as a whole—is destroying religion. On the contrary, evidence reveals that America, compared with other industrialized nations, remains a fairly religious nation.

Religion in America, then, remains a viable topic to explore. Before doing so, we will place religion in its proper relationship to sociology as a field, define religion as an institution, examine religion from the functionalist and conflict perspectives, and explore both religious organization and the ways people express their religious beliefs.

RELIGION AND SOCIOLOGY

The study of religion attracted the attention of sociologists in the nineteenth century. In fact, all the early major sociologists were interested in the place religion holds in social life. Auguste Comte, the man who coined the term *sociology,* saw sociology as a new religion—a religion of science—and sociologists as high priests. Other pioneering sociologists, such as Emile Durkheim, Max Weber, and Karl Marx, viewed religion as a key to some of the mysteries of social life. Religion, they believed, could help explain the presence of social order

and the great cultural variation of societies. Despite this early scholarly interest in the place of religion in social life, the sociological study of religion disappeared during the first half of the twentieth century. It was not until after World War II that social scientists again became interested in the scientific study of religion.

Sociologists of religion neither explain nor endorse particular religions but rather are interested in all sociological aspects of religion. This interest must begin with a meaningful sociological definition of religion.

The Sociological Meaning of Religion

Religion is a unified system of beliefs and practices relative to sacred things. This definition of religion comes from Emile Durkheim, whose work was based on studies of the Australian aborigines in the late nineteenth century. According to Durkheim, every society distinguishes between the **profane**—nonsacred aspects of life—and the **sacred**—things that are set apart and given a special meaning that transcends immediate existence.

Sacred things take on a public or social character that makes them appear important in themselves; the profane do not. Bolivian tin miners attach sacred meaning to figures of the devil, or Tio, and to figures of a bull (Nash, 1972). Because Americans do not share these religious beliefs, these cultural items are part of their nonsacred, or profane, world. Moreover, some nonreligious aspects of culture may take on a sacred character. R. P. Cuzzort and E. W. King have illustrated the difference between the sacred and the profane in the broadest sense.

> When Babe Ruth was a living idol to baseball fans, the bat he used to slug his home runs was definitely a profane object. It was Ruth's personal instrument and had little social value in itself. Today, however, one of Ruth's bats is enshrined in the Baseball Hall of Fame. It is no longer used by anyone. It stands, rather, as an object which in itself represents the values, sentiments, power, and beliefs of all members of the baseball community. What was formerly a profane object is now in the process of gaining some of the qualities of a sacred object (Cuzzort and King, 1976:27).

Babe Ruth's bat illustrates two things about the sociological study of religion. First, a profane object can become sacred, and vice versa. Second, sociologists can deal with religion without becoming involved in theological issues. By focusing on the cultural and social aspects of religion, sociologists can successfully avoid questions about the ultimate validity or invalidity of any particular religion. This point needs to be elaborated.

The Sociological Study of Religion

Religion involves a *transcendent reality*—a set of meanings attached to a world beyond human observation. Because sociologists who study religion must deal with this dimension of unobservability, their task is particularly difficult. The following questions indicate the unique problems sociologists face when studying religion. Can an empirical science actually study the unobservable? Can scientific objectivity be maintained while investigating something as value laden as religion? Can sociologists retain any religious faith after scientifically probing religion? Each of these questions deserves a brief answer.

Obviously, sociologists cannot study the unobservable. As part of an empirical science, sociologists who investigate religion must formulate theories and hypotheses on aspects of religion that can be measured and observed. Consequently, sociologists avoid the strictly spiritual side of religion and focus on religion's socially relevant aspects.

Objectivity, as we saw in Chapter 2 ("Sociologists Doing Research"), is only a goal for which scientists strive. Scientists are not always successful, even in non-controversial areas of research. In an area of study with the emotional overtones of religion, vigilance against bias must be especially sharp. Thus, sociologists investigating religion do not attempt to determine the truth or falseness of any religion. Evaluative stances involving a definition of good or bad, true or false, are avoided; therefore, theological issues are not within the purview of the sociology of religion. Thus, sociologists can point out that conceptions of God, religious beliefs, and religious practices vary from culture to culture without attempting to make judgments of the validity of any of these issues. Because the validity of any religion is founded in faith, sociologists restrict themselves to the social manifestations of religion.

Sociologists, then, are not in the business of determining which religious faith people ought to follow. For this reason, it is perfectly possible for them to retain a religious faith of their own while investigating the social dimensions of religion. In fact, sociologists, like other people, vary in their religious orientations. They have varying degrees of religious commitment, exhibit a myriad of religious behaviors, and belong to quite different religious organizations. (See Figure 14.1.)

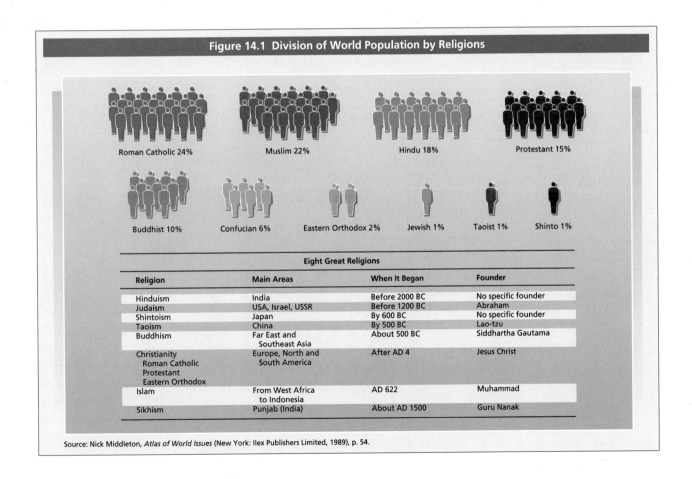

Figure 14.1 Division of World Population by Religions

Roman Catholic 24%   Muslim 22%   Hindu 18%   Protestant 15%

Buddhist 10%   Confucian 6%   Eastern Orthodox 2%   Jewish 1%   Taoist 1%   Shinto 1%

Eight Great Religions

Religion	Main Areas	When It Began	Founder
Hinduism	India	Before 2000 BC	No specific founder
Judaism	USA, Israel, USSR	Before 1200 BC	Abraham
Shintoism	Japan	By 600 BC	No specific founder
Taoism	China	By 500 BC	Lao-tzu
Buddhism	Far East and Southeast Asia	About 500 BC	Siddhartha Gautama

Christianity
  Roman Catholic
  Protestant
  Eastern Orthodox
	Europe, North and South America	After AD 4	Jesus Christ

Islam	From West Africa to Indonesia	AD 622	Muhammad
Sikhism	Punjab (India)	About AD 1500	Guru Nanak

Source: Nick Middleton, *Atlas of World Issues* (New York: Ilex Publishers Limited, 1989), p. 54.

1. A _____ is a unified system of beliefs and practices relative to sacred things.

2. _____ things are set apart and given special meaning that transcends immediate existence.

3. Sociologists who study religion are normally forced to abandon their religious beliefs and convictions. *T or F?*

Answers: 1. religion 2. sacred 3. F

THEORETICAL PERSPECTIVES

The functionalist and conflict perspectives have a quite different view of the sociological dimensions of religion. We begin with functionalism, which examines the role of religion in society in terms of the functions it serves.

Functionalism and Religion

We know from archaeological discoveries and anthropological observations that religion has existed in some form in nearly all societies (Nielsen, 1993; see Figure 14.2 for a global distribution of major religions). The earliest evidence of religion has been traced as far back

as 50,000 B.C., when humans had already begun to bury their dead, a practice that suggests belief in some existence after death. In early Rome there were specific gods for objects and events. There was a god of doors, a god of money, and a goddess of fever. The ancient Hebrews were told by their god—in both the Book of Genesis and the Book of Leviticus—that pigs were unclean animals whose pollution would spread to all who touched or tasted them. To the tribes of New Guinea and Pacific Melanesian Islands, on the other hand, pigs are holy creatures worthy of ancestral sacrifice (Harris, 1974).

Emile Durkheim, the first sociologist to examine religion from a scientific viewpoint, wondered why all societies have some form of religion. In one of his books, *The Elementary Forms of Religious Life* (1995, orig-

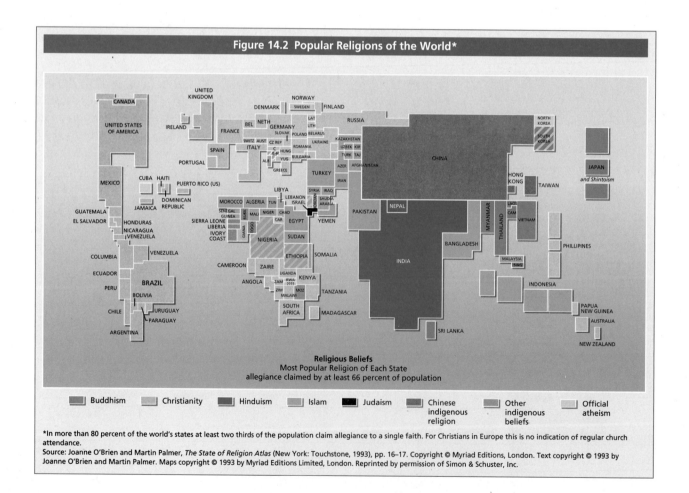

Figure 14.2 Popular Religions of the World*

Religious Beliefs
Most Popular Religion of Each State
allegiance claimed by at least 66 percent of population

Buddhism   Christianity   Hinduism   Islam   Judaism   Chinese indigenous religion   Other indigenous beliefs   Official atheism

*In more than 80 percent of the world's states at least two thirds of the population claim allegiance to a single faith. For Christians in Europe this is no indication of regular church attendance.

Source: Joanne O'Brien and Martin Palmer, *The State of Religion Atlas* (New York: Touchstone, 1993), pp. 16–17. Copyright © Myriad Editions, London. Text copyright © 1993 by Joanne O'Brien and Martin Palmer. Maps copyright © 1993 by Myriad Editions Limited, London. Reprinted by permission of Simon & Schuster, Inc.

inally published in 1915), Durkheim offered an explanation for the ubiquity of religion, an explanation that lies in the function religion performs for society. The essential function of religion, he believed, was to provide through sacred symbols a mirror for members of society to see their common unity. Through religious rituals, people were worshipping their society and thereby reminding themselves of their shared past and future existence. Sociologists have identified the following basic functions of religion: (1) to provide legitimation for social arrangements, (2) to promote a sense of unity, (3) to provide a sense of meaning, and (4) to promote a sense of belonging.

How does religion legitimate social arrangements?
Legitimation refers to explanations of why a society is—and generally should be—the way it is; legitimation helps both to justify and to explain the status quo. Legitimations tell us why some people have power and others do not, why some are rich and others poor, why some are slaves and others masters. Many such legitimations are based on religion. In fact, Emile Durkheim saw legitimation as the central function of religion. He said that the ideas of the sacred that develop within any society represent the society itself. When people think of their souls, Durkheim believed, they are really thinking of the social element within themselves—an element that is superior to the individual and will live on after the death of the individual.

If, as Durkheim believed, religion is a representation of society, its power to provide legitimation for society is considerable. Through its system of beliefs, a religion offers an explanation of the nature of social life, the existence of evil, and the imperfect nature of the world. Religion has been used to justify slavery in America (many ministers in the nineteenth-century South preached that blacks were descendants of Ham, a son of Noah whose ancestors were condemned to slavery) and male superiority (Eve is said to have been created from one of Adam's ribs, and it was her encouragement that led Adam to eat the forbidden apple). One of the most important and widely agreed-on social functions of religion, then, is the legitimation or justification of social arrangements.

How does religion promote social unity? Durkheim pointed out that religion acts as a kind of glue that holds society together. Without religion, society would be chaotic. As Cuzzort and King have stated, Durkheim "provided the greatest justification for religious doctrine ever granted by a social scientist when he claimed that all societies must have religious commitments. Without religious dedication there is no social order" (Cuzzort and King, 1976:43). Not all sociologists would agree that religion is absolutely necessary for every society's existence. Some cite cases—Northern Ireland

Why does religion exist in all societies? According to Emile Durkheim's functionalist explanation, religion is ubiquitous because sacred symbols (such as the cross on top of this church) serve as mirrors through which members of society are able to see their common unity.

being a recent example—in which religion has caused societies to fragment, even to engage in civil war. Thus, it is perhaps more accurate to say that religion is usually a source of social unity.

How does religion provide a sense of meaning?
Religion not only provides explanations for the nature of social life and promotes social unity, but also gives individuals a sense of worth and meaning far beyond day-to-day life. Thus, many people mark important events in life—birth, sexual maturity, marriage, death—with religious ceremonies and explain such events in religious terms. Religion gives believers cosmic significance and gives eternal significance to a short and uncertain earthly existence.

How does religion promote a sense of belonging?
Religious organizations provide opportunities for people to associate with others who share such similarities as ideas, ways of life, and ethnic or racial backgrounds. Religion thus provides a kind of group identity for those who seek it. Because religion is not controlled by government or business in religiously pluralistic societies, people may join religious organizations freely and feel that they have some degree of influence over the direction these organizations will take. For many people in modern society, membership in a religious organization provides a sense of community. This counteracts depersonalization, powerlessness, and rootlessness.

Of course, religious organizations are not unique in providing a sense of belonging. Fraternal groups, clubs, and other voluntary associations do much the same sort of thing. Because of religion's other functions, however, the sense of belonging provided by religious organizations can be particularly appealing.

Conflict Theory and Religion

As previously noted, Durkheim believed that religion supports existing societies by providing legitimation for social arrangements. Durkheim's ideas are still being debated, however, and other sociological theorists have identified quite different relationships between religion and society.

Some of the alternative views can be traced to Georg Friedrich Hegel, an eighteenth-century German philosopher who believed that ideas and beliefs, including religious ones, determine the nature of social life. Ideas and beliefs, Hegel taught, cause social life to be structured in certain ways. Hegel's position was very popular within the religious and intellectual establishment of his day, but it was not without critics. By the early nineteenth century, a group of critics known as the young Hegelians had begun to develop a radical reanalysis of Hegel's position. They claimed to set Hegel "on his head." Social structure determines ideas and beliefs, not the other way around. More specifically, the economic structure of a society determines the nature of the society's religion. Karl Marx decided the group had gone too far the other way and developed a view encompassing both ideas and structure.

What are Marx's major ideas regarding religion? Marx believed that humans create a unified system of beliefs and practices relative to sacred things and then act as if it were something beyond their control. People become "alienated" from the system they have set up. Although humans have the power to change (or, better yet, in Marx's mind, to abandon) the religion they have created, they don't, because they see it as a binding force to which they must conform.

Religion, Marx wrote, is managed by the ruling class to justify its economic, political, and social advantage over the oppressed. Poverty, degradation, and misery are said by those in power to be part of God's plan. To eliminate inequalities and injustices is to tamper with God's will. Change is sacrilegious to the elites. Many illustrations of close ties between elites and religion exist. In fact, a close relationship between the dominant classes and the church has been the rule in human history. The Catholicism of medieval Europe, for example, was so closely identified with the upper classes that the lower and lower-middle classes welcomed the Protestant Reformation. In his classic study *Millhands*

and Preachers, Liston Pope (1942) demonstrated the control of churches in a small North Carolina community (Gastonia) by the managers of the mill.

> *The churches were inextricably bound to mill management by their finances if not by their ideology. Organized religion was a key instrument in creating a manipulable labor force and in resisting the union's effort to break with the exploitative status quo. Ministers not only justified management practices; they were mute in reaction to the violent recriminations exerted by the community against the union organizers (Pope, 1942:xx).*

In a restudy of Gastonia, churches were still providing latent support for long-standing economic relations in the community. Marx contended that religion is used by the oppressed as an explanation and justification for their deplorable existence and to provide hope for a better life after death. Consequently, they can find self-fulfillment in their oppression. Marx thought of religion as a narcotic for the oppressed.

> *Religious suffering is at the same time an expression of real suffering and a protest against real suffering. Religion is the sigh of the oppressed creature, the sentiment of a heartless world, and the soul of soulless condition. It is the opium of the people (quoted in Bottomore, 1963:43–44, originally published in 1844).*

Marx thought that if people abandoned the religion of their oppressors, they would turn from the comforts offered by an afterlife to concentrate on the present. This would lead them to reject their oppressed economic condition and decide to do something about it. Marx was also aware of the difficulty of people gaining a conscious awareness of the power of the religion of the capitalist oppressor. A contemporary scholar of ideology makes the latter point in this way:

> *The most efficient oppressor is the one who persuades his underlings to love, desire and identify with his power, and any practice of political emancipation thus involves that most difficult of all forms of liberation, freeing ourselves from ourselves (Eagleton, 1994:xiii–xiv).*

How did Weber view the place of religion in social life? Whereas Marx contended that religion tends to retard social change because the elites wish to maintain the status quo, Max Weber (1958, originally published in 1904) believed that religion sometimes encourages social change. He looked for a case that would demonstrate his point, and he found it in the relationship between the ideas of early Protestantism and the rise of capitalism.

The Sacred Ganges

This CNN report explores the place of the Ganges River in the religious life of India. Hindus have been taught that the Ganges is sacred. According to the Hindu faith, the Ganges is so spiritually pure that it cannot possibly be polluted. Consequently, despite its polluted state, over a million Hindus enter its waters every day. To die and be cremated on the banks of the Ganges is believed to insure salvation among devout Hindus.

1. Use India's religious view of the Ganges to distinguish between the sacred and the profane.

   _____

   _____

2. State the religious role of the Ganges from the functionalist perspective.

   _____

   _____

For most Americans, there is nothing sacred about river water. Hindus in India, on the other hand, attribute to the Ganges River a special meaning that rises above immediate existence.

Weber began with the fact that capitalism had become widespread only in northwestern Europe and America. Why had this happened? Why had capitalism not emerged as the basic economic system on a large scale in other parts of the world? Through an extensive study of various religions of the world, Weber (1951, 1952, 1958, 1964a) found a possible answer in a basic compatibility between what he termed the spirit of capitalism and the Protestant ethic.

What are the spirit of capitalism and the Protestant ethic?   Weber noted that capitalism involved a radical redefinition of work. With capitalism, work changed from something that had to be done to something that people should do; work became a moral obligation rather than a mere necessity. Capitalism also required the reinvestment of profits if businesses were to grow. Because capitalism involved the reinvestment of capital to increase capital, reinvestment for the future was more important than immediate consumption. All of this Weber called the *spirit of capitalism.*

Most major religions did not define hard work as an obligation or demand the reinvestment of capital for further profits rather than for immediate enjoyment, but particular Protestant sects did. Weber called this cluster of values, norms, beliefs, and attitudes embodied in Protestantism that he thought favored the emergence of modern capitalism the **Protestant ethic.**

What is the nature of the Protestant ethic?   Weber found the roots of the Protestant ethic in the seventeenth-century theology of Calvinism. Of particu-

lar importance was the Calvinist belief that every person's eternal fate was unalterably predetermined by God at birth—predestination. Society, the Calvinists believed, was divided into the few who were elected to salvation and the many who were predestined to spend eternity in hell. Individuals could do nothing to alter their fate.

The many condemned to hell might well be expected to conclude that having a good time was the only way to live. Some other tenets of Calvinism, however, discouraged idleness, pleasure seeking, and living for the present. For one thing, Calvinism taught its followers to look for signs of God's blessing on them. Individuals could do nothing to change their standing with God, but Calvin contended that God would identify his select by rewarding them in this world. This was translated into the belief that the more successful people were in this life, the more sure they were of being a member of God's select few. Second, consumption beyond necessity was considered sinful; those who engaged in self-pleasure were agents of the devil. Finally, Calvinists believed that the chief purpose of human beings was to glorify God on earth through one's occupational calling. Because everyone's material rewards were really God's and the purpose of life was to glorify God, profits should be multiplied (through reinvestment) rather than used in the pursuit of personal pleasures.

Other Protestant religious beliefs may have contributed to the development of capitalism. For example, Protestants endorsed the idea that salvation came from unselfish good works here on earth. Protestants were encouraged to do more and more work while receiving less and less of the material rewards for themselves.

Does the Protestant ethic still exist? The debate over Weber's Protestant ethic continues (Lenski, 1963, 1971; Schuman, 1971; Nelsen, 1973; Hammond and Williams, 1976; Shepard, et al., 1997). However, any link between Protestantism and economic behavior that may have existed two hundred years ago appears to be rather weak today. Surveys show that Protestants do not value success and hard work any more than Catholics do. Moreover, Jews have higher incomes and are more successful occupationally than Protestants. If the Protestant ethic still exists, it does not appear to be the exclusive property of Protestants. As a matter of fact, it may not be the property of any religious group.

Did Protestantism cause capitalism to develop? Weber did not contend that capitalism developed *because* of Protestantism. (Capitalism had actually flourished among Catholics in some parts of the world before the Protestant Reformation.) He did, however, suggest that early Protestantism seems to have provided a social environment *conducive* to the emergence of this type of economic system. He recognized that capitalism could exist apart from Protestant dogma. Weber himself pointed out that capitalism helped destroy many of the foundations of Protestantism. He was well aware that from the seventeenth century on, religion had gradually lost its hold over capitalist society.

An Evaluation

Interpretations of the relationship between religion and social change, as the preceding paragraphs show, have been quite varied. Marx, writing from the conflict perspective, believed that religion retards social change in part because of its ties to an economic elite. Unlike Marx, Weber argued that religion could encourage social change by supporting new social arrangements.

Although the two views appear contradictory, each may be correct under some conditions. Meredith McGuire (1997) has suggested that the relationship between religion and social change depends on the quality of religious beliefs and practices existing in a society, dominant ways of thinking in a culture, the social location of religion within a larger society, and the internal structure of religious organizations and movements. For example, McGuire has argued that religion is likely to promote change if it provides a critical standard against which the established social system can be measured. When religion provides such a standard, it allows the social critic to say, "This is what we say is the right way to act, but look how far our group's actions are from these standards!" Thus, the American civil rights movement produced results partly because of the support of religious leaders who pointed to the incompatibility of racial discrimination with American values favoring equality and opportunity.

The ties that exist between religious organizations and other portions of society also affect religion's role in social change. In many Latin American countries, for example, the Roman Catholic Church plays an important role in social change and protest because it is the sole dissenting grassroots organization with an established network and sympathetic leanings. When other modes of action, such as interest groups exist, religion may play a very minor role in social change.

The relationship of religion to social change, then, is quite variable. Although the interpretations developed by Marx and Weber call attention to specific aspects of

---

FEEDBACK

1. According to Karl Marx, religion is a force working for
   a. the good of all people.
   b. the good of the proletariat.
   c. maintenance of the status quo.
   d. the creation of conflict between the classes.
2. According to Marx, religion is created by the _____ to justify its economic, political, and social advantages over the oppressed.
3. The _____ is the belief that work is an obligation and that capital should be reinvested for further profits rather than for immediate enjoyment.
4. According to Max Weber,
   a. early Protestantism provided a social environment conducive to capitalism.
   b. capitalism developed because of Protestantism.
   c. capitalism strengthened Protestantism.
   d. religion will always be linked with economy.
5. Religion
   a. has been found to promote social change under almost all conditions.
   b. is unrelated to social change.
   c. promotes social change only under certain conditions.
   d. always supports the interests of underprivileged groups in society.

Answers: 1. c 2. elite 3. Protestant ethic 4. a 5. c

that relationship, no single interpretation appears to be applicable to all situations.

RELIGIOUS ORGANIZATION AND RELIGIOSITY

Religious Organization

In Western societies, most people tend to relate to religion through some organizational structure. For this reason, the nature of religious organization is an important component of the sociological study of religion. Early scholars developed four basic types of religious organization: church, denomination, sect, and cult (Troeltsch, 1931; Niebuhr, 1968, originally published in 1929).

How do sociologists distinguish between the basic types of religious organization? A **church** is a life-encompassing religious organization to which all members of a society belong. This type of religious organization exists when religion and the state are closely intertwined. King Henry VIII's divorce troubles stemmed from the Roman Catholic Church's hold on sixteenth-century England. In Elizabethan England, Anglican Archbishop Richard Hooker once wrote that "there is not any man of the Church of England but the same man is also a member of the commonwealth; nor any man a member of the commonwealth which is not also of the Church of England."

A **denomination** is one of several religious organizations that most members of a society accept as legitimate. Because denominations are not tied to the state, membership in them is voluntary and competition among them for members is socially acceptable. Being one religious organization among many, a denomination accepts the values and norms of the secular society and the state, although it may at times be in opposition to them. Most American "churches"—Methodist, Episcopalian, Presbyterian, Baptist, Roman Catholic, and Reformed Jew—are denominations.

A **sect** is an exclusive and cohesive religious organization based on the desire to reform the beliefs and practices of another religious organization. Sects are formed by breaking away from either an established denomination or another sect that is felt to have lost its religious purity. Sect formation usually involves a rejection of the surrounding society that has become too secular. Although withdrawal is usually only psychological, some sects actually form communal groups apart from the larger society. An example of extreme social withdrawal is the Amish (Kraybill and Olshan, 1994). Less extreme sects in existence today include the

Seventh-day Adventists, the Quakers, and the Assemblies of God.

Other types of religious organizations exist that are not included in the types of religious organization just presented. The most important of these is the *cult*. Sects have prior links with another religious organization. A sect, in fact, is formed as a result of a deliberate break with a "parent" religious organization. The rupture occurs, however, not to establish a new religious faith but to redeem traditions that have been lost by the parent organization. Luther, for example, claimed to be rescuing the true church, not creating a new one. A **cult** is a religious organization whose characteristics are not drawn from existing religious traditions within a society. Whether imported from outside the society or created within the society, cults bring something new to the larger religious environment. The world was shocked in 1997 when reports came of the ritualistic suicides of thirty-nine members of the Heaven's Gate cult in California (Thomas, 1997). Cults do not usually appear in such an extreme and bizarre form. More conventional examples of cults are the Unification Church, the Divine Light Mission, and the Church of Scientology (Clark, 1993).

Although the term *cult* covers a wide range of groups and organizations, the religious cults existing in the United States today share several characteristics. At their center are an authoritarian structure, rejection of the secular world's laws and ways, strict discipline of

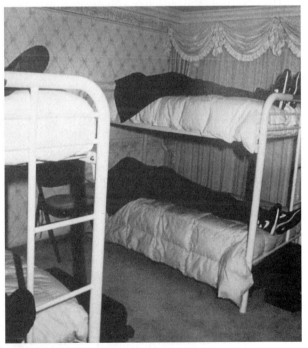

The Heaven's Gate cult in California constituted an extreme religious group. This is a photograph of a few of the thirty-nine members who participated in a mass suicide in 1997.

adherents, rigidity in thinking, conviction of sole possession of truth and wisdom, belief in the group's moral superiority, and discouragement of individualism (Appel, 1983; Collins, 1991). More is said about cults at the end of the section on religion in the United States.

Religiosity

Charles Glock and Rodney Stark (1965, 1968) have spent many years exploring individual religious experiences in terms of **religiosity**—the ways in which people express their religious interests and convictions. Their work focuses on the types of religious attitudes and behavior people display in their everyday lives.

How do people display religiosity?   Glock and Stark have distinguished five dimensions of religiosity. The first dimension is *belief,* or some idea about what is considered to be true. People may, for example, believe that Christ died for us all on the cross, that everyone is created equal, that there is no god but Allah, or that all people of a certain race should be exterminated. Beliefs may relate to some supernatural world above or beyond us or to the ways we perceive the world around us.

Ritual—religious practices expected of members of a religion—is the second dimension of religiosity. Ritual may be private (personal prayer) or public (attending church).

The *intellectual* dimension of religiosity, which refers to the expectation that religious people will be informed about their faith, may involve knowledge of the books of the Bible or an interest in such religious aspects of human existence as evil, suffering, and death.

Experience—certain feelings attached to religious expression—is the fourth dimension of religiosity. This dimension is the hardest to measure, but James Davidson (1975) has contended that it can be probed by asking people whether or not they think specific religious experiences are desirable and whether or not they have had specific religious experiences.

The final dimension of religiosity—*consequences*—refers to things people claim they do or do not do as a result of religious beliefs, rituals, knowledge, or experiences. Consequences may be social (opposing or supporting capital punishment or abortion) or personal (restricting one's sex life to marriage, telling the truth regardless of the cost).

Why use a dimensional approach to religiosity?   The dimensional approach demonstrates that being religious can mean several different things. The dimensions of religiosity are related to one another. Some research has found, in fact, that the relationships are strong enough to cast doubt on the appropriateness of separating religiosity into different dimensions (Clayton and Gladden, 1974). The dimensions may, however, operate independently at times. People who attend church (ritual dimension), for example, may know very little of the church's doctrines (intellectual dimension). People who attend church regularly may very well believe that people who do not go to church will spend eternity in hell (belief dimension).

Exhibiting religious practices (rituals) is an important dimension of religiosity. These Baptists are engaging in the traditional practice of total immersion in baptism.

1. Match the following descriptions and types of religious organizations:

   ____ a. church
   ____ b. sect
   ____ c. denomination
   ____ d. cult

   (1) a religious organization whose characteristics are not drawn from existing religious traditions within a society
   (2) an exclusive and cohesive religious organization based on the desire to reform the beliefs and practices of another religious organization
   (3) a life-encompassing religious organization to which all members of a society belong
   (4) one of several religious organizations within a society that are considered to be legitimate

2. _____ refers to the ways people express their religious interests and convictions.

3. Match the dimensions of religiosity with the examples beside them.

   ____ a. belief
   ____ b. ritual
   ____ c. intellectual
   ____ d. experience
   ____ e. consequences

   (1) prayer in church
   (2) knowing the content of the Koran
   (3) having seen an angel
   (4) conviction that the Bible is divinely inspired
   (5) marching in support of prayer in schools

Answers: 1. a. (3) b. (2) c.(4) d. (1) 2. Religiosity 3. a. (4) b. (1) c. (2) d. (3) e. (5)

RELIGION IN THE UNITED STATES

The nature rites of South Sea Islanders and the well-ordered Church of the Middle Ages were relatively simple and stable forms of religion. By comparison, advanced industrial society presents a special challenge to the sociological study of religion.

The Development of Religion in America

The search for religious freedom was only one among many reasons that colonists came to America, but there was a genuine religious element in both colonization and the American Revolution (Brauer, 1976; Marty, 1985). Robert Bellah has described the American religious connection this way:

In the beginning, and to some extent ever since, Americans have interpreted their history as having religious meaning. They saw themselves as being a "people" in the classical and biblical sense of the word. They hoped they were a people of God (Bellah et al. 1991:2).

Religion in the colonies was highly varied, but beginning in the second half of the eighteenth century, the approach to religion was increasingly influenced by the thinkers of the Enlightenment—Bentham, Voltaire, Diderot, Hume, and many others—in Europe and the British Isles. The same ideas that inspired the questioning of arbitrary colonial government and the writing of the Declaration of Independence also led to a critical examination of religion and its relationship to the state. Although the leaders of the American Revolution seldom raised arguments against religious faith, they were sharply critical of any entanglement of religion with the power of the state. Despite this tradition, the United States has experienced many incidents of religious persecution, including many directed at immigrant groups. Nevertheless, the ideas of separation of church and state and freedom of religious expression have become cornerstones of American life.

Secularization: Real or Apparent?

Religion has been an important influence during most of human history, including the history of the United States. The industrialization (and accompanying secularization) of many nations, however, has had the effect of diminishing the influence of religion. An examination of the role of religion in the United States today must therefore begin with consideration of the process of secularization.

Secularization is the process through which religion loses influence over society as a whole. During this process, other social institutions are emptied of religious content and freed from religious control (Wilson, 1982). Religion itself becomes a specialized, isolated institution. In preindustrial society, religion and social life were inseparable. Hester Prynne, the central character of Nathaniel Hawthorne's novel *The Scarlet Letter*, was forced by both religious leaders and peers in her Puritan community to live forever with shame and ostracism for her act of adultery. Religion and social life were so intertwined in seventeenth-century America that the discovery of her violation of the religious norm prohibiting adultery led to Hester's loss of any right to a normal social existence in her community. As industrialization proceeds, conventional religion loses some of its influence.

Secularization unquestionably dilutes the influence of religion on society because it leads to a separation between the religious institution and other institutions such as the polity, economy, family, and education. Still, the extent of the decline of the influence of religion created by secularization varies from society to society and from time to time (Beckford and Luckmann, 1989).

There are some indications of a diminution of the importance of religion in the United States. The percentage of Americans claiming that religion is very important in their lives declined from 75 percent in 1952 to 58 percent in 1995. (See Figure 14.3.) The Princeton Religion Index, comprising eight leading indicators, has also declined since the 1940s. (See Figure 14.4.) Finally, only 14 percent of the public in 1957 indicated that religion was losing influence on American life; in 1995, 57 percent of the public saw a lessening of religion's influence on American life (Gallup, 1996).

Is secularization destroying religiosity in the United States? In a restudy of Middletown, Theodore Caplow, Howard Bahr, and Bruce Chadwick (1983) conclude that secularization is not destroying religion in the United States. They found the current residents of Middletown (which they take to be representative of the United States) to be as committed to religion as were preceding generations. Whether measured by the number of churches per capita, the proportion of regular churchgoers, or financial support of the churches, Caplow and his colleagues state that the trend is toward greater involvement in religious affairs.

In fact, as stated in the *Sociological Imagination* opening this chapter, compared with other industrialized countries, America still appears to be a fairly religious nation. (See Table 14.1.) Only 8 percent of the American population is without a religious preference. About 88 percent identify themselves as Protestants, Catholics, Jews, or Mormons (Gallup, 1996). There are now over three hundred recognized denominations and sects and thousands of independent congregations in the United States (Jacquet, 1990). About seven in ten Americans belong to some church, and over half of these claim to be active in their congregations. Four Americans in ten state that they have attended a church or synagogue in a typical week. (In England, for example, the average weekly church attendance is 14 percent.) Furthermore, although the proportion of Americans belonging to a church or synagogue has declined slightly from a high of 76 percent in 1947 to 69 percent in 1995, church attendance has changed very little over the years. Since 1939, weekly church or synagogue attendance in the United States has remained relatively stable—from 41 percent in 1939 to 43 percent in 1995. Americans also tend to support traditional religious beliefs. Ninety-six percent of the American population believe in God or a universal spirit, 65 percent believe in life after death, 90 percent believe in heaven, and 73 percent believe in hell. Seventy-two percent believe in the existence of angels (Gallup, 1996).

Why the continuing interest in religion? Caplow and his colleagues suggest that religion is sought as a buffer against the insatiable demands of the state. According

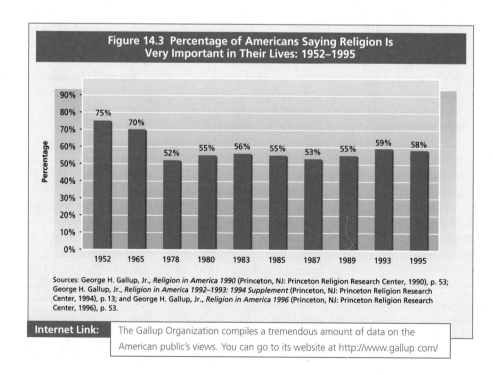

Figure 14.3 Percentage of Americans Saying Religion Is Very Important in Their Lives: 1952–1995

Sources: George H. Gallup, Jr., *Religion in America 1990* (Princeton, NJ: Princeton Religion Research Center, 1990), p. 53; George H. Gallup, Jr., *Religion in America 1992–1993: 1994 Supplement* (Princeton, NJ: Princeton Religion Research Center, 1994), p. 13; and George H. Gallup, Jr., *Religion in America 1996* (Princeton, NJ: Princeton Religion Research Center, 1996), p. 53.

Internet Link: The Gallup Organization compiles a tremendous amount of data on the American public's views. You can go to its website at http://www.gallup.com/

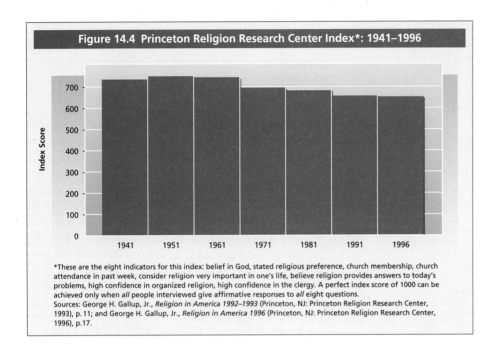

Figure 14.4 Princeton Religion Research Center Index*: 1941–1996

*These are the eight indicators for this index: belief in God, stated religious preference, church membership, church attendance in past week, consider religion very important in one's life, believe religion provides answers to today's problems, high confidence in organized religion, high confidence in the clergy. A perfect index score of 1000 can be achieved only when *all* people interviewed give affirmative responses to *all* eight questions.
Sources: George H. Gallup, Jr., *Religion in America 1992–1993* (Princeton, NJ: Princeton Religion Research Center, 1993), p.11; and George H. Gallup, Jr., *Religion in America 1996* (Princeton, NJ: Princeton Religion Research Center, 1996), p.17.

to Gallup, Americans are searching for spiritual moorings for three basic reasons: the threat of nuclear war, loneliness, and disenchantment with what is perceived as a society without rules and with an anything-goes philosophy. Several sociologists have suggested that religion remains important in helping many Americans deal with important problems such as grief, fear of the unknown, misery, hopelessness, hunger, death, and meaning beyond life in this world (Stark and Bainbridge, 1981, 1985; Hammond, 1985; Wind and Lewis, 1994; Greeley, 1996).

Does this mean that secularization is not an important force in American society today? No, it does not. As Robert Wuthnow (1985) points out, when Weber, Durkheim, and Marx wrote about secularization, they were not thinking of short-term, individual-level religious attitudes and behavior such as belief in God or church attendance. They were observing major historical trends in European history since the Middle Ages, such as the separation of church and state. From this long-term, large-scale perspective, argues Wuthnow, secularization has accompanied modernization throughout the industrial world, including the United States.

In addition to reducing the social influence of religion, secularization promotes the mixing of the sacred and the profane (Fenn, 1978). American churches are becoming more secular, and the American clergy are increasingly involved in community affairs. The churches and church leaders were heavily involved in the civil rights movement and the Vietnam War protests of the 1960s (Hadden and Rymph, 1973), and are still concerned with such political issues as world peace and poverty. Although the clergy's participation

in public affairs has receded in favor of parishioners' inner, spiritual needs, the clergy remain involved in solving such social problems as drug abuse, juvenile delinquency, hunger, malnutrition, homelessness, and school violence. Secular society is entering religion in other ways as well. Churches are now, for example, permitting the use of "worldly" musical instruments such as saxophones and electric guitars in worship services. Jazz and rock liturgies are being performed in church. The televangelists of the "electronic church" use many marketing and show-business techniques to attract viewers and bring in vast amounts of money (Hadden and Swann, 1981; Hoover, 1988).

Robert Wuthnow (1976, 1978, 1990) argues that secularization, like increased religious fervor, should be seen as a process that fluctuates in strength. According to him, the religious revival of the 1950s waned because of opposition to it within the countercultural generation of the 1960s. Wuthnow thinks that a return to religious commitment may occur as successive age groups, unaffected by the counterculture, mature.

According to one group of researchers, secularization may well be lowering commitment to religious institutions in the United States. According to C. Kirk Hadaway, Penny Long Marler, and Mark Chaves (1993), characterizations of religious commitment in America typically rely on poll data. Self-reports of church attendance, they contend, inflate estimates of church attendance. Comparing church attendance rates based on actual counts of those who attend church with self-reported rates of church attendance, these researchers conclude that church attendance rates for Protestants and Catholics in the United States are actually about one-half the generally cited rates.

TABLE 14.1	Global Comparisons in Religiosity

Consider Selves Religious Persons

Italy	83%
United States	81
Ireland	64
Spain	63
Great Britain	58
West Germany	58
Hungary	56
France	51
Non-ethnic Lithuanians	50
Ethnic Lithuanians	45
Czechoslovaks	49
Slovaks only	69
Czechs only	38
Scandinavia	46

Attend Church at Least Weekly

Ireland	82%
United States	43
Spain	41
Italy	36
West Germany	21
Czechoslovakia	17
Ethnic Lithuanians	15
Non-ethnic Lithuanians	12
Great Britain	14
Hungary	13
France	12
Scandinavia	5

Average Ratings of Importance of God in One's Life ("10" is of highest importance)

United States	8.2
Ireland	8.0
Northern Ireland	7.5
Italy	6.9
Spain	6.4
Finland	6.2
Belgium	5.9
Great Britain	5.7
West Germany	5.7
Norway	5.4
Netherlands	5.3
Hungary	4.8
France	4.7
Denmark	4.4

Source: George H. Gallup, Jr., *Religion in America*, 1992–93 (Princeton, NJ: Princeton Religion Research Center, 1993), p. 70.

Civil and Invisible Religion

There is still other evidence of the enduring influence of religion on American society. This can be seen in the existence of *civil religion* and *invisible religion*.

What is civil religion? Every American president's inaugural address (except Washington's second one) has made reference to God. Similar religious references exist throughout American life: Thanksgiving Day is a day of public thanksgiving and prayer; our currency proclaims "In God We Trust"; the pledge of allegiance to our flag contains the phrase "under God"; most formal public occasions open with prayer. Even voluntary organizations such as the Rotary Club and the Boy Scouts carry out this religious theme.

None of these public religious allusions, however, are supposed to involve any specific religion. References are not made to Jesus Christ, Moses, or the Virgin Mary but are made to the general concept of God, a concept compatible with nearly all religions. **Civil religion,** then, is a public religion that expresses a strong tie between a deity and a society's way of life and is broad enough to encompass almost the entire nation. Robert Bellah (1967, 1975; Bellah and Hammond, 1980), the sociologist most closely identified with the concept of civil religion, has summarized it in this way:

> The separation of church and state [in America] has not denied the political realm a religious dimension. Although matters of personal religious beliefs, worship, and association are considered to be strictly private affairs, there are, at the same time, certain common elements of religious orientation that the great majority of Americans share. These have played a crucial role in the development of American institutions and still provide a religious dimension for the whole fabric of American life, including the political sphere. This public religious dimension is expressed in a set of beliefs, symbols, and rituals that I am calling the American civil religion (Bellah and Hammond, 1980:171).

The concept of civil religion continues to generate debate (Gehrig, 1981; Demerath and Williams, 1985; Audi and Wolterstorff, 1996), but many studies support Bellah's claim for the existence of a civil religion in the United States (Wimberley et al., 1976; Christenson and Wimberley, 1978; Wimberley and Christenson, 1980; Kearl and Rinaldi, 1983; Warner, 1994). Martin Marty (1985) points out that civil religion is alive and well today, thanks in part to its adoption by the conservative evangelical churches which in the 1980s linked God and nationalism.

What is invisible religion? According to some social scientists, focusing only on traditional religious beliefs and practices causes us to underestimate the religiosity of Americans (Luckmann, 1967; Yinger, 1969, 1970). These social scientists believe that many Americans who do not belong to churches or sects still practice a type of religion. These individuals, according to Luckmann, practice an **invisible religion**—a private religion that is substituted for formal religious organizations, practices, and beliefs in a secular society.

In a study of 208 households in Baton Rouge, Louisiana, Richard Machalek and Michael Martin (1976) found support for the existence of an invisible religion, even in a region of the United States that is thought to be deeply religious in the conventional sense. Their approach was to ask participants in their survey to list the most important concerns of life beyond immediate daily problems ("ultimate concerns") and to indicate the means they used to cope with them. They found that 18 percent of the respondents cited such traditional religious concerns as life after death, salvation, and the existence of heaven. More significantly, 8 percent listed a variety of less religiously conventional concerns, such as the need for consideration and love for others, the need for less selfishness in the world, national corruption, lack of political leadership, poverty, the need for social equality, and the meaning of life. One-third of the sample relied on conventional religious practices—praying, studying the Bible, formalized church activities, and informal religious study groups—to help them handle their ultimate concerns. For two-thirds of the sample, coping strategies were not related to institutionalized religion. They relied on such humanistic strategies as reflection and meditation, participation in civic organizations, and informal discussion groups composed of family and friends.

Even people who ordinarily remain outside formal religion tend to observe religious ceremony during such rites of passage as birth, marriage, and death. For example, Gorer (1965) found that although 25 percent of his sample did not believe in life after death, 98 percent arranged religious rites when they buried their relatives.

Whether or not the United States is as religious as it has been in the past, it is clear that religion still affects the American way of life. Americans join churches and synagogues and attend services in comparatively large numbers, and civil religion and private religion influence their daily lives. Moreover, there has recently been a revival of religious fundamentalism (Riesebrodt, 1993).

Resurgence of Religious Fundamentalism

Since the late 1960s, most mainline American Protestant denominations—Methodists, Lutherans, Presbyterians, Episcopalians—have either been declining in membership or fighting to hold their own. In contrast, contemporary fundamentalists, the conservatives of Protestantism, have been growing (Bedell,

Despite the declining influence of religion in American life, the United States is relatively religious compared with other industrialized nations. Whereas 40 percent of Americans indicate they normally attend a church or synagogue weekly, only 10 percent of those in England report doing so.

1993, 1997; see Figure 14.5). Fundamentalists exist in all Protestant organizations, but they are predominantly found in such religious bodies as the Mormons, the Assemblies of God, the Seventh-day Adventists, the Southern Baptists, and the Jehovah's Witnesses.

What is the nature of fundamentalism today? **Fundamentalism** is based on the rejection of the effects of modern society on religion and the corresponding desire to adhere closely to traditional beliefs, rituals, and doctrines. The roots of contemporary religious fundamentalism can be found in the latter part of the nineteenth century. Two issues disturbed the early fundamentalists. First, fundamentalists were concerned about the spread of secularism. Science was challenging the Bible as a source of truth, Marxism was portraying religion as an opiate for the masses, Darwinism was challenging the biblical interpretation of creation, and religion was generally losing its traditional influence on all social institutions. Second, fundamentalists rejected the movement away from emphasis on the traditional message of Christianity toward an accent on social service (Johnstone, 1996).

The theological agenda of today's fundamentalists is very close to that of their nineteenth-century forebears. Today's fundamentalists believe in the literal truth of the Scriptures, being "born again" through acceptance of Jesus Christ as the Son of God, the responsibility of all believers to give witness for God, the presence of Satan as an active force for evil, and the destruction of the world prior to the Messiah's second coming (Barnhart, 1987).

Are all fundamentalists alike? Variations exist, of course, among fundamentalists. The term *evangelical,*

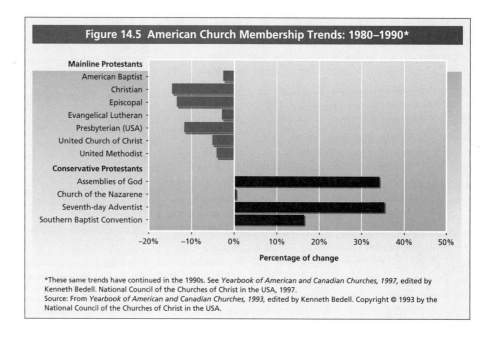

Figure 14.5 American Church Membership Trends: 1980–1990*

*These same trends have continued in the 1990s. See *Yearbook of American and Canadian Churches, 1997,* edited by Kenneth Bedell. National Council of the Churches of Christ in the USA, 1997.
Source: From *Yearbook of American and Canadian Churches, 1993,* edited by Kenneth Bedell. Copyright © 1993 by the National Council of the Churches of Christ in the USA.

for example, is often used as a synonym for fundamentalism. Most fundamentalists are less accommodating in their opposition to the rest of the world and more theologically rigid than evangelicals. Most evangelicals are on the political right or in the center, but some are on the ideological left. Evangelicals share much of the traditional theology of fundamentalism, but often reject its political and social conservatism (Quebedeaux, 1978; Hunter, 1987; Roof, 1994). Other religious organizations that share in much of the fundamentalist theology also have some unique beliefs and practices of their own. Neo-Pentacostalism—or the *charismatic movement,* as it is sometimes called—has occurred for the most part within traditional religious organizations, particularly the Roman Catholic and Episcopal churches. Those involved in this movement often speak of being "born again" or of receiving "the baptism of the Holy Spirit." But central to most neo-Pentecostal groups is the experience of "speaking in tongues" (glossolalia), which believers claim is a direct gift of the Holy Spirit (Cox, 1992).

Why has fundamentalism reappeared? There are several reasons for the revival of fundamentalism. First, many Americans feel that their world is out of control. The social order of the 1950s has been shattered by a string of traumatic events beginning with the civil rights movement and progressing through campus violence, political assassinations, the Vietnam War, and Watergate. Fundamental religion, with its absolute answers and promise of eternal life, provides a strong anchor in a confusing, bewildering world. Second, fundamentalist churches, by placing great emphasis on warmth, love, and caring, provide solace to people who

are witnessing and experiencing the weakening of family and community ties. Mainline churches are more formal and impersonal. Third, fundamentalist churches offer a more purely sacred environment, whereas the religious message of mainline churches has been increasingly diluted by secularization.

If the preceding are some of the reasons that many Americans have turned to fundamentalist churches, it is the electronic church that has articulated the appropriate messages to a receptive nationwide audience. In its role as part of the mass media, the electronic church has been a very important contributing factor in the revival of religious fundamentalism (Hadden and Swann, 1981). The place of the electronic church deserves some elaboration. (See *Doing Research.*)

What has been the place of the electronic church in the resurgence of fundamentalism? In the past, religious evangelists typically traveled from community to community and preached to the faithful in tents, open fields, or rented meeting halls. Modern communications technology has changed all that. Television, radio, and direct mail campaigns via computers have made it possible for religious broadcasters to reach millions of people easily, rapidly, and repetitively. Ministers, particularly from the fundamentalist side of Protestantism, have created vast and profitable television ministries. Prime-time preachers such as Jerry Falwell, Jimmy Swaggart, Oral Roberts, Pat Robertson, and Jim Bakker became among the best-known people in the United States. It is estimated that there are over 1,400 radio stations and 60 television stations totally owned by religious organizations. According to the National Religious Broadcasters Association, there are about 350 television ministries. From the thousands of weekly hours of religious programming come well over 500 million tax-free dollars each year (Johnstone, 1996).

Sexual and financial scandals surrounding Jim Bakker and Jimmy Swaggart have damaged the electronic church. The proportion of Americans indicating they have "a great deal" or "quite a lot" of confidence in religion as an institution decreased from 66 percent in 1985 to 57 percent in 1995, a drop of 9 percentage points (Gallup, 1996).

It remains to be seen whether this drop in public confidence permanently weakens the electronic church and the fundamentalist religions themselves. The outcome may depend partly on the results of a sweeping Internal Revenue Service investigation of nearly every major television ministry in the United States begun in 1987 in the wake of reported widespread financial abuses by former PTL Club hosts Jim and Tammy Bakker. Among others, the IRS examined Jerry Falwell's Liberty University and Pat Robertson's alleged illegal use of millions of dollars from his Christian Broadcasting Network to help underwrite his presidential campaign. Further discouragement came with Jim Bakker's conviction and imprisonment.

There is an important connection in the United States today between religious fundamentalism and politics. We will explore this connection later in this chapter.

New Religious Movements

Many new religious movements have emerged alongside the resurgence of fundamentalism (Robbins, 1988; Beckford, 1991; Heelas, 1996). Although some of these are offshoots of fundamentalism, most are not related. In fact, many of them are cults related to Eastern religions. Speaking broadly, these new religious movements embrace such varied organizations as the neo-Pentecostalists in established churches, the Jesus People on the fringe of the Christian establishment, the Reverend Moon's Unification Church, politically oriented movements such as the Moral Majority, Scientology, the Guru Maharaj Ji, the Hare Krishnas, and the Messianic Jews.

We have already discussed neo-Pentacostalism. Now let's turn to religious movements that can be classified as religious cults. Two examples are the Unification Church and Scientology.

What is the Unification Church? One of the most controversial of the new religious movements is the Reverend Sun Myung Moon's Unification Church, one of the new mystical religions whose adherents are generally (and negatively) referred to as Moonies. Mr. Moon is a Korean industrialist turned prophet-minister-messiah who proclaims a religion that is a combination of Protestantism and anticommunism (Barker, 1984).

Several interesting sociological observations can be made about the Unification Church. First, like some of the other new mystical movements such as the Hare Krishnas, total conformity is expected of its members. Second, in spite of parental fears, its morality is generally conservative, sometimes rigidly so. More often than not it endorses such basic American values as a powerful God, a strong sense of family, and individualism. Third, the more the Unification Church grows, the more it resembles a denomination. For example, it now has its own seminary with faculty members who belong to respected national professional societies.

Although the Unification Church is small in terms of the number of members, it has attracted considerable attention. It forced society to deal with difficult issues regarding the legal rights of religious practitioners in relation to family members who may believe that the church members have been brainwashed (Richardson, 1982; Shepherd, 1982; Bromley and Shupe, 1984). More recently, the Unification Church has been in the public spotlight because of its economic activities. The Reverend Moon's movement now controls more than $10 billion in business assets around the world. These assets include an automobile plant in Asia, the University of Bridgeport (Connecticut), the *Washington Times* newspaper, Atlantic Video, and the Nostalgia television network (Byron, 1993).

What is Scientology? Scientology is a method of spiritual therapy that grew out of the work of the late science fiction writer L. Ron Hubbard. Hubbard claimed to have discovered the therapy and technique of Dianetics, a new approach to mental health that was received with faddish seriousness in the 1950s. Later, Hubbard developed an elaborate system of Scientology that included far more metaphysical speculation,

Contemporary religious fundamentalism in the United States has its roots in the nineteenth century. True to their nineteenth-century antecedents, American religious fundamentalists today interpret the Bible literally, believe in Satan as an active force for evil, and wait for the eventual destruction of the world prior to the Messiah's second coming.

SOCIOLOGY

William Stacey and Anson Shupe—The Electronic Church

In the recent past, religious evangelists typically traveled from community to community and preached to the faithful in tents, open fields, or rented meeting halls. Modern-day communications technology has changed all that. Within the past decade, religious programming on radio and television has expanded greatly.

Although the "electronic church" has attracted considerable attention, disagreement exists as to the actual size of its audience and the extent of its impact. Many evangelists claim to have very large audiences, but most rating services estimate the total religious television audience to be of a rather modest size, approximately seven to twelve million viewers.

William Stacey and Anson Shupe (1982) have advanced the sociological understanding of the influence of the electronic church by examining the characteristics of its viewers. They surveyed residents of the Dallas-Fort Worth metropolitan area. This area, often referred to as the "buckle" of the southern Bible Belt, is certainly not representative of the nation as a whole, but it provides a particularly good place to begin an examination of the electronic church, because such religious programming seems to have wide appeal in the southern United States and because the Dallas-Fort Worth area has access to all of the major syndicated religious programs.

Stacey and Shupe found that regular viewers were likely to have relatively low incomes and to have completed less than a high school education. Viewers were also concentrated among individuals over thirty-five years of age,

Robert Tilton is part of the "electronic church." Controversy exists regarding the effects of televangelists. Although some attribute significant influence to the electronic church, others think its effects are limited.

---

particularly the doctrine of reincarnation. Hubbard and his followers claim to be able to perform mental and physical healing. Thus, Scientology is a combination of medicine and religion. Its practitioners charge for their services; have very little formal religious ritual; and, like the Unification Church and some other new religious movements, demand absolute conformity to stated doctrines. In recent years, however, the requirement that followers sever all non-Scientological connections has been relaxed somewhat.

Are these new religious movements unique? The new religious movements have many connections with older religions. The New Testament, for example, demands that Christians leave family and friends for the sake of "the gospel." And throughout the history of Western Christendom, movements have periodically arisen that demand absolute conformity, self-denial, and separation from family and friends who could not be converted (Beckford, 1985). Similarly, the political activities of the Moral Majority had many parallels in American history, and similar political involvement by fundamentalist religious groups occurs in other nations (Marty, 1980; Yinger and Cutler, 1982). Nor are more exotic cults new to our times. Consider this example from New York State in the early 1880s:

Some frontier religions were mildly eccentric, while others were extremely bizarre. Possibly the most flamboyant messiah of all was Isaac Bullard. Clad in nothing but a bearskin loincloth, he led his troop of adherents from Vermont through New York and Ohio and finally south to Missouri, proclaiming a religion compounded of free love, communism and dirt—Bullard often boasting that he hadn't washed in seven years (Wise, 1976:20).

In late November 1978, news began to arrive in the United States, first slowly and then with a tragic rapidity, that a semireligious, socialistic colony in Guyana, South America, headed by the Reverend Jim Jones—founder of the California-based People's Temple—had been the scene of one of the most shocking suicide-murder rites in recent times. The bizarre accounts appearing in the media prompted many people to ask how anyone could have become involved in the cult.

Some tried to answer the question by dismissing the participants as ignorant or mentally unbalanced individuals. But as more facts were uncovered, people learned that many of the members were fairly well educated young people and that Jones was trusted and respected by part of the California political establishment. They also learned that such events, although rare, have occurred before.

people in households with relatively more children, females, and long-term residents of Texas. Blue-collar workers were more likely than white-collar workers to watch such programs, but retired persons and homemakers were more likely to watch than were people with jobs.

Those who attended church regularly tended more to watch religious programming, which is an important finding because it contradicts the claim that the electronic church is depriving local churches of members. Those with orthodox religious orientations were also more likely to tune in such programs. When the members of specific denominations were surveyed, it was found that among members of fundamentalist denominations (such as Pentecostals and Seventh-day Adventists), those displaying more

church religiosity (church membership, contributions, and so forth) were more likely to view religious programming. Among members of conservative denominations (such as Southern Baptists and Disciples of Christ) and moderate denominations (Methodists, Presbyterians, and Episcopalians), those with more orthodox religious beliefs seemed more inclined to watch these programs. In summary, the electronic church:

preaches to the converted who are already predisposed, or self-selected, to seek out its messages. These are persons who are members of fundamentalist congregations and/or persons with highly orthodox religious beliefs (Stacey and Shupe, 1982:299).

Critical feedback
1. According to Stacey and Shupe's research, what kinds of people are most likely to watch religious programming? How typical are they of the population as a whole?
2. On the basis of their research, Stacey and Shupe conclude that the impact of the electronic church on local churches and politics is quite limited. Others, however, suggest that the impact will become increasingly stronger because television evangelists are gaining increasing access to communications technology that is becoming more instrumental in shaping social consciousness. Would you predict that the electronic church will have greater social impact in the future? Why or why not?

Moreover, it would happen again. Like Jim Jones, David Koresh was a pathological, charismatic cult leader. He and many of his Branch Davidians, some sixty people, including seventeen children, went up in flames in their Waco, Texas, compound in 1993 under FBI assault. Koresh saw himself as Jesus Christ in sinful form, and his devoted followers probably believed they were on their way to join Jesus in heaven after the Apocalypse (Edmonds and Potok, 1993; Gibbs, 1993a).

Why are people willing to go to such extremes? Sociology cannot totally answer this larger question, but it can help us understand why people may be disposed to join the many religious cults now in existence in the United States.

Why have these new religious movements emerged?
Reasons for the emergence of these new religious movements closely related to fundamentalism have already been covered in the discussion of the revival of fundamentalism. Moreover, such movements should not be identified with the likes of the People's Temple of Jim Jones or David Koresh's Branch Davidians. This section looks exclusively at some explanations for the current rise of cults and why people join them.

Thomas Robbins, Dick Anthony, and James Richardson (1978) contend that cults are the product of

a value crisis or normative breakdown in modern industrial society. In this view, some people attempt to escape moral ambiguity—a sense of personal helplessness, doubt, and uncertainty—by grasping onto a new "truth."

According to Charles Glock and Robert Bellah, interest in cults stems from the excessive emphasis on self-interest in American Protestantism. The exaggeration of the parts of the Bible emphasizing self-interest has resulted in a failure to provide either social benefits or a sense of personal worth in complex urban society. The new religious yearning seen in cults is geared toward socialism and mysticism, which are incompatible with American denominations that emerged from an agricultural or frontier environment:

The demand for immediate, powerful, and deep religious experience, which was part of the turn away from future-oriented instrumentalism toward present meaning and fulfillment, could on the whole not be met by the [traditional] religious bodies.... In many ways Asian spirituality provided a more thorough contrast to the rejected utilitarian individualism than did biblical religion. To external achievement it posed inner experience; to the exploitation of nature, harmony with nature; to the impersonal organization, intense relation to a guru (Glock and Bellah, 1976:340-341).

Harvey Cox (1977) has isolated four reasons for Americans' being drawn to Eastern religious cults. First, most of the converts to these cults are looking for friendship, companionship, acceptance, warmth, and recognition. The cult provides a supportive community that helps overcome past loneliness and isolation. The Eastern religious cults provide emotional ties that converts cannot find at home, school, church, or work. Many of them even adopt kinship terms to give recruits new identities that set them apart from their former existence—sister, brother, Hare Krishna. Often a convert is renamed entirely. Entertainer Steve Allen's son Brian joined the Church of Jesus Christ at Armageddon in Seattle and was "reborn" under the name Logic Israel.

Second, most of the Eastern religious cults emphasize immediate experience and emotional gratification rather than deliberation and rational argument. Converts report "experiencing" religion rather than merely thinking about it. Whether by meditation, speaking in tongues, or singing hymns, adherents have the frequent and intense emotional experiences they could not find elsewhere.

Third, Eastern religious cults emphasize authority. By having a firm authority structure and a clear, simple set of beliefs and rules, they offer converts something in which to believe. Converts profess to exchange uncertainty, doubt, and confusion for trust and assurance.

Finally, these cults purport to offer authenticity and naturalness in an otherwise "artificial" world. By emphasizing natural foods, communal living apart from "civilization," a uniform dress code, and sometimes nudity, these groups attempt to show they are not part of the "plastic society."

Social Correlates of Religion

Social class and politics are two important social correlates of religious preference. Before discussing them, however, it will be helpful to provide a picture of religious affiliation in the United States.

What are the religious preferences of Americans? Although there are over three hundred denominations and sects in the United States, Americans are largely Protestant (58 percent) and belong to a few major denominations—Baptist (20 percent), Methodist (10 percent), Lutheran (6 percent), Presbyterian (4 percent), and Episcopalian (4 percent). Fourteen percent prefer other Protestant denominations. Catholics constitute a relatively large proportion of the American population (25 percent) and Jews a small proportion (2 percent). Fewer than one in ten Americans has no religious preference (Gallup, 1996). Table 14.2 lists religious organizations in the United States with memberships above 300,000.

TABLE 14.2	Membership in Selected Religious Organizations in the United States
Religion	**Number of Members**
Roman Catholic	60,280,454
Southern Baptist Convention	15,663,296
United Methodist Church	8,538,662
Jewish	6,840,000
Lutheran Church in America	5,190,489
Muslim	5,000,000
Presbyterian Church (U.S.A.)	3,669,489
Episcopal Church	2,536,550
Assemblies of God	2,387,982
United Church of Christ	1,472,213
Jehovah's Witnesses	966,243
Christian Church (Disciples of Christ)	929,725
Seventh-day Adventists	790,731
Church of the Nazarene	601,900
Salvation Army	453,150
Wisconsin Evangelical Lutheran Synod	412,478
Reformed Church in America	306,312

Sources: Yearbook of American and Canadian Churches, 1997, edited by Kenneth Bedell. Copyright © 1997 by the National Council of the Churches of Christ in the USA. The estimate for Muslims is cited in Azim A. Nanji (Ed.), *The Muslim Almanac*, Detroit, Gale Research, Inc., 1996, p. 141. The estimate for Jewish membership comes from Frank S. Mead and Samuel S. Hill, *Handbook of Denominations in the United States*, Nashville, TN: Abingdon Press, 1995.

What is the relationship between social class and religious characteristics? As Figure 14.6 reveals, there are marked differences in social class (as measured by education and income) among the adherents of various religions in the United States. Generally speaking, Presbyterians, Episcopalians, and Jews are on the top of the stratification structure. Below them are Lutherans, Catholics, and Methodists. When measured by education and income, Baptists, on the average, come out on the bottom. Because these are average figures, there are, of course, many individual exceptions to these rankings.

The social-class differences are partly due to self-selection—people tend to want to belong to churches with members who have socioeconomic characteristics similar to their own. The socioeconomic differences occur partly because religious organizations place varying emphasis on worldly success versus otherworldly rewards (Niebuhr, 1968). Finally, a historical dimension exists. With the exception of the Jewish groups, the higher-status denominations have been at the top for a long time. Thus, members of these denominations have had tremendous socioeconomic advantages. Lutherans and Roman Catholics, in contrast, tend to be later immigrants who have not had as many opportunities to achieve higher status (Johnstone, 1996).

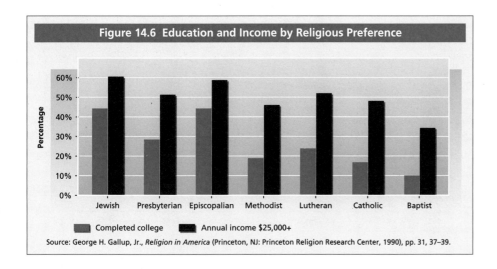

Figure 14.6 Education and Income by Religious Preference

Completed college   Annual income $25,000+

Source: George H. Gallup, Jr., *Religion in America* (Princeton, NJ: Princeton Religion Research Center, 1990), pp. 31, 37–39.

Differences in religiosity exist between the upper and lower classes. Reflecting Marx's view of religion as an opiate of the masses, many early sociologists saw religion primarily as a source of solace for the deprived poor. This entrenched view was challenged by research in the 1940s that showed higher church affiliation among the wealthy. Further research in the 1950s and 1960s revealed that religion is important at both extremes of the stratification structure. It is just that the upper and lower classes express their religious beliefs in different ways. The upper classes display their religiosity through church membership, church attendance, and observance of ritual, whereas lower-class people more often pray privately and have emotional religious experiences (Stark and Bainbridge, 1985).

What is the relationship between religion and politics? American religious groups vary widely in socioeconomic characteristics, so we must be careful in interpreting empirical connections between religion and politics. Some of the apparent influence of religion on political attitudes and behavior must in part be due to social-class factors.

The Democratic Party is endorsed more than the Republican Party by Protestants, Catholics, and Jews. Followers of the Jewish faith are particularly aligned with the Democratic Party, followed in strength of support by Catholics and Protestants. This is predictable, because Protestants generally are more politically conservative than Catholics or Jews. Of the major Protestant denominations, the greatest support for the Republican Party is found among Episcopalians and Presbyterians, which is hardly surprising, because the upper classes are more likely to be identified with the Republican Party. Despite their affiliation with the more conservative Republican Party, however, Episcopalians and Presbyterians are less socially conservative than Baptists, who are the

strongest supporters of the Democratic Party of all Protestant denominations.

At the presidential level, Protestants have been significantly more likely to vote for Republican candidates than have Catholics or Jews. Since 1952, Catholics have voted for Democratic candidates, except in 1972 when they voted for Richard Nixon and in 1980 and 1984 when they voted for Ronald Reagan. Protestants have broken their string of Republican presidential votes only in 1964, when 55 percent voted for Lyndon Johnson.

Religion and the Baby Boomers

What is unique about baby boomers? Two points need to be made immediately about America's baby boomers, the generation born between 1946 and 1964 whose oldest members are now in their fifties. First, there are a lot of them. Their number—seventy-five million—represents almost 40 percent of the adult population and about 28 percent of the entire population of the United States. When baby boomers say "jump," American society generally replies "how high"? Like a python ingesting a pig, America's schools, political system, mass media, and housing industry have adapted to accommodate their needs and take advantage of their consumption potential at every stage of the life cycle. Second, having grown up during a period of startling change since World War II, baby boomers are not the homogeneous generation their collective label implies. Their ranks include all social classes, ethnic groups, and races. They are also separated by gender. Despite this diversity, baby boomers have a shared identity sufficient to view them as a socially meaningful generation.

This generational juggernaut has already influenced the religious institution earlier in their lives. They seem poised to do so again as they approach or enter middle age (Bouvier and De Vita, 1991).

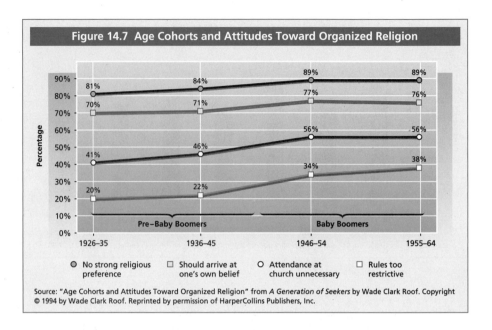

Figure 14.7 Age Cohorts and Attitudes Toward Organized Religion

Source: "Age Cohorts and Attitudes Toward Organized Religion" from *A Generation of Seekers* by Wade Clark Roof. Copyright © 1994 by Wade Clark Roof. Reprinted by permission of HarperCollins Publishers, Inc.

What have been the effects of the baby boomers on religion? The religious institution first felt the effects of the baby boomers when their parents swelled membership rolls by faithfully bringing them to church. The second shock came when the baby boomers dropped out of their churches in record numbers after leaving home. As revealed in Figure 14.7, the two **age cohorts**—persons born during the same time period in a particular population—born prior to and during World War II (1926–1935 and 1936–1945) have a much stronger orientation toward organized religion than the two baby boomers' birth cohorts (1946–1954 and 1955–1964). These differences in religious attitudes help to explain why the bulk of baby boomers left their churches after leaving home and why a substantial proportion still refuses to return.

According to some researchers, large numbers of baby boomers are finding their way back to the spiritual roots they abandoned. If true, American religion is once again about to feel the shock of a shift made by the baby-boom generation. Organized religion may be in for another landscaping.

About one-third of the baby boomers never left the church, and over 40 percent remain dropouts from organized religion. It is the returning one-fourth that is altering the nature of formal religion in America and the religious practices of millions of its inhabitants. These are some of the observations made by Wade Clark Roof, who has conducted an extensive survey on religion among the baby boomers (Roof, 1994).

Why are many baby boomers returning to church? Baby boomers are at the age when preceding generations of Americans have typically become more active in organized religion. According to a Gallup poll, many baby boomers of both genders are following this pattern. (See Table 14.3.)

Although some baby boomers are returning to church for needed social support following severe career disappointments or divorces, most are doing what parents normally do—they show concern about the well-being of their children. Many baby-boomer parents believe it their duty to expose their children to religious tenets that can be embraced or rejected of

Some evidence suggests that many baby boomers, who left their churches in larger numbers in their early twenties, are returning to their religious roots. If this happens, religion in America will feel the shock wave the baby boomers always make when they touch an institution.

TABLE 14.3	Baby Boomers and Religion					
Importance of Religion	**Men**	**Women**		**Church member?**	**Men**	**Women**
Very important	39%	58%		Yes	61%	69%
Fairly important	42	31		No	39	31
Not very important	18	11		**Nonmembers expecting to join in next 5 years**		
Importance next 5 years				Yes	38	39
More important	61	66		No	55	57
Same	25	20		Not sure	7	4
Less important	13	12		**Want children to receive religious education?**		
Time will spend on religious activities next 5 years				Yes	73	78
More time	37	44		No	17	15
Same time	55	52		Not sure/not applicable	10	7
Less time	7	3				

Source: George H. Gallup, Jr., *Religion in America, 1992–93* (Princeton, NJ: Princeton Religion Research Center, 1993) p. 15.

their own choice later. Others see their children failing to be exposed to moral values outside a religious context. For some, the church is an instrument for reinforcing values they are attempting to instill in their children at home. Still others are looking to organized religion for the sense of community it can provide in a highly individualistic world.

According to Roof, the return of baby boomers may even accelerate. In his study, older baby boomers with children were more involved in church than younger members of their generation. As the latter began to have children, many of them may also turn to churches. Whether this happens or not, organized religion will be energized by the baby boomers at midlife.

FEEDBACK

1. _____ is the process whereby religion loses influence over society as a whole.
2. According to Caplow, Bahr, and Chadwick's restudy of Middletown,
   a. American ministers are becoming more socially conscious.
   b. secularization has proceeded too far.
   c. the trend is toward less religious involvement in the United States.
   d. the trend is toward greater religious involvement in the United States.
3. A _____ religion is a public religion that expresses a strong link between God and the American way of life.
4. An _____ religion is a private religion dealing with ultimate concerns that is substituted for formal religious organizations, practices, and beliefs in a secular society.
5. _____ is based on the rejection of the effects of modern society on traditional religious beliefs, rituals, and doctrines.
6. Which of the following is *not* one of the reasons presented for the resurgence of fundamentalism in the United States?
   a. fear that the world is out of control
   b. provision of solace by fundamentalist organizations
   c. the electronic church
   d. the rise of invisible religion
7. The origins of new religious movements appear to lie in
   a. social conditions in modern industrial society.
   b. desecularization.
   c. the rising influence of the Middle East.
   d. the decline of civil religion.
8. Which of these groupings of religious organizations is at the top of America's stratification structure?
   a. Jewish, Presbyterian, Episcopalian
   b. Lutheran, Jewish, Catholic
   c. Catholic, Episcopalian, Methodist
   d. Methodist, Baptist, Presbyterian
9. Which of the following is the major reason many baby boomers are turning to churches?
   a. concern for the ethical and religious training of their children
   b. need for comfort following divorces
   c. guilt over letting their own religiously oriented parents down
   d. fear of dying without practicing a religious faith

Answers: 1. Secularization 2. d 3. civil 4. invisible 5. Fundamentalism 6. d 7. a 8. a 9. a

SUMMARY

1. Religion is a unified system of beliefs and practices about sacred things. From his studies of Australian aborigines, Emile Durkheim concluded that every religion distinguishes between the sacred and the profane.

2. Because religion involves a set of meanings attached to a world beyond human observation, sociologists who study it scientifically face some unique problems. They do not evaluate the validity of various religions, leaving theological issues to theologians and philosophers.

3. According to anthropologists, religion is found in nearly all societies. Its importance to all people is due in part to the functions it serves. Religion offers explanations for the structure of societies, promotes social unity, and provides a sense of meaning and belonging.

4. Writing from the conflict perspective, Karl Marx contended that religion is used by elites to justify and maintain their power and is used by the oppressed to soothe their misery and maintain hope for improvement. Religion, in his view, is the opiate of the masses.

5. According to the conflict interpretation, religion retards social change because those in power wish to maintain the status quo. Max Weber, on the other hand, believed that religion could promote social change. He attempted to support this idea by demonstrating a connection between the Protestant ethic and the rise of capitalism. The Protestant ethic emphasizes hard work and reinvestment for profit, characteristics that were highly compatible with the needs of emerging capitalism.

6. Two of the most important components of religion are religious organization and religiosity. The major forms of religious organization are churches, denominations, sects, and cults. The ways people express their religious interests and convictions (religiosity) can be analyzed in terms of five dimensions: belief, ritual, intellectual, experience, and consequences. These dimensions of religiosity are closely related, but many important studies have focused on a single dimension. Examples include studies of religious conversion (belief dimension), commitment (ritual dimension), and the relationships of religious participation to prejudice and deviant behavior (consequence dimension).

7. Secularization has weakened the control religion had over society in earlier times. The sacred and the profane are increasingly intermixed. The existence of secularization does not mean, however, that Americans are not religious. Although many observers have noted a decline in the role of religion in the United States, others contend that religion may be stronger than has been supposed. For one thing, Americans practice a civil (public) religion and an invisible (private) religion.

8. Moreover, there has been a revival of religious fundamentalism in the United States. This fundamentalism, which is occurring largely outside the mainline religious organizations, shares a theological conservatism with nineteenth-century American fundamentalism. In addition, contemporary fundamentalism has its own conservative social and political agenda.

9. The resurgence of religious fundamentalism is due to several factors, not the least of which is the electronic church. Television, radio, and computers (for direct mail campaigns) have enabled fundamentalist organizations to spread their messages to vast numbers of people on a continuous basis. It remains to be seen whether recent scandals among some televangelists will permanently damage both the electronic church and fundamentalism itself.

10. The rise of new religious movements since the 1960s may also be an indication of a religious revival. Neo-Pentecostalism, the Unification Church, Scientology, and other new religious movements may be a response to contemporary feelings of confusion, doubt, helplessness, and uncertainty.

11. Two important social correlates of religion are social class and politics. The major religious faiths can be ranked in terms of educational level, income, and occupational prestige. Also, the upper and lower classes express their religiosity in quite different ways.

12. The baby boomers, a generation that has influenced all aspects of American social life since they began being born in 1946, are having a new effect on organized religion. Having dropped out of churches early in their lives, many are returning primarily out of concern for the ethical and religious training of their children.

LEARNING OBJECTIVES REVIEW

After careful study of this chapter, you will be able to:

- State the sociological meaning of religion.
- Demonstrate the different views of religion taken by functionalists and conflict theorists.
- Distinguish the basic types of religious organization.
- Discuss the meaning and nature of religiosity.
- Define secularization and describe its relationship to religiosity in the United States.
- Differentiate civil and invisible religion in America.
- Describe the current resurgence of religious fundamentalism in the United States.
- Identify a wide variety of new religious movements in America.
- Outline the relationship of baby boomers with religion.

CONCEPT REVIEW

Match the following concepts with the definitions listed below them:

____ a. secularization   ____ d. age cohort   ____ g. denomination

____ b. invisible religion   ____ e. profane

____ c. sacred   ____ f. civil religion

1. Persons born during the same time period in a particular population.
2. A public religion that expresses a strong tie between a deity and a society's way of life.
3. The process whereby religion loses its influence over society as a whole.
4. One of several religious organizations that members of a society accept as legitimate.

5. A private religion that is substituted for formal religious organizations, practices, and beliefs in a secular society.
6. Nonsacred aspects of our lives.
7. Things that are set apart and given a special meaning that transcends immediate human existence.

CRITICAL THINKING QUESTIONS

1. Do you think the distinction between the sacred and the profane still has validity in America today? Explain why or why not with examples. _____
_____
_____

2. Consider the four functions of religion identified by sociologists. Which of these functions are being performed by religion in the contemporary United States? Illustrate each. _____
_____
_____

3. Describe Weber's Protestant ethic. Explain why Marx could not have endorsed the existence of the relationship between the Protestant ethic and the spirit of capitalism proposed by Weber. _____
_____
_____

4. What do you think is the relationship between secularization and religiosity in the United States at the present time? Develop your answer with material from the text. _____
_____
_____

5. Evaluate the place of civil religion and invisible religion in America today._____
_____
_____

6. Identify what you think are the three most important effects of fundamentalism on religion in contemporary America. Support your case with specifics. _____
_____
_____

MULTIPLE CHOICE QUESTIONS

1. **Which of the following statements is *false*?**
   a. A profane object can become a sacred object.
   b. Profane objects do not take on a public or social character that makes them appear important in themselves.
   c. Religion is a unified system of beliefs and practices relative to sacred things.
   d. In their research, sociologists must avoid the strictly spiritual side of religion.
   e. Sociologists cannot successfully avoid questions about the ultimate validity of any particular religion.

2. **Those aspects of life that are everyday, ordinary, and taken for granted are labeled the**
   a. sacred.
   b. transcendental.
   c. profane.
   d. legitimate.
   e. material.

3. **Archaeological discoveries and anthropological observations have demonstrated that religion**
   a. began around the eleventh century.
   b. tends to dissolve into a limited number of basic denominations.
   c. has been more successful at governing societies than political parties.
   d. always contains an element of salvation and afterlife.
   e. has existed in some form in nearly all known societies.

4. **Emile Durkheim contended that _____ was the central function of religion.**
   a. sacredness
   b. supernaturalism
   c. legitimation
   d. transcendentalism
   e. salvation

5. **Religious ceremonies marking important events in life, such as birth, sexual maturity, marriage, and death, illustrate the contribution of religion to**
   a. a sense of unity.
   b. a sense of anomie.
   c. a sense of time.
   d. a sense of meaning.
   e. the legitimization of social arrangements.

6. **Karl Marx contended that religion _____ social change.**
   a. promotes
   b. retards
   c. refracts
   d. serves no function in
   e. precludes

7. **The Calvinist belief that every person's eternal fate is unalterably predetermined by God at birth is known as**
   a. sacred legitimation.
   b. transcendentalism.
   c. predestination.
   d. deism.
   e. immaculate salvation.

8. **According to Max Weber,**
   a. early Protestantism provided a social environment conducive to capitalism.
   b. religion tends to be linked to the economy in most societies.
   c. capitalism developed as a direct result of Protestantism.
   d. Protestantism developed as a direct result of capitalism.
   e. religion serves as an economic opiate for the masses.

9. **The ways in which people express their religious interests and convictions is called _____ by sociologists.**
   a. piety
   b. religiosity
   c. worship
   d. secularization
   e. ritualism

10. **_____ is the process through which religion loses influence over society as a whole.**
    a. Dereligiosity
    b. Transcendentalism
    c. Segmentation
    d. Secularization
    e. Pragmatism

11. **According to studies of religiosity in the United States,**
    a. secularization is destroying religion in the United States.
    b. only half of all Americans now believe in God or a universal spirit.
    c. only 20 percent of all Americans now believe in life after death.
    d. the proportion of regular churchgoers has declined sharply.
    e. compared with other industrialized countries, America still appears to be a fairly religious nation.

12. **Researchers cited in the text attribute Americans' continuing interest in religion to three basic reasons: the threat of nuclear war, loneliness, and**
    a. the need to impress others with their piety.
    b. disenchantment with what they perceive as an anomic society.
    c. basic spiritual and psychological needs.
    d. the absence of competing institutions to fulfill their needs.
    e. guilt and fear of divine punishment after death.
13. **The involvement of contemporary clergy in community affairs demonstrates**
    a. the mixing of the sacred and the profane.
    b. their resistance to secularization.
    c. the strong tie between a deity and a society's way of life.
    d. the demise of civil religion.
    e. the increasing significance of religion.

14. **Many of the new religious movements in the United States have emerged alongside the resurgence of**
    a. fundamentalism.
    b. escapism.
    c. desecularization.
    d. invisible religion.
    e. civil religion.
15. **Which of the following statements about baby boomers is *incorrect*?**
    a. About one-third of baby boomers never left their churches.
    b. Baby boomers are religiously heterogeneous.
    c. Older baby boomers with children are more involved in church than younger members of their generation.
    d. Baby boomers do not appear to be returning to the spiritual roots they had abandoned.
    e. The bulk of baby boomers abandoned their churches after leaving home.

FEEDBACK REVIEW

True-False
1. Sociologists who study religion are normally forced to abandon their religious beliefs and convictions. *T or F?*

Fill in the Blank
2. According to Marx, religion is created by the _____ to justify its economic, political, and social advantages over the oppressed.
3. _____ is based on the rejection of the effects of modern society on traditional religious beliefs, rituals, and doctrines.

Multiple Choice
4. According to Karl Marx, religion is a force working for
   a. the good of all people.
   b. the good of the proletariat.
   c. maintenance of the status quo.
   d. the creation of conflict between the classes.
5. According to Max Weber,
   a. early Protestantism provided a social environment conducive to capitalism.
   b. capitalism developed because of Protestantism.
   c. capitalism strengthened Protestantism.
   d. religion will always be linked with economy.
6. Which of the following is *not* one of the reasons presented for the resurgence of fundamentalism in the United States?
   a. fear that the world is out of control
   b. provision of solace by fundamentalist organizations
   c. the electronic church
   d. the rise of invisible religion
7. The origins of new religious movements appear to lie in
   a. social conditions in modern industrial society.
   b. desecularization.
   c. the rising influence of the Middle East.
   d. the decline of civil religion.

8. Which of these groupings of religious organizations is at the top of America's stratification structure?
   a. Jewish, Presbyterian, Episcopalian
   b. Lutheran, Jewish, Catholic
   c. Catholic, Episcopalian, Methodist
   d. Methodist, Baptist, Presbyterian

9. Which of the following is the major reason many baby boomers are turning to churches?
   a. concern for the ethical and religious training of their children
   b. need for comfort following divorces
   c. guilt over letting their own religiously oriented parents down
   d. fear of dying without practicing a religious faith

Matching

10. Match the dimensions of religiosity with the examples beside them.
    _____ a. belief   (1) prayer in church
    _____ b. ritual   (2) knowing the content of the Koran
    _____ c. intellectual   (3) having seen an angel
    _____ d. experience   (4) conviction that the Bible is divinely inspired
    _____ e. consequences   (5) marching in support of prayer in school

GRAPHIC REVIEW

- -

Figure 14.6 shows the relationship between education and income by denomination in the United States. If you can answer these questions, you will demonstrate an understanding of these data.

1. Describe the relationship between social class and religious preference among Americans.

   _____

   _____

2. Using these data and an understanding of fundamentalism in America, can one infer any relationship among social class, religious preference, and fundamentalism? Explain.

   _____

   _____

ANSWER KEY

- -

Concept Review	Multiple Choice	......Feedback Review
a. 3	1. e	1. F
b. 5	2. c	2. elite
c. 7	3. e	3. Fundamentalism
d. 1	4. c	4. c
e. 6	5. d	5. a
f. 2	6. b	6. d
g. 4	7. c	7. a
	8. a	8. a
	9. b	9. a
	10. d	10. a. 4
	11. e	b. 1
	12. b	c. 2
	13. a	d. 3
	14. a	e. 5
	15. d	

Chapter Fifteen

Health and Health Care

OUTLINE

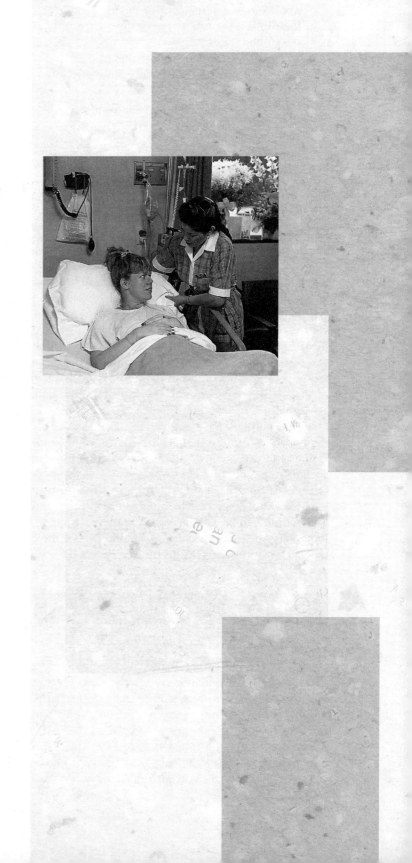

After careful study of this chapter, you will be able to:
- Define the concept of a health-care system and identify its major components.
- Apply functionalism, conflict theory, and symbolic interactionism to the health-care system in the United States.
- Discuss the distribution of diseases in the United States according to age, gender, race and ethnicity, and social class.
- Describe what is meant by looking at the health-care system as a medical-industrial complex.
- Describe the major health-care delivery innovations in America today.
- Discuss health-care reform in the United States today.

SOCIOLOGICAL IMAGINATION

Is Canada the major exception among modern societies in having a national health-care system? Canada is not the only industrialized country with a health-care system organized and operated by the federal government. In fact, the United States is the sole exception among highly developed nations in failing to provide a federally centralized health insurance program as a free or inexpensive service for all its citizens on an equal access basis.

To this point we have explored the most thoroughly developed institutions in modern society—family, education, polity, economy, and religion. Neither modern society nor sociology is static, however. As society changes, sociology expands to encompass the study of new social phenomena. Several additional institutions are beginning to take on increasing importance in contemporary society. This chapter explores one of these emerging institutions—the health-care system.

HEALTH CARE AS A SOCIAL INSTITUTION

Health Care and Society

Health is a major issue in every society. Whether they rely on shamans, barber-surgeons, magic, rituals, herbs, or hospitals, people in all societies develop distinct patterns for coping with sickness and death. In the 1830s, Alexis de Tocqueville noted the concern Americans tend to have for their well-being. Even he might well be overwhelmed by the current mania over physical health among Americans. Industries that supply the needs of joggers, weight lifters, health-food devotees, tennis players, and aerobic dancers are booming.

The responses Americans have made to this heightened interest in their personal health are clearly embedded in culture. Americans tend to respond to caring for their health needs with aggressiveness. This aggressiveness, contends Lynn Payer (1989), is reflected in differences in medical practice in the United States and other industrialized countries. Compared with their European colleagues, American physicians practice extremely aggressive medicine, and their patients expect it. Thus, American physicians are much more likely to perform surgery, run numbers of tests, and prolong life for individuals known to be in pain and facing imminent death.

The Nature of the Health-Care System

The **health-care system** embraces the professional services, organizations, training academies, and technological resources that are committed to the treatment, management, and prevention of disease. The American health-care system constitutes a very large sector of the total productive apparatus of the United States. Its growing claim on national resources shows it to be an institution that is increasing in importance. As shown in Figure 15.1, Americans spent nearly $950 billion on health care in 1994, up from approximately $75 billion in 1970. Since 1980, total health expenditures have increased almost 300 percent (U.S. Bureau of the Census, 1996a). These numbers more clearly reveal the size of the health-care system when considered as a proportion of the gross domestic product (GDP)—the monetary value of all the goods and services produced by the economy during a given year. Health-care expenditures now account for about 14 percent of the GDP; the comparable figure in 1970 was about 7 percent. (See Figure 15.2.)

What are the components of the health-care system? By way of introduction, we will briefly discuss four major components of the health-care system: physicians, nurses, hospitals, and patients. Each of these components is covered in greater depth later, along with some new organizational approaches to the delivery of health care in modern societies.

Although physicians constitute only about 10 percent of health-care workers in the United States, they establish the framework within which everyone else works. Only physicians are authorized by most states to diagnose illness, prescribe medicine, and certify such events as birth and death. Hospitalized patients may be attended by many professionals and other employees, but decisions about their diagnosis and treatment are made by physicians. American physicians' responsibili-

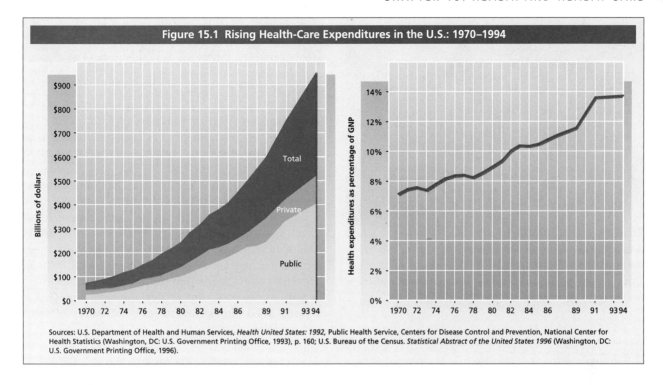

Figure 15.1 Rising Health-Care Expenditures in the U.S.: 1970–1994

Sources: U.S. Department of Health and Human Services, *Health United States: 1992,* Public Health Service, Centers for Disease Control and Prevention, National Center for Health Statistics (Washington, DC: U.S. Government Printing Office, 1993), p. 160; U.S. Bureau of the Census. *Statistical Abstract of the United States 1996* (Washington, DC: U.S. Government Printing Office, 1996).

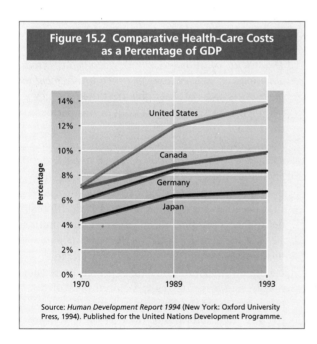

Figure 15.2 Comparative Health-Care Costs as a Percentage of GDP

Source: *Human Development Report 1994* (New York: Oxford University Press, 1994). Published for the United Nations Development Programme.

ties are matched by high levels of social prestige and monetary rewards.

Nursing became a recognized profession only in the late nineteenth century, and its identity continues to undergo change. When Florence Nightingale was sent (at her insistence) with nurses and supplies to help British army doctors in Turkey in 1854, she encountered substantial resistance. One colonel wrote: "the ladies seem to be on a new scheme, bless their hearts... I do not wish to see, neither do I approve of, ladies doing the drudgery of nursing" (quoted in Mumford, 1983:289). After finally winning acceptance, Nightingale influenced the establishment of nursing schools in England and elsewhere. Although this represented a major advance in medical care, Nightingale's vision of the nurse's role was more narrow than most nurses would now accept:

> She insisted that nurses should be "clean, chaste, quiet, and religious." She saw the nurse as providing wifely support, keeping the hospital going so the physician could do his important work (Mumford, 1983:290).

Possibly as a result of this initial orientation, nursing has experienced frequent controversy regarding education, professional roles, and compensation.

Training programs for nurses have gradually shifted from three-year diploma programs in hospitals to associate and baccalaureate programs in colleges and universities. College- and university-based training was slow to develop, despite a need for increasingly sophisticated knowledge on the part of nurses, because hospital administrators enjoyed the plentiful supply of cheap labor generated by training programs within their own hospitals. Though the shift to colleges and universities is well established, the fact that it took more than half a century symbolizes the professional and economic problems of nurses (Lynaugh and Brush, 1996).

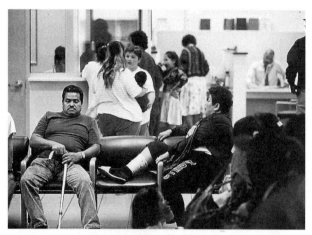

Americans have always been health conscious, a point made by Alexis de Tocqueville in the 1830s. It is not surprising, therefore, that the health-care system constitutes a very large proportion of the total productive apparatus of the U.S. economy.

Hospitals provide specialized medical services to a variety of types of inpatients and outpatients. They range from small facilities specializing in patients who need short-term, uncomplicated care to large medical centers that provide long-term care involving complex technology. Medical centers and some large hospitals typically combine patient care with medical research and training.

People who are not employees or practitioners within the health-care system usually enter the system only because they have been defined by others as ill or injured. Such a definition is based on a complex social process involving many people. Symptoms that may be considered proof of illness in some circumstances are ignored in others. The elderly, for example, are labeled ill far more easily than young adults because of the common assumption that illness is a normal aspect of old age. Whether or not a patient's designation as sick is biologically justified, this social definition places the patient in a complex set of rights and responsibilities that Talcott Parsons calls the *sick role*. This leads us directly into the theoretical perspectives, because the analysis of patients in terms of the sick role is part of the functionalist approach to the health-care system.

THEORETICAL PERSPECTIVES AND THE HEALTH-CARE SYSTEM

Functionalism

Talcott Parsons (1951, 1964a, 1964b, 1975), an eminent proponent of functionalism, proposed a view of sickness that was distinctively sociological rather than merely medical. For the first time, sickness became relevant to sociological theory and research. Parsons began on the assumption that health problems are a threat to society. If people are sick and cannot fulfill their roles, society will not function smoothly. Either functions will not be performed or they will be performed less well by others who fill in for the sick. To prevent people from pretending to be ill, sickness becomes a special form of deviant behavior. This is accomplished through the institutionalization of a sick role in every society.

What is the sick role?   The **sick role** serves to remove people from active involvement in everyday routines, give them special protection and privileges, and set the stage for their return to their normal social roles. Parsons identifies four major aspects of the sick role:

1. *The sick are permitted to withdraw temporarily from other roles or, at least, to reduce their involvement in them.* As long as others agree that they are sick, people can miss work or school or not perform their usual household duties.
2. *It is assumed that the sick, although engaging in deviant behavior, cannot become well through simply willing the sickness away.* Consequently, negative sanctions are avoided because others agree that the illness is not the patient's fault. Illness is defined as a physical rather than a moral matter.
3. *The sick are expected to define their condition as undesirable.* People are expected to want to become well again. They are expected to neither resign themselves to being sick nor take advantage unfairly of the benefits of being ill (relief from normal activities, sympathy, service from others).

1. The _____ embraces the professional services, organizations, training academies, and technological resources that are committed to the treatment, management, and prevention of disease.
2. Health-care expenditures now account for about _____ percent of the American GDP.
   a. 5   b. 10   c. 14   d. 21

Answers: 1. health-care system 2. c

4. *The sick are expected to seek and to follow the advice of competent health-care providers.* Although not blamed for the onset of their illness, the sick are considered responsible for doing whatever is required to improve. Failure to cooperate in efforts to cure themselves results in others defining them as deviants.

If any of these elements are missing, the privileges of the sick role may be withdrawn. For example, the quality of care given to those who are seen as "uncooperative" may be decreased. Some illnesses, including those linked to obesity and AIDS (acquired immunodeficiency syndrome), may even be denied the usual privileges of the sick role by some who view them as the patient's own fault. The existence of the sick role and the relationships formed around it demonstrate the social nature of illness.

What are the criticisms of the concept of the sick role? Like any important theoretical formulation, Parsons's concept of the sick role has been closely scrutinized. Several criticisms have been leveled at it. One criticism depicts the sick role as too narrow in scope. It speaks only to Western society, it does not capture all aspects of illness, and it does not seem to apply to such conditions as pregnancy. According to another objection, the sick role applies primarily to acute illnesses, such as viral infections, which are curable. In the case of an incurable disease, such as inoperable cancer, the patient is not expected to leave the sick role until death. Also, charge some critics, the sick role is not applied to stigmatized illnesses such as AIDS and mental illness. Some have also contended that the sick role refers to the professional health-care system and ignores other means of coping with sickness, such as self-treatment or the use of religious faith. Although analyzing sickness sociologically requires more than the concept of the sick role, Parsons's effort was pioneering and is still quite useful (Twaddle, 1981; Twaddle and Hessler, 1987).

A second issue relevant to the functionalist perspective is an exploration of the medical profession's rise to a position of preeminence in American society. More about the financial and prestige rewards for physicians is discussed under the conflict theory explanation.

Why has the American medical profession risen to its present heights? In his Pulitzer Prize–winning book, *The Social Transformation of American Medicine,* Paul Starr (1982) offers a functionalist explanation for the advancement of American physicians from their traditional lowly status. Using a pluralist orientation, Starr contends that American physicians were able to gain the recognition of medicine as a profession through the efforts of the American Medical Association (AMA),

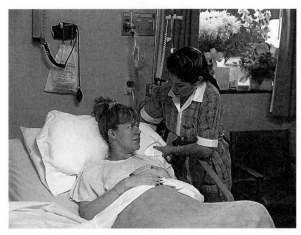

The sick role, which places considerable responsibility on the sick person to return to normal activities, can be applied only to curable illnesses. This sick role does not cover individuals with incurable diseases, such as terminal cancer or AIDS. This woman will be expected to abandon the sick role as soon as she can return to her life as it was before the accident.

especially after its reorganization around the turn of the century. As an interest group, the AMA was able to wield some power vis-à-vis competing interest groups such as patent medicine producers who offered therapy and advice in addition to selling medicine. The patent medicine companies were direct competitors because early physicians, in addition to treating patients, often prepared their own drugs.

According to Starr, Americans were persuaded by the physicians' organization that they "needed" physicians to handle their health problems. The combination of a responsive American public and a convincing lobby elevated physicians in the eyes of the American public. In fact, today the American public accords greater confidence to those people controlling medicine than to the U.S. Supreme Court ("Why Don't Americans Trust the Government?," 1996).

Starr explicitly rejects a Marxian interpretation of the rise of American medicine. It is to that interpretation that we now turn.

Conflict Theory

Several issues are covered here. They include the conflict perspective's alternative explanation for the rise in prestige of the American medical profession, the explanation of the high pay and status of physicians, the professional and economic problems of nurses, and inequality of health care.

How do conflict theorists explain the rise in the prestige of the medical profession in the United States? According to Marxist scholar Vicente Navarro (1976,

1986), Starr's explanation for the way American medicine has evolved lies in the values and beliefs of American society. Although Navarro credits Starr with recognizing that the AMA was successful in persuading the American people of the connection between their beliefs and values and the preeminence of medicine, he offers a very different interpretation.

As a Marxist, Navarro starts with the assumption that the structure of American society has built-in power differentials based on class, race, and gender. The nature of American institutions, including medicine, is the product of conflicts, of which class conflict is central but not alone. Although other nonclass-based types of conflict exist, class conflict determines the boundaries within which all other struggles occur. The success of the medical profession, then, is due to the power it possesses as a result of its combined connections with the dominant capitalist class, the white race, and the male gender.

The same political and economic forces that determine the nature of capitalism determine the nature of the institution of medicine. Thus, it was not merely persuasion that promoted the interests of the physicians but, more importantly, the coercive and repressive power physicians possessed because of the capitalist power structure. Navarro would point, for example, to medicine's attempt to create a monopolization of medical services by regular physicians through repression or control of alternative (competing) approaches to health care such as spiritual healing, acupuncture, osteopathy (based on the idea that disease results from loss of structural integrity), and homeopathy (treatment of disease by administering minute doses of a remedy that would in healthy persons produce symptoms of the disease being treated). In fact, a 1987 U.S. Supreme Court decision ruled that the American Medical Association was guilty of unfair restraint of trade in its efforts to keep osteopaths from practicing.

Navarro recognizes that the existence of conflict implies that the dominant power blocs in capitalist society do not always get their way. The existence of a dominant class, race, and gender means that there are dominated classes and races and a dominated gender. And the dominated do not always passively accept their domination. It is continuous conflicts and struggles that determine alterations in American society and in medicine.

Why are physicians highly rewarded?   Conflict theorists reject the functionalist explanation for the high financial and prestige rewards of physicians. According to functionalist theory, the high rewards accorded physicians are necessary for talented people to consider entering a demanding profession requiring extensive training. According to the conflict perspective, physi-

cians' rewards are high only because the medical establishment keeps medical school enrollments artificially low, thereby limiting price competition among physicians. There is probably some truth in both of these explanations.

Whatever the reason, not all physicians have enjoyed the privileges accorded those in the United States today. In eighteenth-century England and early twentieth-century France, physicians were poorly paid and were socially marginal. In the Soviet Union of 1990, physicians received only about 75 percent as much income as the average industrial worker. Contemporary Great Britain has a few specialists who command very high incomes, but most general practitioners are only moderately well paid. Although physicians in most contemporary societies have higher status than was the case in ancient Rome (where physicians were primarily slaves, freedmen, and foreigners, and medicine was considered a very low occupation), the situation in the United States is unusual (Starr, 1982).

Will physicians be able to maintain their present level of rewards?   Whether physicians in the United States will be able to maintain their privileged and powerful position is not known. For one thing, the number of physicians per 10,000 population in the United States has risen sharply since 1950. (See Figure 15.3.) Also, because of declines in population growth, increasing costs of malpractice insurance, and the movement from general practice to specialized medicine, physicians are becoming less and less likely to work as independent professionals. Increasing numbers of physicians are going into group practice or managed care organizations, where expenses are shared, or are taking salaried positions with organizations. As these changes occur, individual physicians are beginning to lose control over their work.

Perhaps the most subtle loss of autonomy for the profession will take place because of the increasing corporate influence over the rules and standards of medical work. Corporate management is already thinking about the different techniques for modifying the behavior of physicians, getting them to accept management's outlook and integrate it into their everyday work (Starr, 1982:447–448).

Medicine retains enough mystery that physicians are unlikely to lose all their social status. As long as physicians use professional jargon, write illegible prescriptions, maintain control over patients' records, and force patients to wait in a waiting room "because the doctor's time is so valuable" (Mumford, 1983:334), they will be held in awe by many members of society. It appears, however, that their ability to dictate the nature of the medical system is eroding.

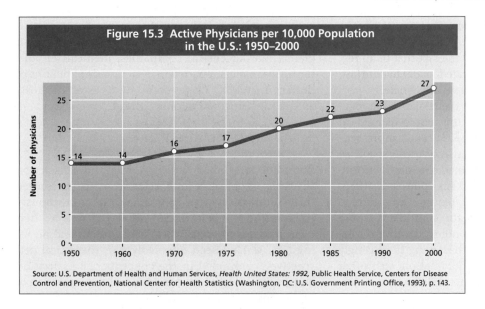

Figure 15.3 Active Physicians per 10,000 Population in the U.S.: 1950–2000

Source: U.S. Department of Health and Human Services, *Health United States: 1992*, Public Health Service, Centers for Disease Control and Prevention, National Center for Health Statistics (Washington, DC: U.S. Government Printing Office, 1993), p. 143.

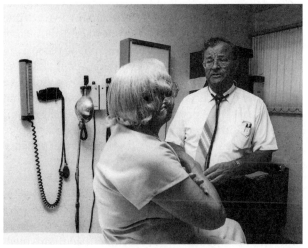

Functionalists and conflict theorists have very different explanations for the high rewards enjoyed by physician specialists in the United States. While functionalists attribute the handsome income of physicians to their talent and training, advocates of conflict theory credit the medical establishment with limiting the number of available physicians.

What are some of the professional and economic problems nurses face?   Role expectations and authority relationships are problematic for nurses. Although nurses are around patients far more than physicians are, they lack the authority to make any but the most routine decisions about patient care. In many situations, nurses must convey recommendations to physicians while appearing to be passive. Leonard Stein illustrates this problem of domination in a hypothetical telephone conversation between a nurse and a physician:

This is Dr. Jones.

(An open and direct communication.)

Dr. Jones, this is Miss Smith of 2W—Mrs. Brown, who learned today of her father's death, is unable to fall asleep.

(This message has two levels. Openly, it describes a set of circumstances, a woman who is unable to sleep and who that morning received word of her father's death. Less openly, but just as directly, it is a diagnostic and recommendation statement; i.e., Mrs. Brown is unable to sleep because of her grief, and she should be given a sedative. Dr. Jones, accepting the diagnostic statement and replying to the recommendation statement, answers.)

What sleeping medication has been helpful to Mrs. Brown in the past?

(Dr. Jones, not knowing the patient, is asking for a recommendation from the nurse, who does know the patient, about what sleeping medication should be prescribed. Note, however, his question does not appear to be asking her for a recommendation. Miss Smith replies.)

Phenobarbital mg 100 was quite effective night before last.

(A disguised recommendation statement. Dr. Jones replies with a note of authority in his voice.)

Phenobarbital mg 100 before bedtime as needed for sleep, got it?

(Miss Smith ends the conversation with the note of a grateful supplicant.)

Yes I have and thank you very much doctor (Stein, 1967:700).

Despite the difficult tasks and inconvenient hours associated with nurses' work, nurses' pay ranks below that of most professionals. As noted in Chapter 10 ("Inequalities of Gender"), women receive substantially lower pay for their work than men. This is certainly the case for nurses compared with physicians. The low pay received by nurses may relate to the fact that of all registered nurses, about 95 percent are female. In contrast, only about 15 percent of physicians are female.

Efforts by nurses to improve their salaries are regularly resisted by hospitals, but the nature of medical work could lead to nursing's being taken more seriously. Medical work is increasingly understood to involve more than prescribing medicine and performing surgery. Anselm Strauss and his colleagues (1985) show that successful treatment requires attention to safety, comfort, emotion, and the coordination of a variety of activities. Nurses, they note, play critical roles in these tasks. Nursing itself is becoming increasingly specialized, and the need for coordination with other professionals is increasing (Kovner, 1990).

How do conflict advocates view inequality in health care in the United States? The concern of conflict advocates for health-care inequality begins with **epidemiology**—the study of the patterns of distribution of diseases within a population. Social epidemiologists relate the occurrence of illness to individuals and their social and physical environments (Rockett, 1994). Conflict theorists are interested in the reason minorities and the poor have lower life expectancies and higher incidence of certain diseases than the general American population. (They are aware that certain illnesses are associated with some minorities more than others. For example, African Americans are more prone to heart disease, and alcoholism is more prevalent among Irish Americans.) Conflict proponents trace the proclivity for ill health among minorities and the poor to the nature of capitalist society. Specifically, they point to the role of the health-care establishment in maintaining unequal access to medical care (Navarro, 1976, 1986; McKinlay, 1985; Waitzkin, 1986; Weitz, 1996).

How does the medical establishment promote unequal access to health care? Inequality in access to medical care is promoted by making medical care a commodity to be bought on the market. Those without the money to purchase health care are likely to be ill more often, have more serious diseases, and die younger. Unable to purchase good health care, they suffer the consequences. A suggested solution to this problem is the institution of a national health-care program. In fact, the topic of health-care reform in the United States will close out this chapter.

Why is affordable health care not provided to the poor in the United States? According to conflict theorists, the answer lies in the commercialization of medicine for the sake of profit. To them, medical care is a capital-istic industry, an industry that brings to its participants—hospitals, nursing care homes, nurses, physicians, pharmaceutical companies, medical schools, medical suppliers, pharmacies—hundreds of billions of dollars annually. These combined vested interests ensure the perpetuation of health-care inequality.

Symbolic Interactionism

According to symbolic interactionism, humans act on the basis of subjective interpretations of reality learned from interaction with others and are continuously interpreting the behavior of others (as well as their own) in terms of socially acquired symbols and meanings. As we have seen repeatedly, this theoretical perspective is grounded in the concepts of socialization and symbols. This section examines the socialization of physicians, the background and socialization of nurses, and the influence of labeling.

The socialization process of most physicians begins early. Those who eventually become doctors tend to have made the decision early in life, many by the time they are sixteen years old. This is not accidental. Early decision occurs in part because many of those who enter medical school have been influenced by relatives who are physicians. This, of course, also helps to explain why physicians tend to come from the higher social classes. This presocialization, however important in the decision to become a physician, does not prepare medical students for the socialization they experience in medical school.

What are the most common socialization experiences of medical students? That medical students acquire technical knowledge and skills and learn to diagnose and treat illness goes without saying. In addition, medical school socializes students to accept the beliefs, norms, values, and attitudes associated with the medical profession. They learn, for example, to be dispassionate and unemotional about illness, suffering, and death; to never criticize another physician publicly; and to avoid admitting to patients their lack of knowledge. They also learn gradually to view themselves as doctors and to display a confident, professional appearance (Colombotos and Kirchner, 1986; Twaddle and Hessler, 1987). Medical school also transforms students from eager neophytes who wish to absorb all medical knowledge to more experienced individuals who realize that their human limitations require a selective learning strategy. Part of this aspect of their socialization is to understand what is considered safe not to learn. This determination is made by students assessing what they think the faculty wants them to learn, a social process involving meanings, symbols, and subjective interpretations (Becker et al., 1961).

How are nurses socialized? Although nurses come from the range of social classes, they tend to be white, middle-class females. Compared to the general female college student population, nursing students are more altruistic, benevolent, and generous; they are less interested in power, controlling others, and advancing themselves. They place emphasis on having the freedom to make their own decisions and on being treated empathetically. These characteristics are supported during the four years of undergraduate nursing education:

> *During their four years of nursing education, students are socialized to a nurse role that values individualized, direct patient care, rational knowledge, and innovation, and they are taught to think of themselves as autonomous, professional persons (Twaddle and Hessler, 1987:215).*

The problem is that upon graduation nurses are faced with a rigid, bureaucratized work environment controlled by physicians. Their work lives must be spent in a setting that contradicts the values they brought to their education and that their socialization in school reinforced. This creates the potential for many conflicts and difficulties. Nurses are constantly attempting to interpret their occupational world with symbols and meanings that do not fit their work situation.

How is the labeling perspective applied to illness? As discussed in Chapter 7 ("Deviance and Social Control"), the labels and stigmas applied to people affect the way in which others behave toward those people and, in turn, affect the behavior of the labeled people. "Sickness" and "illness" are social labels that can be used to stigmatize people.

The nature of health-related labels is based on culture. Because culture varies, definitions associated with health and illness are also diverse. Mark Zborowski (1952, 1969) found Jews and Italians more likely to talk openly about their pain, whereas "old Americans" and the Irish, with objectively the same levels of pain, tended not to complain. Similarly, Irving Zola (1966) reported that although the Irish did not see any relationship between their illnesses and their social relationships with others, the Italians presented problems with social relationships as a significant part of their illnesses.

Given this background, the existence of labels and stigma associated with certain forms of illness is not surprising. The elderly are erroneously labeled "disabled" (for example, forgetful) far more often than young adults because of the common assumption that illness is a normal aspect of aging. People with mental health problems are often stigmatized, which may lead to the adoption of a personal identity based on the properties of the stigma.

There is no more timely illustration of labeling and stigmatizing than the current view of AIDS in the United States. Because AIDS was initially found among homosexual men, who are already stigmatized in American society, people who acquire AIDS from such sources as blood transfusions or from heterosexual partners are also labeled and stigmatized (Altman, 1986). The taint of deviance and immorality is currently attached to AIDS patients. The stigma of AIDS is reflected in many responses. About one-third of Americans believe that AIDS is a punishment for the decline in moral standards (Gallup, 1992). The AIDS-related stigma persists despite the fact that the disease cannot be acquired through casual social interaction.

That the three theoretical perspectives provide very different insights into the health-care system is to be expected. In fact, in the absence of a single unifying sociological theory, the field is fortunate to have three powerful theoretical perspectives. Together, these theoretical approaches unmistakably underscore the social nature of sickness and health care.

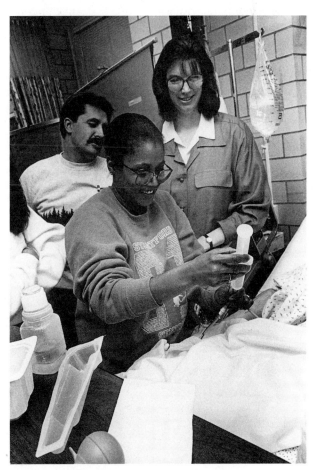

Nursing students place high value on making their own medically related decisions and on being treated with empathy. These expectations usually are not met in the inflexible organizational setting in which they work after their training is completed.

DOING RESEARCH

Howard S. Becker—The Socialization of Physicians

In 1961, Howard Becker and his colleagues published a book entitled *Boys in White*. Using the technique of participant observation, these sociologists examined the socialization of medical students from a symbolic interactionist perspective. The picture Becker painted of the student culture in medical school was not a glamorous one. Medical students are seen grappling with a sense of eroding idealism, diminished concern for patient well-being, and bewilderment regarding the shortcomings of the field of medicine.

Bright college graduates who enter medical school to be doctors face an initial rude awakening. The gratifying thoughts of helping others through the practice of a high-status, honored profession

are in the first year in medical school supplanted by the recognition of unforeseen barriers. Before one can be a physician, one must first learn to be a medical student. Learning how to act as a medical student blots out concern for how one will act as a doctor in the distant future.

Much of *Boys in White* concentrates on the ways in which students recognize and solve the pressing problems posed by their teachers and the work they assign. Medical students turn their eyes from the glorious distant mountain to the swamp that threatens to overwhelm them. Becker and his associates, then, focus on the social interaction between medical students and their instructors as they undergo a prolonged rite of passage into the profession of medicine.

These researchers began their research without firm, testable hypotheses and preset data-gathering research instruments. Because the researchers wished to "discover" how the medical school as an organization influenced the basic perspectives of physicians-

in-the-making—something Becker and his colleagues did not know beforehand—they simply could not design formal research instruments in advance. This is not to say, however, that they lacked any design for their research. The research team recognized the need for some organizing principles for their study, principles they found in symbolic interactionism. Becker and his colleagues, then, chose to view behavior within medical school as the outcome of a process in which the participants control and shape their conduct by taking into account (via role taking) the expectations of other participants.

The authors stated that their commitment to symbolic interactionism led them to use participant observation as their major research technique. Although the researchers did not pose as medical students, they did "participate" in the life of the medical student. The researchers' intense and prolonged periods of observation came as they attended classes and laboratories with students, listened to student conversations outside of class, accompanied stu-

Persons with AIDS are a current example of labeling and stigmatizing. Because AIDS appears among drug users such as these two young men, there is a tendency in the United States to stigmatize all people with this disease.

dents on rounds with attending physicians, ate with students, and took night calls with students.

In the end, Becker and his colleagues challenge the commonly shared view (on the part of both the public and medical school faculty) that medical students shed their idealism for cynicism as they progress through their training. Often cited evidence for this alleged shift from idealism to cynicism is the detached way medical students come to view their patients. Death and disease become depersonalized for medical students. Because medical students view death and disease unemotionally, as a medical responsibility rather than a tragedy, they are viewed by laypeople as cruel and heartless.

The researchers do not deny that medical students appear to become cynical and detached in medical matters. This appearance, however, is attributed not to the loss of an idealistic long-range perspective but to the need to survive the overwhelming demands of the training experience. A cadaver, for example, is not viewed as some-

body's daughter or mother but as a necessary means for gathering information for a forthcoming examination. Successful medical students ask not what they can do for a cadaver but what a cadaver can do for them. It is, in short, manifestations of the urgent and immediate need to get through school that give the appearance of cynicism to outsiders.

As graduation approaches, the researchers contend, another change in medical students occurs. As medical students lose their concern for the immediate situational demands of school, they begin to openly display the general concern for service to others they had upon entrance to medical school. The major difference is that this new idealism is more informed than the original idealism.

This conclusion regarding cynicism is understandable within the framework of symbolic interactionism. Individuals can simultaneously hold more than one set of sometimes conflicting values and rely on those values that seem to them relevant in a particular situa-

tion. As the situation changes, an individual's definition of the situation on which the individual constructs his or her behavior is altered. The dormant values of idealism replace the values of cynicism as the situation of the medical student changes. Corresponding alterations in behavior follow.

Critical feedback

1. Do you think the researchers' method of investigation is appropriate? Why or why not? What other methods could have been used?
2. Does the symbolic interactionist interpretation of medical school socialization ring true to you? Explain.
3. If you think they would be appropriate, explain how functionalism and conflict theory could be brought to bear on the study of the socialization of medical students. If you don't think one or both of these two theoretical perspectives are applicable, please explain why.

---

1. The _____ serves to remove people from active involvement in everyday routines, give them special protection and privileges, and set the stage for their return to their normal social roles.
2. If a sick person in American society does not attempt to recover, he or she will be labeled a _____.
3. According to the _____ perspective, physicians rose to dominance in the American health-care system by taking such actions as attempting to create a monopoly on medical services.
4. AIDS appears most frequently among homosexuals and intravenous drug users. This is an example of _____.
5. According to conflict theorists, affordable health care is not provided to the poor in the United States because of the desire for _____.
6. Nurses are caught in a conflict between their values and their work environment. Which of the following perspectives best helps us understand this situation?
   a. functionalism
   b. conflict theory
   c. symbolic interactionism
   d. exchange theory
7. America's negative reaction to AIDS victims illustrates the process of _____.

Answers: 1. sick role 2. deviant 3. conflict 4. epidemiology 5. profit 6. c 7. labeling

HEALTH IN THE UNITED STATES

Epidemiology, as noted earlier, refers to the study of the patterns of distribution of diseases within a population. Health and sickness, social epidemiologists have demonstrated, are strongly influenced by social factors, as reflected in several demographic characteristics and social-class level. Prominent among these demographic characteristics are income, age, gender, race, and ethnicity.

Age and Health

What is the relationship between illness and age?
Thanks in large part to improvements in medical treatment, working conditions, diet, living conditions, and sanitation, Americans are now living longer. The average life expectancy—average age at death—for a newborn baby in the United States today is about seventy-six years, an increase of almost 50 percent since the turn of the century, when life expectancy was forty-seven years (World Population Data Sheet, 1997). Still, the United States has a relatively high infant mortality rate among more developed countries. (See Table 15.1.)

Life expectancy may well reach eighty-five before long in industrialized countries. If so, more and more American citizens will be fairly healthy until near the end of their lives. Whether or not life expectancy in industrialized countries reaches eighty-five, the health of the aged is going to be better than it was for previous generations.

Causes of death vary with age. Infectious diseases, such as pneumonia and influenza, are the biggest killers of children under six years of age. Adolescents and young adults die more from suicide, murder, and accidents (notably automobile related). Chronic illnesses, such as heart diseases and strokes, are most likely to take the lives of the aged.

Gender and Health

How do health differences vary by gender? Across the world, females have a significantly greater life expectancy than males. In the United States, for example, the average life expectancy for females is 79 years, compared to 73 years for males (*World Population Data Sheet,* 1997). The popular conception that males die earlier than females from the stress of harder work is false. Starting at birth and continuing throughout life, males exhibit higher death rates than females. Except for diabetes, men die at a higher rate than women at all ages for all causes of death. Although women have the advantage of longevity over men, they are at a relative disadvantage in terms of physical sickness. American

TABLE 15.1	Infant Mortality Rates, Selected Countries: 1997
Country	**Deaths per 1,000 Live Births**
Japan	4.0
Singapore	4.0
Switzerland	5.0
France	5.0
Germany	5.1
Czechoslovakia	6.0
Canada	6.2
United Kingdom	6.2
United States	7.3
China	31.0
Nigeria	84.0

Source: *World Population Data Sheet 1997* (Washington, DC: Population Reference Bureau, 1997).

females suffer higher rates of acute illnesses, such as infectious diseases and respiratory ailments. Only in the case of injuries do males have higher rates of acute illness than females. Not surprisingly, women are more involved in the health-care system than men.

Why do women live longer but have more illness?
There is no consensus regarding the irony of men enjoying better health than women but dying earlier. Several explanations, however, are plausible. It may be that the female gender role creates greater stress than the masculine gender role. Because they have a greater interest in health, women may be more alert to illness (Verbrugge, 1986). The nature of the male gender role may also contribute to this situation. First, men do not seem to handle stress as well as women. The "macho" aspect of the male gender role may cause men to avoid asking for help from physicians and following medical advice when given. Males are also not as likely as females to relieve stress through supportive, intimate relationships (Nathanson, 1989; Verbrugge, 1989). As a result, stress may be more likely to lead to death for men than for women. Moreover, the male gender role encourages behavior injurious to one's health. Society encourages the engagement of males in dangerous activities such as drinking, smoking, driving fast, and fighting. Young males are, in fact, killed in automobile accidents at a rate three times higher than females and are twice as likely as females to die of cigarette- and alcohol-related diseases during middle age.

Race, Ethnicity, and Health

What health patterns are associated with race and ethnicity? With the exception of Asian Americans, racial and ethnic minorities in the United States have poorer

Smoking and Health

Until 1997, all representatives of the tobacco industry publicly denied that tobacco was either addictive or harmful to one's health. The CEOs of the American tobacco companies made this assertion in sworn testimony to a Congressional committee as late as 1996. The American Medical Association, as indicated in this CNN report, recognizes smoking as the primary avoidable cause of death in the United States. Although no other single commercial product is as connected with disease as tobacco, cigarette companies continue to target the most vulnerable segments of the population—the young, minorities, and the poor.

Although the American tobacco industry denies it, its critics claim cigarette makers target young people such as these. According to critics, Joe Camel, in his many guises, appears to be aimed at America's youth population.

1. Sociologists are concerned with the distribution of illnesses in the population. When they point out that the poor and minorities have a higher incidence of smoking-related diseases, they are engaged in the study of

   _____

2. Smoking, in recent years, has begun to carry a stigma. Stigmatization of smokers is best understood by which of the three major theoretical perspectives? Explain.

   _____

   _____

health than the general white population. The infant mortality rate for African American infants, for example, is more than double that of white infants. (See Figure 15.4.) Although African Americans have experienced an increase in life expectancy since the turn of the century, white males typically outlive African American males by over six years. White females have a life expectancy over five years longer than African American females. (See Figure 15.5.) African Americans are more likely than whites to die from thirteen of the fifteen major causes of death (O'Hare et al., 1990). Although Hispanic Americans have lower mortality rates from heart disease and cancer than either African Americans or non-Hispanic whites, Hispanic Americans have a shorter life expectancy, a higher infant mortality rate, and higher death rates from diabetes, pneumonia, tuberculosis, and influenza than non-Hispanic white Americans. Despite improvements in health since 1950, Native Americans still suffer disproportionately from diabetes, venereal disease, hepatitis, alcoholism, tuberculosis, malnutrition, and suicide.

Although some diseases such as sickle cell anemia, hypertension, and diabetes seem to be biologically based, most minority group illnesses are traced to the disadvantages associated with poverty. Minorities in the United States die younger and suffer more illness than the white majority, not primarily because of their racial and ethnic characteristics but because of low income. People in poverty get infectious diseases from living in unsanitary, overcrowded conditions; are exposed to extreme and frequent violence; cannot afford to maintain a nutritious diet; and do not receive proper medical treatment, either preventive or after the fact. Consequently, minorities tend to die more from infectious diseases and diseases caused by improper diet rather than from long-term illnesses such as heart disease and cancer.

Another way to demonstrate the connection between illness and poverty is to compare Asian Americans with other minorities. Asian Americans, who are the most well-off minority group in the United States, have the lowest age-adjusted mortality rate of all minority groups in the country. Asian Americans also have the lowest infant mortality rate and have significantly lower suicide and homicide rates.

Worse health among minorities may be due to factors in addition to low income. Even within the same income groups, African Americans have a higher death rate than whites. In fact, lower income whites have a significantly lower death rate from hypertension than the highest income African Americans. One prominent explanation is the physical and psychological stress experienced by minority group members who remain targets of prejudice and discrimination despite social and economic success. (See *Doing Research,* Chapter 9, "Inequalities of Race, Ethnicity, and Age.")

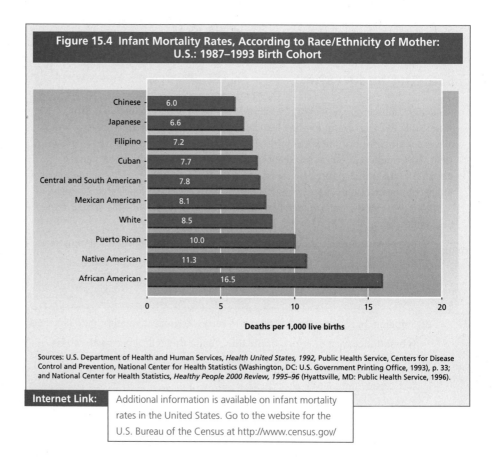

Figure 15.4 Infant Mortality Rates, According to Race/Ethnicity of Mother: U.S.: 1987–1993 Birth Cohort

Deaths per 1,000 live births

Sources: U.S. Department of Health and Human Services, *Health United States, 1992,* Public Health Service, Centers for Disease Control and Prevention, National Center for Health Statistics (Washington, DC: U.S. Government Printing Office, 1993), p. 33; and National Center for Health Statistics, *Healthy People 2000 Review, 1995–96* (Hyattsville, MD: Public Health Service, 1996).

Internet Link: Additional information is available on infant mortality rates in the United States. Go to the website for the U.S. Bureau of the Census at http://www.census.gov/

Social Class and Health

What are the effects of social class on the incidence of illness? For the poor, good health is just one more thing they have less of than the more affluent. Over 30 percent of the poor have no health insurance coverage (U.S. Bureau of the Census, 1996a). The health profile of the poor pretty well mirrors that of minority groups. On nearly all measures—infant mortality, life expectancy, infectious diseases, chronic health problems—members of the lower class are worse off than those in the middle and upper classes. The contributing factors to the poor health of the lower class are quite familiar by now—unsanitary and substandard living

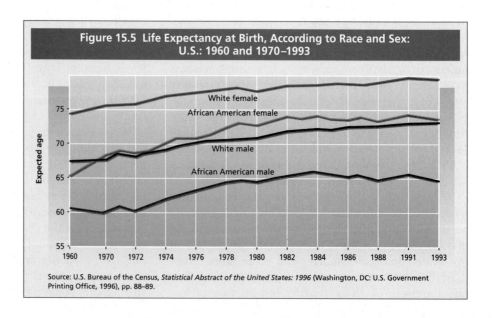

Figure 15.5 Life Expectancy at Birth, According to Race and Sex: U.S.: 1960 and 1970–1993

White female
African American female
White male
African American male

Expected age

Source: U.S. Bureau of the Census, *Statistical Abstract of the United States: 1996* (Washington, DC: U.S. Government Printing Office, 1996), pp. 88–89.

conditions, exposure to violence, inadequate nutrition, alcohol and other drug abuse, lack of medical care.

In itself, access to medical care is not the solution to poor health in the lower class. Equality of medical care alone would not equalize health levels between social classes. Providing quality medical care only addresses the results of poor environmental conditions. Adequate medical care will help only in conjunction with lifestyle changes such as better housing and eating habits (Marmot, Shipley, and Rose, 1984; Cockerham, 1995).

Social-class level has been consistently found to be related to mental illness. A series of early, highly respected studies on social class and mental health established the existence of the highest incidence of mental illness among members of the lower class (Faris and Dunham, 1939; Hollingshead and Redlich, 1958; Srole et al., 1962). Although the lower class has a higher incidence of mental problems, including personality disorders and schizophrenia, the upper classes tend to have more anxiety and mood problems (Cockerham, 1985, 1996; Link, Dohrenwend, and Skodol, 1986).

Why do members of the lower class have a higher incidence of mental illness? Three basic explanations have been offered for the higher incidence of mental illness among the poor (Cockerham, 1995). The *genetic* explanation posits a biological tendency to mental illness among members of the lower social class. Studies of twins indicate that genetics by itself is an insufficient explanation. Social factors also contribute. According to the *social stress* explanation, members of the lower class are more susceptible to mental disorders because of the extra stress they experience from trying to cope with their deprivation. Evidence on this explanation does not speak with a single voice. Other scholars, supporters of the *social selection* explanation, account for the higher rates of mental disorders in the lower class by a mobility process—persons in other social classes with mental disorders tend to end up in the lower class, and mentally healthy persons born in the lower class move up to higher social classes. Although the evidence is inconclusive, it seems likely that social class contributes to poor mental health more than mental illness affects social-class level (Link, Dohrenwend, and Skodol, 1986).

Most social epidemiologists, who think research on the link between social class and mental illness must continue, do not attribute primacy to any single explanation. Most likely, genetic, social selection, and stress factors each contribute to the higher incidence of mental disorders among those at the bottom of the class structure.

THE CHANGING HEALTH-CARE SYSTEM

The health-care system, like other American institutions, is undergoing rapid change. First we'll cover several important areas of change, starting with what has been called the *medicalization* of society. Following this is a discussion of the new medical-industrial complex and organizations delivering health care. Finally, health-care innovations are reviewed.

The Medicalization of Society

What is the process of medical diagnosis today? Within contemporary society, the formal power to define people as ill rests with physicians. On the basis of physicians' diagnoses, students may (or may not) be excused from school and employees may (or may not) be allowed to miss work. Social critic Ivan Illich (1995) contends that such power is potentially dangerous to society. We are, he believes, experiencing the process of **medicalization** in which problems that were once considered matters of individual responsibility and choice are now considered a part of the domain of medicine and physicians:

> In a medicalized society the influence of physicians extends not only to the purse and the medicine chest but also to the categories to which people are assigned. Medical bureaucrats subdivide people into those who may drive a car, those who may stay away from work, those who must be locked up, those who may become soldiers, those who may cross borders, cook, or practice prostitution, those who may not run for the vice-presidency of the United States, those who are dead, those who are competent to commit a crime, and those who are liable to commit one (Illich, 1995:76–77).

1. Although American women have a longer life expectancy than men, women are more involved in the health-care system as patients. *T or F?*
2. With the exception of _____, racial and ethnic minorities in the United States have poorer health than the general white population.
3. Although the upper class has a higher incidence of serious mental problems, the lower class tends to have more anxiety and mood problems. *T or F?*
4. Higher rates of mental illness among the lower class are pretty clearly due to the genetic factor alone. *T or F?*

Answers: 1. T 2. Asian Americans 3. F 4. F

Although the effects of medicalization may or may not be as pervasive as Illich contends, many aspects of life—mental illness, abuse of alcohol and other drugs, hyperactivity and other behavior disorders in children, homosexuality, and some criminal behavior, to name a few—that once were considered evidence of character or spiritual flaws are now interpreted in medical terms. Such interpretations can lead to more humane treatment, but the nature of the treatment may also serve to profit certain groups. Pharmaceutical companies, for example, gain opportunities to sell drugs for everyday anxieties and other problems that individuals may once have handled on their own. Similarly, some categories of physicians may profit from opportunities to diagnose and treat people who would not otherwise be considered ill. The question of whether the treatments that serve the interests of corporations and physicians are necessarily beneficial for those who are now defined as ill is a matter of continuing concern.

The New Medical-Industrial Complex

Paul Starr (1982) writes of the rise of the *medical-industrial complex,* by which he means the network of physicians, medical schools, hospitals, medical equipment suppliers, pharmaceutical companies, health insurance carriers, and other for-profit organizations. This traditional coalition, with its shared interests, was instrumental in the rise to power of the American medical profession. A new force, however, has arrived on the scene. A "new" medical-industrial complex has emerged with the entry of the large health-care corporation. Entry of the for-profit corporation in the health-care system promises another major transformation for the medical profession that is due to the trend toward a higher level of integrated control of the health-care system. With the medical corporation has come for-profit hospitals and the increasing employment of doctors in group or large corporate settings. Both the AMA and physicians are, in fact, abandoning their traditional commitment to the private (solo) practice of medicine (Twaddle and Hessler, 1987).

Although doctors probably will not become "proletarianized," corporate employment will involve a significant loss of autonomy and independence. Physicians' pace of work and practice routines will be more regulated. Choice of retirement time will no longer be the sole prerogative of the physicians. Doctors will increasingly be measured against standardized performance criteria. Mistakes will be monitored carefully to reduce the risk of malpractice losses.

Will the medical profession lose power as a result of the corporatization of medicine? Without question, the new health-care corporate giants will wield considerable power, and the medical profession will have to share its traditionally held power with them. In that sense, the profession of medicine will not continue to be the sole dominating factor in determining and controlling the nature of the health-care system to which it has become accustomed. But as medical sociologist David Mechanic writes:

> . . . it would be naive to assume that the enormous power of the professional, painstakingly and skillfully cultivated over this century, will be quickly or substantially reversed despite changing conditions. . . . Physicians will continue to affect substantially the patterns of medical care, appropriate standards, and the costs associated with varying episodes of illness (Mechanic, 1986: vii).

Organizations Delivering Health Care

Hospitals are organizations that provide specialized medical services to a variety of patients, including some who continue to live in their homes and others who will spend a prolonged period in the facility. Hospitals range from small facilities specializing in patients who need short-term, uncomplicated care to large medical centers that provide long-term care often involving complex technology. Medical centers and some large hospitals typically combine patient care with medical research and training.

How important are hospitals? The United States has become highly dependent on hospitals. Trends in childbirth illustrate this. In 1940 over one-half of American babies were born at home; today, despite some resurgence in interest in home delivery, almost all babies are born in hospitals (Hoff and Schneidermann, 1985).

What are the benefits and negative consequences of hospitals? The increased reliance of people on hospitals and their advanced technology has resulted in a number of tangible benefits. Childbirth has become safer for mothers and infants who experience complications. The ability to treat such diseases as cancer and heart disease has increased dramatically. Other diseases, such as tuberculosis, have declined greatly in their frequency.

Other, more problematic consequences have also been experienced. Because many hospitals involve combinations of highly specialized professionals and equipment, they are contributing to an increasingly specialized and incomplete view of health. Although acknowledging that health care has advanced considerably since physicians had to make diagnoses with the benefit of little information besides what patients told them, medical historian Stanley Joel Reiser sees dangers in the increasing reliance on sophisticated technology:

Machines inexorably direct the attention of both doctor and patient to the measurable aspects of illness, but away from the "human" factors that are at least equally important. Insofar as technological evidence occupies the time and commands the chief allegiance of both doctor and patient, it diminishes the possibility that a close personal relationship will develop between the two. So, without realizing what has happened, the physician in the last two centuries has gradually relinquished...unsatisfactory attachment to subjective evidence—what the patient says—only to substitute a devotion to technological evidence—what the machine says (Reiser, 1978:229–230).

The physician's view of the patient, then, is becoming increasingly incomplete. Physicians, according to Reiser, are likely to become estranged from their patients and their own judgment as a result of their increasing reliance on technological devices.

The lack of a clear-cut relationship between medical practice in hospitals and health has been demonstrated by James James (1979), who found that during a 1976 work slowdown by physicians in Los Angeles, hospital incomes dropped $17.5 million but death rates actually fell below the usual rate. James speculates that the reduced number of elective operations may have lowered the death rate during the period of the slowdown.

How are hospitals contributing to rising health-care costs?   Health-care costs have increased dramatically in recent years. Hospital expenses account for a large proportion of the increase. When patients check into a hospital, they are paying for more than a room and a bed. A large staff of professional and nonprofessional employees and increasingly expensive equipment are underwritten by patients and their insurance companies. Patients are not in a position to form an organized group to fight rising costs (and individual patients who may be concerned about rising health costs in general insist on the very best facilities for themselves and members of their families). Moreover, physicians generally lack expertise in the efficient planning and use of health-care resources (Jonas, 1978). As a result, efforts to contain medical costs are poorly organized.

Are investor-owned hospitals a solution to rising hospital costs?   There is a trend toward for-profit hospitals. On the assumption that corporate efficiency will aid in medical cost containment, more and more local hospitals are being absorbed by huge, privately owned hospital chains, such as the Hospital Corporation of American and Humana. Research, however, runs counter to the early optimism of supporters of this "privatization" movement. An extensive study of investor-owned and nonprofit community hospitals revealed that costs per hospital visit were nearly 20 percent higher in the for-

profit hospitals and that overhead expenses per admission were 4 percent lower in community hospitals (Pattison and Katz, 1983). Two additional studies also show that costs are significantly higher for corporate hospitals than for the nonprofit hospitals (State of Florida, 1981–82, 1982–83). It appears, then, that although chain hospitals are a good investment because of higher costs to patients, they are less cost-effective than community not-for-profit hospitals.

Health-Care Delivery Innovations

Traditionally, Americans have relied on private physicians for their health care. In cases of serious illness, patients were sent to hospitals, which often encouraged physicians to send large numbers of patients to increase hospital revenue. The emphasis throughout this process has been on the treatment of existing diseases rather than on the prevention of disease. The treatment itself has been somewhat narrow. Medical treatment has been emphasized, whereas efforts to deal with patients' emotional needs have been far less systematic. Changes are occurring, however. Health care has become more expensive both for consumers, who must pay for treatment, and for providers, who must pay for malpractice insurance, new equipment, and the services of many specialists and technicians. As life expectancy increases, problems of an emotional nature are becoming as serious as strictly medical problems. Several changes in health-care delivery organizations have been made in efforts to deal with such problems. Health maintenance organizations, physician-hospital organizations, and hospices are representative of the kinds of innovations being made.

What are health maintenance organizations?   **Health maintenance organizations** (HMOs) are health plans in which members pay set fees on a regular basis and in return receive all necessary health care at little or no additional cost. Most HMOs are group practices with physicians representing a variety of specializations. From the point of view of physicians, HMOs provide access to a full range of expertise while saving individual physicians from the need to worry about the mechanics of running a business. From the point of view of medical consumers, HMOs "improve the market" by providing an alternative to health plans in which benefits are received only after the onset of an illness or accident. Although HMOs provide treatment, their major orientation is the prevention of disease through early detection and preventive medicine (Beck, 1993).

What are the pluses and minuses of HMOs?   Because the plan is based on prepayment, members do not avoid necessary health care simply because of a fear

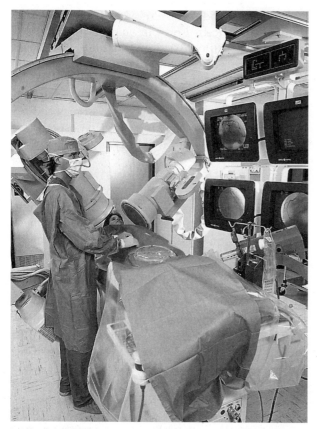

Much of the skyrocketing cost of medical care in the United States is due to the changing nature of hospitals. This operating room scene reflects the sophisticated technology that most hospital administrators and physicians appear to desire. Once obtained, this costly technology is underwritten by patients, insurance companies, and the federal government.

that they will be unable to afford it. Some critics of HMOs believed that the availability of extensive treatment at little or no additional cost would lead to the HMOs' downfall because of members who use their services excessively. It appears, however, that this has not been the case. As Emily Mumford notes:

Many people do not enjoy going to the doctor; some even very rich people avoid doctors. The record of savings in HMO's suggests that most people do not use health services excessively when they are free (Mumford, 1983:384).

If HMOs, with their preventive emphasis, can reduce chronic illness and unhealthy personal habits, such health plans can be expected to reduce the overall cost of medicine. Whether or not HMOs can do so is a matter of serious debate. Aligned on one side are those who see health care as a professional relationship between doctor and patient (Anders, 1996; Relman, 1997). On the other side are supporters of HMOs who see health care as a service industry in which "consumers" rather than "patients" select products on the basis of market-

driven criteria such as quality, convenience, and cost (Herzlinger, 1997). Evidence is not sufficient to support either side in this important controversy.

How prevalent are HMOs? Despite their origins during the 1920s, HMOs are only beginning to operate on a sufficiently large scale for their effects on the quality and cost of health care to be understood completely. In 1995, HMOs had an enrollment of 103 million members. This represents nearly a tenfold increase since 1976 (Spragins, 1996).

What are physician-hospital organizations? Physician-hospital organizations (PHOs) are joint ventures between hospitals and organized groups of physicians to reach predetermined business objectives. Typically, activity is centered on coordinating managed care contracting. They provide the same services as most managed care organizations: claims processing, quality assurance, receipt and distribution of fees, and recruitment and credentialing of health-care personnel. PHOs have been formed mostly in areas where community hospitals tend not to have access to organized medical groups with which they can connect (Friend, 1996).

HMOs and PHOs speak to the need to practice preventive medicine while controlling medical expenses. The medical system is also confronted with the problem of how to provide appropriate care for terminally ill patients. An important response has been the hospice.

What are hospices? Although HMOs speak to the need to practice preventive medicine while controlling medical expenses, the medical system is also confronted with the problem of how to provide appropriate care for the terminally ill. Advances in medical procedures and technology have resulted in the ability to prolong life even in the case of serious illness, but these advances do not necessarily improve the situation faced by patients who are experiencing physical pain, psychological trauma, and the knowledge that their disease will eventually result in their death. **Hospices**—organizations designed to provide support for the dying and their families—are a recent response to this problem.

The hospice movement has its roots in England, but during the past two decades it has gained in importance in the United States. The physical form of hospices varies, with some located in hospitals, some in separate buildings, and some providing services to patients in their homes. Regardless of their location, however, they make the services of physicians, nurses, counseling psychologists, and social workers available to patients and their families.

Whereas hospitals are oriented toward curing diseases, hospices are oriented toward making patients with apparently incurable diseases as comfortable as possible.

These differing orientations result in very different strategies. For example, pain killers may be administered much more freely in a hospice setting. Hospices can do nothing to reverse an unfavorable medical diagnosis. They can, however, assist patients in maintaining their dignity during their final months of life.

Can patients decide when to die?   Hospices provide help for patients whose deaths appear inevitable, but they do not speak to the problems of patients whose lives can be maintained for an extended period of time by advanced technology and treatment but who have no real hope for recovery. Ivan Illich provides a clear illustration of the problem.

> I know of a woman who tried, unsuccessfully, to kill herself. She was brought to the hospital in a coma, with a bullet lodged in her spine. Using heroic measures the surgeon kept her alive, and he considers her case a success: she lives, but she is totally paralyzed; he no longer has to worry about her ever attempting suicide again (Illich, 1995:198).

Because the technology exists for sustaining life even when patients are comatose and improvement is unlikely, serious questions exist as to whether treatment can be withheld under such circumstances. Many states are now dealing with such difficult questions as when a patient can legally be considered dead and whether patients, their families, and their physicians have the right to withhold or to discontinue treatment for patients with no apparent prospect for recovery. Opinions range from those who believe that a duty exists to maintain human life under all circumstances to a contention by Dr. Christiaan Barnard (the pioneer of heart transplant operations), who has been quoted in the press as saying that unplugging machines and ceasing treatment is insufficient for some patients. When suffering is severe and recovery impossible, Barnard says, physicians should practice "active euthanasia," killing patients rather than allowing them to continue enduring pointless suffering (Mumford, 1983). Dr. Jack Kevorkian, who accepts credit for helping ninety to one

hundred individuals commit "assisted suicide," was the first to announce publicly that he was helping terminally ill patients who seek an end to their lives. Thus far, the courts have upheld his right to do this, although his medical license has been revoked by the state of Michigan (Ilka, 1998).

HEALTH-CARE REFORM IN THE UNITED STATES

As you have already learned in the *Sociological Imagination* opening this chapter, most modern societies provide health care as a free or inexpensive service through their governments. The United Kingdom, although it has a system of private medical care, promotes equality of medical care through its National Health Service, which provides the vast majority of its services without charge at the point of delivery. Canada also has a public health-care delivery system committed to the provision of equal access. As in the United Kingdom, health care in Canada is financed through the government. (See Table 15.2.)

Of the highly developed countries, the United States is the only one without national health insurance for all its citizens. Approximately forty-one million Americans are without health insurance. Even with the enactment of Medicaid, about half of America's poor are without medical coverage. Moreover, except for the Veterans Administration hospital system, the military, and the Indian Health Service, the United States is without any direct mechanism for the delivery of health-care services.

In the past, some congressional attempts have been made to enact a national health insurance program. In addition, some senators were proposing legislation in 1989 that would have forced nearly all American businesses to provide minimum health insurance benefits to most workers. This is important because health insurance in the United States is a privilege largely enjoyed, with the exception of the elderly, by those

FEEDBACK

1. The term used to describe the process through which problems that were once considered matters of individual responsibility and choice become a part of the domain of medicine and physicians is called _____.
2. A "new" medical-industrial complex has been created because of the entry of the _____ into the health-care system.
3. Because of the increasing movement from solo practice to group and corporate medicine, doctors will eventually become "proletarianized." *T or F?*
4. Investor-owned hospitals are less efficient than not-for-profit community hospitals. *T or F?*
5. Health plans in which workers pay set fees and receive all necessary health care at no additional cost are called _____.
6. Organizations providing support for the dying and their families are called _____.

Answers: 1. medicalization 2. corporation 3. F 4. T 5. health maintenance organizations 6. hospices

TABLE 15.2	Comparison of U.S. and Canada's Health-Care Systems		
		Canada	**United States**
Population		27.3 million (1991)	263 million (1995)
Provinces/States		10 plus two territories	50 plus D.C.
Total expenditures for health care		$55 billion (1990)	$949 billion (1994)
Health costs as a percentage of GDP		9.0% (1991)	13.7% (1994)
Per capita cost		$1,683 (1989)	$3,510 (1994)
Coverage		Universal	41 million uninsured (1994)
Access		Patients pick their doctor; government pays the bills	Patients pick their doctor; payment by patient, insurer, or government
Physicians who treat patients per 100,000 population		1 per 449 (1989)	1 per 435 (1994)
Hospital beds per capita		1 for 148 people (1988)	1 for 233 people (1994)
Percentage of each U.S. dollar spent on administrative costs		11% (1987)	26% (1994)
Malpractice insurance premiums, annual mean for all physicians		$2,446 (1989)	$15,500 (1989)
Rate of coronary bypass surgery per 100,000 residents		41.8 (1989)	149.3 (1989)
Cesarean section rate per 100 deliveries		19.5 (1989)	22.8 (1993)
Infant mortality per 1,000 live births		7.3 (1990)	8.4 (1993)
Life expectancy in years			
Males		74 (1989)	72 (1993)
Females		81 (1989)	79 (1993)

Sources: *World Almanac 1994;* National Center for Health Statistics; General Accounting Office; Congressional Budget Office; American Medical Association; Canadian Medical Association; Canadian Medical Protective Association; National Commission to Prevent Infant Mortality; Statistics Canada; Health Care Financing Administration; U.S. Department of Health and Human Services, *Health United States 1992,* Public Health Service, Centers for Disease Control and Prevention, National Center for Health Statistics (Washington, DC: U.S. Government Printing Office, 1993); U.S. Bureau of the Census, *Statistical Abstract of the United States 1996* (Washington, DC: U.S. Government Printing Office, 1996).

who have full-time jobs with well-established firms. Those with the power—including business interests and the medical establishment—have thus far successfully frustrated these efforts. By invoking the threats of socialized medicine and profit reduction, contend conflict proponents, elites in the United States have used capitalist ideology to prevent equality of access to medical care across the population. In fact, say some conflict theorists, some form of socialized medicine will be required if equality of health care is to become a reality in the United States.

The Need for Health-Care Reform

Why is there a need for health-care reform? One motivation for health-care reform is economic. America annually devotes a larger share of its GDP on health care than any other nation. (Refer to Figure 15.2.) In 1992, health care accounted for 14 percent of America's gross domestic product or $833 billion. This amounts to over $3,000 annually for every man, woman, and child. If current trends continue, health-care spending in the United States is expected to reach 18 percent of GDP by 2000 (*World Development Report 1993,* 1993). And, as you saw at the beginning of this chapter, the

costs of health care are skyrocketing. The escalating costs of medical care also negatively affect the overall health of the economy. Economic resources of government—in excess of $800 billion annually—desperately needed in other areas of public life are being siphoned off by a health-care system out of control (Marmor and Barr, 1992a).

A second reason for reform is the substantial proportion of Americans who do not have access to medical care except on an emergency basis. (See Figure 15.6.) A large segment of the forty-one million uninsured citizens are the most disadvantaged on the bottom of the stratification structure. Children are also vulnerable. This vulnerability extends beyond the poor. Increasingly, employers are either requiring employees to pay high premiums and deductibles for the coverage of their children and spouses, or their company health plans simply do not cover dependents. Almost one-fourth of uninsured children live with parents who are insured, a proportion expected to rise to one-half by 2000. Nearly ten million children in the United States are without health insurance (U.S. Bureau of the Census, 1996a).

A third reason for health reform relates to quality of life. One obvious aspect of a lowered quality of life is

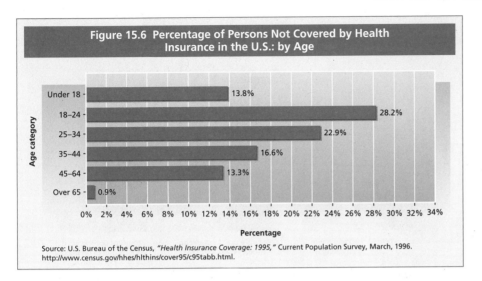

Figure 15.6 Percentage of Persons Not Covered by Health Insurance in the U.S.: by Age

Age category	Percentage
Under 18	13.8%
18–24	28.2%
25–34	22.9%
35–44	16.6%
45–64	13.3%
Over 65	0.9%

Source: U.S. Bureau of the Census, *"Health Insurance Coverage: 1995,"* Current Population Survey, March, 1996. http://www.census.gov/hhes/hlthins/cover95/c95tabb.html.

the lack of health care or inferior health care for millions of Americans. Perhaps less obvious is the fear and desperation the uninsured feel who know that an illness or accident experienced by themselves or a member of their family could either go untreated or absorb whatever economic resources they have. Moreover, even the insured live with the prospect of losing coverage should they lose their jobs. As you saw in Chapter 13 ("Political and Economic Institutions"), more Americans, blue and white collar, are experiencing job loss and lower compensation even if employed.

Health-Care Reform Options

What are the basic health-care reform options? "Modified competitiveness" is based on market principles such as consumer cost sharing. This approach, however, recognizes universal health coverage as a precondition to health-care reform.

A second option, **managed competition,** is related to the first option, but is a combination of free-market forces and government regulation. Under this option, health care would be structured around several large health plans modeled on health maintenance organizations (HMOs). To increase their competitive edge, employers and other consumers would form large purchasing networks, with health-care providers bidding on their business. Individuals would be free to select a plan, and the costs would be paid either through an income or payroll tax. The idea behind this option is to create competition and to reward those health-care providers that have superior performance in cost, quality, and patient satisfaction (*World Development Report 1993,* 1993).

A third option is the **single payer** approach like the Canadian model, under which medical services are financed governmentally. Canadians choose their doctors and hospitals and bill the government according to a standardized fee structure. Contrary to popular opinion in the United States, a public opinion comparison of the U.S. and Canadian health-care systems shows that the Canadian system is perceived to be more successful in providing access to needed services, engenders higher levels of consumer satisfaction with medical services, and is regarded much more favorably than the U.S. system (Taylor, 1994b, see Figure 15.7).

A fourth option—based primarily on the German model—uses a **play or pay** mechanism through which universal coverage is provided by requiring employers either to offer employees health coverage ("play") or to pay into a public fund for covering the uninsured. Access to medical care in Germany, considered among the best in the world, is guaranteed for life (Atkinson, 1994).

A significant proportion of the population in the United States is medically uninsured. Of the 41 million uninsured, a large share are the poorest Americans. Where are the uninsured in poverty to obtain the cash required of them when they need health care?

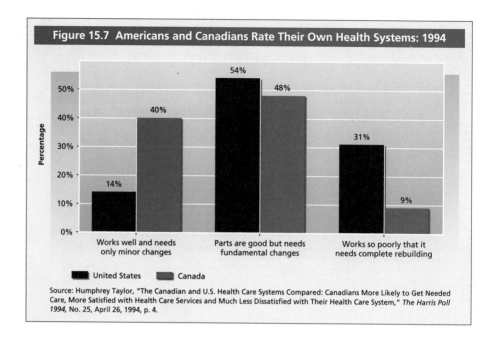

Figure 15.7 Americans and Canadians Rate Their Own Health Systems: 1994

Source: Humphrey Taylor, "The Canadian and U.S. Health Care Systems Compared: Canadians More Likely to Get Needed Care, More Satisfied with Health Care Services and Much Less Dissatisfied with Their Health Care System," *The Harris Poll 1994*, No. 25, April 26, 1994, p. 4.

FEEDBACK

1. Approximately _____ million Americans are without health insurance.
2. Which of the following highly industrialized countries does not have national health-care coverage?
   a. Germany   c. United States
   b. France   d. Canada
3. The major reason the United States does not have a national health-care program is because its citizens want different results from health-care programs than citizens of other developed societies. *T or F?*
4. Of those Americans without health-care protection, the poor and children comprise the majority. *T or F?*
5. The approach to health-care delivery under which medical services are financed governmentally is called
   a. managed competition.   c. all payer.
   b. single payer.   d. play or pay.

Answers: 1. 41 2. c 3. F 4. T 5. b

SUMMARY

1. All societies must develop ways of coping with the issues of health care and dying. Modern societies have evolved large and complex health-care systems. In the United States, for example, 13.7 percent of the gross domestic product is accounted for by health-care expenditures.

2. The major components of the health-care system are physicians, nurses, patients, and hospitals. Physicians control the health-care system in the United States and are highly rewarded in terms of power, prestige, and money. Partly because nursing began as a low-status, female-dominated occupation, it still suffers from low power and pay.

3. Functionalists advocate the concept of the sick role, which carries with it prescriptions for the treatment and behavior of patients. Patients are expected to withdraw from or reduce their involvement in other roles, attempt to recover, and seek the advice of physicians and others

providing care. Failure to cooperate in efforts to recover may result in denial of the usual privileges associated with the sick role.

4. Functionalists and conflict theorists disagree about the medical profession's rise to power in the United States. One functionalist explanation contends that the medical profession has become preeminent because it is consistent with what Americans feel they need. The conflict interpretation attributes the power of the medical profession to such efforts as the attempt to drive out competing approaches to health care. Conflict theorists also see the high rewards of physicians, the professional and economic problems of nurses, and the inequality of health care as products of power differentials.

5. Symbolic interactionism is useful for understanding the socialization of physicians and nurses. This perspective also deals with the labeling process in illness. It explains,

for example, that a stigma is attached to certain illnesses.

6. The distribution of diseases within a population is related to a variety of social factors. Americans have an increasing life expectancy. Women in the United States live longer than men on the average but are more involved in the health-care system as patients. Generally, health is poorest among members of racial and ethnic minorities. Social-class level is positively correlated with both physical and mental illness.

7. The health-care system in the United States is undergoing significant change. Some critics fear that the medicalization of society benefits corporations and physicians too much. This is said to be occurring as health-related matters once thought to be the responsibility of individuals are becoming a part of the province of medicine and physicians. Others point to the entry of the corporation into the health-care system, a movement that threatens the autonomy of physicians and the power of the medical establishment.

8. Americans have become very dependent on hospitals. This dependence has both positive and negative consequences. Childbirth has become safer, and many diseases are more easily treated and controlled. On the negative side, hospitals, through the promotion of specialization, contribute to an incomplete view of health. Hospitals are also contributing to rising health-care costs. Investor-owned hospitals do not appear to be a solution to this problem. Although profitable to investors, for-profit hospitals are more costly to run and deliver more expensive health care.

9. As health care becomes more extensive and expensive, the trend is away from care by private physicians in hospitals toward alternative means of delivering health care. Two alternatives are health maintenance organizations, which are health plans with predetermined fees paid by patients on a regular basis, and hospices, which provide support for dying people and their families.

10. The public and politicians agree that the United States, the only major industrial power in the world without national health insurance for everyone, must undergo major health-care reform. There are several distinct options for such reform, and the political, social, and economic struggles surrounding the issue cloud the path.

LEARNING OBJECTIVES REVIEW

After careful study of this chapter, you will be able to:

- Define the concept of a health-care system and identify its major components.
- Apply functionalism, conflict theory, and symbolic interactionism to the health-care system in the United States.
- Discuss the distribution of diseases in the United States according to age, gender, race and ethnicity, and social class.

- Describe what is meant by looking at the health-care system as a medical-industrial complex.
- Describe the major health-care delivery innovations in America today.
- Discuss health-care reform in the United States today.

CONCEPT REVIEW

Match the following concepts with the definitions listed below them.

____ a. medicalization
____ b. hospitals

____ c. health maintenance organizations (HMOs)

____ d. epidemiology
____ e. sick role

1. Health plans in which members pay set fees on a regular basis and in return receive all necessary health care at little or no additional cost.
2. Organizations that provide specialized medical services to a variety of patients, including some who continue to live in their homes and others who will spend a prolonged period in the facility.
3. A process in which problems that were once considered matters of individual responsibility and choice are now considered a part of the domain of medicine and physicians.
4. A social definition serving to remove people from active involvement in everyday routines, give them special protection and privileges, and set the stage for their return to their normal social roles.
5. The study of the patterns of distribution of diseases within a population.

CRITICAL THINKING QUESTIONS

1. Evaluate the idea that under certain circumstances sickness is seen as a form of deviant behavior. Draw on the concept of the sick role in forming your answer. _____

_____

_____

2. The medical profession in the United States enjoys a lofty position on the stratification structure. Compare and contrast the functionalist and conflict explanations for this fact. _____

_____

_____

3. Members of the lower class in American society have a higher incidence of mental illness than is documented in other social classes. Evaluate the various explanations for this phenomenon. _____

_____

_____

4. Do you think the health-care system in the United States will be reformed in the next ten to fifteen years? Why or why not? If it is reformed, describe the basic direction you think it will take. _____

_____

_____

MULTIPLE CHOICE QUESTIONS

1. **The size of the health-care system in America is**
   a. decreasing slightly.
   b. leveling off.
   c. steady as it has been for the last two decades.
   d. increasing.
   e. decreasing substantially.

2. **Although _____ constitute only about 10 percent of health-care workers in the United States, they establish the framework within which everyone else works.**
   a. hospice personnel
   b. physicians
   c. registered nurses
   d. practical nurses
   e. medical consultants

3. **Talcott Parsons uses the term _____ to describe the complex set of rights and responsibilities associated with biological illness.**
   a. illness syndrome
   b. sick role
   c. doctor-patient nexus
   d. instrumental role nexus
   e. mutual contact

4. **According to Marxist scholar Vicente Navarro, the success of the medical profession in the United States is due to**
   a. clever lobbying by the American Medical Association.
   b. the high quality of American medical schools.
   c. the dedication of nurses and volunteers.
   d. its connection with dominant societal power blocs.
   e. its roots in religion.

5. **The concern of conflict advocates for health-care inequality begins with _____, the study of the patterns of disease distribution within a population.**
   a. demography
   b. epidemiology
   c. immunology
   d. medical geography
   e. structural medicine

6. **Which of the following statements would symbolic interactionists be *most* likely to endorse?**
   a. Medical schools socialize students to accept the beliefs, norms, values, and attitudes associated with the medical profession.
   b. Nurses are constantly attempting to interpret their occupational world with symbols and meanings that fit their work situation.
   c. "Sickness" and "illness" are now seldom used social labels to stigmatize people in the United States.
   d. The experience of pain is the one area of illness that is not affected by cultural differences.
   e. Because it is now known that AIDS cannot be acquired through casual social interaction, this disease is on the way to losing its negative symbolic baggage.

7. Epidemiologists have demonstrated that health and illness are strongly influenced by _____ factors.
   a. social psychological
   b. demographic and social-class
   c. anthropological
   d. genetic
   e. spiritual

8. Which of the following statements is *incorrect*?
   a. The average life expectancy of Americans has increased almost 50 percent since 1900.
   b. Around the world, females have a significantly greater life expectancy than males.
   c. Asian Americans have the lowest age-adjusted mortality rate of all minorities in the United States.
   d. Social-class level has been consistently found to be related to mental illness.
   e. Among highly industrialized countries, the United States has a relatively low infant mortality rate.

9. With the exception of _____, racial and ethnic minorities in the United States have poorer health than the general white population.
   a. Native Americans
   b. Dominicans
   c. Asian Americans
   d. Aleuts
   e. Puerto Ricans

10. Minority groups in the United States die younger and suffer more illness than the white majority. Epidemiologists attribute this to
    a. low income.
    b. genetic diseases.
    c. racial and ethnic characteristics.
    d. poor health habits.
    e. lack of knowledge about preventative health and adequate nutrition.

11. The entry of large health-care organizations into America's health-care system has led to the creation of
    a. an atmosphere inhospitable to health-care reform.
    b. medicalization.
    c. a medical-industrial complex.
    d. the professionalization of nurses.
    e. physicians' assistants.

12. The current major threat to the power of the medical profession lies in
    a. the new medical-industrial complex.
    b. medicalization.
    c. demedicalization.
    d. hospices.
    e. the failure of the U.S. Congress to pass a health-care reform bill.

13. _____ are health plans in which members pay set fees on a regular basis and in return receive all necessary health care at little or no cost.
    a. Cafeteria plans
    b. Payroll deduction plans
    c. Fee-for-service plans
    d. Hospices
    e. Health maintenance organizations

14. Which of the following statements regarding health-care reform in the United States is *accurate*?
    a. Few modern societies provide health care as a free or inexpensive service through their governments.
    b. The United States is one of the few highly developed countries with national health insurance.
    c. Only the poorest Americans are without access to medical care on a nonemergency basis.
    d. Only a little over one million children in the United States are uncovered by health insurance.
    e. A major reason for health-care reform in the United States is that a substantial proportion of Americans do not have access to medical care except on an emergency basis.

15. Which of the following was *not* discussed as a basic health-care reform option for the United States?
    a. modified competitiveness
    b. managed competition
    c. all payer
    d. single payer
    e. play or pay

FEEDBACK REVIEW

True-False

1. Although American women have a longer life expectancy than men, women are more involved in the health-care system as patients. *T or F?*
2. Investor-owned hospitals are less efficient than not-for-profit community hospitals. *T or F?*
3. The major reason the United States does not have a national health-care program is because its citizens want different results from health-care programs than citizens of other developed societies. *T or F?*

Fill in the Blank

4. The _____ embraces the professional services, organizations, training academies, and technological resources that are committed to the treatment, management, and prevention of disease.
5. If a sick person in American society does not attempt to recover, he or she will be labeled a _____.
6. According to conflict theorists, affordable health care is not provided to the poor in the United States because of the desire for _____.
7. A "new" medical-industrial complex has been created because of the entry of the _____ into the health-care system.

Multiple Choice

8. Health-care expenditures now account for about _____ percent of the American GDP.
   a. 5   b. 10   c. 14   d. 21
9. Nurses are caught in a conflict between their values and their work environment. Which of the following perspectives best helps us understand this situation?
   a. functionalism   b. conflict theory   c. symbolic interactionism   d. exchange theory
10. The approach to health-care delivery under which medical services are financed governmentally is called
   a. managed competition   b. single payer   c. all payer   d. play or pay

GRAPHIC REVIEW

Table 15.2 compares the health-care systems of Canada and the United States on a number of important dimensions. Answer these questions to test your understanding of the implications of the data in this table.

1. Which of these two health-care systems appears to be the most effective? Why?

   _____

   _____

2. Why would most Americans favor the current U.S. health-care model over Canada's? Explain.

   _____

   _____

ANSWER KEY

Concept Review	Multiple Choice	Feedback Review
a. 3	1. d	1. T
b. 2	2. b	2. T
c. 1	3. b	3. F
d. 5	4. d	4. health-care system
e. 4	5. b	5. deviant
	6. a	6. profit
	7. b	7. corporation
	8. e	8. c
	9. c	9. c
	10. a	10. b
	11. c	
	12. a	
	13. e	
	14. e	
	15. c	

Chapter Sixteen

Population and Urbanization

After careful study of this chapter, you will be able to:

- Distinguish among the three population processes.
- Discuss the major dimensions of the world population growth problem.
- Relate the ideas of Thomas Malthus to the demographic transition.
- Describe projected world population growth.
- Distinguish between the concepts of community and city and define the terms being used to signify differences in the size of urban areas.
- Trace the historical development of preindustrial and modern cities.
- Describe some of the major consequences of suburbanization.
- Discuss world urbanization.
- Compare and contrast the three theories of city growth developed by urban ecologists.
- Compare and contrast traditional and contemporary views of the quality of urban life.

SOCIOLOGICAL IMAGINATION

Do city dwellers suffer isolation from personal relationships? According to the popular conception of city life, urbanites are part of the "lonely crowd"—they live amid a sea of people whom they do not know personally. This conception of the quality of urban social life was advocated by pioneers in the study of *urbanism*. Subsequent research has shown that urban residents do engage in personal relationships and that these social ties contribute to a positive emotional state of mind. Those without such social relationships tend to be recently relocated residents who have not yet had time to form close human connections.

The quality of urban life is an important issue within urban populations. There are, of course, many other issues. To begin the study of some of these issues, the chapter first discusses the concept of *demography*.

THE DYNAMICS OF DEMOGRAPHY

The Nature of Demography

Demography is the scientific study of population. It encompasses everything that affects or can be affected by population size, distribution, composition, structure, and change. Although basically a social science, demography draws from many disciplines, including biology, geography, mathematics, economics, sociology, and political science. The dynamics of demographic analysis involve three processes: fertility (births), mortality (deaths), and migration (movement from one geographic area to another). From the operation of these three processes comes alterations in population characteristics as well as social, political, and economic changes (Namboodiri, 1996).

Demography can be divided into two subareas. **Formal demography** deals with gathering, collating, analyzing, and presenting population data. Documentation of changes in the American population, for example, is part of formal demography. **Social demography** is the study of population in relation to its social setting. Social demography places population practices and patterns in a cultural context. For example, social demography helps us to understand that infant mortality is higher among the poor than among the rich.

In one sense, all of us are demographers in that we observe and record the conspicuous events of population change—births, deaths, and relocations of individuals and families. The professional demographer considers these events in two ways. First, the demographer gathers, organizes, and analyzes the patterns of population size, structure, composition, and distribution in an area. Second, the demographer attempts to identify and understand how demographic and social processes interrelate. In other words, we must understand that the quantity of people shapes the quality of social life. At the same time, the quality and fabric of social life shape the quantity and character of population processes. Both these points are elaborated in this chapter.

Knowledge of population trends has never been more important than it is today. The study of population gives us information about population growth, the characteristics of population, the location of population, and the probable long- and short-run effects of demographic trends. Population information is used, for example, to plan for the health, education, transportation, and recreation needs of virtually every community in the United States as well as in parts of most other countries of the world. Demography assists government policymakers and decision makers in meeting the needs of various socioeconomic groups, including the young, the poor, the unemployed, and the elderly. Demography provides insights for industry by generating information needed for identifying changing demands for products and services. In short, demography plays a major role in planning, policy formation, and decision making in both public and private sectors of modern economies. If leaders throughout the world use population data to determine the nature and quality of daily life, it is important for us to know something about this area of sociological study.

The three population processes—fertility, mortality, and migration—are responsible for population growth and decline. They are used to chart past population changes and to predict future ones. Fertility and mortality have both biological and social dimensions.

Fertility

What is fertility? **Fertility** is the number of children born to women. Whereas fertility refers to the "actual" number of children women produce, **fecundity** is the maximum rate at which women can physically produce children. At the lower end, infecund (or sterile) women can naturally produce no children. Although we do not know for sure the upper limit of average fecundity (the estimate is fifteen births per woman) the *Guinness Book of World Records* awards the individual fertility championship to an eighteenth-century woman whose twenty-seven pregnancies produced sixty-nine children (Russell, 1987). The record for a group probably is held by the Hutterites, who migrated from Switzerland to North and South Dakota and Canada in the late nineteenth century. Hutterite women in the 1930s were producing an average of more than twelve children each (Westoff and Westoff, 1971). The Hutterites give us a good estimate of fecundity, because they are the best example of *natural fertility*—the number of children born to women in the total absence of conscious birth control. No other known society has exhibited the fertility level of the Hutterites (Weeks, 1996).

How is fertility measured? The **crude birthrate** is the annual number of live births per one thousand members of a population. The birthrate varies considerably from one country to another. (See Table 16.1.) The birthrate for the United States is fifteen per one thousand. Kenya, in East Africa, experiences a very high birthrate of thirty-eight per one thousand, and Germany, a very low birthrate of ten per one thousand (*World Population Data Sheet*, 1997).

The advantage of the crude birthrate is its simplicity. To calculate it, one needs only the number of births in a year and the size of the population. It is, however, "crude" because it fails to target those women in the population most likely to give birth and ignores the age structure of the population, both of which affect the number of live births in any given year. Consequently, demographers also use the **fertility rate**—the annual number of live births per one thousand women ages fifteen to forty-four. **Age-specific fertility** is the number of live births per one thousand women in specific age groups, such as twenty to twenty-four or thirty-five to thirty-nine.

The birthrate is influenced by both biological and social factors. A major biological influence on the birthrate is women's health. For example, widespread

The birthrate of a society is affected by social factors, including societal attitudes toward reproduction. Due to changing attitudes regarding the proper number of children to have in modern societies, fewer women are having children, and more women are having fewer children. This couple, like more and more of their contemporaries, could well decide to have no more than two children.

TABLE 16.1	World Birthrates and Death Rates		
Region	Crude Birthrate*	Crude Death Rate*	Infant Mortality**
World	24	9	59
Africa	40	14	89
Asia	24	8	58
North America	14	9	7
United States	15	9	7
Canada	13	7	6
Latin America	25	7	39
Europe	10	12	10
Russia	9	14	18
Oceania	19	8	24
More developed nations	11	10	9
Less developed nations	27	9	64

*Number of births or deaths per one thousand persons.

**Number of deaths among children under one year of age per one thousand live births.

Source: *World Population Data Sheet*, 1997 (Washington, DC: Population Reference Bureau, 1997).

disease (especially rubella, or German measles) causes the birthrate to decline because many pregnancies end in miscarriages. Fecundity also influences fertility behavior: it still takes an average of nine months to give birth to a baby. Primary social factors affecting the birthrate of a population are the average age at marriage, the level of economic development, the availability and use of contraceptives and abortion, the number of women in the labor force, the educational status of women, and social attitudes toward reproduction.

Attitudes toward having children have changed dramatically in recent years in the United States. The most obvious result of this change has been the steady decline in the birthrate. More Americans are using contraceptives and having abortions. A greater number believe that two children—or even one child—is an acceptable size for a family. And more American women today are postponing having children until their late twenties and early thirties. As a result, fewer women are having children, and more women are having fewer children. (Refer to Chapter 11, "Family.")

Mortality

What is mortality? **Mortality** refers to deaths within a population. The important dimensions of mortality are life span and life expectancy. **Life span**—the most advanced age to which humans can survive—can only be inferred from the person with the greatest authenticated age, a Japanese man who lived nearly 121 years (Russell, 1987). Although this tells us that people are biologically capable of living to be 120 years old, few even approach this age in a world in which the **life expectancy**—the average number of years that persons in a given population born at a particular time can expect to live—is sixty-six years (*World Population Data Sheet,* 1997).

Morbidity refers to rates of disease and illness in a population. The incidence of a disease or illness represents indicators of health status as well as the relative importance of social versus biological causes of death. For example, in turn-of-the-century America, influenza, pneumonia, and tuberculosis accounted for nearly one-quarter of all deaths, but their share of all deaths has now declined to less than 3 percent. As more people survive to older ages, chronic degenerative diseases such as heart disease, stroke, and cancer have come to represent an increasing share of deaths (nearly seven out of ten). These diseases are not age related but represent the consequences of an industrialized lifestyle, including stress, alcohol consumption, cigarette use, and pollution.

How is mortality measured? The **crude death rate** is the annual number of deaths per one thousand members of a population. Like the birthrate, the death rate varies widely throughout the world. (See Table 16.1.) The worldwide average death rate is nine per one thousand persons. Among more developed countries, the death rate is ten per one thousand, whereas among less developed countries, the death rate is nine per one thousand (*World Population Data Sheet,* 1997). Looking at specific regions of the world, the death rate varies from a low of seven per one thousand in Latin America and Canada to a high of fourteen per one thousand in Africa and Russia. The death rate in the United States is about nine per one thousand.

Demographers are also interested in the variations in birthrates and death rates for various groups. They have devised **age-specific death rates** to measure the number of deaths per one thousand persons in a specific age group, such as fifteen to nineteen or sixty to sixty-four. This measure allows one to compare the risk of death to members of different groups. Although death eventually comes to everyone, the rate at which it occurs depends on many factors, including age, sex, race, occupation, social class, standard of living, and health care.

The **infant mortality rate**—the number of deaths among infants under one year of age per one thousand live births—is considered a good indicator of the health status of any group, because infants are extremely susceptible to variations in food consumption, availability of medical care, and public sanitation. Note in Table 16.1 the similarities in the death rates for various regions as opposed to the wide variations in infant mortality. Infants in less developed countries are seven times more likely to die before their first birthday than infants

The infant mortality rate among African children is the highest in the world. These children have beaten the odds. Among African children under one year of age, eighty-nine per one thousand perish. This compares with seven per one thousand in the United States.

in the more developed nations; this fact illustrates the wide variation in living standards among these nations.

Migration

What is migration? **Migration** refers to the movement of people from one geographic area to another for the purpose of establishing a new residence. We can describe migration in terms of a move either within a country or from one country to another. A current example of international migration—movement to a new country—is the resettlement of Asian refugees into a number of countries around the world. Many of the refugees who settle in the United States in one particular city or region later move to another region, thus becoming internal migrants, as when people move from New York state to Arizona.

How is migration measured? The **gross migration rate** is the annual number of persons per one thousand members of a population who enter (immigrants) or leave (emigrants) a geographic area. Net migration is the effect of immigration and emigration on the size of a population. Thus, the **net migration rate** is the annual increase or decrease per one thousand members of the population as a result of people's leaving and entering the population. The United States, for example, has a net migration rate of about 2.4 per one thousand population. That is, 2.4 more persons per one thousand population enter the country than leave (Haupt and Kane, 1991).

When the U.S. Census Bureau reports migration rates, it refers only to the number of legal immigrants. Many people violate immigration quotas in the United States. In the 1970s, illegal entry into the United

States—primarily from Latin American and Caribbean countries—became a major concern. There are no precise statistics on either the illegal immigration rate or the total number of illegal aliens currently living in the United States. Estimates of the current number of illegal aliens range from three million to six million persons.

Why do people migrate? The most general explanation for migration is the so-called "push-pull" theory. That is, people move either because they are attracted elsewhere or because they feel impelled to leave their present location. Often "push and pull" factors exist simultaneously. The United States was settled by people who were fleeing economic, religious, and social circumstances they wished to avoid and who at the same time were attracted by a new land that held some promise for a better future. People, of course, may not migrate despite significant pushing and pulling factors. Certain barriers, such as the cost of moving or bad health, may outweigh the pushing or pulling factors.

In cases of voluntary movement, economic reasons are the most compelling for migrating. Voluntary migration is also associated with stages in the life cycle. The greatest movement is found among young people reaching maturity who leave home to attend college, to take a job, to marry, or to join the military (Weeks, 1996).

Unfortunately, not all migration decisions are freely made. People may be forced to leave an area or prevented from leaving a locale, both of which occurred among Jews in Hitler's Germany. An environmental crisis or an inhospitable climate producing famine or drought may give people no choice but to abandon their place of residence. Retired people may move from a metropolitan area to a rural community or from a colder climate to a warmer one.

America is, of course, a land of immigrants. People came to the United States in massive numbers either because circumstances forced them to leave their homeland or because they envisioned the attainment of a better life. The latter was the case for this immigrant family at Ellis Island at the turn of the century.

FEEDBACK

1. Match the following terms and definitions.

_____ a. fertility
_____ b. crude death rate
_____ c. crude birthrate
_____ d. demography
_____ e. fecundity
_____ f. age-specific fertility
_____ g. age-specific death rate
_____ h. infant mortality rate
_____ i. migration
_____ j. gross migration rate
_____ k. net migration rate
_____ l. morbidity

(1) the number of deaths per one thousand members of a population per year

(2) the movement of people from one geographic area to another for the purpose of establishing a new residence

(3) the increase or decrease per one thousand members of the population per year as a result of people leaving and entering the population

(4) the number of live births per one thousand women aged fifteen to forty-four years

(5) the number of deaths per one thousand persons in a specific age group

(6) the study of the growth, distribution, composition, and change of a population

(7) the number of live births per one thousand women in specific age groups

(8) the number of persons per one thousand members of a population who enter or leave each year

(9) the number of deaths to infants under one year of age per one thousand live births

(10) rates of disease and illness in a population

(11) the maximum rate at which women can physically produce children

(12) the number of live births per one thousand members of a population per year

Answers: 1. a. 4 b. 1 c. 12 d. 6 e. 11 f. 7 g. 5 h. 9 i. 2 j. 8 k. 3 l. 10

WORLD POPULATION

The World Population Growth Problem

The growth and distribution of the world's population—5.8 billion (Haub, 1997)—have never been evenly divided. The population has grown at markedly different rates throughout history and has not been distributed proportionally across the world's large land masses (Livi-Bacci, 1997).

How fast is the world's population growing? When describing the growth of human population, it is impossible to avoid a certain amount of conjecture, because there has never been a complete worldwide counting of the population. Although most countries now take a census, there are still national populations that have not been counted. Furthermore, the quality of census data varies a great deal from country to country. Nevertheless, it is possible to indicate world population growth patterns by using historical information and recently collected data.

Rapid world population growth—now standing at 1.5 percent per year—is a relatively recent event: "Those of us born before 1950 have seen more population growth during our lifetimes than occurred during the preceding 4 million years since our early ancestors first stood upright" (Brown, 1996:3).Only about 250 million people were on the earth in A.D. 1. (See Figure

16.1.) It was not until 1650 that the world's population doubled, to half a billion. Subsequent doublings have taken less and less time. The second doubling occurred in 1850, bringing the world population to one billion. By 1930, only eighty years later, another doubling had taken place. Only forty-five years later, in 1976, a fourth doubling raised the world's population to four billion. At the current growth rate, the world's population is expected to double again in approximately forty-seven years and will exceed eight billion persons by the year 2025 (*World Population Data Sheet,* 1997). (See Table 16.2.) The world's population has been able to increase so dramatically in part because of the mathematical principle by which population increases.

How does the population grow? This question refers, of course, not to the sexual activity that can produce children but to the underlying numerical principle of population growth. We are accustomed to thinking in terms of linear growth—when something increases in a constant amount in a given time period. If you save $100 a year for ten years, you will end up with $1,000, accumulated in ten equal amounts. But population does not grow linearly. It increases according to the principle of **exponential growth:** the absolute growth that occurs within a given time period becomes part of the base on which the growth rate is applied for the next time period. (See Figure 16.2.) This means that if the growth rate of a population remains the same for two successive years, the absolute growth will be larger in the second year.

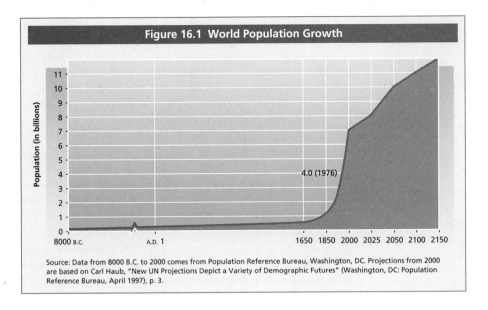

Figure 16.1 World Population Growth

4.0 (1976)

Population (in billions)

8000 B.C. A.D. 1 1650 1850 2000 2025 2050 2100 2150

Source: Data from 8000 B.C. to 2000 comes from Population Reference Bureau, Washington, DC. Projections from 2000 are based on Carl Haub, "New UN Projections Depict a Variety of Demographic Futures" (Washington, DC: Population Reference Bureau, April 1997), p. 3.

Suppose that a city of 100,000 has an annual growth rate of 5 percent for two years. At the end of the first year, its population would have increased by 5,000 (100,000 × 0.05). During the second year, the population would increase by an additional 5,250 (105,000 × 0.05). Consider a concrete case. The world population grew at about 2 percent per year between 1960 and 1970, which produced an increase of 650 million people. At the same rate, by 1980, the world population would have increased by 800 million. This was nearly the case: in 1970 the world population was 3,261 million; in 1980, it stood at 4,414 million, an actual increase of 793 million.

TABLE 16.2	Population Projections by Regions of the World			
Region	Population 1997* (In Millions)	Population Projection for 2025 (In Millions)	Population Increase Between 1997 and 2025 (In Percentages)	Doubling Time (In Years)
Asia	3,552	4,914	38	44
Africa	743	1,313	77	26
Latin America	490	691	41	38
Oceania	29	39	34	63
Europe	729	706	(3)	**
Russia	147	131	(11)	**
United States and Canada	298	372	25	117
World	5,840	8,036	38	47
Less developed areas (Asia excluding Japan, Africa, Latin America, Oceania)	4,666	6,810	46	38
More developed areas (Europe, Russia, United States, Canada, Japan)	1,175	1,226	4	564
World	5,840	8,036	38	47

*Estimated for mid-1997.
**Not projected to occur.
Source: Adapted from *World Population Data Sheet, 1997* (Washington, DC: Population Reference Bureau, 1997).

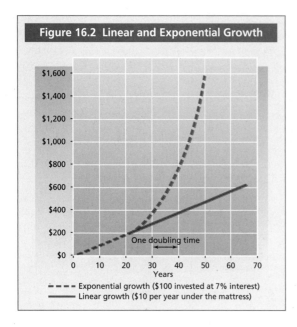

Figure 16.2 Linear and Exponential Growth

- - - - Exponential growth ($100 invested at 7% interest)
——— Linear growth ($10 per year under the mattress)

There is a familiar story that offers a classic example of exponential growth. The story tells of a clever courtier who presented a beautiful chess set to his king and in return asked only that the king give one grain of rice for the first square on the chess board, two grains, or double the amount, for the second square, four (doubling again) for the third, and so forth. The king, not being mathematically inclined, agreed and ordered the rice brought forth. The eighth square required 128 grains, and the twelfth took more than a pound of rice. Long before reaching the sixty-fourth square, the king's coffers were depleted. Even today, the world's richest king could not produce enough rice to fill the final square. It would require more than 200 billion tons, or the equivalent of the total world's production of rice for 653 years.

Because of the principle of exponential growth, a common term used in analyzing population growth is **doubling time**—the number of years needed to double the original population size. If a population is growing at 1 percent per year, it takes only seventy years to double. The number of people added each year becomes part of the total population, which then increases by another 1 percent in the following year.

The current world population growth rate of 1.5 percent may not seem high, but it represents an increase of over 240,000 people every day. At this rate, the world's population grows by the size of the population of Utah every seven days (*World Population Data Sheet*, 1997). To put it another way, the human population is growing at the rate of 167 per minute, over 240,000 per day, over 87 million per year. The pace of growth is so rapid that the additions to the world's population could equal the population of the United States in just over three years. At this rate, more people will be born in the year 2050 than were born in the 1,500 years after the birth of Christ.

On the encouraging side, the rate of world population growth has been declining since the 1970s. Data from fifty-six countries with an estimated population of ten million or more indicate that the average annual rate of increase dropped from 2.1 percent in 1970 to 1.7 percent in the 1980s, to 1.5 percent in 1997 (*World Population Data Sheet*, 1997).

The top half of Table 16.2 shows the population in various regions of the world, along with the projected populations for 2025. Consider the projected percentage of population increase in the less developed regions: Asia, 38 percent; Africa, 77 percent; Latin America, 41 percent; Oceania, 34 percent. Among the more developed regions, the highest percentage of population increase is 25 percent (in the United States and Canada). When the less developed and more developed regions are categorized separately, as they are on the bottom half of Table 16.2, two things become clear: Not only is the population projected to increase by 46 percent in less developed regions between 1997 and 2025, but the rate of increase is over ten times that of more developed regions. Doubling times for less developed regions vary from a low of 26 years (Africa) to a high of 63 years (Oceania). In contrast, doubling times for more developed regions range from 117 years (United States and Canada) to Europe, whose population is not expected to double.

The concentration of the world's population in less developed regions is equally apparent from the bottom half of Table 16.2. About 80 percent of all people lived in the less developed regions of the world in 1997, and 85 percent are projected to do so by 2025. The majority of people live in those nations that are least able to provide for their needs.

What difference does one child make? The importance of limiting family size, even by one child, can be illustrated by population projections for the United States. Even though the United States is not going to a three-child average in the future, the hypothetical American case can help us understand the importance of population control. Figure 16.3 contrasts the projected population of the United States in the year 2070 for an average family size of two children and an average family size of three children. When small decreases in the death rate and net migration at the present level are assumed, an average two-child family size would result in a population of 300 million in 2015. Taking the hypothetical average family size of three children, the U.S. population would grow to 400 million by 2013. As time passed, the difference of only one extra child per family would assume added significance. By 2070, the two-child family would produce a population of 350 million, but the three-child family would push the population to close to a billion. To say it another way, with an average family of two children, the U.S.

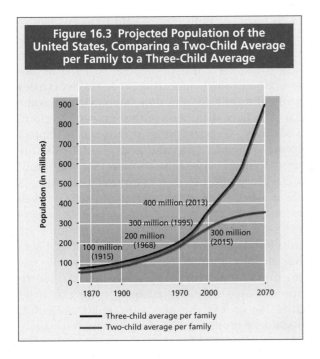

Figure 16.3 Projected Population of the United States, Comparing a Two-Child Average per Family to a Three-Child Average

400 million (2013)

300 million (1995)

200 million (1968)

300 million (2015)

100 million (1915)

Three-child average per family

Two-child average per family

population would not quite double itself between 1970 and 2070. But should the three-child family have been the average, the population would have doubled itself twice during this same period.

The importance of limiting population in developing regions becomes clearer when the effect of even one child added to the average number of children in a fam-

ily is recognized. Moreover, the addition of one child per family has a greater effect as the population base gets larger; not only is one extra person added, but theoretically that one person will be involved with the reproduction of yet another three, and on it goes. The largest populations are found in developing countries, which also have the largest average number of children per family.

Age-sex population pyramids illustrate the dependency burden that results from different rates of population growth. The **dependency ratio** is the ratio of persons in the dependent ages (under fifteen and over sixty-four) to those in the "economically active" ages (fifteen to sixty-four). The dependency ratio reflects the economic burden the productive portion of a population must carry. When these ratios are broken down into the youth dependency and old-age dependency, the social, economic, and political costs of the dependency burden become evident. Developing nations have a higher total dependency ratio, and their youth dependency ratio is more than twice that of the developed nations, whereas the reverse is true for the old-age dependency ratio. Figure 16.4 displays age-sex pyramids for Iran and the United States. After inspecting these contrasting pyramids, you would not be surprised to learn that one individual of working age supports about half the number of dependents in the United States as an individual would in Iran.

For the developing nations, a high youth dependency ratio means that national income is diverted from

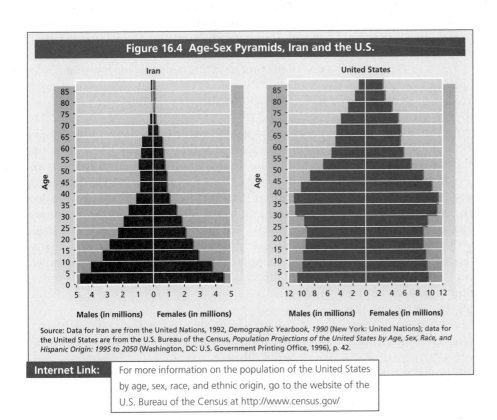

Figure 16.4 Age-Sex Pyramids, Iran and the U.S.

Iran

United States

Males (in millions)   Females (in millions)

Males (in millions)   Females (in millions)

Source: Data for Iran are from the United Nations, 1992, *Demographic Yearbook, 1990* (New York: United Nations); data for the United States are from the U.S. Bureau of the Census, *Population Projections of the United States by Age, Sex, Race, and Hispanic Origin: 1995 to 2050* (Washington, DC: U.S. Government Printing Office, 1996), p. 42.

Internet Link: For more information on the population of the United States by age, sex, race, and ethnic origin, go to the website of the U.S. Bureau of the Census at http://www.census.gov/

savings, and hence economic development, to population maintenance (food, housing, schools, health care). The rising old-age dependency ratio heralds other problems as well, such as the need for increasing health-care services and developing institutional arrangements for the long-term care of the elderly and the emergence of roadblocks to occupational mobility.

Although concern about population growth is deep, it is not new. In 1798, Thomas Robert Malthus, an English minister and economist, published *An Essay on the Principle of Population,* one of the most important works ever written on population growth and its repercussions. Malthus's thesis was so pessimistic that he helped economics earn the label the "dismal science."

The Malthusian Perspective

In his essay, Malthus proposed a set of relationships between population growth and economic development. Here are his central propositions:

1. Population, if left unchecked, will tend to exceed the available food supply.
2. Checks on population can be *positive*—those factors that increase mortality, such as famines, disease, wars—or *preventive*—those that decrease fertility, such as later marriages or abstaining from sexual relations in marriages.
3. For the poor, any improvement in income is lost to additional births; this leads to reduced food consumption and standards of living and eventually to the operation of positive checks.
4. The wealthy and better educated already exercise preventive checks.

This last point leads to some of Malthus's suggestions for solving the "population problem." Because the wealthy and better educated had achieved controlled fertility, the method of extending this to the poor would be through universal education, which would offer all members of the population the chance of improving their situation. "The desire of bettering our condition, and the fear of making it worse, has been constantly in action and has been constantly directing people into the right road" (Malthus, 1798:477). The beneficial effects of education on population control could be enhanced, Malthus suggested, by raising people's aspirations for a higher standard of living. This would be accomplished by raising wages above the minimum required for subsistence, thus providing the poor an opportunity to choose between more children at a minimal standard of living or smaller families with a higher quality of life.

Malthus's writings have profoundly influenced analyses of population change. Although events during the late nineteenth and early twentieth centuries brought his predictions into question, the last half of this century has raised the specter of a Malthusian crisis: population growth outstripping both food resources and the ability of different societies to provide even minimal quantities of basic necessities (housing, health care, education, employment). Some social scientists—the neo-Malthusians—have modified Malthus's propositions in an attempt to explain the current world situation and predict possible futures. First, neo-Malthusians note that the use of contraceptives has not distorted marital relations as Malthus feared that it might. Second, historical developments since Malthus's time indicate that values promoting smaller families and supporting norms are positively related to certain kinds of social and economic changes. In other words, values on family size and norms for fertility regulation are adapted to socioeconomic conditions, enabling society to avoid Malthusian checks through self-regulation. Third, neo-Malthusians argue that many nations have a rate of population growth that overloads this self-regulating process because population growth is so excessive that resources are diverted from socioeconomic change to population maintenance (Humphrey and Buttel, 1986).

A debate is currently taking place between those who believe that the world's population is exploding beyond control and those who are convinced that the population explosion will be defused in the near future. Those predicting a slackening of the current high population growth rate often do so based on the fact that modern methods of birth control are gaining widespread acceptance around the globe. This debate is more easily understood when viewed within the context of the process of the demographic transition.

The Demographic Transition

What is the demographic transition? If developed nations have defied Malthus's theory, they have done so by making what is called the **demographic transition**—the process by which a population gradually moves from high birthrates and death rates to low birthrates and death rates, as a result of economic development. (See Figure 16.5.) The number of stages in the demographic transition ranges from three to five, depending on how finely the process is broken down. What follows is a five-stage model with illustrative examples.

1. Both the birthrate and the death rate are high, and population growth is modest. No countries are at this stage today.
2. The birthrate remains high, but the death rate begins to drop because of modernization. The rate of population growth is very high. Most

Figure 16.5 Stages of the Demographic Transition

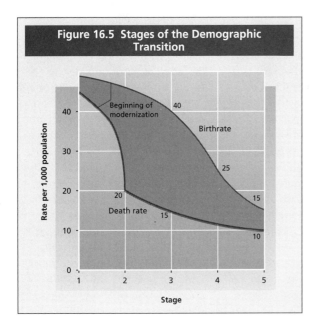

sub-Saharan African countries are presently at this stage.

3. The birthrate begins to decline, but because the death rate continues to go down, population growth is still rapid. Many Latin American countries are currently at this stage.

4. Although the death rate continues to drop, the decline in the birthrate becomes more precipitous. The potential exists for a decline in the rate of population growth. Nations such as Taiwan and South Korea are approaching this stage.

5. Both the birthrate and the death rate are low, and the population grows slowly or not at all. North America, Europe, and Japan are at this stage today. Europe, in fact, is in what is being called the "second demographic transition."

What is the second demographic transition? You may recall from Table 16.2 that the expected population decrease for Europe between 1997 and 2025 is 3 percent and that Europe's population is not expected to double again. This reflects the second demographic transition, which Europe alone is experiencing. Europe's first demographic transition began with a gradual death-rate decline in the early nineteenth century, followed by a birthrate decline beginning around 1880. The transition to both low birthrates and low death rates was completed by the 1930s. Europe's second demographic transition began in the mid-1960s.

The primary demographic characteristic of the second demographic transition, which now must be reached by the less developed countries, is the decline in fertility from slightly above **replacement level**—the birthrate at which a couple replaces itself in the population (2)—to a fertility level of 1.7. Should Europe maintain this lower fertility rate, and if immigration does not interfere, Europe's population will decline, something that has already happened in Austria, Denmark, Germany, Hungary, and Russia. The original formulation of the demographic transition, as noted earlier, ended with stable or nongrowing populations (Van de Kaa, 1987).

Future World Population Growth

World population growth has reached a watershed. After more than two hundred years of acceleration, the annual population growth rate is declining. The current growth rate of 1.5 percent compares favorably with the peak of 2.04 percent in the late 1960s. Moreover, it is projected to reach 1.4 percent to 1.5 percent in 2000 and zero at the end of the next century ("The UN Long-Range Population Projections," 1992). The birthrate is already low in developed countries and is now declining in many developing countries (McNicoll, 1992). Similarly, the population growth rate is declining somewhat in both developed and developing areas of the world.

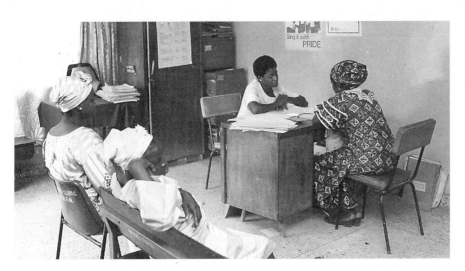

Family planning is crucial to the slowing of population growth. India, a country of about 1 billion whose population is expected to double in thirty-four years, needs more family planning clinics and more people willing to use them.

DOING RESEARCH

Kristin Luker—The Politics of Abortion and Motherhood

In the United States, a debate is currently raging on the issue of abortion. Kristin Luker's (1984) study, a first attempt at understanding the origins of this debate, its meaning, and the perspective of pro-life and pro-choice advocates, is both timely and important.

Luker's book is divided into two parts. First, Luker explores the origins of the abortion debate and its evolution as a public policy issue. In the second part of the book—the part this *Doing Research*

focuses on—Luker outlines the worldviews of pro-life and pro-choice activists. The data for this analysis came from extensive and intensive interviews with 212 abortion activists. Luker interviewed at least one individual from each of the major pro-life and pro-choice groups in California and conducted comparative interviews with abortion activists in six additional states.

According to Luker, abortion per se is only the tip of the iceberg when pro-life and pro-choice activists think about abortion. For these activists, abortion is tied to a number of other key issues, including basic beliefs and feelings about parenthood, the family, religion, roles of the sexes, and the biological nature of men and women.

According to the pro-life view, because men and women are

intrinsically different, they have different roles to play—men are designed to work outside the home, and women are best suited to take care of their homes, children, and husbands. Women are assumed to be the best at expressing tenderness, caring for others, and exhibiting self-sacrifice. When women work outside the home, all members of the family are thought to lose: men lose the nurturing women provide, women must do without the protection and cherishing men offer, and children lose the full-time love and attention of at least one parent. Pro-life people believe that being a parent is a natural rather than a social role and are opposed to the current norm of having two children per family. (One of every five pro-life activists in this study had at least six children.)

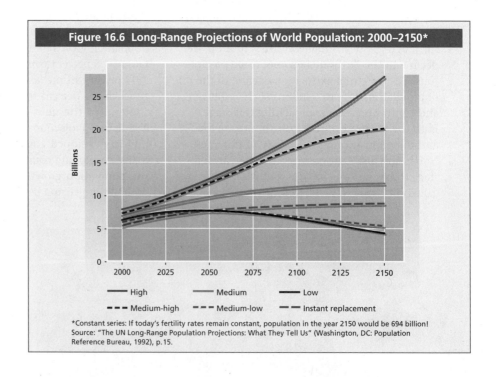

Figure 16.6 Long-Range Projections of World Population: 2000–2150*

*Constant series: If today's fertility rates remain constant, population in the year 2150 would be 694 billion!
Source: "The UN Long-Range Population Projections: What They Tell Us" (Washington, DC: Population Reference Bureau, 1992), p. 15.

Despite the reduction in the annual growth rate and birthrate, the world's population will continue to increase, as we have seen. About seven billion people are expected to inhabit the globe by 2000. Throughout the first half of the twenty-first century, the annual growth rate is expected to decline until world population stabilizes at about twelve billion people. (See Figure 16.6.) At this point, the world will have reached **zero population growth**—when deaths are balanced by births so that the population does not increase.

Dimension for dimension, the pro-choice activists hold directly opposite values and beliefs. Pro-choice people believe in the equality and similarity of the sexes. Not only are family roles not naturally determined, but they stand as roadblocks to full equality between men and women. According to the pro-choice advocates, the number of children one has should be geared to the available emotional, social, and financial resources. If children overtax the available resources, all members of the family suffer, including the children.

According to Luker, these two contrasting orientations rest on two very dissimilar moral foundations. Pro-life activists center their lives around God and believe in "the rightness of His plan for the world." Pro-choice activists take human reason as the bedrock of their worldview. Because they believe that humans can discover truth for themselves, they tend to be optimistic about "human nature."

Concern about abortion in America has in the past, concludes Luker, revolved around the medical profession's right to decide for mothers. The current debate revolves around an entirely new issue—motherhood. Because of new medical technologies, women can now decide, for the first time in history, what their family roles will be and how these roles will fit into their lives. In essence, Luker states, pro-life and pro-choice advocates have squared off in a struggle to define the nature and meaning of motherhood. Pro-life activists hold up motherhood as the central and most rewarding role women can perform. To pro-choice activists, motherhood is only one of many possible roles and can become a detriment when it is offered as the only role for women.

Critical feedback

1. Summarize the perspectives of pro-life and pro-choice activists. Do you agree with Luker's characterization? Which perspective (or parts of each perspective) do you share?
2. Analyze the pro-life and pro-choice views of motherhood in terms of the major theoretical perspectives. Which perspective best fits each view?
3. Analyze the politics of abortion and motherhood within the context of the population problem.

Contrary to popular opinion, limiting the average family size to two children does not immediately produce zero population growth. There is a time lag of sixty to seventy years because of the high proportion of young women of childbearing age in the world's population. Even if each of these women had only two children, the world population would grow because there are so many young women. Zero population growth will be reached only when this disproportionate number of young mothers disappears ("The UN Long-Range Population Projections," 1992).

The time lag is what demographers call "population momentum." The growth of the world's population, like a huge boulder careening down a mountain, cannot be stopped immediately. But the sooner the momentum of current population growth is halted, the better. The sooner the world fertility rate reaches the replacement level, the sooner it will reach zero population growth. The ultimate size of the world's population when it does stop growing depends greatly on when the replacement level is achieved. To state it another way, each decade it takes to reach replacement level fertility will cause the world's population to be 15 percent larger.

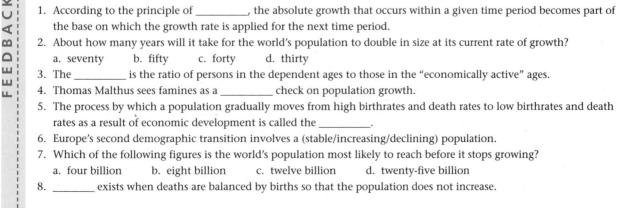

FEEDBACK

1. According to the principle of _____, the absolute growth that occurs within a given time period becomes part of the base on which the growth rate is applied for the next time period.
2. About how many years will it take for the world's population to double in size at its current rate of growth?
   a. seventy   b. fifty   c. forty   d. thirty
3. The _____ is the ratio of persons in the dependent ages to those in the "economically active" ages.
4. Thomas Malthus sees famines as a _____ check on population growth.
5. The process by which a population gradually moves from high birthrates and death rates to low birthrates and death rates as a result of economic development is called the _____.
6. Europe's second demographic transition involves a (stable/increasing/declining) population.
7. Which of the following figures is the world's population most likely to reach before it stops growing?
   a. four billion   b. eight billion   c. twelve billion   d. twenty-five billion
8. _____ exists when deaths are balanced by births so that the population does not increase.

Answers: 1. exponential growth 2. b 3. dependency ratio 4. positive 5. demographic transition 6. declining 7. c 8. Zero population growth

COMMUNITY, CITY, AND METROPOLIS

For centuries, people have lived in large population settlements. The nature of those settlements has undergone important changes, and as the settlements have changed, the language used to discuss them has also been altered. Before you can begin to understand the changing conditions in urban areas, you must examine the terminology used to describe them.

Some Basic Terms

What is a community? Long before the rise of cities, people depended on communities to satisfy their common needs. In fact, a **community** is a concentration of people whose major social and economic needs are satisfied primarily within the area where they live. It is easy to identify the network of social relationships and organizations that fulfill basic social and economic needs in a small community of several hundred people, but the situation is far more complex when several million people live within a given geographic area. Some sociologists even doubt that such large concentrations of people can usefully be described as communities. Consequently, social scientists are now concentrating on cities and metropolitan areas rather than on communities.

What is a city? The census bureaus of most nations define cities as population aggregates exceeding a certain size. In Denmark and Sweden, an area with 200 inhabitants qualifies as a city. Japan uses a much higher cutting point—30,000 inhabitants. The cutting point used by the U.S. Census Bureau is 2,500. These relatively low cutting points were set at a time when urbanization had just begun and population concentrations were small; they are clearly too low for modern times. Moreover, the use of the number of inhabitants to define a city—regardless of the specific number that is used—overlooks other important aspects of cities. Besides containing a reasonably large number of people, cities are characterized by permanence, density, and the heterogeneity and occupational specialization of inhabitants. In brief, a **city** is a dense and permanent concentration of people living in a limited geographic area who earn their living primarily through nonagricultural activities.

Are cities the largest urban units? An individual city may grow to be quite large, but a city's official political boundaries often fail to keep pace with the growth of its population and economic activities. Outlying areas may resist being incorporated into a city because of a fear of higher taxes or loss of political control. Or continued expansion of a city's official boundaries may be impossible because of state or county lines or other nearby cities. A particular urban area, then, may extend far beyond the boundaries of a city.

New terms have had to be coined to identify such enlarged urban areas (Palen, 1997; also see Table 16.3). One such term is **Metropolitan Statistical Area (MSA)**—a grouping of counties that contains at least one city of 50,000 inhabitants or more or an "urbanized area" of at least 50,000 inhabitants and a total metropolitan population of at least 100,000. The establishment of criteria for designating MSAs has allowed the Census Bureau to identify large areas whose populations have interdependent relationships with a central city. This is an important development because the influence of any city extends far beyond its official boundaries. There are 255 MSAs in the United States (U.S. Bureau of the Census, 1997b).

Even the MSA, however, is too limited to identify fully some urban areas. Many MSAs do not exist in isolation. For example, although Newark, New Jersey, qualifies as an MSA, it must be considered in relation to surrounding MSAs, including New York City. Because of such situations, the Census Bureau now identifies a **Primary Metropolitan Statistical Area (PMSA)**—adjacent MSAs that are closely integrated economically and socially. The United States now has 73 PMSAs.

Urban growth has led to yet a third category. A **Consolidated Metropolitan Statistical Area (CMSA)** consists of two or more sets of neighboring PMSAs (also known as a megalopolis). Table 16.4 lists the 18 CMSAs in the United States. At the present time, 13 percent of the American population resides in the top two CMSAs, 21 percent in the top five, 31 percent in the top ten, 38 percent in the entire 18 CMSAs.

TABLE 16.3	U.S. Bureau of the Census Definitions

Metropolitan Statistical Area (MSA): A grouping of counties that contains at least one city of 50,000 inhabitants or more or an "urbanized area" of at least 50,000 inhabitants and a total metropolitan population of at least 100,000.

Primary Metropolitan Statistical Area (PMSA): Adjacent MSAs that are closely integrated economically and socially.

Consolidated Metropolitan Statistical Area (CMSA): Two or more sets of neighboring PMSAs (also known as a *megalopolis*).

Metropolitan Area (MA): The overarching term covering MSAs, PMSAs, and CMSAs.

TABLE 16.4	Consolidated Metropolitan Statistical Areas (CMSAs) in the United States		
CMSA	Population (1,000)	Percent African American	Percent Latino
1. New York-Northern New Jersey-Long Island	19,796	18	15
2. Los Angeles-Riverside-Orange County	15,302	9	33
3. Chicago-Gary-Kenosha	8,527	19	11
4. Washington-Baltimore	7,051	25	4
5. San Francisco-Oakland-San Jose	6,513	9	16
6. Philadelphia-Wilmington-Atlantic City	5,959	18	4
7. Boston-Worcester-Lawrence	5,497	5	4
8. Detroit-Ann Arbor-Flint	5,256	21	2
9. Dallas-Fort Worth	4,362	14	13
10. Houston-Galveston-Brazoria	4,099	18	21
11. Miami-Fort Lauderdale	3,408	19	33
12. Seattle-Tacoma-Bremerton	3,226	5	3
13. Cleveland-Akron	2,899	16	2
14. Denver-Boulder-Greeley	2,190	5	13
15. Portland-Salem	1,982	3	4
16. Cincinnati-Hamilton	1,894	11	5
17. Milwaukee-Racine	1,637	13	4
18. Sacramento-Yolo	1,588	7	12

Source: U.S. Bureau of the Census, *Statistical Abstract of the United States 1996* (Washington, DC: U.S. Government Printing Office, 1996), pp. 40–42.

FEEDBACK

1. A _____ is a concentration of people whose major social and economic needs are satisfied within the area in which they live.
2. Which of the following characteristics is *not* included in the definition of a city?
   a. permanence
   b. dense settlement
   c. advanced culture
   d. occupational specialization in nonagricultural activities
3. A Primary Metropolitan Statistical Area (PMSA) might be contained within a Consolidated Metropolitan Statistical Area. *T or F?*

Answers: 1. community 2. c 3. T

THE URBAN TRANSITION

The nature of the contemporary world has been changed by **urbanization**—the process by which an increasingly larger proportion of the world's population lives in urban areas. Urbanization has been such a prevalent trend that it is now taken for granted in many parts of the world. In fact, this process has resulted in cities so large and interdependent, as you have just seen, that it is now necessary to think in terms of strips of interlocking cities. Near the end of the twentieth century, more people will be living in urban than in rural areas (Linden, 1993b). This is a fairly recent development in human history, however.

The size of the first cities—founded only five thousand or six thousand years ago—was quite small by contemporary standards. Ur, located at the point where the Tigris and Euphrates Rivers meet, was one of the world's first major cities, but at its peak, it held only about 24,000 people. Later, during the Roman Empire, it is unlikely that many cities had populations larger than 33,000. The population of Rome itself was probably under 350,000 (Hawley, 1971).

In addition to their small size, the cities that developed during the ancient and medieval periods contained only a small proportion of the world's population. As recently as 1800, less than 3 percent of the world's population lived in cities of 20,000 or more (Davis, 1955). By contrast, 43 percent of the world's population—and over 75 percent of North America's population—now live in urban areas (*World Population Data Sheet,* 1997). Only about 5 percent of the American population lived in urban areas in 1790 (U.S. Bureau of the Census, 1993j).

What accounts for this increase in urbanization? To find the answer to this question we must take a closer

The lower Pacific Coast of California has developed into a megalopolis. This night shot of the Hollywood Freeway in Los Angeles provides a dramatic hint at the mammoth size of a megalopolis.

look at urbanization before and after the Industrial Revolution. Although rapid urbanization is a fairly recent occurrence, the groundwork for it had been developing for several centuries.

Preindustrial Cities

What enabled the first cities to develop? According to anthropological research, the first urban settlements were located around 3500 B.C. in Mesopotamia (in southwestern Asia between the Tigris and Euphrates Rivers). This location tells us something about a prime condition necessary for the formation of a city—a surplus food supply. The Mesopotamian region is among the world's most fertile areas, and at that time, its people were relatively advanced in the development of plant cultivation. Because people in cities hold primarily nonagricultural jobs, they require an agricultural surplus. Before 3500 B.C., few areas of the world were capable of producing a dependable supply of extra food for nonagriculturally productive people. For this reason, cities were initially slow to develop (Adams, 1965).

The existence of a surplus food supply explains the development of cities but does not explain people's attraction to them. Cities tended to attract four basic types of people: elites, functionaries, craftspeople, and the poor and destitute. For elites, the city provided a setting for consolidating political, military, or religious power. The jewelry and other luxury items found in the tombs of these elites symbolize the benefits that this small segment of the population gained from their consolidation of power and control. Those who lived in cities as political or religious officials (functionaries) received considerably fewer benefits, but their lives were undoubtedly easier than those of the peasant-farmers in the countryside. Craftspeople, still lower on the stratification structure, came to the city to work and sell their products to the elites and functionaries. The poor and destitute, who were lured to the city for economic relief, were seldom able to improve their condition (Gist and Fava, 1974).

Do preindustrial types of cities still exist? Many parts of Africa, Asia, and Latin America are not industrialized, and therefore, still have some preindustrial types of cities. This is particularly true of capital cities, which attract migrants from rural areas because of the symbolic importance of these cities and the promise of a better life. Unfortunately, those who migrate to many of these cities are disappointed because the employment opportunities they expected to find simply do not exist. Also, many cities in developing nations lack the housing, sanitation, transportation, and communication facilities needed to accommodate their residents. As a result, many city dwellers in developing nations are forced to live in makeshift housing with few or none of the facilities that most people in modern societies consider necessities. Although cities in developing nations are not able to fulfill the social and economic needs of their residents, many people are driven to them because their life chances in the overpopulated rural areas and small towns are even worse.

The Rise of the Modern City

As merchants gained power and wealth during the late medieval period, they often established small industries to give themselves a stable supply of goods. As these industries gradually increased in size, the stage was being set for rapid city growth.

Why did the Industrial Revolution encourage the rapid growth of cities? The **Industrial Revolution** was the combination of technological developments beginning in the eighteenth century that created major changes in transportation, agriculture, commerce, and industry. Technological developments during this period allowed more dependable production and transportation with less human labor. As food production and transportation became more dependable, it became possible for more people with nonagricultural jobs to live in cities. The major reason for the growth of cities following the Industrial Revolution, however, was the increased use of factories for manufacturing.

Factories were not established to encourage the growth of cities, but they almost immediately had that effect, because one factory tends to attract others to the same location. Factories located together can share raw materials and transportation costs; and at a time when transportation and communication were slow and undependable, it was desirable for producers of machinery and equipment to be near factories using the equipment. The concentration of industry led to dense population settlements of factory workers and in turn attracted retailers, innkeepers, entertainers, and a wide range of people offering services to city dwellers. Of course, the greater availability of urban services often enticed still more factories to locate in a city, keeping the cycle going. Although subsequent technological and social developments affected the growth of cities in different ways, the overall trend toward urbanization had been set—the industrial world was becoming an urbanized world.

How does modern urban growth differ from life during preindustrial times? Life in preindustrial cities was directed toward the city's center. The walls, moats, and other protective devices found in these cities symbolized their role in defending the cities' inhabitants. It was safer to live in cities than outside them. The importance of the center of the city in preindustrial times is also reflected in residential patterns. Members of the upper class lived near the center of preindustrial cities, which gave them access to the most important government and religious buildings and, as a rule, the main market. The outskirts (including what we would now consider the suburban areas) were populated by the poor and those with low-status craft occupations, such as tanning and butchering. These people lived on the outskirts partly because more privileged residents wanted to have minimal contact with them. More important, though, the outskirts were the most undesirable areas because of their lack of accessibility to the central city (Sjoberg, 1960).

Modern cities, like preindustrial cities, located their major economic and government activities in their central cores. Cities in industrialized nations, however, are connected by sophisticated communication and transportation networks, which have had the opposite effect of the walls and other protective devices of preindustrial cities. Thus, industrial cities tend to have an outward orientation. Most of their development has occurred away from the central city in the surrounding suburban areas. In industrialized nations, population movement from rural to urban areas has been accompanied by suburbanization. In fact, many central cities are now losing population to the areas surrounding them.

Suburbanization and Its Consequences

Metropolitan dispersal—which occurs when central cities lose population to the area surrounding them—is a hallmark of contemporary American urban growth (Baldassare, 1992; Gober, 1993). By 1980, seventeen of the twenty-six central cities (as defined by political boundaries) that had a population of half a million or more in 1970 showed losses in population. But the population outside central cities is expanding several times as fast as the central cities themselves. According to the U.S. Census Bureau, the population of the central cities of the 255 MSAs increased by only .002 percent between 1970 and 1980, whereas the population of MSA areas outside the central cities (almost entirely suburbs) increased by 18.2 percent during the same period (U.S. Bureau of the Census, 1984). During 1991–1992, the percentage of Americans moving from central cities to suburbs was twice the rate of movement from suburbs to central cities (U.S. Bureau of the Census, 1993j). This is a continuation of a longer trend that has involved more than a doubling of the proportion of the population living in suburbs since 1950.

What makes suburbanization possible? Suburbanization has become an important trend partly because of technological developments. Improvements in communication (such as telephones, radios, television, computers, and fax machines) allow people to live away from the central city without losing touch with what is going on there. Developments in transportation (especially highways, automobiles, and trucks) have made it possible both for people to commute to work and for many businesses to leave the central city for suburban locations.

Technology is not the only cause of suburbanization, however. Both cultural and economic pressures have encouraged the development of suburbs. Partly because of America's frontier heritage, American culture has always contained an antiurban strain. Some Americans prefer urban life, but many report that they would rather live in a rural setting. Suburbs, with their low-density housing, have allowed many people to escape what they see as urban congestion without leaving the urban areas completely. Economic pressures favoring suburbs stem from the scarcity and high cost of land in the central city. Developers of new housing, retail, and industrial projects have often found that suburban locations are far less expensive than those near the central city.

Government policy has often increased the impact of economic forces. Federal Housing Administration

In part because of their frontier heritage, many Americans continue to hold a bias against urban living. Suburbs, such as this one in Maryland, permit people to remain near the advantages of urban centers while living in areas with lower density housing.

regulations, for example, have encouraged the financing of new houses (which can be built most cheaply in suburban locations) rather than the refurbishing of older houses in central city areas. As residents have moved to suburban areas, increasing numbers of retail stores and jobs have followed (Steinnes, 1982).

When suburbanization began to become noticeable in the 1930s, only the upper and middle classes could bear the cost. It was not until the 1950s that they were joined by the working class. Racial and ethnic diversity in suburbs has lagged behind social-class diversity. Despite federal legislation prohibiting housing discrimination, many suburbs have remained largely white.

Central-city minorities are beginning to move to the suburbs in greater numbers (Frey, 1984, 1985; Massey and Denton, 1987; Farley, 1997). Although America's suburban counties remain predominantly white, minorities have poured into the suburbs over the past ten years. According to the 1990 census figures, there are 3.7 million more whites in suburban counties than in 1980, compared to 4 million more Hispanics, 2.1 million more Asians, and 1.7 million more African Americans (Vobejda, 1991). The proportion of minorities living in predominantly suburban counties increased by 30 percent, to 25 million people during the past decade. A greater number of Latinos and Asian Americans, in fact, now reside in suburban counties than live in central cities. The percentage of African Americans living in suburbs increased almost 10 percent during the 1980s (Duke and Morin, 1991).

Despite this decline in the degree of residential segregation, African Americans remain the most isolated ethnic or racial minority in the United States (O'Hare et al., 1991; Farley and Frey, 1994; Palen, 1997; South and Crowder, 1997). Although the percentage of African Americans living in suburbs has increased from 16 percent to 26 percent between 1970 and 1990, the percentage of African Americans living in central cities has declined only slightly. (See Table 16.5.) The situation is better for Latinos and Asians, as noted above. Not only are Latinos and Asians more suburbanized—and thus less concentrated in the central cities than African Americans—they are less segregated in suburban areas (Palen, 1997). Even this progress has a downside. As the more successful minority members move to the suburbs, the "underclass" accounts for a larger proportion of the remaining central-city population (Massey, 1990; Massey and Eggers, 1990). (See Chapter 9, "Inequalities of Race, Ethnicity, and Age.")

The concentration of any racial or ethnic minority in a given area does not necessarily mean that slums will develop. African American ghettos have appeared in part because the residents are poor and lack the educational and employment opportunities needed to change their condition. Because of inferior education and continued discrimination in housing and employment, the movement of African Americans out of inner-city slums has been infinitely slower than it was for white minorities. Names can be changed and nationalities can be denied, but skin color is indelible. Even with a reduction in employment and housing discrimination, the foreseeable future of urban African Americans includes living in the central city, most often within ghettos. Thus, a major problem has been created in the United States by the process of suburbanization. As middle-class whites move to the suburbs, the

TABLE 16.5	African American and White Populations by Metropolitan Residence: 1970-1990					
	1970		1980		1990	
	African Americans	Whites	African Americans	Whites	African Americans	Whites
Total number (in thousands)	22,581	177,749	26,495	188,372	29,931	199,827
Percentage	100.0	100.0	100.0	100.0	100.0	100.0
Metropolitan areas	74.3	67.8	81.1	73.3	83.8	75.6
Central cities	58.2	27.8	57.7	24.9	57.3	25.8
Suburbs	16.1	40.0	23.3	48.4	26.4	49.8
Nonmetropolitan areas	25.7	32.2	18.9	26.7	16.2	24.4

Sources: 1970—U.S. Bureau of the Census, *1970 Census of Population*, PC (I)-BI, United States Summary (Washington, DC: U.S. Government Printing Office, 1972), Table 48; 1980—U.S. Bureau of the Census, *1980 Census of Population*, Part B, PC 80-I-BI, Part I, United States Summary (Washington, DC: U.S. Government Printing Office, 1983), Table 38; 1990—U.S. Bureau of the Census, *1990 Census of Population*, *Social and Economic Characteristics, United States*, CP-2-1 (Washington, DC: U.S. Government Printing Office, 1993).

central city is becoming increasingly populated by minorities, the poor, and the elderly. The result is a concentration of problems of economic and social dependency in cities that are financially unprepared to deal with them. This situation has been captured in what is known as the *central-city dilemma*.

What is the central-city dilemma? The creation and persistence of inner-city slums are only partially due to the socioeconomic characteristics of their residents. Actually, their socioeconomic characteristics are intertwined in a much larger picture, of which residential, educational, and employment discrimination are only a part. As the poor, unskilled, and uneducated migrate into the city, they are figuratively passed in the outbound lanes by members of the middle class and by manufacturers and retailers who are headed for the suburbs. It is there that commerce and industry find lower tax rates, less expensive land, less congestion, and many of their customers who have already left the central city. Accompanying the exodus of the middle-class residents, manufacturers, and retailers is the shrinking of the central-city tax base.

The severity of the problem is indicated by the fact that no city in a sample studied by Marshall and Lewis (1983) was able to attract more high-status white males from the suburbs than it lost in either of two periods (1955–1960 and 1965–1970). Cities founded in 1890 or before were able to slow some of their earlier decline, but the decline of cities founded after 1890 actually accelerated during the periods studied.

Services must continue for those remaining in the central city, but many of those requiring municipal services (such as education, sanitation, police and fire protection, and public transportation) cannot financially support the services. Without adequate funding, these services, along with the physical facilities in the central

city, continue to deteriorate. If city government attempts to rally by raising taxes, it runs the risk of driving even more of its tax base to the suburbs. If nothing is done about the decay of physical facilities and services, those who can afford it will escape the mess, taking their tax and consumer dollars with them. And all the while those who have already left the city for suburban living continue to commute to work, using public transportation, streets, and sidewalks without compensating tax support. Thus, suburban dwellers who work in cities derive direct economic advantages without paying sufficient taxes to support facilities that make these advantages possible.

This, then, is the dilemma of the central city— diminishing public revenues are not sufficient to cope with the influx of the poor, unskilled, and uneducated who have difficulty supporting themselves, let alone contributing to the financial health of the inner city. Poor inner-city residents face consequences of this urban dilemma in addition to the deterioration of the physical surroundings and public services.

What are the consequences of the central-city dilemma? Poor inner-city African American children are handicapped by inferior educational facilities and teachers, and the dropout rate among them is considerably higher than among whites (Kozol, 1991). Their unemployment rate is at least twice as high as it is for urban whites during prosperous times, and it skyrockets when the economy falters. Those who are employed generally must take dead-end, low-paying jobs because the more desirable jobs go to the better educated. A catalog of social ills flows from this socioeconomic condition in the inner city. African Americans in big-city slums exist in a world of poverty, congestion, prostitution, drug addiction, broken homes, and brutality. Crime rates, for example, appear to increase in central cities that have been affected by suburbanization, both because suburbanization creates stratification along city-suburban lines and because it contributes to an absolute decline in the socioeconomic status, economic and tax base, and physical structure of the central city (Farley and Hansel, 1981; Ganz, 1985; Yinger, 1995).

World Urbanization

Urbanization is a worldwide movement. From 1800 to the mid-1980s, the number of urban dwellers increased almost one hundred times, whereas the population increased only about fivefold. Over 2.5 billion people—43 percent of the world's population—now live in urban areas. In more developed countries, 74 percent of the population lives in urban areas compared to 36 percent in less developed countries (*World Population Data Sheet,* 1997).

The pattern of urbanization has been different for more developed and less developed countries. Most of the urban growth in less developed countries before the turn of the century occurred through colonial expansion. Western countries, which had been involved in colonial expansion since the late fifteenth century, held half the world under colonial rule by the latter part of the nineteenth century. It has only been since World War II that many of these colonial countries have become independent nations (Bardo and Hartman, 1982).

Since gaining independence, these former colonies have been experiencing rapid urbanization and industrialization. In fact, urbanization in these areas is now proceeding faster than it did in the West during its urban expansion period. The rate of urbanization for major industrial nations in the West was 15 percent each decade throughout the nineteenth century. In the 1960s, the rate of urbanization in major low-economy countries was 20 percent per decade. During the twentieth century, the urban population in these areas has increased seventy times, nine times faster than the urban population of Western countries expanded during the nineteenth century (Light, 1983).

One difference in the pattern of urbanization in less developed countries, then, is the rapidity of the urban growth rate. Other differences are worth noting.

What are some additional differences in the pattern of urbanization in less developed and more developed countries? In the first place, industrial development in less developed countries, unlike the Western experience, has not kept pace with urbanization. Cities of North America and Europe had jobs for all migrants from rural areas. In the cities of less developed countries, the supply of labor from the countryside exceeds the demand for labor. A high rate of unemployment is the obvious result. The term **overurbanization** has been created to describe a situation in which a city is unable to supply adequate jobs and housing for its inhabitants.

Another difference between urbanization in more developed and less developed countries is the number and size of cities. When grouped by size, cities in developed countries form a pyramid: a few large cities at the top, many medium-size cities in the middle, and a large base of small cities (Light, 1983). In the less developed world, in contrast, many countries have one tremendously big city that dwarfs a large number of villages. Calcutta, India, and Mexico City are examples. Of the world's ten largest cities, only two—Shanghai and

The pattern of urbanization is different for high-income and low-income economies. In high-income countries there are a few very large cities, many medium-sized cities, and a much larger number of small cities. In many low-income countries, there is one large city that dominates a myriad of villages.

Calcutta—were in less developed areas in 1950. By 2000, as you can see in Figure 16.7, the United Nations estimates that seven of the top ten largest urban areas will be in less developed countries. By the turn of the century, there will be twenty-one "megacities" with populations of ten million or more. Of these, eighteen will be in developing countries, including the most impoverished societies in the world (Linden, 1993b).

Why do people in developing countries move to large cities with inadequate jobs and housing? Urban soci-ologists see the operation of "push" and "pull" factors in the rural-urban migration in less developed countries. Peasants are pushed out of their villages because expanding rural populations cannot be supported by the existing agricultural economy. Peasants are forced to migrate elsewhere, and cities are at least an alternative. Peasants are also attracted to cities by perceived opportunities for better education, employment, social welfare support, and availability of good medical care, despite the fact that they are likely to be disappointed (Firebaugh, 1979).

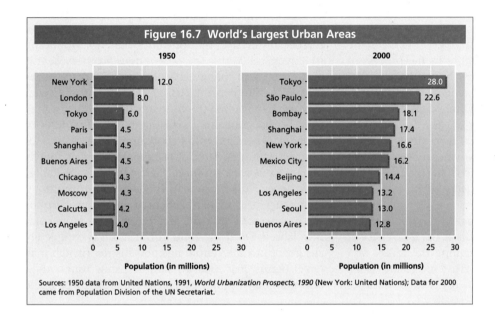

Figure 16.7 World's Largest Urban Areas

1950

City	Population (in millions)
New York	12.0
London	8.0
Tokyo	6.0
Paris	4.5
Shanghai	4.5
Buenos Aires	4.5
Chicago	4.3
Moscow	4.3
Calcutta	4.2
Los Angeles	4.0

2000

City	Population (in millions)
Tokyo	28.0
São Paulo	22.6
Bombay	18.1
Shanghai	17.4
New York	16.6
Mexico City	16.2
Beijing	14.4
Los Angeles	13.2
Seoul	13.0
Buenos Aires	12.8

Sources: 1950 data from United Nations, 1991, *World Urbanization Prospects, 1990* (New York: United Nations); Data for 2000 came from Population Division of the UN Secretariat.

FEEDBACK

1. _____ is the process by which an increasingly larger proportion of the world's population lives in urban areas.
2. Preindustrial cities first began to develop in areas with an _____ surplus.
3. Thanks to worldwide urbanization, preindustrial cities no longer exist. *T or F?*
4. The rate of city growth increased very rapidly following the _____.
5. One of the major trends in urban growth in modern society is _____ dispersal.
6. The _____ exists because diminishing public revenues are not sufficient to cope with the influx of the poor, unskilled, and uneducated.
7. The recent increase in African American suburbanization is finally resolving the central-city dilemma. *T or F?*
8. _____ occurs when a city is unable to supply adequate jobs and housing for its inhabitants.

Answers: 1. Urbanization 2. agricultural 3. F 4. Industrial Revolution 5. metropolitan 6. central-city dilemma 7. F 8. Overurbanization

URBAN ECOLOGY

Although every city is unique in certain respects, the patterns of all cities can be interpreted in terms of efforts by city residents to adapt to one another and to their environment. The processes and results of such adaptations are the subject matter of **urban ecology**—the study of the relationships between humans and their environments within cities.

Sociologists at the University of Chicago in the 1920s and 1930s who were studying how urban residents are affected by living in cities realized that such effects varied in different parts of the city. For example, Harvey Zorbaugh (1929), in a study of continuing importance, found significant differences between

Chicago's wealthy Gold Coast and its slum areas. The realization that such differences existed led to additional questions. Why do differences exist between areas of a city? How do different areas affect one another? What are the processes through which areas are changed? To answer these and other questions, the University of Chicago sociologists developed theories of urban ecology (Flanagan, 1993; Kleniewski, 1997).

Theories of City Growth

Three major theories of city growth have been developed. **Concentric zone theory** describes urban growth in terms of circular areas that grow from the central city outward. **Sector theory** emphasizes the importance of transportation routes in the process of urban growth. **Multiple nuclei theory** focuses specific geographic or historical influences on areas within cities. The three theories lead to quite different images of the use of space in urban areas. (See Figure 16.8.) However, these three theories tell us more when considered together than they do when considered separately. To understand why this is so, we must first examine each theory in greater detail.

What is concentric zone theory? Ernest Burgess (1925), like other early sociologists at the University of Chicago, was interested in explaining the causes and consequences of Chicago's growth. One result of his work was the concentric zone theory, which describes the process of city growth in terms of circles of distinctive zones that develop from the central city outward. As illustrated in Figure 16.8, the innermost circle is the *central business district,* the heart of the city, containing major government and private office buildings, banks, retail and wholesale stores, and entertainment and cultural facilities. Because land values in the central city are high, space is at a premium. The central business district contains a large proportion of a city's important businesses partly because the less important ones are unable to compete for the expensive space in the central business district.

The central business district exerts an especially strong influence on other parts of a city. This influence is especially clear in the zone immediately surrounding it. Burgess called this the *zone in transition* because it is in the process of changing. As new businesses and activities enter the central business district, the district expands by invading the zone next to it. This area may have been a residential area inhabited by middle- or upper-class families, who left because of the invasion of business activities. Most of the property in this zone is bought by those with little interest in the area. Rather than investing money in building maintenance, landowners simply extract rent from the property or sell it at a profit after the area has become more commercialized. Until the zone in transition is completely absorbed into the central business district (which may never occur), it is used for slum housing, warehouses, and marginal businesses that are unable to compete

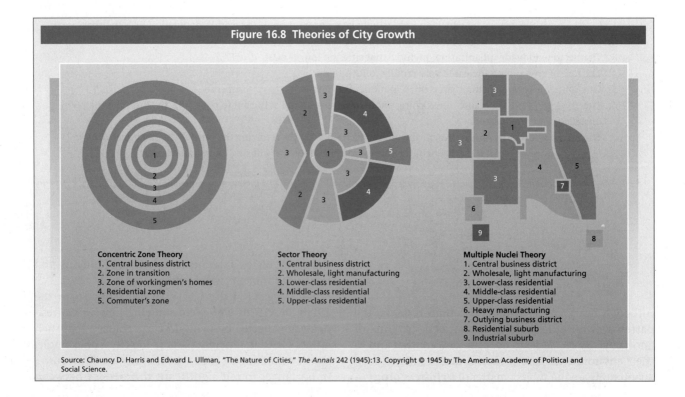

Figure 16.8 Theories of City Growth

Concentric Zone Theory
1. Central business district
2. Zone in transition
3. Zone of workingmen's homes
4. Residential zone
5. Commuter's zone

Sector Theory
1. Central business district
2. Wholesale, light manufacturing
3. Lower-class residential
4. Middle-class residential
5. Upper-class residential

Multiple Nuclei Theory
1. Central business district
2. Wholesale, light manufacturing
3. Lower-class residential
4. Middle-class residential
5. Upper-class residential
6. Heavy manufacturing
7. Outlying business district
8. Residential suburb
9. Industrial suburb

Source: Chauncy D. Harris and Edward L. Ullman, "The Nature of Cities," *The Annals* 242 (1945):13. Copyright © 1945 by The American Academy of Political and Social Science.

economically for space in the central business district itself. In short, the invasion of business activities creates deterioration in the zone in transition.

Surrounding the zone in transition are three zones devoted primarily to housing. The *zone of workingmen's homes* contains modest but stable neighborhoods populated largely by blue-collar workers. In the northern United States, the zone of workingmen's homes is often inhabited by second-generation immigrants who have had enough financial success to leave the deteriorating zone in transition. Next comes a *residential zone* containing mostly middle-class and upper middle-class neighborhoods. Single-family dwellings dominate this zone, which is inhabited by managers, professionals, white-collar workers, and some well-paid factory workers. On the outskirts of the city, often outside the official city limits, is the *commuters' zone,* which contains upper-class and upper middle-class suburbs.

What is sector theory?   Homer Hoyt (1939) developed sector theory on the basis of a study he performed for the Federal Housing Administration. Hoyt's study indicated that contrary to concentric zone theory, patterns of land use do not change according to distance from the central business district. Instead, land is more strongly affected by major transportation routes. As Figure 16.8 shows, sectors tend to be pie-shaped wedges going from the central business district to the city's outskirts. Each sector is organized around a major transportation route. Once a given type of activity is organized around a transportation route, its nature tends to be set. Thus, some sectors will be predominantly industrial, others will contain stores and professional offices, others will be "neon strips" with motels and fast-food restaurants, and still others will be residential sectors, each with its own social class and ethnic composition.

As in concentric zone theory, cities are circular in shape. But because of the importance of transportation routes extending from the central business district, the boundaries of many cities form a starlike pattern rather than a uniformly circular shape. The exact shape of a city, however, is not a major issue in sector theory. What is important is the tendency for patterns of land use to be organized around transportation routes.

What is multiple nuclei theory?   Many cities have areas that cannot be explained by either concentric zone or sector theory. For example, the Beacon Hill area of Boston defies concentric zone theory by remaining an expensive, prestigious neighborhood despite its location near the central business district. Because of such cases, Chauncy Harris and Edward Ullman (1945) have suggested that cities do not necessarily follow any particular pattern. (Figure 16.8 shows one example of a multiple nuclei city pattern.) A city may have several separate centers, some devoted to manufacturing, some to retail trade, some to residential use, and so on. These centers are said to develop with no particular reference to the central business district. Instead, they reflect such factors as geography and the history and traditions of particular areas. Because the factors affecting land use are unique to each city, multiple nuclei theorists have not predicted any particular pattern of land use that would apply to all cities. Of the three theories of city growth, this is by far the most flexible.

Which of these theories of city growth is correct? Although each of these theories calls attention to the dynamics of city growth, none of them applies to all cities. Each theory, however, does emphasize the importance of certain factors that cannot be overlooked by anyone interested in city growth. Concentric zone theory underscores the fact that growth in any one area of a city is part of the processes of segregation, invasion, and succession. This theory also reveals the importance of economic power in city growth: the distribution of space in cities is heavily influenced by those with the money to buy the land they want for the purposes they have in mind. Concentric zone theory is in need of some modification, however, because it is based on the assumption that cities will continue to grow and that change will occur as one area encroaches on another. Many central cities, however, have ceased to grow. In those cases, the "push-from-the-center dynamic no longer applies" (Choldin and Hanson, 1981:561).

Sector theorists have also contributed to an understanding of urban dynamics. As they have noted, transportation routes have a strong influence on cities. Decisions about the placement of railroad lines had important effects on the growth of cities in the nineteenth and early twentieth centuries. Highways and major streets have an even larger impact now.

Although multiple nuclei theory is vague in its predictions, the types of geographic and historical factors it emphasizes are also important for understanding any specific city. Moreover, the apparent decline of many central business districts may increase the importance of specific local factors in the future growth of cities.

No single ecological theory describes the growth pattern of all cities; each is valuable in itself. In recognition of this, more recent attempts to explain patterns of land use have combined the insights of all three theories. Several researchers have reported that people distribute themselves in cities in a way that can be predicted only by combining concentric zone theory and sector theory (Berry, 1965; Timms, 1971; Berry and Kasarda, 1977).

Different sectors, for example, attract people from different social classes. The geography and history associated with some transportation routes out of a city encourage people in higher social classes to settle there.

Other sectors, which are less appealing, are left to lower income people. Within any particular sector, however, family status and lifestyle seem to vary as concentric zone theory predicts. The newer single-family homes at the edge of the city attract different types of people than the apartment complexes and older homes that are likely to be found near the center of the city. Exact patterns of land use in any urban area, however, are

likely to depend on the cultural values that affect that urban area (Abu-Lughod, 1980).

Regardless of the ultimate fate of the concentric zone, sector, and multiple nuclei theories, the theories have performed an important service. By calling attention to the factors influencing land use in urban areas, these theories remind us that cities have not grown in a completely random manner.

1. Match the following zones of the concentric zone theory with the adjacent descriptions:
    ____ a. central business district
    ____ b. commuters' zone
    ____ c. residential zone
    ____ d. zone in transition
    ____ e. zone of workingmen's homes
    (1) one that is the next zone out from the zone in transition
    (2) site of major stores and offices
    (3) zone with physical deterioration stemming from few repairs
    (4) the zone the wealthiest people live in
    (5) the zone that is largely middle-class in composition
2. The theory that emphasizes the importance of transportation routes is called _____ theory.
3. The _____ theory contends that the growth of cities does not necessarily follow any specific pattern.
4. One of the ultimate contributions of the three major theories of city growth is their demonstration that cities do not grow in a random fashion. *T or F?*

Answers: 1. a. (2) b. (4) c. (5) d. (3) e. (1) 2. sector 3. multiple nuclei 4. T

QUALITY OF URBAN LIFE

As just discussed, sociologists from the University of Chicago contributed to our understanding of the pattern of land use in cities. Their interests were not, however, confined simply to the ecological characteristics of cities. These sociologists were also concerned with the effects of cities on people. Their basic conclusion: people in cities live a unique way of life. The distinctive way of life said to be developed in urban areas is referred to as **urbanism.**

Cities have always been thought to produce a wide variety of lifestyles. During medieval times, cities were viewed as places to escape the domination of powerful landowners, and it was said that "city air makes one free." Many of today's cities are also considered to be sources of intellectual and cultural diversity, sophistication, and freedom. Louis Wirth (1938), the University of Chicago sociologist who first developed the idea of urbanism as a way of life, agreed in part with the intellectual tradition associating city life with freedom and tolerance. He believed that because urban residents are confronted with a variety of lifestyles, they are more tolerant of diversity and less ethnocentric than residents of small towns or rural areas (Wilson, 1985, 1991). However, Wirth also drew on an intellectual tradition that saw impersonality and social disorganization in urban areas, a tradition represented in the work of German sociologist Ferdinand Tönnies. Later sociologists, as noted in the *Sociological Imagination* begin-

ning this chapter, have a more positive view of the quality of urban social life.

The Traditional Perspective on Urbanism: Loss of Community

What was Tönnies' viewpoint? Although he did not use the term *urbanism*, Tönnies was the first sociologist to contrast the way of life in rural and urban settings. As noted in Chapter 5 ("Social Structure and Society"), Tönnies distinguished between **gemeinschaft** (community) and **gesellschaft** (society). In the gemeinschaft—the type of society existing before the Industrial Revolution—social relationships were intimate because people knew each other. Daily life revolved around family and was based on tradition. There was a sense of a shared way of life and an interest in the common good of the community. Urbanization changes all that, according to Tönnies. In the gesellschaft, social relationships are impersonal because most people with whom one interacts are strangers. Although the immediate family is still important, the wider kinship network disappears. With the fading of tradition goes the sharing of common values. Interest shifts from the community's welfare to self-interest.

How did Wirth conceptualize urbanism? In Wirth's view, the urban conditions that promote sophistication, freedom, and diversity also involve a high cost. Urbanism, Wirth said, was marked by a depersonalization of human relationships. Rather than involving their total personalities in interaction, urban residents interact on the basis of rigidly defined roles. For example, a customer in a rural store may combine shopping with a lengthy conversation with the store's owner, but interactions in most stores in urban areas are usually confined to whatever is necessary to exchange money and products. This emphasis on rigid role prescriptions, Wirth concluded, makes it difficult for urban residents to maintain a sense of their own individuality and makes them indifferent and blasé. Heavy reliance on roles occurs in the city partly because there are so many people that it is impossible to know everyone personally. Urbanites may actually know more people than do rural residents, but they know fewer people well enough to interact with them on a personal basis.

According to Wirth, urban residents behave as they do because of three demographic characteristics of cities: large size, high density (a large number of residents per square mile), and much diversity within the population. Wirth contended that as cities become larger, more densely settled, and more socially diverse, residents become more sophisticated and more reliant on impersonal relationships to protect themselves from the claims of others.

Contemporary Perspective on Urbanism: Community Sustained

Wirth's theory of urbanism has been criticized on several grounds, the most frequent being that it is an overgeneralization. Some critics charge that Wirth's view of urbanism mistakenly assumes that all urban residents share a common lifestyle (Young, 1990). A leading spokesperson for this line of thought is Herbert Gans (1962b, 1968).

Do all urban residents share the same way of life? Gans has described five basic population groups found within inner-city areas, the first three of which live in the inner city by choice and are protected from the extreme negative consequences depicted by Wirth.

- *Cosmopolites.* People in this category are well educated and have high incomes. They live in inner-city areas because they are attracted to the areas' cultural advantages, such as art museums, theaters, and symphony orchestras.
- *The unmarried and childless.* People in this category also live in the inner city by choice. They may not feel a strong commitment to the inner city, but living there allows them to meet other people, and the area is convenient to jobs and entertainment facilities.
- *Ethnic villagers.* People in this category live in inner-city neighborhoods with strong ethnic identities. Residents of such neighborhoods are suspicious of the city as a whole and participate little in its affairs. The neighborhood itself, however, is an important source of intimate, enduring relationships.
- *The trapped.* Some people—such as the elderly living on pensions, or public assistance—cannot afford to leave the inner city. They may identify with their own neighborhood, but this does not

SOCIOLOGY IN THE NEWS

Urban Asian Americans

The exploding Asian American population in American cities is the topic of this CNN report. Asian Pacific Americans, the report states, is the fastest growing immigrant group in the United States. The 3 million Asian Pacific Americans in California will grow to 8.5 million by 2020; the 3.5 million in mid-Atlantic states will become over 20 million in three decades. Like all recent large immigrant groups, Asian Americans are concentrated in inner cities.

1. Into which of Herbert Gans's five basic population groups would the Asian Americans described in the CNN report most likely fall? Why?

   _____

   _____

2. Would you expect urban Asian Americans to engage in some form of a gemeinschaft-like social life? Explain.

   _____

   _____

Asian Americans immigrating to cities follow a family pattern similar to other urban immigrant groups. They tend to live in extended families, with two or more generations residing together.

All urban residents do not share the same way of life. Many of the people pictured here are "cosmopolites"— they are well-educated, have high incomes, and are drawn to the inner city because of its cultural advantages.

prevent them from becoming upset over changes in the central-city area, which they consider undesirable.

• *The deprived.* Poor minorities, the psychologically and physically challenged, some divorced mothers, and other deprived people see the city as a source of increased opportunities for employment or welfare assistance. These people's poverty, however, prevents such individuals from choosing where they will live in the city. These people are therefore exposed to the most extreme effects of population size, density, and diversity.

According to Gans, then, no blanket statement (such as that formulated by Wirth) applies to everyone living in the inner city. Although the trapped and the deprived are unable to escape many of the adverse effects described by Wirth, members of the other three categories have chosen to live in the inner city. They would not have made this choice if they did not see many advantages to living there.

Are city dwellers isolated from personal relationships?
According to Wirth's theory, urban residents have relatively few opportunities for participation in informal relationships involving family or friends. Subsequent research shows, however, that urban residents do participate in informal relationships and such relationships play an important role in maintaining a positive emo-

tional state (Kadushin, 1983; Creekmore, 1985; Fischer, 1982, 1984, 1995). In an intentional test of Wirth's theory, John Kasarda and Morris Janowitz (1974) found that population size and density do not necessarily produce a lack of involvement in local friendship, kinship, or other social relationships. Those who are not highly involved in such relationships in a large city tend to be those who have not lived there long enough to become a part of it. John Esposito and John Fiorillo (1974) found essentially the same thing in a study of working-class neighborhoods in New York City.

Other evidence indicates that families continue to provide urban residents with emotional support and mutual aid. (Refer to Chapter 11, "Family.") Studies based on observations of actual behavior in areas as diverse as lower-class slums and middle-class suburbs are showing the importance of personal relationships among family members and friends. For example, Gans (1962a) observed the West End of Boston in which many aspects of Italian rural life continue to be honored. Personal relationships are encouraged there, some have argued, because the area is dominated by one ethnic group. However, neighborhoods in general appear to be important sources of interaction and support. For example, a study of the Chicago slum of Addams by Gerald Suttles (1968) has shown that even in a low-income area populated by four different ethnic groups, a social order based on informal and personal relationships can develop.

FEEDBACK

1. The distinctive way of life allegedly developed in urban areas is referred to as _____.
2. According to Ferdinand Tönnies, urbanism corresponds to the (gemeinschaft/gesellschaft).
3. Louis Wirth relied on three characteristics to explain the development of urbanism as a way of life. What were they?
   a. age of city, population density, political structure
   b. population size, economic base, political structure
   c. population size, population density, population diversity
   d. economic base, political structure, population structure
4. According to Herbert Gans, which of the following has relatively little choice within the inner city?
   a. cosmopolites, the unmarried and childless
   b. ethnic villagers, the deprived
   c. the unmarried and childless, ethnic villagers
   d. the trapped, the deprived
5. Research shows that urban residents are largely isolated from family and friends. *T or F?*

Answers: 1. urbanism 2. gesellschaft 3. c 4. d 5. F

SUMMARY

1. Demography is the scientific study of population. Population data are very important today in part because of their use by government and industry.
2. Demographers use three population processes to account for population change. These processes, which are measured in various ways, are fertility, mortality, and migration.
3. Because population grows exponentially, it increases more each year in absolute numbers even if the rate of growth remains stable. This principle of population growth caused Thomas Malthus to predict that population size would ultimately outstrip the food supply, resulting in mass starvation and death.
4. More developed nations have avoided fulfilling Malthus's dire predictions largely through undergoing what is known as the demographic transition. This is the process accompanying economic development in which a population gradually moves from high birthrates and death rates to low birthrates and death rates. Europe seems to be experiencing a second demographic transition, which involves an eventual decline in population size.
5. In more developed societies, economic growth has led to a lower fertility rate for a number of reasons. But these societies' type of economic development, along with continued inequalities, has not led to a reduction in fertility in less developed countries. These latter countries may or may not pass through the demographic transition.
6. Although the world's population growth rate is declining after two hundred years of acceleration, the world's population will continue to grow. It is projected that world population growth will stop at just over twelve billion by the end of the twenty-first century.
7. As urban areas have increased in size, sociologists have gradually shifted their attention from communities to larger units, such as Metropolitan Statistical Areas and Consolidated Metropolitan Statistical Areas. However, the basic unit of study is the city, which is characterized by a large number of inhabitants, permanence, density, and occupational specialization of inhabitants.
8. Although the process of urbanization is now commonplace throughout the world, the first cities developed only five thousand to six thousand years ago. The first preindustrial cities developed in fertile areas where plant cultivation was sufficiently advanced to support a nonagricultural population. A surplus food supply permitted relatively large numbers of people to perform nonagricultural work while living in cities. Preindustrial types of cities still exist in underdeveloped parts of the world.
9. The Industrial Revolution, with its technological innovations and alterations in social structure, caused a major increase in the rate of urbanization. The development of factories was an especially important influence on the location of cities. Although cities and metropolitan areas have become quite important in industrialized societies, they are being strongly affected by suburbanization and a recent tendency for medium-sized metropolitan and nonmetropolitan areas to grow more rapidly than large metropolitan areas.
10. Suburbanization has created severe problems for central cities and their residents. The problems are particularly serious for minorities, the poor, and the elderly in the inner city.
11. Less developed countries are having a different experience with urbanization than did the West. In less developed countries, the urban growth rate has been faster, industrialization has been slower, and a base of small and middle-sized cities has not developed. Still, peasants in less developed countries migrate to cities both because they are forced off the land and because cities appear to provide unique opportunities.

REVIEW GUIDE

12. Urban ecologists are concerned with studying the processes human populations use to adjust to their environments. Urban ecologists have developed three major models of spatial development in cities: concentric zone theory, sector theory, and multiple nuclei theory. Each of the three theories has received some empirical support, but it appears that combining insights from all three will be the most useful.

13. According to the traditional view of urbanism, cities provide a free and sophisticated environment but are characterized by impersonality and social disorganization. Empirical research indicates that this description is not applicable to all urban residents and that many people in urban areas are involved in personal and social relationships.

LEARNING OBJECTIVES REVIEW

After careful study of this chapter, you will be able to:

- Distinguish among the three population processes.
- Discuss the major dimensions of the world population growth problem.
- Relate the ideas of Thomas Malthus to the demographic transition.
- Describe projected world population growth.
- Distinguish between the concepts of community and city and define the terms being used to signify differences in the size of urban areas.

- Trace the historical development of preindustrial and modern cities.
- Describe some of the major consequences of suburbanization.
- Discuss world urbanization.
- Compare and contrast the three theories of city growth developed by urban ecologists.
- Compare and contrast traditional and contemporary views of the quality of urban life.

CONCEPT REVIEW

Match the following concepts with the definitions listed below them:

____ a. gemeinschaft
____ b. multiple nuclei theory
____ c. crude death rate
____ d. urban ecology
____ e. fecundity
____ f. replacement level
____ g. sector theory
____ h. age-specific fertility

____ i. social demography
____ j. doubling time
____ k. urbanization
____ l. community
____ m. gross migration rate
____ n. Industrial Revolution
____ o. Consolidated Metropolitan Statistical Area (CMSA)

____ p. morbidity
____ q. demography
____ r. net migration rate
____ s. Metropolitan Statistical Area (MSA)
____ t. fertility rate
____ u. life span

1. The birthrate level at which married couples replace themselves in the population.
2. Rates of disease and illness in a population.
3. The maximum rate at which women can physically produce children.
4. A description of the process of urban growth emphasizing the importance of transportation routes.
5. The branch of demography that studies population in relation to its social setting.
6. A grouping of counties that contains at least one city of 50,000 inhabitants or more or an "urbanized area" of at least 50,000 inhabitants and a total metropolitan population of at least 100,000.
7. Tönnies's term for the type of society based on tradition, kinship, and intimate social relationships.

8. A social group whose members live within a limited geographic area and whose social relationships fulfill major social and economic needs.
9. The annual number of live births per one thousand women ages fifteen to forty-four.
10. The annual number of deaths per one thousand members of a population.
11. The annual increase or decrease per one thousand members of a population resulting from people leaving and entering the population.
12. The scientific study of population.
13. A description of the process of urban growth that emphasizes specific and historical influences on areas within cities.

14. The most advanced age to which humans can survive.
15. The combination of technological developments beginning in the eighteenth century that created major changes in transportation, agriculture, commerce, and industry.
16. The number of live births per one thousand women in a specific age group.
17. The study of the relationships between humans and their environments within cities.
18. An urban area composed of two or more sets of neighboring Primary Metropolitan Statistical Areas (PMSAs).
19. The number of years needed to double the size of a population.
20. The process by which an increasingly larger proportion of the world's population lives in urban areas.
21. The annual number of persons per one thousand members of a population who enter or leave a geographic area.

CRITICAL THINKING QUESTIONS

1. Suppose that the average number of children per family in the United States increased from 1.8 to 3. Discuss the short- and long-term effects this would have on population growth._____
_____

2. Should the slow-growing nations, such as the United States, have immigration policies that help relieve the population pressures on countries with high-growing populations? Defend your position using sociological concepts. _____
_____
_____

3. Explain and evaluate the usefulness of the following theories of city growth: concentric zone theory, sector theory, and multiple nuclei theory. _____
_____
_____

4. According to Louis Wirth, what are the major characteristics of urbanism as a way of life? Evaluate the accuracy of Wirth's explanation in view of more recent research findings._____
_____
_____

MULTIPLE CHOICE QUESTIONS

1. **The three population processes that are responsible for population growth and decline are**
   a. fecundity, morality, and migration.
   b. fecundity, morbidity, and migration.
   c. fertility, morbidity, and migration.
   d. fertility, mortality, and migration.
   e. fecundity, mortality, and morbidity.

2. **Which of the following *best* describes contemporary Americans' behavior regarding reproduction?**
   a. Fewer American women are having abortions.
   b. Fewer Americans are using contraceptives.
   c. More American women are having children in their late teens and early twenties.
   d. More American women are now beginning to have larger families.
   e. Fewer American women are having children, and more American women are having fewer children.

3. **The demographic term for deaths within a population is**
   a. morbidity.
   b. mortality.
   c. terminality.
   d. finality.
   e. turnover.

4. **Rates of disease and illness in a population are known as**
   a. life expectancy.
   b. mortality.
   c. morbidity.
   d. life span.
   e. social pathology.

5. **The movement of people from one geographic area to another for the purpose of establishing a new residence is known as**
   a. fecundity.
   b. exponential transition.
   c. migration.
   d. morbidity.
   e. the demographic transition.

6. **_____ is the principle by which a population increases.**
   a. Linear growth
   b. Natural logarithms
   c. Multiplication
   d. Exponential growth
   e. Arithmetic explosion

7. **For Thomas Malthus, the solution to the population problem was**
   a. early marriage.
   b. improved contraception.
   c. universal education.
   d. federally financed abortions.
   e. infanticide.

8. **_____ is the process by which a population gradually moves from high birthrates and death rates to low birthrates and death rates as a result of economic development.**
   a. Exponential growth
   b. The demographic transition
   c. The dependency ratio
   d. The fertility-morbidity ratio
   e. The net migration rate

9. **According to the text, there are three areas of the world presently in the fifth stage of the demographic transition. These areas are**
   a. China, Japan, and North America.
   b. Australia, Iran, and Europe.
   c. South America, Europe, and North America.
   d. Iran, North America, and Europe.
   e. Europe, North America, and Japan.

10. **The primary characteristic of the "second demographic transition" is**
    a. a decline in fertility from slightly above replacement level to slightly below.
    b. a reduction in the net migration rate.
    c. an increase in mortality.
    d. the progress achieved regarding adult literacy.
    e. the shift from gesellschaft to gemeinschaft patterns of social organization.

11. **_____ has been reached when deaths are balanced by births so that a population does not increase.**
    a. Zero population growth
    b. Exponential growth
    c. Equilibrium ratio
    d. The fecundity level
    e. Demographic transition

12. **The primary resource needed by a preindustrial city was**
    a. a strong trading center.
    b. a surplus food supply.
    c. adequate employment opportunities.
    d. a good system of waterways.
    e. a completed demographic transition.

13. **Metropolitan dispersal refers to**
    a. the spreading out of industry throughout a city.
    b. the dispersal of retail outlets outward due to inner-city deterioration.
    c. middle- and upper-class whites buying up downtown property after urban renewal.
    d. cities losing population to the areas surrounding them.
    e. cities gaining population over surrounding areas.

14. **Which of the following theories of city growth explicitly deals with the "zone in transition"?**
    a. concentric zone theory
    b. overurbanization theory
    c. sector theory
    d. ecological invasion theory
    e. multiple nuclei theory

15. **Which of the following statements *best* describes the personal relationships of city dwellers?**
    a. Urban residents conduct their social relationships on the basis of rigidly defined role prescriptions.
    b. Urban dwellers rely on impersonal relationships to protect themselves from the social claims of others.
    c. Kinship networks have disappeared from large cities.
    d. Urban residents know few people well enough to interact on a personal basis.
    e. Large population size does not necessarily produce a lack of involvement in local friendship, kinship, and other social relationships.

FEEDBACK REVIEW

True-False

1. Thanks to worldwide urbanization, preindustrial cities no longer exist. *T or F?*
2. One of the ultimate contributions of the three major theories of city growth is their demonstration that cities do not grow in a random fashion. *T or F?*

Fill in the Blank

3. Thomas Malthus sees famine as a _____ check on population growth.
4. Preindustrial cities first began to develop in areas with an _____ surplus.
5. _____ occurs when a city is unable to supply adequate jobs and housing for its inhabitants.
6. The distinctive way of life allegedly developed in urban areas is referred to as _____.

Multiple Choice

7. Which of the following figures is the world's population most likely to reach before it stops growing?
   a. four billion
   b. eight billion
   c. twelve billion
   d. twenty-five billion
8. Which of the following characteristics is *not* included in the definition of a city?
   a. permanence
   b. dense settlement
   c. advanced culture
   d. occupational specialization in nonagricultural activities
9. According to Herbert Gans, which of the following has relatively little choice within the inner city?
   a. cosmopolites, the unmarried and childless
   b. ethnic villagers, the deprived
   c. the unmarried and childless, ethnic villagers
   d. the trapped, the deprived

Matching

10. Match the following zones of the concentric zone theory with the adjacent descriptions:

    ____ a. central business district
    ____ b. commuters' zone
    ____ c. residential zone
    ____ d. zone in transition
    ____ e. zone of workingmen's homes

    (1) one that is the next zone out from the zone in transition
    (2) site of major stores and offices
    (3) zone with physical deterioration stemming from few repairs
    (4) the zone the wealthiest people live in
    (5) the zone that is largely middle-class in composition

GRAPHIC REVIEW

Metropolitan residence by race since 1970 is displayed in Table 16.5. Answer these questions to test your grasp of the meaning of these data.

1. What conclusions about race and residence can be drawn from these data? Speak to both the situation within decades as well as across decades.

    _____

    _____

2. Describe the central-city dilemma. Relate the data in this table to the creation and possible amelioration of this problem.

    _____

    _____

ANSWER KEY

Concept Review				Multiple Choice	Feedback Review
a.	7	l.	8	1. d	1. F
b.	13	m.	21	2. e	2. T
c.	10	n.	15	3. b	3. positive
d.	17	o.	18	4. c	4. agricultural
e.	3	p.	2	5. c	5. Overurbanization
f.	1	q.	12	6. d	6. urbanism
g.	4	r.	11	7. c	7. c
h.	16	s.	6	8. b	8. c
i.	5	t.	9	9. e	9. d
j.	19	u.	14	10. a	10. a. 2
k.	20			11. a	b. 4
				12. b	c. 5
				13. d	d. 3
				14. a	e. 1
				15. e	

Chapter Seventeen

Collective Behavior and Social Movements

After careful study of this chapter, you will be able to:

- Describe the various social activities engaged in by dispersed collectivities.
- Describe the basic nature of a crowd and identify the basic types of crowds.
- Compare and contrast contagion theory and emergent norm theory.
- Define the concept of social movement and identify the primary types of social movements.
- Apply the concepts of relative deprivation and unfulfilled rising expectations to the emergence of social movements.
- Compare and contrast relative deprivation theory with value-added theory and resource mobilization theory.
- Trace the phases in the life course of social movements.

SOCIOLOGICAL IMAGINATION

Is crowd behavior basically emotional and irrational? An affirmative answer to this question seems to be the obvious response. Earlier, even most sociologists assumed crowd behavior to be without reason or structure. Sociologists today, however, recognize the structure and rationality inherent in crowd behavior, even in a revolutionary mob toppling a government. Such revolutionaries have a common purpose and organize their behavior to achieve their shared objective.

Mob behavior does not fit in with the normal flow of events in social life. Sociologists, however, are not without an explanation for even the most apparently unstructured social phenomenon. These social events come under the heading of *collective behavior.*

DEFINING COLLECTIVE BEHAVIOR

Collective behavior refers to the relatively spontaneous and unstructured social behavior of people who are responding to similar stimuli. It is *collective* because it usually takes place among a relatively large number of people. The phrase "responding to similar stimuli" means that collective behavior is a reaction on the part of people to some person or event outside themselves. Episodes of collective behavior involve social interaction in which participants influence one another's behavior.

The study of collective behavior poses some particular problems for sociology. In the first place, sociologists are accustomed to studying structured behavior.

Second, how are researchers going to investigate a social phenomenon that occurs spontaneously? Despite these difficulties, sociologists have managed to conduct some interesting research and formulate some useful theories on types of collective behavior. This is partly because there is more structure and rationality to collective behavior than appears on the surface.

Presentation of the forms of collective behavior starts with the more disorganized, unplanned, and short-term forms of collective behavior, including rumors, mass hysteria, panics, fads, fashions, and publics. Next crowds and riots are covered. Social movements, the most highly structured, enduring, and rational form of collective behavior, closes the chapter.

DISPERSED COLLECTIVITIES

In the more structured forms of collective behavior, such as crowds and social movements, people are in physical contact with one another. Other forms of collective behavior occur among dispersed members of a mass society. These *dispersed collectivities* engage in the less structured forms of collective behavior—rumors, mass hysteria, panics, fads, fashions, and publics. Ralph Turner and Lewis Killian point out that the behavior occurring among members of dispersed collectivities is not highly individualized or atomistic. Dispersed collectivities display a uniformity of response to some common object of attention and act in awareness of membership in a collectivity:

> When people are scattered about, they can communicate with one another in small clusters of people; all of the members of a public need not hear or see what every other member is saying or doing. And they can communicate in a variety of ways—by telephone, letter, Fax machine, computer linkup, as well as through second-, or third-, or fourth-hand talk in a gossip or rumor network (Goode, 1992:255).

Rumors

In the *Aeneid,* Virgil wrote these lines: "Rumor! What evil can surpass her speed? In movement she grows mighty, and achieves strength and dominion as she swifter flies." Rumor, as Virgil's words underscore, has a very bad reputation. Rumors may be benign, as in the case of continual Elvis Presley sightings, or they can do considerable damage, as you will see. They are often communicated as the truth when in fact they may be false. At best, rumors are usually inaccurate and misleading. In any event, the likelihood of a rumor being spread depends in part on the degree of anxiety a person feels, the extent of uncertainty the person is experiencing about events, the credibility of the person

Because rumors may be true or false, they should be viewed skeptically. For example, you would have been misled had you believed that the crash of TWA Flight 800 was due to "friendly" fire from a U.S. warship.

passing on the rumor, and the relevance of the rumor to the person hearing it (Rosnow, 1988, 1991).

What is a rumor?   A **rumor** is a widely circulating story whose truth is questionable. Rumors usually focus on people or events that are of great interest to others. Part of the mass media makes its fortune on the public's attraction to rumors. There are magazines devoted exclusively to the private lives and loves of rock idols and movie stars; tabloid newspapers filled with titillating stories based on guesswork, half-truths, and innuendos; and more "respectable" national publications that cater to the public's desire to learn about the private lives of the rich, famous, and offbeat. As these examples suggest, rumors and gossip are closely related.

Three days after the news of the nuclear accident at Three Mile Island near Harrisburg, Pennsylvania, a rumor flashed throughout the surrounding area. The plant was said to be about to explode, which would devastate everything nearby and spread radioactive fallout for miles around. Another rumor had it that McDonald's restaurants increased the protein content of their hamburgers with ground worms. According to another rumor, the combination of a soft drink and Pop Rocks candy would cause stomachs to explode. A rumor circulated after Saddam Hussein invaded Kuwait in the summer of 1990 that Iraqi troops had crossed the Kuwaiti border into Saudi Arabia. The rumor drove up the price of the dollar for several hours on the world foreign exchange markets (Frankel, 1990). Rumors swirled around Michael Dukakis, the then-governor of Massachusetts, during his 1988 presidential campaign against George Bush. One of the most widely dispersed

rumors was that Dukakis had twice been under psychiatric care for depression. Within a matter of days, Dukakis dropped eight points in the national public opinion polls (Johnson, 1991). The crash of TWA Flight 800 was attributed by some to the accidental firing of a surface-to-air missile from an American warship (Hosenball, 1996). None of these rumors proved to be true, but they were spread and believed in part because they touched on people's insecurities, uncertainties, and anxieties (Rosnow, 1991).

The effects of rumors are not always frivolous. The damage caused by rumors is revealed in this statement from the *Report of the National Advisory Commission on Civil Disorders,* which focused on the urban riots and violence in large American cities during the summer of 1967.

> *Rumors significantly aggravated tension and disorder in more than 65 percent of the disorders studied by the Commission. Sometimes, as in Tampa and New Haven, rumor served as the spark which turned an incident into a civil disorder. Elsewhere, notably Detroit and Newark, even where they were not precipitating or motivating factors, inflaming rumors made the job of police and community leaders far more difficult (Report of the* National Advisory Commission on Civil Disorders, *1968:326).*

Potentially harmful rumors persisted into the winter of 1967–1968 following the "long, hot summer" of 1967. One of these rumors was the Detroit castration rumor.

> *A mother and her young son are shopping at a large department store. At one point the boy goes to the lavatory. He is a*

long time returning, and the mother asks the floor supervisor to get him. The man discovers the boy lying unconscious on the floor. He has been castrated. Nearby salesclerks recall that several teenage boys were seen entering the lavatory just before the young boy and leaving shortly after he was discovered (Rosenthal, 1971:36).

In the white version, the castrated youth was white and his assailants black; in the black version, the colors were reversed. In a city as volatile and frightened as Detroit at that time, this rumor seemed quite believable and could have led to further rioting and violence.

Akin to rumors are what Jan Harold Brunvand calls *urban legends* (Brunvand, 1981, 1984, 1986, 1989). Although urban legends may incorporate current rumors, they tend to have a longer life and wider acceptance. Consider the case of "The Baby in the Oven," in which a babysitter gets high on drugs while babysitting and does a tragic thing:

This couple with a teenage son and a little baby left the baby with this hippie-type girl who was a friend of the son's. They went to a dinner party or something, and the mother called in the middle of the evening to see if everything was all right. "Sure," the girl says. "Everything's fine. I just stuffed the turkey and put it in the oven." Well, the lady couldn't remember having a turkey, so she figured something was wrong. She and her husband went home and they found the girl had stuffed the baby and put it in the oven. Now the son used a lot of drugs; she was his friend, so I guess they figure she took them, too... (Brunvand, 1980:55).

Like rumors, urban legends permit us to play out some of our hidden fears and guilt feelings by being shocked and horrified when we hear the story involving other people. "The Baby in the Oven" story may represent guilt feelings for sometimes leaving our children with strangers or a deep fear of outsiders entering our homes.

Mass Hysteria and Panics

What is mass hysteria? **Mass hysteria** exists when collective anxiety is created by acceptance of one or more false beliefs. Orson Welles's famous "Men from Mars" radio broadcast in 1938, though based entirely on H. G. Welles's novel *The War of the Worlds,* caused nationwide hysteria. About one million listeners became frightened or disturbed and thousands of Americans hit the road to avoid the invading Martians, and telephone lines were jammed as people shared rumors, anxieties, fears, and escape plans (Houseman, 1948; Cantril, 1982; Barron, 1988).

Other examples of mass hysteria are the response to imagined witches in seventeenth-century Salem, Massachusetts (refer to *Doing Research* in Chapter 7,

"Deviance and Social Control"), and the reaction of many Americans to AIDS following the death of actor Rock Hudson in the mid-1980s. In Salem, twenty-two people labeled witches died—twenty by execution—before the false testimony of several young girls began to be questioned. The mass hysteria dissipated only after the false beliefs were discredited. There has been some hysteria in the United States regarding AIDS. A 1987 Gallup poll showed that a substantial proportion of Americans held to false beliefs regarding the spread of AIDS—30 percent believed insect bites can spread the disease, 26 percent related the spread to food handling or preparation, 26 percent thought AIDS could be transmitted via drinking glasses, 25 percent saw a risk in being coughed or sneezed upon, and 18 percent believed that AIDS could be contracted from toilet seats (Gallup, 1988). These mistaken ideas persisted on a widespread basis for a while despite the medical community's conclusion that AIDS is spread through sexual contact, by sharing hypodermic needles, and by transfusion of infected blood. As of the late 1990s, toleration, compassion, and understanding of AIDS had increased substantially.

What is a panic? A **panic** occurs when people react to a real threat in fearful, anxious, and often self-damaging ways. Panics usually occur in response to such unexpected events as fires, invasions, and ship sinkings. Over 150 people, for example, died in the Kentucky Beverley Hills Supper Club in 1977 when a panic reaction to a fire caused a jamming of the escape routes.

Interestingly enough, people often do not panic after natural disasters such as earthquakes and floods. Although panics may occur at the outset, major natural catastrophes usually lead to highly structured behavior (Erikson, 1976; Dynes and Tierney, 1994).

Fads and Fashions

What are fads? **Fads** are unusual patterns of behavior that spread rapidly, are embraced zealously, and disappear after a short time. The widespread popularity of a fad rests largely on its novelty. In the 1950s, college students held contests to see who could get the most people inside a telephone booth or automobile. Somewhat more novel was "streaking" (running naked across college grounds or through occupied classrooms) which delighted students in the early 1970s (Aguirre, Quarantelli, and Mendoza, 1988). A reminder of that fad occurred in the summer of 1991 when a sixteen-year-old female, wearing only her eyeshadow, ran down the first fairway at the British Open golf tournament in Southport, England. Ironically, on that same day, two male fans celebrated the Atlanta Braves' five-run sixth inning performance by baring it all on the field. One of these streakers impressed spectators and players alike

A fad is a form of collective behavior that is unusual, spreads quickly, is zealously adopted, and is short-lived. The rage over Beanie Babies is a recent example of a fad.

with a head-first slide into homeplate. More recent fads include Beanie Babies, body tattoos, "washboard abs," the Macarena, and cigar smoking (Hamilton, 1997).

What are fashions?   Fashions are patterns of behavior that are widely approved but expected to change periodically. Although the most widely recognized examples are related to appearance (clothing, jewelry, hairstyles), fashions also come and go in such diverse areas as automobile design, home decorating, architecture, and politics. As societies modernize, fashion becomes more important and changes more rapidly. Traditional dress for both sexes changed very little among premodern Native Americans and Africans. In contemporary American society, clothing styles change every year. Women can hardly keep up with the proper dress or skirt length, and men are having a harder time determining whether cuffs on pants are fashionable or passé. By the early 1990s piercing had become a fashion rage. Americans, from teenagers to adults, were adorning themselves with a variety of objects inserted in holes in their tongues, noses, navels, and nipples (Rogers, 1993).

Fashion goes beyond clothing and other aspects of personal appearance. Slang goes in and out of favor (Lofland, 1993). What self-respecting teenager today would be caught calling something "neat"? "Awesome" would seem appropriate as a one-word description of something impressive.

Why is fashion more important in modern society?   In the first place, modern societies are based on growing economies fueled by mass consumption. Styles in clothing, automobiles, eyewear, and sporting equipment must change if profits are to be made and people employed. Second, without traditions to supply the brakes, people in modern societies are eager to respond to new fashions constantly being created by entrepreneurs and established corporations. Finally, the relative affluence in modern societies provides enough disposable income for people to indulge their desire for novelty and change in fashion.

Publics and Public Opinion

Politicians are constantly referring to the "American public." What do they mean by this? Do they suppose that all Americans think alike? They certainly seem to when they make statements such as "The American public does not want tax increases" or "The American public cares about the environment." Actually, the opinions of Americans are not the same; opinions vary with the issue involved, and people disagree on any given issue. Therefore, it clouds understanding to think of anything so broad as *the* American public. If, however, it is not meaningful to speak of the American public as a total entity, it is proper to think of the existence of many publics in American society. Sociologists are careful to distinguish between publics and public opinion.

What is the difference between a public and public opinion?   A **public** is composed of people who are concerned about an issue on which opinions differ. Publics form around such issues as drug abuse, drunk driving, civil rights, women's rights, environmental protection, gun control, prayer in schools, and the waging of war. People in a public not only disagree on an issue, but also recognize the right to disagree. Because this makes debate legitimate, people concerned about an issue feel free to express their opinions on it. The public's expression of attitudes and beliefs about an issue is called **public opinion.**

How are publics and public opinion formed?   The process of forming publics and the opinions publics express does not usually begin with a clear-cut issue. Before what Ralph Turner and Lewis Killian (1987) call the "crystallization" phase, an issue around which a public eventually forms is unfocused. Unrest and discussion exist, but clear lines of demarcation have not been drawn. Although families, friends, and small discussion groups may express themselves on the issue, public debate has not yet begun. The issue is crystallized only when some galvanizing event occurs or when some public figure is able to transform widespread but unfocused concern into a sharply defined issue.

An example of creating a public through the crystallization of an issue is provided by reactions to a disease new to the United States in the 1980s. AIDS, a fatal disease with no known cure, was discovered in the homosexual population in 1981. Although unrest and concern existed, they were not focused, because it seemed that only homosexuals gave the disease to one another. Panic ensued, however, as AIDS showed up in heterosexual married women and in children. A public still was not formed until the latter part of 1985, when this question arose: Should a child either with AIDS or simply carrying the virus be allowed to attend public school? A national debate, complete with emotional demonstrations, ensued. According to civil liberties advocates, denying such a child the right to attend public school is discriminatory, in part because AIDS is not passed through casual contact. Parents citing the obligation of public schools to provide children with a safe learning environment expressed fear that the disease could be contracted nonsexually, as through blood from playground accidents.

The crystallization process is aided by certain techniques. One is the dissemination of propaganda; another is the formation of interest groups.

Propaganda is information consciously and systematically disseminated to influence the decisions, actions, and beliefs of people on an issue. Although the information may be true or false, propagandists attempt to manipulate public opinion by camouflaging the real purpose of their campaign and by presenting only one side of the issue. In the AIDS case, all sorts of false information was spread, including the idea that the disease could be picked up from toilet seats.

As noted in Chapter 13 ("Political and Economic Institutions"), an **interest group** is composed of people who organize to achieve some specific shared goal by influencing political decision making. Publics usually form interest groups during the crystallization of an issue, interest groups that continue to function long after the issue has ceased to be at the top of the national agenda. Thus, the National Organization of Women (NOW), the National Rifle Association (NRA), the National Association for the Advancement of Colored People (NAACP), and the American Association of Retired Persons (AARP) continue to press for the issues of concern to their publics in the political process.

CROWDS

The Nature and Types of Crowds

The best-known and most dramatic form of collective behavior is the crowd. Crowd behavior is interesting to most of us because it involves intense emotions. If the content of local and national television news and our long-standing interest in disaster movies are any indication, people are alternately fascinated and horrified by the unleashing of basic passions that often occurs in crowds. We are engrossed by reports of mobs, rioters, and crowds celebrating a joyous event or lamenting a sad one.

What is a crowd? A **crowd** is a temporary collection of people who share a common point of interest. The temporary residents of a large campground, each occupied with his or her own activities and interests, constitute an aggregate. But if some event, such as the landing of a hot-air balloon or the appearance of a bear, serves as a common stimulus to draw the campers together, a crowd exists. What will happen next is highly unpredictable. The specific subject of mutual interest—the common stimulus—is not important. What is significant is the fact that a number of individuals have been drawn together in a situation that has the potential to stimulate an emotional reaction.

A crowd situation involves ambiguity and uncertainty; participants have no definite preconceived ideas about the way they should behave toward one another or toward some target on which their attention is converging. Members of a crowd, however, are certain about one thing—they share the urgent feeling that something either is about to happen or should be made to happen (Marx and McAdam, 1994; McPhail, 1994).

What are the basic types of crowds? Although it is accurate to make generalizations about crowds, not all crowds are alike. Herbert Blumer (1969a) has distinguished four basic types of crowds. A *casual crowd* is the least organized, least emotional, and most temporary

1. A _____ is a widely circulating story where truth is questionable.
2. _____ exists when collective anxiety is created by acceptance of one or more false beliefs.
3. A _____ occurs when people react to a genuine threat in fearful, anxious, and often self-damaging ways.
4. Unusual patterns of behavior that spread rapidly, are embraced zealously, and disappear after a short time are called _____.
5. _____ are patterns of behavior that are widely approved but expected to change periodically.
6. A _____ is composed of people who are concerned about an issue on which opinions vary.
7. _____ refers to information consciously and systematically disseminated to influence the decisions, actions, and beliefs of people on an issue.

Answers: 1. rumor 2. Mass hysteria 3. panic 4. fads 5. Fashions 6. public 7. Propaganda

A "conventional" crowd gathers for a particular purpose and follows some predetermined norms defining appropriate behavior. It is expected that the Armenian and American fans at this soccer match would prominently display their respective national flags.

type of crowd. Although a casual crowd shares some point of interest, it is minor and fades quickly. Members of a casual crowd may gather with others to observe the aftermath of an accident, to watch someone threatening to jump from a building, or to observe an energetic child tap-dancing in the street in the French Quarter of New Orleans.

A *conventional crowd* has a specific purpose and follows some existing guidelines defining appropriate behavior. People watching a film, flying on a chartered flight to their university's ballgame, or observing a tennis match are in conventional crowds. As in casual crowds, there is little interaction among members of conventional crowds. The fact that the activity of the conventional crowd follows some established procedures distinguishes this type of crowd from a casual crowd.

Expressive crowds have no purpose beyond unleashing emotion. Their members are collectively caught up in a dominating, all-encompassing mood of the moment. Free expression of emotion—yelling, crying, laughing, jumping—is the main characteristic of this type of crowd. Religiously fervent fundamentalist worshipers, hysterical fans at a rock concert, the multitude gathered at Times Square on New Year's Eve, and the some 250,000 Americans at Woodstock 1994 are all examples of expressive crowds.

Finally, a crowd that takes some action toward a target is an *acting crowd*. This type of crowd concentrates intensely on some objective and engages in aggressive behavior to achieve it. A conventional crowd may become an acting crowd, as when a crowd of European soccer fans abandons the guidelines for watching the match in order to attack the officials. Similarly, an expressive crowd may become an acting one, as in the case of fans celebrating in the streets after their city's

team has won the Super Bowl who end up overturning cars and destroying other types of property. Mobs and riots are two major types of acting crowds.

What are mobs and riots?   A **mob** is an emotionally stimulated, disorderly crowd that is ready to use destructiveness and violence to achieve a specific purpose. Through a group understanding reached via verbal and nonverbal behavior, a mob knows what it wants to do and considers all other things as distractions. In fact, individuals who are tempted to deviate from the mob's purpose are pressured to conform. If a mob is storming a seat of government, for example, it is inappropriate for participants to waste their time raping or stealing, although that may come later. Concentration on the main event is maintained by strong leadership.

Mobs have a long and violent history. The French Revolution is a classic example of mob action. With the cry "To the Bastille! To the Bastille!" a Parisian mob stormed this symbol of their oppression on July 14, 1789.

The population of Paris was aroused, the unruly element of the city was in the streets, their wrath directed against the prison-fortress, the bulwark of feudalism, the stronghold of oppression, the infamous keeper of the dark secrets of the kings of France. The people had always feared, always hated it, and now against its sullen walls was directed the torrent of their wrath (Morris, 1893:269).

The overthrow of royalty and the desire for political freedom were the larger objectives of the French Revolution, but this mob's more specific and limited goal was to destroy the symbol of tyranny, oppression, and fear. As in all revolutions, there were other mob actions (Tilly, 1992). In October 1789, a largely female mob forcibly entered the royal palace at Versailles, and peasants throughout France overthrew their feudal masters. Ultimately, King Louis XVI and Queen Marie Antoinette were escorted to Paris for execution by a mob.

The formation of mobs is not limited to revolutions. During the mid-eighteenth century, American colonists mobbed tax collectors as well as other political officials appointed by the British. During the Civil War, more than one thousand people were killed or injured as armed mobs protested against the Union Army's draft. Mobs in the American South have acted as judges, juries, and executioners in the lynching of African Americans (as well as some whites) since the end of the nineteenth century. (It is estimated that approximately five thousand lynchings occurred between 1882 and 1968; no such illegal hangings have been recorded since 1968.) Early labor-management relations in the United States were marked with mob violence. The 1960s saw mass violence in urban ghettos and on college campuses. In 1989 mob violence occurred in response to Iran's Ayatollah Khomeini's death threat to Salman Rushdie, author of *The Satanic Verses*. At least thirty-five people, including fifteen policemen, were injured in rampaging mob action in New Delhi, India. Later six anti-Rushdie demonstrators were killed by police in Pakistan, and ten protestors were killed in clashes with police in Bombay, India.

Some acting crowds, although engaged in deliberate destructiveness and violence, do not have the mob's sense of common purpose. These episodes of crowd destructiveness and violence are **riots.** Riots involve a much wider range of activities than mob action. Whereas a mob surges to burn a particular building, to lynch an individual, or to throw bombs at a government official's car, rioters often direct their violence and destructiveness at targets simply because they are convenient. The ghetto riots in many large American cities during the summer of 1967 occurred against a background of massive unemployment, uncaring slum landlords, poverty, discrimination, and charges of excessive police brutality. The actions of the rioters in the twenty-odd cities involved did not, however, lash out at the underlying cause of the riots—the enduring gap between black and white America (*Report of the National Advisory Commission on Civil Disorders*, 1968). Although the rioters were protesting against discrimination and deprivation and although white-owned businesses were damaged more frequently than those owned by African Americans, the fact remains that African Americans did more damage to themselves (the overwhelming majority of the dead were African American) and their neighborhoods than they did to the establishment. As further evidence of some rather random behavior in riots, many participants saw the riots as a way to loot stores or to have some destructive, violent fun.

Riots are not just something that happened in the 1960s. A vicious race riot occurred in Miami in 1980 (Porter and Dunn, 1984). Several rioting incidents occurred in Europe in 1985 (Moody, 1985). In West Germany, thousands of youths rampaged in sixteen cities; five days of violence (smashed windows, looted shops, burned cars) resulted in dozens of injured people and about five hundred persons being detained by the police. Violence also occurred in south London in 1985 in an area hard-hit by race riots in 1981. After two nights of arson and looting, seventy-four people were injured and over two hundred arrested. Later in that same week, rioting broke out in Liverpool, England, an area that had also seen racial disturbances in 1981. In the aftermath of the 1992 acquittals of the police officers charged in the brutal beating of Rodney King, Los Angeles experienced America's deadliest riots in twenty-five years. Two days of rioting following the acquittals left the City of Angels with at least fifty-three dead, 2,300 injured, over 16,000 arrested, and an estimated $800 million in damage from looting and burning (Duke and Escobar, 1992; Mathews, 1992).

Theories of Crowd Behavior

Some theories have been developed to explain crowd behavior, including contagion theory and emergent norm theory.

What is contagion theory? **Contagion theory** emphasizes the irrationality of crowds that is created by participants stimulating one another to higher and higher levels of emotional intensity. This theory has its roots in the classic work of Le Bon (1960, originally published in 1895), a French aristocrat who disdained crowds composed of the masses. People in crowds, Le Bon thought, were reduced to a nearly subhuman level:

> By the mere fact that he forms part of an organized crowd, a man descends several rungs in the ladder of civilization. Isolated, he may be a cultivated individual; in a crowd, he is a barbarian—that is, a creature acting by instinct. He possesses the spontaneity, the violence, the ferocity, and also the enthusiasm and heroism of primitive beings... (Le Bon, 1960:32).

Herbert Blumer (1969a) has attempted to avoid Le Bon's elitist biases in reformulating contagion theory. For Blumer, the basic process in crowds is "circular reaction"—people mutually stimulating one another. There are three stages to this process. In *milling,* the first stage, people move around in an aimless and random fashion, much like excited herds of cattle or sheep. The primary effect of milling is the establishment of rapport.

Through milling, people become increasingly aware of and sensitive to one another; they enter something akin to a hypnotic trance. All of this prepares the crowd to act in a concerted and spontaneous way.

The second stage, *collective excitement,* is a more intense form of milling. At this stage, crowd members become impulsive, unstable, and highly responsive to the actions and suggestions of others. Individuals begin to lose their personal identities and take on the identity of the crowd.

The last stage, *social contagion,* is an extension of the other stages. Behavior in this stage involves rigid, unthinking, and nonrational transmission of a mood, impulse, or form of behavior (Blumer, 1969a). Fans at soccer games in Europe have launched attacks on the referees of such proportion that games have been interrupted and people killed or injured. Taking a less extreme case, people who intend to spend a few minutes observing an auction end up buying white elephants because they become caught up in the excitement and competition of bidding.

Although Blumer's theory is more refined than Le Bon's, it still implies that people in crowds are irrational and out of control. Sociologists today know that much of crowd behavior, even within acting crowds such as a mob, is actually very rational (McPhail, 1991). That is, even mobs can be shown to pursue their objective by doing certain predictable and rational things. As noted earlier, the leadership of a mob taking control of a nation's capital buildings pressures its members to forego distractions such as theft and rape until the takeover is completed. Emergent norm theory illustrates the structure and rationality of crowd behavior.

What is emergent norm theory?   Emergent norm theory stresses the similarity between typical, everyday social behavior and crowd behavior. In both situations, norms guide behavior (Turner, 1964; Turner and Killian, 1987). According to emergent norm theory, rules develop within crowds to tell people how they are expected to behave. These rules, of course, are emergent norms, because the crowd participants are not aware of them until they find themselves in the situation. These norms are developed on the spot as crowd participants

pick up cues as to expected behavior in the situation. Emergent norm theory contends, in short, that crowd behavior is no different from noncrowd behavior, except that crowds do not have ready-made norms.

This explanation of crowd behavior is quite different from contagion theory. According to contagion theory, all individuals in crowds conform because of the emotions they share with the others involved. According to emergent norm theory, people in crowds conform because of group pressure to do so. This conformity may be active or passive. Some people in a riot may take home as many watches and rings as they can carry; others may simply not interfere with the looters, although they take nothing for themselves. In Nazi Germany, some people destroyed the stores of Jewish merchants, while others watched silently, afraid to disagree for fear that others would ridicule or hurt them. According to emergent norm theory, then, people in a crowd may be present for a variety of reasons, and they do not all behave in the same way (McPhail and Wohlstein, 1983; Zucher and Snow, 1990).

Contemporary sociologists view crowd behavior as a more structured and rational phenomenon than is apparent on the surface. Sociologists also agree that a social movement is the most highly structured, rational, and enduring form of collective behavior (Goode, 1992; Marx and McAdam, 1994).

SOCIAL MOVEMENTS

The Nature of Social Movements

A **social movement** involves a large number of people acting together with some degree of leadership and organization to promote or prevent social change. It is the form of collective behavior that has the most structure, lasts the longest, and is the most likely to create social change (Lofland, 1996). Most social movements are mounted to bring about a desired change of some sort. This was as much the case for Nazism and the American Revolution as it is for the gay liberation movement. Still, as the definition indicates, social movements can take a

FEEDBACK

1. A _____ is a temporary group of people who are reacting to the same event or individual.
2. An _____ crowd has no purpose or direction beyond the unleashing of emotions.
3. Mob and riot are simply two terms for the same type of crowd. *T or F?*
4. Much crowd behavior is structured and rational. *T or F?*
5. Some individuals at a lynching do not participate, or give only verbal support, but do not attempt to stop it. Which of the following theories of crowd behavior best explains this?
   - a. contagion theory
   - b. crowd decision theory
   - c. emergent norm theory
   - d. casual crowd theory

Answers: 1. crowd 2. expressive 3. F 4. T 5. c

A social movement involves a large number of people acting together with some degree of leadership and organization to promote or to prevent social change. Most social movements, such as the movement to end apartheid in South Africa, seek social change. South African blacks at the ballot box is one important indicator of change.

stand against social change. The movements that brought Prohibition to early twentieth-century America, the movement against abortion, the senior rights movement, and the new feminist movement are examples (Blanchard, 1994; Ferree and Hess, 1994; Powell, Williamson, and Branco, 1996).

What are the primary types of social movements? David Aberle (1991) has identified four basic types of social movements. A **revolutionary movement** attempts to change a society totally. Examples are revolutionary movements like the one led by Mao Zedong that gave China a socialist form of government. A **reformative movement** aims to bring partial change to a society. The Women's Christian Temperance Union and the antiwar movement of the 1960s illustrate this type of social movement. A **redemptive movement** focuses on totally changing individuals rather than society. The religious cult of David Koresh, which resulted in the deaths of many people in Waco, Texas, was a redemptive movement. Finally, an **alternative movement** seeks only limited changes in individuals. Zero Population Growth was an example of an alternative movement because it attempted to persuade people to limit the size of their families but did not advocate

sweeping changes in lifestyle or the establishment of legal penalties for having large families.

Sociologists have been successful in formulating explanations of social movements largely because a social movement is the most highly structured form of collective behavior. The major theories of social movements are relative deprivation theory, value-added theory, and resource mobilization theory.

Relative Deprivation Theory

Satisfaction with present conditions does not fire people to push, shove, and die for change. It is the frustrated and discontented who want change and who are the most willing to fight for it. Discontent with present conditions, in short, is necessary for collective action. People must see existing conditions as unfair and unjust (Rose, 1982). Thus, American colonists protested taxation without representation; Fidel Castro pointed to the vast gap between rich and poor Cubans; and Iranian revolutionaries in 1979 felt strongly that the shah had gone too far with the processes of modernization, Westernization, and secularization. Discontent is more likely to lead to a social movement if it is linked with relative deprivation and unfulfilled rising expectations.

What is relative deprivation? **Relative deprivation** is felt when people compare themselves with others and believe that they should have as much as those others have. Women's liberationists compare the situation of women to men, and gays in the United States underscore the penalties they suffer when they reveal their sexual preference. Government statistics indicate that African Americans receive less income than whites of comparable educational background. African Americans who are aware of this fact are likely to experience relative deprivation. Because a comparison is made between one's own situation and the situation of others, deprivation of this type is purely relative. There is no absolute standard for comparison—only the conviction among certain people that they wrongfully have less than some specific others have.

What are unfulfilled rising expectations? **Unfulfilled rising expectations** occur when newly raised hopes for a better life either are not satisfied at all or are not satisfied as rapidly as people had expected. For example, newly industrializing countries are likely to experience some revolutionary discontent when people who have been poor all their lives are suddenly promised a better life. They revolt not just because of their poverty but because their expectations about their material well-being have changed more rapidly than their actual material condition. The phenomenon of unfulfilled rising expectations helps explain why many revolutionary situations arise only after people have

Mounting a Social Movement

As you know from Chapter 8 ("Social Stratification"), this CNN report highlights the lower-class status of blacks in Brazil. Until recently, Afro Brazilians have not rebelled against the institutionalized discrimination they face. Black leaders have now formed the Afro Brazilian Movement to raise the consciousness of their people. They are promoting self-acceptance and self-esteem in order to create a demand for a better place in Brazilian society.

This woman is part of the Afro Brazilian Movement. She is distributing literature designed to inform others of the institutionalized discrimination Afro Brazilians face.

1. Does this report show that collective behavior is socially structured? Explain.

   _____

   _____

2. Does relative deprivation theory apply to the Afro Brazilians? Why or why not?

   _____

   _____

experienced some economic and social improvement. Alexis de Tocqueville (1955, originally published in 1835) observed improvement in the French peasant's economic situation:

> It is a singular fact that this steadily increasing prosperity, far from tranquilizing the population, everywhere promoted a spirit of unrest. The general public became more and more hostile to every ancient institution, more and more discontented; indeed, it was increasingly obvious that the nation was heading for a revolution.
>
> Moreover, those parts of France in which the improvement in the standard of living was most pronounced were the chief centers of the revolutionary movement. Such records of the Ile-de-France region as have survived prove clearly that it was in the districts in the vicinity of Paris that the old order was soonest and most drastically superseded. In these parts the freedom and wealth of the peasant had long been better assured than in any other pays d'élection (Tocqueville, 1955:175).

It will be interesting to observe events in the new Russian Federation and China subsequent to the changes occurring in those societies.

James Davies (1979) has linked unfulfilled rising expectations to revolutionary social movements. According to Davies's J-curve theory, a revolutionary movement is most probable when a period of rising expectations accompanied by actual economic improvement is followed by a decline in the fortunes of the masses. (See Figure 17.1.) According to this model, once expectations begin to rise, they continue to do so. As long as expected need satisfaction and actual need satisfaction are reasonably close, people will tolerate the gap between what they want and what they are getting. It is when actual need satisfaction falls off sharply (note the upside-down J formed by the curve in Figure 17.1) that the gap between what people want and what they have becomes intolerable. At this point, a revolutionary social movement is most likely to occur.

What is the major shortcoming of relative deprivation theory? The primary difficulty with relative deprivation theory is the theory's inability to explain the existence of social discontent without the development of a subsequent social movement. Although African Americans had long been discontented and suffered from relative deprivation, a social movement was not mounted until the 1950s. We must conclude, then, that although discontent and deprivation are necessary conditions of social movements, they are not sufficient ones. Discontent and deprivation must precede a movement, but they cannot produce one alone. The other two theories of social movements point to other factors.

Value-Added Theory

One of the strengths of Neil Smelser's *value-added theory* is the theory's applicability to many forms of collective behavior. Although the theory can be applied to such types of collective behavior as panics and riots, we describe it here only within the context of social movements.

What is value-added theory? Smelser's theory is based on a concept borrowed from the field of economics. In the value-added process, each step in the creation of something contributes (adds value) to the final

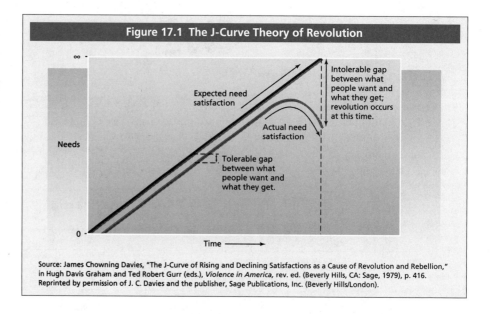

Figure 17.1 The J-Curve Theory of Revolution

Source: James Chowning Davies, "The J-Curve of Rising and Declining Satisfactions as a Cause of Revolution and Rebellion," in Hugh Davis Graham and Ted Robert Gurr (eds.), *Violence in America*, rev. ed. (Beverly Hills, CA: Sage, 1979), p. 416. Reprinted by permission of J. C. Davies and the publisher, Sage Publications, Inc. (Beverly Hills/London).

product. Smelser gives an example involving automobile production:

> An example of [the value-added process] is the conversion of iron ore into finished automobiles by a number of stages of processing. Relevant stages would be mining, smelting, tempering, shaping, and combining the steel with other parts, painting, delivering to retailer, and selling. Each stage "adds its value" to the final cost of the finished product. The key element in this example is that the earlier stages must combine according to a certain pattern before the next stage can contribute its particular value to the finished product, an automobile. Painting, in order to be effective as a "determinant" in shaping the product, has to "wait" for the completion of the earlier processes. Every stage in the value-added process, therefore, is a necessary condition for the appropriate and effective condition of value in the next stage. The sufficient condition for final production, moreover, is the combination of every necessary condition, according to a definite pattern (Smelser, 1971:13–14).

Smelser's **value-added theory** outlines six conditions that are necessary and sufficient for the development of a social movement. That is, Smelser specifies conditions that must exist if a social movement is to occur and that will lead to a social movement if they are present.

What are the conditions for the development of a social movement? *Structural conduciveness* refers to the presence of a context for the development of a social movement. The college student demonstrations and violence in the 1960s and 1970s occurred both because there was a war that students could protest against and because most college campuses had places where rallies and protest meetings could be held. Without the war

and without ready access to places for demonstrations, the antiwar movement could not have taken place.

A second condition promoting the emergence of a social movement is the presence of *structural strains*—conflicts, ambiguities, and discrepancies within a society. The conflict between the belief in political self-determination for all countries and the federal government's desire to save Southeast Asia from communism contributed to the antiwar movement in the 1960s. Probably the key structural strain in this case was the government's continued stance that there was no war (no legal war had been declared) despite the vast resources being devoted to battle and the obvious casualties that combat was producing.

The combination of structural conduciveness and structural strains increases the probability that a social movement will occur. When a third condition—*generalized beliefs*—is also present, a social movement is even more likely. Generalized beliefs indicate the sources of structural strains, the characteristics of those sources, and what should be done to reduce them. The civil rights movement, for example, was guided by the general belief that racial inequality should be eradicated.

Even when structural conduciveness, structural strains, and generalized beliefs exist, a social movement might not occur. One or more *precipitating factors* must occur to place the wheels of collective behavior into action. President Nixon's announcement on April 30, 1970, that American troops had been sent into Cambodia began a chain of events that left four Kent State University students dead and nine others wounded at the hands of National Guardsmen (Lewis, 1972).

Once these first four conditions exist, the only block to the emergence of a social movement is the *mobilization of participants for action.* It is at this point that leaders become very important. Martin Luther King, Jr., seized the opportunity provided by Rosa Parks, the black woman who refused to move to the back of the bus in Montgomery, Alabama, in 1955. This incident was the precipitating factor for a massive boycott that led to the desegregation of Montgomery's bus system.

The sixth determinant of a social movement is *social control*—efforts on the part of society (media, police, courts, community leaders, political officials) to prevent, minimize, or interrupt the momentum created by the existence of the first five conditions. If appropriate techniques of social control are applied, a potential social movement may be prevented, even though the first five determinants are present. Sometimes, social controls can be applied after a panic, riot, or mob action has already started. At that point, control efforts may block the social movement, minimize its effects, or make matters worse.

What are the strengths and weaknesses of value-added theory?   According to value-added theory, more than discontent and deprivation are necessary for a social movement to emerge. Beyond structural strains there must be structural conduciveness, generalized beliefs, precipitating factors, and the mobilization of participants for action. Value-added theory has been criticized, however, for failing to consider the importance of resources—funds, people, abilities—for the emergence and success of social movements (Olzak and West, 1991). The critics of the theory reject the traditional view of social movements as irrational, spontaneous, and initially unstructured (Oberschall, 1973; Tilly, 1978; Opp, 1989; Gamson, 1990; Opp and Roehl, 1990). According to other critics, the value-added approach works best for crowd behavior, but even in that type of collective behavior all five factors are not always present (Goode, 1992).

Resource Mobilization Theory

What is involved in resource mobilization theory? According to resource mobilization theory, a key to galvanizing people for collective action is the mobilization of resources. **Resource mobilization** is the process through which members of a social movement secure and use the resources needed to advance their cause. Resources encompass human skills such as leadership, organizational ability, and labor power as well as material goods such as money, property, and equipment (Cress and Snow, 1996; McCarthy and Wolfson, 1996).

John Lofland (1979) writes of the "white-hot" mobilization efforts of the Unification Church ("Moonies") in the 1970s. Upon taking up residence in the United States in 1971, the Reverend Sun Myung Moon was shocked by the scant resource mobilization work that had been done. He set out to change the situation through a series of moves. He began by establishing a set of long- and short-term goals for the movement, including new-member and fund-raising quotas. Elaborate publicity campaigns and events were staged, including speaking tours and rallies. An organizational structure was established and a national training center created. The trainees were subsequently sent across the country in evangelistic teams. The United States was divided into ten regions, each headed by a regional director. Each regional director supervised several state directors who were in charge of local center directors. These efforts required that about $15 million be generated each year, some of it coming from Korean and Japanese branches of the movement. (Allegations were made that some of the money came from the South Koreans and American CIA.) Many local groups made money from their own businesses, such as house-cleaning services, gas stations, and restaurants. Between 1971 and 1974,

One of the conditions promoting the development of a social movement is the existence of conflict within a society. In 1997, these Mexican teachers mounted a protest movement against the ruling Institutional Revolutionary Party (PRI), burning campaign propaganda, blocking city streets, and clashing with police.

DOING RESEARCH

Frances Fox Piven and Richard Cloward—The Success and Failure of Poor People's Movements

Frances Fox Piven and Richard Cloward researched four poor people's social movements, two originating during the Great Depression of the 1930s and two spawned after World War II. During the 1930s, the Workers' Alliance of America emerged from the protest of the unemployed, and the Congress of Industrial Organizations rose from the insurgency of industrial workers. The post–World War II era brought two poor people's movements—the civil rights movement and the movement of welfare recipients that created the National Welfare Rights Organization.

Little work has been done on the movement of the unemployed in the Great Depression, so Piven and Cloward used both secondary and primary sources. Because much historical analysis has been done on the movement of industrial workers and the civil rights movement, the researchers used secondary sources. Analysis of the civil rights movement was based on the direct knowledge of and involvement in the movement on the part of Piven and Cloward, along with the few existing studies of the movement.

All of these protest movements were researched from a conflict perspective. Piven and Cloward contend that power is held by those who control the means of physical coercion and the means of creating wealth. Because institutions, which are in the hands of the elite who hold the power, are extremely efficient in maintaining political docility, Piven and Cloward argue, the lower classes have the chance to push for their own class interests only on rare occasions. Thus, on those rare occasions, the forms the protests take and the effects they can have are largely shaped by the elites. This is what Piven and Cloward mean when they depict social protest as a socially structured phenomenon. This influence of the larger social context through elites, the researchers argue, normally limits the size and force of the poor people's protests.

According to Piven and Cloward, leaders of lower class protest movements have generally relied on the creation of formal organizations with a mass membership composed of the lower classes themselves. This approach, contend the authors, diminishes the probability of success for two reasons, one external to movements and one internal to them.

Externally, the creation of organizations does not successfully force the elites to make the concessions necessary to generate the resources needed to sustain protest organizations over a sufficiently extended period of time. Elites, because of fear of the excessive political energy of the masses, seem to side with the insurgents, asking for their views and encouraging them to state their grievances to appropriate governmental bodies. The elites, say Piven and Cloward, are not actually responding to the protest

membership increased from five hundred to two thousand. (For more detail on the current resources of the Unification Church, see Chapter 14, "Religion.")

The case of the Unification Church makes a point central to resource mobilization theory: collective action is organizationally based and led by rational people calculating the likelihood of achieving their ends (Morris, 1981, 1984). Resource mobilization advocates contend that while other social movement theories see the organization as something that emerges as the movement develops, resource mobilization theory sees preexisting organizational structure as central to getting a social movement off the ground. Organizational structure (and associated resources) is seen as a basic instrument for initiating a social movement.

Several scholars have emphasized the role of "outsiders" in the creation of modern social movements (McCarthy and Zald, 1977; Zald and McCarthy, 1987; Burstein, 1991). Outsiders may be volunteers who care about the plight of a category of people or professionals knowledgeable about the organization and management of social movements. For example, white students and adults from the North brought commitment and organizational skills to the civil rights movement in the 1960s.

How has resource mobilization theory been criticized? Although critics praise resource mobilization theory for shedding light on the importance of resources in mounting and sustaining social movements, they fault it for deemphasizing the necessary social discontent and strain needed for a social movement to emerge. Resources are of little use if people are not sufficiently dissatisfied with present conditions (Jenkins, 1983; Klandermans, 1984). Although recognizing the importance of organization and planning, some social movement theorists foresee a danger in dismissing the important role of spontaneity in the emergence and development of social movements. Social movement theory, critics write, must not organization but are responding rather to the fear of insurgency itself. Once insurgency subsides, as it usually does, most of the organizations based on it disappear. The protest-based organizations that do survive are those that are useful to the elites who control the resources on which their existence depends. Surviving organizations must abandon their politics of opposition.

Unwittingly, then, protest organizers help the elites channel the protest of the insurgent masses into normal political channels. As a result, two things tend to occur. After the heat of insurgency has subsided, either the organizations fade away for lack of elite-supplied resources, or they survive because of their increasing usefulness to the elites on whom they are dependent for resources. The poor, of course, lose with either alternative.

Not only are protest organizations not as successful as they might be because of the influence of elites, assert Piven and Cloward, but the energy spent on building organizations channels resources away from activities that would produce better results. Instead of

mobilizing the protest sentiments of the poor to create major change, leaders of the movement are preoccupied with organization building. As a result, the potential momentum of the poor people's willingness to protest is diminished. Piven and Cloward put it this way:

When workers erupted in strikes, organizers collected dues cards; when tenants refused to pay rent and stood off marshals, organizers formed building committees; when people were burning and looting, organizers used that "moment of madness" to draft constitutions (Piven and Cloward, 1979:xii).

Piven and Cloward see a lesson for leaders of future poor people's protests. Future organizers and leaders should understand that past protest efforts have failed because their leaders did not understand the influence of existing social structure on the forms of political activity available to the poor. Future lower class protest mobilizations should not be squandered out of ignorance of political and social tendencies.

Critical feedback

1. Which of the four types of social movements described in this chapter do you think most poor people's social movements fit? Explain.

2. Explain why you think the conflict perspective might underlie this research. Are there alternative theoretical perspectives that could have been used? Why or why not?

3. Do you think the research methods used by Piven and Cloward are appropriate? Explain your position.

4. If the leaders of a poor people's protest movement knew and believed Piven and Cloward's conclusions, could they use the conclusions to enhance their degree of success? Why or why not?

lose sight of the emotional factors that lead people to join a movement whose likelihood of success is small and must consider the ways in which organization leaders are affected by the unpredictable, spontaneous actions of the people involved (Killian, 1984; Opp, 1988; Rule, 1989; Scott, 1995).

The Future Direction of Social Movement Theory

The three theories of social movements as just discussed are more complementary than they are mutually exclusive. Relative deprivation theory provides an emphasis on discontent from a social-psychological, or micro, viewpoint. Value-added theory implies discontent (within the concept of structural strain) but focuses on the operation of a number of factors at the macro level of analysis. Value-added theory implies the need for resources through its mobilization factor, but

resource mobilization theory is needed to spell out this contributing factor to the rise of social movements.

Social movement theory in the future will likely encompass both preexisting structure and spontaneity, both rationality and irrationality. It seems inevitable that the viewpoints of resource mobilization theory and other social movement theories will be considered in the future by those on both sides of the debate.

There is currently a distinction being made between "old" and "new" social movements (Eyerman, 1992; Haferkamp and Smelser, 1992). Old social movements, such as the labor movement, are class-based. They are based on the struggle for power and control over economic conditions within the context of industrial capitalism. New social movements, such as the women's movement, the ecology movement, the peace movement, the gay movement, and the animal-rights movement, are not embedded in economics and opposition to capitalism. Rather than expressing conflicts of industrial society and industrialization, new social movements rest on conflicts thought to be appropriate to postindustrial society. The new social movements are fueled less by economic than by cultural conflicts. They are aimed at redefinitions of norms and values rather than at questions of economics and who gets what. New social movements are more global in focus and tend to center on quality-of-life issues (Melucci, 1980; McAdam, McCarthy, and Zald, 1988; Kriesi, 1989; Scott, 1995).

Life Course of Social Movements

What is the life-course approach to social movements? There are several related models outlining phases in the life course of social movements (Blumer, 1974; Mauss, 1975; Tilly, 1978). Despite disagreement among sociologists, five stages in the birth, life, and death of social movement can be identified: turbulence, organization, institutionalization, fracturing, deterioration. (See Figure 17.2.)

During the *turbulence* stage, people who feel a mild threat to something they value begin to read and write for the media, hold informal meetings, and write letters to political officials and others. If these activities are not threatening to the society, they are indulgently accepted as legitimate. Various institutions and agencies of the society make efforts at conciliation, compromise, and absorption to restore a basic consensus. These attempts to quiet things down usually prompt movement leaders to push harder. Societal hostility may follow. A movement can die at this point, or it can progress to the next stage. An example of a social movement in the turbulence stage was the antinuclear power movement, which gained momentum as a result of the near disaster at the Three Mile Island nuclear plant in April 1979. Shortly after the near meltdown at the Harrisburg plant, an attempt was made to organize a nationwide march to the White House behind such slogans as "The People Put Nuclear Power on Trial!" "No More Harrisburgs!" and "March to Stop Nuclear Power!"

The *organization* stage is marked by the formation of ad hoc committees, caucuses, and formal organizations by those segments of the public who have come to recognize a threat to one of their vital interests. This jelling of the movement can occur because of repressive actions by the government or other institutions, or it can result from disappointment when nothing has been done about the problem. Although the movement is not yet nationally organized, it is formally organized on the local and regional levels. Only massive repression or giving in to the movement's demands can stop it at this point. Leaders of the antinuclear war, or peace, movement failed to coalesce (Silber, 1982; Kleidman, 1993; Lofland, 1993).

Institutionalization occurs when political or other institutions recognize that a problem exists and attempt to create mechanisms to cope with it. The movement is at the height of its public acceptance, power, support, and respectability at this point. It is nationally organized, has many members and lots of money, influences political processes, and has the attention of the media. Examples of institutionalized movements are those related to women's liberation, ecology and the environment, obscenity and pornography, and alcohol and other drugs.

Ironically, the very success of a social movement leads to the movement's *fracturing*. As a result of the movement's success, acceptance, and respectability, the movement accommodates itself to society. Many active supporters as well as the sympathetic public come to believe that the situation has improved enough that their vital interests are no longer seriously threatened. Those who remain in the movement begin to fight among themselves about future strategies and tactics. Some want to continue until the problem is completely eliminated, others call for modification of the original

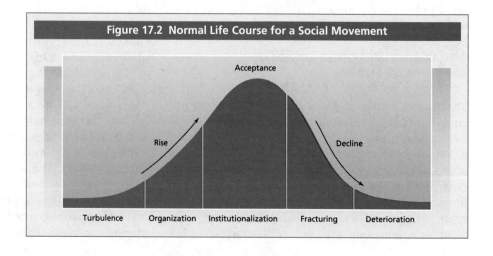

Figure 17.2 Normal Life Course for a Social Movement

Acceptance

Rise

Decline

Turbulence   Organization   Institutionalization   Fracturing   Deterioration goals, and still others advocate the pursuit of new objectives. The civil rights movement and the War on Poverty, both of which were institutionalized in the 1960s, are now fragmented.

The last stage is *deterioration,* which ironically is often brought on by the success of the movement (Tucker, 1976; von Eschen, Kirk, and Pinard, 1976). The most important aspects of the movement's program have become part of society, many of the movement's leaders have joined the establishment, and public sup-

port has evaporated. "True believers" may turn in desperation to violence or terrorism to keep the movement alive, as the Weathermen did at the end of the New Left movement of the 1960s. This behavior serves to alienate the core of the movement from former leaders and the larger society.

The life-course approach clearly illustrates the involvement of social movements in change. Social movements work for change; they themselves change as time passes (Cummins, 1993).

FEEDBACK

1. A _____ is the form of collective behavior that has the most structure, lasts the longest, and is the most likely to create social change.
2. Which of the following is an example of a reformative social movement?
   a. the French Revolution
   b. Zero Population Growth
   c. the Jesus People
   d. Women's Christian Temperance Union
3. _____ is felt when people compare themselves with others and believe that they should have as much as those others have.
4. _____ occur when newly raised hopes for a better life either are not satisfied at all or are not satisfied as rapidly as people expect them to be.
5. Once widespread discontent exists within a society, a social movement is bound to occur. *T or F?*
6. According to the _____ theory, several conditions exist that are necessary and sufficient for the emergence of a social movement.
7. _____ is the process through which members of a social movement secure and use the resources needed to press for social change.
8. Which of the following is *not* one of the stages in the life course of a social movement?
   a. institutionalization   b. regeneration   c. deterioration   d. organization   e. turbulence

Answers: 1. social movement 2. d 3. Relative deprivation 4. Unfulfilled rising expectations 5. F 6. value-added 7. Resource mobilization 8. b

SUMMARY

1. Most areas of sociological study are based on the assumption that social life is predictable, orderly, and recurrent. Collective behavior is an important exception because much of it is spontaneous, short term, and relatively unstructured. Some forms of collective behavior, however, are more planned, structured, and enduring.
2. On the more unstructured end of the collective behavior continuum are dispersed collectivities—rumors, mass hysteria, panics, fads, fashions, and publics. Even these forms of collective behavior are structured to some degree.
3. Rumors have a very bad reputation, usually deserved, because they are composed of inaccurate, distorted, or false information. Rumors can be playful or harmless, or they can be damaging and hurtful.
4. Mass hysteria occurs when people become anxious because of acceptance of false beliefs. Panics take place when there is a collective reaction to a real threat.
5. Unusual patterns of behavior that are adopted quickly, accepted enthusiastically, and disappear soon are called

fads. Fashions are patterns of behavior that are widely approved but expected to change periodically. Fashions are more central to modern than premodern societies.
6. A public is composed of people who are concerned about an issue on which opinions differ. When attitudes and beliefs about an issue are expressed, public opinion exists. For an issue to become important enough for a public to emerge, a process of crystallization must take place. This usually happens when a dramatic event occurs or when some public figure is able to influence enough people. Use of propaganda and the formation of interest groups are two major means of creating and sustaining a public.
7. Crowd behavior is fascinating to most people because it usually involves intense feelings and sometimes outrageous behavior. There are casual, conventional, expressive, and acting crowds. Mobs and riots are the two best-known examples of acting crowds.
8. Two quite different theories attempt to explain crowd behavior. Contagion theory stresses the irrationality of

REVIEW GUIDE

crowds and the build-up of intense emotion produced by social interaction within a large collection of people. Emergent norm theory depicts crowd behavior as being more rational. According to emergent norm theory, crowd behavior is guided by norms that arise spontaneously.

9. Social movements are closer to conventional social behavior than is crowd behavior because they are more permanent and more organized. Yet, they are not so permanent and structured as most aspects of social life.

10. There are three major explanations of social movements. According to relative deprivation theory, discontent and unfulfilled rising expectations create a breeding ground for social movements. Value-added theory outlines six necessary and sufficient conditions for the development of a social movement: structural conduciveness, structural strains, generalized beliefs, precipitating factors, mobilization of participants for action, and social con-

trol. According to the resource mobilization theory, resources are a crucial ingredient in the mounting of a social movement. Resources include human skills and economic assets, both of which often come from people outside the movement itself.

11. The debate continues over the relative importance of pre-existing organizational structure and rationality on the one hand and the gradual emergence of organizational structure and spontaneity on the other. Some advocates of resource mobilization theory and other social movement theories are recognizing the validity of one another's viewpoint.

12. The temporary nature of social movements is reflected in a movement's life course. Social movements that survive beyond the beginning, or stage of turbulence, usually progress through organization, institutionalization, fracturing, and deterioration. Oddly enough, the success of a social movement ensures the movement's death.

LEARNING OBJECTIVES REVIEW

After careful study of this chapter, you will be able to:

- Describe the various social activities engaged in by dispersed collectivities.
- Describe the basic nature of a crowd and identify the basic types of crowds.
- Compare and contrast contagion theory and emergent norm theory.
- Define the concept of social movement and identify the primary types of social movements.

- Apply the concepts of relative deprivation and unfulfilled rising expectations to the emergence of social movements.
- Compare and contrast relative deprivation theory with value-added theory and resource mobilization theory.
- Trace the phases in the life course of social movements.

CONCEPT REVIEW

Match the following concepts with the definitions listed below them:

____ a. revolutionary movement
____ b. unfulfilled rising expectations
____ c. collective behavior
____ d. mass hysteria

____ e. emergent norm theory
____ f. relative deprivation
____ g. panic
____ h. redemptive movement

____ i. fashions
____ j. rumor
____ k. public
____ l. crowd

1. The type of social movement that focuses on totally changing individuals rather than society.
2. Patterns of behavior that are widely approved but expected to change periodically.
3. The type of social movement that attempts to change society totally.
4. A condition that exists when people compare themselves with others and believe that they should have as much as those others have.

5. The relatively spontaneous and unstructured social behavior of people who are responding to similar stimuli.
6. The form of collective behavior in which general social anxiety is created by acceptance of one or more false beliefs.
7. A widely circulating story whose truth is questionable.
8. People concerned about an issue on which opinions differ.

9. The theory of crowd behavior that stresses the similarity between typical, everyday social behavior and crowd behavior.

10. When newly raised hopes for a better life either are not satisfied at all or are not satisfied as rapidly as people had expected.

11. Collective behavior that occurs when people react to a genuine threat in fearful, anxious, and often self-damaging ways.

12. A temporary collection of people who share a common point of interest.

CRITICAL THINKING QUESTIONS

1. Sociologists take collective behavior to be a legitimate area of study on the grounds that it is more structured than it appears at first glance. Develop three examples of collective behavior of which you have some knowledge to support this claim by sociologists. _____

_____

_____

2. You have participated in crowd behavior at some time in your life. Think of an instance and identify it as one of the four types of crowds described in the text. Using your personal experience, provide examples of behavior within the crowd that illustrate why you think it was a particular type of crowd. _____

_____

_____

3. Consider the women's movement you have witnessed in your lifetime. Which theory of social movements do you think best explains it? Defend your answer with specific links between the theory you have selected and the nature of the women's movement as you understand it. _____

_____

_____

4. Use your knowledge of the civil rights movement that began in the 1960s to demonstrate your understanding of the life-course approach to social movements. Identify the stage at which you think this movement now stands. Explain your choice with specific illustrations of the current state of the civil rights movement. _____

_____

_____

MULTIPLE CHOICE QUESTIONS

1. **The study of collective behavior poses some particular problems for sociology because**
   a. sociologists are accustomed to studying social interaction.
   b. collective behavior is difficult to define in sociological terms.
   c. sociologists are accustomed to studying structured behavior.
   d. sociological theories do not account for collective behavior.
   e. sociological research methods are inappropriate for the study of collective behavior.

2. Jan Harold Brunvand argues that _____, such as "The Baby in the Oven," permits us to play out hidden fears and guilt feelings by being shocked and horrified when we hear the story involving other people.
   a. fads
   b. fables
   c. fairy tales
   d. urban legends
   e. riddles

3. Orson Welles's famous "Men from Mars" radio broadcast produced
   a. a riot.
   b. a revolution.
   c. a redemptive social movement.
   d. a fashion.
   e. mass hysteria.

4. When parents in a local school rebel against an AIDS victim attending school with their children, they are said to be exhibiting
   a. faddish behavior.
   b. mass hysteria.
   c. victimization.
   d. a public interest reaction.
   e. collective hypocrisy.

5. Items such as body tattoos, "washboard abs," and the Macarena are examples of
   a. advertising propaganda.
   b. social movements.
   c. fashions.
   d. fads.
   e. crowd behavior.

6. The crystallization phase of an issue of concern to the public is aided by propaganda and by
   a. the formation of interest groups.
   b. rumor control centers.
   c. the development of casual crowds.
   d. riots.
   e. social contagion.

7. Propagandists attempt to
   a. engage in a debate with individuals or groups that disagree with their agenda.
   b. manipulate public opinion by camouflaging the real purpose and presenting one side of an issue.
   c. inform public opinion.
   d. set the agenda for public debate by providing issues to think about.
   e. articulate the public's attitudes and beliefs about one or more issues.

8. Organizations such as the National Rifle Association, the National Organization of Women, and the National Association of Retired Persons are examples of
   a. peer groups.
   b. primary groups.
   c. social categories.
   d. interest groups.
   e. social groups.

9. Fans at a rock concert are an example of a _____ crowd.
   a. casual
   b. acting
   c. expressive
   d. conventional
   e. instrumental

10. When a crowd of soccer fans abandons the guidelines for watching the match in order to attack the officials, a/an _____ crowd has become a/an _____ crowd.
    a. casual; acting
    b. conventional; acting
    c. expressive; acting
    d. conventional; expressive
    e. casual; expressive

11. An emotionally stimulated, disorderly crowd that is ready to use destructiveness and violence to achieve a specific purpose is known as a/an
    a. mob.
    b. expressive crowd.
    c. unconventional crowd.
    d. revolution.
    e. interest group.

12. In his reformation of contagion theory, Herbert Blumer defined _____ as the stage in which people become increasingly aware and sensitive to one another and enter something akin to a hypnotic trance.
    a. rising expectations
    b. social contagion
    c. milling
    d. collective excitement
    e. crowd decision

13. According to _____ theory, rules develop within crowds to tell people how they are expected to behave.
    a. emergent norm
    b. crowd decision
    c. value-added
    d. value-clarification
    e. contagion

14. The Zero Population Growth movement constitutes an example of a/an _____ social movement.
    a. redemptive
    b. reformative
    c. evolutionary
    d. revolutionary
    e. alternative

15. The role of "outsiders" in the creation of modern social movements is *most* prominent in _____ theory.
    a. value-added
    b. relative deprivation
    c. emergent norm
    d. contagion
    e. resource mobilization

FEEDBACK REVIEW

True-False

1. Mob and riot are simply two terms for the same type of crowd. *T or F?*
2. Much crowd behavior is structured and rational. *T or F?*

Fill in the Blank

3. Unusual patterns of behavior that spread rapidly, are embraced zealously, and disappear after a short time are called

   _____.
4. _____ refers to information consciously and systematically disseminated to influence the decisions, actions, and beliefs of people on an issue.
5. An _____ crowd has no purpose or direction beyond the unleashing of emotions.
6. A _____ is the form of collective behavior that has the most structure, lasts the longest, and is the most likely to create social change.
7. _____ is the process through which members of a social movement secure and use the resources needed to press for social change.

Multiple Choice

8. Some individuals at a lynching do not participate, or give only verbal support, but do not attempt to stop it. Which of the following theories of crowd behavior best explains this?
   a. contagion theory
   b. crowd decision theory
   c. emergent norm theory
   d. casual crowd theory
9. Which of the following is an example of a reformative social movement?
   a. the French Revolution
   b. Zero Population Growth
   c. the Jesus People
   d. Women's Christian Temperance Union
10. Which of the following is *not* one of the stages in the life course of a social movement?
    a. institutionalization
    b. regeneration
    c. deterioration
    d. organization
    e. turbulence

ANSWER KEY

Concept Review	Multiple Choice	Feedback Review
a. 3	1. c	1. F
b. 10	2. d	2. T
c. 5	3. e	3. fads
d. 6	4. b	4. Propaganda
e. 9	5. d	5. expressive
f. 4	6. a	6. social movement
g. 11	7. b	7. Resource mobilization
h. 1	8. d	8. c
i. 2	9. c	9. d
j. 7	10. b	10. b
k. 8	11. a	
l. 12	12. c	
	13. a	
	14. e	
	15. e	

Chapter Eighteen

Social Change in Modern Society

After careful study of this chapter, you will be able to:

- Define social change and illustrate the three social processes contributing to social change.
- Discuss the role of technology, population, the natural environment, conflict, and ideas as sources of social change.
- Compare and contrast the cyclical and evolutionary perspectives on social change.
- Describe the major features of the functionalist and conflict perspectives on social change.
- Define modernization and explain some of its major consequences.
- Define world-system theory and summarize some of the consequences flowing from its description of the structure of the world economy.
- Explain the major points of the pessimists and optimists regarding the future of the physical environment.
- Discuss the possible directions human community may take in the twenty-first century.

SOCIOLOGICAL IMAGINATION

Do successful revolutions bring about rapid and radical social change?   It seems definitional to conclude that radical alterations in a society follow a successful revolutionary movement. There are dramatic examples of truly revolutionary changes in the nature of a society: France (eighteenth century), China (twentieth century), Russia (early twentieth century). Sometimes, however, victory celebrations are not followed by a new social order. Change subsequent to the American Revolution was quite gradual and long-term. Although America bears little resemblance to its nature in 1776, the changes cannot be linked to any specific event.

Revolutions are only part of one factor contributing to the alteration of social structure. After defining *social change,* we turn immediately to several of its sources.

DEFINITION OF SOCIAL CHANGE

The speed with which change is occurring in the modern world is startling. Suppose for a moment that the history of the planet Earth were measured by a 365-day period, with midnight January 1 marking the starting point and December 31 being today. Under this time span, each Earth "day" would represent twelve million years of actual time. This would mean that the first form of life, a simple bacterium, would have appeared

in February. More complex life, such as fish, would appear about November 20. On December 10, the dinosaurs appeared; by Christmas they were extinct. The first recognizable human beings would not appear until the afternoon of December 31. Our species, *homo sapiens,* would emerge shortly before midnight of the last day of the year. All of recorded history is encapsulated in the last sixty seconds of the year (Ornstein and Ehrlich, 1991).

It takes no Rip Van Winkle to tell us that we are in a period of bewildering and devastating change. Indeed, it has become a cliché to say that change is one of the most constant features of American society. In a very short time, we have gone from such questions as Does a lady kiss in public? to questions such as Should mothers give their daughters the pill? Should prisons issue condoms to inmates (due to the AIDS scare)? Should a university or college make contraceptives available to single students? In fact, condom machines have now been installed in restrooms at some universities. For spring break, some American universities have distributed "survival kits," including such necessities as toothpaste and condoms. There is now a rule in college football motivated by fear of the spread of AIDS. A player must miss at least one play or his team call a timeout if a wound is oozing or bleeding or if a player's uniform has blood on it. The player must have the cut closed or the clothing cleaned or changed. A college basketball player must come out of the game when blood is involved. A fear of AIDS is now emerging in boxing (Gorman, 1996).

Social change refers to alterations in social structures that have long-term and relatively important consequences. All social structures change. The speed of change may vary from the glacial to the mercurial, but the existence of social change is a constant. This chapter first examines the major sources of social change and several theories of social change. A discussion of modernization and world-system theory is followed by an examination of the future of society.

SOURCES OF SOCIAL CHANGE

Predicting the precise changes that will occur is difficult at best. (See Table 18.1.) This is partly because each society absorbs forces for change in ways at least somewhat consistent with its traditional culture and social structures. In addition, we are capable of modifying our behavior to avoid a predicted state of affairs or deliberately create a given condition (Caplow, 1991). This fact should not, however, discourage the attempt to understand the sources of social change. Sociologists have not been discouraged; they have identified several major sources of social change.

TABLE 18.1	Key Assumptions in Predicting Social Change in America

The most accurate prognosticator of trends in American society has been the Frenchman, Alexis de Tocqueville. Tocqueville's two volumes published in the 1830s entitled *Democracy in America* either correctly predicted many of the paths America subsequently followed or, when wrong, displayed an amazing grasp of the nature of American society. Tocqueville's success has been attributed to several key assumptions he made:

- The persistence of major social institutions is taken for granted. Unlike many of his contemporaries—and many of ours—Tocqueville did not expect the family, religion, or the state to disappear or to be rapidly transformed.
- The stability of human nature is taken for granted. Tocqueville did not expect men and women to become much better or worse or different than history had shown them to be.
- Two very powerful, long-term trends are expected to continue: a trend towards equality and a trend towards administrative centralization.
- Social change is always channeled and circumscribed by the limited availability of material resources.
- The future is not fully determined by the past. The future, viewed from the present, is an amalgam of probabilities based on past experience and of contingencies arising from accident and free will.
- There are no social forces aside from human actions. Historical events are not foreordained by factors beyond human control.

Source: Adapted from Theodore Caplow, *American Social Trends*, New York: Harcourt Brace Jovanovich, 1991, p. 216.

Social Processes

Three interrelated social processes contribute to social change: discovery, invention, and diffusion.

How do these three social processes promote social change? **Discovery** involves either creating something or reinterpreting something already in existence. Early navigators, fortified by their skill and courage, persevered to discover the roundness of the earth. Many worldwide changes followed this pivotal discovery, not the least of which were new patterns of migration, intercontinental commerce, and colonization. Salt, first used as a flavor enhancer for food, was subsequently used as money in Africa and as a religious offering among early Greeks and Romans. More significantly, fire, which had been used for warmth and cooking by prehistoric peoples for an undetermined period of time, was used by the first agriculturists in about 7000 B.C. to clear fields and to create ash for fer-

tilizer. Fire was central to the ensuing transition to agricultural society.

Invention is the creation of a new element by combining two or more already existing elements and the creation of new rules for their use as a unique combination. Examples of invention on the material side of culture come easily to mind. It was not so much the materials Orville and Wilbur Wright fashioned into an airplane—most of the parts were available to many who had vainly tried to fly before them—but the way in which the Wright Brothers combined and manipulated the elements.

The pace of social change through invention is closely tied to the complexity of the cultural and social base. As the base becomes more complex and varied, the number of elements as well as the number of ways the elements can be combined into inventions increases at a fantastic rate. Therefore, the more complex and varied a society, the more rapidly it will change. This fact helps to explain why, even though several million years passed between the evolution of the human species and the invention of the automobile and the airplane, we reached the moon in the same century that Henry Ford and the Wright Brothers came up with their inventions.

When one group borrows something from another group—whether it be norms, values, roles, or styles of architecture—change occurs through the process of **diffusion.** The extent and rate of invention depend on the complexity of a society, but the extent and rate of diffusion depend on contact between members of different groups. Borrowing from others may involve entire societies, as in the American importation of cotton growing from India. Or diffusion may take place between groups within the same society, as in the social and cultural spinoffs of the jazz subculture created by African American musicians in New Orleans which include "funkiness" in behavior, dress, and speech.

To gain acceptance, a borrowed element to some degree must fit into the group to which it is being introduced. Despite the upsurge of unisex fashions in America today, wearing a Scottish kilt could get a construction worker laughed off the top of an unfinished skyscraper. Wearing kilts still clashes too violently with the American definition of manhood. If skirts are ever to become as acceptable for American men as pants are for women, either their form will have to be modified or our concept of masculinity will have to change.

Integrating a borrowed element into a new social setting may also involve selectivity of only certain aspects. The Japanese, for example, can accept capitalism but resist the American form of government, style of conducting business, and family structure. Diffusion, in short, usually involves modification and selectivity.

Most of the elements of a group are borrowed rather than created. The process of invention is important, but usually far more elements are borrowed from others than are created within the group, particularly in modern societies.

Technology

Technology is the part of culture, including ideas and hardware, that is used to reach practical goals. Concentration on technology as a source of social change has a long history in sociology (Blumer, 1990; Teich, 1993; Volti, 1995). Karl Marx placed the importance of technology for social change in sharp perspective when he wrote: "The windmill gives you society with the feudal lord; the steammill, society with the industrial capitalist" (Marx, 1920:119). Although the relationship between technology and social change is not this simple and direct—and Marx knew it—such a relationship is easily established.

Consider the social effects of what appear to be minor technological changes. The invention and diffusion of the stirrup altered medieval warfare. Soldiers on foot were easily defeated by knights on horseback because the stirrup permitted knights to continuously hold their lances rather than throw them; the knights did not have to reload nearly as often. This new military technology—a combination of the knight, the horse, and the lance—helped to engender the feudal system, with the knightly class at the top (White, 1972). Although originally thought of as a mere means of transportation, the automobile has had uncounted effects on the nature of modern society. For example, it has changed dating and courtship patterns, contributed to a new sexual morality, created whole new industries, promoted industrial expansion, produced suburbanization, and contributed immeasurably to environmental pollution. The creation of the silicon chip has introduced technological change at an unprecedented rate. Although it took more than a century for telephone ownership in the United States to reach 94 percent, the personal computer market reached its present 40 percent household penetration in just over twenty years. In less than fifteen years, almost 25 percent of American households had a cell phone; it has taken the Internet less than five years to reach a 26 percent household penetration ("The Silent Boom," 1997). The societal transformations wrought by the microprocessor are many and far-reaching. The workplace is undergoing a transformation, with effects as far-reaching as the Industrial Revolution. Telecommunications technology will allow many workers to work from their homes, but will also mean far less human interaction (McGinn and Raymond, 1997–98). The microprocessor has radically changed many surgical techniques, from microsurgeries to radio wave therapy (Cowley and Underwood, 1997–98). Perhaps the most profound effect of the microprocessor has been its contribution

The relationship between technology and social change is firmly established. Compare the pattern of family activities in homes with television sets to the activities of this family with a radio in the 1920s.

to the end of the cold war. Soviet military personnel recognized the potential of computer guidance systems for long-range munitions: "The realization that the Soviet Union did not, and would not, possess the computer or sensor capabilities to produce such weapons—and so, in the long run, must inevitably face defeat at Western hands—was a significant factor behind Mikhail Gorbachev's decision to sue for peace" (Watson and Barry, 1997–98). Now that scientists have produced Dolly, a sheep clone, who can know where this capability will lead if medical science applies it to human embryos (Elmer-DeWitt, 1993b; "Hello, Dolly," 1997).

It is pointless to detail here the contribution of technology to major societal shifts such as the rise of agriculture and industrialization. (Refer to Chapter 5, "Social Structure and Society.") More importantly, the role of technology in social change should be placed in perspective. Specifically, *technological determinism* needs to be considered.

Why should we be careful of technological determinism? **Technological determinism** is the theory of social change that assumes technology to be the primary cause of the nature of social structure. This is a seductive viewpoint. By use of dramatic examples, it is easy to conclude that the nature of a society can be traced directly to the society's technological base. Actually, the relationship between technology and social change is more complicated (Lauer, 1991). First, social change can occur without technological developments. Ancient Greek commercial development was not based on improved production methods. Second, the introduction of technology does not necessarily lead to social change. A South Indian village did not change its social structures following a switch from dry

Technology is a major contributor to the process of social change the world is currently experiencing. We are only beginning to see some of the social changes that will follow in the wake of the World Wide Web. The demand for capacity to carry information traffic on the Web now doubles every one hundred days.

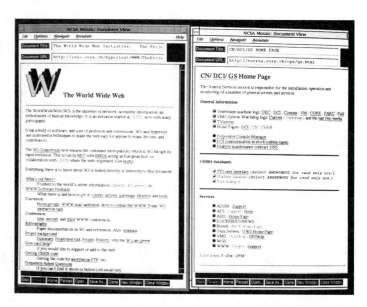

to wet cultivation of crops made in the community twenty-five years earlier (Epstein, 1962). Third, as just implied in the discussion of diffusion, the particular effects of technology will vary from society to society because the adoption and use of technology is always filtered through the nature of a given society. Although Russia, for example, has computer technology, it places much tighter controls on the use of the technology by individual citizens.

Although it seems inadvisable to see technology as a necessary and sufficient cause of social change, the contribution of technology to change should not be underestimated. It is best to view technology as an extremely important factor in a network of causes of social change. Nor should we ignore the fact that like population, technological innovation grows at an exponential rate. David Freeman (1974) makes this point by compressing the total lifetime of the earth (some five billion years) into an eighty-day period and calculating when some significant technological changes occurred: the Stone Age began six minutes ago, the agricultural revolution took place fifteen seconds ago, metals were first used ten seconds ago, and the Industrial Revolution started three-tenths of one second ago.

Population

Alterations in population size and composition, as seen in Chapter 16 ("Population and Urbanization"), can have significant rippling effects within and between societies. Population pressures in China have led to a concern for the conservation of resources. This in turn has produced a state policy of replacing traditional, but more expensive, burial rites with cremation. Conflicts and strains within families are one consequence. Concern for its burgeoning population has also produced a national population-control program in China with incentives for the one-child family. This is already affecting Chinese social structure, as in the dishonesty engendered in anxious local officials who underreport birth statistics to meet government quotas. Less developed countries are undergoing many dramatic changes because their declining mortality is not being matched by lowered fertility.

America's baby boom following World War II is a prime illustration of a changing population structure's effects on an entire society. Americans born between 1946 and 1964 expanded medical personnel and facilities for child health care and created the need for more teachers and schools in the 1950s and 1960s. Those thirty-year-olds now in the labor market are experiencing increased competition for jobs and fewer opportunities to move up the career ladder. When the baby boomers retire, problems of health care and Social Security support by the working population (which will be smaller because of the declining birthrate since the 1960s) will loom large. Longer working hours, retraining programs, and reeducation for older people will probably become prevailing patterns. As America's population continues to age, more attention will be paid to the elderly. For example, there will be more extended-care homes for the aged, increased emphasis in medicine on geriatrics, and more television advertising and programming targeting the elderly.

The Natural Environment

The natural environment has changed the way Americans live many times. The existence of the vast territory west of the original thirteen colonies led to the great expansion that ultimately stretched to the Pacific Ocean. This western movement wrought untold changes, not the least of which was the destruction of many Native American cultures. The Great Depression of the 1930s was brought on in part by the dust bowl that had been the great agricultural Midwest. Americans who experienced economic inflation in the 1970s and early 1980s are now driving high-priced, smaller automobiles, in large part because of the shock waves from the Arab oil embargo in the early 1970s.

There are plenty of intimations of the close connections between population, the natural environment, and social change in Chapter 16. Moreover, this theme is developed more explicitly in the closing section of this chapter.

Conflict: Revolution and War

Thinkers have long recognized the presence of conflict in human interaction. From the Greek historian Polybius to Karl Marx, conflict has been viewed by many early thinkers as the central form of human interaction. Marx and more contemporary conflict theorists who see social conflict as the prime mover in social change are covered in the next major section. For now, let's focus on revolution and war as sources of social change involving conflict.

What is a revolution?   A **revolution** is the type of social movement that involves the toppling of a political regime through violent means. According to Charles Tilly, a revolution results in the replacement of one set of power holders by another (Tilly, 1978, 1997). Most revolutionaries, of course, expect more fundamental changes than this. Marx, for example, expected workers' revolutions to eliminate class-based inequality and therefore to have a profound effect on the social and economic structures of the societies in which they occurred.

Are revolutions normally followed by such radical changes?   As suggested in the *Sociological Imagination* opening this chapter, they are not. In Crane Brinton's view, the changes introduced by revolutionary leaders are followed eventually by restoration of a social order similar to prerevolutionary days (1990). In other words, even successful revolutions fail to tear down the old order and bring in an entirely new one. After the victory celebrations are over, a society operates with a great deal of continuity with the past. In an important sense, things return to "normal" sometime after the revolution has been pronounced a success.

Do revolutions bring about no change?   There is a middle ground between Marx's position that revolutions produce major changes and Brinton's position that revolutions have little lasting effect. Tilly argues that revolutions do have lasting effects if there is a genuine transfer of power. Some revolutions, Tilly writes, have not had lasting effects, because revolutionaries seized control of a state too weak to be used as an instrument for the creation of change. Eighteenth-century French revolutionaries, in contrast, took control of a powerful, centralized state, allowing them to seize and redistribute the property of aristocrats and churches. Despite later reactions against the leaders of the French Revolution, such actions had lasting effects. Similarly, the Chinese communists created a strong central state of their own, allowing them to reorganize village structure in a way that permanently altered Chinese society.

In most cases, the new social order created by a successful revolution is likely to be a compromise between the new and the old. Contemporary China illustrates that revolutions do not result in the wholesale changes promised by their leaders. A cornerstone of the communist revolution in China was sexual equality; liberation from sexism was a revolutionary plank. The situation for Chinese women has improved, but complete sexual equality still does not exist.

The extent of change created by a revolution clearly varies from the great to the small. The effects of a revolution depend on the nature of the situation in which that revolution occurs. One thing is certain, however. The potential importance of revolution as a means for creating social change should not be overlooked or minimized. Revolutions can be a powerful mechanism for social change, whether or not they create the total change promised by those who lead them and expected by those who participate in them.

A revolution is, of course, only one type of social movement. Social change can also be brought about, though not always, by other types of social movements. The civil rights movement and the women's movement have both produced fundamental changes in American society (Curtis and Aguirre, 1993). Poor people's movements have not been so successful. (Refer to *Doing Research* in Chapter 17, "Collective Behavior and Social Movements.") Whether successful or not, social movements involve conflict and, when successful, leave society different than before they occurred.

Social change, then, is brought about by revolutionary conflict within societies. Change is also created by conflict between societies in the form of war. War and change are closely intertwined. The golden age of Athens (fifth century B.C.) started with the Persian War and ended with the Peloponnesian War. Rome's flowering in the first and second centuries cannot be separated from wars in Europe and Asia. War was a backdrop to the Age of Enlightenment. The founding of the United States is no exception; it required a war. The Civil War was instrumental in America's transition from an agricultural to an industrial society. Momentous technological, economic, and social change also followed World War I, World War II, and the Vietnam War.

How does war promote social change?   Robert Nisbet (1989) writes eloquently on the connection between war and change. Change is created through diffusion as war breaks down insulating barriers between societies. The crossbreeding and intermingling of cultures during and following a war leave all societies involved different than they were before the conflict. Wars also promote invention and discovery. During the first World War (1914–1918), the U.S. government was able, because of the pressure of war, to promote and finance the development of such technologies as the airplane, automobile, and radio, each of which contributed to the social and cultural revolution that followed the war. And America's culture, both during and after World War I, was exported to societies all over the world.

Ideas

Although Marx did not overlook the importance of ideas in social change, he did believe that the determining forces in historical change lay in social and economic conditions. Max Weber, partially in response to Marx, emphasized the role of ideas in social change. This position clearly undergirds Weber's work on the relationship between eighteenth-century Calvinism, a primary part of Protestantism, and the development of industrial capitalism. Weber's thesis was explicated in Chapter 14 ("Religion"), so it is sufficient to say here that Weber believed that the ideas inherent in Calvinism produced behavior needed for industrial capitalism to flourish.

Alfred North Whitehead wrote that "a general idea is always a danger to the existing order" (Whitehead, 1933:22). We have seen the validity of this observation in the analysis of social movements and the previous discussion of revolution. Recall that two of the major theories of social movements emphasize the crucial place of ideas in generating social change. In relative deprivation theory, people are said to be ready to challenge the establishment on the belief that they have less of some valued thing (freedom, wealth, equality) than they should have. According to Neil Smelser's value-added theory, a necessary condition for the emergence of a social movement is the widespread belief that something is wrong and that some action should be taken to rectify it.

Do ideas always lead to social change? Sometimes ideas retard social change. During the Middle Ages, Christian religious ideas prevented change (Gay, 1977). In fact, the Catholic Church based its strength on total resistance to change (Manchester, 1993). For example, economic development was slowed in part because usury (charging interest on loans) was considered a sin.

Slavery in the United States was successfully defended for a long time on the basis of false ideas that depicted African Americans as inferior. Similar false ideas have been used to deny women equal opportunity in many societies, including the United States.

Connections Among the Sources of Change

A careful reading of this section on the sources of social change reveals their interrelatedness. Several, or all, of these factors can be operating in combination. Consider abortion. Abortions are possible because of medical inventions that have been widely diffused. Underlying the ability to perform abortions is medical technology. This technology affects the population by lowering the birthrate, which, in turn, affects family size and a number of other social conditions (child-free marriages, dual-employed couples, sexual freedom). A lowered birthrate also reduces demand on limited natural resources. Conflict between pro-life and pro-choice forces in the United States has existed ever since abortion was legalized by the Supreme Court in 1973 in the *Roe v. Wade* decision. This conflict, of course, has been based on differing ideas regarding when a fetus constitutes a human being.

THEORETICAL PERSPECTIVES

It is important to understand the sources of social change. Description aids our understanding in any area of sociological study. Sociologists, however, wish to delve more deeply into the phenomena associated with social change. It is through theory that we can go beyond description toward explanation. With the help of anthropologists and historians, sociologists depend on three major theoretical perspectives for understanding why and how social change is part of social life.

A brief examination of some cyclical theories, used early in the study of social change, underscores the dif-

---

FEEDBACK

1. _____ refers to alterations in social structures that have long-term and relatively important consequences.
2. The more complex a society, the more rapidly it will change through invention. *T or F?*
3. The process of _____ usually involves modification and selectivity.
4. The theory of social change that assumes technology to be the primary cause of the nature of social structure is called _____.
5. The baby boom in the United States is a good example of the effects of _____ on an entire society.
6. _____ is a type of social movement that involves the toppling of a political regime through violent means.
7. Contrary to conventional wisdom, war retards social change. *T or F?*
8. Was it Karl Marx or Max Weber who emphasized the role of ideas in creating social change? _____

Answers: 1. Social change 2. T 3. diffusion 4. technological determinism 5. population 6. Revolution 7. F 8. Weber ference between description and explanation. This background makes the three more contemporary theoretical approaches to social change more meaningful.

The Cyclical Perspective

Scholars were baffled. Why would the most developed nations in the world spend $337 billion between 1914 and 1918 (World War I) to kill over eight million soldiers? This was very poor evidence indeed for a belief in the natural progression of civilization. Was change necessarily progress, scholars began to ask? Was Western civilization on the road to further improvement or on the way down?

The dramatic upheaval during the early part of the twentieth century led scholars to consider the possibility that civilizations rise and fall rather than develop in a straight line. The rise and fall of Rome, after all, had been documented in six large volumes by British historian Edward Gibbon between 1776 and 1788. Although many scholars disagreed with Gibbon's conclusions, they did not miss the point that social change may very well follow a pattern of growth and decay. Three of the most prominent advocates of the cyclical perspective were historians Oswald Spengler (1880–1936), Arnold Toynbee (1889–1975), and sociologist Pitirum Sorokin (1889–1968).

In 1918, Spengler published *The Decline of the West,* in which he compared societies to living things, as Spencer did, but with a vital difference. Whereas Spencer saw advanced societies as superior survivors of the struggle of the fittest, Spengler believed that civilizations are born, ripen, decay, and perish. Specifically, he predicted that Western civilization—which had traded zest for materialism—was going to be replaced by a civilization from Asia. Spengler's work is now considered more literature than scientific explanation.

Almost thirty years later, British historian Arnold Toynbee (1946) published *A Study of History,* in which he presented his challenge-response theory of social change. According to Toynbee, civilizations rise only when a proper challenge in the environment is presented to a people who can respond successfully to it. If a challenge from the environment is not issued or if the society fails to meet a challenge, that society will either fail to rise in the first place or decline.

Like Spengler and Toynbee, Sorokin did not see in history any linear pattern of change. To him, sociocultural history involves a cyclical alternation among three reasonably homogeneous cultural types. *Ideational cultures* are based on the conviction that truth and value come from God. Emphasis is on the spiritual and the next world. In *sensate cultures,* truth is discovered through empirical observation by the senses. It is experience in the current world—gained through

touching, smelling, hearing, seeing, and tasting—that reveals truth and reality. Sensate cultures are not absolutist but allow for reinterpretation of reality. People in ideational cultures focus on spiritual needs; members of sensate cultures place top priority on physical needs and gratification of the senses. Whereas ideational cultures tend to be spiritual, sensate cultures are characterized by pleasure seeking and materialism. Sorokin's third cultural type, the *idealistic culture,* is a blend of the ideational and the sensate, with somewhat more accent on the ideational.

For Sorokin, then, history is best described as continuous movement from the predominance of one type of culture to another. According to Sorokin, sensate culture dominated in Christ's time. Christian medieval Europe typified the shift from the sensate to the ideational. Idealistic culture emerged in thirteenth- and fourteenth-century Europe. Modern Western society is again grounded in sensate culture.

Because history tells us that societies do, in fact, rise and fall, cyclical perspectives on social change are especially appealing. The proof is there for anyone who looks. Cyclical perspectives have a major problem, however. They are not so much explanations for change as descriptions of change. Spengler does not indicate why societies mature and decay; he just tries to show that they do. Toynbee does not state why a formerly successful society fails to meet a new challenge; he just writes that societies failing to respond to a challenge will decline. Sorokin provides no explanation for the cyclical pattern he describes; he just attempts to substantiate its existence with an amazing amount of historical information.

Is the cyclical perspective even worth considering? The cyclical perspective is worthy of consideration for several reasons. First, it represents an early, historically important attempt to understand social change. Second, it helps to legitimate the scientific study of social change. Third, it is useful as a backdrop for distinguishing between description and explanation. Each of three contemporary theoretical perspectives—evolutionary theory, functionalist theory, and conflict theory—attempts to explain why or how societies change.

The Evolutionary Perspective

What was the nature of early evolutionary theory? "Every day in every way, things get better and better." In simple terms, this idea of progress, of societies constantly moving toward improvement, was at the heart of the evolutionary perspective that dominated nineteenth-century social thought. In its nineteenth-century form, the evolutionary perspective held that societies must pass through a set series of stages, each of which results in a more complex and advanced stage of development.

The evolutionary perspective became popular during the nineteenth century in part because of the nature of those times. During the middle and late nineteenth century, Western Europeans prided themselves on their superiority over other peoples. Progress and territorial expansion were watchwords of the times. Contact with preliterate people made through colonial expansion only offered ethnocentric Europeans further proof that their societies were the most highly civilized ever to exist. "Primitives" simply represented early stages of social evolution.

Another important influence on the rise of the evolutionary perspective of social change was the work of Charles Darwin (1809–1882), which provided an intellectual framework for interpreting the place of primitive and advanced societies in the world. Darwin's idea that all living things tend to improve and become more complex as they develop seemed to fit the fact that some societies were more advanced than others.

In the hands of the English sociologist Herbert Spencer, Darwin's biological ideas were transformed into social Darwinism. Spencer drew a parallel between living organisms and societies and coined the expression "survival of the fittest" to explain why some peoples had become "civilized" whereas others remained "savages" and "barbarians." Social Darwinism was so widely accepted in the Western world that governments in Western Europe and America believed that they had the right to dominate, protect, and tutor less developed societies. Such was thought to be the destiny and burden of nature's hardiest people.

Social Darwinism and classical evolutionary perspectives on change survived into the first part of the twentieth century, but they are no longer considered valid. Social scientists no longer believe that societies necessarily improve as they change or that developing societies are destined to follow the Western model. Nevertheless, more recent perspectives on social change have overcome some of the faults of the classical evolutionary perspective (Sanderson, 1990; Chirot, 1994).

What is the viewpoint of modern evolutionary theorists? The contemporary evolutionary perspective rejects the unilinear, or single-direction, assumption of earlier perspectives. Given the social and cultural diversity around the world, modern evolutionary theorists ask, How can we believe that all societies develop in a single direction? Because we know that societies can develop in many ways—some even toward greater unhappiness and deterioration—evolution must be multilinear, or multidirectional. Deliberately missing from the current evolutionary perspective, then, are the ideas of definite and orderly stages of development and of change inevitably producing

progress and greater happiness. There are many paths and directions to evolutionary change (Steward, 1979).

The most recent sociocultural evolutionary theory comes from Gerhard Lenski, Patrick Nolan, and Jean Lenski (1995), whose initial assumption is that both change and stability must be considered if human societies are to be understood. Despite the dramatic changes that have taken place in the world over the past ten thousand years, they argue, most individual societies have successfully resisted change. This, of course, is a paradox: How can rapid social change occur when individual societies successfully resist change? To resolve this apparent paradox, Lenski, Nolan, and Lenski offer a proposition: a total system can change despite resistance to change in a majority of its parts, providing the parts that fail to change do not survive. This, they assert, is precisely what has happened in the "world system of societies." Almost none of those societies that changed little while they existed are still around today. Conversely, nearly all surviving societies are those that have been greatly altered. Thus, say Lenski, Nolan, and Lenski, a process of natural selection has operated in the world system of societies which favors innovative societies over those that resist change the most successfully.

For Lenski, Nolan, and Lenski, the key to societal survival is the accumulation of information, particularly information relevant to subsistence. In fact, improvements in subsistence technology are said to be a "necessary" precondition for a society to grow in complexity, size, wealth, or power. This is because subsistence technology is the corpus of information that provides the energy members of a society need to maintain their endeavors. As available energy increases, activities of members increase. To say it another way, subsistence technology sets limits for activity and change within a society. Subsistence technology is also important because it stimulates improvement in other types of technology, as in production, transportation, communication, and war. These further technological advances promote additional advances in subsistence technology. Lenski, Nolan, and Lenski are not technological determinists. For example, they think that a society's beliefs and values are crucial because, in part, they affect the extent to which a people are open to innovation and change.

What does modern evolutionary theory explain? Sociocultural evolutionary theory helps us understand how societal changes are made. It shows that as subsistence technologies have improved, societies have had the means for gradually progressing from a hunting and gathering to a postindustrial form. As we shall see, modern evolutionary theory is informed, in part, by both functionalist and conflict theory.

The Functionalist Perspective

As we have shown repeatedly, functionalism emphasizes social stability and continuity over social change. Because the central question of functionalism focuses on the contributions various social aspects of society make to society's maintenance as a whole, it may seem contradictory to speak of a functionalist theory of social change. Nevertheless, it is not inappropriate. Two functionalist theories of social change—that proposed by William Ogburn and that of Talcott Parsons—are especially interesting. Both of these theories are based on the concept of *equilibrium.*

What is the connection between functionalism and the concept of equilibrium? The word *equilibrium* implies balance and consistency, for it connotes attempts to reestablish stability after some internal or external disturbance. An ordinary room thermostat for temperature control is the most common physical analogue. When applied to social life, the concept of equilibrium holds that societies are inherently stable and that any changes within them are eventually assimilated to achieve a new state of equilibrium. A society in change, then, moves from stability to temporary instability to stability once again. Sociologists refer to this as a "dynamic," or "moving," equilibrium. For example, in 1972, a broken dam led to the destruction of the community of Buffalo Creek, West Virginia. The physical destruction of the community was accompanied by death and the loss of the old fabric of life. Despite the ensuing chaos, residents of the community managed to pull their lives together again. Things were not the same as they had been before, but a new equilibrium was built out of the physical, social, and human wreckage (Erikson, 1976).

This application of the concept of equilibrium should remind you of functionalism. In fact, many sociologists believe that functionalism is based on the concept of equilibrium. Kingsley Davis, for example, writes that "it is only in terms of equilibrium that most sociological concepts make sense. Either tacitly or explicitly, anyone who thinks about society tends to use the notion. The functional-structural approach to sociological analysis is basically an equilibrium theory" (Davis, 1949:634). Although not all sociologists would agree that most sociological concepts make sense only in terms of equilibrium, most would concur that functionalism involves the concept of equilibrium.

Before turning to the two specific functionalist theories of social change, we should consider the similarity of and difference between the evolutionary and functionalist approaches to the explanation of social change, because both of the functionalist theories of change we will describe draw on evolutionary theory. Both the evolutionary and functionalist perspectives view society as composed of many highly differentiated parts, all of which contribute to the maintenance of a smooth-running, stable society. However, whereas the evolutionary perspective emphasizes a constant forward direction spurred by change, the functionalist perspective places emphasis on society's ability to recover its balance if equilibrium is upset by change. In functionalism, the various parts of society are seen as highly integrated, so that if a change occurs in one part, other parts are affected. The chain reaction is supposed to eventually restore balance because change is absorbed and distributed among a society's elements until equilibrium is reestablished. The emphasis in the functionalist approach is always on a return to stability and order after some restructuring has taken place.

What is Ogburn's theory of social change? If a new equilibrium is achieved, the parts of society, asserted William Ogburn (1964, originally published in 1922), do not all reach the new balance at the same time. Some parts lag in time behind others. Ogburn applied the term **cultural lag** to any situation in which disequilibrium is caused by one aspect of a society failing to change at the same rate as an interrelated aspect. More specifically, Ogburn believed that changes in the nonmaterial aspects of culture (norms, values, beliefs) lag behind alterations in the material culture (technology, inventions). Significant social change occurs when the nonmaterial culture is forced to change because of a prior change in the material culture. Ogburn's well-known example is the lag between the technological ability to cut down entire forests and the subsequent emergence of the conservationist movement. More recently, we can point to the sexual norms, values, and beliefs now in the process of attempting to catch up with the widespread distribution of birth-control technology that occurred years ago. Cultural lag still exists. The Roman Catholic Church officially continues to oppose birth control, while many members of the church are waging personal struggles over contraception. The continuing conflict (sometimes violent) between pro-choice and pro-life advocates over the issue of abortion is a prime illustration of cultural lag. Finally, in a related area, consider the legal, ethical, and social dilemmas yet to be resolved flowing from the technology permitting surrogate motherhood.

How does Parsons approach social change? The concept of equilibrium is at the heart of Talcott Parsons's theory of social change. In his early work, Parsons (1937, 1951) did not emphasize social change. Societies were depicted as systems attempting to resist change in order to maintain their current state of equilibrium. It was only later that Parsons (1966, 1971, 1977) began to depict change as contributing to the creation of a new state of equilibrium with characteristics different than before change was introduced.

Consistent with his roots in evolutionary theory, Parsons was interested in the processes by which societies become more complex. The first process is *differentiation,* by which aspects of a society are broken into separate parts. As we saw earlier, in simple societies the family was responsible for nearly all functions—economic, educational, medical, emotional, and recreational. As societies became more complex, these functions began to break off (differentiate) from the family. Jobs were found in factories, education occurred in schools, doctors and hospitals cared for people's medical needs, people outside the home were used for emotional support, and amusement was supplied by a variety of nonfamilial sources.

Differentiation brought on the need for the second process—*integration.* All the newly evolving social units had to form workable links for a new equilibrium—different from the previous one—to be established. Ways had to be developed for schools and families to mesh, for parents to accept their children's leaving the land for work in the factories, and for parents to cope with their offspring's establishing their own lives outside their local communities. The process of integration not only leads to a new state of equilibrium, but also, in conjunction with the process of differentiation, helps to produce a much more complex type of society.

What are the contributions and criticisms of the functionalist perspective on social change? A major contribution of the functionalist perspective is its attempt to explain both stability and change. Ogburn tries to show that a society's ways of thinking, feeling, and behaving are constantly attempting to catch up with prior technological change. Technology, for him, is an independent variable leading to further social change. Parsons identifies differentiation and integration as processes integral to the maintenance of a moving equilibrium. His work is also valuable in emphasizing that change does not imply total change; continuity as well as change exists.

Functionalists have been criticized for having a conservative bias. The concept of equilibrium, argue some critics, assumes internal societal resistance to change. Also, the equilibrium model of functionalism is charged with depicting change as external to societies: societies are changing not because of their own internal dynamics but as a result of being forced to make adjustments. By failing to explore internal sources of change, it is argued, functionalists ignore the many forces for change that influence a society. Finally, functionalists have been faulted for focusing on gradual change and failing to consider radical change (Collins, 1977; Giddens, 1979, 1987, 1997). The conflict perspective addresses these criticisms.

The Conflict Perspective

Whereas the evolutionary and functionalist perspectives are based on the assumption that society is inherently stable, the conflict perspective depicts society as unstable, with ever-present conflict and inconsistency. In addition, the evolutionary and functionalist perspectives view society as an integrated whole whose various parts work harmoniously to achieve balance, whereas the conflict perspective emphasizes the separate parts and the conflict that occurs among them. According to the conflict perspective, social and cultural change occurs as a result of the struggles and conflicts among groups representing different segments of a society. The resolution of conflict among various segments of a society results in social change. The conflict perspective sees scarcity of desired resources—over which groups are in conflict—as the major source of the social instability that characterizes all societies.

What are the origins of the conflict perspective? Many of the basic assumptions of the conflict perspective emerge from the writings of Karl Marx, who wrote that "without conflict, no progress: this is the law which civilization has followed to the present day" (Marx, in Feuer, 1959:7). Ralf Dahrendorf, a modern advocate of the conflict perspective, summarizes Marx's view of society in this succinct passage:

For Marx, society is not primarily a smoothly functioning order of the form of a social organism, a social system, or a static social fabric. Its dominant characteristic is, rather, the continuous change of not only its elements, but its very structural form. This change in turn bears witness to the presence of conflicts as an essential feature of every society. Conflicts are not random; they are a systematic product of the structure of society itself. According to this image, there is no order except in the regularity of change (Dahrendorf, 1959:28).

More specifically, Marx believed that the struggle for scarce economic resources, particularly property, is the primary stimulus for change. Change in capitalist society is authored by class struggle. As a capitalist society develops, according to Marx, it begins to divide into two classes: those who possess property and own the means for production (the bourgeoisie) and the exploited workers who sell their labor merely to survive (the proletariat). As time passes, these two classes continue to polarize as the rich get richer and the poor get poorer. As polarization continues, members of society increasingly tend to become members of the proletariat or of the bourgeoisie; in-between classes begin to disappear. At some point, a revolution permits the oppressed proletariat to seize power. With the proletarian revolution comes the development of a classless society. The absence of classes eliminates the source of conflict.

This closed factory represents a significant loss in jobs for individuals and loss in tax revenue for the local community and state. According to conflict theorists, the ability and willingness of a corporation to move away despite these negative consequences lies in its superior control over resources.

What are the contributions and criticisms of Marx's approach to social change?   Defenders of Marx contend that contrary to the charge, Marx was able to avoid economic determinism. Although Marx contended that the economic base of any society is the ultimate determinant of the society's nature and course, he also allowed for the influence of the noneconomic aspects of a society on the society's economic base. Marx is credited with effectively showing that how people think, feel, and behave reflects the basic underlying economic foundation of their society. In addition, Marx refined the view that social conflict is built into all stratification structures and that the distribution of power is rooted in the economic system (Lauer, 1991).

According to critics, Marx placed too much emphasis on economic factors as the determinants of social change while downplaying relevant social and cultural forces. Others criticized Marx for failing to see that conflict can integrate societies as well as tear them apart and for not recognizing the prevalence of cooperation over conflict within societies. Marx's views on revolution have also been challenged. Contrary to Marx's prediction, the critics assert, most revolutions of this century have come from the middle class rather than the working class. Also, the critics continue, communist revolutions did not occur in highly industrialized Western societies such as the United States, Great Britain, or Germany but in the agrarian societies of China, Russia, and Cuba. Finally, non-Marxists note the lack of a polarization of capitalist societies into ruling classes and working classes. They point to the emergence of a large middle class—composed of neither workers nor owners—in modern capitalist societies (Vago, 1996).

Although the conflict perspective on social change almost disappeared from the sociological scene during the first half of the twentieth century, it has been revitalized. This reemergence is well illustrated in the work of Ralf Dahrendorf, who follows some of Marx's ideas on conflict but rejects others (1958a, 1958b, 1959). Though Dahrendorf sees instability and change as results of struggles over resources, he disagrees with Marx on the type of resources involved. Marx envisioned a struggle over material resources; Dahrendorf views change as resulting from conflict over power—over who controls whom. In addition to their differences over the type of resource scarcity that provokes conflict, Marx and Dahrendorf also disagree on the number of societal segments involved in conflict. Marx saw society as formed of two opposing social classes, those who own the means of production and those who don't. Dahrendorf, seeing power as the source of conflict, proposes that conflict can occur among groups at all levels of society. Rather than one grand conflict, Dahrendorf envisions a multitude of simultaneous conflicts among interest groups in all segments and at all levels of society, the resolution of which results in change. By rejecting Marx's contention that history is created by class conflict alone, Dahrendorf is able to give prominence to conflicts among all types of interest groups—political, economic, religious, racial, sexual. Societies change as power relationships among interest groups change.

To the present time, historical events offer more support for Dahrendorf's interpretation of the conflict perspective than to Marx's. Class conflict on the order described by Marx has not occurred in any capitalist society. Social classes have not been polarized into major warring factions. Rather, capitalist societies have become composed of countless competing groups. In America, racial groups struggle over the issue of equal economic opportunity, and environmentalists and industrialists fight over the proper balance in environmental protection and economic development.

Some sociologists have concluded that no single perspective on social change is clearly superior to others in all respects. They assert that social and cultural change is too complex to be captured by any single theory, given our present state of knowledge. They believe that because each perspective has its strengths and weaknesses, the best way to explain social change is through the combination of various aspects of each theory. These efforts have been called *functionalist conflict* theory (Collins, 1994).

Attempts to Reconcile the Functionalist and Conflict Perspectives

There has been no fully successful synthesis of the functionalist and the conflict perspectives on social change, but some interesting attempts have been made. Although he does not reconcile all the differences between these two perspectives, Pierre van den Berghe (1978) is able to point out some commonalities. First, both perspectives observe society as systems with interrelated parts. Second, both perspectives are capable of viewing conflict as a contributor to social integration and integration as a producer of conflict. Third, both theoretical perspectives assume an evolutionary view of social change. Finally, van den Berghe sees both perspectives as equilibrium models.

Like Marx, Lewis Coser (1964, 1991) sees the pressure for social change coming from conflict occurring as competing groups and classes rise and fall in the struggle for power and the protection of their interests. Unlike Marx, however, Coser does not believe that cataclysmic social change necessarily follows social conflict or that change produced by conflict must necessarily come about suddenly. In fact, Coser distinguishes between change *within a system* and change *of a system*. A society flexible enough to adjust to a new environment will experience change while maintaining its basic structure. Examples of change *within* a society are the recent civil rights and women's movements in the United States. If a society is not flexible enough, a change *of* the system may occur as a result of conflict. The Chinese Communist Revolution illustrates change *of* a society.

According to Coser, the behavior of elites strongly influences whether conflict will produce a readjustment within the society or a breakdown and formation of a new society. If the dominant groups are flexible enough to allow free expression of complaints and make appropriate adjustments, change within the society is more likely to occur. Should those with the power choose to protect their interests by resisting change and stifling grievances, they run the risk of intensifying conflict and producing a change of the society.

Radical change of a society may occur suddenly, but it does not necessarily have to, according to Coser.

Can We All Get Along?

Modern American society is changing toward such ethnic and racial diversity, some in this CNN report argue, that the accompanying social conflict among minority groups will tear it apart. The conflict in such places as Lebanon, Yugoslavia, and the former Soviet Union are offered as examples of diversity leading to social disunity. Other observers see the increasing racial and ethnic variety as a source of positive social change consistent with the historically diverse nature of American society. These individuals point to a changing social world whose hallmark is globalism.

1. Which theoretical perspective—functionalism or conflict theory—would be endorsed by those concerned about diversity leading to social fragmentation? Explain.

   _____

   _____

2. Why would advocates of the other theoretical perspective not share this concern?

   _____

   _____

Violence, as in the looting pictured above, is cited as supporting evidence by those persons who see ethnic and racial diversity as a source of social fragmentation.

Although basic institutions, values, and social relationships did change immediately following the Chinese Communist Revolution, change following the American war for independence was gradual. The United States is a fundamentally different society than it was before 1776, but this restructuring cannot be linked to any specific event. In other words, for Coser, the fundamental restructuring of a society can take place as a result of gradual, cumulative changes within it.

Lenski, Nolan, and Lenski (1995) also attempt to combine the functionalist and conflict perspectives. In the first place, they give equal weight to stability and change. Second, they incorporate conflict explicitly into their theory. For example, Lenski, Nolan, and Lenski contend that when societies have come into conflict with one another for territory and other vital resources, those that have been technologically more advanced have usually been dominant. They make ample provision for conflict between societies, pointing out the need for societies to defend themselves and their territories from the invasion of outsiders. Nor do they ignore conflict within societies, highlighting, for example, the struggle in industrializing agrarian societies between the elites in control of their society's resources and the masses who supply the labor for subsistence-level returns.

Do these attempts at reconciliation mean that the functionalist and conflict perspectives can be abandoned in the area of social change? Although these attempts at reconciliation may be interesting, they do not eliminate the need for separate functionalist and conflict perspectives on social change. Continued work by functionalists will tell us more about the processes involved in the maintenance of a dynamic equilibrium—how societies maintain stability and order while undergoing change. Conflict theorists should continue to provide insights into how change occurs as a result of the struggles among those with differential power and opposed interests.

MODERNIZATION AND WORLD-SYSTEM THEORY

The Nature of Modernization

It seems as if the entire world is either already modernized or attempting to move in that direction. This move toward modernization is, in fact, one of the most significant trends in human history.

What is modernization? **Modernization** is a process involving all those social and cultural changes accompanying economic development. As a society modernizes, it moves from being a traditional, undeveloped society toward becoming an economically developing society. This transition from traditional to developing society involves a host of changes (Haferkamp and Smelser, 1992).

What are some of the major consequences of modernization? Some dramatic demographic changes are associated with modernization. Population growth occurs as the death rate declines and life expectancy increases. Between the thirteenth and seventeenth centuries, for example, Europe had a life expectancy as low as twenty years; by 1930, life expectancy had increased to over sixty years and now stands at seventy-three years (*World Population Data Sheet*, 1997). A second demographic change is the movement of the population from rural to urban areas. Whereas most members of traditional societies work the land, most members of modern societies live and work in towns and cities, because that is where industry and jobs are located.

Stratification structures are altered by modernization. Traditional society is characterized by a bipolar stratification structure—the wealthy at one end, the poor masses at the other. Modernization brings about an expansion of the middle and upper classes as the sharing of wealth becomes more widespread. As more

FEEDBACK

1. According to the cyclical perspective on social change, societies rise and fall rather than move continuously toward improvement. *T or F?*
2. Cyclical theories explain why social change occurs. *T or F?*
3. Classical evolutionary theory assumed that change leads to improvement. *T or F?*
4. According to modern evolutionary theory, is social change unilinear or multilinear? _____
5. According to the _____ perspective, society is inherently stable, and any change that occurs is eventually assimilated so that a new state of equilibrium is achieved.
6. _____ refers to any situation in which disequilibrium is caused by one aspect of a society failing to change at the same rate as an interrelated aspect.
7. According to the _____ perspective, social and cultural change occurs as a result of the struggles among groups representing different segments of a society.
8. An attempt to synthesize the functionalist and conflict perspectives has come through _____ theory.

Answers: 1. T 2. F 3. T 4. multilinear 5. functionalist 6. Cultural lag 7. conflict 8. evolutionary emphasis is placed on personal effort and achievement, social mobility increases and inequality generally declines. Actually, although status increases for the population in general—and women in particular—it declines for the elderly.

The political institution is greatly affected by modernization. The role of the state expands as it becomes more centralized and more involved in social and economic affairs. At the same time, modernization generally promotes political democracy. Although political power is not totally equalized as a result of modernization—there are always powerful elites—power is more widely dispersed. Moreover, the extent of democratization varies considerably from society to society.

Modernization moves education from the family to a formal system of educational organizations. Education is designed not just for the privileged few but for the entire society; primary education, in fact, is intended for all members of a modernizing society. This is because a literate population is necessary for the development of an adequate workforce in an industrialized economy. The educational curriculum must also be changed to become more technical and more secular.

The last major consequence of modernization relates to family life. The nuclear family tends to replace the extended family. Because the economy is not based on a familial division of labor, people must move to cities where the jobs are located. This does not mean that extended family ties are obliterated; they simply become much harder to maintain, and much of their tightness is lost. One universal consequence of modernization is that other social units take over functions that were formerly the family's domain. For example, government assumes more responsibility for the elderly, schools take care of most of the children's educational needs, and the locus of entertainment shifts to the mass media.

These social and cultural changes are usually associated with modernization. Does this mean that all of them will occur in every society that is undergoing modernization? Will they be expressed in the same specific ways in all societies? These questions have been debated by sociologists.

Are the social and cultural characteristics of modernizing societies becoming similar?   This question has received conflicting answers. Advocates of the pattern of *convergence* foresee the development of social and cultural similarity among modernizing nations. They contend that as societies adopt the newest technological arrangements they tend to develop social and cultural commonalities. This is said to occur because a particular technology carries in its wake a particular occupational structure, which in turn exercises similar wide-ranging effects on various social structures and social relationships. The necessity for an industrializing society to develop an educated populace is another major unifying force. Other pressures for uniformity include the enlarged role of government, the influence of multinational corporations within modernizing countries, and the accelerated growth of large-scale organizations.

Supporters of the pattern of *divergence* do not see social and cultural homogeneity as an inevitable result of modernization. Other sociologists emphasize the effects of idiosyncratic social and cultural forces to explain the emergence of new social patterns in modernizing societies. Advocates of divergence offer as evidence less developed countries that are now undergoing modernization along lines somewhat different from the path taken by Western societies. Many less developed countries who chose the Soviet economic model from the 1950s onwards are, along with former Communist countries in Eastern Europe, now adopting the capitalist model of the West (*World Development Report,* 1991; Schnitzer, 1997). Could this signal greater convergence in the future?

As in most complex debates, there is a middle ground that is probably closer to the truth. Social and cultural elements of traditional societies do not necessarily vanish in the process of modernization; traditional elements may blend with the new. Changes in the family during modernization, for example, depend in part on the type of traditional family a society has. Some traditional family structures are more compatible with modernization than others. To cite another example, democratization usually accompanies modernization, but its extent and form may vary because of the specific social and cultural traditions of a society. Although modern Japan is more democratic than it was in premodern times, it is less democratic than most Western societies. And although the Soviet Communist Party leaders voted overwhelmingly in July 1991 to endorse the principle of private property, freedom of religion, and a pluralistic political system, serious political opposition has since arisen. It remains to be seen how such legislative changes will be reflected in democratic forms.

World-System Theory

To this point, the process of modernization and its consequences have been shown from the point of view of events *within* countries. When modernization is considered from an international perspective, quite different theories emerge. One of these is world-system theory, which also depicts social and cultural diversity among modernizing nations. Its approach is unique (VanRossem, 1996; Chase-Dunn and Hall, 1997).

What is world-system theory?   According to **world-system theory,** the pattern of a nation's development largely depends on the nation's location in the world economy. According to conflict theorist Immanuel Wallerstein (1979b, 1989) and some other world-system theorists, the world economy is divided into segments. Core nations, such as the United States, dominate the world economy through the provision of managerial expertise and technological innovation. This domination leads to the control and exploitation of the rest of the world; the core nations use the cheap labor and natural resources of peripheral nations (such as less developed nations) to benefit themselves.

What are the consequences of this arrangement? As a result of the world-system arrangement, the standard of living is much higher in core nations, whose skilled workers provide labor of their own free will. Workers in peripheral nations who provide unskilled, coerced labor suffer a low standard of living. In addition, peripheral nations do not develop as much as they could if they were not dominated and exploited by the core nations. In fact, world-system advocates contend, the prime mechanism in perpetuating this arrangement is the prevention of full development in peripheral nations. Consequently, the economic, social, and cultural gap between core and peripheral nations is increasing, thus ensuring that developing and developed societies will not converge (Rossides, 1990b).

THE FUTURE OF SOCIETY

The question of the future of modern society is absurdly broad and complex. In this concluding section, we will explore only two aspects of the future. First, we look at the environment, the physical context that will have an overriding influence on the nature of human existence in the future. Second, the social environment of the future is discussed within the context of the call for a renewed sense of community.

The Physical Environment

As shown in Chapter 16 ("Population and Urbanization"), eighteenth-century economist Thomas Malthus believed that the crush of overpopulation would be infinitely greater than our ability to produce enough food. In the short run, at least, science and technology have so far forestalled the dire consequences envisioned by Malthus. However, Malthus's emphasis on the earth's limitations has not been forgotten. **Carrying capacity** refers to the number of organisms of a given species that can be sustained in a given geographic area. This term captures the idea that there is a limit to the population size of any species that can be maintained in an area without depleting the natural resources for the next generation (Postel, 1994). In recent years, the industrialized nations of the world have become painfully aware of the limitations of their natural resources, particularly fossil fuels (oil, natural gas, and coal). The carrying capacity of developed countries is being tested by the growing population and high demand for energy to support an industrial economy.

Some current descriptions of the limits of the earth's resources and existing environmental pollution offer little encouragement to future generations. Scholars do not agree, however, on what the future holds (Pearlstein, 1997). Some believe that if resource depletion, pollution, and population growth continue at the present rate, world society is seriously threatened (Flavin, 1997). Others believe that technology can save the day (French, 1997). Let's examine these two viewpoints more closely.

What is the negative view of the future? Those who fear the breakdown of industrial society see themselves as realists. If they are gloomy about our prospects, they argue, it is only because they are facing the facts regarding the planet's physical condition. (See Table 18.2.) They point to recent scientific reports claiming that the protective ozone layer over densely populated areas of

FEEDBACK

1. _____ is a process involving all those social and cultural changes accompanying economic development.
2. Which of the following is not one of the major consequences of modernization discussed in the text?
    a. the social and cultural convergence of modernizing countries
    b. increased urbanization
    c. greater equality
    d. more political democracy
    e. widespread development of the nuclear family
3. According to _____ theory, the pattern of a nation's development largely depends on the nation's location in the world economy.
4. World-system theory would predict a (convergence/divergence) between core and peripheral nations.

Answers: 1. Modernization 2. a 3. world-system 4. divergence

始#

510 SOCIOLOGY

TABLE 18.2	Bad News: Changes in the Earth's Physical Condition
Forests	Each year the Earth's tree cover diminishes by some 17 million hectares, an area the size of Austria. Forests are cleared for farming, harvests of lumber and firewood exceed sustainable yields, and air pollution and acid rain take a growing toll on every continent.
Land	Annual losses of topsoil from cropland are estimated at 24 billion tons, roughly the amount on Australia's wheatland. Degradation of grazing land is widespread throughout the developing world, North America, and Australia.
Climate System	The amount of carbon dioxide, the principal greenhouse gas in the atmosphere, is now rising 0.4 percent per year from fossil-fuel burning and deforestation. Record hot summers of the eighties may well be exceeded during the nineties.
Air Quality	Air pollution reached health-threatening levels in hundreds of cities and crop-damaging levels in scores of countries.
Plant and Animal Life	As the number of humans inhabiting the planet rises, the number of plant and animal species drops. Habitat destruction and pollution are reducing the earth's biological diversity. Rising temperatures and ozone layer depletion could add to losses.

Source: Reprinted from *State of the World, 1991*, A Worldwatch Institute Report on Progress Toward a Sustainable Society, Project Director: Lester R. Brown. By permission of W. W. Norton & Company, Inc. Copyright © 1991 by Worldwatch Institute.

before they can dry on the line. For miles around Copsa Mica, the Romanian town dubbed "black town," the trees and grass are so saturated with soot that they appear to have been covered with ink (Painton, 1990). Pollution is also an environmental problem in the People's Republic of China. Coal soot from factories wraps Peking in a thick black haze much like that in Pittsburgh in the earlier days of unregulated industrial production (Baumol and Blinder, 1996; see Figure 18.1).

Some important research projects support this unpleasant portrait of the future. Under the auspices of the Club of Rome, a prestigious international group formed out of concern for the world's future, a group of researchers from the Massachusetts Institute of Technology launched a sophisticated computer-simulation study of the future of industrial society, using population growth, resource supply, food supply, industrial output, and pollution as key variables. No matter what solutions they entered into their model—"unlimited" resources, pollution control, birth control, increased food yield—the result of their projections was the same: industrial society will collapse within a hundred years because population and the world economy are growing too fast. The assumption that technology can save us, these researchers conclude, is a false one. All that technological advances can do is delay the inevitable end, because dominant cultural patterns and institutional arrangements perpetuate behaviors that are now socially and environmentally destructive.

Another study of the future, an 800-page report on the earth's population, natural resources, and environmental problems, takes a very pessimistic view of the world in 2000.

If present trends continue, the world in 2000 will be more crowded, more polluted, less stable ecologically, and more vulnerable to disruption than the world we live in now.

Pessimists regarding the sustainability of the Earth underscore the fact that Eastern Europe is the most polluted area on the planet. The heavy pollution from this factory in former East Germany is graphic testimony to the problem.

the United States has thinned twice as fast as previously predicted. Ultraviolet rays, a cause of skin cancer, are also said to be reaching the ground for longer periods of the year (Snider, 1993). If this report of the accelerating depletion of the ozone layer is accurate, the Environmental Protection Agency states that the number of Americans dying of skin cancer over the next fifty years will not be the 9,300 predicted earlier but will reach 200,000. Those concerned about the sustainability of the planet also point to what has been called Communism's dirtiest secret—the fact that Eastern Europe is the world's most polluted region. In parts of Poland and Czechoslovakia, the air is so polluted by factories and power plants that washed clothes are dirty

Figure 18.1 More Bad News: Current Worldwide Carbon Dioxide Emissions

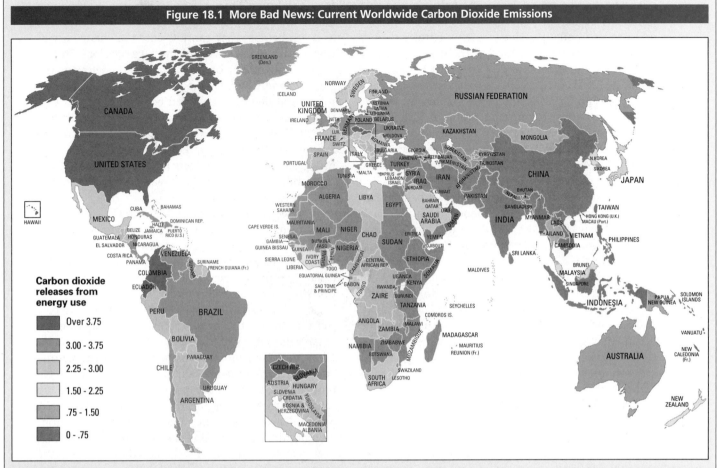

Carbon dioxide releases from energy use

- Over 3.75
- 3.00 - 3.75
- 2.25 - 3.00
- 1.50 - 2.25
- .75 - 1.50
- 0 - .75

This figure shows the carbon dioxide releases from energy use, industrial activities, and deforestation, expressed as tons of carbon per person per year. Countries with the highest emissions are Germany and the United States. Carbon dioxide, created primarily from the burning of fossil fuels such as oil, natural gas, and coal, is a major contributor to the buildup of "greenhouse gases" that climatologists fear will create a long-term warming of the Earth's atmosphere.

Source: Adapted from William C. Clark, "Managing Planet Earth." Copyright © 1989 by *Scientific American, Inc.* All rights reserved.

Internet Link: For more information on international ecology, you can go to the United Nations Environment Programme/Industry and Environment at http://www.unepie.org/home.html

Serious stresses involving population, resources, and environment are clearly visible ahead. Despite greater material output, the world's people will be poorer in many ways than they are today.

For hundreds of millions of the desperately poor, the outlook for food and other necessities of life will be no better. For many it will be worse. Barring revolutionary advances in technology, life for most people on earth will be more precarious in 2000 than it is now—unless the nations of the world act decisively to alter current trends (The Global 2000 Report to the President, *1980, Vol. 1:1).*

By the time you read the quote above, nearly twenty years will have passed from the date of its publication. Thoughtful people who worry about the future of the environment do not believe the warning of *The Global 2000 Report* was heeded. Their fears are heightened as

the world stands on the threshold of the twenty-first century.

What is the positive view of the future? As you might suspect, these reports have attracted many critics. Some have criticized the research methods and assumptions of the studies. They believe that the extreme case presented by those predicting disaster does more harm than good (Budiansky, 1993). It alienates less developed countries that do not yet have the luxury of worrying about pollution and the depletion of natural resources. And by crying wolf too loudly, the environmentalists anesthetize people to real problems that can and should be solved. Some critics of the Global 2000 study believe that doomsday prophecies may even become self-fulfilling because they may lead us to reduce efforts to improve the situation. Others are convinced that birthrates will be forced to decline before any breakdown can occur. Still others contend that technology will prevent collapse. Those in this latter category count on breakthroughs such as the one claimed by researchers Stanley Pons and Martin Fleischmann who announced in April 1989 that they had sustained nuclear fusion (creating energy by forcing two atomic nuclei together) at room temperature. If such fusion proves to be possible, the potential exists to provide an inexhaustible source of energy that, unlike nuclear fission (breaking apart atomic nuclei), yields very little radioactive waste (Lemonick, 1993).

Those who see a brighter future do not believe that continued population and economic growth will cause the world to reach the limits of its natural resources and lead to environmental deterioration. They believe that there is no fixed or predictable connection between economic growth, resource depletion, and pollution. In their view, economic growth can continue without depleting our resources or destroying the environment. If economic growth is harming us, they conclude, it is because we are not allocating our resources and protecting our environment properly. Stopping growth, in other words, will not necessarily halt resource depletion and pollution.

Some observers are convinced that a more rational approach is to weigh the cost of solving environmental problems against benefits obtained. Our high priority on pollution problems, they say, has sometimes cost us more than it has been worth; we have spent too much money attempting to reach higher standards of cleanliness than is practical or affordable. Besides, they contend, our city air has less carbon monoxide in it now than it had fifty years ago from the burning of coal, coke, and charcoal without adequate oxygen flow. (See Table 18.3.) These observers predict that industrial growth in less developed countries will eventually slow down population growth. However, industrial development requires greater food production, which requires the application of fertilizers and pesticides. If environmentalists insist on stricter controls on the use of these chemicals for food production, the developing countries' economic development will be retarded—and population growth will not be slowed down.

So what does the future hold environmentally?   If you want to know what the future holds, take the word of a palmist you trust. In other words, it is impossible to know at the present time what the future holds. However, most social scientists probably agree with environmentalist Paul Ehrlich: "the real question, for those concerned about realistic solutions, is whether researchers and decision-makers can devise ways to bring human behavior into harmony with physical reality in time" (Ehrlich, Ehrlich, and Holdren, 1977:957). More recently, Ehrlich wrote of the "boiled

frog" syndrome. Frogs placed in a pan of water whose temperature is slowly increased will sit until they boil to death because they cannot sense the change occurring in their lethal environment. Like the frogs, Ehrlich writes, humans do not appear capable of detecting the equally deadly trends in population expansion, environmental deterioration, and economic growth that threatens to boil the planet (Ornstein and Ehrlich, 1991). This conclusion is, of course, based on the assumption that world resources are limited, that technology cannot solve all our problems, and that the environmental deterioration created by technology is difficult to overestimate.

If the consequences associated with the earth's population, natural resources, and environmental problems are to be handled successfully, it is necessary for the public to become concerned, for government leaders to genuinely seek solutions, and for long-term thinking to replace short-term thinking (Clark, 1989; Regan, 1990; Bright, 1997). Scientist Lester Brown, project director for the Worldwatch Institute, a respected organization devoted to raising public and governmental awareness of problems associated with the natural environment, contends that the past five years have brought heightened interest in such problems. In the fourteenth annual publication of *State of the World*, Brown, Flavin, and French (1997) argue that environmental problems are being taken more seriously than ever. Concern, Brown says, is also building over the apparent increase in the earth's temperature and in the potential loss of the earth's tropical rain forests. Brown points to steps being taken in Russia and China as well as by the United Nations and the World Bank to build a more sustainable world. It does seem that problems of the environment are becoming more personalized. Americans, for example, are sufficiently worried about the dramatic increase in skin cancer associated with the depletion of the earth's ozone layer to create a boom in the sun block industry.

In the eleventh annual examination of the global environment done by Brown and the Worldwatch Institute, little reason for optimism about the future is offered. Deterioration of the physical environment is said to continue:

Signs of environmental constraints are now pervasive. Cropland is scarcely expanding any more, and a good portion of existing agricultural land is losing fertility. Grasslands have been overgrazed and fisheries over harvested, limiting the amount of additional food from these sources. Water bodies have suffered extensive depletion and pollution, severely restricting future food production and urban expansion. And natural forests—which help stabilize the climate, moderate water supplies, and harbor a majority of the planet's terrestrial biodiversity—continue to recede (Postel, 1994:3).

TABLE 18.3	Mixed News: Annual Emissions of Air Pollutants in the United States: 1940 to 1991 (Millions of metric tons per year)					
Year	Suspended Particulate Matter	Sulfur Oxides	Nitrogen Oxides	Volatile Organics	Carbon Monoxide	Lead*
1940	22.8	18.0	6.8	18.5	81.6	n/a
1950	24.5	20.3	9.3	20.8	86.3	n/a
1960	21.1	20.0	12.8	23.6	88.4	n/a
1970	18.1	28.2	18.1	27.1	98.8	203.8
1980	8.5	23.2	20.4	22.7	76.2	70.6
1985	7.3	20.7	20.0	21.3	67.5	21.0
1991	7.4	20.7	18.8	16.8	62.1	5.0

n/a = not available
* Lead is measured in thousands of metric tons per year.

Definitions of Air Pollutants

Suspended particulate matter—minute dust particles—results primarily from industrial processes and fuel combustion. Smaller particulates can carry toxic substances, be toxic themselves, and imbed themselves in lung tissue. Suspended particulate matter can aggravate respiratory illnesses.

Sulfur oxides largely originate from the combustion of coal and oil by electrical utilities and industrial processes. Sulfur dioxide is a main contributor, along with nitrogen oxides, to acid deposition.

Nitrogen oxides are caused by the combustion of fuel by industry, automobiles, and electrical utilities. Nitrogen oxides contribute to photochemical smog and ozone, which corrode wood and stone and threaten the health of humans and animals. Nitrogen oxides aggravate respiratory illnesses and combine with water in the atmosphere to form acid deposition.

Volatile organic compounds originate from the combustion of fossil fuels by automobiles and power stations, industrial processes, refineries, and volatilization of organic solvents and fuels. In the presence of sunlight, these organic compounds contribute to the formation of ozone. Ozone damages plant and animal tissue, prematurely ages the lungs, and causes other respiratory damage.

Carbon monoxide is formed from combustion of fossil fuels, mostly gasoline and diesel fuel. Exposure to carbon monoxide is greatest in urban areas. If present at high concentrations, carbon monoxide can cause drowsiness, slowed reflexes, and possibly death.

Lead in the atmosphere results mainly from the combustion of lead-containing gasoline by automobiles. Chronic exposure to lead—a heavy metal—can lead to anemia, convulsions, and kidney and brain damage.

Sources: U.S. Environmental Protection Agency, *National Air Pollutant Emission Estimates*, 1980–1984, EPA-450/4-85/014, 1986; *National Air Quality and Emission Trends Report*, 1985, EPA-450/4-87-001, 1987. Adapted from Susan Weber (ed.), *USA by Numbers*, Washington, DC: Zero Population Growth, Inc., 1988, pp. 133–134. Zero Population Growth, 1400 16th St. NW, Suite 320, Washington, D.C. 20036. The 1991 data came from Dale Curtis and Barry Walden Walsh, *Environmental Quality*, Council on Environmental Quality, Washington, DC: U.S. Government Printing Office, January 1993.

The situation may not be this dire. Some encouragement can be taken from the 1989 agreement of several European nations to eliminate by 2000 all production and use of chemical compounds (chlorofluorocarbons) harmful to the atmosphere's ozone protective shield. Seventy countries have signed an agreement committing industrialized nations to eliminate all chlorofluorocarbons by 2000 and committing developing nations to such elimination by 2010. The United States, oil companies, clean-fuels manufacturers, environmental and consumer groups, automakers, and state governments reached a rare accord in 1991 on a program to supply cleaner-burning gasoline to America's smoggiest cities beginning in 1995. In a more recent pact, 159 nations agreed to the Kyoto Protocol, which requires signers to reduce emissions of six greenhouse gases from 6 to 8 percent below 1990 levels (Bartholet and Breslau, 1997). It remains to be seen whether these apparent signs of concern for the planet and its inhabitants are permanent and sufficiently strong to prompt serious action on the part of citizens, government officials, and business leaders ("The Carrying Capacity Debate," 1996; Sagoff, 1997).

DOING RESEARCH

Katherine S. Newman—The Withering of the American Dream?

Katherine Newman's *Declining Fortunes* does not deliver an uplifting message. After the economic recovery from the Great Depression in the 1930s, each generation of Americans expected hard work to bring a standard of living higher than their parents'. A boost to this expectation of an increasingly more prosperous life was intensified by the unprecedented economic expansion following World War II. Newman's message runs counter to the experience of Americans after 1945 and their enormous progeny, the seventy-six million baby boomers born between 1946 and 1964. Both generations lived in a culture full of optimism about the future. Upwardly spiraling affluence was assumed to be part of the destiny of American citizens.

What is Newman's message? Social and economic change are placing the American Dream in jeopardy. The downscaling of jobs and pay that occurred during the 1980s and 1990s have replaced the earlier optimism with anger, doubt, and fear. The traditional middle-class way of life is threatened by recession, slow economic growth, high unemployment, declining work compensation, and a staggering national debt.

Newman set out to interpret the meaning of this new reality among middle-class Americans. She wished to understand the effects on family life, political attitudes, and personal identities as well as the "rage, disappointment, and . . . sense of drift in communities across the land" (Newman, 1993:x).

Bypassing the "bloodless" statistics, Newman decided to talk with some of the Americans behind the graphs and dire popular media accounts. She conducted personal interviews with 150 Americans living in a suburban community representative of much of America. "Pleasanton" is populated by the mix of skilled blue-collar workers and white-collar professionals from a variety of ethnic and religious origins that has been the hallmark of suburbia. Newman spent two years studying these people who are attempting to make sense of the unexpected wrong turn the promise of America has taken. Her respondents were schoolteachers, guidance counselors, and sixty families whose children are now grown.

As an anthropologist, Newman approached this exploration of downward mobility through attempting to understand the way residents of Pleasanton were viewing what is happening to the American Dream. She looked for hidden meanings of culture as much as the more apparent behavior and norms. The stresses associated with changing economic conditions, she believed, would be more likely to bring cultural expectations, disappointments, and conflicts close enough to the surface for a trained social scientist to see. In fact, she saw intergenerational conflict, racial and ethnic division, and marital discord that might be expected from a generation that will not lose all it has achieved but will not get what it had been led to believe would be its due. When reading the following statement by Lauren Caulder who was born in the 1950s when hard work promised success, the shattering of confidence in the American Dream

The Social Environment

An attempt to cover all aspects of the possible nature of the social environment in the twenty-first century and beyond would be futile. Even the nature of community in the future is too complicated for just treatment. The nature of community in the future, however, is so central to human existence that it must be introduced.

As you know from Chapter 1 ("The Sociological Perspective"), the founders of sociology were preoccupied with the negative effects of modernity on human community. Each of these early sociological thinkers grappled with the question of the maintenance of social and cultural meaning in the increasingly fragmented and individualistic social context they saw being created by industrialization. An early German sociologist, Ferdinand Tönnies (introduced in Chapter 5, "Social Structure and Society"), shared with these founders a concern for the loss of community in the wake of modernization. You may recall that in *gemeinschaft,* or preindustrial societies, members are oriented toward the values of the community as a whole. Kinship, tradition, cooperation, and intimate social relations fostered a high degree of social integration. *Gesellschaft* (industrial) societies, in contrast, are characterized by individualism, social fragmentation, weak family ties, competition, and impersonal social relationships. According to Tönnies, the transition from preindustrial to industrial society created a shift from among baby boomers will be readily apparent.

I'll never have what my parents had. I can't even dream of that. I'm living a lifestyle that's way lower than it was when I was growing up and it's depressing. You know it's a rude awakening when you're out in the world on your own . . . I took what was given to me and tried to use it the best way I could. Even if you are a hard worker and you never skipped a beat, you followed all the rules, did everything they told you you were supposed to do, it's still horrendous. They lied to me. You don't get where you were supposed to wind up. At the end of the road it isn't there. I worked all those years and then I didn't get to candy land. The prize wasn't there, damn it (Newman, 1993:3).

After a detailed and often personal exploration of what Newman calls the "withering American Dream," she turns to the larger social and political implications of a society in transition from upward mobility based on effort and merit to one in which the social class of birth promises increasingly to determine one's future social and economic situation.

According to Newman, the soul of America is at stake. She raises these kinds of questions: Will Americans turn to exclusive self-interest or care for others as well as themselves? Will suburbanites turn a blind eye to the rapidly deteriorating inner cities? Will the generations and racial and ethnic groups turn inward or attempt to bridge the divides that threaten to separate them further? A partial answer to these questions is reflected in public opinion regarding the use of federal, state, and local tax revenues.

In this regard, Newman does not paint a pretty picture. If the residents of Pleasanton are any guide, Americans do not wish to invest in the future for the common good. Public and secondary schools, colleges and universities, and inner cities, for example, are receiving a rapidly declining share of public economic support. In conclusion, Newman states:

This does not augur well for the soul of the country in the twenty-first century. Every great nation draws its strength from a social contract, an unspoken agreement to provide for one another, to reach across the narrow self-interests of generations, ethnic groups, races, classes, and genders toward some vision of the common good. Taxes and budgets—the mundane preoccupations of city hall—express this commitment, or lack of it, in the bluntest *fashion. Through these mechanistic devices, we are forced to confront some of the most searching philosophical questions that face any country: What do we owe one another as members of a society? Can we sustain a collective sense of purpose in the face of the declining fortunes that are tearing us apart, leaving those who are able to scramble for advantage and those who are not to suffer out of sight? (Newman, 1993:221).*

Critical feedback

1. Think about your experiences at home and in other social institutions (schools, churches). State the conception of the American Dream these experiences gave you. Critically analyze the ways in which society shaped your conception.
2. If the American Dream is withering, many social changes are in store. Describe the major outlines of change that you foresee.
3. Suppose Katherine Newman had decided to place her study in the context of sociological theory. Write a conclusion to her book from the theoretical perspective—functionalism, conflict, symbolic interactionism—that you think is most appropriate.

community interest to self-interest, from integrated social institutions to an atomistic individualism existing within a society composed of fragmented and isolated social institutions.

Tönnies did not believe that the social fragmentation of the industrial age could last over the long term. People, he argued, would eventually find the impersonality and social fragmentation so undesirable that a society combining aspects of gemeinschaft and gesellschaft would eventually emerge.

Recently, a number of prominent sociologists have argued that members of modern society are not isolated individuals; people are participants in a complex social web of institutional commitments. In fact, the argument goes, members of a society can only live their lives through social institutions—education, polity, economy, family, religion. Although those sociologists do not believe that we can return to the traditional small communities of Tönnies's gemeinschaft, they think we can recognize and take responsibility for the social institutions that connect Americans to each other and connect the United States to the rest of the world. Without nostalgically longing for small, isolated communities of the past, we can question our apparent exclusive commitment to autonomy and individualism in the light of our need to share in the effort to rejuvenate our corporations, families, schools, and cities. In this view, we all share in responsibility for cultivating the common good (Etzioni, 1988, 1993, 1995; Selznick, 1992; Wilson, 1993; Bellah, 1991, 1996).

A number of prominent sociologists have recently contended that we should pay attention to our shared responsibility for cultivating the common good. Habitat for Humanity volunteers, who are building this home, are one answer to this call.

These American sociologists are clearly emphasizing community and social integration over extreme individualism and social fragmentation. Is this expression of concern a reflection of genuine social change? Will society be remolded along the lines Tönnies predicted? Are these social scientists giving—"philosophical voice to the yearnings of ordinary folks who wish to preserve their liberties while reclaiming their vision of a decent community, one in which the moral senses will become as evident in public life as they now are in private life" (Wilson, 1993:248)? No firm answers are available for these questions. (See *Doing Research* for a pessimistic assessment). There is another view of society, *postmodernism,* that offers some stimulating speculation on the nature of community in contemporary society.

What is the postmodern view of community?
Postmodernism is a generic term applied to the postindustrial stage of social development in which people are dominated by media images and are connected through rapidly expanding electronic sources of information. In its most extreme form, postmodernism contends that contemporary society is composed of fragmented social institutions and isolated individuals more closely connected to their technological devices (computers, CD players, televisions) than to society or to other people.

Radical postmodernism embraces nihilism—a rejection of all traditional principles, beliefs, values, and social institutions. A more moderate variant of postmodernism not only rejects nihilism, but envisions the possibility of a renewed sense of community and social connectedness based on a technologically sophisticated system of national and international communications (Harvey, 1989; Griswold, 1994).

This view of postmodernism rests on the creation and dissemination of a variety of technological marvels (Elmer-DeWitt, 1993a, Powell, 1993). Some of these are already in the hands of consumers: electronic mail; wireless telephones; global telephone communication; cable and satellite television; interactive video games; information data banks on CD ROMs; and on-line television programming. The Internet, a global computer network, will probably soon be accessible through cable television (Elmer-DeWitt, 1994; Hafner, 1996). Other technological innovations that appear to be on the way to the mass market are interactive television and interactive movies.

As the electronic information superhighway or World Wide Web is extended to an ever-widening audience, contend the more optimistic postmodernists, human beings within societies and around the globe will be in touch with each other in an unprecedented way. Wendy Griswold makes this summary statement:

[I]f electronic communications make intimate interactions possible for more and more people regardless of where they are physically located, it may lead to a greater communion among them, a greater sense of what they have in common as people. Recall that when Americans and Europeans watched the CNN reports of the Gulf War in 1991, they shared a heightened sense of all being "in this together." Could communications foster a global sense of "all being in this together"? (Griswold, 1994:149).

A moderate view of postmodernism has envisioned the electronic information superhighway. This bank of television monitors suggests the distance along this road we are at the present time.

Griswold is not saying that a sense of community will come from technology. She is wondering about it. Moreover, this form of community is not what Tönnies and the founders of sociology had in mind when they thought about the need for community. How could it be; they could not imagine the world we live in today. We will have to wait to see, as Griswold notes, if the new technology produces a postmodern society culturally connected electronically within and between societies or if the new technology leads people to live in isolated worlds of their own choosing. Will people retreat into their secluded worlds of computers, CD ROMs, Walkmans, and videos, or will they achieve some sense of national and international integration? The answer to this question is likely to come in the relatively near future.

FEEDBACK

1. _____ refers to the number of organisms of a given species capable of being sustained in a given geographic area.
2. Thanks to modern technology, the earth clearly has a limitless capacity to support its population at its current rate of population and economic growth. *T or F?*
3. Postmodernists agree that the electronic information superhighway will produce a greater sense of national and international community. *T or F?*

Answers: 1. Carrying capacity 2. F 3. F

SUMMARY

1. Social change refers to alterations in social structures that have long-term and relatively important consequences. The pace of social change over most of the world is accelerating at a dramatic rate.

2. Although predicting the precise nature of social change within a society is hazardous, several sources of social change are pretty well understood. Discovery, invention, and diffusion are the major social processes through which social change occurs. Other important sources of social change are technology, population, the natural environment, conflict, and ideas. There are many interrelationships among these sources of social change.

3. Sociologists have gone beyond identifying sources of social change to develop theoretical perspectives on social change. The cyclical, evolutionary, functionalist, and conflict perspectives view social change in very different ways. The cyclical perspective, short on explanatory power, envisions societies as changing through the process of growth and decay. Contemporary evolutionary theory emphasizes multilinear change (that is, various societies may evolve in many diverse directions). The functionalist perspective depicts societies as relatively stable, integrated systems that are constantly seeking a new equilibrium after some major change has occurred. According to the conflict perspective, societies are unstable systems that are constantly undergoing change. Attempts to combine the insights of the functionalist and conflict perspectives, although interesting, have not eliminated the need to retain separate perspectives.

4. Modernization, the process that involves numerous social and cultural changes accompanying economic develop-

ment, is a worldwide phenomenon. Consequences of modernization include rapid population growth, expanding urbanization, increased equality, greater political democracy, the push for universal education, and a trend toward the nuclear family.

5. The existence of modernization does not necessarily mean that all modernizing societies are on the road to social and cultural convergence. This is partly because modernization is filtered through existing social and cultural arrangements.

6. According to world-system theory, the pattern of a nation's development largely rests on the nation's place in the world economy. So long as core nations control and exploit peripheral nations, the development gap between the former and the latter will increase. This gap works against convergence.

7. The future of contemporary society has both a physical and a social side. Many social and biological scientists are concerned that resource depletion, overpopulation, and pollution have set the world on a path to destruction. Some critics even claim that industrial society will collapse within a hundred years unless population growth is stabilized and economic growth is stopped. More optimistic observers contend that technology and the laws of supply and demand will see us through.

8. On the social side, many sociologists are currently calling for a renewed sense of community. They are also debating whether new technology, embodied in the electronic information superhighway, will lead to further fragmentation of society or encourage the development of a sense of community within societies and around the globe.

LEARNING OBJECTIVES REVIEW

After careful study of this chapter, you will be able to:

- Define social change and illustrate the three social processes contributing to social change.
- Discuss the role of technology, population, the natural environment, conflict, and ideas as sources of social change.
- Compare and contrast the cyclical and evolutionary perspectives on social change.
- Describe the major features of the functionalist and conflict perspectives on social change.

- Define modernization and explain some of its major consequences.
- Define world-system theory and summarize some of the consequences flowing from its description of the structure of the world economy.
- Explain the major points of the pessimists and optimists regarding the future of the physical environment.
- Discuss the possible directions human community may take in the twenty-first century.

CONCEPT REVIEW

Match the following concepts with the definitions listed below them:

____ a. diffusion   ____ c. invention   ____ e. carrying capacity

____ b. technology   ____ d. social change   ____ f. postmodernism

1. The number of organisms of a given species capable of being sustained in a given geographic area.
2. The social process that occurs when members of one group borrow social and cultural elements from another group.
3. The creation of a new element by combining two or more already existing elements and creating new rules for their use as a unique combination.
4. A generic term applied to the postindustrial stage of social development in which people are dominated by media images and are connected through electronic channels of information.
5. Alterations in social structures that have long-term and relatively important consequences.
6. The part of culture, including ideas and hardware, that is used to reach practical goals.

CRITICAL THINKING QUESTIONS

1. What do you think is the proper perspective on technological determinism? In the process of answering this question, be sure to incorporate the perspective of Karl Marx. _____

_____

_____

2. Identify a major social change that has occurred in your lifetime. What do you think are the major sources of this change? Be careful to relate each source of change to the nature of the change itself. _____

_____

_____

3. Is evolutionary theory compatible or incompatible with functionalism and conflict theory? You should make clear the nature of each of these perspectives on social change in the process of formulating your answer. _____

_____

_____

4. Does modernization lead to convergence or divergence among societies? Use examples. _____

_____

_____

5. If you were asked by your junior high school to deliver a speech to current students on the future of the world's physical environment, what would you write as a one-page summary of your talk? _____

_____

_____

_____

MULTIPLE CHOICE QUESTIONS

1. Alterations in social structures that have long-term and relatively important consequences are known as
   a. modernization.
   b. demographic change.
   c. cultural lag.
   d. social change.
   e. social movements.

2. Three interrelated processes that contribute to social change are discovery, invention, and
   a. innovation.
   b. confiscation.
   c. migration.
   d. fragmentation.
   e. diffusion.

3. **That part of culture, including ideas and hardware, that is used to reach practical goals is known as**
   a. science.
   b. technology.
   c. morality.
   d. social structure.
   e. invention.

4. **Which of the following statements *best* describes the relationship between technology and social change?**
   a. Social change cannot occur without technological developments.
   b. Technology does not necessarily lead to social change.
   c. The effects of technology are similar in all societies.
   d. Technology can be viewed as a single cause of social change.
   e. Technology has little effect on social change.

5. **According to historian Crane Brinton, the changes introduced by revolutionary leaders are**
   a. usually tinged with sacred significance.
   b. best thought of as discoveries.
   c. followed eventually by restoration of a social order similar to prerevolutionary days.
   d. widely adopted by the people.
   e. simply another example of cultural lag.

6. **Max Weber, partially in response to Marx, emphasized the role of _____ in social change.**
   a. technology
   b. revolution
   c. ideas
   d. inventions
   e. territorial discoveries

7. **A major criticism of the cyclical perspective is that it**
   a. assumes society is inherently stable.
   b. assumes that society must pass through a set of progressive stages.
   c. fails to take into account the history of societies.
   d. describes rather than explains change.
   e. emphasizes inventions rather than discoveries.

8. **A _____ equilibrium exists when a society moves from stability to temporary instability to stability once again.**
   a. sporadic
   b. dynamic
   c. diffused
   d. calibrated
   e. homeostatic

9. **According to the conflict theory, _____ is the major source of social instability that characterizes all societies.**
   a. technology
   b. revolution
   c. ideas
   d. politics
   e. scarcity of desired resources

10. **_____ is the process involving all those social and cultural changes accompanying economic development.**
    a. Modernization
    b. Postindustrialization
    c. Evolution
    d. Convergence
    e. Technological determinism

11. **The development of social and cultural similarity among modernizing nations is known as**
    a. convergence.
    b. world-system theory.
    c. congruence.
    d. consistency.
    e. modernization.

12. **According to _____ theory, a nation's ability to compete in the worldwide economic system depends on that nation's location in the world economy.**
    a. conflict
    b. social-economic
    c. world-system
    d. evolutionary
    e. cyclical

13. **Which of the following statements *best* fits with world-system theory?**
    a. The standard of living is much lower in core nations.
    b. Workers in peripheral nations experience a high standard of living.
    c. The economic gap between core and peripheral nations is increasing.
    d. Peripheral nations develop rapidly due to the aid given by core nations.
    e. The process of integration comes before the process of differentiation.

14. **The number of organisms of a given species that can be sustained in a given geographic area is known as the**
    a. life expectancy ratio.
    b. cultural lag ratio.
    c. dependency ratio.
    d. carrying capacity.
    e. ecological quota.

15. **Which of the following statements is *not* consistent with postmodernism?**
    a. In the most highly developed societies, people are connected through rapidly expanding electronic sources of information.
    b. Contemporary society is composed of individuals isolated from others.
    c. Radical postmodernism rejects all traditional principles, beliefs, values, and social institutions.
    d. Contemporary society is composed of fragmented social institutions.
    e. Technology is now producing a sense of community in modern society.

FEEDBACK REVIEW

True-False

1. According to the cyclical perspective on social change, societies rise and fall rather than move continuously toward improvement. *T or F?*
2. Cyclical theories explain why social change occurs. *T or F?*
3. Classical evolutionary theory assumed that change leads to improvement. *T or F?*
4. Postmodernists agree that the electronic information superhighway will produce a greater sense of national and international community. *T or F?*

Fill in the Blank

5. The theory of social change that assumes technology to be the primary cause of the nature of social structure is called _____.
6. The baby boom in the United States is a good example of the effects of _____ on an entire society.
7. Was it Karl Marx or Max Weber who emphasized the role of ideas in creating social change? _____
8. According to modern evolutionary theory, is social change unilinear or multilinear? _____
9. According to the _____ perspective, social and cultural change occurs as a result of the struggles among groups representing different segments of a society.

Multiple Choice

10. Which of the following is *not* one of the major consequences of modernization discussed in the text?
    a. the social and cultural convergence of modernizing countries
    b. increased urbanization
    c. greater equality
    d. more political democracy
    e. widespread development of the nuclear family

ANSWER KEY

Concept Review	Multiple Choice	Feedback Review
a. 2	1. d	1. T
b. 6	2. e	2. F
c. 3	3. b	3. T
d. 5	4. b	4. F
e. 1	5. c	5. technological determinism
f. 4	6. c	6. population
	7. d	7. Weber
	8. b	8. multilinear
	9. e	9. conflict
	10. a	10. a
	11. a	
	12. c	
	13. c	
	14. d	
	15. e	

Glossary

Absolute poverty   The absence of enough money to secure life's necessities.

Achieved status   A slot within a social structure occupied because of an individual's efforts.

Age cohorts   Persons born during the same time period in a particular population.

Ageism   A set of beliefs, norms, and values used to justify age-based prejudice and discrimination.

Age-specific death rate   The number of deaths per one thousand persons in a specific age group.

Age-specific fertility   The number of live births per one thousand women in a specific age group.

Age stratification   The unequal distribution of scarce desirables in a society based on chronological age.

Agricultural society   A society whose subsistence relies primarily on the cultivation of crops with plows drawn by animals.

Alternative movement   The type of social movement that seeks only limited changes in individuals.

Anomie   A social condition in which norms are weak, conflicting, or absent.

Anticipatory socialization   The process of preparing oneself for learning new norms, values, attitudes, and behaviors.

Ascribed status   A slot within a social structure that is not earned or chosen.

Assimilation   Those processes whereby groups with distinctive identities become culturally and socially fused.

Authoritarian personality   The type of personality characterized by excessive conformity; submissiveness to authority figures; inflexibility; repression of impulses, desires, and ideas; fearfulness; and arrogance toward persons or groups thought to be inferior.

Authoritarianism   The type of political system controlled by nonelected rulers who generally permit some degree of individual freedom.

Authority   Power accepted as legitimate by those subjected to it.

Beliefs   Ideas concerning the nature of reality.

Bilateral descent   The familial arrangement in which inheritance and descent are passed equally through both parents.

Blended family   Composed of at least one remarried man or woman with a minimum of one child from a previous marriage.

Bourgeoisie   Members of a society who own the means for producing wealth.

Bureaucracy   A type of formal organization based on rationality and efficiency.

Capitalism   A system founded on beliefs in private property and the pursuit of profit.

Carrying capacity   The number of organisms of a given species capable of being sustained in a given geographic area.

Case study   A thorough, recorded investigation of a small group, incident, or community.

Caste system   The type of stratification structure in which there is no social mobility because social status is inherited and cannot subsequently be changed.

Causation   The idea that events occur in a predictable, nonrandom way and that one event leads to another.

Charismatic authority   Legitimate power based on an individual's personal characteristics.

Church   A life-encompassing religious organization to which all members of a society belong.

City   A dense and permanent concentration in a limited geographic area of people who gain their living primarily through nonagricultural activities.

Civil religion   A public religion that expresses a strong tie between a deity and a society's way of life.

Class conflict   Marx's idea that the key to the unfolding of history is the conflict between those controlling the means for producing wealth and those laboring for them.

Class consciousness   A sense of identification with the goals and interests of the members of one's social class.

Coercion   The type of social interaction in which an individual is compelled to behave in certain ways by another individual or a group.

Cognition   The mental process that enables humans to think, remember, recognize, and imagine.

Cohabitation   Living with someone in a marriagelike arrangement without legal obligations and responsibilities of formal marriage.

Collective behavior   Elementary patterns of social interaction among people who are responding to similar stimuli.

Communitarian capitalism   The type of capitalism that emphasizes the interests of employees, customers, and society.

Community   A social group whose members live within a limited geographic area and whose social relationships fulfill major social and economic needs.

Comparable worth   The idea that similar wages should be paid for jobs requiring equivalent skills, responsibility, and effort, even if the duties performed are dissimilar.

Concentric zone theory   A description of the process of urban growth emphasizing circular areas that develop from the central city outward.

Conflict   When individuals or groups work against one another to obtain a larger share of the valuables.

Conflict perspective   The theoretical perspective that emphasizes conflict, competition, change, and constraint within a society.

Conformity   A type of social interaction in which an individual behaves toward others in ways expected by the group.

Consolidated Metropolitan Statistical Area (CMSA)   An urban area composed of two or more sets of neighboring PMSAs.

Conspicuous consumption   The consumption of goods and services in order to display one's wealth to others.

Contagion theory   The theory of crowd behavior that emphasizes the irrationality of crowds that is created by participants stimulating one another to higher and higher levels of emotional intensity.

Contingency approach   An approach to organization theory in which the structure of organizations is related to the situations faced by organizations.

Contingent employees   Individuals hired by companies on a part-time, short-term, or contract basis to replace full-time employees.

Control group   The group in an experiment that is not exposed to the experimental variable.

Control theory   The conception that conformity is based on the existence of a strong bond between individuals and society.

Controlled experiment   A laboratory experiment that attempts to eliminate all possible contaminating influences on the variables being studied.

Cooperation   A form of social interaction in which individuals or groups combine their efforts to reach some common goal.

Cooperative learning   A nonbureaucratic classroom structure in which students study in groups with teachers as guides rather than as the controlling agents determining all activities.

Corporation   An organization owned by shareholders who have limited liability and limited control over organizational affairs.

Correlation   A statistical measure in which a change in one variable is associated with change in another variable.

Counterculture   A subculture that is deliberately and consciously opposed to certain central aspects of the dominant culture.

Credentialism   The idea that credentials (educational degrees) are unnecessarily required for many jobs.

Criminal justice system   A system for controlling crime comprised of police, courts, and corrections.

Crowd   A temporary collection of people who share a common point of interest.

Crude birthrate   The annual number of live births per one thousand members of a population.

Crude death rate   The annual number of deaths per one thousand members of a population.

Cult   A religious organization whose characteristics are not drawn from existing religious traditions within a society.

Cultural lag   Any situation in which disequilibrium is caused by one aspect of a society's failing to change at the same rate as an interrelated aspect.

Cultural particulars   The widely varying, often distinctive ways societies may use to handle common problems presented by the existence of cultural universals.

Cultural relativism   The idea that any given aspect of a particular culture should be evaluated in terms of its place within the larger cultural context of which it is a part rather than in terms of some alleged universal standard that is applied across all cultures.

Cultural transmission theory   The theory that deviance is a part of a subculture transmitted through socialization.

Cultural universals   General cultural traits thought to exist in all known cultures.

Culture   The material objects as well as the patterns for thinking, feeling, and behaving that are passed from generation to generation among members of a society.

Culture shock   The psychological and social stress one may experience when confronted with a radically different cultural environment.

De facto subjugation   Subjugation based on common, everyday social practices.

De jure subjugation   Subjugation based on the law.

Democracy   The type of political system in which elected officials are held responsible for fulfilling the goals of the majority of the electorate.

Democratic control   The form of control in which authority is split evenly between husband and wife.

Demographic transition   The process by which a population gradually moves from high birthrates and death rates to low birthrates and death rates as a result of economic development.

Demography   The scientific study of population.

Denomination   One of several religious organizations that members of a society accept as legitimate.

Dependency ratio   The ratio of persons in the dependent ages (under fifteen and over sixty-four) to those in the "economically active" ages (fifteen to sixty-four).

Dependent variable   A variable in which a change (or effect) can be observed.

Desocialization   The process of relinquishing old norms, values, attitudes, and behaviors.

Deterrence   Intimidation of members of society into compliance with requirements of the legal system.

Deviance   Behavior that violates significant social norms.

Deviant   A person who has violated one or more of society's most highly valued norms, especially those valued by the elites.

Differential association theory   The theory that crime and delinquency are more likely to occur among individuals who have been exposed to more unfavorable attitudes toward the law than to favorable ones.

Differential power   A theory according to which a majority uses prejudice and discrimination as weapons of power in the domination of one or more subordinate minorities.

Diffusion   Occurs when members of one group borrow social and cultural elements from another group.

Direct institutionalized discrimination   Organizational or community actions intended to deprive a racial or ethnic minority of its rights.

Discovery   The social process of creating something or reinterpreting an existing thing.

Discrimination   Unequal treatment of individuals based on their minority membership.

Divorce rate   The number of divorces in a particular year for every one thousand members of the total population.

Divorce ratio   The number of divorced persons per one thousand persons who are married and living with their spouses.

Doubling time   The number of years needed to double the size of a population.

Downsizing   The decision on the part of top management to reduce its workforce in order to cut costs, increase profits, and enhance stock values.

Dramaturgy   The symbolic interactionist approach that depicts social life as theater.

Drives   Impulses to do something to reduce discomfort.

Dual labor market   The existence of a split between core and peripheral segments of the economy and the division of the labor force into preferred and marginalized workers.

Dynamic equilibrium   The assumption on the part of functionalists that a society both changes and maintains most of its original structure over time.

Dysfunction   A negative consequence of some element of a society.

Economic determinism   The idea that the nature of a society is based on the society's economy.

Economy   The institution designed for the production and distribution of goods and services.

Educational equality   Occurs when schooling produces the same results (achievement and attitudes) for lower-class and minority children as it does for other children.

Ego   The conscious, rational part of the personality.

Elitism   The theory of power distribution that sees society in the control of a few individuals and organizations.

Emergent norm theory   The theory of crowd behavior that stresses the similarity between typical, everyday social behavior and crowd behavior.

Endogamy   Mate-selection norms requiring individuals to marry within their kind.

Epidemiology   The study of the patterns of distribution of diseases within a population.

Ethnic minority   A people that is socially identified and set apart by others and by itself on the basis of its unique cultural or nationality characteristics.

Ethnocentrism   The tendency to judge others in terms of one's own cultural standards.

Ethnomethodology   The study of the processes people develop and use in understanding the routine behaviors expected of themselves and others in everyday life.

Exogamy   Mate-selection norms requiring individuals to marry someone outside their kind.

Experimental group   The group in an experiment exposed to the experimental variable.

Exponential growth   When the absolute growth of a population that occurs within a given time period becomes part of the base on which the growth rate is applied for the next time period.

Extended family   A family consisting of two or more adult generations that share a common household and economic resources.

Fads   Unusual patterns of behavior that spread rapidly, are embraced zealously, and disappear after a short time.

False consciousness   Acceptance of a system that works against one's own interests.

Family   A group of people related by marriage, blood, or adoption.

Family of orientation   The family into which an individual is born.

Family of procreation   The family group established upon marriage.

Fashions   Patterns of behavior that are widely approved but expected to change periodically.

Fecundity   The maximum rate at which women can physically produce children.

Feminism   A social movement aimed at the achievement of sexual equality—socially, legally, politically, and economically.

Feminist theory   A theoretical perspective that links the lives of women and men to the structure of gender relationships within society.

Feminization of poverty   The trend toward more of the poor in the United States being women and children.

Fertility   The number of children born to women.

Fertility rate   The annual number of live births per one thousand women age fifteen to forty-four.

Field research   A research approach for studying aspects of social life that cannot be measured quantitatively and that are best understood within a natural setting.

Folk society   A society that rests on tradition, cultural and social consensus, family, personal ties, little division of labor, and an emphasis on the sacred.

Folkways   Norms without moral overtones.

Formal demography   The branch of demography that deals with gathering, collating, analyzing, and presenting population data.

Formal organization   A social structure deliberately created for the achievement of one or more specific goals.

Formal sanctions   Rewards and punishments that may be given only by officially designated persons.

Functionalism   The theoretical perspective that emphasizes the contribution (functions) made by each part of a society.

Fundamentalism   A rejection of the effects of modern society on religion and the corresponding desire to adhere closely to traditional beliefs, rituals, and doctrines.

Game stage   According to Mead, the stage of development during which children learn to consider the roles of several people at the same time.

Gemeinschaft   Tönnies's term for the type of society based on tradition, kinship, and intimate social relationships.

Gender   The expectations and behaviors associated with a sex category within a society.

Gender identity   An awareness of being masculine or feminine.

Generalized other   An integrated conception of the norms, values, and beliefs of one's community or society.

Gesellshaft Tönnies's term for the type of society characterized by weak family ties, competition, and impersonal social relationships.

Goal displacement Occurs when organizational rules and regulations become more important than organizational goals.

Government The political structure that rules a nation.

Gross migration rate The annual number of persons per one thousand members of a population who enter or leave a geographic area.

Group Several people who are in contact with one another; share some ways of thinking, feeling, and behaving; take one another's behavior into account; and have one or more interests or goals in common.

Groupthink A situation in which pressure toward uniformity discourages members of a group from expressing their reservations about group decisions.

Hawthorne effect When unintentional behavior on the part of researchers influences the results they obtain from those they are studying.

Health-care system The combination of professional services, organizations, training academies, and technological resources that are committed to the treatment, management, and prevention of disease.

Health maintenance organizations (HMOs) Health plans in which members pay set fees on a regular basis and in return receive all necessary health care at little or no additional cost.

Hidden curriculum The nonacademic agenda in schools that teaches children skills thought to be necessary for success in modern bureaucratic society.

Homogamy Mate-selection norms requiring individuals to marry someone with similar social characteristics.

Horizontal mobility A change from one occupation to another at the same general status level.

Horticultural society A society that solves the subsistence problem primarily through the domestication of plants.

Hospices Organizations designed to provide support for the dying and their families.

Hospitals Organizations providing specialized medical services to a variety of patients, including some who continue to live in their homes and others who will spend a prolonged period in the facility.

Humanist sociology The theoretical perspective that places human needs and goals at the center of sociology.

Hunting and gathering society A society that solves the subsistence problem through hunting animals and gathering edible fruits and vegetables.

Hypothesis A tentative, testable statement of a relationship between particular variables.

"I" The spontaneous and unpredictable part of the self.

Id The part of the personality containing biologically inherited urges, impulses, and desires.

Ideal culture Aspects of culture publicly embraced by members of a society.

Ideal-type method A method that involves isolating the most basic characteristics of some social entity.

Ideology A set of ideas used to justify and defend the interest and actions of those in power in a society.

Imitation stage The developmental stage during which, according to Mead, a child imitates the physical and verbal behavior of a significant other without comprehending the meaning of what is being imitated.

Incapacitation The removal of criminals from society.

Income The amount of money received by an individual or group.

Independent variable A variable that causes something to happen.

Indirect institutionalized discrimination Unintentional organizational or community actions that negatively affect a racial or ethnic minority.

Individualistic capitalism The type of capitalism founded on the principles of self-interest, the free market, profit maximization, and the highest return possible on stockholder investment.

Industrial Revolution The combination of technological developments beginning in the eighteenth century that created major changes in transportation, agriculture, commerce, and industry.

Industrial society A society whose subsistence is based primarily on the application of science and technology to the production of goods and services.

Infant mortality rate The number of deaths among infants under one year of age per one thousand live births.

Informal organization A group in which personal relationships are guided by rules, rituals, and sentiments not provided for by the formal organization.

Informal sanctions Rewards and punishments that may be applied by most members of a group.

In-group A group toward which one feels intense identification and loyalty.

Instincts   Genetically inherited, complex patterns of behavior that always appear among members of a particular species under appropriate environmental conditions.

Institution   A cluster of social structures that fulfills one or more of the fundamental needs of a society.

Interest group   A group organized to achieve some specific shared goal by influencing political decision making.

Intergenerational mobility   Social mobility that takes place from one generation to the next.

Interorganizational relationship   A pattern of interaction among authorized representatives of two or more formally independent organizations.

Intragenerational mobility   Social mobility that occurs within the career of an individual.

Invention   The creation of a new element by combining two or more already existing elements and creating new rules for their use as a unique combination.

Invisible religion   A private religion that is substituted for formal religious organizations, practices, and beliefs in a secular society.

Iron law of oligarchy   The principle stating that power tends to become concentrated in the hands of a few members of any organized group.

Labeling theory   The theory that deviance exists when some members of a group or society label others as deviants.

Latent function   An unintended and unrecognized consequence of some element of a society.

Laws   Norms that are formally defined and enforced by designated persons.

Life chances   The likelihood of securing the "good things in life," such as housing, education, good health, and food.

Life expectancy   The average number of years that persons in a given population born at a particular time can be expected to live.

Life span   The most advanced age to which humans can survive.

Looking-glass self   One's self-concept based on perceptions of others' judgments.

Macrosociology   The level of analysis that focuses on relationships among social structures without reference to the interaction of the people involved.

Managed competition   An approach to health-care delivery structured around incentives for large health-care providers to compete on the basis of cost, quality, and patient satisfaction.

Manifest function   An intended and recognized consequence of some element of a society.

Marriage   A legal union in which public approval is given to sexual activity, having children, and assuming mutual rights and obligations.

Mass hysteria   The form of collective behavior in which general social anxiety is created by acceptance of one or more false beliefs.

Mass media   Those means of communication that reach large heterogeneous audiences without any personal interaction between the senders and the receivers of messages.

Master status   A status that affects most other aspects of a person's life.

Material culture   The concrete, tangible objects within a culture.

Matriarchal control   The form of control in which the oldest female living in the household has the authority.

Matrilineal descent   The familial arrangement in which descent and inheritance are passed from the mother to her female descendants.

Matrilocal residence   When a married couple is expected to live with the wife's parents.

"Me"   The socialized part of the self.

Mechanical solidarity   Social unity based on a consensus of values and norms, strong social pressure for conformity, and dependence on tradition and family.

Medicalization   A process in which problems that were once considered matters of individual responsibility and choice are now considered a part of the domain of medicine and physicians.

Meritocracy   A society in which social status is based on ability and achievement rather than background or parental status.

Metropolitan dispersal   The process by which central cities lose population to the areas surrounding them.

Metropolitan Statistical Area (MSA)   A grouping of counties that contains at least one city of fifty thousand inhabitants or more or an "urbanized area" of at least fifty thousand inhabitants and a total metropolitan population of at least one hundred thousand.

Microsociology   The level of analysis concerned with the study of people as they interact in daily life.

Migration   Movement of people from one geographic area to another for the purpose of establishing a new residence.

Minority   A people who possess some distinctive physical or cultural characteristics, are dominated by the majority, and are denied equal treatment.

Mob   An emotionally stimulated, disorderly crowd that is ready to use destructiveness and violence to achieve a specific purpose.

Modernization   A process involving all those social and cultural changes accompanying economic development.

Monogamy   The form of marriage in which one man is married to only one woman at a time.

Monopoly   Situation in which a single company controls a market.

Morbidity   Disease and illness in a population.

Mores   Norms of great moral significance thought to be vital to the well-being of a society.

Mortality   Deaths within a population.

Multiculturalism   An approach to education in which the curriculum accents the viewpoints, experiences, and contributions of women and diverse ethnic and racial minorities.

Multinational corporations   Firms in highly industrialized societies with operating facilities throughout the world.

Multiple causation   The idea that an event occurs as a result of several factors operating in combination.

Multiple nuclei theory   A description of the process of urban growth that emphasizes specific geographic and historical influences on areas within cities.

Nation-state   The political entity that holds authority over a specified territory.

Neolocal residence   When a newly married couple establishes a residence separate from either of their parents.

Net migration rate   The annual increase or decrease per one thousand members of a population resulting from people's leaving and entering the population.

Norms   Rules defining appropriate and inappropriate ways of behaving for specific situations.

Nuclear family   The smallest group of individuals (mother, father, and children) that can be called a family.

Objectivity   The principle of science stating that scientists are expected to prevent their personal biases from influencing their results and their interpretation of the results.

Obligations   Roles informing individuals of the behavior others expect from them.

Occupational sex segregation   The concentration of men and women in different occupations.

Oligopoly   A situation in which a combination of companies controls a market.

Open classroom   A nonbureaucratic philosophy of education that advocates elimination of the sharp line of authority between teachers and students, the predetermined curriculum for all children of a given age, the constant comparison of students' performance, the use of competition as a motivator, and the grouping of children according to performance and ability.

Open class system   The type of stratification structure in which an individual's social status is based on merit and individual efforts.

Operational definition   A definition of an abstract concept in terms of simpler, observable procedures.

Organic-adaptive systems   Organizations based on rapid response to change rather than on the continuing implementation of established administrative principles.

Organic solidarity   Social unity based on a complex of highly specialized statuses that makes members of a society dependent on one another.

Organizational environment   All the forces outside an organization that exert an actual or a potential influence on the organization.

Out-group   A group toward which one feels opposition, antagonism, and competition.

Overurbanization   A situation in which a city is unable to supply adequate jobs and housing for its inhabitants.

Panic   Collective behavior that occurs when people react to a genuine threat in fearful, anxious, and often self-damaging ways.

Participant observation   The type of field research technique in which a researcher becomes a member of the group being studied.

Patriarchal control   The form of control in which the oldest male living in the household has the authority.

Patriarchal society   The type of society that is controlled by males who use their power, prestige, and economic advantage to dominate females.

Patriarchy   A hierarchical social structure in which women are dominated by men.

Patrilineal descent   The familial arrangement in which descent and inheritance are passed from the father to the male descendants.

Patrilocal residence   When a married couple is expected to live with the husband's parents.

Peer group   A group composed of individuals of roughly the same age and interests.

Personality   A relatively organized complex of attitudes, beliefs, values, and behaviors associated with an individual.

Physician-hospital organizations   Joint ventures between hospitals and organized groups of physicians to reach predetermined business objectives.

Play or pay   An approach to health-care delivery in which universal coverage is provided by requiring employers either to offer employees coverage ("play") or to pay into a public fund for covering the uninsured ("pay").

Play stage   According to Mead, the stage of development during which children take on the roles of individuals, one at a time.

Pluralism   The theory of power distribution that sees decision making as the result of competition, bargaining, and compromise among diverse special interest groups.

Political action committees   Organizations established by interest groups for the purpose of raising and distributing funds to selected political candidates.

Political institution   The institution within which political power is obtained and exercised.

Political party   An organization designed to gain control of government through the election of candidates to public office.

Polyandry   The form of marriage in which one woman is married to two or more men at the same time.

Polygyny   The form of marriage in which one man is married to two or more women at the same time.

Population   All those people with the characteristics a researcher wants to study within the context of a particular research question.

Positivism   The use of observation, experimentation, and other methods of the physical sciences in the study of social life.

Postmodernism   A generic term applied to the postindustrial stage of social development in which people are dominated by media images and are connected globally through electronic channels of information.

Power   The ability to control the behavior of others even against their will.

Power elite   A unified coalition of top military, corporate, and government leaders.

Prejudice   Negative attitudes toward some minority or toward individual members of a minority.

Presentation of self   The ways that we, in a variety of social situations, attempt to create a favorable evaluation of ourselves in the minds of others.

Prestige   Social recognition, respect, and admiration that a society attaches to a particular status.

Primary deviance   Isolated acts of norm violation.

Primary group   Several people who are emotionally close, know one another well, seek one another's company because they enjoy being together, and have a "we" feeling.

Primary Metropolitan Statistical Area (PMSA)   An urban area consisting of adjacent metropolitan areas that are closely integrated economically and socially.

Primary relationship   A relationship that is intimate, personal, based on a genuine concern for another's total personality, and fulfilling in itself.

Profane   Nonsacred aspects of our lives.

Proletariat   Members of a society who labor for the bourgeoisie at subsistence wages.

Propaganda   Information consciously and systematically disseminated to influence the decisions, actions, and beliefs of people on an issue.

Protestant ethic   The cluster of values, norms, beliefs, and attitudes embodied in Protestantism that Max Weber contended favored the emergence of modern capitalism.

Public   People concerned about an issue on which opinions differ.

Public opinion   The public's expression of attitudes and beliefs about an issue.

Race   A distinct category of people who are alleged to share certain biologically inherited physical characteristics.

Racism   An ideology that links people's physical characteristics with their psychological or intellectual superiority or inferiority.

Random sample   A sample selected on the basis of chance so that each member of a population has an equal opportunity of being selected.

Rationalism   The solution of problems on the basis of logic, data, and planning rather than tradition or superstition.

Rational-legal authority Authority based on rules and procedures associated with political offices.

Real culture Subterranean patterns for thinking, feeling, and behaving.

Recidivism A return to crime after prior conviction.

Redemptive movement The type of social movement that focuses on totally changing individuals rather than society.

Reference group A group we use to evaluate ourselves and to acquire attitudes, beliefs, values, and norms.

Reflexes Simple, biologically inherited, automatic reactions to physical stimuli.

Reformative movement The type of social movement that aims to bring partial change to a society.

Rehabilitation The resocialization of criminals into conformity with the legal code of society.

Relative deprivation A condition that exists when people compare themselves with others and believe that they should have as much as those others have.

Relative poverty The measurement of poverty by a comparison of the economic condition of those at the bottom of a society with other segments of the population.

Reliability Exists when a measurement technique yields consistent results on repeated applications.

Religion A unified system of beliefs and practices relating to sacred things.

Religiosity The ways in which people express their religious interests and convictions.

Replacement level The birthrate level at which married couples replace themselves in the population.

Replication The duplication of the same study to ascertain its accuracy.

Resocialization The process of learning to adopt new norms, values, attitudes, and behaviors.

Resource mobilization The process through which members of a social movement secure and utilize the resources needed to press for social change.

Retribution The demand from criminals compensation equal to their offense against society.

Revolution The type of social movement that involves the toppling of a political regime through violent means.

Revolutionary movement The type of social movement that attempts to change society totally.

Rights Roles informing individuals of the behavior that can be expected from others.

Riot A crowd that engages in episodes of destructiveness and violence without a mob's sense of common purpose.

Role conflict Occurs when the performance of a role in one status clashes with the performance of a role in another status.

Role distancing The process of mentally separating oneself from a role.

Role performance The actual conduct involved in putting roles into action.

Role strain Occurs when the roles of a single status are inconsistent with one another.

Role taking The process of mentally assuming the viewpoint of another individual and then responding to oneself from that imagined viewpoint.

Roles Culturally defined rights and obligations attached to social statuses indicating the behavior expected of individuals holding them.

Rumor Inaccurate, distorted, or erroneous information spread by word of mouth or by the mass media.

Sacred Things that are set apart and given a special meaning that transcends immediate human existence.

Sample A limited number of cases drawn from a population.

Sanctions Rewards and punishments used to encourage desired behavior.

Secondary analysis The use of information already collected by someone else for another purpose.

Secondary deviance Deviance as a lifestyle and personal identity.

Secondary group A group that is impersonal and task oriented and involves only a segment of the lives and personalities of its members.

Secondary relationship A relationship that is impersonal, involves only limited aspects of another's personality, and exists to accomplish a specific purpose beyond the relationship itself.

Sect An exclusive and cohesive religious organization based on the desire to reform the beliefs and practices of another religious organization.

Sector theory A description of the process of urban growth emphasizing the importance of transportation routes.

Secularization The process whereby religion loses its influence over society as a whole.

Self-concept An individual's image of herself or himself as an entity separate from other people.

Self-fulfilling prophecy   When an expectation leads to behavior that causes the expectation to become a reality.

Sex   The biological distinction between male and female.

Sexism   A set of beliefs, norms, and values used to justify sexual inequality.

Sexual harassment   The use of one's superior power in making unwelcomed sexual advances.

Sick role   A social definition serving to remove people from active involvement in everyday routines, give them special protection and privileges, and set the stage for their return to their normal social roles.

Significant others   Those persons whose judgments are the most important to an individual's self-concept.

Single payer   An approach to health-care delivery under which medical services are financed governmentally.

Social aggregate   A number of people who happen to be physically located together.

Social category   A number of persons who share a social characteristic.

Social change   Alterations in social structures that have long-term and relatively important consequences.

Social class   A segment of a population whose members have a relatively similar share of the desirable things and who share attitudes, values, norms, and an identifiable lifestyle.

Social control   Means for promoting conformity to a group's or society's rules.

Social demography   The branch of demography that studies population in relation to its social setting.

Social dynamics   The study of social change.

Social exchange   When one person voluntarily does something for another expecting a reward in return.

Social gerontology   The scientific study of the social dimensions of aging.

Social interaction   The process in which people mutually influence one another's behavior.

Social mobility   Movement of individuals or groups within a stratification structure.

Social movement   A large number of people acting together with some degree of leadership and organization to promote or prevent social change.

Social network   A web of social relationships that joins one directly to other people and groups and, through those individuals and groups, indirectly to additional parties.

Social statics   The study of stability and order in society.

Social stratification   The creation of layers of people possessing unequal shares of scarce desirables.

Social structure   Patterned, recurring social relationships.

Socialism   An economic system founded on the belief that the means of production should be owned by the people as a whole and that government should plan and control the economy.

Socialization   The process through which people learn to participate in group life through the acquisition of culture.

Society   People who live within defined territorial borders and who share a common culture.

Sociobiology   The study of the biological basis of human behavior.

Sociological imagination   The set of mind that allows individuals to see the relationship between events in their personal lives and events in their society.

Sociology   The scientific study of social structure.

Spurious correlation   When the apparent relationship between two variables is actually produced by a third variable that affects both of the original two variables.

Strain theory   The theory that deviance is most likely to occur when there is a discrepancy between culturally prescribed goals and legitimate (socially approved) means of obtaining them.

Status   The position that a person occupies within a group.

Status set   The total number of statuses that an individual occupies at any particular time.

Stereotype   A set of ideas based on distortion, exaggeration, and oversimplification that is applied to all members of a social category.

Stigma   An undesirable characteristic used by others to deny the deviant full social acceptance.

Structural differentiation   When one social structure divides into two or more social structures that operate more successfully separately than the one alone would under the new historical circumstances.

Structural mobility   Mobility that occurs because of changes in the distribution of occupational opportunities.

Subculture   A group that is part of the dominant culture but differs from it in some important respects.

Subjective approach   A research method in which the aim is to understand some aspect of social reality through the study of the subjective interpretations of the participants themselves.

Superego   That part of the personality containing all the ideas of what is right and wrong that have been learned from those close to individuals, particularly parents.

Survey   A research method in which people are asked to answer a series of questions.

Symbol   Something that stands for or represents something else.

Symbolic interactionism   The type of interactionism that focuses on interaction among people based on mutually understood symbols.

Taboo   A norm so strong that its violation is thought to be punishable by the group or society or even by some supernatural force.

Technological determinism   The theory of social change that assumes technology to be the primary cause of the nature of social structure.

Technology   The part of culture, including ideas and hardware, that is used to reach practical goals.

Theoretical perspective   A set of assumptions accepted as true by its advocates.

Theory   A logically integrated set of propositions that can be tested against reality.

Total institution   A place in which residents are separated from the rest of society and are controlled and manipulated by those in charge.

Totalitarianism   The type of political system in which a ruler with absolute power attempts to control all phases of social life.

Tracking   A process used in schools to place individual students in educational programs consistent with the school's expectations of a student's eventual occupation.

Traditional authority   Legitimate power rooted in custom.

Trained incapacity   Occurs when previous training prevents someone from learning to adapt to new situations.

Underclass   People in poverty who are either continuously unemployed or underemployed because of the absence of job opportunities and/or required job skills.

Unfulfilled rising expectations   When newly raised hopes for a better life are either not satisfied at all or not satisfied as rapidly as people had expected.

Urban ecology   The study of the relationship between humans and their environment within cities.

Urban society   A society in which social relationships are impersonal and contractual, kinship is de-emphasized, cultural and social consensus are not complete, the division of labor is complex, and secular concerns outweigh sacred ones.

Urbanism   A distinctive way of life supposedly experienced by residents of cities and characterized by impersonality, a lack of primary relationships, tolerance, and sophistication.

Urbanization   The process by which an increasingly large proportion of a population comes to live in urban areas.

Validity   Exists when a research instrument actually measures what it was designed to measure.

Value-added theory   The theory outlining six conditions that are necessary and sufficient for the emergence of a social movement.

Value-free research   Research in which personal biases are not allowed to affect its conduct and outcomes.

Values   Broad cultural principles embodying ideas about what most people in a society consider to be desirable.

Variable   Something that occurs in different degrees among individuals, groups, objects, and events.

Verifiability   A principle of science by which any given piece of research can be duplicated (replicated) by other scientists.

Verstehen   The method of understanding the behavior of others by putting oneself mentally in another's place.

Vertical mobility   An upward or downward change in status level.

Voucher   A tax allowance granted to cover part of the costs for an academic year at the parental school of choice.

Wealth   All the economic resources possessed by an individual or group.

White-collar crime   A crime committed by respectable and high-status people in the course of their occupations.

World-system theory   A theory of modernization that sees the pattern of a nation's development as largely dependent on the nation's location in the world economy.

Zero population growth   When deaths are balanced by births so that a population does not increase.

References

Aberle, David. *The Peyote Religion Among the Navaho.* Norman: University of Oklahoma Press, 1991.

Abu-Lughod, Janet. *Rabat.* Princeton, NJ: Princeton University Press, 1980.

Achenbaum, W. Andrew. *Old Age in the New Land.* Baltimore: Johns Hopkins University Press, 1978.

Acker, Joan. *Doing Comparable Worth.* Philadelphia: Temple University Press, 1988.

Adams, Bert N. *Kinship in an Urban Setting.* Chicago: Markham, 1968.

____. "Isolation, Function, and Beyond: American Kinship in the 1960s." *Journal of Marriage and the Family* 32 (November 1970):575–597.

Adams, Robert. *The Evolution of Urban Society.* Chicago: Aldine, 1965.

Adelmann, P. K., T. C. Antonucci, S. E. Crohan, and L. M. Coleman. "Empty Nest, Cohort, and Employment in the Well-Being of Midlife Women." *Sex Roles* (1989):173–189.

Adelson, Joseph. "Why the Schools May Not Improve." *Commentary* 78 (October 1984):40–45.

Adorno, T. W., Else Frenkel-Brunswik, Daniel J. Levinson, and R. Nevitt Sanford. *The Authoritarian Personality.* New York: Harper, 1950.

Aguirre, B. E., E. L. Quarantelli, and Jorge L. Mendoza. "The Collective Behavior of Fads: The Characteristics, Effects, and Career of Streaking." *American Sociological Review* 53 (1988):569–584.

Ahlburg, Dennis A., and Carol J. De Vita. "New Realities of the American Family." *Population Bulletin* 47 (August 1992): 1–44.

Alba, Richard D. "The Twilight of Americans of European Ancestry: The Case of Italians." *Ethnic and Racial Studies* 8 (January 1985):134–158.

____. *The Transformation of White America.* New Haven, CT: Yale University Press, 1990.

Albert, Michel. *Capitalism vs. Capitalism.* New York: Four Walls Eight Windows, 1993.

Alexander, Jeffrey C., Bernard Giesen, Richard Munch, and Neil J. Smelser (eds.). *The Micro-Macro Link.* Berkeley: University of California Press, 1987.

Alford, Robert R., and Roger Friedland. *Power of Theory.* Cambridge, England: Cambridge University Press, 1985.

Allen, Francis A., et al. *The Decline of the Rehabilitative Ideal.* New Haven, CT: Yale University Press, 1981.

Allport, Gordon. *The Nature of Prejudice.* Garden City, NY: Doubleday, 1958.

Alter, Catherine, and Jerald Hage. *Organizations Working Together.* Newbury Park, CA: Sage, 1992.

Alter, Jonathan. "How America's Meanest Lobby Ran Out of Ammo." *Newsweek* (May 16, 1994):24–25.

Altman, Dennis. *AIDS in the Mind of America.* Garden City, NY: Anchor Books, 1986.

American 2000: An Educational Strategy. Washington, DC: U.S. Department of Education, 1991.

American Association of School Administrators. *Sex Equality in Educational Materials.* Executive Handbook Series, Vol. IV. Washington, DC: American Association of School Administrators, 1974.

American Association of University Women. *How Schools Shortchange Girls.* Washington, DC: AAUW Educational Foundation, 1992.

American Sociological Association. "Embarking Upon a Career with an Undergraduate Degree in Sociology." Washington, DC, 1993.

____. "Careers in Sociology." Washington, DC, 1995.

____. "Code of Ethics." Washington, DC, 1997.

Anders, George. *Health Against Wealth.* Boston: Houghton Mifflin, 1996.

Andersen, Kurt. "An Eye for an Eye." *Time* (January 24, 1983):28–39.

Andersen, Margaret L. *Thinking About Women.* 4th ed. Boston: Allyn and Bacon, 1997.

Anderson, Jack. *The Anderson Papers.* New York: Ballantine Books, 1974.

Anderson, T. B. "Widowhood as a Life Transition: Its Impact on Kinship Ties." *Journal of Marriage and the Family* 46 (February 1984):105–114.

Andrews, Lori B. *Black Power, White Blood.* New York: Pantheon Books, 1996.

Appel, Willa. *Cults in America*. New York: Holt, Rinehart and Winston, 1983.

Aramoni, Aniceto. "Machismo." *Psychology Today* 5 (January 1972):69–72.

Aronowitz, Stanley, and Henry A. Giroux. *Education Under Siege*. South Hadley, MA: Bergin and Garvey Publishers, 1985.

Aronson, E., and A. Gonzalez. "Desegregation, Jigsaw, and the Mexican American Experience." In P. A. Katz and D. A. Taylor (eds.), *Eliminating Racism*. New York: Plenum Press, 1988, pp. 301–314.

Asch, Solomon E. "Opinions and Social Pressure." *Scientific American* 193 (November 1955):31–35.

Ashour, Ahmen Sakr. "The Contingency Model of Leadership Effectiveness: An Evaluation." *Organizational Behavior and Human Performance* 9 (June 1973):339–355.

Atchley, Robert C. *Social Forces and Aging*. 8th ed. Belmont, CA: Wadsworth, 1997.

Atchley, Robert C., and S. Miller. "Types of Elderly Couples." In T. H. Brubaker (ed.), *Family Relationships in Later Life*. Beverly Hills, CA: Sage, 1983, pp. 77–90.

Atkinson, Paul. "Ethnomethodology: A Critical Review." In W. Richard Scott and Judith Blake (eds.), *Annual Review of Sociology* 14 (1988):441–465.

Atkinson, Rick. "Germany's Health-Care System Such a Success It Needs Reform." *Washington Post* (June 22, 1994):A17.

Audi, Robert, and Nicholas Wolterstorff. *Religion in the Public Square: Debating Church and State*. Lantham, MD: Rowan and Littlefield, 1996.

Babbie, Earl R. *The Practice of Social Research*. 7th ed. Belmont, CA: Wadsworth, 1995.

Bailey, William C., and Ruth D. Peterson. "Murder and Capital Punishment: A Monthly Time-Series Analysis of Execution Publicity." *American Sociological Review* 54 (October 1989):722–743.

Baldassare, Mark. "Suburban Communities." *Annual Review of Sociology* 18 (1992):475–494.

Bales, Robert F. *Interaction Process Analysis*. Reading, MA: Addison-Wesley, 1950.

Ball, Richard A. "A Poverty Case: The Analgesic Subculture of the Southern Appalachians." *American Sociological Review* 33 (December 1968):885–895.

Ballantine, Jeanne H. *The Sociology of Education*. 3d ed. Englewood Cliffs, NJ: Prentice Hall, 1993.

Balrig, Flemming. *The Snow-White Image*. Oslo: Norwegian University Press, 1988.

Bandura, Albert. "Self-Efficacy Mechanism in Human Agency." *American Psychologist* 37 (February 1982):122–147.

_____. *Social Foundations of Thought and Action*. Englewood Cliffs, NJ: Prentice Hall, 1986.

Banfield, Edward C. *The Unheavenly City*. Boston: Little, Brown, 1970.

Bardo, John W., and John J. Hartman. *Urban Sociology*. Itasca, IL: Peacock, 1982.

Barker, Eileen. *The Making of a Moonie*. London: Basil Blackwell, 1984.

Barlett, Donald L., and James B. Steele. *America: Who Stole the Dream?* Kansas City, KA: Andrews and McMeel, 1996.

Barlow, Hugh D. *Introduction to Criminology*. 6th ed. Glenview, IL: Scott, Foresman, 1992.

Barnett, Ola W., Cindy L. Miller-Perrin, and Robin D. Perrin. *Family Violence Across the Lifespan*. Newbury Park, CA: Sage, 1997.

Barnhart, Joe Edward. *The Southern Baptist Holy War*. Austin: Texas Monthly Press, 1987.

Barrett, Laurence I. "Fighting for God and the Right Wing." 1993 (September 13, 1993):58–60.

Barrett, Michael J. "The Case for More School Days." *The Atlantic Monthly* (November 1990):78–106.

Barron, James. "Fondly Recalling a Martian Invasion." *New York Times* (September 16, 1988):B1, B3.

Barron, Milton L. "Minority Group Characteristics of the Aged in American Society." *Journal of Gerontology* 8 (October 1953):477–482.

Barrow, Georgia M. *Aging, the Individual, and Society*. 6th ed. St. Paul, MN: West, 1996.

Barry, Herbert III, Margaret K. Bacon, and Irvin L. Child. "A Cross-Cultural Survey of Some Sex Differences in Socialization." *Journal of Abnormal and Social Psychology* 55 (November 1957):327–332.

Bartholet, Jeffrey, and Karen Breslau. "Wake up Call." *Newsweek* (December 22, 1997):66–68.

Baruch, Grace, and Jeanne Brooks-Gunn (eds.). *Women in Midlife*. New York: Plenum, 1984.

Baumohl, Bernard. "When Downsizing Becomes Dumbsizing." *Time* (March 15, 1993):55.

Baumol, William J., and Alan S. Blinder. *Economics*. 7th ed. Fort Worth: Dryden Press, 1996.

Baxter, Janeen. "Is Husband's Class Enough? Class Location and Class Identity in the United States, Sweden, Norway, and Australia." *American Sociological Review* 59 (April 1994):220–235.

Beals, Ralph L., Harry Hoijer, and Alan R. Beals. *An Introduction to Anthropology*. 5th ed. New York: Macmillan, 1977.

Beck, Melinda. "Doctors Under the Knife." *Newsweek* (April 5, 1993):28–33.

Becker, Howard S. *Art Worlds*. Berkeley: University of California Press, 1982.

____. *Outsiders*. New York: Free Press, 1991.

Becker, Howard S., et al. *Boys in White*. Chicago: University of Chicago Press, 1961.

Beckford, James A. *Cult Controversies*. London: Tavistock, 1985.

____. *New Religious Movements and Rapid Social Change*. London: Sage, 1991.

Beckford, James A., and Thomas Luckmann (eds.). *The Changing Face of Religion*. Newbury Park, CA: Sage, 1989.

Bedell, Kenneth B. *Yearbook of American and Canadian Churches 1993*. Nashville, TN: Abingdon Press, 1993.

____. Yearbook of American and Canadian Churches 1997. Nashville, TN: Abingdon Press, 1997.

Begley, Sharon. "Not Just a Pretty Face." *Newsweek* (November 1, 1993):63–67.

____. "The 3-Million-Year-Old Man." *Newsweek* (April 11, 1994):84.

____. "The Science Wars." *Newsweek* (April 21, 1997):54–57.

Bell, Daniel. *The Coming of Post-Industrial Society*. New York: Basic Books, 1976.

____. "Downfall of the Business Giants." *Dissent* (Summer 1993):316–323.

Bellah, Robert N. "Civil Religion in America." *Daedalus* 96 (Winter 1967):1–21.

____. *The Broken Covenant*. New York: Seabury Press, 1975.

____. *Beyond Belief*. Berkeley: University of California Press, 1991.

Bellah, Robert N., and Phillip E. Hammond. *Varieties of Civil Religion*. New York: Harper & Row, 1980.

Bellah, Robert N., et al. *The Good Society*. New York: Knopf, 1991.

Bellah, Robert N., et al. *Habits of the Heart*. (Updated edition.) Berkeley: University of California Press, 1996.

Bellas, Marcia L. "Comparable Worth in Academia: The Effects on Faculty Salaries of the Sex Composition and Labor-Market Conditions of Academic Disciplines." *American Sociological Review 59* (December 1994):807–821.

Belsky, Gary. "Escape from America." *Money* (July 1994):60–70.

Bem, Sandra. "Androgyny Versus Tight Little Lives of Fluffy Women and Chesty Men." *Psychology Today* 9 (September 1975):58–62.

Bendix, Reinhard. *Max Weber*. Garden City, NY: Doubleday, 1962.

Benedict, Ruth. *Patterns of Culture*. New York: Mentor, 1946.

Bennett, William J. "American Education: Making It Work." *Chronicle of Higher Education* 34 (May 4, 1988):29–41.

Bennis, Warren G. "Beyond Bureaucracy." *Transaction* 2 (July/August 1965):31–35.

____. "Response to Shariff: Beyond Bureaucracy Baiting." *Social Science Quarterly* 60 (June 1979):20–24.

Benokraitis, Nijole V. *Marriages and Family*. 2d ed. Upper Saddle River, NJ: Prentice Hall, 1996.

Bensman, Joseph, and Israel Gerver. "Crime and Punishment in the Factory: The Function of Deviancy in Maintaining the Social System." *American Sociological Review* 28 (August 1963):588–598.

Berardo, D. H., C. L. Shehan, and G. R. Leslie. "A Residue of Transition: Jobs, Careers, and Spouses' Time in Housework." *Journal of Marriage and the Family* (May 1987):381–390.

Berger, Bennett M. *The Survival of a Counterculture*. Berkeley: University of California Press, 1981.

Berger, Bennett M., Bruce Hacket, and R. Mervyn Millar. "The Communal Family." *The Family Coordinator* 21 (October 1972):419–427.

Berle, Adolph A., Jr., and Gardiner C. Means. *The Modern Corporation and Private Property*. New York: Harcourt Brace Jovanovich, 1968.

Berlin, Brent, and Paul Kay. *Basic Color Terms*. Berkeley: University of California Press, 1969.

Bernard, Thomas J. "Structure and Control: Reconsidering Hirschi's Concept of Commitment." *Justice Quarterly* 4 (1987):409–424.

Bernstein, Aaron. "Inequality: How the Gap Between the Rich and Poor Hurts the Economy." *Business Week* (August 15, 1994):78–83.

Bernstein, Paul. *American Work Values*. Albany: State University of New York Press, 1997.

Berry, Brian J. L. "Internal Structure of the City." *Law and Contemporary Problems* 30 (Winter 1965):111–119.

Berry, Brian J. L., and John D. Kasarda. *Contemporary Urban Ecology*. New York: Macmillan, 1977.

Bianchi, Suzanne M., "America's Children: Mixed Prospects." *Population Bulletin* 45 (June 1990):3–41.

Bianchi, Suzanne M., and Daphne Spain. *Women, Work, and Family in America*. Washington, DC: Population Reference Bureau, 1996.

Bickman, Leonard, and Debra J. Rog (eds.). *Handbook of Applied Social Research Methods*. Thousand Oaks, CA: Sage, 1997.

Bielby, William T., and Denise D. Bielby. "Family Ties: Balancing Commitments to Work and Family in Dual Earner Households." *American Sociological Review* 54 (October 1989):776–789.

Bierstedt, Robert. "Sociology and General Education." In Charles H. Page (ed.), *Sociology and Contemporary Education*. New York: Random House, 1963, pp. 40–55.

Bigler, Rebecca, and Lynn Liber. "The Role of Attitudes and Interventions in Gender-Schematic Processing." *Child Development* 61 (1990):1440–1452.

Billings, Dwight B., and Thomas Urban. "The Socio-Medical Construction of Transsexualism: An Interpretation and Critique." *Social Problems* 29 (February 1982):266–282.

Billson, Janet Mancini, and Bettina Huber. "Embarking Upon a Career with an Undergraduate Degree in Sociology." Washington, DC: American Sociological Association, 1993.

Binstock, R. H. "Federal Policy Toward the Aging—Its Inadequacies and Its Politics." *Natural Journal* 11 (November 1978):1838–1845.

____. "The Aged as Scapegoat." *Gerontologist* 23 (1983):136–143.

Birnbaum, Jeffry H., and Alan S. Murray. *Showdown at Gucci Gulch*. New York: Vintage, 1988.

Black Elected Officials: A National Roster, 1993. Washington, DC: Joint Center for Political and Economic Studies Press, 1994.

Black, Clifford. "Clinical Sociology and Criminal Justice Professions." *Free Inquiry in Creative Sociology* 12 (1984):117–120.

"Black/White Relations in the United States." *The Gallup Poll Social Audit*. Princeton, NJ: The Gallup Organization, 1997.

Blake, Robert, and Wayne Dennis. "The Development of Stereotypes Concerning the Negro." *Journal of Abnormal and Social Psychology* 38 (October 1943):525–531.

Blanchard, Dallas A. *The Anti-Abortion Movement and the Rise of the Religious Right*. Old Tappan, NJ: Twayne Publishers, 1994.

Blau, Peter M. *Exchange and Power in Social Life*. New York: Wiley, 1964.

Blau, Peter M., and Otis Dudley Duncan. *The American Occupational Structure*. New York: Wiley, 1967.

Blau, Peter M., and Marshall W. Meyer. *Bureaucracy in Modern Society*. 3d ed. New York: Random House, 1987.

Blau, Peter M., and W. Richard Scott. *Formal Organizations*. San Francisco: Chandler, 1962.

Block, Fred, Richard A. Cloward, Barbara Ehrenreich, and Frances Fox Piven. *The Mean Season*. New York: Pantheon, 1987.

Block, Jeanne H. "Conceptions of Sex Role: Some Cross-Cultural and Longitudinal Perspectives." *American Psychologist* 28 (June 1973):512–526.

Block, Jeanne H., Jack Block, and Per F. Gjerde. "The Personality of Children Prior to Divorce." *Child Development* 57(1986):827–840.

Bloom, Allan. *The Closing of the American Mind*. New York: Simon & Schuster, 1987.

Bluestone, Barry, and Bennett Harrison. *The Deindustrialization of America*. New York: Basic Books, 1984.

Blum, Linda M. *Between Feminism and Labor*. Berkeley: University of California Press, 1991.

Blumer, Herbert. "Collective Behavior." In Alfred M. Lee (ed.), *Principles of Sociology*. 3d ed. New York: Barnes & Noble, 1969a, pp. 65–121.

____. *Symbolic Interactionism*. Englewood Cliffs, NJ: Prentice Hall, 1969b.

____. "Social Movements." In R. Serge Denisoff (ed.), *The Sociology of Dissent*. New York: Harcourt Brace Jovanovich, 1974, pp. 74–90.

____. (ed.). *Industrialization as an Agent of Social Change*. Hawthorne, NY: Aldine de Gruyter, 1990.

Blumstein, Philip, and Pepper Schwartz. *American Couples*. New York: Pocket Books, 1985.

Bonacich, Edna. "Advanced Capitalism and Black/White Race Relations in the United States: A Split Labor Market Interpretation." *American Sociological Review* 41 (February 1976):34–51.

Booth, William. "1990's Social Reformer Now Toes the Line." *Washington Post* (July 12, 1994):A1, A4.

Boroughs, Don L. "Winter of Discontent." *U.S. News & World Report* (January 22, 1996):47–54.

Bose, Christine E., and Peter H. Rossi. "Prestige Standings of Occupations as Affected by Gender." *American Sociological Review* 48 (June 1983):331–342.

Bottomore, Tom B. *Karl Marx*. New York: McGraw-Hill, 1963.

____. *Political Sociology*. 2d ed. Minneapolis: University of Minnesota Press, 1993.

Bouvier, Leon F., and Carol J. De Vita. "The Baby Boom—Entering Midlife." *Population Bulletin* 46 (November 1991):1–35.

Bouza, Anthony. *How to Stop Crime*. New York: Plenum, 1993.

Bowles, Samuel, and Herbert Gintis. *Schooling in Capitalist America*. New York: Basic Books, 1976.

Box, Steven. *Deviance, Reality and Society*. 2d ed. New York: Holt, Rinehart and Winston, 1981.

Brace, Kimball W. (ed.). *The Election Data Book: A Statistical Portrait of Voting in America*, 1992. Lanham, MD: Bernan Press, 1993.

Bradley, Harriet. *Men's Work, Women's Work: A Sociological History of the Sexual Division of Labor in Employment*. Minneapolis: University of Minnesota Press, 1989.

Brajuha, Mario, and Lyle Hallowell. "Legal Intrusion and the Politics of Fieldwork: The Impact of the Brajuha Case." *Urban Life* 14 (January 1986):454–478.

Brand, David. "It Was Too Good to Be True." *Time* (June 1, 1987):59.

Brauer, Jerald C. (ed.). *Religion and the American Revolution.* Philadelphia: Fortress Press, 1976.

Braus, Patricia. "What Workers Want." *American Demographics* (August 1992):30–37.

Bridges, George S., and Robert D. Crutchfield. "Law, Social Standing, and Imprisonment." *Social Forces* 66 (March 1988):699–724.

Bright, Chris. "Tracking the Ecology of Climate Change." In Lester R. Brown, Christopher Flavin, and Hilary French (eds.), *State of the World 1997.* New York: Norton, 1997.

Brint, Steven, and Jerome Karabel. *The Diverted Dream.* New York: Oxford University Press, 1989.

Brinton, Crane. *Ideas and Men.* 2d ed. Englewood Cliffs, NJ: Prentice Hall, 1963.

____. *The Anatomy of Revolution.* New York: Knopf, 1990.

Brissett, Dennis, and Charles Edgley. *Life as Theater.* 2d ed. Hawthorne, NY: Aldine de Gruyter, 1990.

Brody, William. "Building a New Kind of Academy." *Planning for Higher Education* 25 (Summer 1997):72–74.

Bromley, David G., and Anson D. Shupe, Jr. *New Christian Politics.* Macon, GA: Mercer University Press, 1984.

Bronfenbrenner, Urie. *Two Worlds of Childhood.* New York: Russell Sage Foundation, 1970.

Brouilette, John R. "The Liberalizing Effect of Taking Introductory Sociology." *Teaching Sociology* 12 (January 1985):131–148.

Broverman, Inge K., et al. "Sex-Role Stereotypes and Clinical Judgments of Mental Health." *Journal of Consulting Psychology* 34 (February 1970):1–7.

Brown, Lester R. *State of the World 1996.* New York: Norton, 1996.

Brown, Lester R., Christopher Flavin, and Hilary French (eds.). *State of the World 1997.* New York: Norton, 1997.

Browne, M. Neil, and Stuart M. Keeley. *Asking the Right Questions.* 3d ed. Englewood Cliffs, NJ: Prentice Hall, 1990.

Brubaker, T. H. "Family in Later Life: A Burgeoning Research Area." *Journal of Marriage and the Family* (November 1990):959–981.

____. "Families in Later Life: A Burgeoning Research Area." In A. Booth (ed.), *Contemporary Families.* Minneapolis: National Council on Family Relations, 1991, pp. 226–248.

Bruner, Jerome. "Schooling Children in a Nasty Climate." *Psychology Today* (January 1982):57–63.

Brunner, H. G., M. Nelsen, X. O. Breakefield, H. H. Ropers, and B. A. van Oost. "Abnormal Behavior Associated with a Point Mutation in the Structural Gene for Monoamine Oxidase A." *Science* 262 (October 22, 1993):578–580.

Brunvand, Jan Harold. "Urban Legends: Folklore for Today." *Psychology Today* 14 (June 1980):50ff.

____. *The Vanishing Hitchhiker.* New York: Norton, 1981.

____. *The Choking Doberman and Other "New" Urban Legends.* New York: Norton, 1984.

____. *The Mexican Pet.* New York: Norton, 1986.

____. *Curses! Broiled Again!* New York: Norton, 1989.

Buckley, Stephen. "Shrugging Off the Burden of a Brainy Image." *Washington Post* (June 17, 1991):Dl, D6.

Budiansky, Stephen. "The Doomsday Myths." *U.S. News & World Report* (December 13, 1993):81–91.

Bulmer, Martin. *The Chicago School of Sociology.* Chicago: University of Chicago Press, 1984.

Bunzel, John H. (ed.). *Challenge to American Schools.* New York: Oxford University Press, 1985.

Burgess, Ernest W. "The Growth of the City." In Robert E. Park, Ernest W. Burgess, and Robert D. McKenzie (eds.), *The City.* Chicago: University of Chicago Press, 1925, pp. 47–62.

Burgess, J. K. "Widowers." In C. S. Chilman, E. W. Nunnally, and F. M. Cox (eds.), *Variant Family Forms.* Beverly Hills, CA: Sage, 1988, pp. 150–164.

Burns, T., and G. M. Stalker. *The Management of Innovation.* 2d ed. London: Tavistock, 1966.

Burros, Marian. "Even Women at the Top Still Have Floors to Do." *New York Times* (May 31, 1993):A1, A11.

Burstein, Paul. "Legal Mobilization as a Social Movement Tactic: The Struggle for Equal Employment Opportunity." *American Journal of Sociology* 96 (March 1991):1201–1225.

Butler, Joseph. *Five Sermons.* Indianapolis: Hackett, 1983.

Butler, Robert N. *Why Survive?* New York: Harper & Row, 1975.

Byron, Christopher. "Seems Like Old Times." *New York* (September 27, 1993):22–23.

Cafferty, Pastora San Juan, and William McCready (eds.). *Hispanics in the United States.* New Brunswick, NJ: Transaction Books, 1985.

Calasanti, Toni M. "Participation in a Dual Economy and Adjustment to Retirement." *International Journal of Aging and Human Development* 26 (1988):13–27.

Califano, Joseph A., Jr. *America's Health Care Revolution.* New York: Random House, 1985.

____. "Overview: Welfare Reform, Introduction." *Yale Law & Policy Review* 11(1993):109–112.

Callaghan, Karen A. (ed.). *Ideals of Feminine Beauty.* Westport, CT: Greenwood, 1994.

Callan, Victor J. "The Voluntarily Childless and Their Perceptions of Parenthood and Childlessness." *Journal of Comparative Family Studies* 14 (Spring 1983):87–96.

Camic, Charles. *Reclaiming the Sociological Classics.* Malden, MA: Blackwell Publishers, 1998.

Campbell, Frances A., and Craig T. Ramey. "Mid-Adolescent Outcomes for High Risk Students: An Examination of the Continuing Effects of Early Intervention." Paper Presented at the Biennial Meeting of the Society for Research in Child Development, 1993.

____. "Effects of Early Intervention on Intellectual and Academic Achievement: A Follow-up Study of Children from Low-Income Families." *Child Development* 65 (1994):684–698.

Cancio, A. Silvia, T. David Evans, and David J. Maume, Jr. "Reconsidering the Declining Significance of Race: Racial Differences in Early Career Wages." *American Sociological Review* 61 (August 1996):541–556.

Cantril, Hadley. *The Invasion from Mars.* Princeton, NJ: Princeton University Press, 1982.

Caplan, Nathan, Marcella H. Choy, and John K. Whitmore. "Indochinese Refugee Families and Academic Achievement." *Scientific American* 266 (February 1992):36–42.

Caplovitz, David. *The Poor Pay More.* New York: Free Press, 1963.

Caplow, Theodore. *American Social Trends.* New York: Harcourt Brace Jovanovich, 1991.

Caplow, Theodore, Howard M. Bahr, and Bruce A. Chadwick. *All Faithful People.* Minneapolis: University of Minnesota Press, 1983.

Caplow, Theodore, Howard M. Bahr, Bruce A. Chadwick, Rueben Hill, and Margaret Holmes Williamson. *Middletown Families.* Minneapolis: University of Minnesota Press, 1982.

Cappelli, Peter, et al. *Change at Work.* New York: Oxford, 1997.

Carlson, Eliot. "Negroes, Jews Press Efforts to Join Groups That Now Refuse Them." *Wall Street Journal* (September 10, 1969).

Carnoy, Martin. *The Politics and Economics of Race in America.* Cambridge: Cambridge University Press, 1994.

Carnoy, Martin, and Henry Levin. *The Limits of Educational Reform.* New York: McKay, 1976.

"The Carrying Capacity Debate." *Population Today* (April 1996):3.

Carter, Stephen L. *The Culture of Disbelief.* New York: Basic Books, 1993.

Cashion, Barbara G. "Female-Headed Families: Effects on Children and Clinical Implications." *Journal of Marital and Family Therapy* (April 1982):77–85.

Casler, Lawrence. "The Effects of Extra Tactile Stimulation on a Group of Institutionalized Infants." *Genetic Psychology Monographs* 71 (1965):137–175.

Cassirer, Ernst. *The Philosophy of the Enlightenment.* Princeton, NJ: Princeton University Press, 1951.

Castro, Janice. "Disposable Workers." *Time* (March 15, 1993):47.

Center for the American Woman and Politics. "Fact Sheet." April 1997.

Centers, Richard. *The Psychology of Social Classes.* New Haven, CT: Yale University Press, 1949.

Chafetz, Janet Saltzman. *Sex and Advantage.* Totowa, NJ: Rowmann and Allanheld, 1984.

____. *Gender Equity.* Newbury Park, CA: Sage, 1990.

Chagnon, Napoleon A. *Yanomamö.* 5th ed. Fort Worth: Harcourt Brace College Publishers, 1997.

Chambliss, William J. "The Saints and the Roughnecks." *Society* 11 (November/December 1973):24–31.

Chambliss, William J., and Robert Seidman. *Law, Order, and Power.* 2d ed. Reading, MA: Vantage, 1982.

Charles, Maria. "Cross-National Variation in Occupational Sex Segregation." *American Sociological Review* 57 (August 1992):483–502.

Chase-Dunn, Christopher, and Thomas D. Hall. *Rise and Demise: Comparing World Systems.* Boulder, CO: Westview Press, 1997.

Cherlin, Andrew J., et al. "Longitudinal Studies of Effects of Divorce on Children in Great Britain and the United States." *Science* 7 (June 1991):1386–1389.

Children's Defense Fund. *The State of America's Children—1991.* Washington, DC: Children's Defense Fund, 1991.

Chilman, Catherine, Elam Nunnally, and Fred Cox (eds.). *Variant Family Forms.* Newbury Park, CA: Sage, 1988.

Chirot, Daniel. *How Societies Change.* Thousand Oaks, CA: Pine Forge Press, 1994.

Choi, N. G. "Correlates of the Economic Status of Widowed and Divorced Elderly Men." *Journal of Family Issues* 12 (March 1992):38–54.

Choldin, Harvey M., and Claudine Hanson. "Subcommunity and Change in a Changing Metropolis." *Sociological Quarterly* 22 (Autumn 1981):549–564.

Christenson, James A., and Ronald C. Wimberley. "Who Is Civil Religious?" *Sociological Analysis* 39 (Spring 1978):77–83.

Christie, Nils. *Crime Control as Industry.* 2d and enlarged ed. London: Routledge, 1994.

Chriswick, Barry R. "The Skills and Economic Status of American Jewry: Trends over the Last Half-Century." *Journal of Labor Economics* 11(1993):229–242.

Cicourel, Aaron, and John Kitsuse. *The Educational Decision Makers.* New York: Bobbs-Merrill, 1963.

Clabes, J. "Caught in the Middle: Sandwich Generation." *Santa Barbara News Press* (May 1, 1989):B-10.

Clark, Burton. "The Cooling Out Function of Higher Education." *American Journal of Sociology* 65 (May 1960):569–576.

____. "Cooling Out Function Revisited." *New Directions for Community Colleges* 32 (1980):15–31.

Clark, Charles S. "Cults in America." *C Q Researcher* 3 (May 7, 1993):385–408.

Clark, Elizabeth J. "Intervention for Cancer Patients: A Clinical Approach to Program Planning." *Journal of Applied Sociology* 1 (1984):83–96.

Clark, William C. "Managing Planet Earth." *Scientific American* 261 (September 1989):47–54.

Clausen, John A. *The Life Course.* Englewood Cliffs, NJ: Prentice Hall, 1986.

Clayton, Obie, Jr. (ed.). *An American Dilemma Revisited.* New York: Russell Sage Foundation, 1996.

Clayton, Richard R., and James W. Gladden. "The Five Dimensions of Religiosity: Toward Demythologizing a Sacred Artifact." *Journal for the Scientific Study of Religion* 13 (June 1974):135–143.

Clayton, Richard R., and Harwin L. Voss. "Shacking Up: Cohabitation in the 1970s." *Journal of Marriage and the Family* 39 (May 1977):272–284.

Cleiren, Marc. *Bereavement and Adaptation.* Bristol, PA: Taylor and Francis, 1992.

Clinard, Marshall B., and Robert F. Meier. *Sociology of Deviant Behavior.* 8th ed. New York: Harcourt Brace, 1992.

Cloward, Richard A., and Lloyd E. Ohlin. *Delinquency and Opportunity.* New York: Free Press, 1966.

Coakley, Jay J. "The Sociological Perspective: Alternate Causations of Violence in Sport." In D. Stanley Eitzen (ed.), *Sport in Contemporary Society.* 4th ed. New York: St. Martin's Press, 1993, pp. 98–111.

Cockburn, Alexander. "Beat the Devil." *The Nation* (January 17, 1994):42–43.

Cockburn, C. "The Relations of Technology." In R. Crompton, and M. Mann (eds.), *Gender and Stratification.* Cambridge, England: Polity, 1986.

Cockerham, William C. "Sociology and Psychiatry." In H. Kaplan and B. Sadeck (eds.), *Comprehensive Textbook of Psychiatry,* Vol. 1, 4th ed., Baltimore: Williams and Wilkins, 1985, pp. 265–273.

____. *Medical Sociology.* 6th ed. Englewood Cliffs, NJ: Prentice Hall, 1995.

____. *Sociology of Mental Disorder.* 4th ed. Upper Saddle River, NJ: Prentice Hall, 1996.

Cohen, Albert K. *Delinquent Boys.* New York: Free Press, 1977.

Cohen, David, and Marvin Lazerson. "Education and the Corporate Order." *Socialist Revolution* 2 (March/April 1972):47–72.

Coleman, James S. "The Concept of Equality of Educational Opportunity." *Harvard Educational Review* 38 (Winter 1968):7–22.

____. *Equal Educational Opportunity.* Cambridge, MA: Harvard University Press, 1969.

____. *Equality and Achievement in Education.* Boulder, CO: Westview Press, 1990.

Coleman, James S., et al. *Equality of Educational Opportunity.* Washington, DC: U.S. Government Printing Office, 1966.

Coleman, James S., Sally B. Kilgore, and Thomas Hoffer. "Public and Private Schools." *Society* 19 (January/February 1982a):4–9.

Coleman, James S., et al. "Cognitive Outcomes in Public and Private Schools." *Sociology of Education* 55 (April/July 1982b):65–76.

Coleman, James S., et al. "Achievement and Segregation in Secondary Schools: A Further Look at Public and Private School Differences." *Sociology of Education* 55 (April/July 1982c):162–182.

Coleman, James S., Thomas Hoffer, and Sally B. Kilgore. "Questions and Answers: Our Response." *Harvard Educational Review* 51 (November 1981):526–545.

Coleman, James W. "Toward an Integrated Theory of White-Collar Crime." *American Journal of Sociology* 93 (September 1987):406–439.

____. *The Criminal Elite.* 4th ed. New York: St. Martin's Press, 1997.

Coleman, L. M., T. C. Antoncucci, P. K. Adelmann, and S. E. Crohan. "Social Roles in the Lives of Middle-Aged and Older Black Women." *Journal of Marriage and the Family* 49 (November 1987):761–771.

Collier, Theresa J. "The Stigma of Mental Illness." *Newsweek* (April 26, 1993):16.

Collins, John J. *The Cult Experience.* New York: Charles C. Thomas, 1991.

Collins, Randall. "Functional and Conflict Theories of Educational Stratification." *American Sociological Review* 36 (December 1971):1002–1019.

____. *Conflict Sociology.* New York: Academic Press, 1975.

____. *The Credential Society.* New York: Academic Press, 1979.

____. *Four Sociological Traditions.* New York: Oxford University Press, 1994.

Colombotos, John, and Corinne Kirchner. *Physicians and Social Change*. New York: Oxford University Press, 1986.

Comfort, Alex. "Age Prejudice in America." *Social Policy* 17 (November/December 1976):3–8.

Common Cause, "Common Cause Guide to Democratic Party Soft Money Donors." July, 1997a.

Common Cause, "Common Cause Guide to Republican Party Soft Money Donors." July, 1997b.

Comte, Auguste. *The Positive Philosophy*. Translated and edited by Harriet Martineau. London: Bell, 1915.

Connor, Walter D. *Socialism, Politics, and Equality*. New York: Columbia University Press, 1979.

Conrad, Peter, and Rochelle Kern (eds.). *Sociology of Health and Illness*. 3d ed. New York: St. Martin's Press, 1990.

Cook, Karen S. "Exchange and Power in Networks of Interorganizational Relations." *Sociological Quarterly* 18 (Winter 1977):62–82.

Cooley, Charles Horton. *Human Nature and the Social Order*. New York: Scribner's, 1902.

Coon, Dennis. *Essentials of Psychology*. 7th ed. Pacific Grove, CA: Brooks/Cole, 1997.

Corsaro, William, and Thomas Rizzo. "Discussions and Friendship: Socialization Processes in the Peer Culture of Italian Nursery School Children." *American Sociological Review* 53 (December 1988):879–894.

____. "An Interpretive Approach to Childhood Socialization." *American Sociological Review* 55 (June 1990):466–468.

Coser, Lewis A. *The Functions of Social Conflict*. New York: Free Press, 1964.

Coser, Lewis A., et al. *Introduction to Sociology*. 3d ed. San Diego: Harcourt Brace Jovanovich, 1991.

Cowgill, Donald O., and Lowell D. Holmes (eds.). *Aging and Modernization*. New York: Appleton-Century-Crofts, 1972.

Cowley, Geoffrey, and Anne Underwood. "Surgeon, Drop that Scalpel." *Newsweek* (Winter, 1997–98):77–78.

Cox, Frank D. *Human Intimacy*. 6th ed. St. Paul, MN: West, 1993.

Cox, Harold G. *Later Life*. 2d ed. Englewood Cliffs, NJ: Prentice Hall, 1988.

Cox, Harvey. "Eastern Cults and Western Culture." *Psychology Today* (July 1977):36ff.

____. *Religion in the Secular City*. New York: Simon & Schuster, 1992.

Craig, Grace. *Human Development*. 6th ed. Englewood Cliffs, NJ: Prentice Hall, 1992.

Crain, Robert L. "School Integration and Occupational Achievement of Negroes." *American Journal of Sociology* 75 (January 1970):593–606.

Crain, Robert L., and Carol S. Weisman. *Discrimination, Personality and Achievements*. New York: Seminar Press, 1972.

Creekmore, C. R. "Cities Won't Drive You Crazy." *Psychology Today* (January 1985):46–53.

Cress, Daniel M., and David A. Snow. "Mobilization at the Margins: Resources, Benefactors, and the Viability of Homeless Social Movement Organizations." *American Sociological Review* 61 (December 1996):1089–1109.

Crewdson, John. "Fraud in Breast Cancer Study." *Chicago Tribune* (March 13, 1994):1.

Crosby, F. E. *Juggling*. New York: Free Press, 1993.

Cullen, John B., and Shelley M. Novick. "The Davis-Moore Theory of Stratification: A Further Examination and Extension." *American Journal of Sociology* 84 (May 1979):1424–1437.

Cummings, Scott. "White Ethnics, Racial Prejudice, and Labor Market Segmentation." *American Journal of Sociology* 95 (January 1980):938–950.

Cummins, Eric. *The Rise and Fall of California's Radical Prison Movement*. Stanford, CA: Stanford University Press, 1993.

Curley, Tom. "Hawaii Gay Marriage Case has National Implications." *USA Today* (December 5, 1996):4A.

Curtis, Russell L., and Benigno E. Aguirre (eds.). *Collective Behavior and Social Movements*. Boston: Allyn and Bacon, 1993.

Curtiss, Susan. *Genie*. New York: Academic Press, 1977.

Cushman, Thomas. *Notes from Underground: Rock Music Counterculture in Russia*. New York: St. Martin's Press, 1995.

Cuzzort, R. P., and E. W. King. *Humanity and Modern Social Thought*. 2d ed. Hinsdale, IL: Dryden Press, 1976.

____. *Twentieth Century Social Thought*. 4th ed. New York: Holt, Rinehart and Winston, 1989.

Cyert, Richard M., and James G. March. *A Behavioral Theory of the Firm*. Englewood Cliffs, NJ: Prentice Hall, 1963.

Dahl, Robert. *Who Governs?* New Haven, CT: Yale University Press, 1961.

Dahrendorf, Ralf. "Out of Utopia: Toward a Reorientation of Sociological Analysis." *American Journal of Sociology* 64 (September 1958a):115–127.

____. "Toward a Theory of Social Conflict." *Journal of Conflict Resolution* 2 (June 1958b):170–183.

____. *Class and Class Conflict in Industrial Society*. Stanford, CA: Stanford University Press, 1959.

Danziger, Sheldon H., and Daniel H. Weinberg (eds.). *Fighting Poverty*. Cambridge, MA: Harvard University Press, 1986.

Darder, Antonia, and Rodolfo D. Torres. *The Latino Reader*. Malden, MA: Blackwell, 1997.

Davidson, James D. "Glock's Model of Religious Commitment: Assessing Some Different Approaches and Results." *Review of Religious Research* 16 (Winter 1975):83–93.

____. "Social Differentiation and Sports Participation: The Case of Golf." In Robert M. Pankin (ed.), *Social Approaches to Sport*. Rutherford, NJ: Fairleigh Dickinson University Press, 1982, pp. 181–224.

Davies, Bronwyn. *Frogs and Snails and Feminist Tales*. New York: Pandora Press, 1990.

Davies, Christie. "Sexual Taboos and Social Boundaries." *American Journal of Sociology* 87 (March 1982):1032–1063.

Davies, James Chowning. "The J-Curve of Rising and Declining Satisfactions as a Cause of Revolution and Rebellion." In Hugh Davis Graham and Ted Robert Gurr (eds.), *Violence in America* (rev. ed.). Beverly Hills, CA: Sage, 1979.

Davis, F. James. *Minority-Dominant Relations*. Arlington Heights, IL: AHM Publishing Corporation, 1978.

____. "Up and Down Opportunity's Ladder." *Public Opinion* 5 (June/July 1982):11ff.

Davis, Kingsley. "Final Note on a Case of Extreme Isolation." *American Journal of Sociology* 52 (March 1947):232–247.

____. *Human Society*. New York: Macmillan, 1949.

____. "The Origin and Growth of Urbanization in the World." *American Journal of Sociology* 60 (March 1955):429–437.

Davis, Kingsley, and D. Rowland. "Old and Poor: Policy Challenges in the 1990's." 2 (1991):37–59.

Davis, Kingsley, and Wilbert Moore. "Some Principles of Stratification." *American Sociological Review* 10 (April 1945):242–249.

Dawkins, Richard. *The Selfish Gene*. New York: Oxford University Press, 1976.

de Witt, K. "Battle is Looming on U.S. College Aid to Poor Students." *New York Times* (May 28, 1991):1,9.

Deegan, Mary Jo. *American Ritual Dramas*. New York: Greenwood, 1989.

Deegan, Mary Jo, and Michael Stein. "American Drama and Ritual: Nebraska Football." *International Review of Sport Sociology* 3 (1978):31–44.

DeFleur, Melvin L., and Sandra Ball-Rokeach. *Theories of Mass Communication*. 4th ed. New York: Longman, 1982.

Degler, Carl N. *In Search of Human Nature*. New York: Oxford University Press, 1991.

del Pinal, Jorge, and Audrey Singer. "Generations of Diversity: Latinos in the United States." *Population Bulletin* 52 (October 1997):1–48.

Deloria, Vine, Jr., and Clifford Lytle. *The Nations Within*. New York: Pantheon Books, 1984.

Demerath, N. J., III, and Rhys H. Williams. "Civil Religion in an Uncivil Society." *The Annals* 480 (July 1985):154–166.

Demo, D. H., and A. C. Acock. "The Impact of Divorce on Children." In A. Booth (ed.), *Contemporary Families*. Minneapolis: National Council on Family Relations, 1991, pp. 162–191.

Dentler, Robert A. "The Political Sociology of the African American Situation: Gunnar Myrdal's Era and Today." *Daedalus* 124 (Winter 1995):15–36.

Denzin, Norman K., and Yvonna S. Lincoln (eds.). *Handbook of Qualitative Research*. Thousand Oaks, CA: Sage, 1993.

DeSpelder, L. A., and D. L. Strickland. *The Last Dance*. Mountain View, CA: Mayfield Press, 1991.

Deutschman, Alan. "Pioneers of the New Balance." *Fortune* (May 20, 1991):60–68.

Dewar, Helen, and Barbara Vobejda. "Clinton Signs Head Start Expansion." *Washington Post* (May 19, 1994):A1.

DeYoung, Alan J. *Economics and American Education*. White Plains, NY: Longman, 1989.

DiIulio, John J., Jr., and Anne Morrison Piehl. "Does Prison Pay?: The Stormy National Debate Over the Cost-Effectiveness of Imprisonment." *The Brookings Review* (Fall 1991):28–35.

Directory of Corporate Affiliations, 1996. New Providence, NJ: National Register Publishing Company, 1996.

Dobzhansky, Theodosius. *Mankind Evolving*. New Haven, CT: Yale University Press, 1962.

Doerner, William R. "To America with Skills." *Time* (July 8, 1985):46ff.

Dohrenwend, Bruce P., and Barbara Snell Dohrenwend. "Sex Differences and Psychiatric Disorders." *American Journal of Sociology* 81 (May 1977):1447–1454.

Dollard, John, et al. *Frustration and Aggression*. New Haven, CT: Yale University Press, 1939.

Domhoff, G. William. *Who Rules America Now?* New York: Touchstone Books,1986.

____. *The Power Elite and the State*. Hawthorne, NY: Aldine de Gruyter, 1990.

____. *State Autonomy or Class Dominance*. New York: Aldine de Gruyter, 1996.

Dornbush, Sanford M., et al. "Single Parents, Extended Households, and the Control of Adolescents." *Child Development* 56 (1985):326–341.

Douglas, Mary. "Accounting for Taste." *Psychology Today* 13 (July 1979):44ff.

Doyle, James A. *The Male Experience*. 3d ed. Madison, WI: Brown & Benchmark, 1995.

Dreeben, Robert. *On What Is Learned in School.* Reading, MA: Addison-Wesley, 1968.

Dreeban, Robert, and A. Gamoran. "Race, Instruction, and Learning." *American Sociological Review* 51 (1986):660–669.

Drucker, Peter F. "The Coming of the New Organization." *Harvard Business Review* 66 (January/February 1988):45–53.

Duff, Christina. "Profiling the Aged: Fat Cats or Hungry Victims." *Wall Street Journal* (September 28, 1995): B1,B16.

Duke, Lynne, and Gabriel Escobar. "A Looting Binge Born of Necessity, Opportunity." *Washington Post* (May 10, 1992):Al, A23.

Duke, Lynne, and Richard Morin. "Demographic Shift Reshaping Politics." *Washington Post* (August 13, 1991):A7.

Dunbar, Leslie (ed.). *Minority Report.* New York: Pantheon, 1984.

Durkheim, Emile. *The Division of Labor in Society.* New York: Free Press, 1964a.

____. *Suicide.* Translated by John A. Spaulding and George Simpson. Edited by George Simpson. New York: Free Press, 1964b.

____. *The Rules of Sociological Method.* Translated by Sarah A. Solovay and John H. Mueller. Edited by George E. G. Catlin. New York: Free Press, 1966.

____. *The Elementary Forms of the Religious Life.* New York: Free Press, 1995.

Dychtwald, Ken. *Age Wave.* New York: Bantam Books, 1990.

Dye, Thomas R. *Who's Running America?* 5th ed. Englewood Cliffs, NJ: Prentice Hall, 1990.

Dynes, Russell R., and Kathleen J. Tierney (eds.). *Disasters, Collective Behavior, and Social Organization.* Newark: University of Delaware Press, 1994.

Eagleton, Terry. *Ideology.* London: Longman, 1994.

Eaton, Judith S. "Minorities, Transfer, and High Education." *Peabody Journal of Education* (Fall 1990).

____. "Encouraging Transfer: The Impact on Community Colleges." *Educational Record* 72 (Spring 1991):34–39.

Ebenstein, William, and Edwin Fogelman. *Today's Isms.* 9th ed. Englewood Cliffs, NJ: Prentice Hall, 1985.

Edelman, Marian Wright. "Children." In Mark Green (ed.), *Changing America.* New York: Newmarket Press, 1992, pp. 431–442.

Edelman, Peter. "The Worst Thing Bill Clinton Has Done." *The Atlantic Monthly* (March, 1997):43–58.

Edmonds, Patricia, and Mark Potok. "Followers 'See Themselves on Side of Jesus'." *USA Today* (April 20, 1993):A1, A2.

Ehrenreich, Barbara. "Helping the Rich Stay That Way." *Time* (April 18, 1994):86.

Ehrlich, Paul R., Anne H. Ehrlich, and John P. Holdren. *Ecoscience.* 3d ed. San Francisco: Freeman, 1977.

Eisenstock, Barbara. "Sex-Role Differences in Children's Identification with Counterstereotypic Televised Portrayals." *Sex Roles* 10 (1984):417–430.

Eismeier, Theodore J., and Philip H. Pollock, III. *Business, Money, and the Rise of Corporate PACs in American Elections.* Westport, CT: Greenwood Press, 1988.

Eitzen, D. Stanley, and Norman R. Yetman. "Racial Dynamics in American Sports: Continuity and Change." In Robert M. Pankin (ed.), *Social Approaches to Sport.* Rutherford, NJ: Fairleigh Dickinson University Press, 1982, pp. 156–180.

El Nasser, Haya. "Judges Say 'Scarlet Letter' Angle Works." *USA Today* (June 25, 1996):1A–2A.

Elikann, Peter T. *The Tough-On-Crime Myth: Real Solutions to Cut Crime.* New York: Plenum, 1996.

Elkin, Frederick, and Gerald Handel. *The Child and Society.* 5th ed. New York: McGraw-Hill, 1991.

Elliott, John E. *Comparative Economic Systems.* 2d ed. Belmont, CA: Wadsworth, 1985.

Elliott, Michael. "Crime and Punishment." *Newsweek* (April 18, 1994):18–22.

Ellwood, David T. *Poor Support.* New York: Basic Books, 1989.

Elmer-DeWitt, Philip. "Take a Trip into the Future on the Electronic Superhighway." *Time* (April 12, 1993a):50–54.

____. "Cloning—Where Do We Draw the Line?" *Time* (November 8, 1993b): 61–70.

____. "Battle for the Soul of the Internet." *Time* (July 25, 1994): 50–56.

Empey, LaMar T. *American Delinquency.* 2d ed. Homewood, IL: Dorsey, 1982.

Employment and Training Report of the President. Washington, DC: U.S. Government Printing Office, 1978.

Engels, Friedrich. *The Origin of the Family, Private Property, and the State.* New York: International Publishing, 1942.

England, Paula. *Comparable Worth.* Hawthorne, NY: Aldine de Gruyter, 1989.

Entwisle, Doris R., and Karl L. Alexander. "Early Schooling as a 'Critical Period' Phenomenon." In K. Namboodiri and R.G. Corwin (eds.), *Sociology of Education and Socialization.* Greenwich, CT: JAI Press, 1989, pp. 27–55.

____. "Entry Into School: The Beginning School Transition and Educational Stratification in the United States." *Annual Review of Sociology* 19 (1993):401–423.

Epstein, Cynthia Fuchs, and Rose Laud Coser (eds.). *Access to Power*. Boston: Allen and Unwin, 1981.

Epstein, T. Scarlett. *Economic Development and Social Change in South India*. New York: Humanities Press, 1962.

Erickson, Robert, and John H. Goldthrope. *The Constant Flux: A Study of Class Mobility in Industrial Societies*. Oxford, U.K.: Clarendon Press, 1992.

Erikson, Erik H. *Childhood and Society*. 2d ed. New York: Norton, 1964.

____. *The Life Cycle Completed*. New York: Norton, 1982.

Erikson, Kai T. *Wayward Puritans*. New York: Wiley, 1966.

____. "A Comment on Disguised Observation in Sociology." *Social Problems* 14 (Spring 1967):366–373.

____. *Everything in Its Path*. New York: Simon & Schuster, 1976.

Ermann, M. David, and Richard J. Lundman. *Corporate and Governmental Deviance*. 5th ed. New York: Oxford University Press, 1996.

Eshleman, J. Ross. *The Family*. 8th ed. Boston: Allyn and Bacon, 1997.

Esping-Andersen, Gosta. *Welfare States in Transition*. Thousand Oaks, CA: Sage, 1996.

Esposito, John, and John Fiorillo. "Who's Left on the Block? New York City's Working Class Neighborhoods." In Hans Spiegel (ed.), *Citizen Participation in Urban Development*. Fairfax, VA: Learning Resources, 1974, pp. 307–334.

Etzioni, Amitai. *A Comparative Analysis of Complex Organizations* (rev. ed.). New York: Free Press, 1975.

____. "Education for Mutuality and Civility." *Futurist* 16 (October 1982):4–7.

____. *The Spirit of Community*. New York: Crown Publishers, 1993.

____. *Rights and the Common Good*. New York: St. Martin's Press, 1995.

Evans, Jean. *Three Men*. New York: Knopf, 1954.

Evans, Sara M. *Personal Politics*. New York: Vintage, 1980.

____. *Born for Liberty: A History of Women in America*. New York: Free Press, 1989.

Evans, Sara M., and Barbara J. Nelson. *Wage Justice*. Chicago: University of Chicago Press, 1989.

Eyerman, Ron. *Studying Collective Action*. Newbury Park, CA: Sage, 1992.

Eysenck, Hans. *Crime and Personality*. 3d ed. London: Routledge and Kegan Paul, 1977.

Fallows, James. "The Case Against Credentialism." *The Atlantic* (December 1985):49–67.

Faludi, Susan. *Backlash*. New York: Anchor Books, 1992.

"Falwell May Revive Moral Majority Group." *The Plain Dealer* (November 8, 1992).

Farhi, Paul. "TV Ratings Agreement Reached." *Washington Post* (July 10, 1997):A1,A16.

Faris, Robert E., and H. Warren Dunham. *Mental Disorders in Urban Areas*. Chicago: University of Chicago Press, 1939.

Farley, Christopher John. "The Butt Stops Here." *Time* (April 18, 1994):58–64.

Farley, John E. *Majority-Minority Relations*. 3d ed. Englewood Cliffs, NJ: Prentice Hall, 1995.

Farley, John E., and Mark Hansel. "The Ecological Context of Urban Crime: A Further Exploration." *Urban Affairs Quarterly* 17 (September 1981):37–54.

Farley, Reynolds. *Blacks and Whites*. Cambridge, MA: Harvard University Press, 1984.

____. "Three Steps Forward and Two Back? Recent Changes in the Social and Economic Status of Blacks." *Ethnic and Racial Studies* 8 (January 1985):4–28.

____. *The New American Reality: Who We Are, How We Got There, Where We Are Going*. New York: Russell Sage Foundation, 1996.

____. "Modest Declines in U. S. Residential Segregation Observed." *Population Today* 25 (February 1997):1–2.

Farley, Reynolds, and Walter R. Allen. *The Color Line and the Quality of American Life*. New York: Oxford University Press, 1989.

Farley, Reynolds, and William H. Frey. "Changes in the Segregation of Whites from Blacks During the 1980's: Small Steps Toward a More Integrated Society." *American Sociological Review* 59 (February 1994):23–45.

Farrell, Christopher. "Time to Prune the Ivy." *Business Week* (May 24, 1993):112–118.

Farrell, Ronald A., and Victoria Lynn Swigert (eds.). *Social Deviance*. 3d ed. Belmont, CA: Wadsworth, 1988.

Fasteau, Marc. "The Male Machine: The High Price of Macho." *Psychology Today* 8 (September 1975):60.

Fattah, Ezzat A. "Is Capital Punishment a Unique Deterrent? A Dispassionate Review of Old and New Evidence." *Journal of Criminology* 23 (July 1981):291–311.

Fausto-Sterling, Anne. *Myths of Gender*. New York: Basic Books, 1987.

Feagin, Joe R. *Subordinating the Poor*. Englewood Cliffs, NJ: Prentice Hall, 1975.

____. "The Continuing Significance of Race: Antiblack Discrimination in Public Places." *American Sociological Review* 56 (February 1991):101–115.

____. *Racial and Ethnic Relations*. 5th ed. Englewood Cliffs, NJ: Prentice Hall, 1996.

Feagin, Joe R. and Clairece Booker Feagin. *Discrimination American Style*. Melbourne, FL: Krieger, 1984.

Feagin, Joe R., and Hernan, Vera. *White Racism*. New York: Routledge, 1995.

Featherman, David L. "The Socio-Economic Achievement of White Religio-Ethnic Sub-groups." *American Sociological Review* 36 (April 1971):207–222.

Fedarko, Kevin. "Bodies of Evidence." *Time* (December 6, 1993):70.

Federal Election Commission. "Top 50 PAC's—Contributions to Candidates January 1, 1995–December 31, 1996." (http://www.fec.gov/finance/paccnye.htm) July 1, 1997.

Feeley, Malcolm, and Jonathon Simon. "The New Penology: Notes on the Emerging Strategy of Corrections and Implications." *Criminology* 30 (1992):449–474.

Fein, Helen. *Genocide*. Newbury Park, CA: Sage, 1993.

Fenn, Richard. *Toward a Theory of Secularization*. Storrs, CT: Society for the Scientific Study of Religion, 1978.

Fenwick, Charles R. "Crime and Justice in Japan: Implications for the United States." *International Journal of Comparative and Applied Criminal Justice* 6 (1982):62–71.

Ferguson, Ronald F. "Shifting Challenges: Fifty Years of Economic Change Toward Black-White Earnings Equality." *Daedalus* 124 (Winter 1995):37–76.

Fergusson, D. M., L. J. Horwood, and F. T. Shannon. "A Proportional Hazards Model of Family Breakdown." *Journal of Marriage and Family* 46 (August 1984):539–549.

Ferrar, Jane W. "Marriages in Urban Communal Households: Comparing the Spiritual and the Secular." In Peter J. Stein, Judith Richman, and Natalie Hannon (eds.), *The Family*. Reading, MA: Addison-Wesley, 1977, pp. 409–419.

Ferree, M., and B. Hess. *Controversy and Coalition: The New Feminist Movement Across Three Decades of Change*. Old Tappan, NJ: Twayne Publishers, 1994.

Feuer, Lewis S. (ed.). *Marx and Engels*. New York: Doubleday, 1959.

Fiedler, Fred E. *A Theory of Leadership Effectiveness*. New York: McGraw-Hill, 1967.

Fierman, J. "Jobs in America: When Will You Get a Raise?" *Fortune* (July 12, 1993):34–36.

Fine, Gary Alan. "The Sad Demise, Mysterious Disappearance, and Glorious Triumph of Symbolic Interactionism." *Annual Review of Sociology* 19 (1993):61–87.

_____. (ed.). *A Second Chicago School*. Chicago: University of Chicago Press, 1995.

_____. *The Culture of Restaurant Work*. Berkeley and Los Angeles: University of California Press, 1996.

Fineman, Howard. "Leadership? Don't Ask Us." *Newsweek* (May 11, 1992):42.

_____. "God and the Grass Roots." *Newsweek* (November 8, 1993):42–46.

_____. "Shifting Racial Lines." *Newsweek* (July 10, 1995):38–39.

Finke, Roger, and Rodney Stark. *The Churching of America*. New Brunswick, NJ: Rutgers University Press, 1992.

Firebaugh, Glenn. "Structural Determinants of Urbanization in Asia and Latin America." *American Sociological Review* 44 (April 1979):199–215.

Firebaugh, Glenn, and Kenneth E. Davis. "Trends in Antiblack Prejudice, 1972–1984: Region and Cohort Effects." *American Journal of Sociology* 94 (September 1988):251–272.

Fischer, Claude S. *To Dwell Among Friends*. Chicago: University of Chicago Press, 1982.

_____. *The Urban Experience*. 2d ed. New York: Harcourt Brace Jovanovich, 1984.

_____. "The Subcultural Theory of Urbanism: A Twentieth-Year Assessment." *American Journal of Sociology* 101 (November 1995):543–577.

Fischer, David Hackett. *Growing Old in America*. New York: Oxford University Press, 1977.

Flanagan, William G. *Contemporary Urban Sociology*. New York: Cambridge University Press, 1993.

Flavin, Christopher. "The Legacy of Rio." In Lester R. Brown, Christopher Flavin, and Hilary French (eds.), *State of the World 1997*. New York: Norton, 1997.

Fletcher, Max E. "Harriet Martineau and Ayn Rand: Economics in the Guise of Fiction." *American Journal of Economics and Sociology* 33 (October 1974):367–379.

Florida, Richard, and Martin Kenney. "Transplanted Organizations: The Transfer of Japanese Industrial Organization to the U.S." *American Sociological Review* 56 (June 1991):381–398.

Flynn, C. P. "Relationship Violence by Women: Issues and Implications." *Family Relations* 39 (1990):194–198.

Forbush, William Byron (ed.). *Fox's Book of Martyrs*. Philadelphia: John C. Winston Company, 1926.

Frankel, Glenn. "Persian Gulf Region Rife with Rumors Amid News Blackout." *Washington Post* (August 6, 1990):A16.

Freeman, David M. *Technology and Society*. Chicago: Rand McNally, 1974.

Freeman, Jo. "The Origins of the Women's Liberation Movement." *American Journal of Sociology* 78 (January 1973):792–811.

Freeman, Linton C. "The Sociological Concept of 'Group': An Empirical Test of Two Models." *American Journal of Sociology* 98 (July 1992):152–166.

Freeman, Richard B. *Black Elite*. New York: McGraw-Hill, 1977.

French, Hilary F. "Learning from the Ozone Experience." In Lester R. Brown, Christopher Flavin, and Hilary French (eds.), *State of the World 1997*. New York: Norton, 1997, pp. 151–171.

French, Marilyn. *The Women's Room*. New York: Ballantine, 1988.

Freud, Sigmund. *Civilization and Its Discontents*. London: Hogarth, 1930.

Frey, William H. "Lifecourse Migration and Metropolitan Whites and Blacks and the Structure of Demographic Change in Large Central Cities." *American Sociological Review* 49 (December 1984):803–827.

Frey, William H. "Mover Destination Selectivity and the Changing Suburbanization of Metropolitan Blacks and Whites." *Demography* 22 (May 1985):223–243.

Friedan, Betty. *The Feminine Mystique*. New York: Dell, 1963.

____. *The Fountain of Age*. New York: Simon and Schuster, 1993.

Friedman, Milton. *Capitalism and Freedom*. Chicago: University of Chicago Press, 1982.

Friedrich, Carl J., and Zbigniew K. Brzezinski. *Totalitarian Dictatorship and Autocracy*. 2d ed. Cambridge, MA: Harvard University Press, 1965.

Friedson, Eliot. *Medical Work in America*. New Haven, CT: Yale University Press, 1989.

Friend, Peter M. "PHO Growing Pains." *Healthcare Executive* (May/June 1996):12–16.

Frieze, Irene H., Jacquelynne E. Parsons, Paula B. Johnson, Diane N. Rable, and Gail L. Zellman. *Women and Sex Roles*. New York: Norton, 1978.

Fritz, Jan M. *Clinical Sociology Handbook*. New York: Garland, 1985.

Fritz, Jan M., and Elizabeth J. Clark. *Sociological Practice: The Development of Clinical and Applied Sociology*. East Lansing: Michigan State University Press, 1989.

Fritz, Jan M., and Elizabeth J. Clark (eds.). *Sociological Practice: Health Sociology*. East Lansing: Michigan State University Press, 1991.

Fruch, Terry, and Paul McGhee. "Traditional Sex Role Development and Amount of Time Watching Television." *Developmental Psychology* 11 (1975):109ff.

Fukuyama, Francis. *The End of History and The Last Man*. New York: Basic Books, 1991.

Fuller, John L., and William R. Thompson. *Behavior Genetics*. St. Louis: Mosby, 1978.

Furstenberg, Frank F., Jr., and Andrew J. Cherlin. *Divided Families*. Cambridge, MA: Harvard University Press, 1991.

Gallup, Alec, and Frank Newport. "Death Penalty Support Remains Strong, But Most Feel Unfairly Applied." *The Gallup Poll News Service* 56 (June 26, 1991):1.

Gallup, George H., Jr. *The Gallup Poll*. Wilmington, DE: Scholarly Resources, 1988.

____. *Religion in America 1990*. Princeton, NJ: Princeton Religion Research Center, 1990.

____. *The Gallup Poll, Public Opinion 1991*. Wilmington, DE: Scholarly Resources, 1992.

____. *Religion in America, 1992–93*. Princeton, NJ: Princeton Religion Research Center, 1993.

____. "Crime." *The Gallup Poll: Public Opinion 1993*. Wilmington, DE: Scholarly Resources, 1994.

____. *The Gallup Poll, Public Opinion 1993*. Wilmington, DE: Scholarly Resources, 1994a.

____. *Religion in America 1992–1993, 1994 Supplement*. Princeton, NJ: Princeton Religion Research Center, 1994b.

____. *Religion in America 1996*. Princeton, NJ: The Princeton Religion Research Center, 1996.

Gallup, George H., Jr., and Larry Hugick. "Racial Tolerance Grows, Progress on Racial Equality Less Evident." *The Gallup Poll Monthly* 297 (June 1990):23–25.

Gallup, George H., Jr., and Frank Newport. "Time at a Premium for Many Americans." *Gallup Poll Monthly* (November 1990):43–49.

Game, Ann, and Andrew W. Metcalfe. *Passionate Sociology*. Thousand Oaks, CA: Sage, 1996.

Gamoran, Adam. "Instructional and Institutional Effects of Ability Grouping ." *Sociology and Education* 59 (1986):185–198.

____. "The Variable Effects of High School Tracking." *American Sociological Review* 57 (December 1992):812–828.

Gamson, William. *The Strategy of Social Protest*. 2d ed. Belmont, CA: Wadsworth, 1990.

Gann, L. H., and Peter J. Duignan. *The Hispanics in the United States*. Boulder, CO: Westview Press, 1986.

Ganong, Lawrence, and Marilyn Coleman. *Remarried Family Relationships*. Thousand Oaks, CA: Sage, 1994.

Gans, Herbert J. *The Urban Villagers*. New York: Free Press, 1962a.

____. "Urbanism and Suburbanism as Ways of Life." In Arnold M. Rose (ed.), *Human Behavior and Social Processes*. Boston: Houghton Mifflin, 1962b, pp. 625–648.

____. *People and Plans*. New York: Basic Books, 1968.

____. "The Uses of Poverty: The Poor Pay All." *Social Policy* 2 (July/August 1971):20–24

____. "Reconstructing the Underclass: The Term's Danger as a Planning Concept." *Journal of the American Planning Association* 56 (Summer 1990):271–277.

Ganz, Alexander. "Where Has the Urban Crisis Gone?" *Urban Affairs Quarterly* 20 (June 1985):449–468.

Garbarino, J. "The Incidence and Prevalence of Child Maltreatment." In L. Ohlin and M. Tonry (eds.), *Family Violence*. Chicago: University of Chicago Press, 1989, pp. 219–261.

Gardner, David Pierpont. *A Nation at Risk: The Imperative for Educational Reform*. Report of the National Commission on Excellence in Education. Washington, DC: U.S. Government Printing Office, 1983.

Gardner, Jennifer M. "Recession Swells Count of Displaced Workers." *Displaced Workers, 1987–91*. U.S. Department of Labor. Bureau of Labor Statistics. Bulletin 2427. Washington, DC: U.S. Government Printing Office, 1993.

Gardner, Lytt. "Deprivation Dwarfism." *Scientific American* 227 (July 1972):76–82.

Garfinkel, Harold. *Studies in Ethnomethodology*. Cambridge, UK: Polity Press, 1984.

Gay, Peter. *The Enlightenment: The Science of Freedom*. New York: Norton, 1977.

Gecas, Viktor. "The Self-Concept." In Ralph H. Turner and James F. Short, Jr. (eds.), *Annual Review of Sociology*. Vol. 8. Palo Alto, CA: Annual Reviews Inc., 1982, pp. 1–33.

Gecas, Viktor, and Michael L. Schwalbe. "Beyond the Looking-Glass Self: Social Structure and Efficacy-Based Self-Esteem." *Social Psychology Quarterly* 46 (1983):77–88.

Gehrig, Gail. "The American Civil Religion Debate: A Source for Theory Construction." *Journal for the Scientific Study of Religion* 20 (March 1981):51–63.

Geis, Gilbert, Robert Meier, and Lawrence Salinger (eds.). *White-Collar Crime: Offenses in Business, Politics and the Professions*, 3d ed. New York: Free Press, 1995.

Gelles, Richard J. *Intimate Violence in Families*. 3d ed. Thousand Oaks, CA: Sage, 1997.

Gelles, Richard J., and Donileen Loseke (eds.). *Current Controversies on Family Violence*. Newbury Park, CA: Sage, 1993.

Gelpi, Barbara C., et al. (eds.). *Women and Poverty*. Chicago: University of Chicago Press, 1986.

George, Linda K. "Sociological Perspectives on Life Transitions." *Annual Review of Sociology* 19 (1993):353–373.

George, Linda K., and George L. Maddox. "Subjective Adaptation to Loss of the Work Role: A Longitudinal Study." In Jon Hendricks and C. Davis Hendricks (eds.), *Dimensions of Aging*. Cambridge, MA: Winthrop, 1979, pp. 331–338.

Gerber, Jurg. "Heidi and Imelda: The Changing Image of Crime in Switzerland." *Criminology* 8 (March 1991):121–128.

Gersoni-Stavn, Deane. *Sexism and Youth*. New York: Bowker, 1974.

Gerth, H. H., and C. Wright Mills (eds.). *From Max Weber*. New York: Oxford University Press, 1958.

Gest, Ted. "Are White-Collar Crooks Getting Off Too Easy?" *U.S. News & World Report* 99 (July 1985):43.

Gibbons, Tom. "Justice Not Equal for Poor Here." *Chicago Sun-Times* (February 24, 1985):1,18.

Gibbs, Nancy. "Oh My God, They're Killing Themselves." *Time* (May 3, 1993a):27–43.

____. "Angels Among Us." *Time* (December 27, 1993b):56–65.

____. "The Vicious Cycle." *Time* (June 20, 1994a)

Gibbs, R., and M. Straus. "'Til Death Do Us Part." *Time* (January 18, 1993):38–45.

Gibson, Margaret A., and John V. Ogbu. *Minority Status and Schooling*. New York: Garland Publishing, 1991.

Giddens, Anthony. *Studies in Social and Political Theory*. London: Hutchinson, 1979.

____. *Profiles and Critiques in Social Theory*. London: Macmillan, 1982.

____. *The Constitution of Society*. Berkeley: University of California Press, 1984.

____. *Social Theory and Modern Sociology*. Stanford, CA: Stanford University Press, 1987.

____. *Introduction to Sociology*. 2d ed. New York: Norton, 1997.

Gilbert, Dennis A., and Joseph A. Kahl. *The American Class Structure*. 4th ed. Belmont, CA: Wadsworth, 1993.

Gilbert, Lucia Albino. *Two Careers/One Family*. Newbury Park, CA: Sage, 1993.

Gilder, George. *Wealth and Poverty*. New York: Basic Books, 1981.

Giles, Jeff. "Generation X." *Newsweek* (June 6, 1994):62–72.

Gilmore, David D. *Manhood in the Making*. New Haven, CT: Yale University Press, 1991.

Giroux, Henry A. "Theories of Reproduction and Resistance in the New Sociology of Education: A Critical Analysis." *Harvard Educational Review* 53 (August 1983):257–293.

Giroux, Henry A., and David Purpel (eds.). *The Hidden Curriculum and Moral Education*. Berkeley, CA: McCutchan, 1983.

Gist, Noel P., and Sylvia Fleis Fava. *Urban Society*. 6th ed. New York: Crowell, 1974.

Glaser, Daniel, and Edna Erez. *Evaluation Research and Decision Guidance*. New Brunswick, NJ: Transaction Books, 1988.

Glasgow, Douglas G. *The Black Underclass*. New York: Vintage, 1981.

Gleckman, H., J. Carey, R. Mitchell, T. Smart, and C. Pousch. "The Technology Payoff." *Business Week* (June 14, 1993):51ff.

Glick, P., and S. L. Lin. "More Young Adults Are Living with Their Parents: Who Are They?" *Journal of Marriage and the Family* (February 1986):107–112.

Glock, Charles Y., and Robert N. Bellah (eds.). *The New Religious Consciousness.* Berkeley: University of California Press, 1976.

Glock, Charles Y., and Rodney Stark. *Religion and Society in Tension.* Chicago: Rand McNally, 1965.

____. *American Piety.* Berkeley: University of California Press, 1968.

Glueck, Sheldon, and Eleanor Glueck. *Unraveling Juvenile Delinquency.* Cambridge, MA: Harvard University Press, 1950.

Gober, Patricia. "Americans on the Move." *Population Bulletin* 48 (November 1993):1–40.

Goff, D., L. Goff, and S. Lehrer. "Sex-Role Portrayals of Selected Female Television Characters." *Journal of Broadcasting* 24 (1980):466–478.

Goffman, Erving. *Encounters.* Indianapolis: Bobbs-Merrill, 1961a.

____. *Asylums.* Garden City, NY: Anchor Books, 1961b.

____. *Stigma.* Englewood Cliffs, NJ: Prentice Hall, 1963.

____. *The Presentation of Self in Everyday Life.* New York: Overlook Press, 1974.

____. *Gender Advertisements.* New York: Harper & Row, 1979.

____. "The Interaction Order." *American Sociological Review* 48 (February 1983):1–17.

Goldberg, Gertrude Schaffner, and Eleanor Kremen (eds.). *The Feminization of Poverty.* Westport, CT: Greenwood, 1990.

Goldberg, Herb. *The Hazards of Being Male.* New York: New American Library, 1987.

Goldberg, Philip. "Are Women Prejudiced Against Women?" *Transaction* 5 (April 1968):28–30.

Goldberger, Arthur S., and Glen G. Cain. "The Causal Analysis of Cognitive Outcomes in the Coleman Report." Madison, WI: Institute for Research on Poverty. University of Wisconsin, December 1981.

____. "The Causal Analysis of Cognitive Outcomes in the Coleman, Hoffer, and Kilgore Report." *Sociology of Education* 55 (April/July 1982):103–122.

Goldfarb, William. "Psychological Privation in Infancy and Subsequent Adjustment." *American Journal of Orthopsychiatry* 15 (April 1945): 247–255.

Goldman, Ari L. "Religion Notes: Choosing Jews." *New York Times* (October 23, 1993):8.

Goldscheider, Frances K., and Calvin Goldscheider. "Leaving and Returning Home in 20th Century America." *Population Bulletin* 48 (March 1994):1–35.

Goldscheider, Frances K., and Linda J. Waite. *New Families, No Families?* Chapel Hill: University of North Carolina Press, 1991.

Goleman, D. "An Emerging Theory on Blacks' IQ Scores." *New York Times Education Supplement* (April 10, 1988):22–24.

Golf World. (July 4, 1997):30,33.

Good, Mary-Jo DelVecchio, Paul E. Brodwin, Byron J. Good, and Arthur Kleinman (eds.). *Pain as Human Experience: An Anthropological Perspective.* Berkeley: University of California Press, 1994.

Goode, Erich. *Collective Behavior.* Fort Worth, TX: Sanders College Publishing, 1992.

Goode, William J. *World Revolution and Family Patterns.* New York: Free Press, 1970.

Goodman, Ann B., Carole Siegel, Thomas J. Craig, and Shang P. Lin. "The Relationship Between Socioeconomic Class and Prevalence of Schizophrenia, Alcoholism, and Affective Disorders Treated by Inpatient Care in a Suburban Area." *American Journal of Psychiatry* 140 (1983):166–170.

Goodwin, Leonard. *Do the Poor Want to Work?* Washington, DC: The Brookings Institution, 1972.

Gordon, Edmund W. "Toward Defining Equality of Educational Opportunity." In Frederick Mosteller and Daniel P. Moynihan (eds.), *On Equality of Educational Opportunity.* New York: Vintage, 1972, pp. 423–426.

Gordon, Milton M. *Assimilation in American Life.* New York: Oxford University Press, 1964.

____. *Human Nature, Class, and Ethnicity.* New York: Oxford University Press, 1978.

Gordon, Robert A. "SES Versus IQ in the Race-IQ-Delinquency Model." *International Journal of Sociology and Social Policy* 7 (1987):30–96.

Gorer, Geofrey. *Grief and Mourning in Contemporary Britain.* London: Cresset Press, 1965.

Gorman, Christine. "Blood, Sweat and Tears." *Time* (February 1996):59.

Gornick, Vivian, and Barbara K. Moran (eds.). *Women in Sexist Society.* New York: New American Library, 1974.

Gottfredson, Michael R., and Travis Hirschi. *A General Theory of Crime.* Stanford, CA: Stanford University Press, 1990.

Gould, R. "Growth Toward Self Tolerance." *Psychology Today* (February 1975).

Gould, Stephen Jay. "The Piltdown Conspiracy." *Natural History* 89 (August 1980):8–28.

____. *The Mismeasurement of Man.* New York: Norton, 1981.

Gouldner, Alvin N. "Metaphysical Pathos and the Theory of Bureaucracy." *American Political Science Review* 49 (June 1955):496–507.

Gracey, Harry L. "Learning the Student Role: Kindergarten as Academic Boot Camp." In Dennis H. Wrong and Harry L. Gracey (eds.), *Readings in Introductory Sociology.* 3d ed. New York: Macmillan, 1977, pp. 215–226.

"Graduates Take Rites of Passage into Japanese Corporate Life." *Financial Times* (April 8, 1991):4.

Grandjean, B. "An Economic Analysis of the Davis-Moore Theory of Stratification." *Social Forces* 53 (June 1975):543–552.

Granovetter, Mark. "The Strength of Weak Ties." *American Journal of Sociology* 78 (May 1973):1360–1380.

Graubard, Allen. "The Free School Movement." *Harvard Educational Review* 42 (August 1972):351–373.

Greeley, Andrew M. "Political Attitudes Among American White Ethnics." In Charles H. Anderson (ed.), *Sociological Essays and Research.* (rev. ed.). Homewood, IL: Dorsey, 1974, pp. 202–209.

____. *Ethnicity, Denomination, and Inequality.* Beverly Hills, CA: Sage, 1976.

____. *Religious Change in America.* Cambridge, MA: Harvard University Press, 1996.

Green, Gary S. *Occupational Crime.* 2d ed. Chicago: Nelson-Hall Publishers, 1997.

Greenberg, David F., and Ronald C. Kessler. "The Effects of Arrests on Crime: A Mulivariate Panel Analysis." *Social Forces* 60 (March 1982):771–790.

Greenfeld, Lawrence. "Examining Recidivism." Bureau of Justice Statistics, U.S. Department of Justice. Washington, DC: U.S. Government Printing Office, 1985.

Greenhouse, Linda. "No Help for Dying." *New York Times* (June 27, 1997):A1,A15.

Greenhouse, Steven. "Strikers A+U.P.S. Backed by Public." *New York Times* (August 17, 1997):A1,A16.

Gregory, Paul R., and Robert C. Stuart. *Comparative Economic Systems.* 3d ed. Boston: Houghton Mifflin, 1989.

Greider, William. *Who Will Tell the People?* New York: Simon & Schuster, 1992.

Griffin, John Howard. *Black Like Me.* Boston: Houghton Mifflin, 1961.

Griswold, Wendy. *Cultures and Societies in a Changing World.* Thousand Oaks, CA: Pine Forge Press, 1994.

Gross, Beatrice, and Ronald Gross. *The Great School Debate.* New York: Simon & Schuster, 1985.

Gross, L., and S. Jeffries-Fox. "What Do You Want to Be When You Grow Up, Little Girl?" In Gary Tuchman, Arlene Kaplan Daniels, and James Benet (eds.), *Hearth and Home.* New York: Oxford University Press, 1978, pp. 240–265.

Grubb, W. Norton. "The Decline of Community College Transfer Rates: Evidence from National Longitudinal Surveys." *Journal of Higher Education* (March/April 1991):194–217.

Grubb, W. Norton, and Robert H. Wilson. "Trends in Wage and Salary Inequality, 1967–1988." *Monthly Labor Review* 115 (June 1992):23–39.

Grunwald, Michael. "How She Got a Job: Welfare-to-Work Isn't Cheap." *The American Prospect* (July/August, 1997):25–29.

Grusky, David B., and Robert M. Hauser. "Comparative Social Mobility Revisited: Models of Convergence and Divergence in Sixteen Countries." *American Sociological Review* 49 (February 1984):19–38.

Gueron, Judith M. "Welfare and Poverty: The Elements of Reform." *Yale Law & Policy Review* 11 (1993):113–125.

Guinzburg, Suzanne. "Education's Earning Power." *Psychology Today* 17 (October 1983):20–21.

Gur, R. C., et al. "Sex Differences in Regional Cerebral Glucose Metabolism During a Resting State." *Science* 267 (January 27, 1995):528–531.

Gutman, Henry G. *The Black Family in Slavery and Freedom, 1750–1925.* New York: Pantheon, 1983.

Guttentag, Marcia, and Helen Bray. "Teachers as Mediators of Sex Role Standards." In Alice G. Sargent (ed.), *Beyond Sex Role.* St. Paul: West, 1977, pp. 395–411.

Hacker, Andrew. *Two Nations: Black and White, Separate, Hostile, and Unequal.* Expanded and updated edition. New York: Ballantine Books, 1995.

Hacker, Helen Mayer. "Women as a Minority Group." *Social Forces* 30 (October 1951):60–69.

Hadaway, C. Kirk, Penny Long Marler, and Mark Chaves. "What the Polls Don't Show: A Closer Look at U.S. Church Attendance." *American Sociological Review* 58 (December, 1993):741–752.

Hadden, Jeffrey K., and Raymond R. Rymph. "The Marching Ministers." In Jeffrey K. Hadden (ed.), *Religion in Radical Transition.* 2d ed. New Brunswick, NJ: Transaction Books, 1973, pp. 99–109.

Hadden, Jeffrey K., and Charles E. Swann. *Prime Time Preachers.* Reading, MA: Addison-Wesley, 1981.

Haferkamp, Hans, and Neil J. Smelser (eds.). *Social Change and Modernity.* Chapel Hill: University of North Carolina Press, 1992.

Hafner, Katie. "Technology: Log On and Shoot." *Newsweek* 128:7 (August 12, 1996):58.

Hagan, Frank E. *Introduction to Criminology.* 3d ed. Chicago: Nelson-Hall Publishers, 1994a.

Hagan, John. "The Social Embeddedness in Crime and Unemployment." *Criminology* 31 (November 1993):464–492.

____. *Crime and Disrepute.* Thousand, Oaks, CA: Pine Forge Press, 1994b.

Hagan, John, and Alberto Palloni. "Toward Structural Criminology: Method and Theory in Criminological Research." In Ralph H. Turner and James F. Short (eds.), *Annual Review of Sociology* 12 (1986):431–449.

Hager, Mary G. "The Myth of Objectivity." *American Psychologist* 37 (May 1982):576–579.

Hall, Edward T. *The Silent Language.* Garden City, NY: Doubleday,1966.

____. "How Cultures Collide." *Psychology Today* 10 (July 1976):66ff.

____. "Learning the Arabs' Silent Language." *Psychology Today* 13 (August 1979):45ff.

Hall, John. "President's Remarks Leave Settlement Up in the Air." *American Business Review* (September 18, 1997):1.

Hall, Richard H. "Intraorganizational Structural Variation: Application of the Bureaucratic Model." *Administrative Science Quarterly* 7 (December 1962):295–308.

____. *Organizations.* 6th ed. Englewood Cliffs, NJ: Prentice Hall, 1996.

Hallinan, Maureen T. "Equality of Educational Opportunity." In W. Richard Scott and Judith Blake (eds.), *Annual Review of Sociology* 14 (1988):249–268.

____. (ed.). *Restructuring Schools: Promising Practices and Policies.* South Bend, IN: University of Notre Dame, 1995.

Halpern, Diane F. *Critical Thinking Across the Curriculum: A Brief Edition of Thought and Knowledge.* Mahwah, NJ: Lawrence Erlbaum Associates, 1997.

Hamilton, Kendall. "Blowing Smoke." *Newsweek* (July 21, 1997):54–60.

Hammer, Joshua. "Inside a War Zone: 'The Situation is Desperate'." *Newsweek* (June 20, 1994):44–46.

Hammond, Phillip E. *The Sacred in a Secular Age.* Berkeley: University of California Press, 1985.

Hammond, Phillip E., and Kirk R. Williams. "The Protestant Ethic Thesis: A Social-Psychological Assessment." *Social Forces* 54 (March 1976):579–589.

Hampden-Turner, Charles, and Alfons Trompenaars. *The Seven Cultures of Capitalism.* New York: Currency Doubleday, 1993.

Hampton, Robert L., et al. (eds.). *Family Violence.* Newbury Park, CA: Sage, 1993.

Hancock, LynNell, and Pat Wingert. "A Mixed Report Card." *Newsweek* (November 13, 1995):69.

Handel, Gerald (ed.). *Childhood Socialization.* Hawthorne, NY: Aldine de Gruyter, 1988.

____. "Revising Socialization Theory." *American Sociological Review* 55 (June 1990):463–466.

Handel, Warren. *Ethnomethodology.* Englewood Cliffs, NJ: Prentice Hall, 1982.

Hannan, Michael T., and John Freeman. *Organizational Ecology.* Cambridge, MA: Harvard University Press, 1989.

Harlow, Harry F. "The Young Monkeys." *Psychology Today* 5 (1967):40–47.

Harlow, Harry F., and Margaret Harlow. "Social Deprivation in Monkeys." *Scientific American* 207 (November 1962):137–146.

Harlow, Harry F., and Robert R. Zimmerman. "Affectional Responses in the Infant Monkey." *Science* 21 (August 1959):421–432.

Harrington, Michael J. *The Other America.* New York: Penguin, 1971.

____. *The New American Poverty.* New York: Penguin, 1984.

Harris, Chauncy D., and Edward L. Ullman. "The Nature of Cities." *Annals of the American Academy of Political and Social Science* 242 (November 1945):7–17.

Harris, Fred R., and Roger W. Wilkens (eds.). *Quiet Riots.* New York: Pantheon, 1988.

Harris, John F., and Judith Havemann. "Welfare Rolls Continue Sharp Decline." *Washington Post* (August 13, 1997):A1,A6.

Harris, Kathleen Mullan. "Work and Welfare Among Single Mothers in Poverty." *American Journal of Sociology* 99 (September 1993):317–352.

Harris, Marvin. *Cows, Pigs, Wars, and Witches.* New York: Random House, 1974.

____. "Sociobiology and Biological Reductionism." In Ashley Montagu (ed.), *Sociobiology Examined.* New York: Oxford University Press, 1980, pp. 311–335.

Harrison, Bennett. "Corporate Restructuring, National Economic Policy and the Worsening Distribution of Labor Income in the United States Since the 1970's." Paper prepared for the Solutions for the New Workforce Conference, Washington, DC, September 28, 1987.

Harrison, Bennett. *Lean and Mean.* New York: Basic Books, 1994.

"Harvard Delays in Reporting Fraud." *Science* 215 (January 29, 1982):478.

Harvey, David. *The Condition of Postmodernity.* Oxford: Basil Blackwell, 1989.

Hasegawa, Keitaro. *Japanese Style Management.* Tokyo: Kodansha International Ltd., 1986.

Hatch, Nathan O. *The Democratization of American Christianity.* New Haven, CT: Yale University Press, 1989.

Haub, Carl. "World Population Rises to 5.840 Billion in 1997." *Population Today* 25 (May 1997):1–2.

Haupt, Arthur, and Thomas T. Kane. *Population Handbook.* Washington, DC: Population Reference Bureau, 1991.

Hawley, Amos H. "Human Ecology." In David L. Sills (ed.), *International Encyclopedia of the Social Sciences.* New York: Macmillan, 1968, pp. 329–337.

____. *Urban Society.* 2d ed. New York: Wiley, 1971.

____. "The Logic of Macrosociology." *Annual Review of Sociology* 18 (1992):1–14.

Hawley, W. D. "Achieving Quality Integrated Education— With or Without Federal Help." In F. Schultz (ed.), *Annual Editions: Education 85/86.* Guilford, CT: Dushkin Publishing, 1985, pp. 142–145.

Hawley, W. D., and M. A. Smylie. "The Contribution of School Desegregation to Academic Achievement and Racial Integration." In P. A. Katz and D. A. Taylor (eds.), *Eliminating Racism.* New York: Plenum Press, 1988, pp. 281–297.

Hays, Kim. *Practicing Virtues: Moral Traditions at Quaker and Military Boarding Schools.* Berkeley and Los Angeles: University of California Press, 1994.

Heckscher, Charles. *White-Collar Blues.* New York: Basic Books, 1995.

Heelas, Paul. *The New Age Movement.* Cambridge, MA: Blackwell Publishers, 1996.

Hefner, Robert, and Rebecca Meda. "The Future of Sex Roles." In Marie Richmond-Abbott (ed.), *The American Woman.* New York: Holt, Rinehart and Winston, 1979, pp. 243–264.

Heilbroner, Robert L. *The Making of Economic Society.* 9th ed. Englewood Cliffs, NJ: Prentice Hall, 1993.

Helle, H. J., and S. N. Eisenstadt (eds.). *Macro-Sociological Theory.* Vol. 1. Beverly Hills, CA: Sage, 1985a.

____. *Micro-Sociological Theory.* Vol. 2. Beverly Hills, CA: Sage, 1985b.

"Hello, Dolly." *The Economist* 18 (March 1, 1997):17–18.

Hendricks, Jon, and C. Davis Hendricks. *Aging in Mass Society.* 3d ed. New York: HarpCollege, 1987.

Henig, Jeffrey R. *Rethinking School Choice.* Princeton, NJ: Princeton University Press, 1994.

Henry, William A., III. "News as Entertainment: The Search for Dramatic Unity." In Elie Abel (ed.), *What's News.* San Francisco: Institute for Contemporary Studies, 1981, pp. 133–158.

Heritage, John. *Garfinkel and Ethnomethodology.* Cambridge, UK: Polity, 1984.

Herrnstein, Richard J., and Charles Murray. *The Bell Curve: Intelligence and Class Structure in American Life.* New York: Free Press, 1994.

Herzlinger, Regina. *Market-Driven Health Care.* Reading, MA: Addison-Wesley, 1997.

Hesiod. *Theogony, Works and Days, Shield.* Translation, Introduction, and Notes by Apostolos N. Athanassakis. Baltimore: Johns Hopkins University Press, 1983.

Hessler, Richard M. *Social Research Methods.* St. Paul, MN: West, 1992.

Hickey, Tom, Louise A. Hickey, and Richard A. Kalish. "Children's Perceptions of the Elderly." *Journal of Genetic Psychology* 112 (June 1968):227–235.

Hilbert, Richard A. "Ethnomethodology and the Micro-Macro Order." *American Sociological Review* 55 (December 1990):794–808.

Hill, Patrice. "Clinton May Trim Some Budget Pork." *Washington Times* (August 5, 1997):A1,A14.

Hinkle, Roscoe C. *Developments in American Sociological Theory.* Albany: State University of New York Press, 1994.

Hirsch, E. D. *The Schools We Need and Why We Don't Have Them.* New York: Doubleday, 1996.

Hirsch, E. D., Jr. *Cultural Literacy.* Boston: Houghton Mifflin, 1987.

Hirschi, Travis. *Causes of Delinquency.* Berkeley: University of California Press, 1972.

Hirschi, Travis, and Michael R. Gottfredson. "Substantive Positivism and the Idea of Crime." *Rationality and Society* 2 (October 1990):412–428.

Hirschi, Travis, and Hanan Selvin. *Principles of Survey Analysis.* New York: Free Press, 1973.

Hobbes, Thomas. *Leviathan.* Edited by Herbert W. Schneider. New York: Liberal Arts Press, 1958.

Hochschild, Arlie R. *The Managed Heart: Commercialization of Feeling.* Berkeley: University of California Press, 1983.

____. *The Time Bind: When Work Becomes Home and Home Becomes Work.* New York: Henry Holt, 1997.

Hochschild, Arlie, with Anne Machung. *The Second Shift.* New York: Avon, 1997.

Hodge, Robert W., Paul M. Siegel, and Peter H. Rossi. "Occupational Prestige in the United States, 1925–1963." *American Journal of Sociology* 70 (November 1964):286–302.

Hodge, Robert W., and Donald J. Treiman. "Class Identification in the U.S." *American Journal of Sociology* 73 (March 1968):535–547.

Hodge, Robert W., Donald J. Treiman, and Peter H. Rossi. "A Comparative Study of Occupational Prestige." In Reinhard Bendix and Seymour Martin Lipset (eds.), *Class, Status, and Power.* 2d ed. New York: Free Press, 1966, pp. 309–321.

Hodson, Randy, and Robert L. Kaufman. "Economic Dualism: A Critical Review." *American Sociological Review* 47 (December 1982):727–739.

Hodson, Randy, and Teresa A. Sullivan. *The Social Organization of Work.* Belmont, CA: Wadsworth, 1990.

Hoebel, E. Adamson. *The Law of Primitive Man.* New York: Atheneum, 1983.

Hoecker-Drysdale, Susan. *Harriet Martineau.* Oxford: Berg Publishers, 1994.

Hoerr, John. "Sharpening Minds for a Competitive Edge." *Business Week* (December 17, 1990):72–78.

Hoff, Gerard Alan, and Lawrence J. Schneidermann. "Having Babies at Home: Is It Safe? Is It Ethical?" *Hastings Center Report* 15 (December 1985):19–27.

Hoffman, Lois W., and Francis I. Nye. *Working Mothers.* San Francisco: Jossey-Bass, 1975.

Hofstadter, Richard. *America at 1750.* New York: Vintage, 1973.

Hogan, Dennis P., and Nan Marie Astone. "Transition to Adulthood." In Ralph H. Turner and James F. Short, Jr. (eds.), *Annual Review of Sociology* 12 (1986):109–130.

Hollingshead, August B. *Elmtown's Youth.* New York: Wiley, 1949.

Hollingshead, August B., and Frederick C. Redlich. *Social Class and Mental Illness.* New York: Wiley, 1958.

Holmberg, Allan R. *Nomads of the Long Bow.* Garden City, NY: Natural History Press, 1969.

Holt, John. *How Children Fail.* New York: Dell, 1967.

Hook, Donald D. *Death in the Balance.* Lexington, MA: Heath, 1989.

Hoover, Kenneth R., and Todd Donovan. *The Elements of Social Scientific Thinking.* 6th ed. New York: St. Martin's Press, 1995.

Hoover, Stewart M. *The Social Sources of the Electronic Church.* Newbury Park, CA: Sage, 1988.

Hooyman, Nancy R., and H. Asuman Kiyak. *Social Gerontology.* 4th ed. Boston: Allyn and Bacon, 1996.

Hope, Keith. "Vertical and Nonvertical Class Mobility in Three Countries." *American Sociological Review* 47 (February 1982):99–113.

Horwitz, Allan V. *The Logic of Social Control.* New York: Plenum, 1990.

Horwitz, Robert A. "Psychological Effects of the Open Classroom." *Review of Educational Research* 49 (Winter 1979):71–86.

Hosenball, Mark. "The Anatomy of a Rumor." *Newsweek* (November 23, 1996):43.

Hotaling, G. T., et al. (eds.). *Family Abuse and Its Consequences.* Beverly Hills, CA: Sage, 1988.

Hougland, James G., Jr., and Willis A. Sutton, Jr. "Factors Influencing Degree of Involvement in Interorganizational Relationships in a Rural County." *Rural Sociology* 43 (Winter 1978):649–670.

Houseman, John. "The Men from Mars." *Harper's Magazine* 197 (December 1948):74–82.

Howe, Neil, and William Strauss. "America's 13th Generation." *New York Times* (April 16, 1991).

Hoyt, Homer. *The Structure and Growth of Residential Neighborhoods in American Cities.* Washington, DC: Federal Housing Authority, 1939.

Huber, Peter, and Jessica Korn. "The Plug-and-Play Economy." *Forbes* (July 7, 1997):268–272.

Hudson, R. B., and J. B. Gonyea. "Political Mobilization and Older Women." *Generations* 14 (Summer 1990): 67–72.

Hughes, John A., Peter J. Martin, and W. W. Sharrock. *Understanding Classical Sociology.* Thousand Oaks, CA: Sage, 1995.

Huizinga, David, and Delbert S. Elliot. "Juvenile Offenders: Prevalence, Offender Incidence, and Arrest Rates by Race." *Crime and Delinquency* 33 (April, 1987):206–223.

Hull, J. B. "Women Find Parents Need Them Just When Careers Are Resuming." *Wall Street Journal* (September 9, 1985).

Human Development Report 1997. New York: Oxford University Press, 1997.

Hummel, Ralph P. *The Bureaucratic Experience.* 3d ed. New York: St. Martin's Press, 1987.

Humphrey, Craig R., and Frederick H. Buttel. *Environment, Energy and Society.* Malabar, FL: Krieger Publishing, 1986.

Humphreys, Laud. *Tearoom Trade.* New York: Aldine, 1979.

Hunter, Floyd. *Community Power Structure.* Chapel Hill: University of North Carolina Press, 1953.

____. *Community Power Succession.* Chapel Hill: University of North Carolina Press, 1980.

Hunter, James Davison. *Evangelicism.* Chicago: University of Chicago Press, 1987.

Hurn, Christopher J. *The Limits and Possibilities of Schooling.* 3d ed. Boston: Allyn and Bacon, 1993.

Iacocca, Lee, and William Novak. *Iacocca.* New York: Bantam, 1984.

Ilka, Douglas. "Kevorkian Drops Off Body of Massachusetts Man." *The Detroit News* (February 4, 1998):1

Illich, Ivan. *Medical Nemesis.* London: M. Boyars, 1995.

Inciardi, James A. *Criminal Justice.* 5th ed. Forth Worth, TX: Harcourt Brace College Publishers, 1996.

Infante, Esme M. "Panel: Nicotine Addictive." *USA Today* (August 13, 1994):A1.

Ingrassia, Michele, and Melinda Beck. "Patterns of Abuse." *Newsweek* (July 4, 1994):26–33.

Isikoff, Michael. "Christian Coalition Steps Boldly Into Politics." *Washington Post* (September 10, 1992):A1, A14.

Jackman, Mary R., and Robert W. Jackman. *Class Awareness in the United States.* Berkeley: University of California Press, 1983.

Jackson, Elton F., and Harry J. Crockett, Jr. "Occupational Mobility in the United States: A Point Estimate and Trend Comparison." *American Sociological Review* 29 (February 1964):5–15.

Jackson, Philip W. *Life in the Classrooms*. New York: Holt, Rinehart and Winston, 1968.

Jacob, John E. "Black America, 1993: An Overview." In Billy J. Tidwell (ed.), *The State of Black America 1994*. New York: National Urban League, 1994, pp. 1–9.

Jacobs, H. "Aging and Politics." In R. H. Binstock and L. K. George (eds.), *Handbook of Aging and the Social Sciences*. New York: Academic Press, 1990.

Jacquet, Constant H., Jr. (ed.). *Yearbook of American and Canadian Churches, 1990*. Nashville: Abingdon Press, 1990.

Jaggar, Alison M., and Paula S. Rothenberg. *Feminist Frameworks*. 3d ed. New York: McGraw-Hill, 1993.

James, James J. "Impact of the Medical Malpractice Slowdown in Los Angeles County: January 1976." *American Journal of Public Health* 69 (May 1979):437–443.

James, Paul. *Nation Formation*. Thousand Oaks, CA: Sage, 1996.

Janis, Irving L. *Groupthink*. Boston: Houghton Mifflin, 1982.

Janowitz, Morris. *Sociology and the Military Establishment*. 3d ed. Beverly Hills, CA: Sage, 1974.

Janssen-Jurreit, Marie Louise. *Sexism*. New York: Farrar Strauss Giroux, 1982.

Jencks, Christopher. *Inequality*. New York: Basic Books, 1972.

Jenkins, J. Craig. "Resource Mobilization Theory and the Study of Social Movements." In Ralph H. Turner and James F. Short, Jr. (eds.), *Annual Review of Sociology* 9 (1983):527–553.

Jensen, Arthur. "How Much Can We Boost IQ and Scholastic Achievement?" *Harvard Educational Review* 39 (Winter 1969):1–123.

Johnson, David W., and Frank P. Johnson. *Joining Together*. 5th ed. Boston: Allyn and Bacon, 1994.

Johnson, Haynes. *Sleepwalking Through History*. New York: Norton, 1991.

____. *Divided We Fall*. New York: Norton, 1995.

Johnson, William B., and Arnold H. Packer. *Workforce 2000*. Indianapolis: Hudson Institute, 1987.

Johnstone, Ronald L. *Religion in Society*. 5th ed. Upper Saddle River, NJ: Prentice Hall, 1996.

Jonas, Steven. *Medical Mystery*. New York: Norton, 1978.

Jones, A. H. M. *The Decline of the Ancient World*. New York: Holt, Rinehart and Winston, 1966.

Jones, James H. *Alfred C. Kinsey: A Public/Private Life*. New York: Norton, 1997.

Jones, James M. *Prejudice and Racism*. Reading, MA: Addison-Wesley, 1972.

____. *Bad Blood*. (expanded ed.) New York: Free Press, 1993.

Jones, Landon Y. *Great Expectations*. New York: Ballantine, 1986.

Jost, Kenneth. "Downward Mobility." *CQ Researcher* 3 (July 23, 1993):627–647.

Joy, L. A., M. M. Kimball, and M. L. Zabrack. "Television and Aggressive Behavior." In T. M. Williams (ed.), *The Impact of Television*, New York: Academic Press, 1986.

Kadushin, Charles. "Mental Health and the Interpersonal Environment: A Reexamination of Some Effects of Social Structure on Mental Health." *American Sociological Review* 48 (April 1983):188–198.

Kaelble, Hartmut. *Social Mobility in the 19th and 20th Centuries*. New York: St. Martin's Press, 1986.

Kagan, Jerome. *The Nature of the Child*. New York: Basic Books, 1986.

Kain, Edward L. *The Myth of Family Decline*. Lexington, MA: Lexington Books, 1990.

Kalisch, Philip A., and Beatrice J. Kalisch. "Sex-Role Stereotyping of Nurses and Physicians on Prime-Time Television: A Dichotomy of Occupational Portrayals." *Sex Roles* 10 (1984):533–553.

Kalish, Carol B. *International Crime Rates*. U.S. Bureau of Justice Statistics. Washington, DC: U.S. Government Printing Office, 1988.

Kalmijn, Matthijs. "Mother's Occupational Status and Children's Schooling." *American Sociological Review* 59(April, 1994):257–275.

Kamarck, Elaine Ciulla. "Family Policy." In Mark Green (ed.), *Changing America*. New York: Newmarket Press, 1992, pp. 443–456.

Kanter, Rosabeth Moss. *Commitment and Community*. Cambridge, MA: Harvard University Press, 1972.

____. *Men and Women of the Corporation*. New York: Basic Books, 1993.

Kantrowitz, Barbara. "A Nation Still at Risk." *Newsweek* (April 19, 1993a):46–49.

____. "The Group Classroom." *Newsweek* (May 10, 1993b):73.

Kaplan, David A. "Bobbitt Fever." *Newsweek* (January 24, 1994):52–55.

Karatnycky, Adrian. "Democracies on the Rise, Democracies at Risk." *Freedom Review* 26 (January/February 1995):5–10.

Karp, David A., Gregory Stone, and William C. Yoels. *Being Urban*. 2d ed. Westport, CT: Greenwood, 1991.

Kart, C. S. *The Realities of Aging*. Boston: Allyn and Bacon, 1990.

Kasarda, John D., and Morris Janowitz. "Community Attachment in Mass Society." *American Sociological Review* 39 (June 1974):328–339.

Kasindorf, Martin, and Daniel Shapiro. "Asian-Americans: A 'Model Minority'." *Newsweek* (December 6, 1982):39–42.

Kass, Drora, and Seymour Martin Lipset. "America's Newest Wave of Jewish Immigrants." *New York Times Magazine* (December 7, 1980).

Katchadourian, Herant, and John Boli. *Cream of the Crop: The Impact of Elite Education in the Decade After College.* New York: Basic Books, 1994.

Katel, Peter. "The Bust in Boot Camps." *Newsweek* (February 21, 1994):26.

Katz, Irwin. "Desegregation or Integration in Public Schools?: The Policy Implications of Research." *Integrated Education* 5 (December 1967/January 1968):15–28.

Katz, Michael B. *Class, Bureaucracy and Schools.* New York: Praeger, 1975.

____. (ed.). *The "Underclass" Debate.* Princeton, NJ: Princeton University Press, 1993.

Kearl, Michael C. *Endings.* New York: Oxford University Press, 1989.

Kearl, Michael C., and Anoel Rinaldi. "The Political Uses of the Dead as Symbols in Contemporary Civil Religions." *Social Forces* 61 (March 1983):693–708.

Kellam, Sheppard G., Margaret E. Ensminger, and R. Jay Turner. "Family Structure and the Mental Health of Children." *Archives of General Psychiatry* 34 (September 1977):1012–1022.

Keller, Evelyn Fox. *Reflections on Gender and Science.* New Haven, CT: Yale University Press, 1985.

Kelley, Harold H., and John W. Thibaut. "Group Problem Solving." In Gardner Lindzey and Elliot Aronson (eds.), *The Handbook of Social Psychology.* Vol. 4. Reading, MA: Addison-Wesley, 1969, pp. 735–785.

Kelley, Maryellen R. "New Process Technology, Job Design, and Work Organization: A Contingency Model." *American Sociological Review* 55 (1990):191–208.

Kellough, Richard D. *A Resource Guide for Teaching: K-12.* 2d ed. Upper Saddle River, NJ: Merrill, 1997.

Kelly, Delos H. (ed.). *Deviant Behavior.* 5th ed. New York: St. Martin's Press, 1996.

Kephart, William M., and Davor Jedlicka. *The Family, Society, and the Individual.* 7th ed. New York: HarperCollins, 1990.

Kephart, William M., and William Zellner. *Extraordinary Groups.* 5th ed. New York: St. Martin's Press, 1994.

Kertzer, David I., and Peter Laslett (eds.). *Aging in the Past.* Berkeley: University of California Press, 1995.

Kiechel, W. "How Will We Work in the Year 2000?" *Fortune* (May 17, 1993):38.

Kilgore, Sally B. "The Organizational Context of Tracking in Schools." *American Sociological Review* 56 (April 1991):189–203.

Killian, Lewis M. "Organization, Rationality and Spontaneity in the Civil Rights Movement." *American Sociological Review* 49 (December 1984):770–783.

Kinder, Donald R. "The Continuing American Dilemma: White Resistance to Racial Change 40 Years After Myrdal." *Journal of Social Issues* 42 (1986):15lff.

Kinsey, Alfred C., and Paul Gebhard. *Sexual Behavior in the Human Female.* Philadelphia: Saunders, 1953.

Kinsey, Alfred C., Wardell Pomeroy, and Clyde Martin. *Sexual Behavior in the Human Male.* Philadelphia: Saunders, 1948.

Kirkpatrick, David. "Smart New Ways to Use Temps." *Fortune* (February 15, 1988):110–116.

Kissman, Kris, and Jo Ann Allen. *Single-Parent Families.* Newbury Park, CA: Sage, 1993.

Kitano, Harry H. *Japanese Americans.* New York: Chelsea House, 1993.

Kitano, Harry H., and Roger Daniels. *Asian Americans.* Englewood Cliffs, NJ: Prentice Hall, 1988.

Kitson, Gay C., and Helen J. Raschke. "Divorce Research: What We Know; What We Need to Know." *Journal of Divorce* 4 (Spring 1981):1–37.

Klandermans, Bert. "Mobilization and Participation: Social Psychological Explanations of Resource Mobilization Theory." *American Sociological Review* 49 (1984):583–600.

Kleidman, Robert. *Organizing for Peace.* Syracuse, NY: Syracuse University Press, 1993.

Klein, Stephen P., Susan Turner, and Joan Petersilia. *Racial Equity in Sentencing.* Santa Monica, CA: Rand, 1988.

Kleniewski, Nancy. *Cities, Change, and Conflict.* Belmont, CA: Wadsworth, 1997.

Klineberg, Otto. "Race Differences: The Present Position of the Problem." *International Social Science Bulletin* 2 (Autumn 1950):460–466.

Kluegel, James R. "Trends in Whites' Explanations of the Black-White Gap in Socioeconomic Status, 1977–1989." *American Sociological Review* 55 (August 1990):512–525.

Kluegel, James R., and Eliot R. Smith. "Whites' Beliefs About Blacks' Opportunity." *American Sociological Review* 47 (August 1982):518–532.

____. *Beliefs About Inequality.* Hawthorne, NY: Aldine de Gruyter, 1986.

Knox, David, and Caroline Schacht. *Choices in Relationships.* 5th ed. Belmont, CA: Wadsworth, 1997.

Kohl, Herbert. *Growing Minds.* New York: Harper & Row, 1984.

Kohn, Melvin L. "Social Class and Parent-Child Relationships: An Interpretation." *American Sociological Review* 68 (January 1963):471–480.

____. *Class and Conformity.* 2d ed. Chicago: University of Chicago Press, 1977.

____. "Unresolved Issues in the Relationship Between Work and Personality." In Kai Erikson and Steven Peter Vallas (eds.), *The Nature of Work.* New Haven, CT: Yale University Press, 1990.

Kohn, Melvin L., and Carmi Schooler. "Occupational Experience and Psychological Functioning: An Assessment of Reciprocal Effects." *American Sociological Review* 38 (February 1973):97–118.

____. "The Reciprocal Effects of the Substantive Complexity of Work and Intellectual Flexibility: A Longitudinal Assessment." *American Journal of Sociology* 84 (July 1978):24–52.

____. "Job Conditions and Personality: A Longtitudinal Assessment of Their Reciprocal Effects." *American Journal of Sociology* 87 (May 1982):1257–1286.

____. *Work and Personality: An Inquiry into the Impact of Social Stratification.* Norwood, NJ: Ablex, 1983.

Kolbe, Richard, and Joseph C. LaVoie. "Sex-Role Stereotyping in Preschool Children's Picture Books." *Social Psychology Quarterly* 44 (December 1981):369–374.

Komarovsky, Mirra. "Cultural Contradictions and Sex Roles: The Masculine Case." *American Journal of Sociology* 78 (January 1973):873–885.

Kovner, Christine. "Nursing." In Anthony R. Kovner (ed.), *Health Care Delivery in the United States.* 4th ed. New York: Springer, 1990, pp. 87–105.

Kowalewski, David. *Transnational Corporations and Caribbean Inequalities.* New York: Praeger, 1982.

Kozol, Jonathan. *Death at an Early Age.* Boston: Houghton Mifflin, 1968.

____. *Savage Inequalities.* New York: Crown, 1991.

Kraar, Louis. "The Real Threat to China's Hong Kong." *Fortune* 135 (May 26, 1997):84.

Kraditor, Aileen S. *The Ideas of the Women Suffrage Movement 1890–1920.* Garden City, NY: Anchor Books, 1971.

Krause, Tadeusz, and Kazimierz M. Slomczynski. "How Far to Meritocracy? Empirical Tests of a Controversial Thesis." *Social Forces* 63 (March 1985):623–642.

Kraybill, Donald B., and Marc A. Olshan (eds.). *The Amish Struggle with Modernity.* Hanover, NH: University Press of New England, 1994.

Kriesberg, Louis. *Social Inequality.* Englewood Cliffs, NJ: Prentice Hall, 1979.

Kriesi, Hanspeter. "New Social Movements and the New Class in the Netherlands." *American Journal of Sociology* 94 (March 1989):1078–1116.

Krohn, Marvin D., and James L. Massey. "Social Control and Delinquent Behavior: An Examination of the Elements of the Social Bond." *Sociological Quarterly* 21 (Autumn 1980):529–543.

Kruschwitz, Robert B., and Robert C. Roberts. *The Virtues.* Belmont, CA: Wadsworth, 1987.

Kübler-Ross, Elizabeth. *Death.* Englewood Cliffs, NJ: Prentice Hall, 1975.

____. *Death's Final Stage.* New York: Simon and Schuster, 1986.

____. *On Death and Dying.* New York: Collier Books, 1993.

Kuhn, Thomas S. *The Structure of Scientific Revolutions.* 3d ed. Chicago: University of Chicago Press, 1996.

Kuttner, R. "Competitiveness Craze." *Washington Post* (August 12, 1993a):A17.

____. "The Corporation in America, Is It Socially Redeemable?" *Dissent* (Winter 1993b):35–49.

Kuznik, Frank. "Fraud Busters." *Washington Post Magazine* (April 14, 1991):22–31.

Labich, Kenneth. "Can Your Career Hurt Your Kids?" *Fortune* (May 20, 1991):38–56.

Lacayo, Richard. "Wounding the Gun Lobby." *Time* (March 29, 1993):29–30.

Ladd, Everett C. "The Prejudices of a Tolerant Society." *Public Opinion* 10 (July/August 1987):2–3, 56.

Lamb, Michael B., M. Ann Easterbrooks, and George W. Holden. "Reinforcement and Punishment Among Preschoolers: Characteristics, Effects, and Correlates." *Child Development* 51 (1980):1230–1236.

Lampert, Leslie. "Fat Like Me." *Ladies Home Journal* (May 1993):154ff.

Landes, David S., and Charles Tilly (eds.). *History as Social Science.* Englewood Cliffs, NJ: Prentice Hall, 1991.

Landry, Bart. *The New Black Middle Class.* Berkeley: University of California Press, 1988.

Lane, Robert E. *The Market Experience.* New York: Cambridge University Press, 1991.

Lasswell, Harold. "The Structure and Function of Communication in Society." In Lyman Bryson (ed.), *The Communication of Ideas.* New York: Harper & Row, 1948.

Lauer, Robert H. *Perspectives on Social Change.* 5th ed. Boston: Allyn and Bacon, 1991.

Lawrence, Paul R., and Jay W. Lorsch. *Organization and Environment.* Boston: Division of Research, Graduate School of Business Administration, Harvard University, 1967.

Lazarsfeld, Paul. *Survey Design and Analysis.* New York: Free Press, 1955.

Le Bon, Gustave. *The Crowd.* New York: Viking, 1960.

Leach, Edmond. *Social Anthropology.* New York: Oxford University Press, 1982.

Leacock, Eleanor Burke. *Teaching and Learning in City Schools.* New York: Basic Books, 1969.

Lee, Alfred McClung. *Toward Humanist Sociology.* New York: Oxford University Press, 1978.

____. *Sociology for People.* Syracuse, NY: Syracuse University Press, 1990.

Lee, Gary R. *Family Structure and Interaction.* Philadelphia: Lippincott, 1977.

Lee, Valerie E., and Kenneth Frank. "Students' Characteristics that Facilitate the Transfer from Two-Year to Four-Year Colleges." *Sociology of Education* (July 1990):178–193.

"Left Out." *Newsweek* (March 21, 1983):26–35.

Leibenstein, Harvey. *Inside the Firm.* Cambridge, MA: Harvard University Press, 1987.

Lemann, Nicholas. *The Promised Land.* New York: Knopf, 1991.

LeMasters, E. E. *Blue-Collar Aristocrats.* Madison: University of Wisconsin Press, 1975.

Lemert, Charles, and Ann Branaman. (eds.). *The Goffman Reader.* Malden, MA: Blackwell Publishers, 1997.

Lemert, Edwin M. *Human Deviance, Social Problems, and Social Control.* 2d ed. Englewood Cliffs, NJ: Prentice Hall, 1972.

Leming, Michael R., and George E. Dickinson. *Understanding Dying, Death, and Bereavement.* 2d ed. New York: Holt, Rinehart and Winston, 1990.

Lemonick, Michael D. "Blinded by the Light." *Time* (December 20, 1993):54.

____. "How Man Began." *Time* (March 14, 1994):80–87.

Lengermann, Patricia Madoo. "Robert E. Park and the Theoretical Content of Chicago Sociology: 1920–1940." *Sociological Inquiry* 58 (Fall 1988):361–377.

Lenski, Gerhard E. *The Religious Factor.* (rev. ed.). Garden City, NY: Doubleday, 1963.

____. "The Religious Factor in Detroit, Revisited." *American Sociological Review* 36 (February 1971):48–50.

____. *Power and Privilege.* Chapel Hill: University of North Carolina Press, 1984.

____. "Rethinking Macrosociological Theory." *American Sociological Review* 53 (April 1988):163–171.

Lenski, Gerhard E., Patrick Nolan, and Jean Lenski. *Human Societies.* 7th ed. New York: McGraw-Hill, 1995.

Leo, John. "A No-Fault Holocaust." *U. S. News & World Report* (July 21, 1997):14.

Leonard, Wilbert Marcellus, II. *A Sociological Perspective of Sport.* 4th ed. New York: Macmillan, 1993.

Lerner, Jacqueline. *Working Women and Their Families.* Newbury Park, CA: Sage, 1993.

Leslie, Gerald R., and Sheila K. Korman. *The Family in Social Context.* 7th ed. New York: Oxford University Press, 1989.

Levin, Jack, and William C. Levin. *Ageism.* Belmont, CA: Wadsworth, 1980.

____. *The Functions of Discrimination and Prejudice.* 2d ed. New York: Harper & Row, 1994.

Levine, Daniel U., and Robert J. Havighurst. *Society and Education.* 8th ed. Boston: Allyn and Bacon, 1992.

Levine, David I., and Laura D'Andrea Tyson. "Participation, Productivity, and the Firm's Environment." In Alan S. Blinder (ed.), *Paying for Productivity.* Washington, DC: The Brookings Institution, 1990, pp. 183–237.

Levine, Sol, and Paul E. White. "Exchange as a Conceptual Framework for the Study of Interorganizational Relationships." *Administrative Science Quarterly* 5 (March 1961):583–610.

Levinger, G. "Editor's Page." *Journal of Social Issues* 42 (1986).

Levinson, Daniel J. *Seasons of a Man's Life.* New York: Random House, 1978.

____. "A Conception of Adult Development." *American Psychologist* 41 (January 1986):3–13.

Levitan, Sar A. *Programs in Aid of the Poor.* Ann Arbor: University of Michigan Institute of Labor, 1990.

Levy, Frank. *Dollars and Dreams.* Beverly Hills, CA: Sage, 1987.

Lewellen, Ted C. "Political Anthropology." In Samuel L. Long (ed.), *The Handbook of Political Behavior.* Vol. 3. New York: Plenum Press, 1981, pp. 333–392.

Lewin, Arie Y., and Carroll U. Stephens. "Designing Postindustrial Organizations: Combining Theory and Practice." In George P. Huber and William H. Glick (eds.), *Organizational Change and Redesign.* New York: Oxford University Press, 1993, pp. 393–409.

Lewis, Gordon H. "Role Differentiation." *American Sociological Review* 37 (August 1972):424–434.

Lewis, I. A., and William Schneider. "Hard Times: The Public on Poverty." *Public Opinion* 8 (June/July 1985):2ff.

Lewis, Jerry. "A Study of the Kent State Incident Using Smelser's Theory of Collective Behavior." *Sociological Inquiry* 42 (1972):87–96.

Lewis, Lionel. "Working at Leisure." *Society* (July/August 1982):27–32.

Lewis, Michael. *The Culture of Personality.* Amherst: University of Massachusetts Press, 1978.

Lewis, Oscar. *The Study of Slum Culture.* New York: Random House, 1978.

Liazos, Alexander. "The Poverty of the Sociology of Deviance: Nuts, Sluts, and Preverts." *Social Problems* 20 (Summer, 1972):103–120.

Lieberman, Janet. "A Plan for High School/Community College Collaboration." *The College Board Review* (Fall 1989):14–19.

Lieberman, Myron. *Public Education.* Cambridge, MA: Harvard University Press, 1993.

Liebow, Elliot. *Talley's Corner.* Boston: Little, Brown, 1967.

Light, Ivan. *Cities in World Perspective.* New York: Macmillan, 1983.

Likert, Rensis. *New Patterns of Management.* New York: McGraw-Hill, 1961.

Lin, Nan, Walter M. Ensel, and John C. Vaughn. "Social Resources and Strength of Ties: Structural Factors in Occupational Status Attainment." *American Sociological Review* 46 (August 1981):382–399.

Lincoln, James R., and Arne L. Kalleberg. *Culture, Control, and Commitment.* New York: Cambridge University Press, 1990.

Lincoln, James R., and Kerry McBride. "Japanese Industrial Organization in Comparative Perspective." In W. Richard Scott and James F. Short, Jr. (eds.), *Annual Review of Sociology* 13 (1987):289–312.

Linden, Eugene. "Can Animals Think?" *Time* (March 22, 1993a):55–61.

____. "Megacities." *Time* (January 11, 1993b):28–38.

Link, Bruce G. "Understanding Labeling Effects in the Area of Mental Disorders." *American Sociological Review* 52 (February 1987):96–112.

Link, Bruce G., and Francis Cullen. "Reconsidering the Social Rejection of Ex-Mental Patients: Levels of Attitudinal Response." *American Journal of Community Psychology* 11 (1983):261–273.

Link, Bruce G., Bruce P. Dohrenwend, and Andrew E. Skodol. "Socio-Economic Status and Schizophrenia: Noisome Occupational Characteristics as a Risk Factor." *American Sociological Review* 51 (April 1986):242–258.

Link, Bruce G., Francis T. Cullen, James Frank, and John Wozniak. "The Social Rejection of Former Mental Patients: Understanding Why Labels Matter." *American Journal of Sociology* 92 (May 1987):1461–1500.

Linton, Ralph. "The One Hundred Percent American." *The American Mercury* 40 (April 1937):427–429.

Linz, Juan J. "An Authoritarian Regime: Spain." In Erik Allardt and Yrjo Littunen (eds.), *Cleavages, Ideologies and Party Systems.* Helsinki: The Academic Bookstore, 1964.

Lipset, Seymour Martin. "Harriet Martineau's America." In Seymour Martin Lipset (ed.), *Harriet Martineau, Society in America.* Garden City, NY: Doubleday, 1962, pp. 5–41.

____. "Social Mobility in Industrial Societies." *Public Opinion* 5 (June/July 1982):41–44.

____. "Blacks and Jews: How Much Bias?" *Public Opinion* 10 (July/August 1987):4–5, 57–58.

____. *Continental Divide.* New York: Routledge, 1990.

Lipset, Seymour M., and Hans L. Zetterberg. "Social Mobility in Industrial Societies." In Seymour M. Lipset and Reinhard Bendix (eds.), *Social Mobility in Industrial Society.* Berkeley: University of California Press, 1964, pp. 11–76.

Lipset, Seymour Martin, and Reinhard Bendix (eds.). *Social Mobility in Industrial Society.* Berkeley: University of California Press, 1964.

Lipset, Seymour Martin, Martin Trow, and James Coleman. *Union Democracy.* New York: Free Press, 1956.

Lipsey, Mark. "Juvenile Delinquency Treatment: A Meta-Analytic Inquiry into the Variability of Effects." In Mark Lipsey (ed.), *Meta-Analysis for Explanation: A Casebook.* New York: Russell Sage Foundation, 1991.

Little, Suzanne. "Sex Roles in Faraway Places." *Ms.* (February 1975):77ff.

Litwak, Eugene. "Occupational Mobility and Extended Family Cohesion." *American Sociological Review* 25 (February 1960):9–21.

Litwak, Eugene. "Models of Bureaucracy Which Permit Conflict." *American Journal of Sociology* 67 (September 1961):177–184.

Livi-Bacci, Massimo. *A Concise History of World Population.* 2d ed. Malden, MA: Blackwell Publishers, 1997.

Livingston, E. *Making Sense of Ethnomethodology.* London: Routledge and Kegan Paul, 1987.

Locke, John. *Second Treatise of Government.* Edited by C.B. Macpherson. Indianapolis, IN: Hackett, 1980.

Lodge, George C. *The New American Ideology.* New York: Knopf, 1975.

____. *The American Disease.* Washington Square: New York University Press, 1986.

Lodge, George C., and Ezra F. Vogel (eds.). *Ideology and National Competitiveness.* Cambridge, MA: Harvard Business School, 1987.

Lofland, John. "White-Hat Mobilization: Strategies of a Millenarian Movement." In Mayer N. Zald and John D. McCarthy (eds.), *The Dynamics of Social Movements.* Cambridge, MA: Winthrop, 1979, pp. 157–166.

____. *Protest.* New Brunswick, NJ: Transaction Books, 1991.

____. *Polite Protesters.* Syracuse, NY: Syracuse University Press, 1993.

____. *Social Movement Organizations*. Hawthorne, NY: Aldine de Gruyter, 1996.

Logan, Charles H. "Problems in Ratio Correlation: The Case of Deterrence Research." *Social Forces* 60 (March 1982):791–810.

Logan, John R., and Harvey L. Molotch. *Urban Fortunes*. Berkeley: University of California Press, 1987.

Lombroso, Cesare. *Crime*. Boston: Little, Brown, 1918.

Longmire, Dennis R. "American Attitudes Among the Ultimate Weapon: Capital Punishment." In Timothy J. Flanagan and Dennis Longmire (eds.), *Americans View Crime and Justice*. Thousand Oaks, CA: Sage, 1996, pp. 93–108.

Longshore, Douglas. "School Racial Composition and Blacks' Attitudes Toward Desegregation: The Problem of Control in Desegregated Schools." *Social Science Quarterly* 63 (December 1982a):674–687.

____. "Race Composition and White Hostility: A Research Note on the Problem of Control in Desegregated Schools." *Social Forces* 61 (September 1982b):73–78.

Lopreato, Joseph. "From Social Evolutionism to Biocultural Evolutionism." *Sociological Forum* 5 (1990): 187–212.

Lord, George F., III, and William W. Falk. "Hidden Income and Segmentation: Structural Determinants of Fringe Benefits." *Social Science Quarterly* 63 (June 1982):208–224.

Lowe, M. "The Dialectic of Biology and Culture." In M. Lowe and R. Hubbard (eds.), *Woman's Nature*. New York: Pergamon, 1983, pp. 39–62.

Luckmann, Thomas. *The Invisible Religion*. New York: Macmillan, 1967.

Luker, Kristin. *Abortion and the Politics of Motherhood*. Berkeley: University of California Press, 1984.

Lund, Caroline. Introduction to *Women and the Family* by Leon Trotsky. New York: Pathfinder Press, 1970, pp. 7–18.

Lux, Kenneth. *Adam Smith's Mistake*. Boston: Shambhala, 1990.

Lyerly, Robert, and James K. Skipper, Jr. "Differential Rates of Rural-Urban Delinquency: A Social Control Approach." *Criminology* 19 (February 1982):385–394.

Lyman, Stanford M. *Chinese Americans*. New York: Random House, 1974.

Lynaugh, Joan E., and Barbara L. Brush. *American Nursing*. Malden, MA: Blackwell Publisher, 1996.

Lynch, David J. "Dying Dreams, Dead-End Streets." *USA Today* (September 20, 1996):1B-2B.

Lynd, Robert S., and Helen M. Lynd. *Middletown*. New York: Harcourt, Brace & World, 1929.

____. *Middletown in Transition*. New York: Harcourt, Brace & World, 1937.

Machalek, Richard, and Michael Martin. "'Invisible' Religions: Some Preliminary Evidence." *Journal for the Scientific Study of Religion* 15 (December 1976):311–321.

MacIntyre, Alasdair. *A Short History of Ethics*. 1st Touchstone ed. New York: Simon and Schuster, 1997.

Macionis, John J. *Sociology*. 6th ed. Upper Saddle River, NJ: Prentice Hall, 1997.

MacKellar, F. Landis, and Machiko Yanagishita. "Homicide in the United States: Who's at Risk." Washington, D.C.: Population Reference Bureau, 1995.

Maddox, Brenda. "Homosexual Parents." *Psychology Today* 16 (February 1982):62, 66–69.

Majority Staff of the Senate Judiciary Committee. *Violence Against Women*. Washington, DC, 1992.

Malinowski, Bronislaw. *The Sexual Life of Savages*. New York: Liveright, 1929.

Malthus, Thomas. *An Essay on the Principle of Population*. London: Reeves and Turner, 1798.

Manchester, William. *A World Lit Only by Fire*. Boston: Little, Brown, 1993.

Manski, Charles F. "Educational Choice (Vouchers) and Social Mobility." *Institute for Research on Poverty*. Madison: University of Wisconsin, June 1992.

Marcus, Ruth, and Dan Balz. "Clinton Outlines Plan to Break Welfare Cycle." *Washington Post* (June 5, 1994):A1, A18.

Markle, Gerald E., and Ronald J. Troyer. "Cigarette Smoking as Deviant Behavior." In Ronald A. Farrell and Victoria Lynn Swigert (eds.), *Social Deviance*. 3d ed. Belmont, CA: Wadsworth, 1988, pp. 82–91.

Marmor, Theodore R., and Michael S. Barr. "Health." In Mark Green (ed.), *Changing America*. New York: Newmarket Press, 1992a, pp. 399–414.

Marmor, Theodore R., Jerry L. Mashaw, and Philip L. Harvey. *America's Misunderstood Welfare State*. New York: Basic Books, 1992.

Marmot, M. G., M. J. Shipley, and Geoffrey Rose. "Inequalities in Death-Specific Explanations of a General Pattern." *Lancet* 83 (May 5, 1984):1003–1006.

Marriott, M. "Afrocentrism: Balancing or Skewing History?" *New York Times* (August 11, 1991):1,18.

Marsden, Peter V. "Core Discussion Networks of Americans." *American Sociological Review* 52 (February 1987):122–131.

Marshall, Harvey, and Bonnie L. Lewis. "City-Suburb Competition for High Status Males: Age and Size as Determinants." *Sociology and Social Research* 67 (January 1983):129–145.

Martineau, Harriet. *Society in America*. Paris: Bandry's European Library, 1837.

____. *How to Observe Manners and Morals.* London: C. Knight, 1838.

Martocchio, B. "Grief and Bereavement." *Nursing Clinics of North America* 20 (1985):327–346.

Marty, Martin E. "Fundamentalism Reborn: Faith and Fanaticism." *Saturday Review* (May 1980):37–42.

____. *Pilgrims in Their Own Land.* New York: Penguin, 1985.

Marx, Gary T., and Douglas McAdam. *Collective Behavior and Social Movements.* Englewood Cliffs, NJ: Prentice Hall, 1994.

Marx, Karl. *The Poverty of Philosophy.* Translated by H. Quelch. Chicago: Charles H. Kerr, 1920.

____. *Capital.* Edited by Friedrich Engels. New York: International, 1967.

Marx, Karl, and Friedrich Engels. *The Communist Manifesto.* New York: Penguin, 1969.

Mason, Philip. *Patterns of Dominance.* New York: Oxford University Press, 1970.

Massey, Douglas S. "American Apartheid: Segregation and the Making of the Underclass." *American Journal of Sociology* 96 (September 1990):329–357.

Massey, Douglas S., and Nancy A. Denton. "Trends in the Residential Segregation of Blacks, Hispanics, and Asians: 1970–1980." *American Sociological Review* 52 (December 1987):802–825.

____. *American Apartheid.* Cambridge, MA: Harvard University Press, 1996.

Massey, Douglas S., and Mitchell L. Eggers. "The Ecology of Inequality: Minorities and the Concentration of Poverty, 1970–1980." *American Journal of Sociology* 95 (March 1990):1153–1188.

Mathews, Tom. "The Siege of L.A." *Newsweek* (May 11, 1992):30–38.

Maun, Michael. *The Sources of Social Power.* New York: Cambridge University Press, 1986.

____ (ed.). *The Rise and Decline of the Nation State.* New York: Basil Blackwell, 1990.

Mauss, Armand L. *Social Problems as Social Movements.* Philadelphia: Lippincott, 1975.

Maxwell, Gerald. "Why Ascription: Parts of a More-or-Less Formal Theory of the Functions and Dysfunctions of Sex Roles." *American Sociological Review* 40 (1975):445–455.

Maynard, Douglas W. "On the Functions of Social Conflict Among Children." *American Sociological Review* 50 (April 1985):207–223.

Maynard, Micheline. "Female Treasurers Crack Corporate Glass Ceiling." *USA Today* (July 11, 1994):6B.

Mazlish, Bruce. *A New Science.* University Park, PA: Penn State Press, 1993.

McAdam, Doug, John D. McCarthy, and Mayer N. Zald. "Social Movements." In Neil J. Smelser (ed.), *Handbook of Sociology.* Newbury Park, CA: Sage, 1988, pp. 695–737.

McAllister, Bill. *Washington Post.* (October 29, 1992):A1,A4.

____. "Census Calls Sampling Essential." *Washington Post* (July 15, 1997):A1,A4.

McCaghy, Charles H., and Timothy A. Capron. *Deviant Behavior.* 4th ed. Boston: Allyn and Bacon, 1997.

McCarthy, John D., and Mark Wolfson. "Resource Mobilization by Local Social Movement Organizations: Agency, Strategy, and Organization in the Movement Against Drinking and Driving." *American Sociological Review* 61 (December 1996):1070–1088.

McCarthy, John D., and Mayer N. Zald. "Resource Mobilization and Social Movements: A Partial Theory." *American Journal of Sociology* 82 (1977):1212–1241.

McCormack, Gavan, and Yoshio Sugimoto. *The Japanese Trajectory.* Cambridge, MA: Cambridge University Press, 1988.

McDonald, Scott. "Sex Bias in the Representation of Male and Female Characters in Children's Picture Books." *Journal of Genetic Psychology* 150 (December 1989):389–401.

McGinn, Daniel, and Joan Raymond. "Workers of the World, Get Online." *Newsweek* (Winter 1997–98):32–33.

McGrath, Ellie. "Confucian Work Ethic." *Time* (March 28, 1983):52.

McGuire, Meredith B. *Religion, The Social Context.* 4th ed. Belmont, CA: Wadsworth, 1997.

McKinlay, John B. (ed.). *Issues in the Political Economy of Health Care.* New York: Tavistock, 1985.

McLanahan, Sara, and Gary Fam Sandefur. *Growing Up with a Single Parent.* Cambridge, MA: Harvard University Press, 1996.

McLaughlin, Steven, and Associates. *The Changing Lives of Women.* Chapel Hill: University of North Carolina Press, 1988.

McManus, Ed. "The Death Penalty and the Race Factor." *Illinois Issues* 11 (March 1985):47.

McNicoll, Geoffrey. "Changing Fertility Patterns and Politics in the Third World." *Annual Review of Sociology* 18 (1992):85–108.

McPhail, Clark. *Myth of the Madding Crowd.* Hawthorne, NY: Aldine de Gruyter, 1994.

McPhail, Clark, and Ronald T. Wohlstein. "Individual and Collective Behaviors Within Gatherings, Demonstrations, and Riots." In Ralph H. Turner and James F. Short, Jr. (eds.), *Annual Review of Sociology* 9 (1983):579–600.

McQuail, Denis. *Mass Communication Theory*. 2d ed. Beverly Hills, CA: Sage, 1987.

Mead, George Herbert. *Mind, Self and Society*. Chicago: University of Chicago Press, 1934.

Mead, Margaret. *Sex and Temperament in Three Primitive Societies*. New York: Mentor Books, 1950.

Mechanic, David. "Foreword." In John Colombotos and Corinne Kirchner, *Physicians and Social Change*. New York: Oxford University Press, 1986, pp. vii–x.

Mednick, Sarnoff A., William J. Gabrielli, and Barry Hutchings. "Genetic Influences in Criminal Convictions: Evidence from an Adoption Cohort." *Science* 224 (1984):891–894.

Melucci, Alberto. "The New Social Movements: A Theoretical Approach." *Social Science Information* 19 (May 1980):199–226.

Menaghan, E. G., and T. L. Parcel. "Parental Employment and Family Life: Research in the 1980's." In A. Booth (ed.), *Contemporary Families*. Minneapolis: National Council on Family Relations, 1991, pp. 361–380.

Merris, Martina, Annette D. Bernhardt, and Mark S. Handcock. "Economic Inequality: New Methods and New Trends." *American Sociological Review* 59 (April 1994):205–219.

Merton, Robert K. *Social Theory and Social Structure*. (enlarged ed.). New York: Free Press, 1968.

____. *On Social Structure and Science*. Chicago: University of Chicago Press, 1996.

Meyer, Daniel R., and Maria Cancian. "Life After Welfare." *Population Today* (July/August, 1997):4–5.

Meyer, John W., and W. Richard Scott. *Organizational Environments*. Beverly Hills, CA: Sage, 1985.

Michaels, Marguerite. "Sorry, Wrong Country." *Time* (June 6, 1994):34.

Michels, Robert. *Political Parties*. New York: Free Press, 1949.

Miles, Robert. *Racism*. London: Tavistock/Routledge, 1989.

Milgram, Stanley. "Behavioral Study of Obedience." *Journal of Abnormal and Social Psychology* 67 (1963):371–378.

____. "Some Conditions of Obedience and Disobedience to Authority." *Human Relations* 18 (1965):57–76.

____. "The Small World Problem." *Psychology Today* 1 (1967):61–67.

____. *Obedience to Authority*. New York: Harper & Row, 1974.

Milkman, Ruth. *Farewell to the Factory*. Berkeley: University of California Press, 1997.

Mill, John Stuart. *Representative Government*. London: Everyman, 1910.

____. *On Liberty and Utilitarianism*. New York: Bantam Books, 1993.

Miller, Delbert C. "Sociologists in the Corporate World: Academic, Research, and Practice Roles in Business and Industry." Washington, DC: American Sociological Association, 1994.

Miller, Joanne, and Howard H. Garrison. "Sex Roles: The Division of Labor at Home and in the Workplace." *Annual Review of Sociology* 8 (1982):237–262.

Millicent, Lawton. "Schools' 'Glass Ceiling' Imperils Girls, Study Says." *Education Week* (February 12, 1992):17.

Mills, C. Wright. *The Power Elite*. New York: Oxford University Press, 1956.

____. *The Sociological Imagination*. New York: Oxford University Press, 1959.

Mintz, Beth, and Michael Schwartz. "The Structure of Intercorporate Unity in American Business." *Social Problems* 29 (December 1981a):87–103.

Mintz, Beth, and Michael Schwartz. "Interlocking Directorates and Interest Group Formation." *American Sociological Review* 46 (December 1981b):851–869.

____. *The Power Structure of American Business*. Chicago: University of Chicago Press, 1985.

Mintz, Morton, and Jerry Cohen. *America, Inc.* New York: Dial Press, 1972.

Mintzberg, Henry. *The Structuring of Organizations*. Englewood Cliffs, NJ: Prentice Hall, 1979.

Mirowsky, J., and C. E. Ross. "Social Patterns of Distress." *Annual Review of Sociology* 12 (1986):23–45.

Moen, Phyllis. "The Family Impact of Maternal Employment: Its Effects on Children." In Sameha Peterson, Judy Richardson, and Gretchen Kreuter (eds.), *The Two-Career Family*. Washington, DC: University Press, 1978, pp. 182–197.

____. *Women's Two Roles*. New York: Auburn House, 1992.

Moir, Anne. *Brain Sex: The Real Difference Between Men and Women*. New York: Lyle Stuart, 1991.

Moller, David Wendell. *Confronting Death*. New York: Oxford University Press, 1995a.

____. *Death and Dying*. New York: Oxford University Press, 1995b.

"Money Kills." *The Economist* 341 (November 16, 1996):51.

Monk-Turner, Elizabeth. "The Occupational Achievements of Community and Four-Year College Entrants." *American Sociological Review* 55 (October 1990):719–725.

Montagu, Ashley. "Don't Be Adultish." *Psychology Today* 11 (August 1977):46–50, 55.

Montero, Darrel. "The Japanese Americans: Changing Patterns of Assimilation Over Three Generations." *American Sociological Review* 46 (December 1981):829–839.

Moody, John. "Street Wars: Youths Vent Their Rage." *Time* (October 14, 1985):5l.

Moore, Joan, and Harry Pachon. *Hispanics in the United States*. Englewood Cliffs, NJ: Prentice Hall, 1985.

Moore, Thomas S. *The Disposable Work Force*. Hawthorne, NY: Aldine de Gruyter, 1996.

Mor, V., et al. (eds.). *The Hospice Experiment*. Baltimore, MD: Johns Hopkins Press, 1988.

Morgan, C. "Adjusting to Widowhood: Do Social Networks Make It Easier?" *Gerontologist* 29 (1989):101–107.

Morganthau, Tom. "The Price of Neglect." *Newsweek* (May 11, 1992):54–55.

____. "IQ: Is It Destiny?" *Newsweek* (October 24, 1994):53–55.

Morganthau, Tom, and Seema Nayyar. "Those Scary College Costs." *Newsweek* (April 29, 1996):52–56.

Morin, Richard. "America's Middle-Class Meltdown." *Washington Post* (December, 1991):C1–C2.

____. "Crime Time: The Fear, The Facts." *Washington Post* (January 30, 1994):C1.

Morris, Aldon D. "Black Southern Sit-in Movement: An Analysis of Internal Organization." *American Sociological Review* 46 (December 1981):744–767.

____. *The Origins of the Civil Rights Movement*. New York: Free Press, 1984.

Morris, Charles. *Historical Tales*. Atlanta: Martin and Hoyt, 1893.

Morris, Norval, and Michael Tonry. *Between Prison and Probation*. New York: Oxford University Press, 1990.

Moskos, Charles C., and Frank R. Wood (eds.). *The Military*. Washington, DC: Pergamon-Brassey, 1988.

Mueller, D. P., and P. W. Cooper. "Children of Single Parent Families: How They Fare as Young Adults." *Family Relations* 35 (January 1986):169–176.

Muller, Jerry Z. *Adam Smith in His Time and Ours*. New York: Free Press, 1993.

Mumford, Emily. *Medical Sociology*. New York: McGraw-Hill, 1983.

Mumford, Lewis. *The City in History*. New York: Harcourt Brace Jovanovich, 1968.

Murdock, George P. *Our Primitive Contemporaries*. New York: Macmillan, 1935.

____. "The Common Denominator of Cultures." In Ralph Linton (ed.), *The Science of Man in the World Crisis*. New York: Columbia University Press, 1945, pp. 123–142.

____. "World Ethnographic Sample." *American Anthropologist* 59 (August 1957):664–687.

Murray, Charles. *Losing Ground*. New York: Basic Books, 1984.

Murray, Stephen O. *American Gay*. Chicago: University of Chicago Press, 1996.

Musheno, Michael C. "Criminal Diversion and Social Control: A Process Evaluation." *Social Science Quarterly* 63 (June 1982):280–292.

Mussen, Paul H. "Early Sex-Role Development." In David Goslin (ed.), *Handbook of Socialization Theory and Research*. Chicago: Rand McNally, 1969, pp. 707–732.

Myrdal, Gunnar. *An American Dilemma*. New York: Harper & Row, 1944.

____. *Objectivity in Social Research*. New York: Pantheon, 1969.

Nader, Ralph, and Mark Green. "Crime in the Suites." *The New Republic* (April 29, 1972):17–19.

Nagata, Donna K. *Legacy of Injustice*. New York: Plenum, 1993.

Nakamura, David. "Equity Leaves Its Mark on Male Athletes." *Washington Post* (July 7, 1997):A1,A10.

Nakao, Keiko, and Judith Treas. "Computing 1989 Occupational Prestige Scores." General Social Survey Methodological Report Number 70, 1990.

Nakao, Keiko, Robert W. Hodge, and Judith Treas. "On Revising Prestige Scores for All Occupations." General Social Survey Methodological Report Number 69, October 1990.

Namboodiri, Krishman. *A Primer of Population Dynamics*. New York: Plenum, 1996.

Nanji, Azim A. *The Muslim Almanac*. Detroit: Gale Research, 1996.

Nash, June. "Devils, Witches, and Sudden Death." *Natural History* 81 (March 1972):52ff.

Nathanson, C. A. "Social Roles and Health Status Among Women and the Significance of Employment." *Social Science and Medicine* 14 (1980):463–471.

____. "Sex Differences in Mortality." *Annual Review of Sociology* 10 (1989):191–213.

Navarro, Vincente. *Medicine Under Capitalism*. New York: Prodist, 1976.

____. *Crisis, Health, and Medicine*. New York: Tavistock, 1986.

Nee, Victor, and David Stark (eds.). *Remaking the Economic Institutions of Socialism*. Stanford, CA: Stanford University Press, 1989.

Nelan, Bruce W. "More Harm Than Good." *Time* (March 15, 1993a):40–45.

____. "The Dark Side of Islam." *Time* (October 4, 1993b):62–64.

Nelsen, Benjamin. "Weber's Protestant Ethic: Its Orgins, Wanderings, and Foreseeable Futures." In Charles Y. Glock and Philip E. Hammonds (eds.), *Beyond the Classics?* New York: Harper & Row, 1973, pp. 71–130.

Nettler, Gwynn. *Explaining Crime.* 3d ed. New York: McGraw-Hill, 1984.

Neugarten, B. L. "Adult Personality: Toward a Psychology of the Life Cycle." In B. L. Neugarten (ed.), *Middle Age and Aging.* Chicago: University of Chicago Press, 1968, pp. 137–147.

"The New World of Work." *Business Week* (October 17, 1994):76–86.

Newman, Katherine S. *Falling from Grace.* New York: Free Press, 1988.

____. *Declining Fortunes.* New York: Basic Books, 1993.

Niebuhr, H. Richard. *The Social Sources of Denominationalism.* New York: World, 1968.

Nielsen, Niels, C., Jr., et al. *Religions of the World.* 3d ed. New York: St. Martin's Press, 1993.

Nietzsche, Friedrich. *Beyond Good and Evil.* New York: Penguin, 1987.

Nilson, Linda Burzotta. "Reconsidering Ideological Lines: Beliefs About Poverty in America." *Sociological Quarterly* 22 (Autumn 1981):531–548.

Nimkoff, Meyer F. (ed.). *Comparative Family Systems.* Boston: Houghton Mifflin, 1965.

Nisbet, Robert A. *The Sociological Tradition.* New York: Basic Books,1966.

____. *The Social Bond.* New York: Knopf, 1970.

____. *The Present Age.* New York: Harper & Row, 1989.

Novak, Michael. *The Unmeltable Ethnics.* 2d ed. New Brunswick: Transaction, 1996.

Nydeggar, Corinne N. "Family Ties of the Aged in Cross-Cultural Perspective." In Beth B. Hess and Elizabeth W. Markson (eds.), *Growing Old in America.* New Brunswick, NJ: Transaction Press, 1985.

Oakes, Jeannie. *Keeping Track.* New Haven, CT: Yale University Press, 1985.

Oates, Stephen B. *Malice Toward None.* New York: Mentor Books,1977.

Oberschall, Anthony. *Social Conflict and Social Movements.* Englewood Cliffs, NJ: Prentice Hall, 1973.

O'Brien, Joanne, and Martin Palmer. *The State of Religion Atlas.* New York: Touchstone, 1993.

O'Connell, Martin. "Where's Papa? Fathers' Role in Child Care." *Population Bulletin* 20 (September 1993):1–20.

Ogburn, William F. *On Culture and Social Change.* Chicago: University of Chicago Press, 1964.

O'Hare, William P. *America's Welfare Population: Who Gets What?* Washington, DC: Population Reference Bureau, 1987.

____. "Can the Underclass Concept Be Applied to Rural Areas?" Population Reference Bureau. Staff Working Papers, January 1992a.

____. "America's Minorities—The Demographics of Diversity." *Population Bulletin* 47 (December 1992b):1–47.

____. "A New Look at Poverty in America." *Population Bulletin* 51 (1996):1–48.

____. "African Americans in the 1990s." Washington, DC: Population Reference Bureau, 1991.

O'Hare, William P., and Brenda Curry-White. "The Rural Underclass: Examination of Multiple-Problem Populations in Urban and Rural Settings." Population Reference Bureau. Staff Working Papers, January 1992.

O'Hare, William P., and Judy C. Felt. "Asian Americans: America's Fastest Growing Minority Group." *Population Trends and Public Policy.* Washington, DC: Population Reference Bureau, 1991.

O'Hare William P., et al. *Real Life Poverty in America: Where the American Public Would Set the Poverty Line.* Washington, DC: A Center on Budget and Policy Priorities and Families USA Foundation Report, 1990.

Ohlin, Lloyd, and Michael Tonry (eds.). *Family Violence.* Chicago: University of Chicago Press, 1989.

Olneck, Michael R., and James Crouse. "The IQ Meritocracy Reconsidered." *American Journal of Education* 88 (November 1979):1–31.

Olzak, Susan, and Joane Nagel (eds.). *Competitive Ethnic Relations.* San Diego, CA: Academic Press, 1986.

Olzak, Susan, and Elizabeth West. "Ethnic Conflict and the Rise and Fall of Ethnic Newspapers." *American Sociological Review* 56 (August 1991):458–474.

Omi, Michael, and Howard Winant. *Racial Formation in the United States.* New York: Routledge & Kegan Paul, 1994.

O'Neill, Hugh M., and D. Jeffrey Lenn. "Voices of Survivors: Words that Downsizing CEOs Should Hear." *Academy of Management Executive* 9 (November 1995):23–34.

Oommen, T. K. *Protest and Change.* Newbury Park, CA: Sage, 1990.

Opp, Karl-Dieter. "Grievances and Participation in Social Movements." *American Sociological Review* 53 (1988):853–864.

____. *The Rationality of Political Protest.* Boulder, CO: Westview, 1989.

Opp, Karl-Dieter, and Wolfgang Roehl. "Repression, Micromobilization, and Political Protest." *Social Forces* 69 (1990):521–547.

Orfield, Gary A. "The Growth of Segregation in American Schools: Changing Patterns of Separation and Poverty Since 1968." A Report of the Harvard Project on School Desegregation, 1993.

Orfield, Gary A., et al. "Status of School Desegregation: The Next Generation." Report to the National School Board Association. Alexandria, VA: National School Board Association, 1992.

Ornstein, N., T. Mann, and M. Malbin. "Vital Statistics on Congress 1993–1994." Washington, DC: *Congressional Quarterly,* 1994.

Ornstein, Robert, and Paul Ehrlich. *New World New Mind.* London: Paladin, 1991.

Orum, Anthony M. *Introduction to Political Sociology.* 2d ed. Englewood Cliffs, NJ: Prentice Hall, 1983.

Ouchi, William G. *Theory* Z. New York: Avon Books, 1993.

Padan-Eisenstark, Dorit D. "Are Israeli Women Really Equals? Trends and Patterns of Israeli Women's Labor Force Participation: A Comparative Analysis." *Journal of Marriage and the Family* 35 (August 1973):538–545.

Painton, Frederick. "There the Sky Stays Dark." *Time* (May 28, 1990):40–42.

Palen, John J. *The Urban World.* 5th ed. New York: McGraw-Hill, 1997.

Palmore, Erdman. "Are the Aged a Minority Group?" *Journal of the American Geriatrics Society* 26 (May 1978):214–217.

_____. "The Facts on Aging Quiz: A Review of Findings." *The Gerontologist* 20 (December 1980):669–672.

_____. "More on Palmore's Facts on Aging Quiz." *The Gerontologist* 21 (April 1981):115–116.

Parrillo, Vincent N. *Diversity in America.* Thousand Oaks, CA: Pine Forge Press, 1996.

_____. *Strangers to These Shores.* 5th ed. Boston: Allyn and Bacon, 1997.

Parsons, Talcott. *The Structure of Social Action.* New York: McGraw-Hill, 1937.

_____. *The Social System.* New York: Free Press, 1951.

_____. *Politics and Social Structure.* New York: Free Press, 1959.

_____. "A Functional Theory of Change." In Amitai Etzioni and Eva Etzioni (eds.), *Social Change.* New York: Basic Books,1964a, pp. 83–97.

_____. "Definitions of Health and Illness in the Light of American Values and Social Structure." In Talcott Parsons, *Social Structure and Personality.* New York: Free Press, 1964b, pp. 257–291.

_____. *Societies: Evolutionary and Comparative Perspectives.* Englewood Cliffs, NJ: Prentice Hall, 1966.

_____. *The System of Modern Societies.* Englewood Cliffs, NJ: Prentice Hall, 1971.

_____. "The Sick Role and the Role of the Physician Reconsidered." *Health and Society* (Summer 1975):257–278.

_____. *The Evolution of Societies.* Englewood Cliffs, NJ: Prentice Hall, 1977.

Pascale, Richard Tanner, and Anthony G. Athos. *The Art of Japanese Management.* New York: Simon & Schuster, 1981.

Patchen, Martin. *Black-White Contact in Schools.* West Lafayette, IN: Purdue University Press, 1982.

Patterson, James T. *America's Struggle Against Poverty: 1900–1985.* Cambridge, MA: Harvard University Press, 1986.

Patterson, James T., and Peter Kim. *The Day America Told the Truth.* Englewood Cliffs, NJ: Prentice Hall, 1991.

Patterson, Thomas E. *The American Democracy.* 2d ed. New York: McGraw-Hill, 1993.

Pattison, Robert V., and Hallie M. Katz. "Investor-Owned and Not-for-Profit Hospitals: A Comparison Based on California Data." *New England Journal of Medicine* 309 (August 1983):347–353.

Paul, F. *Declining Access to Educational Opportunities in Metropolitan Chicago 1980–1985.* Chicago: Metropolitan Opportunity Project, University of Chicago, 1987.

Payer, Lynn. *Medicine and Culture.* New York: Henry Holt, 1989.

Pearlstein, Steven. "Layoffs Become a Lasting Reality." *Washington Post* (November 6, 1993):D1.

_____. "Studies on Cost of Global Warming Don't Clear the Air." *Washington Post* (June 12, 1997):E1.

Pearlstein, Steven, and DeNeen L. Brown. "Black Teenagers Facing Worse Job Prospects." *Washington Post* (June 4, 1994):A1.

Peck, Dennis L., and J. Selwyn Hollingsworth (eds.). *Demographic and Structural Change: The Effects of the 1980s on American Society.* Westport, CT: Greenwood, 1996.

Pedersen, Daniel. "Southern Discomfort." *Newsweek* (December 9, 1996):36.

Pelfrey, William V. *The Evolution of Criminology.* Cincinnati, OH: Anderson, 1980.

Pelikan, Jaroslav. *The Idea of the University.* New Haven, CT: Yale University Press, 1992.

Pendleton, Brian F., Margaret M. Poloma, and T. Neal Garland. "Scales for Investigation of the Dual Career Family." *Journal of Marriage and the Family* 42 (May 1980):269–275.

Pennings, Johannes M. "The Relevance of the Structural-Contingency Model for Organizational Effectiveness." *Administrative Science Quarterly* 20 (September 1975):393–410.

Peplau, L. "Roles and Gender." In Harold H. Kelly et al. (eds.), *Close Relationships.* New York: Freeman, 1983, pp. 220–264.

Perrow, Charles. *Complex Organizations.* 3d ed. New York: McGraw-Hill, 1986.

Pessen, Edward. *The Log Cabin Myth.* New Haven, CT: Yale University Press, 1984.

Peters, E. L. "Aspects of Family Life Among the Bedouin of Cyrenaica." In Meyer F. Nimkoff (ed.), *Comparative Family Systems*. Boston: Houghton Mifflin, 1965, pp. 121–146.

Peters, Tom. "Restoring American Competitiveness: Looking for New Models of Organizations." *Academy of Management Executive* 2 (May 1988):103–109.

Peyser, Marc. "Tyranny of the Red Ribbon." *Newsweek* (June 28, 1993):61.

Pfeffer, Jeffrey. "The Micropolitics of Organizations." In Marshall W. Meyer and Associates (eds.), *Environments and Organizations*. San Francisco: Jossey-Bass, 1978, pp. 29–50.

____. *Power in Organizations*. Marshfield, MA: Pitman, 1981.

Phillips, Derek. "Rejection: A Possible Consequence of Seeking Help for Mental Disorders." *American Sociological Review* 28 (December 1963):963–972.

____. "Rejection of the Mentally Ill: The Influence of Behavior and Sex." *American Sociological Review* 29 (October 1964):679–687.

Phillips, John L. *The Origins of the Intellect*. 2d ed. San Francisco: Freeman, 1975.

Phillips, Kevin. *The Politics of the Rich and Poor*. New York: Harper Perennial, 1991.

____. *Boiling Point*. New York: Random House, 1993.

Piaget, Jean. *The Psychology of Intelligence*. Boston: Routledge & Kegan Paul, 1950.

Piaget, Jean, and Barbel Inhelder. *The Psychology of the Child*. New York: Basic Books, 1969.

Pichanik, V. K. *Harriet Martineau*. Ann Arbor: University of Michigan Press, 1980.

Pillemer, K., and D. Finkelhor. "The Prevalence of Elder Abuse: A Random Sample Survey." *The Gerontologist* 28 (1988):51–57.

Pillemer, V., and J. Suiter. "Will I Ever Escape My Children's Problems? Effects of Adult Children's Problems on Elderly Parents." *Journal of Marriage and the Family* (August 1991):585–594.

Pincus, Fred L. "The False Promises of Community Colleges: Class Conflict and Vocational Education." *Harvard Education Review* 50 (August 1980):332–361.

____. "Contradictory Effects of Customized Contract Training in Community Colleges." *Critical Sociology* (Spring 1989):77–91.

Pines, Maya. "The Civilizing of Genie." *Psychology Today* 15 (September 1981):28ff.

Piven, Frances Fox, and Richard A. Cloward. *Poor People's Movements*. New York: Vintage Books, 1979.

Plank, William. *Gulag 65*. New York: Peter Lang, 1989.

Plog, Fred, and Daniel G. Bates. *Cultural Anthropology*. 3d ed. New York: McGraw-Hill, 1990.

Poddel, Lawrence. "Welfare History and Expectancy." *Families on Welfare in New York City*. Preliminary Report No. 5. New York: The Center for Social Research, City University of New York, 1968.

Polk, Kenneth. Book Review of Michael R. Gottfredson and Travis Hirschi, *A General Theory of Crime* (Stanford, CA: Stanford University Press, 1990). In *Crime and Delinquency* 37 (October 1991):575–579.

Pollner, Melvin. "Left of Ethnomethodology: The Rise and Decline of Radical Reflexivity." *American Sociological Review* 56 (June 1991):370–380.

Polonko, Karen A., John Scanzoni, and Jay D. Teachman. "Childlessness and Marital Satisfaction: A Further Assessment." *Journal of Family Issues* 3 (December 1982):545–573.

Ponnuru, Ramesh. "Affirmative Reaction." *National Review* 43 (October 13, 1997):52–56.

Pontell, Henry N. *A Capacity to Punish*. Bloomington: Indiana University Press, 1984.

Pope, Liston. *Millhands and Preachers*. New Haven, CT: Yale University Press, 1942.

Popenoe, David, Jean Bethke Elshtain, and David Blankenhorn. *Promises to Keep: Decline and Renewal of Marriage in America*. Lantham, MD: Rowman and Littlefield Publishers, 1996.

Popenoe, David. *Life Without Father*. New York: Free Press, 1996.

Popkin, James, and Katia Hetter. "America's Gambling Game." *U.S. News & World Report* (March 14, 1994):42–46.

Porter, Bruce, and Marvin Dunn. *The Miami Riot of 1980*. Lexington, MA: Lexington Books, 1984.

Portes, Alejandro, and Cynthia Truelove. "Making Sense of Diversity: Recent Research on Hispanic Minorities in the United States." *Annual Review of Sociology* 3 (1987):359–416.

Posner, Richard A. *Aging and Old Age*. Chicago: University of Chicago Press, 1996.

Postel, Sandra. "Carrying Capacity: Earth's Bottom Line." In Lester R. Brown (ed.), *State of the World*. New York: Norton, 1994, pp. 3–21.

Powell, Bill. "Eyes on the Future." *Newsweek* (May 31, 1993):39–41.

Powell, Brian, and Jerry A. Jacobs. "Gender Differences in the Evaluation of Prestige." *Sociological Quarterly* 25 (Spring 1984a):173–190.

____. "The Prestige Gap: Differential Evaluations of Male and Female Workers." *Work and Occupations* 11 (August 1984b):283–308.

Powell, Lawrence Alfred, John B. Williamson, and Kenneth J. Branco. *Senior Rights Movement*. Old Tappan, NJ: Twayne Publishers, 1996.

Puente, Maria. "Election of Governor a Sign of Growing Political Clout." *USA Today* (November 19, 1996):1A.

Purcell, Piper, and Lara Stewart. "Dick and Jane in 1989." *Sex Roles* 22 (1990):177–185.

Quarantelli, E. L., and Joseph Cooper. "Self-Conceptions and Others: A Further Test of the Meadian Hypothesis." *Sociological Quarterly* 7 (Summer 1966):281–297.

Quebedeaux, Richard. *The Worldly Evangelicals*. New York: Harper & Row, 1978.

Queen, Stuart A., Robert W. Habenstein, and Jill S. Quadagno. *The Family in Various Cultures*. 5th ed. New York: Harper & Row, 1985.

Quinn, Jane Bryant. "What's for Dinner, Mom?" *Newsweek* (April 5, 1993a):68.

____. "The Taxpayers vs. Higher Education." *Newsweek* (November 15, 1993b):51.

Quinney, Richard. *Class, State and Crime*. 2d ed. New York: Longman, 1980.

Radin, Paul. *The World of Primitive Man*. New York: Henry Schuman, 1953.

Raine, Adrian (ed.). *The Psychopathology of Crime*. San Diego, CA: Academic Press, 1993.

Rainwater, Lee. "The Lesson of Pruitt-Igoe." *The Public Interest* 8 (Summer 1967):116–126.

Ramey, Craig T., and Frances Campbell. "Poverty, Early Childhood Education, and Academic Competence: The Abecedarian Experiment." In Aletha C. Huston (ed.), *Children in Poverty*. New York: Cambridge University Press, 1991, pp. 190–221.

Rankin, John. "A Call for 'Higher Order' Thinking." *Washington Post Education Review* (August 4, 1991):19–22.

Ravitch, Diane. "What Makes a Good School?" *Society* 19 (January/February 1982): 10–11.

____. *The Troubled Crusade*. New York: Basic Books, 1984.

Rebach, Howard M., and John G. Bruhn. *Handbook of Clinical Sociology*. New York: Plenum, 1991.

Redfield, Robert. *The Folk Culture of Yucatan*. Chicago: University of Chicago Press, 1941.

Reed, Ralph. *Active Faith*. New York: Free Press, 1996.

Regan, Tom (ed.). *Earthbound*. Prospect Heights, IL: Waveland Press, 1990.

Reibstein, Larry, and Nancy Hass. "Rupert's Power Play." *Newsweek* (June 6, 1994):46–48.

Reich, Robert B. "As the World Turns." *The New Republic* (May 1, 1989):23, 26–28.

____. *The Work of Nations*. New York: Knopf, 1993.

____. *Locked in the Cabinet*. New York: Knopf, 1997.

Reich, Robert B., and John D. Donahue. *New Deals*. New York: Penguin, 1986.

Reilly, Bernard, and Joseph A. DiAngelo. "A Look at Job Redesign." *Personnel* (February 1988):61–65.

Reiman, Jeffrey H. *The Rich Get Richer and the Poor Get Prison*. 3d ed. New York: Wiley, 1990.

Reinharz, Shulamit. *On Becoming a Social Scientist*. New Brunswick, NJ: Transaction Books, 1984.

Reischauer, Edwin O. *The Japanese Today*. Cambridge, MA: Harvard University Press, 1988.

Reiser, Stanley Joel. *Medicine and the Reign of Technology*. New York: Cambridge University Press, 1978.

Reiss, Ira L. *Family Systems in America*. 2d ed. Hinsdale, IL: Dryden Press, 1976.

"The Religion Vote in '92." Princeton, NJ: Princeton Religion Research Center (November 1992):3–4.

Relman, Arnold S. "Dr. Business." *American Prospect* (September/October, 1997):91–95.

Remnick, David. "Soviets Pass Law on Privatization." *Washington Post* (July 2, 1991):A1, A15.

Renne, Karen. "Correlates of Dissatisfaction in Marriage." *Journal of Marriage and the Family* 32 (February 1970):54–67.

____. "Childlessness, Health, and Marital Satisfaction." *Social Biology* 23 (Fall 1976):183–197.

Report of the National Advisory Commission on Civil Disorders. Washington, DC: U.S. Government Printing Office, 1968.

Report on Evaluation Studies of Project Head Start. Office of Child Development. U.S. Department of Health, Education, and Welfare, 1970.

Reskin, Barbara. "Sex Segregation in the Workplace." *Annual Review of Sociology* 19 (1993):241–270.

Reskin, Barbara, and Irene Padavic. *Women and Men at Work*. Thousand Oaks, CA: Pine Forge Press, 1994.

Rezvin, Philip. "Ventures in Hungary Test Theory That West Can Uplift East Bloc." *Wall Street Journal* (April 5, 1990):1.

Rheingold, Harriet L., and Kay V. Cook. "The Content of Boys' and Girls' Rooms as an Index of Parents' Behaviors." *Child Development* 46 (1975):459–463.

Richardson, James T. "Conversion, Brainwashing, and Deprogramming." *The Center Magazine* 15 (March/April 1982):18–24.

Richardson, Laurel, and Verta Taylor (eds.). *Feminist Frontiers*. Reading, MA: Addison-Wesley, 1983.

"The 400 Richest People in America." *Forbes* (October 24, 1988):142ff.

Ridgeway, Cecelia L. *The Dynamics of Small Groups*. New York: St. Martin's Press, 1983.

Ridley, Matt. *The Origins of Virtue*. New York: Viking, 1996.

Riedesel, Paul L. "Who *Was* Harriet Martineau?" *Journal of the History of Sociology* 3 (Spring/Summer 1981):63–80.

Riesebrodt, Martin. *Pious Passion.* Chapel Hill: University of North Carolina Press, 1993.

Riesman, David. *The Lonely Crowd.* New Haven, CT: Yale University Press, 1961.

____. *On Higher Education.* San Francisco: Jossey-Bass, 1980.

Rifkin, Jeremy. *The End of Work.* New York: Tarcher/Putnam, 1995.

Riley, Nancy E. *Gender, Power, and Population Change.* Washington, DC: Population Reference Bureau, 1997.

Rist, Ray C. "On Understanding the Process of Schooling: The Contributions of Labeling Theory." In Jerome Kagan and A. H. Halsey (eds.), *Power and Ideology in Education.* New York: Oxford University Press, 1977, pp. 292–305.

Ritzer, George. *The McDonaldization of Society.* Thousand Oaks, CA: Pine Forge Press, 1993.

____. *Sociological Theory.* 4th ed. New York: McGraw-Hill, 1996.

Ritzer, George, and David Walczak. *Working: Conflict and Change.* 3d ed. Englewood Cliffs, NJ: Prentice Hall, 1986.

Robbins, Stephen P. *Organization Theory.* 3d ed. Englewood Cliffs, NJ: Prentice Hall, 1990.

Robbins, Thomas. *Cults, Converts and Charisma.* London and Newbury Park, CA: Sage, 1988.

Robbins, Thomas, Dick Anthony, and James Richardson. "Theory and Research on Today's New Religions." *Sociological Analysis* 39 (Summer 1978):95–122.

Rockett, Ian R. H. "Population and Health: An Introduction to Epidemiology." *Population Bulletin* 49 (November 1994):1–47.

Rodman, Hyman. "The Lower-Class Value Stretch." *Social Forces* 42 (December 1963):205–215.

Roethlisberger, F. J., and William J. Dickson. *Management and the Worker.* New York: Wiley, 1964.

Rogers, Patrick. "Think of It as Therapy." *Newsweek* (May 31, 1993):65.

Rohlen, Thomas P. *For Harmony and Strength.* Berkeley: University of California Press, 1974.

Rojek, Dean G., and Maynard L. Erickson. "Reforming the Juvenile Justice System: The Diversion of Status Offenders." *Law and Society Review* 16 (1981–82):241–264.

Roof, Wade Clark. *A Generation of Seekers.* San Francisco: Harper San Francisco, 1994.

Roper Organization. *Virginia Slims Poll.* New York: Richard Weiner, 1985.

Rose, Jerry D. *Outbreaks.* New York: Free Press, 1982.

Rose, Peter I. *They and We.* New York: McGraw-Hill, 1980.

Rosenbaum, Jill. "Social Control, Gender and Delinquency: An Analysis of Drug, Property and Violent Offenders." *Justice Quarterly* 4 (1987):117–132.

Rosenfeld, Anne, and Elizabeth Stark. "The Prime of Our Lives." *Psychology Today* (May, 1987):62–72.

Rosenfield, Sarah. "The Effects of Women's Employment: Personal Control and Sex Differences in Mental Health." *Journal of Health and Social Behavior* 30 (March 1989):77–91.

Rosenthal, Marilynn. "Where Rumors Raged." *Transaction* 8 (February 1971):34–43.

Rosenthal, Robert, and Lenore Jacobson. *Pygmalion and the Classroom.* New York: Irvington Publishers, 1989.

Rosenthal, Robert, and D. B. Rubin. "Interpersonal Expectancy Effects: The First 345 Studies." *The Behavioral and Brain Sciences* 3 (1978):377–415.

Rosnow, I. "Status and Role Change Through the Life Cycle." In R. Binstock and E. Shanas (eds.), *Handbook of Aging and the Social Sciences.* New York: Van Nostrand, 1985, pp. 62–93.

Rosnow, Ralph L. "Rumor as Communication: A Contextualist Approach." *Journal of Communication* 38 (Winter 1988):12–28.

____. "Inside Rumor: A Personal Journey." *American Psychologist* 46 (May 1991):484–496.

Rosnow, Ralph L., and Gary Alan Fine. *Rumor and Gossip.* New York: Elsevier, 1976.

Rossi, Alice S. (ed.). *The Feminist Papers.* New York: Bantam, 1974.

Rossides, Daniel W. *The American Class System.* Washington, DC: University Press of America, 1980.

____. *Social Stratification.* Englewood Cliffs, NJ: Prentice Hall, 1990a.

____. *Comparative Societies.* Englewood Cliffs, NJ: Prentice Hall, 1990b.

Roszak, Theodore. *The Making of a Counter Culture.* Berkeley and Los Angeles: University of California Press, 1995.

Rothschild, Joyce. "Alternatives to Bureaucracy: Democratic Participation in the Economy." *Annual Review of Sociology* 12 (1986):307–328.

Rothschild, Joyce, and J. Allen Whitt. *The Cooperative Workplace.* New York: Cambridge University Press, 1987.

Rothschild-Whitt, Joyce. "The Collectivistic Organization: An Alternative to Rational Bureaucratic Models." *American Sociological Review* 44 (August 1979):509–527.

Rottenberg, Linda D. (ed.). "America's Health Care: Which Road to Reform?" *Yale Law & Policy Review* 10 (1992):205–541.

Rotundo, E. Anthony. *American Manhood*. New York: Basic Books, 1993.

Rousseau, Jean-Jacques. *The Social Contract and Discourses*. Translation and Introduction by G.D.H. Cole. London: Everyman's Library, 1983.

Rowe, David C. "Biometrical Models of Self-Reported Delinquent Behavior: A Twin Study." *Behavior Genetics* 13 (1983):473–489.

____. "Genetic and Environmental Components of Antisocial Behavior: A Study of 265 Twin Pairs." *Criminology* 24 (August 1986):513–532.

Rowen, Hobart. "Scandal, Japanese Style: Looking the Other Way." *Washington Post* (June 30, 1991):Hl, H6.

Rubenstein, Carin. "Guilty or Not Guilty." *Working Mother* (May 1991):53–56.

Rubin, Jeffrey Z., Frank J. Provenzano, and Zella Luria. "The Eye of the Beholder: Parents' Views on Sex of Newborns." *American Journal of Orthopsychiatry* 44 (1974):512–519.

Rubin, Lillian B. *Worlds of Pain*. New York: Basic Books, 1992.

____. *Families on the Faultline*. New York: HarperCollins, 1994.

Ruesch, Hans. *Top of the World*. New York: Pocket Books, 1959.

Ruggiero, Josephine A., and Louise C. Weston. "Marketing the B.A. Sociologist: Implications from Research on Graduates, Employers, and Sociology Departments." *Teaching Sociology* 14 (October 1986):224–233.

Ruggiero, Vincent Ryan. *A Guide to Sociological Thinking*. Thousand Oaks, CA: Sage, 1995.

Rule, James B. "Rationality and Nonrationality in Militant Collective Action." *Sociological Theory* 7 (1989):145–160.

Russell, Alan. *Guiness Book of World Records*. New York: Sterling, 1987.

Russell, Anne. "The Twelfth Annual Working Woman Salary Survey." *Working Woman* (January 1991):68.

Ryan, William. *Blaming the Victim*. (rev. and updated ed.). New York: Vintage, 1976.

Sachs, Jeffrey. "Management Positions Available: Call Warsaw." *Wall Street Journal* (March 21, 1990).

Sadker, Myra, and David Sadker. "Sexism in the Schoolroom of the 80's." *Psychology Today* 19 (March 1985):54–57.

____. *Failing at Fairness*. New York: Simon & Schuster, 1995.

Sagoff, Mark. "Do We Consume Too Much?" *The Atlantic Monthly* (June 1997):80–96.

Sahlins, Marshall D. *Tribesmen*. Englewood Cliffs, NJ: Prentice Hall, 1968.

Sakamoto, Arthur, and Meichu D. Chen. "Inequality and Attainment in a Dual Labor Market." *American Sociological Review* 56 (1991):295–308.

Salamans, S. "The Tracking Controversy." *New York Times Education Supplement* (April 10, 1988):56–62.

Salholz, Eloise. "A Conflict of the Have-Nots." *Newsweek* (December 12, 1988):28–29.

Sampson, Robert J., and John H. Laub. "Crime and Deviance Over the Life Course: The Salience of Adult Social Bonds." *American Sociological Review* 55 (October 1990):609–627.

Samuda, Ronald. *The Psychological Testing of American Minorities*. New York: Dodd, Mead, 1975.

Samuelson, Paul A., and William D. Nordhaus. *Microeconomics*. 15th ed. New York: McGraw-Hill, 1995.

Sanders, Catherine M. *Surviving Grief*. New York: Wiley, 1992.

Sanderson, Stephen K. *Social Evolutionism*. New York: Basil Blackwell, 1990.

Sanoff, Alvin P. "Jews Find New Solace in the Old Traditions." *U.S. News & World Report* (April 4, 1983).

Sapir, Edward. "The Status of Linguistics as a Science." *Language* 5 (1929):207–214.

Sapiro, Virginia (ed.). *Women, Biology, and Public Policy*. Beverly Hills, CA: Sage, 1985.

Sapolsky, Robert. "A Gene for Nothing." *Discover* (October 1997):40–46.

Sayles, Leonard R., and George Strauss. *The Local Union*. (rev. ed.). New York: Harcourt Brace Jovanovich, 1967.

Scarf, Maggie. "The More Sorrowful Sex." *Psychology Today* 12 (April 1979):45–52.

Schaefer, Richard T. *Racial and Ethnic Groups*. 6th ed. New York: Longman, 1993.

Schaeffer, Robert K. *Understanding Globalization*. Lanham, MD: Rowman & Littlefield Publishers, 1997.

Scheff, Thomas. "The Labeling Theory of Mental Illness." *American Sociological Review* 39 (June 1974):444–452.

____. *Being Mentally Ill: A Sociological Theory*. 2d ed. New York: Aldine, 1984.

____. *Microsociology*. Chicago: University of Chicago Press, 1990.

Schell, Orville. "Deng's Revolution." *Newsweek* (August 22, 1997):21–27.

Schellenberg, James. A. *The Science of Conflict*. New York: Oxford University Press, 1982.

Scheper-Hughes, Nancy. "Deposed Kings: The Demise of the Rural Irish Gerontocracy." In Jay Sokolorsky (ed.), *Growing Old in Different Societies*. Belmont, CA: Wadsworth, 1983, pp. 130–146.

Schiff, Michel, and Richard Lewontin. *Education and Class.* New York: Oxford University Press, 1987.

Schiller, Bradley R. "Empirical Studies of Welfare Dependency: A Survey." *The Journal of Human Resources* 8 (Supplement 1973):19–32.

Schlegel, Kip, and David Weisburd (eds.). *White-Collar Crime Reconsidered.* Boston, MA: Northeastern University Press, 1994.

Schnitzer, Martin C. *Comparative Economic Systems.* 7th ed. Cincinnati, OH: South-Western College Publishing, 1997.

"Schooling and Success." *Focus* 3 (Summer 1982):9–14.

Schrag, Peter. "The Forgotten American." In John Walton and Donald E. Carns (eds.), *Cities in Change.* 2d ed. Boston: Allyn and Bacon, 1977, pp. 129–137.

Schriesheim, Chester A., and Steven Kerr. "Theories and Measures of Leadership: A Critical Appraisal of Current and Future Directions." In James G. Hunt and Lars L. Larson (eds.), *Leadership.* Carbondale: Southern Illinois University Press, 1977, pp. 9–45.

Schuman, Howard. "The Religious Factor in Detroit: Review, Replication and Reanalysis." *American Sociological Review* 36 (February 1971):30–48.

Schutte, N. S., J. M. Malouff, J. C. Post-Corden, and A. L. Rodasta. "Effects of Playing Videogames on Children's Aggressive and Other Behaviors." *Journal of Applied Psychology* 18 (1988):454–460.

Schwandt, Thomas A. *Qualitative Inquiry: A Dictionary of Terms.* Thousand Oaks, CA: Sage, 1997.

Schwartz, Barry. *The Battle for Human Nature.* New York: Norton, 1987.

Schwarz, John E., and Thomas J. Volgy. *The Forgotten Americans.* New York: Norton, 1992.

Schweinhart, Lawrence J. "How Much Do Good Early Childhood Programs Cost?" *Early Childhood Education and the Public Schools* 3 (April 1992):115–127.

Scimecca, Joseph A. "Humanist Sociological Theory: The State of the Art." *Humanity and Society* 11 (1987):335–352.

Scott, Alan. *Ideology and the New Social Movements.* London: Routledge, 1995.

Scott, W. Richard. *Organizations.* 3d ed. Englewood Cliffs, NJ: Prentice Hall, 1992.

Seefeldt, Carol, Richard K. Jantz, Alice Galper, and Kathy Serock. "Using Pictures to Explore Children's Attitudes Toward the Elderly." *The Gerontologist* 17 (December 1977):506–512.

Sellin, Thorsten. *The Penalty of Death.* Beverly Hills, CA: Sage, 1991.

Selznick, Philip. *The Moral Commonwealth.* Berkeley: University of California Press, 1992.

Sennett, Richard, and Jonathan Cobb. *The Hidden Injuries of Class.* New York: Vintage, 1973.

Service, Elman R. *Origins of the State and Civilization.* New York: Norton, 1975.

____. *The Hunters.* 2d ed. Englewood Cliffs, NJ: Prentice Hall, 1979.

Sewell, William H. "Inequality of Opportunity for Higher Education." *American Sociological Review* 36 (October 1971):793–809.

Shaller, Elliot H., and Mary K. Qualiana. "The Family and Medical Leave Act—Key Provisions and Potential Problems." *Employee Relations Law Journal* 19 (Summer 1993):5–22.

Shanas, Ethel, et al. *Old People in Three Industrial Societies.* New York: Atherton, 1968.

Shapiro, Laura. "Guns and Dolls." *Newsweek* (May 28, 1990):54–65.

____. "Denying the Holocaust." *Newsweek* (December 20, 1993):120.

Shapiro, Margaret. "A Glimmer of Hope for Russia's Economy." *Washington Post* (June 21, 1994):A11.

Sharrock, Wes, and Bob Anderson. *The Ethnomethodologists.* New York: Tavistock, 1986.

Shaw, Clifford, and Henry McKay. *Delinquency Areas.* Chicago: University of Chicago Press, 1929.

Sheldon, William H. *Varieties of Delinquent Youth.* New York: Harper, 1949.

Shepard, Jon M. *Automation and Alienation.* Cambridge, MA: MIT Press, 1971.

____. "Technology, Alienation, and Job Satisfaction." In Alex Inkeles, James Coleman, and Neil Smelser (eds.), *Annual Review of Sociology* 3 (1977):1–21.

Shepard, Jon M., and James G. Hougland, Jr. "Contingency Theory: 'Complex Man' or 'Complex Organization'?" *Academy of Management Review* 3 (July 1978):413–427.

Shepard, Jon M., Jon Shepard, and Richard E. Wokutch. "The Problem of Business Ethics: Oxymoron or Inadequate Vocabulary." *Journal of Business and Psychology* 6 (1991):9–23.

Shepard, Jon M., Carroll U. Stephens, Virginia W. Gerde, and Michael Goldsby. "The Protestant Ethics and the Spirits of Capitalism: An Extension of Max Weber's Contribution to Economic Sociology." Unpublished manuscript, 1997.

Shepherd, William. "Legal Protection for Freedom of Religion." *The Center Magazine* (March/April 1982):30–37.

Sherman, Lawrence. "Defiance, Deference, and Irrelevance: A Theory of the Criminal Sanction." *Journal of Research in Crime and Delinquency* 30 (November 1993a):445–473.

____. "A Brave New Darwinian Workplace." *Fortune* (February 25, 1993b):50.

Shichor, David, and Dale K. Sechrest (eds.). *Three Strikes and You're Out.* Newbury Park, CA: Sage, 1996.

Shils, Edward A., and Morris Janowitz. "Cohesion and Disintegration in the Wehrmacht in World War II." *Public Opinion Quarterly* 12 (Summer 1948):280–315.

Sidel, Ruth. *Keeping Women and Children Last*. New York: Penguin, 1996.

Sifford, Darrell. *The Only Child*. New York: G. P. Putnam's Sons, 1989.

Silber, John R. "Apocalypses Then and Now: The Peace Movement and the Antinuclear Crusade." *Public Opinion* 5 (August/September 1982):42–46.

Silberman, Charles E. *Crisis in the Classroom*. New York: Vintage, 1971.

"The Silent Boom." *Forbes* (July 7, 1997):170–171.

Simmel, Georg. *Conflict and the Web of Group Affiliation*. Translated by Kurt H. Wolff. New York: Free Press, 1964.

Simmons, J. L. *Deviants*. Berkeley, CA: Glendessary Press, 1969.

Simon, David R., and D. Stanley Eitzen. *Elite Deviance*. Boston: Allyn and Bacon, 1982.

Simon, Herbert A. *Administrative Behavior*. 3d ed. New York: Free Press, 1976.

Simpson, George Eaton. *Emile Durkheim*. New York: Crowell, 1963.

Simpson, George Eaton, and J. Milton Yinger. *Racial and Cultural Minorities*. 5th ed. New York: Harper & Row, 1985.

Sinclair, Karen. "Cross-Cultural Perspectives on American Sex Roles." In Marie Richmond-Abbott (ed.), *The American Woman*. New York: Holt, Rinehart and Winston, 1979, pp. 28–47.

Singh, B. Krishna, and J. Sherwood Williams. "Childlessness and Family Satisfaction." *Research on Aging* 3 (June 1981):218–227.

Singleton, Royce A. Jr., Bruce C. Straits, and Margaret Miller Straits. *Approaches to Social Research*. 2d ed. New York: Oxford University Press, 1993.

Siranni, Carmen (ed.). *Worker Participation and the Politics of Reform*. Philadelphia: Temple University Press, 1987.

Sizer, Theodore. *Horace's Hope: What Works for the American High School*. Boston: Houghton Mifflin, 1996.

Sjoberg, Gideon. *The Preindustrial City*. New York: Free Press, 1960.

Skinner, Denise A. "Dual-Career Families: Strains of Sharing." In Hamilton I. McCubbin and Charles R. Figley (eds.), *Stress and the Family*. Vol. 1. New York: Brunner/Mazel, 1983, pp. 90–101.

Skocpol, Theda. "Bringing the State Back In: Strategies of Analysis in Current Research." In Peter B. Evans, Dietrich Rueschemeyer, and Theda Skocpol (eds.), *Bringing the State Back In*. Cambridge, England: Cambridge University Press, 1985, pp. 3–37.

Skocpol, Theda, and Edwin Amenta. "States and Social Policies." In Ralph H. Turner and James F. Short, Jr. (eds.), *Annual Review of Sociology* 12 (1986):131–157.

____. *Disorder and Crime*. Berkeley: University of California Press, 1992.

Skolnick, Arlene S., and Jerome H. Skolnick. *Family in Transition*. 8th ed. Boston: Little, Brown, 1994.

Sloan, Allan. "The Hit Men." *Newsweek* (February 26, 1996):44–48.

____. "The New Rich." *Newsweek* (August 4, 1997):48–55.

Smelser, Neil J. *Theory of Collective Behavior*. New York: Free Press, 1971.

____. *The Sociology of Economic Life*. 2d ed. Englewood Cliffs, NJ: Prentice Hall, 1976.

Smith, Adam. *An Inquiry into the Nature and Causes of the Wealth of Nations*. Edwin Cannan (ed.). New York: Modern Library, 1965.

Smith, Hedrick. *The Russians*. New York: Quadrangle, 1976.

Smith, Lee. "Burned-Out Bosses." *Fortune* (July 25, 1994):44–52.

Smith, Tom W. "Ethnic Images." General Social Survey Project Topical Report No. 19, December 1990.

Smolowe, Jill. "Sex with a Scorecard." *Time* (April 5, 1993):41.

____. "When Violence Hits Home." *Time* (July 4, 1994):18–25.

Snarr, Brian B. "The Family and Medical Leave Act of 1993." *Compensation and Benefits Review* 25 (May/June 1993):6–9.

Snider, Mike. "Ozone Loss Measured Over North America." *USA Today* (April 22, 1993):1A.

Snipp, C. Matthew. "Sociological Perspectives on American Indians." *Annual Review of Sociology* 18 (1992):351–371.

____. "A Demographic Comeback for American Indians?" *Population Today* 24 (November 1996):4–5.

Snizek, William. "Survivors as Victims: Some Little Publicized Consequences of Corporate Downsizing." Paper Presented to the Ministerie van Binnenlandse Zaken Den Haag, The Netherlands, July 1994.

Snyder, M., and W. B. Swann, Jr. "Hypothesis-Testing Processes in Social Interaction." *Journal of Personality and Social Psychology* 36 (1978):1202–1212.

Sokoloff, Natalie J. *Between Money and Love*. New York: Praeger, 1980.

Sokolovsky, Jay (ed.). *Growing Old in Different Societies*. Belmont, CA: Wadsworth, 1983.

Solomon, Jolie. "Smoke From Washington." *Newsweek* (April 4, 1994):45.

____. "Texaco's Troubles." *Newsweek* (November 25, 1996):48–50.

Sorensen, Elaine. *Comparable Worth*. Princeton, NJ: Princeton University Press, 1994.

Sorokin, Pitirim A. *Social Mobility*. New York: Harper, 1927.

____. *Social and Cultural Dynamics*. New York: American Book Company, 1941.

South, Scott J., and Kyle D. Crowder. "Escaping Distressed Neighborhoods: Individual, Community, and Metropolitan Influences." *American Journal of Sociology* 102 (January 1997):1040–1084.

Spector, Malcolm, and John I. Kitsuse. *Constructing Social Problems*. Menlo Park, CA: Cummings, 1977.

Speizer, Jeanne J. "Education." In Barbara Haber (ed.), *The Women's Annual 1982–1983*. Boston: G. K. Hall, 1983.

Spencer, Herbert. *The Principles of Sociology*. New York: Appleton, 1898.

Spencer, Jon Michael. *The New Colored People*. New York: New York University Press, 1997.

Spengler, Oswald. *The Decline of the West*. New York: Knopf, 1926–1928.

Spickard, Paul R. "The Illogic of American Racial Categories." In Maria P. P. Root (ed.), *Racially Mixed People in America*. London: Sage, 1992.

Spindler, George D., and Louise Spindler. "Anthropologists View American Culture." *Annual Review of Anthropology* 12 (1983):49–78.

Spitz, René A. "Hospitalism." In Anna Freud et al. (eds.), *The Psychoanalytic Study of the Child*. Vol. 2. New York: International Universities Press, 1946a, pp. 52–74.

____. "Hospitalism: A Follow-Up Report." In Anna Freud et al. (eds.), *The Psychoanalytic Study of the Child*. Vol. 2. New York: International Universities Press, 1946b, pp. 113–117.

Spitze, B., and J. Logan. "Sons, Daughters, and Intergenerational Social Support." *Journal of Marriage and the Family* (May 1990):420–430.

Spitze, G. "Women's Employment and Family Relations: A Review." In A. Booth (ed.), *Contemporary Families*. Minneapolis: National Council on Family Relations, 1991, pp. 381–404.

Spitzer, Steven. "Toward a Marxian Theory of Deviance." In Delos H. Kelly (ed.), *Criminal Behavior*. New York: St. Martin's Press 1980, pp. 175–191.

Spohn, Cassia C. "Courts, Sentences, and Prisons." *Daedalus* 124 (Winter 1995):119–143.

Spragins, Ellyn. "Does Your HMO Stack Up?" *Newsweek* (June 24, 1996):56–63.

Srole, Leo T., et al. *Mental Health in the Metropolis*. New York: McGraw-Hill, 1962.

Stacey, Judith. *Brave New Families*. New York: Basic Books, 1990.

Stacey, William, and Anson Shupe. "Correlates of Support for the Electronic Church." *Journal for the Scientific Study of Religion* 21 (December 1982):291–303.

Stanley, Thomas J., and William D. Danko. *The Millionaire Next Door*. Atlanta, GA: Longstreet Press, 1996.

Stark, Rodney, and William Sims Bainbridge. "Secularization and Cult Formation in the Jazz Age." *Journal for the Scientific Study of Religion* 20 (December 1981):360–373.

____. *The Future of Religion*. Berkeley: University of California Press, 1985.

Stark, Rodney, Lori Kent, and Daniel P. Doyle. "Religion and the Ecology of a 'Lost' Relationship." *Journal of Research in Crime and Delinquency* 19–20 (January 1982):4–24.

Starr, Mark. "I'm So Scared." *Newsweek* (January 17, 1994a):40–43.

Starr, Mark, and Allison Samuels. "Ear Today, But Gone Tomorrow." *Newsweek* (July 14, 1997):58–60.

Starr, Paul E. *The Social Transformation of American Medicine*. New York: Basic Books, 1982.

Starting Points. The Report of the Carniege Task Force on Meeting the Needs of Young Children. Carnegie Corporation of New York, April, 1994.

State of Florida Hospital Cost Containment Board. *Hospital Cost Containment Board: Annual Reports*. Tallahassee: Florida Cost Containment Board, 1981–1982, 1982–1983.

"Statement on Race and Intelligence." *Journal of Social Issues* 25 (Summer 1969):1–3.

Stavans, Ilan. *The Hispanic Condition*. New York: Harper Collins, 1996.

Steedley, Gilbert. "The Forbes 500." *Forbes* 153 (April 25, 1994):195–230.

Stein, Leonard I. "The Doctor-Nurse Game." *Archives of General Psychiatry* 16 (June 1967):699–700.

Steinmetz, Suzanne K. *Duty Bound: Elder Abuse and Family Care*. Newbury Park, CA: Sage, 1988.

Steinnes. Donald N. "Suburbanization and the 'Malling of America': A Time-Series Approach." *Urban Affairs Quarterly* 17 (June 1982):401–418.

Stern, Philip M. *The Best Congress Money Can Buy*. New York: Pantheon, 1988.

Stern, Robert F. "Participation by Representation." *Work and Occupations* 15 (November 1988):396–422.

Steward, Julian H. *Theory of Culture Change*. Urbana: University of Illinois Press, 1979.

Stinchcombe, Arthur L., et al. *Crime and Punishment-Changing Attitudes in America*. San Francisco: Jossey-Bass, 1980.

Stoessinger, John. "The Anatomy of the Nation-State and the Nature of Power." In Michael Smith, Richard Little, and Michael Shackleton (eds.), *Perspectives on World Politics*. London: Croom Helm, 1981, pp. 25–26.

Stogdill, Ralph M. *Handbook of Leadership*. New York: Free Press, 1974.

Stoltenberg, John, and Hanna McCrum. "The Seventh Annual Working Woman Salary Survey." *Working Woman* (January 1986):73–82.

Stone, Robyn. *The Feminization of Poverty and Older Women*. Washington, DC: U.S. Government Printing Office, 1986.

Stranger, Janice, Nicole C. Batchelder, William Brossman, and Gerald L. Uslarder. "What the Family and Medical Leave Act Means for Employers." *Journal of Compensation and Benefits* 9 (July/August, 1993):12–19.

Strasburger, Victor C. *Adolescents and the Media: Medical and Psychological Impact*. Newbury Park, CA: Sage, 1995.

Straus, Murray A., Richard J. Gelles, and Suzanne K. Steinmetz. *Behind Closed Doors*. Garden City, NY: Doubleday, 1980.

Straus, Roger A. *Using Sociology*. 2d ed. Dix Hills, NY: General Hall, 1994.

Strauss, Anselm, Shizuko Fagerhaugh, Barbara Suczek, and Carolyn Wiener. *Social Organization of Medical Work*. Chicago: University of Chicago Press, 1985.

Strieb, Gordon F. "Are the Aged a Minority Group?" In Bernice L. Neugarten (ed.), *Middle Age and Aging*. Chicago: University of Chicago Press, 1968, pp. 35–46.

Suchman, Edward A. "The 'Hang-Loose' Ethic and the Spirit of Drug Use." *Journal of Health and Social Behavior* 9 (June 1968):146–155.

Sullivan, Stacy. "Genocide Without Corpses." *Newsweek* (November 4, 1996):37.

Sumner, William Graham. *Folkways*. Boston: Ginn, 1906.

Susser, Ida. *Norman Street*. New York: Oxford University Press, 1982.

Sussman, Marvin B. "The Help Pattern in the Middle Class Family." *American Sociological Review* 18 (February 1953):22–28.

Sutherland, Edwin H. "White Collar Criminality." *American Sociological Review* 5 (1940):1–12.

____. *White-Collar Crime*. New Haven, CT: Yale University Press, 1983.

Sutherland, Edwin H., and Donald R. Cressey. *Principles of Criminology*. 11th ed. Dix Hills, NY: General Hall, 1992.

Suttles, Gerald D. *The Social Order of the Slum*. Chicago: University of Chicago Press, 1968.

Swartz, Thomas R., and Kathleen Maas Weigert (eds.). *America's Working Poor*. Notre Dame, IN: University of Notre Dame Press, 1996.

Sweet, Jones, Larry Bumpass, and Vaugn Call. *National Survey of Families and Households*. Madison: University of Wisconsin, Center for Demography and Ecology, 1988.

Swidler, Ann. "Culture in Action: Symbols and Strategies." *American Sociological Review* 51 (April 1986):273–286.

Swinton, David H. "Economic Status of Blacks." In *The State of Black America 1989*. New York: National Urban League, 1989, pp. 129–152.

Szasz, Thomas. *The Myth of Mental Illness*. (rev. ed.). New York: Harper & Row, 1986.

Szymanski, Albert. "Racial Discrimination and White Gain." *American Sociological Review* 41 (June 1976):403–414.

____. *Class Structure*. New York: Praeger, 1983.

Takaki, T. Ronald. *Strangers from a Different Shore*. 2d ed. Boston: Oxford University Press, 1994.

Tannenbaum, Frank. *Slave and Citizen*. New York: Knopf, 1947.

Tavris, Carol. "Men and Women Report Their Views on Masculinity." *Psychology Today* 10 (January 1977):35ff.

Taylor, Alexander L. "The Growing Gap in Retraining." *Time* (March 28, 1983):50–51.

Taylor, Humphrey. "Greatest Hopes and Fears for 1994." *The Harris Poll* (January 3, 1994a):1–3.

____. "The Canadian and U.S. Health Care Systems Compared." *The Harris Poll* (April 26, 1994b):1–6.

Taylor, P. "Therapists Rethink Attitudes on Divorce: New Movement to Save Marriages Focuses on Impact on Children." *Washington Post* (January 29, 1991):A1, A6.

Taylor, Verta. "Social Movement Continuity: The Women's Movement in Abeyance." *American Sociological Review* 54 (October 1989):761–775.

Teich, Albert H. (ed.). *Technology and the Future*. 6th ed. New York: St. Martin's Press, 1993.

Tellegen, Auke, D. T. Lykken, T. J. Bouchard, Jr., and M. McGue. "Heritability of Interests: A Twin Study." *Journal of Applied Psychology* (August 1993):649–661.

Terkel, Studs. *Working*. New York: Ballantine Books, 1985.

Terry, James L. "Bringing Women In: A Modest Proposal." *Teaching Sociology* 10 (January 1983):251–261.

The Carnegie Commission on Higher Education. *Priorities for Action*. New York: McGraw-Hill, 1973.

The Global 2000 Report to the President. Vols. 1, 2, and 3. A report prepared by the Council on Environmental Quality and the Department of State. Washington, DC: U.S. Government Printing Office, 1980.

The Impact of Head Start. Westinghouse Learning Corporation, Ohio University, July 12, 1969.

"The 1997 Salary Report." *Working Woman* (January 1997):31–76.

The World Almanac and Book of Facts, 1997. Mahwah, NJ: K-111 Reference Corporation, 1996.

The World Almanac of U.S. Politics. Mahwah, NJ: World Almanac Books, 1995.

The World Bank Atlas 1997. Washington, DC: The World Bank, 1997.

Therborn, Goran. "The Prospects of Labor and the Transformation of Advanced Capitalism." *New Left Review* 145 (1984):5–38.

____. *The Power of Ideology and the Ideology of Power.* London: New Left Books, 1982.

Thomas, Evan. "The Next Level." *Newsweek* (April 7, 1997):28–36.

Thomas, Evan, and Gregory L. Vistica. "A Question of Consent." *Newsweek* (April 28, 1977):41.

Thomas, Judy Lundstrom. "Religious Right Infiltrates GOP." *New Haven Register* (September 27, 1992):A11.

Thomas, Melvin E., and Michael Hughes. "The Continuing Significance of Race: A Study of Race, Class, and Quality of Life in America, 1972–1985." *American Sociological Review* 51 (December 1986):830–841.

Thomas, Susan L. *Gender and Poverty.* New York: Garland, 1994.

Thompson, E., and V. Colella. "Cohabitation and Marital Stability: Quality or Commitment." *Journal of Marriage and the Family* (May 1992):259–268.

Thompson, Kenneth. "The Organizational Society." In Graeme Salaman and Kenneth Thompson (eds.), *Control and Ideology in Organizations.* Cambridge, MA: MIT Press, 1980, pp. 3–23.

Thompson, L., and A. J. Walker. "Gender in Families." In A. Booth (ed.), *Contemporary Families.* Minneapolis: National Council on Family Relations, 1991, pp. 76–102.

Thompson, Randall J., and Matthew T. Zingraff. "Detecting Sentencing Disparity: Some Problems and Evidence." *American Journal of Sociology* 86 (1981):869–880.

Thurow, Lester. *Head to Head.* New York: William Morrow, 1992.

Tidwell, Billy J. *The State of Black America, 1994.* Washington, DC: National Urban League, 1994.

Tilly, Charles. *From Mobilization to Revolution.* Reading, MA: Addison-Wesley, 1978.

____. "Does Modernization Breed Revolution?" In Jack A. Gladstone (ed.), *Revolutions.* New York: Harcourt Brace Jovanovich, 1986, pp. 47–57.

____. Book Review of Michael Gottfredson and Travis Hirschi, *A General Theory of Crime* (Stanford University Press, 1990). In *American Journal of Sociology* 96 (May 1991):1609–1611.

____. *Coercion, Capital, and European States, AD 990–1992.* Cambridge, MA: Basil Blackwell, 1992.

____. *Social Processes.* Lantham, MD: Rowan and Littlefield, 1997.

Tilly, Louise, and Joan Scott. *Women, Work, and Family.* New York: Holt, Rinehart and Winston, 1978.

Timms, Duncan. *The Urban Mosaic.* Cambridge, England: Cambridge University Press, 1971.

Tobias, Shella. *Sexual Politics.* Boulder, CO: Westview Press, 1997.

Toch, Thomas, and Ted Slafsky. "The Great College Tumble." *U.S. News & World Report* (June 13, 1991):50.

Tocqueville, Alexis de. *The Old Regime and the French Revolution.* Translated by Stuart Gilbert. Garden City, NY: Doubleday, 1955.

Toffler, Alvin. *The Third Wave.* New York: Bantam Books, 1984.

Toland, John. *Adolph Hitler.* Garden City, NY: Doubleday, 1976.

Tomaskovic-Devey, Donald. *Gender and Racial Inequality at Work.* Ithaca, NY: ILR Press, 1993.

Tönnies, Ferdinand. *Community and Society.* Translated and edited by Charles P. Loomis. East Lansing: Michigan State University Press, 1957.

Towers Perrin. "Workforce 2000: A Bottom Line Concern." Indianapolis: Hudson Institute, 1992.

"Towers Perrin Workplace Index." Boston, MA: Towers Perrin, 1997.

Toynbee, Arnold. *A Study of History.* New York: Oxford University Press, 1946.

"Transfer: Making It Work." New York: Ford Foundation, 1987.

Trattner, Walter I. *From Poor Law to Welfare State.* New York: Free Press, 1989.

Treiman, Donald J. *Occupational Prestige in Comparative Perspective.* New York: Academic Press, 1977.

Trimble, J. E. "Stereotypical Images, American Indians, and Prejudice." In P. A. Katz and D. A. Taylor (eds.), *Eliminating Racism.* New York: Plenum Press, 1988, pp. 181–202.

Troeltsch, Ernst. *The Social Teachings of the Christian Churches.* New York: Macmillan, 1931.

Troll, Lillian E. "The Family of Later Life: A Decade Review." *Journal of Marriage and the Family* 33 (May 1971):263–290.

Tuchman, Gaye, Alene Kaplan Daniels, and James Benét (eds.). *Hearth and Home.* New York: Oxford University Press, 1978.

Tucker, Robert C. (ed.). *The Marx-Engels Reader.* New York: Norton, 1972.

____. "The Deradicalization of Marxist Movements." In Robert H. Lauer (ed.), *Social Movements and Social Change.* Carbondale: Southern Illinois University Press, 1976, pp. 227–255.

Tumin, Melvin M. "Some Principles of Stratification: A Critical Analysis." *American Sociological Review* 18 (August 1953):387–394.

Turnbull, Colin. *The Mountain People.* New York: Simon & Schuster, 1972.

Turner, Jonathan H. *American Society.* 2d ed. New York: Harper & Row, 1976.

____. *The Structure of Sociological Theory.* 5th ed. Belmont, CA: Wadsworth, 1991.

____. *Classical Sociological Theory.* Chicago: Nelson-Hall, 1993.

Turner, Jonathan H., Leonard Beeghley, and Charles H. Powers. *The Emergence of Sociological Theory.* 2d ed. Chicago: Dorsey Press, 1989.

Turner, Margery Austin, Raymond J. Struyk, and John Yinger. *Housing Discrimination Study: Synthesis.* U.S. Department of Housing and Urban Development. Washington, DC: U.S. Government Printing Office, 1991.

Turner, Ralph H. "Collective Behavior." In Robert E. L. Faris (ed.), *Handbook of Modern Sociology.* Chicago: Rand McNally, 1964, pp. 382–425.

Turner, Ralph H., and Lewis M. Killian. *Collective Behavior.* 3d ed. Englewood Cliffs, NJ: Prentice Hall, 1987.

Twaddle, Andrew C. *Sickness Behavior and the Sick Role.* Cambridge, MA: Schenkman, 1981.

Twaddle, Andrew C., and Richard M. Hessler. *A Sociology of Health.* 2d ed. New York: Macmillan, 1987.

Tyree, Andrea, Moshe Semyonov, and Robert W. Hodge. "Gaps and Glissandos: Inequality, Economic Development, and Social Mobility in 24 Countries." *American Sociological Review* 44 (June 1979):410–424.

"The UN Long-Range Population Projections: What They Tell Us." Washington, DC: Population Reference Bureau, 1992.

U.S. Bureau of the Census. *Geographical Mobility: March 1982 to March 1983.* Current Population Reports. Series P-20. No. 393. Washington, DC: U.S. Government Printing Office, 1984.

____. *We, The First Americans.* Washington, DC: U.S. Government Printing Office, 1988.

——. *Statistical Abstract of the United States: 1990.* Washington, DC: U.S. Government Printing Office, 1990a.

____. *Household Wealth and Asset Ownership: 1988.* Current Population Reports. Series P-70. No. 22. Washington, DC: U.S. Government Printing Office, 1990b.

____. *Population Projections of the United States, by Age, Sex, Race, and Hispanic Origin: 1992–2050.* Series P-25. No.1092. Washington, DC: U.S. Government Printing Office, 1992a.

____. *Marital Status and Living Arrangements: March 1992.* Current Population Reports. Series P-20. No. 468. Washington, DC: U.S. Government Printing Office, 1992b.

____. *Money Income of Households, Families, and Persons in the United States: 1992.* Current Population Reports. Series P-60. No. 184. Washington, DC: U.S. Government Printing Office, 1993a.

____. *The Black Population in the United States: March 1992.* Current Population Reports. Series P-20. No. 471. Washington, DC: U.S. Government Printing Office, 1993b.

____. *Hispanic Americans Today.* Current Population Reports. Series P-23. No. 183. Washington, DC: U.S. Government Printing Office, 1993c.

____. *The Hispanic Population in the United States.* Current Population Reports. Series P-20. No. 465RV. Washington, DC: U.S. Government Printing Office, 1993d.

____. *1990 Census of Population, Social and Economic Characteristics, American Indian and Alaska Native Areas.* Section 1 of 2. CP-2–1A. Washington, DC: U.S. Government Printing Office, 1993e.

____. *Fertility of American Women: June 1992.* Current Population Reports. Series P-20. No. 470. Washington, DC: U.S. Government Printing Office, 1993f.

____. 1990 Census of Population. *Social and Economic Characteristics, United States.* CP-2–1. November 1993. Washington, DC: U.S. Government Printing Office, 1993g.

____. *Population Projections of the United States by Age, Sex, Race, and Hispanic Origin: 1993 to 2050.* Current Population Reports. Series P-25. No. 1104. Washington, DC: U.S. Government Printing Office, 1993h.

____. *School Enrollment, Social and Economic Characteristics of Students: October 1992.* Current Population Reports. Series P-20. No. 474. Washington, DC: U.S. Government Printing Office, 1993i.

____. *Population Profile of the United States 1993.* Current Population Reports. Special Studies Series P-23. No. 185. Washington, DC: U.S. Government Printing Office, 1993n.

____. *Household Wealth and Asset Ownership: 1991.* Current Population Reports. Series P-70. No. 34. Washington, DC: U.S. Government Printing Office, 1994a.

____. *How We're Changing, Demographic State of the Nation: 1994.* Current Population Reports. Series P-23. No. 187. Washington, DC: U.S. Government Printing Office, 1994b.

____. Unpublished data: "Table 15. Educational Attainment—Total Money Earnings in 1995 of Persons 25 Years Old and Over, by Age, Race, Hispanic Origin, Sex, and Work Experience in 1995." 1995a.

____. *Marital Status and Living Arrangements: March, 1995 (Update).* Current Population Reports. Series P-20. No. 491. Washington, DC: U. S. Government Printing Office, 1995b. http://www.census.gov/population/socdemo/ms-la/95his07.txt

____. *Statistical Abstract of the United States: 1996.* Washington, DC: U.S. Government Printing Office, 1996a.

____. *Poverty in the United States: 1995.* Current Population Reports. Series P-60. No. 194. Washington, DC: U.S. Government Printing Office, 1996b.

____. "Health Insurance Coverage: 1995." *Current Population Survey,* March 1996c. http://www.census.gov/hhes/hlthins/cover95/c95tabb.html.

____. "Health Insurance Coverage: 1995." *Current Population Survey,* March 1996d. http://www.census.gov/hhes/hlthins/cover95/c95tabc.html.

____. *Voting and Registration in the Election of November 1996.* Current Population Reports. Series P20. No. 504. Washington, DC: U.S. Government Printing Office, 1997a.

____. "Metropolitan Areas." 1997b. http://www.census.gov/population/www/estimates/metrodef.html.

U.S. Bureau of Labor Statistics. *Employment and Earnings.* Vol. 38, No. 1. Washington, DC: U.S. Government Printing Office, 1991.

U.S. Commission on Civil Rights. *Racial Isolation in the Public Schools.* Vol. II. Washington, DC: U.S. Government Printing Office, 1967.

____. *Success of Asian Americans: Fact or Fiction?* Clearinghouse Publication 64. Washington, DC: U.S. Government Printing Office, 1980.

U.S. Commission on Wartime Relocation and Internment of Civilians. *Personal Justice Denied.* Washington, DC: U.S. Government Printing Office, 1983.

U.S. Congress. Joint Economic Committee. *The Concentration of Wealth in the United States: Trends in the Distribution of Wealth Among American Families,* July 1986.

U.S. Department of Health and Human Services. National Institute of Mental Health. *Television and Behavior: Ten Years of Scientific Progress and Implications for the Eighties.* Vol. 1: Summary Report. Washington, DC: U.S. Government Printing Office, 1982a.

____. National Institute of Mental Health. *Television and Behavior: Ten Years of Scientific Progress and Implications for the Eighties.* Vol. 2: Technical Reviews. Washington, DC: U.S. Government Printing Office, 1982b.

____. "Improving Minority Health Statistics." U.S. Public Health Service. Office of Minority Health. Washington, DC: U.S. Government Printing Office, 1992.

U.S. Department of Justice. *White Collar Crime.* Washington, DC: U.S. Government Printing Office, 1987.

____. *Advance Report.* Immigration and Naturalization Service. Washington, DC: U.S. Government Printing Office, 1993a.

____. Federal Bureau of Investigation. *Uniform Crime Reports, 1995.* Washington, DC: U.S.Government Printing Office, 1996.

____. Bureau of Justice Statistics. *Criminal Victimization in the United States, 1995.* Washington, DC: U.S. Government Printing Office, 1997.

U.S. Department of Labor. Employment Standards Administration. *Opportunity 2000.* Washington, DC: U.S. Government Printing Office, 1988.

____. Bureau of Labor Statistics. *Employment in Perspective: Women in the Labor Force.* Report 860. Washington, DC: U.S. Government Printing Office, 1997a.

____. Bureau of Labor Statistics. "Issues in Labor Statistics: Who's Coming into the Workplace, and Who's Leaving?" Washington, DC: U.S. Government Printing Office, June 1992b.

____. Bureau of Labor Statistics. *Employment in Perspective: Women in the Labor Force.* No. 865. Washington, DC: U.S. Government Printing Office, 1993.

____. Bureau of Labor Statistics. *Employment and Earnings.* Washington, DC: U.S. Government Printing Office, January 1997b.

U.S. Merit Systems Protection Board. "A Question of Equity: Women and the Glass Ceiling in the Federal Government." Washington, DC: U.S. Government Printing Office, 1992.

U.S. Metro Data Sheet. 2d ed. Washington, DC: Population Reference Bureau, 1993.

Ulbrich, Patricia M. "The Determinants of Depression in Two-Income Marriages." *Journal of Marriage and the Family* 50 (February 1988):121–131.

Useem, Michael. "The Social Organization of the American Business Elite and Participation of Corporation Directors in the Governance of American Institutions." *American Sociological Review* 44 (August 1979):553–572.

____. *The Inner Circle.* New York: Oxford University Press, 1984.

Vago, Steven. *Social Change.* 3d ed. Upper Saddle River, NJ: Prentice Hall, 1996.

Valdivieso, Rafael, and Cary Davis. "U.S. Hispanics: Challenging Issues for the 1990s." *Population Trends and Public Policy.* Washington, DC: Population Reference Bureau, 1988.

Van de Kaa, Dirk J. "Europe's Second Demographic Transition." *Population Bulletin* 42 (March 1987):1–59.

van den Berghe, Pierre L. "Dialectic and Functionalism: Toward a Theoretical Synthesis." *American Sociological Review* 28 (October 1963):695–705.

____. *Race and Racism.* 2d ed. New York: Wiley, 1978.

Van Fleet, David D. *Behavior in Organizations.* Boston: Houghton Mifflin, 1991.

van Gennep, A. *The Rites of Passage.* Chicago: University of Chicago Press, 1960.

Van Zelst, Ramond H. "Sociometrically Selected Work Teams Increase Production." *Personnel Psychology* 5 (Autumn 1952):175–185.

Vander Zanden, James W. *American Minority Relations.* New York: McGraw-Hill, 1990.

VanRossem, Ronan. "The World System Paradigm as General Theory of Development: A Cross-National Test." *American Sociological Review* 61 (June 1996):508–527.

Vaughn, George B. "Institutions on the Edge: America's Community Colleges." *Educational Record* (Spring 1991):30–33.

Veblen, Thorstein. *The Engineers and the Price System.* New York: Viking, 1933.

____. *Theory of the Leisure Class.* New York: Pengiun, 1995.

Verbrugge, Lois. "From Sneezes to Adieux." *American Demographics* (May 1986):35–38, 53–54.

____. "The Twain Meet: Empirical Explanations of Sex Differences in Health and Mortality." *Journal of Health and Social Behavior* 30 (September 1989):282–304.

Villers Foundation. *On the Other Side of East Street.* Washington, DC: Villers Foundation, 1987.

Vobejda, Barbara. "The Heartland Pulses with New Blood." *Washington Post* (August 11, 1991):Al, A18.

____. "Wedding Welfare and Motherhood." *Washington Post* (June 2, 1994):A1, A12.

Volti, Rudi. *Society and Technological Change.* 3d ed. New York: St. Martin's Press, 1995.

von Eschen, Donald, Jerome Kirk, and Maurice Pinard. "The Disintegration of the Negro Non-Violent Movement." In Robert H. Lauer (ed.), *Social Movements and Social Change.* Carbondale: Southern Illinois University Press, 1976, pp. 203–226.

"Voter Turnout from 1945 to 1997: A Global Report on Political Participation." Stockholm, Sweden: International Institute for Democracy and Electoral Assistance, 1997.

Wagster, Emily. "Many Workers Hurt Abroad by Culture Shock." *USA Today* (December 17, 1993):1B.

Waitzkin, Howard. *The Second Sickness.* Chicago: University of Chicago Press, 1986.

Waldman, Steven. "The Stingy Politics of Head Start." *Newsweek.* Special Issue. Education: A Consumer's Handbook (Fall/Winter 1990–1991):78–79.

Walker, A., C. Pratt, and N. Oppy. "Perceived Reciprocity in Family Caregiving." *Family Relations* (January 1992):82–85.

Wallechinsky, David, and Irving Wallace. *The People's Almanac #3.* New York: Bantam, 1981.

Waller, Bruce N. *Critical Thinking: Consider the Verdict.* 2nd ed. Englewood Cliffs, NJ: Prentice Hall, 1994.

Waller, Williard, and Reuben Hill. *The Family.* (rev. ed.). New York: Dryden Press, 1951.

Wallerstein, Immanuel. *The Capitalist World Economy.* New York: Cambridge University Press, 1979b.

____. *The Politics of the World Economy.* Cambridge, England: Cambridge University Press, 1984.

____. *The Modern World-System.* London: Academic Press, 1989.

Wallich, Paul, and Elizabeth Corcoran. "The Discreet Disappearance of the Bourgeoisie." *Scientific American* 226 (February 1992):111.

Wallis, W. Allen, and Harry V. Roberts. *The Nature of Statistics.* New York: Free Press, 1962.

Ward, R., J. Logan, and G. Spitze. "The Influence of Parent and Child Needs on Coresidence in Middle and Later Years." *Journal of Marriage and the Family* (February 1992):209–221.

Ward, Russell B. "The Impact of Subjective Age and Stigma on Older Persons." *Journal of Gerontology* 32 (March 1977):227–232.

Warner, R. Stephen. *Public Religions in the Modern World.* Chicago: University of Chicago Press, 1994.

Warner, W. Lloyd. *Social Class in America.* New York: Harper & Row, 1960.

Warner, W. Lloyd, and Paul S. Lunt. *The Social Life of a Modern Community.* New Haven, CT: Yale University Press, 1941.

Warren, Martin, and Sheldon Berkowitz. "The Employability of AFDC Mothers and Fathers." *Welfare in Review* 7 (July/August 1969): 1–7.

Wasserman, Gary. *The Basics of American Politics.* 5th ed. Glenview, IL: Scott, Foresman, 1988.

Waters, Mary C. *Ethic Options.* Berkeley: University of California Press, 1990.

Watson, George W., and Jon M. Shepard. "The Disaffected Professional in a Changing Contractual Environment." Unpublished manuscript, 1997.

Watson, Russell, and John Barry. "Tomorrow's New Face of Battle." *Newsweek* (Winter, 1997–98):66–67.

Watterson, T. "Social Security: Many Widows Don't Get Benefits." *Baltimore Sun* (June 13, 1990):C24.

Watzman, Nancy, James Youngclaus, and Jennifer Shecter. *Cashing In.* Washington, DC: Center for Responsive Politics, 1997.

Waxman, Chaim I. *The Stigma of Poverty.* New York: Pergamon, 1983.

Weakliem, David L. "Relative Wages and the Radical Theory of Economic Segmentation." *American Sociological Review* 55 (1990):574–590.

Webb, R.K. *Harriet Martineau, A Radical Victorian.* New York: Columbia University Press, 1960.

Weber, Max. *From Max Weber.* Edited by H. H. Gerth and C. Wright Mills. New York: Oxford University Press, 1946.

____. *The Religion of China.* New York: Free Press, 1951.

____. *Ancient Judaism.* New York: Free Press, 1952.

____. *The Religion of India.* New York: Free Press, 1958.

____. *The Sociology of Religion.* Boston: Beacon Press, 1964a.

____. *The Theory of Social and Economic Organization.* Translated by A. N. Henderson and Talcott Parsons. New York: Free Press, 1964b.

Weeks, John R. *Population.* 6th ed. Belmont, CA: Wadsworth, 1996.

Weinberg, Meyer. *Desegregation Research.* 2d ed. Bloomington, IN: Phi Delta Kappan, 1970.

Weisberg, Herbert F., and Jon A. Krosnick. *An Introduction to Survey Research, Polling, and Data Analysis.* Thousand Oaks, CA: Sage, 1996.

Weitz, Rose. *The Sociology of Health, Illness, and Health Care: A Critical Approach.* Belmont, CA: Wadsworth, 1996.

Weitzman, Leonore J., and Deborah Eifler. "Sex Role Socialization in Picture Books for Preschool Children." *American Journal of Sociology* 77 (May 1972):1125–1150.

Weller, Jack E. *Yesterday's People.* Lexington: University Press of Kentucky, 1965.

Werhane, Patricia H. *Adam Smith and His Legacy for Modern Capitalism.* New York: Oxford University Press, 1991.

West, Guida, and Rhoda Lois Blumberg (eds.). *Women and Social Protest.* New York: Oxford University Press, 1990.

Westoff, Leslie Aldridge, and Charles F. Westoff. *From Now to Zero.* Boston: Little, Brown, 1971.

Wetcher, Kenneth. *Save the Males.* Summit, NJ: P.I.A. Press, 1991.

Whitaker, Mark. "Getting Tough at Last." *Newsweek* (May 10, 1993):22.

White, Leslie A. *The Science of Culture.* 2d ed. New York: Farrar, Straus & Giroux, 1969.

White, Lynn. *Medieval Technology and Social Change.* New York: Oxford University Press, 1972.

White, Lynn, and David B. Brinkerhoff. "The Sexual Division of Labor: Evidence from Childhood." *Social Forces* 60 (September 1981):170–181.

White, Ralph K., and Ronald O. Lippitt. *Autocracy and Democracy.* New York: Harper & Row, 1960.

Whitehead, Alfred North. *Adventures of Ideas.* New York: Mentor Books, 1933.

Who Owns Whom 1993. Vol. 2. United Kingdom: Dun and Bradstreet Limited, 1993.

Whorf, Benjamin Lee. *Language, Thought and Reality.* Edited by John B. Carroll. Cambridge, MA: MIT Press, 1956.

"Why Don't Americans Trust the Government?" Menlo Park, CA: *The Washington Post*/Kaiser Family Foundation/Harvard University Survey Project, 1996.

Whyte, William Foote. *Learning from the Field.* Beverly Hills, CA: Sage, 1984.

____. *Street Corner Society.* 4th ed. Chicago: University of Chicago Press, 1993.

Wiatrowski, Michael D., David B. Griswold, and Mary K. Roberts. "Social Control Theory and Delinquency." *American Sociological Review* 46 (October 1981):525–542.

Wiehe, Vernon R. *Sibling Abuse.* Lexington, MA: Heath, 1990.

Wieviorka, Michel. *The Arena of Racism.* Newbury Park, CA: Sage, 1995.

Wiley, Norbert (ed.). *The Marx-Weber Debate.* Beverly Hills, CA: Sage, 1987.

Wilkinson, Doris Y. "Gender and Social Inequality: The Prevailing Significance of Race." *Daedalus* 124 (Winter 1995):167–178.

Wilkinson, Richard (ed.). *Class and Health.* New York: Tavistock, 1986.

Will, Jerrie Ann, Patricia A. Self, and Nancy Datan. "Maternal Behavior and Perceived Sex of Infant." *American Journal of Orthopsychiatry* 46 (January 1976):135–139.

Williams, Christine L. *Gender Differences in Work.* Berkeley: University of California Press, 1989.

Williams, J. Allen, Joetta A. Vernon, Martha C. Williams, and Karen Malecha. "Sex Role Socialization in Picture Books: An Update." *Social Science Quarterly* 68 (March, 1987):148–156.

Williams, John E., and Deborah L. Best. *Measuring Sex Stereotypes.* (rev. ed.). Newbury Park, CA: Sage, 1990.

Williams, Robert L. "Scientific Racism and IQ: The Silent Mugging of the Black Community." *Psychology Today* 7 (May 1974):32–41, 101.

Williams, Robin M., Jr. *American Society.* 3d ed. New York: Knopf, 1970.

Williamson, Robert C. "A Partial Replication of the Kohn-Gecas-Nye Thesis in a German Sample." *Journal of Marriage and the Family* 46 (November 1984):971–979.

Willwerth, James. "The Man from Outer Space." *Time* (April 25, 1994):74–75.

Wilson, Bryan. *Religion in Sociological Perspective.* New York: Oxford University Press, 1982.

Wilson, Edward O. *On Human Nature.* Cambridge, MA: Harvard University Press, 1978.

____. *Biophilia.* Cambridge, MA: Harvard University Press, 1986.

Wilson, James Q. *The Moral Sense.* New York: Free Press, 1993.

____. *Political Organizations.* Princeton, NJ: Princeton University Press, 1995.

Wilson, James Q., and Richard J. Herrnstein. *Crime and Human Nature.* New York: Simon & Schuster, 1985.

Wilson, Kenneth L., and W. Allen Martin. "Ethnic Enclaves: A Comparison of the Cuban and Black Economies in Miami." *American Journal of Sociology* 88 (July 1982):135–160.

Wilson, Stephen. *Informal Groups.* Englewood Cliffs, NJ: Prentice Hall, 1978.

Wilson, Thomas C. "Urbanism and Tolerance: A Test of Some Hypotheses Drawn from Wirth and Stouffer." *American Sociological Review* 50 (February 1985):117–123.

____. "Urbanism, Migration, and Tolerance: A Reassessment." *American Sociological Review* 56 (February 1991):117–123.

Wilson, William Julius. *Power, Racism, and Privilege.* New York: Macmillan, 1973.

____. *The Declining Significance of Race.* 2d ed. Chicago: University of Chicago Press, 1980.

____. "The Urban Underclass." In Leslie W. Dunbar (ed.), *Minority Report.* New York: Pantheon, 1984, pp. 75–117.

____. *The Truly Disadvantaged.* Chicago: University of Chicago Press, 1987.

____. "Studying Inner-City Social Dislocations: The Challenge of Public Agenda Research." *American Sociological Review* 56 (February 1991):1–14.

____ (ed.). *The Ghetto Underclass.* Newbury Park, CA: Sage, 1993.

____. *When Work Disappears: The World of the New Urban Poor.* New York: Knopf, 1997.

Wilson, William Julius, et al. "The Ghetto Underclass and the Changing Structure of Urban Poverty." In Fred R. Harris and Roger W. Wilkins (eds.), *Quiet Riots.* New York: Pantheon, 1988, pp. 123–154.

Wimberley, Ronald C., and James A. Christenson. "Civil Religion and Church and State." *The Sociological Quarterly* 21 (Spring 1980):35–40.

Wimberley, Ronald C., Donald A. Clelland, Thomas C. Hood, and C. M. Lipsey. "The Civil Religious Dimension: Is It There?" *Social Forces* 54 (June 1976):890–900.

Wind, James P., and James W. Lewis (eds.). *American Congregations.* Vol. 2, Chicago: University of Chicago Press, 1994.

Wingert, Pat. "The Sum of Mediocrity." *Newsweek* (December 2, 1996):96.

Winship, Christopher, and Robert D. Mare. "Models for Sample Selection Bias." *Annual Review of Sociology* 18 (1992):327–350.

Wirth, Louis. "Clinical Sociology." *American Journal of Sociology* 37 (July 1931):49–66.

____. "Urbanism as a Way of Life." *American Journal of Sociology* 44 (July 1938):1–24.

____. "The Problem of Minority Groups." In Ralph Linton (ed.), *The Science of Man in the World Crisis.* New York: Columbia University Press, 1945, pp. 347–372.

Wise, David. "Cloak and Dagger Operations: An Overview." In Jerome H. Skolnick and Elliott Currie (eds.), *Crisis in American Institutions.* 3d ed. Boston: Little, Brown, 1976, pp. 88–101.

Wise, William. *Massacre at Mountain Meadows.* New York: Thomas Y. Crowell, 1976.

Wokutch, Richard E. *Worker Protection, Japanese Style.* Ithaca, NY: ILR Press, 1992.

Wolfe, Alan. "The Moral Meanings of Work." *The American Prospect* (September/October 1997): 82–90.

Wolfinger, Raymond E., and Steven J. Rosenstone. *Who Votes?* New Haven, CT: Yale University Press, 1980.

Wolff, Edward N. "The Rich Get Increasingly Richer: Latest Data on Household Wealth During the 1980s." Briefing Paper. Economic Policy Institute. Washington, DC, November 1992.

Wong, Sin-Kwok. "Understanding Cross-National Variation in Occupational Mobility." *American Sociological Review* 55 (August 1990):560–573.

Wood, James R. *Leadership in Voluntary Organizations.* New Brunswick, NJ: Rutgers University Press, 1981.

World Development Report 1991. New York: Oxford University Press, 1991.

World Development Report 1993. New York: Oxford University Press, 1993.

World Development Report 1994. New York: Oxford University Press, 1994.

World Population Data Sheet 1997. Washington, DC: Population Reference Bureau, 1997.

Wortzel, Larry M. *Class in China.* Westport, CT: Greenwood Press, 1987.

Wozniak, Paul R. "Making Sociobiological Sense Out of Sociology." *The Sociological Quarterly* 25 (Spring 1984): 191–204.

Wright, Charles R. "Functional Analysis and Mass Communication." *Public Opinion Quarterly* 24 (Winter 1960):606–620.

____. "Functional Analysis and Mass Communication Revisited." In Jay G. Blumler and Elihu Katz (eds.), *The Uses of Mass Communication.* Beverly Hills, CA: Sage, 1974, pp. 197–212.

Wright, Erik Olin. *Classes.* London: Verse, 1985.

Wright, James D., and Sonia R. Wright, "Social Class and Parental Values for Children: A Partial Replication and Extension of the Kohn Thesis." *American Sociological Review* 4 (June 1976):527–537.

Wright, Kevin N. *The Great American Crime Myth.* New York: Praeger, 1987.

Wright, Lawrence. "Double Mystery." *The New Yorker* (August 7, 1995):45–62.

Wuthnow, Robert. "Recent Patterns of Secularization: A Problem of Generations?" *American Sociological Review* 41 (October 1976):850–867.

____. *Experimentation in American Religion.* Berkeley: University of California Press, 1978.

____. "Review of All Faithful People." *Society* 22 (March/April 1985):87–88.

____. *The Restructuring of American Religion.* Princeton, NJ: Princeton University Press, 1990.

____. *Poor Richard's Principle: Rediscovering the American Dream Through the Moral Dimensions of Work, Business, and Money.* Princeton, NJ: Princeton University Press, 1996.

Yinger, John. *Closed Doors, Opportunities Lost: The Continuing Costs of Housing Discrimination.* New York: Russell Sage Foundation, 1995.

Yinger, Milton J. "A Structural Examination of Religion." *Journal of the Scientific Study of Religion* 8 (Spring 1969):88–89.

____. *The Scientific Study of Religion.* New York: Macmillan, 1970.

____. *Countercultures.* New York: Free Press, 1984.

Yinger, Milton J., and Stephen J. Cutler. "The Moral Majority Viewed Sociologically." *Sociological Focus* 15 (October 1982):289–306.

Yochelson, Samuel, and Stanton E. Samenow. *The Criminal Personality.* Vols. 1 and 2. London: James Aronson, 1994.

Young, Iris Marion. *Justice and the Politics of Difference.* Princeton, NJ: Princeton University Press, 1990.

Young, Michael. *The Rise of the Meritocracy.* New York: Penguin, 1967.

Young, T. R. *The Drama of Social Life.* New Brunswick, NJ: Transaction Books, 1990.

Zald, Mayer N., and John D. McCarthy. "Introduction." In Mayer N. Zald and John D. McCarthy (eds.), *Social Movements in an Organizational Society.* New Brunswick, NJ: Transaction Books, 1987.

Zangwill, Israel. *The Melting Pot.* New York: Macmillan, 1933.

Zawitz, Marianne W. *Report to the Nation on Crime and Justice.* 2d ed. U.S. Department of Justice, Bureau of Justice Statistics, July 1988.

Zborowski, Mark. "Cultural Components in Response to Pain." *Journal of Social Issues* 8 (1952):16–30.

____. *People in Pain.* San Francisco: Jossey-Bass, 1969.

Zelditch, Morris, Jr. "Role Differentiation in the Nuclear Family: A Comparative Study." In Talcott Parsons and Robert F. Bales (eds.), *Family, Socialization and Interaction Process.* New York: Free Press, 1955, pp. 307–352.

Zellner, William W. *Countercultures: A Sociological Analysis.* New York: St. Martin's Press, 1995.

Zigler, Edward, and S. Muenchow. *Head Start.* New York: Basic Books, 1992.

Zigler, Edward, and Sally J. Styfco (eds.). *Head Start and Beyond.* New Haven, CT: Yale University Press, 1993.

Zimbardo, Philip G., S. M. Anderson, and L. G. Kabat. "Induced Hearing Deficit Generates Experimental Paranoia." *Science* (June 26, 1981):1529–1531.

Zoglin, Richard. "Murdock's Biggest Score." *Time* (June 6, 1994):54–56.

Zola, Irving K. "Culture and Symptoms—An Analysis of Patients Presenting Complaints." *American Sociological Review* 31 (October 1966):615–630.

Zopf, Paul E. *American Women in Poverty.* Westport, CT: Greenwood Press, 1989.

Zorbaugh, Harvey. *The Gold Coast and the Slum.* Chicago: University of Chicago Press, 1929.

Zucher, Louis A. *Social Roles.* Beverly Hills, CA: Sage, 1983.

Zucher, Louis A., and David A. Snow. "Collective Behavior: Social Movements." In Morris Rosenberg and Ralph H. Turner (eds.), *Social Psychology.* New Brunswick, NJ: Transaction Books, 1990, pp. 447–482.

Zweigenhaft, Richard L., and G. William Domhoff. *Jews in the Protestant Establishment.* New York: Praeger, 1982.

Name Index

Subject Index

Photo Credits